London
MACMILLAN AND CO.

PUBLISHERS TO THE UNIVERSITY OF
Oxford

A TREATISE

ON

INFINITESIMAL CALCULUS;

CONTAINING

DIFFERENTIAL AND INTEGRAL CALCULUS,
CALCULUS OF VARIATIONS, APPLICATIONS TO ALGEBRA AND GEOMETRY,
AND ANALYTICAL MECHANICS.

BY

BARTHOLOMEW PRICE, M.A., F.R.S., F.R.A.S.,

SEDLEIAN PROFESSOR OF NATURAL PHILOSOPHY, OXFORD.

VOL. II.

INTEGRAL CALCULUS, CALCULUS OF VARIATIONS,
AND
DIFFERENTIAL EQUATIONS.

SECOND EDITION.

"Les progrès de la science ne sont vraiment fructueux, que quand ils amènent aussi le progrès des Traités élémentaires."—CH. DUPIN.

Oxford:

AT THE CLARENDON PRESS.

M.DCCC.LXV.

Pt.

A TREATISE

ON

INTEGRAL CALCULUS

AND

CALCULUS OF VARIATIONS.

BY

BARTHOLOMEW PRICE, M.A., F.R.S., F.R.A.S.,

SEDLEIAN PROFESSOR OF NATURAL PHILOSOPHY, OXFORD.

SECOND EDITION.

"Les progrès de la science ne sont vraiment fructueux, que quand ils amènent aussi le progrès des Traités élémentaires."—CH. DUPIN.

Oxford:

AT THE CLARENDON PRESS.

M.DCCC.LXV.

PREFACE TO THE SECOND EDITION.

In the interval which has elapsed since the publication of the first edition of this work, considerable progress has been made in the Integral Calculus; and much matter of the science which was at that time beyond the range and scope of a didactic treatise has been brought within its limits. I thought fit in that edition to omit the whole theory of Definite Integrals except the barest outline of some of its elementary theorems which were required for subsequent applications; and the Gamma-function and its allied transcendants did not seem to me to be within the scope of an elementary work. Now, on the contrary, no treatise of that character would be complete unless these subjects were discussed at considerable length, and the useful theorems arising out of them, and due to Dirichlet and Liouville, were explained and applied. At that time also it was unnecessary to give any general process for evaluating Definite Integrals; now Cauchy's theory has been received by mathematicians, and it has become necessary to insert it as the nearest approach to a general method of evaluation of these transcendants. The process of evaluating Definite Integrals has been treated much more fully than heretofore; and the application of the Integral Calculus to the Theory of Series, to the peculiarities of Periodic Series, and to the Calculus of Probabilities has been discussed at considerable length, although the higher parts of these subjects are omitted because they are not suited to a treatise intended mainly for educational use. The several parts of the treatise have been enlarged; of the more difficult portions fuller explanations have, been given, and many illustrative examples have been added. The symbols of Determinants have been employed freely in many places; and light is cast by a symmetrical notation on theorems which would be made obscure by a cumbrous symbolism. I have not ventured on the theory of doubly Periodic Functions; for they are too important for cursory and superficial treatment; and a full discussion of their properties would require more space than could be given to it, in a volume which has already exceeded the usual limits of similar works. The latest treatises in which these important transcendants are discussed are mentioned in page 204; and the student will in them find much of the information he is in quest of, although for complete information he must have recourse to the original Memoirs of Abel and Jacobi, and other writers of their School.

I have even more consistently than in the first edition constructed the Calculus on its own basis. It has been hitherto for the most part established on an inversion of the rules of the Differential Calculus; it has had scarcely any principles of its own, and of these none independent of those of the Differential Calculus; the student has been obliged to burden his memory with certain rules which he mechanically applies; he has not been taught to deduce them from first principles, because he has had no principles pregnant with such rules; and of them, at least in the early stages of his knowledge, he can give neither intelligible account nor interpretation; and it is only when he arrives at the first geometrical application that he gets an insight into the meaning of the processes; and his view is even then obscured by an expansion into a series, which he no sooner obtains than he omits all terms, save one, of it.

Now in a science replete with applications so large and so important as those of the Integral Calculus, such a method is unsatisfactory, not to say unphilosophical; and it is neither desirable nor necessary to leave it in this state. Most foreign mathematicians have been alive to the defects, and have succeeded in remedying them: why then should Englishmen be behind? Professor De Morgan is, as far as I know, the first English author who constructed the science on a more philosophical basis; and in his large Treatise on the Differential and Integral Calculus, the Integral Calculus is established on sound principles, and placed early in the course. For purely scientific reasons such an arrangement may be the best, but it may fairly be questioned whether it is convenient for didactic purposes: I have chosen to place it after the Differential Calculus.

In the following treatise the Integral Calculus is considered a part of Infinitesimal Calculus, and as such, is founded on an intelligible conception of Infinitesimals; it is thus a branch of the science of continuous number; its principles are involved in, and effluent from, that fundamental idea; it assumes the existence of an infinitesimal element-function, formed according to an assigned law, the law being involved in the symbolical form of the infinitesimal; and the primary problem is, to determine the finite number or function of number of which the given infinitesimal is the constituent elemental part; that is, Given the infinitesimal element, to find the finite quantity of which it is the infinitesimal element. The required result can evidently only be definite, when the sum of the infinitesimal elements is to be taken between certain fixed limits, which are at a finite distance apart. Thus the primary problem is one of summation of a series, of which the law is given, (for the symbolical form of the element-function, or type-term, determines that,) and the first and the last terms are given, and the sum of these infinitesimal element-functions or differentials is called the Definite Integral. The notion of a Definite Integral is therefore the fundamental one of the Integral Calculus; and the work of the Calculus is, to discover rules for the formation of these, to construct the code of laws which they are subject to, and to investigate the conditions necessary for their application to other subject-matter. Hence it is that the Definite Integrals of simple element-functions are investigated in the early part of the treatise from first principles, and it is only when I have rigorously proved in the most general case that the Definite Integral may be found by an inversion of the process of Differentiation that I have considered myself free to make use of the knowledge of the Differential Calculus, which has been (usually) previously acquired. By these means our labour is diminished, and nothing of principle is lost, because the rules thus found might have been discovered directly from the peculiar principles of the Integral Calculus.

In support of the view of the subject here taken, I allege that on this conception of Infinitesimal elements, and on this conception only, is the Integral Calculus applied to the problems of Rectification, Quadrature and Cubature, and in proof of this allegation I appeal to the processes of Chaps. VI, IX and X; in them the infinitesimal element-function exists previous to the finite function, and the latter is found by the summation of an infinite number of the former. And this is undoubtedly the process, and the only intelligible process, of determining the finite results of an ever-varying law: it is, I assert, on the notion of Infinitesimals only, that the problems of varying velocity can be intelligibly treated by the Integral Calculus.

The following is consequently the outline of the treatise. It consists of three parts; viz. Integral Calculus, commonly so called; the Calculus of Variations; and Differential Equations. The notion of a Definite Integral is stated in its fundamental and most comprehensive form; and the first four chapters are occupied with theorems, evaluations, and other properties of these Integrals. In the VIth and VIIth chapters, the Theory of Single Integration is applied to certain geometrical problems and to the theory of series. In the VIIIth chapter a large extension is given to the subject by means of Multiple Integration, the Elementary Theorems and the Theory of Transformation of Multiple Integrals occupying that chapter. In the three following chapters important applications of this theory to Geometry and to the Calculus of Probabilities are explained; and in the XIIth chapter various methods are discussed which mathematicians have devised for the Reduction of the Order of Integration.

In the course of the Calculus of Variations I have taken the opportunity of expounding at some length the properties of geodesic lines, and in an especial manner those of geodesics on an ellipsoid.

The third part in which Differential Equations, that is element-functions involving two or more dependent variables, are discussed is necessarily imperfect; the subject is surrounded with difficulties and is close on the present boundaries of our knowledge; I can do little else than exhibit such detached portions of it as have yielded to the powers of Analysis.

I am, as in the first Volume, under obligation to many friends for assistance and advice; to Professor Stokes of Pembroke College, Cambridge, to Mr. W. Spottiswoode, M. A., of Oxford, to Mr. H. J. S. Smith, Savilian Professor of Geometry, Oxford; to Professor De Morgan, to M. Moigno, to M. Duhamel; and to many others whose contributions are acknowledged in various parts of the Treatise. And I am also bound to express my sense of obligation to M. Liouville, and M. Crelle, on account of their valuable Journals.

The Chapters mark the salient divisions of the matter; the Articles are numbered continuously throughout the Volume, and their numerals are placed in the inner corners on the top of the pages. Bracketed numerals are also attached to the more important equations and are separate for each Chapter; and reference is for the most part made to the numbers of the Article and of the equation.

The references throughout are made to the second edition of Vol. I.

11, St. Giles', Oxford.
July 1st, 1865.

ANALYTICAL TABLE OF CONTENTS.

INTEGRAL CALCULUS.

CHAPTER I.

THE THEORY OF DEFINITE AND INDEFINITE INTEGRATION.

CHAP. II.

RULES FOR THE INTEGRATION OF ALGEBRAICAL FUNCTIONS.

SECTION 1.—*Integration of Fundamental Algebraical Functions.*

SECTION 2.—*Integration of Rational Fractions.*

18. Definition of rational fractions, and simplification 24
19, 20. Decomposition into partial fractions, when the roots of the denominator are all unequal 25
21, 22. Decomposition into partial fractions when there are in the denominator sets of equal roots 30
23, 24. Integration of $\frac{dx}{x^n \pm 1}$ 35
25. Integration of $\frac{x^m\,dx}{x^n \pm 1}$ 41
26. Integration of rational fractions by various artifices 41
27–30. Integration by reduction of
$\frac{dx}{(a^2+x^2)^n}$, $\frac{dx}{(a^2-x^2)^n}$, $\frac{x^m\,dx}{(a^2 \pm x^2)^n}$ 42

SECTION 3.—*Integration of Irrational Algebraical Functions.*

31–35. Integration of
$\frac{\pm dx}{(a^2-x^2)^{\frac{1}{2}}}$, $\frac{dx}{(2ax-x^2)^{\frac{1}{2}}}$, $\frac{dx}{(a^2+x^2)^{\frac{1}{2}}}$, $\frac{dx}{(2ax+x^2)^{\frac{1}{2}}}$ .. 45
36. Proof of identity of results apparently inconsistent 47
37, 38. Integration of $\frac{dx}{(a+bx\pm cx^2)^{\frac{1}{2}}}$, $\frac{dx}{x(a+bx+cx^2)^{\frac{1}{2}}}$ 48
39, 40. Integration of
$(a^2-x^2)^{\frac{1}{2}}dx$, $(2ax-x^2)^{\frac{1}{2}}dx$, $(a^2+x^2)^{\frac{1}{2}}dx$, $(2ax+x^2)^{\frac{1}{2}}dx$ 49
41. Integration of $\frac{dx}{(a^2 \pm x^2)^{\frac{3}{2}}}$, $\frac{dx}{(a+bx+cx^2)^{\frac{3}{2}}}$ 51
42. Examples in illustration 52

SECTION 4.—*Integration of Irrational Functions by Rationalization.*

43–45. Rationalization of $x^m(a+bx^n)^{\frac{p}{q}}dx$, and examples 53
46, 47. Other forms admitting of rationalization 55

SECTION 5.—*Integration of Irrational Functions by Reduction.*

48. General remarks on the process 56
49. Integration of $\frac{x^n\,dx}{(a^2-x^2)^{\frac{1}{2}}}$ 57
50. Integration of $\frac{x^n\,dx}{(a^2+x^2)^{\frac{1}{2}}}$ 58

51, 52. Integration of $\dfrac{x^n dx}{(2ax \pm x^2)^{\frac{1}{2}}}$ 58
53, 54. Integration of $(a^2 \pm x^2)^{\frac{n}{2}} dx$ 59
55, 56. Integration of $\dfrac{dx}{(a^2 - x^2)^{\frac{n}{2}}}, \dfrac{dx}{x^n (a^2 - x^2)^{\frac{1}{2}}}$ 60
57. Integration of $\dfrac{x^n dx}{(a + bx)^{\frac{1}{2}}}$ 61
58. Integration of $\dfrac{x^n dx}{(a + bx + cx^2)^{\frac{1}{2}}}$ 62

CHAP. III.

INTEGRATION OF LOGARITHMIC AND CIRCULAR FUNCTIONS.

SECTION 1.—*Exponential and Logarithmic Functions.*

59. Integration of $a^x dx$ and $e^{mx} dx$ 64
60, 61. Integration of $x^n e^{ax} dx$, and $x^{-n} e^{ax} dx$ 64
62. Various examples of integration of exponential functions .. 65
63, 64. Integration of logarithmic functions 66

SECTION 2.—*Circular Functions.*

65–67. Integration of fundamental circular functions 67
68, 69. Integration of $(\sin x)^n dx$ and $(\cos x)^n dx$ 71
70. Integration of $(\sin x)^{-n} dx$ and of $(\cos x)^{-n} dx$ 74
71, 72. Integration of $(\sin x)^m (\cos x)^n dx$ 75
73. Integration of $(\tan x)^n dx$ and of $(\cot x)^n dx$ 77
74. Integration of $x^n \sin x\, dx$, and of $x^n \cos x\, dx$ 78
75, 76. Integration of $e^{ax} (\cos x)^n dx$, and of $e^{ax} (\sin x)^n dx$ 78
77. Integration of $f(x) \sin^{-1} x\, dx$, $f(x) \tan^{-1} x\, dx$, &c. 80
78. Integration of $\dfrac{dx}{(a + b \cos x)^n}$ 80
79. Integration by substitution 81
80. Reason why the number of indefinite integrals is so small .. 83

CHAP. IV.

ON DEFINITE INTEGRATION AND ON DEFINITE INTEGRALS.

SECTION 1.—*Definite Integrals determined by Indefinite Integration.*

81. Great importance of definite integration 85
82. Examples of definite integrals determined by indefinite integration 86

SECTION 2.—*The Change of Limits in Definite Integrals.*

SECTION 3.—*The Transformation of a Definite Integral by a change of variable.*

SECTION 4.—*The Differentiation and Integration of a Definite Integral with respect to a Variable Parameter.*

SECTION 5.—*Definite Integrals involving Impossible Quantities. Cauchy's Method of Evaluation.*

Section 6.—*Methods of Approximating to the values of a Definite Integral.*

Section 7.—*The Gamma-Function, and other Allied Integrals.*

SECTION 8.—*The Logarithm-Integral.*

CHAP. V.

SUCCESSIVE INTEGRATION OF AN EXPLICIT FUNCTION OF ONE VARIABLE.

CHAP. VI.

THE APPLICATION OF SINGLE INTEGRATION TO QUESTIONS OF GEOMETRY.

SECTION 1.—*Rectification of Plane Curves; Rectangular Coordinates.*

SECTION 2.—*Rectification of Plane Curves; Polar Coordinates.*

CHAP. VII.

THE APPLICATION OF SINGLE DEFINITE INTEGRATION TO THE THEORY OF SERIES.

SECTION 1.—*The Convergence and Divergence of Series.*

SECTION 2.—*The series of Taylor and Maclaurin.*

SECTION 3.—*The Development of Series by means of Single Definite Integration.*

SECTION 4.—*On Periodic Series, and on Fourier's Integral.*

CHAP. VIII.

ON MULTIPLE INTEGRATION, AND THE TRANSFORMATION OF MULTIPLE INTEGRALS.

SECTION 1.—*On Double, Triple, and Multiple Integration.*

SECTION 2.—*The Transformation of Multiple Integrals.*

SECTION 3.—*The Differentiation of a Multiple Integral with respect to a Variable Parameter.*

CHAP. IX.

QUADRATURE OF SURFACES, PLANE AND CURVED.

SECTION 1.—*Quadrature of Plane Surfaces; Cartesian Coordinates.*

SECTION 2.—*Quadrature of Plane Surfaces; Polar Coordinates.*

SECTION 3.—*Quadrature of Surfaces of Revolution.*

SECTION 4.—*Quadrature of Curved Surfaces.*

SECTION 5.—*Gauss' system of Curvilinear Coordinates.*

CHAP. X.

CUBATURE OF SOLIDS.

SECTION 1.—*Cubature of Solids of Revolution.*

SECTION 2.—*Cubature of Solids bounded by any curved Surface.*

CHAP. XI.

ON SOME QUESTIONS IN THE CALCULUS OF PROBABILITIES, AND ON THE DETERMINATION OF MEAN VALUES.

SECTION 1.—*On the Calculus of Probabilities.*

SECTION 2.—*The Determination of Mean Values.*

CHAP. XII.

REDUCTION OF MULTIPLE INTEGRALS.

SECTION 1.—*Reduction of Multiple Integrals by simple application of the Gamma-function.*

SECTION 2.—*Dirichlet's Method of Reduction by means of a Factor of Discontinuity. Fourier's Integral.*

CALCULUS OF VARIATIONS.

CHAPTER XIII.

EXPOSITION OF THE PRINCIPLES OF THE CALCULUS OF VARIATIONS.

CHAP. XIV.

APPLICATIONS OF THE CALCULUS OF VARIATIONS TO PROBLEMS OF MAXIMA AND MINIMA.

SECTION 1.—*Critical Values of Definite Integrals, whose Element-Functions involve Variables and their Differentials.*

SECTION 2.—*On Geodesic Lines.*

Section 3.—*Investigation of Critical Values of a Definite Integral, whose Element-Function involves Derived-Functions.*

Section 4.—*Discriminating Conditions of Maxima and Minima.*

Section 5.—*Investigation of the Critical Values of a Double Definite Integral.*

DIFFERENTIAL EQUATIONS,

OR

INTEGRATION OF DIFFERENTIAL FUNCTIONS OF TWO AND MORE VARIABLES.

CHAPTER XV.

THE INTEGRATION OF DIFFERENTIAL EQUATIONS OF THE FIRST ORDER.

SECTION 1.—*General Considerations on Differential Equations.*

SECTION 2.—*Exact Total Differential Equations.*

SECTION 3.—*Homogeneous Equations of two Variables.*

SECTION 4.—*The First Linear Differential Equation.*

SECTION 5.—*Partial Differential Equations of the First Order and First Degree.*

SECTION 6.—*Integrating Factors of Differential Equations.*

SECTION 7.—*Singular Solutions of Differential Equations.*

SECTION 8.—*Differential Equations of first order and of any degree.*

SECTION 9.—*Partial Differential Equations of the First Order and Higher Degrees.*

SECTION 10.—*Various Theorems and Applications.*

SECTION 11.—*Solution of Geometrical Problems, dependent on Differential Equations.*

CHAP. XVI.

INTEGRATION OF DIFFERENTIAL EQUATIONS OF ORDERS HIGHER THAN THE FIRST.

SECTION 1.—*General Properties.*

SECTION 2.—*Linear Differential Equations.*

SECTION 3.—*Linear Differential Equations with Constant Coefficients.*

SECTION 6.—*Integration of Partial Differential Equations of Higher Orders.*

CHAP. XVII.

THE SOLUTION OF DIFFERENTIAL EQUATIONS BY THE CALCULUS OF OPERATIONS.

SECTION 1.—*The Solution of Total Differential Equations by Symbolical Methods.*

SECTION 2.—*The Solution of Partial Differential Equations by Symbolical Methods.*

CHAP. XVIII.

INTEGRATION OF SIMULTANEOUS DIFFERENTIAL EQUATIONS.

CHAP. XIX.

INTEGRATION OF DIFFERENTIAL EQUATIONS BY SERIES.

INTEGRAL CALCULUS.

CHAPTER I.

THE THEORY OF DEFINITE AND INDEFINITE INTEGRATION.

ARTICLE I.] The primary problem of the Integral Calculus is the summation of a series of which each term is an infinitesimal, and the number of terms is infinite; all the terms being infinitesimals of the same order, and the difference between two consecutive terms being an infinitesimal of an order higher than that of each term. Thus, according to the doctrine of infinitesimals and infinities, which has been established in Vol. I, if relatively to a given base the orders of infinity and of infinitesimal are the same, the sum will be finite; and as the order of infinity is higher or lower than that of the infinitesimal, so will the sum be infinite or infinitesimal. In most of the cases which will hereafter come under discussion, the sum will be finite; yet not in all: and the quality and the circumstances of those infinitesimals, the sum of an infinite number of which is not finite, will require most cautious and careful consideration.

The infinitesimal term of the series will be expressed as a function of one or more variables; and the variation of the variables will give the several and successive terms of the series. Thus, $f(x)\,dx$, $f(x, y)\,dx\,dy$,...... may be general terms of such a series; and the successive terms will be given by means of the continuous variations of the variables. These are the most general forms of the infinitesimal terms; $x^n dx$, $\cos x\, dx$, $e^{ax+by} dx\,dy$ are particular forms. Such infinitesimal terms are called *infini-*

tesimal elements, being the elements or infinitesimal parts of the whole sum; and the finite factor in each is called the *element-function*. The form of these latter factors evidently assigns the law of the series. In this respect these series are analogous to those ordinary series each of whose terms is a finite quantity, and in which the general term is a type-term, and assigns the law of the series.

The problem of the summation of a series is generally indefinite; but becomes definite when two terms are assigned, the sum of the terms within which, either inclusively or exclusively, is to be found. These assigned terms are called the *limiting terms*, or the limits of the series; the first and last terms being called respectively the inferior and the superior limit. The excess of the superior over the inferior limit is called the *range* of summation. Although the sums of some series can be found for any general limits, say the mth and the nth terms, yet the sums of others can be found only for certain specified limits. These peculiarities depend on the law of the series, and many instances of them will occur hereafter.

Although the number of terms in the series, the sum of which is required to be found in the problem of the Integral Calculus, is infinite; yet as they are infinitesimal, and vary by infinitesimal variations of the subject variables which express their general terms, the difference between the values of the variables at the limits will be a finite quantity; so that the problem is definite as to its limits, and the distance between the limits is finite.

The process by which such sums are found is termed *Integration*, being, as it is, the putting together the parts of which a whole is composed; and the sum of the series of infinitesimal elements between given terms is called a *Definite Integral*, the values of the variable which assign the first and last terms being called the *limits of Integration*; and the excess of the superior limit over the inferior is called the *range* of integration. As the form of each term of the series is the same, if a general term is given, the general form of the sum of a series of terms can in many cases be found, although the first and last terms may not be given; this general sum is called the *Indefinite Integral*.

The *Integral Calculus* is the aggregate of the rules by which Integrals are determined, and the code of laws subject to which Differentials and Integrals in their mutual relations may be applied to questions of Geometry and Physics.

In the early part of the treatise I shall consider functions of one variable; and I shall assume the infinitesimal element to be of the form $f(x)dx$, and the superior and inferior limits to be respectively x_n and x_0; so that $x_n - x_0$ is the range of integration. I shall also assume $f(x)$ to be finite and continuous for all values of x between x_n and x_0: we shall hereby avoid difficulties as to discontinuity and infinite values of functions. Thus the simple problem of the Integral Calculus is to find the sum of an infinite number of infinitesimal elements as x increases by infinitesimal increments from x_0 to x_n; and it takes the following form;

Let $x_n - x_0$ be divided into n infinitesimal parts, and let x_1, $x_2, \ldots x_{n-1}$ be the values of x corresponding to the points of partition. Let S be the definite integral; then observing that $f(x)$ is multiplied by the increment immediately succeeding x, we have

$$S = f(x_0)(x_1 - x_0) + f(x_1)(x_2 - x_1) + \ldots + f(x_{n-1})(x_n - x_{n-1}); \quad (1)$$

and we have to find the value of S in terms of x_0 and x_n.

2.] Let us look at the theory from another point of view, and consider the genesis or origin of element-functions as it is presented to us in the Differential Calculus.

Let us take the following problem; let x be the length of a line OP (see Fig. 1) which varies continuously from $\text{OP}_0 = x_0$ up to $\text{OP}_n = x_n$; on OP_0, OP, OP_n let squares be described, viz. OR_0, OR, OR_n, so that $\text{OR} = x^2$; let OP be increased by an infinitesimal $\text{PQ} = dx$, and on OQ let a square be described; then the increase of x^2 due to the infinitesimal increase of x is $2x\,dx$: suppose a similar process of augmentation to be performed on all values of x from x_0 up to x_n; the effect of this will be that the square x_0^2 will grow into the square x_n^2 by infinitesimal augments, each of which is of the form $2x\,dx$, wherein x receives the successively-increased values. From another point of view however the effect of such a process is, to resolve the finite gnomonic area $\text{OR}_n - \text{OR}_0$ into infinitesimal elements, which are infinitesimal gnomons, each being of the form PR′s, which is expressed by $2x\,dx$; thus x^2 will be resolved into elements $2x\,dx$, corresponding to values of x from $x = x_0$ up to $x = x_n$; and if $x_0 = 0$, the whole square OR_n will be resolved into its gnomonic infinitesimal elements.

Or consider the following more general problem: Let $\text{F}(x)$ be a function of x finite and continuous for all values of x between x_n and x_0; and let the difference $x_n - x_0$ be divided into n equal and finite parts each of which is equal to Δx, so that $x_n - x_0 = n\Delta x$; then by equation (21), Art. 116, of Vol. I,

$$\left.\begin{aligned} F(x_0+\Delta x)-F(x_0) &= \Delta x\, F'(x_0+\theta\Delta x) \\ F(x_0+2\Delta x)-F(x_0+\Delta x) &= \Delta x\, F'(x_0+\Delta x+\theta\Delta x) \\ \cdots\cdots\cdots\cdots \\ F(x_0+n\Delta x)-F\{x_0+(n-1)\Delta x\} &= \Delta x\, F'\{x_0+(n-1)\Delta x+\theta\Delta x\}. \end{aligned}\right\} \quad (2)$$

Let Δx become infinitesimal, that is, become dx; then adding the members of (2), and bearing in mind that $x_n = x_0 + n\,dx$,

$$F(x_n)-F(x_0) = F'(x_0)dx + F'(x_0+dx)dx + F'(x_0+2dx)dx + \ldots \\ \ldots + F'(x_n - dx)dx; \quad (3)$$

that is, the process of growth by infinitesimal increase, on which principle equations (2) are constructed, is equivalent to the resolution of $F(x_n)-F(x_0)$ into infinitesimal elements, as exhibited in the right-hand member of equation (3).

Thus the Differential Calculus is a method by which a given finite function is resolved into its infinitesimal component elements; these being of such a nature, that the aggregate of an infinity of them is required to constitute the finite quantity. Or if the original function is an infinitesimal function, the elements into which it is resolved are infinitesimals of a higher order. The general form of all the elements is the same, as appears from the above examples, and therefore any one is a type of all, and expresses all: but the form of the typical element varies, as the function varies, the law of connexion depending on the process of Differentiation. The subject then on which Differentiation is performed is the function, and the result is the resolution of that function into its elements. The process of Differentiation is therefore one of *Disintegration.*

3.] In the Integral Calculus the data are changed; an infinitesimal element is given, which is the type of all; and the sum of these between certain given limits is to be determined. The process then is the reverse of differentiation, and is that of summation, as before observed: I propose to illustrate it at first by two or three simple examples.

Let us suppose the element to be $2x\,dx$; and the sum of all such to be required, as x continuously increases from x_0 to x_n; for the sake of simplicity, let $x_n - x_0$ be divided into n equal infinitesimal parts, each of which $= i$; so that

$$x_n - x_0 = ni; \qquad \therefore\; i = \frac{x_n - x_0}{n};$$

in the result, $n = \infty$; then the sum

$$= 2\left\{x_0 i + (x_0+i)i + (x_0+2i)i + \ldots + \{x_0+(n-1)i\}i\right\}$$
$$= 2\left\{nx_0 + \frac{n(n-1)}{2}i\right\}i$$
$$= 2\left\{nx_0 + \frac{n(n-1)}{2}\frac{x_n - x_0}{n}\right\}\frac{x_n - x_0}{n}$$
$$= 2\left\{x_0(x_n - x_0) + \left(1-\frac{1}{n}\right)\frac{(x_n-x_0)^2}{2}\right\}$$
$$= x_n^2 - x_0^2, \text{ when } n = \infty;$$

so that the definite integral of $2x\,dx$ is $x_n^2 - x_0^2$.

The problem given in the preceding Article exhibits the meaning of the process from a geometrical point of view. $2x\,dx$ expresses the infinitesimal gnomon contained between the squares described on OP and on OQ, and is consequently the infinitesimal element of x^2: and the sum of all such gnomons between the limits x_n and x_0 is evidently $x_n^2 - x_0^2$. If the inferior limit $= 0$, then the sum $= x_n^2$. This latter result is also true, whatever is the value of x_n; consequently if x_n has the general value x, x^2 is the indefinite integral of $2x\,dx$. As this mode of interpretation is important in giving body to our thoughts, let us take another example; in fig. 16, E is a point (x, y) in the plane of xy referred to the rectangular axes, Ox and Oy. H is a point $(x+dx, y+dy)$ infinitesimally near to E, so that EG $= dx$, EF $=$ GH $= dy$; consequently the infinitesimal rectangular area EH $= dx\,dy$. Now this rectangle is an infinitesimal element of the plane superficies; and is an element of a plane area bounded by any lines in the plane of xy. Thus the whole area will be the integral of these infinitesimal area-elements. If no limits of the area are given, the problem is evidently indefinite. Suppose however the area to be limited; and that it is required to express, say, the area $M_0P_0P_nM_n$ in terms of x_n and x_0, where $OM_0 = x_0$, $OM_n = x_n$, and the equation to the bounding curve is $y = f(x)$. We must find the sum of all the elements, similar to EH, contained between the two ordinates MP and NQ. Now as MP $= f(x)$, NQ $= f(x+dx)$, MN $= dx$; consequently MPQN $= \dfrac{dx}{2}\{f(x)+f(x+dx)\}$, which $= f(x)\,dx$, since dx is infinitesimal; and the whole area is the sum of all the similar elemental slices which are contained between the two ordinates M_0P_0 and M_nP_n. And to find the sum of $f(x)\,dx$ between these given limits is a problem of the Integral Calculus.

Thus the problem requires two processes of integration; (1) we must integrate dy, or, as we say, find the y-integral between the limits 0 and $f(x)$, the former value being the inferior, and the latter the superior limit of y; the result of this integration is evidently $f(x)$; (2) we must integrate $f(x)\,dx$, or, as we say, find the x-integral, from $x = x_0$ to $x = x_n$; the result of this double definite integration will give the area $M_0 P_0 P_n M_n$ in terms of x_0 and x_n.

It will be observed that in the last example the infinitesimal element is an infinitesimal of the second order; that after the y-integration it is an infinitesimal of the first order; and that the final integral, which expresses the area, is finite. This is in exact accordance with the geometrical quantities expressed by the several symbols.

4.] I return now to the general problem of integration of functions of one variable as expressed in equation (1); and to the creation of a convenient system of symbols.

Let S as heretofore represent the definite integral of $f(x)\,dx$ between the limits x_0 and x_n; and let $x_n - x_0$ be divided into n infinitesimal parts; and let $x_1, x_2, \dots x_{n-1}$ correspond to the $(n-1)$ points of partition; then by (1),

$$S = f(x_0)(x_1 - x_0) + f(x_1)(x_2 - x_1) + \dots + f(x_{n-1})(x_n - x_{n-1}). \quad (4)$$

Now on referring to Vol. I, Art. 18, and the mathematical definition of a derived function there given, it appears that if $F(x)$ is a function of x whose derived function is $f(x)$, and x_1 and x_0 are two values of x differing by an infinitesimal, $F(x)$ and $f(x)$ being finite and continuous for all the employed values of their subject-variables,

$$F(x_1) - F(x_0) = f(x_0)(x_1 - x_0). \quad (5)$$

Similarly, if $F(x)$ and $f(x)$ are finite and continuous for all the values of their subject-variables employed in the following equations,

$$\left.\begin{array}{l} F(x_2) - F(x_1) = f(x_1)(x_2 - x_1), \\ \dots\dots\dots\dots \\ F(x_n) - F(x_{n-1}) = f(x_{n-1})(x_n - x_{n-1}). \end{array}\right\} \quad (6)$$

Let all the right-hand members and all the left-hand members of these several equations be added; then by (4) we have

$$S = F(x_n) - F(x_0). \quad (7)$$

We require a symbol to express the relation of F to f; as d or D expresses the differential or infinitesimal element of a variable or of a function; so we employ $\int$ (a long S) to denote the general sum of an infinite number of terms, each of which is an infinitesimal. Thus if $f(x)\,dx$ is the type-element, the sum of an infinite number of which is to be determined, that sum is represented by $\int f(x)\,dx$. Also as thus far the limits of integration are not introduced, this symbol is used to represent the indefinite integral. And thus as d or D is the symbol of differentiation, so is $\int$ the symbol of integration.

The definite integral is conveniently expressed as follows: If x_n and x_0 are the limits of the Integral, x_n being the last or the *superior*, and x_0 the first or the *inferior* limit, then these symbols may be placed at the top and the bottom of $\int$ respectively; so that the definite integral thus determined is expressed by the symbol

$$\int_{x_0}^{x_n} f(x)\,dx.$$

Also since $\mathrm{F}(x)$ is that function whose derived function is $f(x)$, let us represent, as in the Differential Calculus, $f(x)$ by $\mathrm{F}'(x)$; so that in equation (4), S is equal to the sum of infinitesimal elements of which $\mathrm{F}'(x)\,dx$ is the type; and therefore (7) becomes

$$\int_{x_0}^{x_n} \mathrm{F}'(x)\,dx = \mathrm{F}(x_n) - \mathrm{F}(x_0). \tag{8}$$

5.] If the superior limit is x, x being a general value of the variable, subject to the condition that $\mathrm{F}'(x)$ is finite and continuous for all values of x between x_0 and x, then

$$\int_{x_0}^{x} \mathrm{F}'(x)\,dx = \mathrm{F}(x) - \mathrm{F}(x_0); \tag{9}$$

and omitting $\mathrm{F}(x_0)$, which is constant, the indefinite Integral of $\mathrm{F}'(x)\,dx$ is $\mathrm{F}(x)$; and we have

$$\int \mathrm{F}'(x)\,dx = \mathrm{F}(x). \tag{10}$$

Hence it follows that the definite integral of $\mathrm{F}'(x)\,dx$ between the limits x_n and x_0 is the value of the indefinite integral when $x = x_n$, less its value when $x = x_0$; on this account it is frequently and conveniently expressed as follows

$$\int_{x_0}^{x_n} \mathrm{F}'(x)\,dx = \Big[\mathrm{F}(x)\Big]_{x_0}^{x_n} = \mathrm{F}(x_n) - \mathrm{F}(x_0). \tag{11}$$

6.] Perhaps it may be supposed that the value of the definite integral depends on the number and magnitude of the elements $x_1 - x_0$, $x_2 - x_1$, ... $x_n - x_{n-1}$, or on the mode of partition of

$x_n - x_0$ into its parts; if the elements however are infinitesimal, and their number consequently infinite, whatever is the mode of partition, the value of the indefinite integral is the same, as may thus be shewn:

Whatever another mode is, we may consider it to be a subdivision of the first, and thus its elements to be parts of the former elements. Suppose then $x_1 - x_0$ to be divided into n parts, and $\xi_1, \xi_2, \xi_3, \dots \xi_{n-1}$ to be the values of x corresponding to the points of partition, and $\text{F}'(x)dx$ to be the infinitesimal element: then the sum of all the infinitesimal elements corresponding to the successive values of x between x_0 and x_1 is

$$\text{F}'(x_0)(\xi_1 - x_0) + \text{F}'(\xi_1)(\xi_2 - \xi_1) + \dots + \text{F}'(\xi_{n-1})(x_1 - \xi_{n-1});$$

the value of which is, by (4) and (7), $\text{F}(x_1) - \text{F}(x_0)$. And as analogous results are true for each of the other elements $x_2 - x_1$, $\dots x_n - x_{n-1}$, so will the sum be true; and therefore equation (11) is true, independently of the particular mode of partition by which the elements are formed.

Hence we have finally, subject to the condition that $\text{F}'(x)$ is finite and continuous within the limits x_n and x_0,

$$\int_{x_0}^{x_n} \text{F}'(x)\,dx = \text{F}'(x_0)(x_1 - x_0) + \text{F}'(x_1)(x_2 - x_1) + \dots$$
$$\dots + \text{F}'(x_{n-1})(x_n - x_{n-1}) \quad (12)$$
$$= \text{F}(x_1) - \text{F}(x_0) + \text{F}(x_2) - \text{F}(x_1) + \dots + \text{F}(x_n) - \text{F}(x_{n-1})$$
$$= \text{F}(x_n) - \text{F}(x_0); \quad (13)$$

the sum given in the right-hand member of (12) being expressed by either the left-hand member of (12) or by the right-hand member of (13). It will be observed that in the series (12), the terms do not go as far as $\text{F}'(x_n)$; in the definite integral therefore expressed by (13) the value of the element-function at the inferior limit is included, and that at the superior limit is excluded.

7.] To return to the consideration of the indefinite integral: by equation (10)

$$\int \text{F}'(x)\,dx = \text{F}(x);$$

that is, the operation symbolized by $\int dx$, performed on $\text{F}'(x)$, changes it into $\text{F}(x)$; but by the Differential Calculus $\frac{d}{dx}$ is the symbol of an operation which being performed on $\text{F}(x)$ changes it into $\text{F}'(x)$; therefore $\int dx$ and $\frac{d}{dx}$ are so related that one represents a process the reverse of that represented by the other; that is, according to the index law which the symbol $\frac{d}{dx}$ is subject to,

$$\int dx = \left(\frac{d}{dx}\right)^{-1} = d^{-1}dx; \tag{14}$$

$$\therefore \int = d^{-1}; \tag{15}$$

and $\int$ represents an operation which is the reverse of differentiation *.

Hence also $\int$ and d are symbols of operations which destroy each other; and consequently

$$\int d = d\int = d^0 = \int^0 = 1;$$

unity being used as a symbol of an operation which operating on a function leaves it unaltered.

Hence according to the notation of derived functions,

$$\int F'(x)\,dx = F(x),$$

$$\int F''(x)\,dx = F'(x),$$

$$\ldots\ldots\ldots$$

$$\int F^n(x)\,dx = F^{n-1}(x);$$

and in this mode of viewing the subject, the symbol $\int dx$ must be considered as a complex character, indicative of a certain analytical process to be performed on a certain function; the analytical process being the reverse of Derivation.

Hence the problem of Integration resolves itself into this; viz. to determine the function which, when differentiated, produces the infinitesimal element expressing the general term of the series; and therefore as this is a process the reverse of Differentiation, we may make use of our knowledge of the Differential Calculus, and as far as possible invert its rules; for these will thereby become those of the Integral Calculus; such processes we shall enter on in the next Chapter, and thereby obtain indefinite integrals, from which definite integrals may be deduced by means of equation (11).

In this aspect of the Calculus another point requires explana-

* See Vol. I. Art. 420.

tion. Since an arbitrary constant connected with a function of x by addition or subtraction disappears in Differentiation, so in the reverse process such a constant may be introduced; and thereby we have

$$\int F'(x)\,dx = F(x) + C.$$

The same result also follows from equation (9), wherein $-F(x_0)$ being independent of x is constant with respect to it. But as the Integral Calculus might exist previously to the Differential, for the infinitesimal element may exist previously to and independently of the finite function, so its principles ought to have an independent basis. We shall therefore in the first place investigate certain properties of definite integrals, which will be required in the sequel, and also integrate from first principles some infinitesimal elements.

8.] THEOREM I.—If an infinitesimal element has a constant quantity as a factor, the definite integral will also have the same constant factor.

Let the infinitesimal element be $a\,F'(x)\,dx$, wherein a is a constant quantity; then

$$\begin{aligned}\int_{x_0}^{x_n} a\,F'(x)\,dx &= a\,F(x_n) - a\,F(x_0),\\ &= a\,\{F(x_n) - F(x_0)\},\\ &= a\int_{x_0}^{x_n} F'(x)\,dx. \qquad (16)\end{aligned}$$

A constant factor therefore may be taken outside the sign of integration; and similarly may, if required, be removed from without to within the sign of Integration.

The following are particular cases:

$$\int_{x_0}^{x_n} -F'(x)\,dx = -\int_{x_0}^{x_n} F'(x)\,dx. \qquad (17)$$

$$\int_{x_0}^{x_n} \frac{F'(x)\,dx}{C} = \frac{1}{C}\int_{x_0}^{x_n} F'(x)\,dx. \qquad (18)$$

The same theorem is of course true of an indefinite integral.

THEOREM II.—The integral of the algebraic sum of any number of infinitesimal elements is equal to the algebraic sum of the integrals of the same elements.

Let $F'(x)\,dx, f'(x)\,dx, \phi'(x)\,dx, \ldots$ be any infinitesimal elements finite and continuous between the limits x_n and x_0; then

$$\int_{x_0}^{x_n}\{\mathrm{F}'(x)\pm f'(x)\pm\phi'(x)\pm\ldots\}\,dx = \Big[\mathrm{F}(x)\pm f(x)\pm\phi(x)\pm\ldots\Big]_{x_0}^{x_n}$$
$$=\{\mathrm{F}(x_n)-\mathrm{F}(x_0)\}\pm\{f(x_n)-f(x_0)\}\pm\{\phi(x_n)-\phi(x_0)\}\pm\ldots$$
$$=\int_{x_0}^{x_n}\mathrm{F}'(x)\,dx\pm\int_{x_0}^{x_n}f'(x)\,dx\pm\int_{x_0}^{x_n}\phi'(x)\,dx\pm\ldots \quad (19)$$

The same theorem is also true of indefinite integrals.

Hence, and by means of the former theorem,

$$\int_{x_0}^{x_n}\{\mathrm{F}'(x)+\sqrt{-1}f'(x)\}\,dx = \int_{x_0}^{x_n}\mathrm{F}'(x)\,dx+\sqrt{-1}\int_{x_0}^{x_n}f'(x)\,dx. \quad (20)$$

THEOREM III.—If the infinitesimal element is of the form $f(x)\times\mathrm{F}'(x)\,dx$, then

$$\int_{x_0}^{x_n}f(x)\times\mathrm{F}'(x)\,dx = \Big[f(x)\times\mathrm{F}(x)\Big]_{x_0}^{x_n}-\int_{x_0}^{x_n}\mathrm{F}(x)\times f'(x)\,dx.$$

For convenience of notation, let $f(x)=u$, $\mathrm{F}(x)=v$, $\mathrm{F}'(x)\,dx = dv$, and let $u_0, u_1, u_2 \ldots u_n, v_0, v_1, v_2, \ldots v_n$ be the several values of u and v corresponding to $x_0, x_1, x_2, \ldots x_n$ respectively; then

$$\int_{x_0}^{x_n}u\,dv = u_0(v_1-v_0)+u_1(v_2-v_1)+u_2(v_3-v_2)+\ldots+u_{n-1}(v_n-v_{n-1})$$
$$=-u_0v_0-v_1(u_1-u_0)-v_2(u_2-u_1)-\ldots-v_n(u_n-u_{n-1})+u_nv_n$$
$$=u_nv_n-u_0v_0-\{v_1(u_1-u_0)+v_2(u_2-u_1)+\ldots+v_n(u_n-u_{n-1})\};$$

and since the differences between $v_0, v_1, v_2, \ldots v_n$ are infinitesimal, as are also the differences between $u_0, u_1, \ldots u_n$, if we take i to be the *general* symbol of an infinitesimal, we have

$$\begin{aligned} v_1 &= v_0+i_1, & u_1 &= u_0+i'_1,\\ v_2 &= v_1+i_2, & u_2 &= u_1+i'_2,\\ &\cdots & &\cdots\\ v_n &= v_{n-1}+i_n; & u_n &= u_{n-1}+i'_n; \end{aligned}$$

so that

$$\int_{x_0}^{x_n}u\,dv = u_nv_n-u_0v_0-\{v_0(u_1-u_0)+v_1(u_2-u_1)+\ldots+v_{n-1}(u_n-u_{n-1})\}$$
$$-\{i_1(u_1-u_0)+i_2(u_2-u_1)+\ldots+i_n(u_n-u_{n-1})\};$$

the last term of which equality must be neglected, because it contains infinitesimals of a higher order than those of the preceding term, and the preceding term is $\int_{x_0}^{x_n}v\,du$; therefore

$$\int_{x_0}^{x_n}u\,dv = u_nv_n-u_0v_0-\int_{x_0}^{x_n}v\,du \quad (21)$$
$$=\Big[uv\Big]_{x_0}^{x_n}-\int_{x_0}^{x_n}v\,du. \quad (22)$$

And therefore resubstituting

$$\int_{x_0}^{x_n} f(x) \times F'(x)\,dx = \Big[f(x) \times F(x)\Big]_{x_0}^{x_n} - \int_{x_0}^{x_n} F(x) \times f'(x)\,dx. \quad (23)$$

This theorem is known by the name of integration by parts, and is of very frequent use; the form which it assumes in the case of an indefinite integral is

$$\int f(x) \times F'(x)\,dx = f(x) \times F(x) - \int F(x) \times f'(x)\,dx; \quad (24)$$

or if u and v are two functions of x, then

$$\int u\,dv = uv - \int v\,du. \quad (25)$$

THEOREM IV.—If, in order to determine the Integral of $F'(x)\,dx$, it is convenient to introduce another variable z, related to x by the equation $z = \phi(x)$, so that

$$x = f(z),\ dx = f'(z)\,dz,\ F'(x) = F'(f(z));$$

and if z_n and z_0 are the values of z corresponding to x_n and x_0, then

$$\int_{x_0}^{x_n} F'(x)\,dx = \int_{z_0}^{z_n} F'(f(z))f'(z)\,dz;$$

that is, for the definite integral determined by means of x, we may take as its equivalent the other definite integral determined by means of z.

Let $z_1, z_2, \ldots z_{n-1}$ be the values of z corresponding to $x_1, x_2, \ldots x_{n-1}$; then, the elements of x being infinitesimal, we have

$$x_1 - x_0 = f(z_1) - f(z_0) = f'(z_0)(z_1 - z_0),$$

$$x_2 - x_1 = \;.\;.\;.\;.\;.\; = f'(z_1)(z_2 - z_1),$$

$$.\;.\;.\;.\;.\;.\;.\;.\;.\;.\;.\;.\;.\;.$$

$$x_n - x_{n-1} = f(z_n) - f(z_{n-1}) = f'(z_{n-1})(z_n - z_{n-1}).$$

$$\begin{aligned}\int_{x_0}^{x_n} F'(x)\,dx &= (x_1 - x_0)F'(x_0) + (x_2 - x_1)F'(x_1) + \ldots + (x_n - x_{n-1})F'(x_{n-1}) \\ &= F'(f(z_0))f'(z_0)(z_1 - z_0) + F'(f(z_1))f'(z_1)(z_2 - z_1) + \ldots \\ &\qquad + F'(f(z_{n-1}))f'(z_{n-1})(z_n - z_{n-1}) \\ &= \int_{z_0}^{z_n} F'(f(z))f'(z)\,dz; \end{aligned} \quad (26)$$

that is, the two definite integrals are equal; and the latter therefore may be used for the former; and *vice versâ*.

This method is called Integration by substitution, and is of course true for the indefinite integrals as far as variable quantities enter into the functions; the arbitrary constants will however frequently assume different although equivalent forms.

Other Theorems on definite integrals we shall reserve to Chapter IV.

9.] The determination of definite integrals from first principles.

Ex. 1. To determine $\int_{x_0}^{x_n} dx$.

Let, as in Art. 4, $x_n - x_0$ be divided into n infinitesimal parts, and let $x_1, x_2, \ldots x_{n-1}$ be the values of x corresponding to the points of partition; then

$$\int_{x_0}^{x_n} dx = x_1 - x_0 + x_2 - x_1 + x_3 - x_2 + \ldots + x_n - x_{n-1}$$
$$= x_n - x_0.$$

Ex. 2. To determine $\int_{x_0}^{x_n} x^a\, dx$.

Let, as heretofore, $x_1, x_2, x_3, \ldots x_{n-1}$ be the values of x corresponding to the $(n-1)$ points of partition of $x_n - x_0$; and as the mode of partition does not affect the result provided that the elements are infinitesimal, let us suppose that $x_0, x_1, \ldots x_n$ form a geometrical progression whose common ratio differs infinitesimally from unity: that is, let

$$x_1 = x_0(1+i), \qquad \therefore \quad x_1 - x_0 = x_0 i;$$
$$x_2 = x_1(1+i), \qquad x_2 - x_1 = x_1 i;$$
$$\cdots \qquad \cdots$$
$$x_n = x_{n-1}(1+i), \qquad x_n - x_{n-1} = x_{n-1} i;$$

wherein i is infinitesimal; then we have

$$(1+i)^n = \frac{x_n}{x_0}; \text{ thus}$$

$$\int_{x_0}^{x_n} x^a\, dx = x_0^a(x_1 - x_0) + x_1^a(x_2 - x_1) + \ldots + x_{n-1}^a(x_n - x_{n-1})$$
$$= x_0^{a+1} i + x_1^{a+1} i + \ldots + x_{n-1}^{a+1} i$$
$$= x_0^{a+1}\{1 + (1+i)^{a+1} + (1+i)^{2(a+1)} + \ldots$$
$$+ (1+i)^{(n-1)(a+1)}\} i$$
$$= x_0^{a+1} \frac{(1+i)^{n(a+1)} - 1}{(1+i)^{a+1} - 1} i$$
$$= x_0^{a+1} \left\{ \left(\frac{x_n}{x_0}\right)^{a+1} - 1 \right\} \frac{i}{(1+i)^{a+1} - 1}$$
$$= \frac{x_n^{a+1} - x_0^{a+1}}{a+1};$$

for if i is infinitesimal, $(1+i)^{a+1} - 1 = (a+1)i$.

Ex. 3. Determine $\int_{x_0}^{x_n} \frac{dx}{x}$.

Let $x_n - x_0$ be divided into infinitesimal elements, as in the last example; then

$$\int_{x_0}^{x_n} \frac{dx}{x} = \frac{x_1 - x_0}{x_0} + \frac{x_2 - x_1}{x_1} + \dots + \frac{x_n - x_{n-1}}{x_{n-1}}$$

$$= i + i + \dots + i$$

$$= ni;$$

and since $\frac{x_n}{x_0} = (1+i)^n$, $n \log(1+i) = \log x_n - \log x_0$;

$$\therefore \int_{x_0}^{x_n} \frac{dx}{x} = (\log x_n - \log x_0) \frac{i}{\log(1+i)},$$

$$= \log x_n - \log x_0;$$

since $\log(1+i) = i$, when i is infinitesimal.

Ex. 4. Determine $\int_{x_0}^{x_n} a^x dx$.

Let $x_n - x_0$ be divided into n equal parts, each of which is equal to i; so that

$$x_n - x_0 = ni; \text{ then}$$

$$\int_{x_0}^{x_n} a^x dx = a^{x_0} i + a^{x_0+i} i + \dots + a^{x_0+(n-1)i} i$$

$$= a^{x_0} i \{1 + a^i + a^{2i} + \dots + a^{(n-1)i}\}$$

$$= a^{x_0} i \frac{a^{ni} - 1}{a^i - 1}$$

$$= a^{x_0} (a^{x_n - x_0} - 1) \frac{i}{a^i - 1}$$

$$= \frac{a^{x_n} - a^{x_0}}{\log a};$$

since $a^i - 1 = i \log a$, when i is infinitesimal.

Hence we have

$$\int_{x_0}^{x_n} e^x dx = e^{x_n} - e^{x_0}.$$

Ex. 5. Determine $\int_{x_0}^{x_n} \cos x \, dx$.

Let $x_n - x_0$ be divided into n equal elements, each of which is equal to i; so that $x_n - x_0 = ni$;

$$\int_{x_0}^{x_n} \cos x \, dx = i \cos x_0 + i \cos(x_0 + i) + \dots + i \cos\{x_0 + (n-1)i\}$$

$$= i \left\{ \cos x_0 + \cos(x_0 + i) + \dots + \cos\{x_0 + (n-1)i\} \right\}$$

$$\int_{x_0}^{x_n} \cos x\,dx = i\,\frac{\cos\left(x_0+\frac{n-1}{2}i\right)\sin\frac{ni}{2}}{\sin\frac{i}{2}}$$

$$= i\,\frac{\cos\left(x_0+\frac{n-1}{2}\,\frac{x_n-x_0}{n}\right)\sin\frac{x_n-x_0}{2}}{\sin\frac{i}{2}};$$

and therefore, if $n = \infty$, and i is infinitesimal,

$$\int_{x_0}^{x_n} \cos x\,dx = 2\cos\frac{x_n+x_0}{2}\sin\frac{x_n-x_0}{2}$$

$$= \sin x_n - \sin x_0.$$

The preceding examples have given us the following indefinite Integrals; c being an arbitrary constant in each case;

$$\int dx = x+c;$$

$$\int x^a\,dx = \frac{x^{a+1}}{a+1}+c;$$

$$\int \frac{dx}{x} = \log x + c;$$

$$\int a^x\,dx = \frac{a^x}{\log a}+c;$$

$$\int e^x\,dx = e^x + c;$$

$$\int \cos x\,dx = \sin x + c.$$

CHAPTER II.

CONSTRUCTION OF RULES FOR INTEGRATION OF ALGEBRAICAL FUNCTIONS.

10.] We propose in the present and next Chapters to construct the rules of the Integral by inverting those of the Differential Calculus: and first we shall from this point of view exhibit the forms of indefinite integrals which correspond to those theorems on definite integrals which have been proved in Article 8.

Theorem I.—Since $d.a\text{F}(x) = a\text{F}'(x)dx$;

$$\therefore \int a\text{F}'(x)dx = a\text{F}(x),$$

$$= a\int \text{F}'(x)dx; \qquad (1)$$

that is, in the integration of an infinitesimal element one of whose factors is a constant, the constant may be placed outside the sign of integration.

Theorem II.—Since

$$d.\{\text{F}(x) \pm f(x) \pm \dots\} = d.\text{F}(x) \pm d.f(x) \pm \dots$$

$$= \text{F}'(x)dx \pm f'(x)dx \pm \dots;$$

$$\therefore \int \{\text{F}'(x)dx \pm f'(x)dx \pm \dots\} = \text{F}(x) \pm f(x) \pm \dots \qquad (2)$$

that is, the integral of an algebraic sum of infinitesimal elements is equal to the sum of the integrals of the several infinitesimal elements.

$$\therefore \int \{a\text{F}'(x)dx \pm cf'(x)dx\} = a\text{F}(x) \pm cf(x).$$

$$\int \{\text{F}'(x) + \sqrt{-1}f'(x)\}dx = \text{F}(x) + \sqrt{-1}\,f(x).$$

Theorem III.—Since

$$d.\text{F}(x) \times f(x) = f(x) \times \text{F}'(x)dx + \text{F}(x) \times f'(x)dx;$$

$$f(x) \times \text{F}'(x)dx = d.(\text{F}(x) \times f(x)) - \text{F}(x) \times f'(x)dx;$$

$$\therefore \int f(x) \times \text{F}'(x)dx = \text{F}(x) \times f(x) - \int \text{F}(x) \times f'(x)dx; \qquad (3)$$

and therefore if $f(x) =$ [illegible]

$$\int [illegible] = [illegible]$$ [illegible]

THEOREM IV.—Since [illegible]

$$\therefore \int [illegible] = [illegible]$$ [illegible]

which is the theorem in the integral calculus [illegible] to that of the differentiation of [illegible] functions [illegible] Art. 31 of Vol. I. and which may [illegible] prove.

Let $f(x) = x$; [illegible]

$$\therefore \int [illegible] = [illegible]$$
$$= [illegible]$$
$$= [illegible]$$ [illegible]

Integration by this last process [illegible] to the integration of a compound function [illegible] for hence it follows that [illegible] true for x and simple functions of [illegible] functions.

SECTION I.—*Integration of Fundamental Algebraical Functions.*

11.] Integration of $x^n\,dx$.

Since $\qquad d.x^m = m x^{m-1}\,dx$;

$$\therefore \int m x^{m-1}\,dx = x^m;$$

$$\int x^{m-1}\,dx = \frac{x^m}{m}.$$

Let n be substituted for $m-1$; then $m = n+1$; and

$$\int x^n\,dx = \frac{x^{n+1}}{n+1}. \qquad (7)$$

Therefore to integrate $x^n\,dx$, add unity to the index, divide by the index so increased, and by dx.

Of this result the following are particular cases:

(1.) Let n be negative; that is, for n substitute $-n$.

$$\int x^{-n}\,dx = \int \frac{dx}{x^n} = -\frac{x^{-n+1}}{n-1} = -\frac{1}{(n-1)x^{n-1}}.$$

(2.) Let n be fractional, $n = \frac{p}{q}$;

$$\int x^{\frac{p}{q}}\,dx = \frac{q}{p+q}\,x^{\frac{p+q}{q}};$$

therefore
$$\int \frac{dx}{x^2} = -\frac{1}{x};$$

$$\int \frac{dx}{x^3} = -\frac{1}{2x^2};$$

$$\int x^{\frac{1}{2}}\,dx = \frac{2}{3}x^{\frac{3}{2}};$$

$$\int \frac{dx}{x^{\frac{1}{2}}} = 2x^{\frac{1}{2}}.$$

12.] The formula (7) is true for all integral and fractional, positive and negative, values of n, with the exception of, $n = -1$; in which case the right-hand member becomes ∞, and the formula ceases to give an intelligible result: we must therefore return to the principles of definite integration, for they are exact, and by means of them obtain the correct integral.

$$\int_{x_0}^{x} x^n\,dx = \left[\frac{x^{n+1}}{n+1}\right]_{x_0}^{x} = \frac{x^{n+1}-x_0^{\,n+1}}{n+1} = \frac{0}{0}, \text{ when } n = -1;$$

evaluating therefore the indeterminate fraction by the rules of Chapter V, Vol. I; and observing that n is the variable,

$$\int_{x_0}^{x} \frac{dx}{x} = \frac{\log x \,.\, x^{n+1} - \log x_0 \,.\, x_0^{\,n+1}}{1}$$

$$= \log x - \log x_0, \text{ when } n = -1,$$

$$= \log\left(\frac{x}{x_0}\right); \tag{8}$$

a result identical with that of Ex. 3, Art. 9; and therefore

$$\int \frac{dx}{x} = \log x. \tag{9}$$

13.] Extending the results of Art. 11 and 12 to Compound Functions, as we are authorized to do by Theorem IV, Art. 10, we have

$$\int \{f(x)\}^n f'(x)\,dx = \frac{\{f(x)\}^{n+1}}{n+1}; \tag{10}$$

$$\int \frac{f'(x)\,dx}{f(x)} = \log\{f(x)\}. \tag{11}$$

Hence the integral of a fraction, whose numerator is the differential of the denominator, is the Napierian logarithm of the denominator.

Ex. 1. $$\int (a+bx)^n dx = \frac{1}{b}\int (a+bx)^n d(a+bx)$$
$$= \frac{(a+bx)^{n+1}}{b(n+1)}.$$

Ex. 2. $$\int (a+bx^2)^n x\,dx = \frac{1}{2b}\int (a+bx^2)^n d(a+bx^2)$$
$$= \frac{(a+bx^2)^{n+1}}{2b(n+1)}.$$

Ex. 3. $$\int (a^m-x^m)^n x^{m-1} dx = -\frac{1}{m}\int (a^m-x^m)^n d(a^m-x^m)$$
$$= -\frac{(a^m-x^m)^{n+1}}{m(n+1)}.$$

Ex. 4. $$\int (a+bx+cx^2)^n (b+2cx)\,dx$$
$$= \int (a+bx+cx^2)^n d(a+bx+cx^2)$$
$$= \frac{(a+bx+cx^2)^{n+1}}{n+1}.$$

Ex. 5. $$\int \frac{x^2 dx}{(a^3-x^3)^{\frac{1}{2}}} = -\frac{1}{3}\int (a^3-x^3)^{-\frac{1}{2}} d(a^3-x^3)$$
$$= -\frac{2}{3}(a^3-x^3)^{\frac{1}{2}}.$$

Ex. 6. $$\int \frac{dx}{a+bx} = \frac{1}{b}\int \frac{d(a+bx)}{a+bx} = \frac{1}{b}\log(a+bx).$$

Ex. 7. $$\int \frac{(b+2cx)\,dx}{a+bx+cx^2} = \int \frac{d(a+bx+cx^2)}{a+bx+cx^2} = \log(a+bx+cx^2).$$

Ex. 8. $$\int \frac{x^2 dx}{a^3-x^3} = -\frac{1}{3}\int \frac{d(a^3-x^3)}{a^3-x^3} = -\frac{1}{3}\log(a^3-x^3).$$

Ex. 9. $$\int \frac{1-x^n}{1-x}dx = \int \{1+x+x^2+\ldots+x^{n-1}\}\,dx$$
$$= x+\frac{x^2}{2}+\ldots\ldots+\frac{x^n}{n}.$$

14.] Integration of $\frac{dx}{a^2+x^2}$, and of similar forms.

Since $$d.\tan^{-1}\frac{x}{a} = \frac{a\,dx}{a^2+x^2};$$
$$\therefore\quad \tan^{-1}\frac{x}{a} = \int \frac{a\,dx}{a^2+x^2};$$
$$\int \frac{dx}{a^2+x^2} = \frac{1}{a}\tan^{-1}\frac{x}{a}. \qquad (12)$$

And therefore by Theorem IV, Art. 10,

$$\int \frac{f'(x)\,dx}{a^2 + \{f(x)\}^2} = \frac{1}{a} \tan^{-1} \frac{f(x)}{a}. \tag{13}$$

The following is a general form, which admits of being reduced to (12), when the roots of the denominator are impossible:

$$\int \frac{dx}{a + bx + cx^2} = \frac{1}{c} \int \frac{dx}{\frac{a}{c} + \frac{b}{c}x + x^2}$$

$$= \frac{1}{c} \int \frac{d\left(x + \frac{b}{2c}\right)}{\frac{4ac - b^2}{4c^2} + \left(x + \frac{b}{2c}\right)^2}$$

$$= \frac{1}{c} \frac{2c}{(4ac - b^2)^{\frac{1}{2}}} \tan^{-1} \frac{2cx + b}{(4ac - b^2)^{\frac{1}{2}}}$$

$$= \frac{2}{(4ac - b^2)^{\frac{1}{2}}} \tan^{-1} \frac{2cx + b}{(4ac - b^2)^{\frac{1}{2}}}, \tag{14}$$

since $4ac - b^2$ is positive, when the roots of the denominator are impossible.

[redacted]

therefore $$\int \frac{dx}{x^2-a^2} = \frac{1}{2a}\left\{\int \frac{dx}{x-a} - \int \frac{dx}{x+a}\right\}$$

$$= \frac{1}{2a}\left\{\int \frac{d(x-a)}{x-a} - \int \frac{d(x+a)}{x+a}\right\}$$

$$= \frac{1}{2a}\left\{\log(x-a) - \log(x+a)\right\}$$

$$= \frac{1}{2a}\log\frac{x-a}{x+a}. \tag{15}$$

Hence, by virtue of Theorem IV, Art. 10,

$$\int \frac{f'(x)\,dx}{\{f(x)\}^2 - a^2} = \frac{1}{2a}\log\frac{f(x)-a}{f(x)+a}.$$

Again, since $\frac{dx}{a^2-x^2} = \frac{1}{2a}\left\{\frac{dx}{a+x} + \frac{dx}{a-x}\right\}$;

$$\therefore \int \frac{dx}{a^2-x^2} = \frac{1}{2a}\left\{\int \frac{dx}{a+x} + \int \frac{dx}{a-x}\right\}$$

$$= \frac{1}{2a}\left\{\int \frac{d(a+x)}{a+x} - \int \frac{d(a-x)}{a-x}\right\}$$

$$= \frac{1}{2a}\left\{\log(a+x) - \log(a-x)\right\}$$

$$= \frac{1}{2a}\log\frac{a+x}{a-x}. \tag{16}$$

Also when the roots of the denominators of the following form are real and unequal it admits of being reduced to one or other of the above forms:

$$\int \frac{dx}{a+bx+cx^2} = \frac{1}{c}\int \frac{d\left(x+\frac{b}{2c}\right)}{\left(x+\frac{b}{2c}\right)^2 - \frac{b^2-4ac}{4c^2}}$$

$$= \frac{1}{(b^2-4ac)^{\frac{1}{2}}}\log\frac{2cx+b-(b^2-4ac)^{\frac{1}{2}}}{2cx+b+(b^2-4ac)^{\frac{1}{2}}}. \tag{17}$$

Ex. 1. $$\int \frac{dx}{a-bx^2} = \frac{1}{b}\int \frac{dx}{\frac{a}{b}-x^2}$$

$$= \frac{1}{2(ab)^{\frac{1}{2}}}\log\frac{a^{\frac{1}{2}}+xb^{\frac{1}{2}}}{a^{\frac{1}{2}}-xb^{\frac{1}{2}}}.$$

Ex. 2. $\int \frac{dx}{1-2x-x^2} = \int \frac{dx}{2-(x+1)^2}$

$$= \int \frac{d(x+1)}{2-(x+1)^2}$$

$$= \frac{1}{2^{\frac{3}{2}}} \log \frac{2^{\frac{1}{2}}+x+1}{2^{\frac{1}{2}}-x-1}.$$

A comparison of (15) and (16) leads to a result which requires a few words of explanation. For since $a^2-x^2 = -(x^2-a^2)$;

$$\therefore \int \frac{dx}{a^2-x^2} = -\int \frac{dx}{x^2-a^2}$$

$$= -\frac{1}{2a} \log \frac{x-a}{x+a}, \text{ by (15)},$$

$$= \frac{1}{2a} \log \frac{x+a}{x-a};$$

whereas the value of the left-hand member, which is given by (16), is $\frac{1}{2a} \log \frac{a+x}{a-x}$. Thus the subjects of the Logarithm in the two cases are the same quantity affected with different signs; consequently the two integrals differ by only $\log(-1)$, which, by Art. 67, Vol. I, $= \pi\sqrt{-1}$, and is a constant; and either form of the indefinite integral may be taken. If however the integral is definite, both expressions give the same result; for suppose the limits to be x_n and x_0; then either form gives

$$\frac{1}{2a} \log \frac{(x_n+a)(x_0-a)}{(x_n-a)(x_0+a)},$$

and the results are identical. In the exact process of definite integration then the apparent inconsistency disappears.

16.] Integration of $\frac{x^n dx}{(a+bx)^m}$, and of $\frac{dx}{x^n(a+bx)^m}$, m and n being integers.

To integrate $\frac{x^n dx}{(a+bx)^m}$.

$$\text{Let } a+bx = z; \quad \therefore \quad x = \frac{z-a}{b}.$$

$$\therefore \quad b\,dx = dz; \qquad dx = \frac{dz}{b}.$$

$$\therefore \int \frac{x^n dx}{(a+bx)^m} = \int \left(\frac{z-a}{b}\right)^n \frac{dz}{bz^m}$$

$$= \frac{1}{b^{n+1}} \int \frac{(z-a)^n}{z^m} dz. \qquad (18)$$

to integrate which, $(z-a)^n$ must be expanded by the Binomial Theorem, and each term of the expansion, having been divided by z^m, must be integrated separately; and the substitution of $a+bx$ for z made subsequently.

To integrate $\dfrac{dx}{x^n(a+bx)^m}$.

Let $x = \dfrac{1}{z}$; $\quad\therefore\quad dx = -\dfrac{dz}{z^2}$; whereby we have

$$\int \frac{dx}{x^n(a+bx)^m} = -\int \frac{z^{n+m-2}\,dz}{(b+az)^m};$$

which is of the form (18), and may be integrated accordingly.

Ex. 1. $\displaystyle\int \frac{x^2\,dx}{(a+bx)^3}$.

Let $a+bx = z$; $\qquad \therefore\quad x = \dfrac{z-a}{b}$.

$bdx = dz$; $\qquad dx = \dfrac{dz}{b}$.

$$\begin{aligned}
\therefore\quad \int \frac{x^2\,dx}{(a+bx)^3} &= \int \frac{(z-a)^2}{b^2}\,\frac{dz}{bz^3} \\
&= \frac{1}{b^3}\int \frac{(z-a)^2\,dz}{z^3} \\
&= \frac{1}{b^3}\int \frac{z^2-2az+a^2}{z^3}\,dz \\
&= \frac{1}{b^3}\int \left\{\frac{1}{z}-\frac{2a}{z^2}+\frac{a^2}{z^3}\right\} dz \\
&= \frac{1}{b^3}\left\{\log z+\frac{2a}{z}-\frac{a^2}{2z^2}\right\} \\
&= \frac{1}{b^3}\left\{\log(a+bx)+\frac{3a^2+4abx}{2(a+bx)^2}\right\}.
\end{aligned}$$

Ex. 2. $\displaystyle\int \frac{dx}{x^3(a+bx)^2}$.

Let $x = \dfrac{1}{z}$; $\qquad \therefore\quad dx = -\dfrac{dz}{z^2}$.

$$\int \frac{dx}{x^3(a+bx)^2} = -\int \frac{z^3\,dz}{(b+az)^2}.$$

Let $b+az = y$; $\qquad \therefore\quad z = \dfrac{y-b}{a}$;

$$dz = \frac{dy}{a}.$$

$$\therefore \int \frac{dx}{x^2(a+bx)^2} = -\int \frac{(y-b)^2}{a^3}\frac{dy}{y^2}$$

$$= -\frac{1}{a^3}\int \frac{(y-b)^2}{y^2}\,dy$$

$$= -\frac{1}{a^3}\int \left\{1 - \frac{2b}{y} + \frac{b^2}{y^2}\right\} dy$$

$$= -\frac{1}{a^3}\left\{y - 2b\log y - \frac{b^2}{y}\right\}$$

$$= \frac{2b}{a^3}\log\frac{a+bx}{x} - \frac{a+2bx}{a^2x(a+bx)}.$$

17.] Examples of the preceding methods of integration.

Ex. 1. $\int \left(a - \frac{b}{x^3} + cx^{\frac{3}{2}}\right) dx = ax + \frac{b}{2x^2} + \frac{2c}{5}x^{\frac{5}{2}}.$

Ex. 2. $\int (1+x)(1-x^2)x\,dx = \frac{x^2}{2} + \frac{x^3}{3} - \frac{x^4}{4} - \frac{x^5}{5}.$

Ex. 3. $\int \frac{x^4\,dx}{x^2+1} = \frac{x^3}{3} - x + \tan^{-1}x.$

Ex. 4. $\int \frac{x^{n-1}\,dx}{(a+bx^n)^n} = \frac{-1}{n(n-1)b}\frac{1}{(a+bx^n)^{n-1}}.$

Ex. 5. $\int \frac{(a-x)\,dx}{(2ax-x^2)^{\frac{1}{2}}} = (2ax-x^2)^{\frac{1}{2}}.$

Ex. 6. $\int \frac{5x^3\,dx}{3x^4+7} = \frac{5}{12}\log(3x^4+7).$

Ex. 7. $\int \frac{dx}{1+x+x^2} = \frac{2}{3^{\frac{1}{2}}}\tan^{-1}\frac{2x+1}{3^{\frac{1}{2}}}.$

Ex. 8. $\int \frac{(m+nx)\,dx}{a+bx+cx^2} = \frac{n}{2c}\log(a+bx+cx^2)$

$$+ \frac{2mc-nb}{2c}\int \frac{dx}{a+bx+cx^2}.$$

Section 2.—*Integration of Rational Fractions by decomposition into partial Fractions, and by formulæ of reduction.*

18.] A rational fraction is of the form

$$\frac{q_0x^m - q_1x^{m-1} + q_2x^{m-2} - \dots\dots \pm q_{m-1}x \mp q_m}{x^n - p_1x^{n-1} + \dots\dots\dots\dots \pm p_{n-1}x \mp p_n}, \tag{19}$$

the numerator and denominator being algebraical expressions involving only positive and integral powers of x, and $q_0, q_1, \dots q_m, p_1, p_2, \dots p_n$ being constants.

When m is greater than, or equal to n, (19) may by common division be reduced to the sum of an integral algebraical expression, and of a fraction whose denominator will be the same as that of (19), and whose numerator will be of dimensions less by at least unity than the denominator. The integral part may be integrated by the methods of the preceding section, and the fractional part by the method which we now proceed to explain.

The most general form therefore of such a function is

$$\frac{A_1 x^{n-1} - A_2 x^{n-2} + \dots \mp A_{n-1} x \pm A_n}{x^n - p_1 x^{n-1} + p_2 x^{n-2} - \dots \mp p_{n-1} x \pm p_n} \qquad (20)$$

which, for convenience of reference, we shall symbolize by

$$\frac{F(x)}{f(x)}. \qquad (21)$$

Suppose the n roots of $f(x)$ to be $a_1, a_2, \dots a_n$, which may be either real or impossible; and all may be unequal, or there may be one or more systems of equal roots. With a view to the subsequent integration it is necessary to explain a method of resolving a fraction such as (20) into other and more simple fractions capable of immediate integration; and as different processes must be applied, according as all the roots are unequal, or as some (not all) are equal to each other, we shall divide the inquiry into two parts. It will appear that the processes are equally applicable, whether the roots are real or impossible; and if there are systems of equal roots, a process similar to that which is applied to one set must be applied to each of the others.

19.] Let all the roots of $f(x)$ be unequal, so that

$$f(x) = (x-a_1)(x-a_2)\dots(x-a_n). \qquad (22)$$

Then $\frac{F(x)}{f(x)}$ may be resolved into a series of fractions of the form

$$\frac{F(x)}{f(x)} = \frac{N_1}{x-a_1} + \frac{N_2}{x-a_2} + \dots + \frac{N_n}{x-a_n}; \qquad (23)$$

where $N_1, N_2, \dots N_n$ are constants, and to be determined.

The possibility and legitimacy of such an identity as (23) may thus be shown:

Let the right-hand member of (23) be reduced to a common denominator; this will be $f(x)$; then as the identity (not equality only) of the two members of the equation is to be perfect, the two numerators must be identical; and by the hypothesis they may be so: for as $F(x)$ is a function of not more than $(n-1)$ dimensions, it may have n terms, but cannot have more; and therefore involves n coefficients, some of which however may in certain cases have zero values: and the numerator of the right-hand member will be also of $(n-1)$ dimensions and will have n coefficients, involving n undetermined constants $N_1, N_2, \ldots N_n$; by equating, therefore, the coefficients of the same powers of x on both sides of the equation, there will be n different equations, whereby $N_1, N_2, \ldots N_n$ may be determined; and this of course it is possible to do.

Multiplying both sides of (23) by $f(x)$ we have

$$F(x) = N_1 \frac{f(x)}{x-a_1} + N_2 \frac{f(x)}{x-a_2} + \ldots + N_n \frac{f(x)}{x-a_n}. \tag{24}$$

As the two sides are identical, they are the same for *all* values of x; let therefore $x = a_1$; and since $x-a_1$ is a factor of $f(x)$, all the terms of the right-hand side vanish except the first; and that becomes $\frac{0}{0}$, and must therefore be evaluated by the method of Chapter V. Vol. I; whence we have, when $x = a_1$,

$$F(a_1) = N_1 \frac{f'(a_1)}{1}; \qquad \therefore \quad N_1 = \frac{F(a_1)}{f'(a_1)}. \tag{25}$$

$$\left.\begin{array}{lll} \text{Similarly if} & x = a_2, & N_2 = \dfrac{F(a_2)}{f'(a_2)}; \\ & \cdot \;\; \cdot \;\; \cdot & \cdot \;\; \cdot \;\; \cdot \;\; \cdot \\ & x = a_n, & N_n = \dfrac{F(a_n)}{f'(a_n)}. \end{array}\right\} \tag{26}$$

Whence we have

$$\frac{F(x)}{f(x)} = \frac{F(a_1)}{f'(a_1)} \frac{1}{x-a_1} + \frac{F(a_2)}{f'(a_2)} \frac{1}{x-a_2} + \ldots + \frac{F(a_n)}{f'(a_n)} \frac{1}{x-a_n}; \tag{27}$$

and therefore

$$\int \frac{F(x)}{f(x)} dx = \frac{F(a_1)}{f'(a_1)} \int \frac{dx}{x-a_1} + \frac{F(a_2)}{f'(a_2)} \int \frac{dx}{x-a_2} + \ldots$$

$$\ldots + \frac{F(a_n)}{f'(a_n)} \int \frac{dx}{x-a_n}. \tag{28}$$

Thus we have determined in definite forms the values of the n numerators $N_1, N_2, \ldots N_n$, and the decomposition of the rational fraction is complete *.

Now for any integral of the form $\int \frac{dx}{x-a_k}$, we have, by equation (11),

$$\int \frac{dx}{x-a_k} = \log(x-a_k);$$

* Further remarks on the theory of this subject will be found in Leçons 5, 6, 7, "Cours d'Algèbre Supérieure," by J. A. Serret, 2nde ed. Paris, 1854.

which form is, when a_k is real, as convenient as the result admits of: but if a_k is impossible, then, to avoid the Logarithms of impossible quantities, we reduce as follows:

Let $a+\beta\sqrt{-1}$, $a-\beta\sqrt{-1}$ be a pair of conjugate roots, and let the coefficients of the partial fractions corresponding to these roots, and found as above, be $P+Q\sqrt{-1}$, $P-Q\sqrt{-1}$; so that the two partial fractions are

$$\frac{P+Q\sqrt{-1}}{x-(a+\beta\sqrt{-1})} \text{ and } \frac{P-Q\sqrt{-1}}{x-(a-\beta\sqrt{-1})};$$

let these be compounded into a single fraction with a quadratic denominator; whence we have

$$\frac{2P(x-a)-2Q\beta}{(x-a)^2+\beta^2}; \tag{29}$$

and

$$\int \frac{2P(x-a)-2Q\beta}{(x-a)^2+\beta^2}\,dx = P\log\{(x-a)^2+\beta^2\}-2Q\tan^{-1}\frac{x-a}{\beta}. \tag{30}$$

20.] Examples of the preceding method:

Ex. 1. $\displaystyle\int\frac{x\,dx}{x^2-5x+6}$.

To decompose into its partial fractions $\dfrac{x}{x^2-5x+6}=\dfrac{F(x)}{f(x)}$;

the roots of $f(x)$ are 2 and 3; and since $\dfrac{F(x)}{f'(x)}=\dfrac{x}{2x-5}$, the coefficients of $\dfrac{1}{x-2}$ and of $\dfrac{1}{x-3}$ are respectively -2, and 3;

$$\therefore \int\frac{x\,dx}{x^2-5x+6}=\int\left\{\frac{-2\,dx}{x-2}+\frac{3\,dx}{x-3}\right\}$$

$$= -2\log(x-2)+3\log(x-3)$$

$$= \log\frac{(x-3)^3}{(x-2)^2}.$$

Ex. 2. $\displaystyle\int\frac{(x^2+1)\,dx}{x^3+6x^2+11x+6}$.

In this example the roots of the denominator are -1, -2, -3; and since $\dfrac{F(x)}{f'(x)}=\dfrac{x^2+1}{3x^2+12x+11}$, the coefficients of $\dfrac{1}{x+1}$, $\dfrac{1}{x+2}$, and $\dfrac{1}{x+3}$ are respectively 1, -5, and 5;

$$\therefore \int \frac{(x^2+1)\,dx}{x^3+6x^2+11x+6} = \int \frac{dx}{x+1} - 5\int \frac{dx}{x+2} + 5\int \frac{dx}{x+3}$$

$$= \log(x+1) - 5\log(x+2) + 5\log(x+3)$$

$$= \log \frac{(x+1)(x+3)^5}{(x+2)^5}.$$

Ex. 3. $\displaystyle\int \frac{(5x-2)\,dx}{x^3+6x^2+8x}.$

The roots of the denominator are 0, -2, -4; and since $\dfrac{F(x)}{f'(x)} = \dfrac{5x-2}{3x^2+12x+8}$, the coefficients of $\dfrac{1}{x}$, $\dfrac{1}{x+2}$, and $\dfrac{1}{x+4}$ are respectively $-\dfrac{1}{4}$, 3, $-\dfrac{11}{4}$;

$$\therefore \int \frac{(5x-2)\,dx}{x^3+6x^2+8x} = -\frac{1}{4}\int \frac{dx}{x} + 3\int \frac{dx}{x+2} - \frac{11}{4}\int \frac{dx}{x+4}$$

$$= -\frac{1}{4}\log x + 3\log(x+2) - \frac{11}{4}\log(x+4)$$

$$= \log \frac{(x+2)^3}{x^{\frac{1}{4}}(x+4)^{\frac{11}{4}}}.$$

Ex. 4. $\displaystyle\int \frac{dx}{x^3-x^2+2x-2}.$

The roots of the denominator are 1, $+\sqrt{-2}$, $-\sqrt{-2}$; and since $\dfrac{F(x)}{f'(x)} = \dfrac{1}{3x^2-2x+2}$, the coefficients of $\dfrac{1}{x-1}$, $\dfrac{1}{x-\sqrt{-2}}$, and $\dfrac{1}{x+\sqrt{-2}}$ are respectively $\dfrac{1}{3}$, $-\dfrac{2-\sqrt{-2}}{12}$, $-\dfrac{2+\sqrt{-2}}{12}$;

$$\therefore \int \frac{dx}{x^3-x^2+2x-2} = \frac{1}{3}\int \frac{dx}{x-1}$$

$$-\frac{1}{12}\int \left\{ \frac{2-\sqrt{-2}}{x-\sqrt{-2}} + \frac{2+\sqrt{-2}}{x+\sqrt{-2}} \right\} dx$$

$$= \frac{1}{3}\log(x-1) - \frac{1}{3}\int \frac{x+1}{x^2+2}\,dx$$

$$= \quad - \quad - \quad - \quad - \quad - \frac{1}{6}\log(x^2+2) - \frac{1}{3\sqrt{2}}\tan^{-1}\frac{x}{\sqrt{2}}$$

$$= \log \frac{(x-1)^{\frac{1}{3}}}{(x^2+2)^{\frac{1}{6}}} - \frac{1}{3\sqrt{2}}\tan^{-1}\frac{x}{\sqrt{2}}.$$

Ex. 5. $\int \frac{dx}{x^5+5x^3+4x}$.

The roots of the denominator are $0, +\sqrt{-1}, -\sqrt{-1}, +\sqrt{-4}, -\sqrt{-4}$; and since $\frac{F(x)}{f'(x)} = \frac{1}{5x^4+15x^2+4}$; the coefficients of $\frac{1}{x}, \frac{1}{x-\sqrt{-1}}, \frac{1}{x+\sqrt{-1}}, \frac{1}{x-\sqrt{-4}}, \frac{1}{x+\sqrt{-4}}$ are respectively $\frac{1}{4}, -\frac{1}{6}, -\frac{1}{6}, \frac{1}{24}, \frac{1}{24}$;

$$\therefore \int \frac{dx}{x^5+5x^3+4x} = \frac{1}{4}\int\frac{dx}{x} - \frac{1}{6}\int\left\{\frac{1}{x-\sqrt{-1}} + \frac{1}{x+\sqrt{-1}}\right\}dx$$
$$+ \frac{1}{24}\int\left\{\frac{1}{x-\sqrt{-4}} + \frac{1}{x+\sqrt{-4}}\right\}dx$$
$$= \frac{1}{4}\log x - \frac{1}{6}\int\frac{2x\,dx}{x^2+1} + \frac{1}{24}\int\frac{2x\,dx}{x^2+4}$$
$$= \frac{1}{4}\log x - \frac{1}{6}\log(x^2+1) + \frac{1}{24}\log(x^2+4).$$

Ex. 6. $\int \frac{dx}{x^3-1}$.

The roots of the denominator are $1, \frac{-1+\sqrt{-3}}{2}, \frac{-1-\sqrt{-3}}{2}$; and since $\frac{F(x)}{f'(x)} = \frac{1}{3x^2} = \frac{x}{3}$, because $x^3 = 1$; therefore the coefficients of $\frac{1}{x-1}$, $\frac{1}{x-\frac{-1+\sqrt{-3}}{2}}$, and $\frac{1}{x-\frac{-1-\sqrt{-3}}{2}}$ are respectively $\frac{1}{3}, \frac{-1+\sqrt{-3}}{6}, \frac{-1-\sqrt{-3}}{6}$.

$$\therefore \int\frac{dx}{x^3-1} = \frac{1}{3}\int\frac{dx}{x-1}$$
$$+ \frac{1}{6}\int\left\{\frac{-1+\sqrt{-3}}{x-\frac{-1+\sqrt{-3}}{2}} + \frac{-1-\sqrt{-3}}{x-\frac{-1-\sqrt{-3}}{2}}\right\}dx$$
$$= \frac{1}{3}\log(x-1) - \frac{1}{6}\int\frac{2\left(x+\frac{1}{2}\right)+3}{\left(x+\frac{1}{2}\right)^2+\frac{3}{4}}dx$$
$$= \frac{1}{3}\log(x-1) - \frac{1}{6}\log(x^2+x+1) - \frac{1}{\sqrt{3}}\tan^{-1}\frac{2x+1}{\sqrt{3}}.$$

It will be observed that the coefficient of the second fraction corresponding to a pair of conjugate roots is deduced from that of the first by changing the sign of the impossible part.

21.] Let two or more of the roots of $f(x)$ be equal; then the preceding process of resolution does not admit of being applied directly, because in the right-hand member of (23) there will not be n undetermined constants; and consequently the number of unknown constants is not sufficient to render the two numerators identical; in this case we proceed as follows:

Suppose m roots of $f(x)$ to be equal to a_1, and the other roots to be $a_{m+1}, a_{m+2}, \ldots a_n$; so that

$$f(x) = (x-a_1)^m (x-a_{m+1})(x-a_{m+2})\ldots(x-a_n); \qquad (31)$$

then if $\dfrac{F(x)}{f(x)}$ is resolved into a series of fractions of the form (23), the numerator of that which has $(x-a_1)^m$ in the denominator must be of $(m-1)$ dimensions, and involve m undetermined constants; otherwise the equation cannot be an identity; we must therefore suppose

$$\frac{F(x)}{f(x)} = \frac{B_1 x^{m-1} + B_2 x^{m-2} + \ldots + B_{m-1} x + B_m}{(x-a_1)^m} + \frac{N_{m+1}}{x-a_{m+1}} + \ldots$$
$$\ldots + \frac{N_n}{x-a_n}. \qquad (32)$$

But, for the purposes of integration, it is more convenient, and it is allowable, to assume the numerator of the first partial fraction in the form

$$M_1 + M_2(x-a_1) + M_3(x-a_1)^2 + \ldots + M_m(x-a_1)^{m-1}; \qquad (33)$$

so that

$$\frac{F(x)}{f(x)} = \frac{M_1}{(x-a_1)^m} + \frac{M_2}{(x-a_1)^{m-1}} + \ldots + \frac{M_m}{x-a_1} + \frac{N_{m+1}}{x-a_{m+1}} + \ldots$$
$$\ldots + \frac{N_n}{x-a_n}. \qquad (34)$$

As to the numerators of these partial fractions, those corresponding to simple factors in the denominators may be determined by the method of the preceding articles; and for the determination of $M_1, M_2, \ldots M_m$, let

$$(x-a_{m+1})(x-a_{m+2})\ldots(x-a_n) = \phi(x);$$

and let (33), which is the numerator of the fraction which has $(x-a_1)^m$ in the denominator, be symbolized by $\psi(x)$, so that

$$\frac{F(x)}{f(x)} = \frac{\psi(x)}{(x-a_1)^m} + \frac{Q}{\phi(x)}; \qquad (35)$$

Q being a function of x of $n-m-1$ dimensions, which however it will not be requisite to determine; and $\phi(x)$ being the product of all the factors in the denominator short of the set of equal factors; so that

$$f(x) = (x-a_1)^m \phi(x). \tag{36}$$

By the Theorem in Vol. I, Art. 74, Equation (84),

$$\psi(x) = \psi(a_1) + \psi'(a_1)\frac{x-a_1}{1} + \psi''(a_1)\frac{(x-a_1)^2}{1.2} + \dots$$
$$\dots + \psi^{m-1}(a_1)\frac{(x-a_1)^{m-1}}{1.2.3\dots(m-1)}; \tag{37}$$

observing that, as $\psi(x)$ is rational and involves only positive powers of the variable and is of $(m-1)$ dimensions, all derived functions of it after the $(m-1)$th vanish, and that therefore the series (37) has only m terms. Hence, dividing both sides of (37) by $(x-a_1)^m$,

$$\frac{\psi(x)}{(x-a_1)^m} = \frac{\psi(a_1)}{(x-a_1)^m} + \frac{\psi'(a_1)}{1}\frac{1}{(x-a_1)^{m-1}} + \frac{\psi''(a_1)}{1.2}\frac{1}{(x-a_1)^{m-2}}$$
$$+ \dots\dots + \frac{\psi^{m-1}(a_1)}{1.2.3\dots(m-1)}\frac{1}{x-a_1} \tag{38}$$

As to $\psi(x)$ let it be observed that, equating the numerators in equation (35), we have

$$F(x) = \psi(x) \times \phi(x) + Q(x-a_1)^m; \tag{39}$$

$$\therefore \psi(x) = \frac{F(x)}{\phi(x)} - \frac{Q(x-a_1)^m}{\phi(x)}. \tag{40}$$

But as (38) involves $\psi(x)$ and its derived functions up to the $(m-1)$th order, and these when $x = a_1$, the latter term in the right-hand member of (40) will vanish for all values for which $\psi(x)$ is used; we may therefore, for all purposes for which we shall have to use $\psi(x)$, employ the equation,

$$\psi(x) = \frac{F(x)}{\phi(x)}; \tag{41}$$

that is $\psi(x)$ is equal to the numerator of the original fraction, divided by its denominator short of the set of equal factors.

Substituting therefore in (34),

$$\frac{F(x)}{f(x)} = \frac{\psi(a_1)}{(x-a_1)^m} + \frac{\psi'(a_1)}{1}\frac{1}{(x-a_1)^{m-1}} + \frac{\psi''(a_1)}{1.2}\frac{1}{(x-a_1)^{m-2}} + \dots$$
$$\dots + \frac{\psi^{m-1}(a_1)}{1.2.3\dots(m-1)}\frac{1}{x-a_1} + \frac{F(a_{m+1})}{f'(a_{m+1})}\frac{1}{x-a_{m+1}} + \dots$$
$$\dots + \frac{F(a_n)}{f'(a_n)}\frac{1}{x-a_n}; \tag{42}$$

and therefore

$$\int \frac{F(x)}{f(x)} dx = \psi(a_1) \int \frac{dx}{(x-a_1)^m} + \frac{\psi'(a_1)}{1} \int \frac{dx}{(x-a_1)^{m-1}} + \ldots$$
$$\ldots + \frac{\psi^{m-1}(a_1)}{1.2.3\ldots(m-1)} \int \frac{dx}{x-a_1} + \frac{F(a_{m+1})}{f'(a_{m+1})} \int \frac{dx}{x-a_{m+1}} + \ldots$$
$$\ldots + \frac{F(a_n)}{f'(a_n)} \int \frac{dx}{x-a_n}. \quad (43)$$

But
$$\int \frac{dx}{(x-a_1)^r} = -\frac{1}{(r-1)} \frac{1}{(x-a_1)^{r-1}};$$
$$\therefore \int \frac{F(x)}{f(x)} dx = -\frac{\psi(a_1)}{(m-1)} \frac{1}{(x-a_1)^{m-1}} - \frac{\psi'(a_1)}{(m-2)} \frac{1}{(x-a_1)^{m-2}} +$$
$$\ldots\ldots + \frac{\psi^{m-1}(a_1)}{1.2.3\ldots(m-1)} \log(x-a_1) + \ldots. \quad (44)$$

If the denominator contains other sets of equal factors, the series of partial fractions corresponding to them must be determined in a manner precisely analogous to that applied above.

The method also is applicable to sets of equal factors involving impossible roots, in which case we may combine terms of the series (44) corresponding to conjugate factors: for suppose $\alpha+\beta\sqrt{-1}$, $\alpha-\beta\sqrt{-1}$ to be a pair of conjugate roots; and let m equal factors corresponding to each enter into the denominator of the original fraction; then the first terms of the series (44) will be

$$-\frac{1}{m-1} \frac{\psi(\alpha+\beta\sqrt{-1})}{(x-\alpha-\beta\sqrt{-1})^{m-1}} \text{ and } -\frac{1}{m-1} \frac{\psi(\alpha-\beta\sqrt{-1})}{(x-\alpha+\beta\sqrt{-1})^{m-1}}; \quad (45)$$

the sum of which two fractions, short of the common coefficient $\left(-\frac{1}{(m-1)}\right)$, is

$$\frac{\psi(\alpha+\beta\sqrt{-1})(x-\alpha+\beta\sqrt{-1})^{m-1}+\psi(\alpha-\beta\sqrt{-1})(x-\alpha-\beta\sqrt{-1})^{m-1}}{\{(x-\alpha)^2+\beta^2\}^{m-1}}; \quad (4$$

and similarly may the other results be combined; but in order to avoid the logarithms of impossible quantities, the last terms of the series corresponding to a pair of conjugate roots are

$$\frac{\psi^{m-1}(\alpha+\beta\sqrt{-1})}{1.2.3\ldots(m-1)} \frac{1}{x-\alpha-\beta\sqrt{-1}} \text{ and } \frac{\psi^{m-1}(\alpha-\beta\sqrt{-1})}{1.2.3\ldots(m-1)} \frac{1}{x-\alpha+\beta\sqrt{-1}},$$

which may be combined into a single fraction; and of which the numerator is $\{\psi^{m-1}(a+\beta\sqrt{-1})+\psi^{m-1}(a-\beta\sqrt{-1})\}(x-a)$ $+\{\psi^{m-1}(a+\beta\sqrt{-1})-\psi^{m-1}(a-\beta\sqrt{-1})\}\beta\sqrt{-1}$, and whose denominator is $(x-a)^2+\beta^2$, and of which the coefficient is

$$\frac{1}{1.2\ldots(m-1)};$$

and thus the integral of the corresponding element will be of the form

$$\frac{1}{1.2.3\ldots(m-1)}\left\{\nu\log\{(x-a)^2+\beta^2\}-\nu'\tan^{-1}\frac{x-a}{\beta}\right\}. \qquad (47)$$

22.] Examples illustrative of the preceding:

Ex. 1. $\displaystyle\int\frac{x^3+x^2+2}{(x+1)^2(x-1)^2x}dx.$

To determine the coefficient of $\dfrac{1}{x}$;

$$\frac{F(x)}{f'(x)}=\frac{x^3+x^2+2}{5x^4-6x^2+1}; \qquad \text{the coefficient } \therefore = 2.$$

To determine the terms of the series (42) corresponding to $(x-1)^2$;

$$\psi(x)=\frac{x^3+x^2+2}{x(x+1)^2}, \qquad \psi'(x)=\frac{x^3+x^2-6x-2}{x^2(x+1)^3};$$

$$\therefore\ \psi(1)=1; \qquad \psi'(1)=-\frac{3}{4}.$$

To determine the terms of the series (42) corresponding to $(x+1)^2$;

$$\psi(x)=\frac{x^3+x^2+2}{x(x-1)^2}, \qquad \psi'(x)=\frac{-3x^3-x^2-6x+2}{x^2(x-1)^3};$$

$$\therefore\ \psi(-1)=-\frac{1}{2}; \qquad \psi'(-1)=-\frac{5}{4};$$

therefore by series (43),

$$\int\frac{(x^3+x^2+2)dx}{x(x-1)^2(x+1)^2}=2\int\frac{dx}{x}+\int\left\{\frac{1}{(x-1)^2}-\frac{3}{4}\frac{1}{x-1}\right\}dx$$
$$-\int\left\{\frac{1}{2}\frac{1}{(x+1)^2}+\frac{5}{4}\frac{1}{x+1}\right\}dx$$

$$=2\log x-\frac{1}{x-1}-\frac{3}{4}\log(x-1)+\frac{1}{2(x+1)}-\frac{5}{4}\log(x+1),$$

$$=-\frac{3+x}{2(x^2-1)}+\log\frac{x^2}{(x-1)^{\frac{3}{4}}(x+1)^{\frac{5}{4}}}.$$

Ex. 2. $\displaystyle\int\frac{dx}{(x-1)^2(x^2+1)}.$

To determine the coefficients of $\frac{1}{x-\sqrt{-1}}$ and $\frac{1}{x+\sqrt{-1}}$;

since $\frac{F(x)}{f'(x)} = \frac{1}{(x-1)(4x^3-2x+2)}$; the coefficients of $\frac{1}{x-\sqrt{-1}}$ and $\frac{1}{x+\sqrt{-1}}$ are respectively $\frac{1}{4}, \frac{1}{4}$.

To determine the terms of the series (42) corresponding to $(x-1)^2$;

$$\psi(x) = \frac{1}{x^2+1}; \qquad \psi'(x) = \frac{-2x}{(x^2+1)^2};$$

$$\therefore \quad \psi(1) = \frac{1}{2}; \qquad \psi'(1) = -\frac{1}{2};$$

$$\therefore \int \frac{dx}{(x-1)^2(x^2+1)} = \frac{1}{2}\int \frac{dx}{(x-1)^2} - \frac{1}{2}\int \frac{dx}{x-1}$$

$$+\frac{1}{4}\int \left\{ \frac{1}{x-\sqrt{-1}} + \frac{1}{x+\sqrt{-1}} \right\} dx$$

$$= -\frac{1}{2(x-1)} - \frac{1}{2}\log(x-1) + \frac{1}{4}\log(x^2+1)$$

$$= -\frac{1}{2(x-1)} + \log\frac{(x^2+1)^{\frac{1}{4}}}{(x-1)^{\frac{1}{2}}}.$$

Ex. 8. $\int \frac{x^3+1}{(x-1)^4(x^2+1)}\,dx.$

To determine the coefficients of $\frac{1}{x-\sqrt{-1}}$ and of $\frac{1}{x+\sqrt{-1}}$;

since $\frac{F(x)}{f'(x)} = \frac{x^3+1}{(x-1)^3(6x^2-2x+4)}$; the coefficients of $\frac{1}{x-\sqrt{-1}}$ and $\frac{1}{x+\sqrt{-1}}$ are respectively $\frac{1+\sqrt{-1}}{8}, \frac{1-\sqrt{-1}}{8}$.

To determine the terms of the series (42) corresponding to $(x-1)^4$;

$$\psi(x) = \frac{x^3+1}{x^2+1}; \qquad \psi'(x) = \frac{x^4+3x^2-2x}{(x^2+1)^2};$$

$$\psi''(x) = \frac{-2x^3+6x^2+6x-2}{(x^2+1)^3};$$

$$\psi'''(x) = \frac{6x^4-24x^3-36x^2+24x+6}{(x^2+1)^4};$$

$$\therefore \quad \psi(1) = 1; \quad \psi'(1) = \frac{1}{2}; \quad \psi''(1) = 1; \quad \psi'''(1) = -\frac{3}{2};$$

$$\therefore \int \frac{(x^3+1)\,dx}{(x-1)^4(x^2+1)} = \int \frac{dx}{(x-1)^4} + \frac{1}{2}\int \frac{dx}{(x-1)^3} + \frac{1}{1.2}\int \frac{dx}{(x-1)^2}$$

$$-\frac{3}{2.1.2.3}\int \frac{dx}{x-1} + \frac{1}{8}\int \left\{ \frac{1+\sqrt{-1}}{x-\sqrt{-1}} + \frac{1-\sqrt{-1}}{x+\sqrt{-1}} \right\} dx$$

$$= -\frac{1}{3(x-1)^3} - \frac{1}{4(x-1)^2} - \frac{1}{2}\frac{1}{x-1} - \frac{1}{4}\log(x-1)$$

$$+\frac{1}{8}\log(x^2+1) - \frac{1}{4}\tan^{-1}x.$$

As the process of decomposition and subsequent integration is the same in all cases, it is unnecessary here to work out other examples: the student however must exercise himself in them; and the ordinary collections will yield a copious supply.

There are however two particular cases of similar decomposition which exhibit remarkable peculiarities; viz. those in which the denominators are of the forms x^n-1, and x^n+1; and first we will take the simplest forms wherein the numerators are unity.

23.] To determine the integral of $\dfrac{dx}{x^n-1}$.

First let n be odd; then, in Art. 64, Vol. I. it is proved that the roots of $x^n-1=0$ are,

$$1,\ \cos\frac{2\pi}{n} \pm \sqrt{-1}\sin\frac{2\pi}{n},\ \cos\frac{4\pi}{n} \pm \sqrt{-1}\sin\frac{4\pi}{n}, \ldots$$

$$\ldots \cos\frac{n-1}{n}\pi \pm \sqrt{-1}\sin\frac{n-1}{n}\pi;$$

Now $\dfrac{F(x)}{f'(x)} = \dfrac{1}{nx^{n-1}} = \dfrac{x}{nx^n} = \dfrac{x}{n}$, because $x^n=1$ for all the roots of x^n-1, and these are the only values of x for which we have to consider the function.

Therefore the coefficient of $\dfrac{1}{x-1}$ is $\dfrac{1}{n}$;

of $\dfrac{1}{x-\cos\frac{2\pi}{n}-\sqrt{-1}\sin\frac{2\pi}{n}}$ is $\dfrac{1}{n}\left\{\cos\frac{2\pi}{n}+\sqrt{-1}\sin\frac{2\pi}{n}\right\}$;

of $\dfrac{1}{x-\cos\frac{2\pi}{n}+\sqrt{-1}\sin\frac{2\pi}{n}}$ is $\dfrac{1}{n}\left\{\cos\frac{2\pi}{n}-\sqrt{-1}\sin\frac{2\pi}{n}\right\}$;

. . . .

Combining the pairs of conjugate partial fractions according to equation (29), the first pair becomes

$$\frac{2x\cos\frac{2\pi}{n}-2}{n\left\{x^2-2x\cos\frac{2\pi}{n}+1\right\}};$$

and similarly will the other pairs of conjugate partial fractions be compounded; so that the following series will be formed,

$$\frac{1}{x^n-1}=\frac{1}{n}\left\{\frac{1}{x-1}+\frac{2x\cos\frac{2\pi}{n}-2}{x^2-2x\cos\frac{2\pi}{n}+1}+\frac{2x\cos\frac{4\pi}{n}-2}{x^2-2x\cos\frac{4\pi}{n}+1}\right.$$

$$\left.+\ \ldots\ldots+\frac{2x\cos\frac{n-1}{n}\pi-2}{x^2-2x\cos\frac{n-1}{n}\pi+1}\right\}. \quad (48)$$

$$\therefore \int\frac{dx}{x^n-1}=\frac{1}{n}\int\frac{dx}{x-1}+\frac{\cos\frac{2\pi}{n}}{n}\int\frac{\left(2x-2\cos\frac{2\pi}{n}\right)dx}{x^2-2x\cos\frac{2\pi}{n}+1}$$

$$-\frac{2\left(\sin\frac{2\pi}{n}\right)^2}{n}\int\frac{d\left(x-\cos\frac{2\pi}{n}\right)}{\left(x-\cos\frac{2\pi}{n}\right)^2+\left(\sin\frac{2\pi}{n}\right)^2}+\ldots \quad (49)$$

$$=\frac{1}{n}\log(x-1)+\frac{\cos\frac{2\pi}{n}}{n}\log\left(x^2-2x\cos\frac{2\pi}{n}+1\right)$$

$$-\frac{2\sin\frac{2\pi}{n}}{n}\tan^{-1}\frac{x-\cos\frac{2\pi}{n}}{\sin\frac{2\pi}{n}}+\ \ldots\ldots$$

$$\ldots+\frac{1}{n}\cos\frac{n-1}{n}\pi\log\left(x^2-2x\cos\frac{n-1}{n}\pi+1\right)$$

$$-\frac{2}{n}\sin\frac{n-1}{n}\pi\tan^{-1}\frac{x-\cos\frac{n-1}{n}\pi}{\sin\frac{n-1}{n}\pi}. \quad (50)$$

If n is even, then equation (48), by means of Art. 64, Vol. I, becomes

$$\frac{1}{x^n-1}=\frac{1}{n}\left\{\frac{1}{x-1}+\frac{2x\cos\frac{2\pi}{n}-2}{x^2-2x\cos\frac{2\pi}{n}+1}+\cdots\right.$$

$$\left.\cdots+\frac{2x\cos\frac{n-2}{n}\pi-2}{x^2-2x\cos\frac{n-2}{n}\pi+1}-\frac{1}{x+1}\right\}. \quad (51)$$

$$\therefore \int\frac{dx}{x^n-1}=\frac{1}{n}\int\frac{dx}{x-1}+\ldots\ldots-\frac{1}{n}\int\frac{dx}{x+1}$$

$$=\frac{1}{n}\log\frac{x-1}{x+1}+\ldots\ldots$$

$$\ldots+\frac{1}{n}\cos\frac{n-2}{n}\pi\log\left(x^2-2x\cos\frac{n-2}{n}\pi+1\right)$$

$$-\frac{2}{n}\sin\frac{n-2}{n}\pi\tan^{-1}\frac{x-\cos\frac{n-2}{n}\pi}{\sin\frac{n-2}{n}\pi}. \quad (52)$$

Ex. 1. $\int\frac{dx}{x^3-1}$.

The roots of x^3-1 are, 1, $\cos\frac{2\pi}{3}\pm\sqrt{-1}\sin\frac{2\pi}{3}$;

$$\frac{F(x)}{f'(x)}=\frac{1}{3x^2}=\frac{x}{3x^3}=\frac{x}{3};$$

therefore the coefficient of $\frac{1}{x-1}$ is $\frac{1}{3}$,

of $\dfrac{1}{x-\left(\cos\frac{2\pi}{3}+\sqrt{-1}\sin\frac{2\pi}{3}\right)}$ is $\frac{1}{3}\left\{\cos\frac{2\pi}{3}+\sqrt{-1}\sin\frac{2\pi}{3}\right\}$,

of $\dfrac{1}{x-\left(\cos\frac{2\pi}{3}-\sqrt{-1}\sin\frac{2\pi}{3}\right)}$ is $\frac{1}{3}\left\{\cos\frac{2\pi}{3}-\sqrt{-1}\sin\frac{2\pi}{3}\right\}$;

$$\therefore \int\frac{dx}{x^3-1}=\frac{1}{3}\int\frac{dx}{x-1}+\frac{1}{3}\int\frac{2x\cos\frac{2\pi}{3}-2}{x^2-2x\cos\frac{2\pi}{3}+1}dx$$

$$\int\frac{dx}{x^3-1} = \frac{1}{3}\log(x-1)+\frac{1}{3}\cos\frac{2\pi}{3}\log\left(x^2-2x\cos\frac{2\pi}{3}+1\right)$$

$$-\frac{2}{3}\sin\frac{2\pi}{3}\tan^{-1}\left\{\frac{x-\cos\frac{2\pi}{3}}{\sin\frac{2\pi}{3}}\right\}$$

$$=\frac{1}{3}\log(x-1)-\frac{1}{6}\log(x^2+x+1)-\frac{1}{\sqrt{3}}\tan^{-1}\frac{2x+1}{\sqrt{3}}$$

$$=\frac{1}{6}\log\frac{(x-1)^2}{x^2+x+1}-\frac{1}{\sqrt{3}}\tan^{-1}\frac{2x+1}{\sqrt{3}}.$$

24.] To determine $\int\frac{dx}{x^n+1}$.

Let n be even; then, in Art. 65, Vol. I, it is proved that the roots of x^n+1 are $\cos\frac{\pi}{n}\pm\sqrt{-1}\sin\frac{\pi}{n}$, $\cos\frac{3\pi}{n}\pm\sqrt{-1}\sin\frac{3\pi}{n}$, ... $\cos\frac{n-1}{n}\pi\pm\sqrt{-1}\sin\frac{n-1}{n}\pi$.

Now $\frac{F(x)}{f'(x)}=\frac{1}{nx^{n-1}}=\frac{x}{nx^n}=-\frac{x}{n}$, since $x^n=-1$ for all the roots of $x^n+1=0$; therefore the coefficient

of $\frac{1}{x-\cos\frac{\pi}{n}-\sqrt{-1}\sin\frac{\pi}{n}}$ is $-\frac{1}{n}\left\{\cos\frac{\pi}{n}+\sqrt{-1}\sin\frac{\pi}{n}\right\}$;

of $\frac{1}{x-\cos\frac{\pi}{n}+\sqrt{-1}\sin\frac{\pi}{n}}$ is $-\frac{1}{n}\left\{\cos\frac{\pi}{n}-\sqrt{-1}\sin\frac{\pi}{n}\right\}$;

. . . . ;

and combining the pairs of conjugate partial fractions, according to equation (29), the first pair is

$$-\frac{1}{n}\left\{\frac{\cos\frac{\pi}{n}+\sqrt{-1}\sin\frac{\pi}{n}}{x-\left(\cos\frac{\pi}{n}+\sqrt{-1}\sin\frac{\pi}{n}\right)}+\frac{\cos\frac{\pi}{n}-\sqrt{-1}\sin\frac{\pi}{n}}{x-\left(\cos\frac{\pi}{n}-\sqrt{-1}\sin\frac{\pi}{n}\right)}\right\};$$

which becomes $-\frac{1}{n}\frac{2x\cos\frac{\pi}{n}-2}{x^2-2x\cos\frac{\pi}{n}+1}$;

and the other pairs give similar results; so that

$$\int\frac{dx}{x^n+1} = -\frac{1}{n}\int\frac{\left(2x\cos\frac{\pi}{n}-2\right)dx}{x^2-2x\cos\frac{\pi}{n}+1} - \frac{1}{n}\int\frac{\left(2x\cos\frac{3\pi}{n}-2\right)dx}{x^2-2x\cos\frac{3\pi}{n}+1}$$

$$\ldots - \frac{1}{n}\int\frac{\left(2x\cos\frac{n-1}{n}\pi-2\right)dx}{x^2-2x\cos\frac{n-1}{n}\pi+1}, \quad (28)$$

each of which must be integrated according to the process indicated in the preceding article.

Again, let n be odd: then the roots of $x^n+1=0$ are

$$\cos\frac{\pi}{n}\pm\sqrt{-1}\sin\frac{\pi}{n},\quad \cos\frac{3\pi}{n}\pm\sqrt{-1}\sin\frac{3\pi}{n},$$

$$\cos\frac{n-2}{n}\pi\pm\sqrt{-1}\sin\frac{n-2}{n}\pi,\quad -1;$$

so that if the conjugate partial fractions are combined by a process similar to that employed when n is even, the last pair becomes

$$-\frac{1}{n}\,\frac{2x\cos\frac{n-2}{n}\pi-2}{x^2-2x\cos\frac{n-2}{n}\pi+1},$$

and therefore, when n is odd,

$$\int\frac{dx}{x^n+1} = -\frac{1}{n}\int\frac{\left(2x\cos\frac{\pi}{n}-2\right)dx}{x^2-2x\cos\frac{\pi}{n}+1} - \ldots$$

$$-\frac{1}{n}\int\frac{\left(2x\cos\frac{n-2}{n}\pi-2\right)dx}{x^2-2x\cos\frac{n-2}{n}\pi+1} + \frac{1}{n}\int\frac{dx}{x+1}. \quad (29)$$

Ex. 1. $\displaystyle\int\frac{dx}{x^4+1}$.

The roots of $x^4+1=0$ are

$$\cos\frac{\pi}{4}\pm\sqrt{-1}\sin\frac{\pi}{4},\quad \cos\frac{3\pi}{4}\pm\sqrt{-1}\sin\frac{3\pi}{4},$$

$$\frac{F(x)}{f'(x)} = \frac{1}{4x^3} = \frac{x}{4x^4} = -\frac{x}{4};$$

therefore the coefficient

$$\text{of } \frac{1}{x-\cos\frac{\pi}{4}-\sqrt{-1}\sin\frac{\pi}{4}} \text{ is } -\frac{1}{4}\left(\cos\frac{\pi}{4}+\sqrt{-1}\sin\frac{\pi}{4}\right);$$

$$\text{of } \frac{1}{x-\cos\frac{\pi}{4}+\sqrt{-1}\sin\frac{\pi}{4}} \text{ is } -\frac{1}{4}\left(\cos\frac{\pi}{4}-\sqrt{-1}\sin\frac{\pi}{4}\right);$$

. . . . ;

and the pairs of conjugate partial fractions compound into

$$-\frac{1}{4}\frac{2x\cos\frac{\pi}{4}-2}{x^2-2x\cos\frac{\pi}{4}+1}, \text{ and } -\frac{1}{4}\frac{2x\cos\frac{3\pi}{4}-2}{x^2-2x\cos\frac{3\pi}{4}+1};$$

$$\int\frac{dx}{x^4+1} = -\frac{1}{4}\int\frac{\left(2x\cos\frac{\pi}{4}-2\right)dx}{x^2-2x\cos\frac{\pi}{4}+1} - \frac{1}{4}\int\frac{\left(2x\cos\frac{3\pi}{4}-2\right)dx}{x^2-2x\cos\frac{3\pi}{4}+1}$$

$$= -\frac{1}{4}\cos\frac{\pi}{4}\log\left(x^2-2x\cos\frac{\pi}{4}+1\right)+\frac{1}{2}\sin\frac{\pi}{4}\tan^{-1}\left\{\frac{x-\cos\frac{\pi}{4}}{\sin\frac{\pi}{4}}\right\}$$

$$-\frac{1}{4}\cos\frac{3\pi}{4}\log\left(x^2-2x\cos\frac{3\pi}{4}+1\right)+\frac{1}{2}\sin\frac{3\pi}{4}\tan^{-1}\left\{\frac{x-\cos\frac{3\pi}{4}}{\sin\frac{3\pi}{4}}\right\}$$

$$= \frac{1}{4\sqrt{2}}\log\frac{x^2+x\sqrt{2}+1}{x^2-x\sqrt{2}+1}+\frac{1}{2\sqrt{2}}\tan^{-1}\frac{x\sqrt{2}}{1-x^2}.$$

Ex. 2. $\int\frac{dx}{x^5+1}$.

$$\int\frac{dx}{x^5+1} = -\frac{1}{5}\int\frac{\left(2x\cos\frac{\pi}{5}-2\right)dx}{x^2-2x\cos\frac{\pi}{5}+1} - \frac{1}{5}\int\frac{\left(2x\cos\frac{3\pi}{5}-2\right)dx}{x^2-2x\cos\frac{3\pi}{5}+1}$$

$$+\frac{1}{5}\int\frac{dx}{x+1}$$

$$\therefore \int \frac{dx}{x^5+1} = \frac{1}{5}\log(x+1)$$

$$-\frac{1}{5}\cos\frac{\pi}{5}\log\left(x^2-2x\cos\frac{\pi}{5}+1\right)+\frac{2}{5}\sin\frac{\pi}{5}\tan^{-1}\left\{\frac{x-\cos\frac{\pi}{5}}{\sin\frac{\pi}{5}}\right\}$$

$$-\frac{1}{5}\cos\frac{3\pi}{5}\log\left(x^2-2x\cos\frac{3\pi}{5}+1\right)+\frac{2}{5}\sin\frac{3\pi}{5}\tan^{-1}\left\{\frac{x-\cos\frac{3\pi}{5}}{\sin\frac{3\pi}{5}}\right\}.$$

25.] To determine the integrals of $\frac{x^m\,dx}{x^n-1}$ and of $\frac{x^m\,dx}{x^n+1}$, where m is less than n.

To determine $\int \frac{x^m\,dx}{x^n-1}$.

The roots of the denominator are those given in Art. 23; and since
$$\frac{\text{F}(x)}{f'(x)} = \frac{x^m}{nx^{n-1}} = \frac{x^{m+1}}{nx^n} = \frac{x^{m+1}}{n},$$
we may determine without difficulty the coefficients of the several partial fractions, and the form which a pair of conjugate partial fractions assumes when they are combined.

By a similar process may we integrate $\frac{x^m\,dx}{x^n+1}$, and $\frac{\text{F}(x)\,dx}{x^n \pm 1}$, when $\text{F}(x)$ is integral and rational and of not more than $(n-1)$ dimensions.

To the general forms of Arts. 23 and 24 may also be reduced
$$\int \frac{x^m\,dx}{a \pm bx^n};$$
for if we replace bx^n by az^n, the integral becomes
$$\frac{a^{\frac{m-n+1}{n}}}{b^{\frac{m+1}{n}}}\int \frac{z^m\,dz}{z^n \pm 1}.$$

26.] The process however for obtaining the integrals of many infinitesimal-elements of the preceding forms may oftentimes be much simplified by a judicious substitution; the selection of which must be left to the ingenuity of the student, because no general rules can be given. A careful study of the following cases will probably indicate the course whereby he may be led to such a simplification.

Ex. 1. $\int \frac{dx}{x(a^3+x^3)} = \int \frac{x^2 dx}{x^3(a^3+x^3)} = \frac{1}{3}\int \frac{d.x^3}{x^3(a^3+x^3)};$

which, if $x^3 = z$, becomes $\frac{1}{3}\int \frac{dz}{z(a^3+z)}$; and the integral may be determined by the method of Art. 19.

Ex. 2. $\int \frac{x^2 dx}{a^6+x^6} = \frac{1}{3}\int \frac{d.x^3}{(a^3)^2+(x^3)^2} = \frac{1}{3a^3}\tan^{-1}\frac{x^3}{a^3},$

by means of equation (13), Art. 14.

Ex. 3. $\int \frac{dx}{x(a+bx^3)^2} = \frac{1}{b^2}\int \frac{x^2 dx}{x^3(k+x^3)^2}$, if $\frac{a}{b} = k$,

$$= \frac{1}{3b^2}\int \frac{d.x^3}{x^3(k+x^3)^2}$$

$$= \frac{1}{3b^2}\int \frac{dz}{z(k+z)^2}, \text{ if } x^3 = z;$$

and which last integral may be found by the method of Art. 21.

Ex. 4. $\int \frac{x\,dx}{x^4-a^4} = \frac{1}{2}\int \frac{d.x^2}{(x^2)^2-(a^2)^2} = \frac{1}{4a^2}\log\frac{x^2-a^2}{x^2+a^2},$

by means of equation (15), Art. 15.

Ex. 5. $\int \frac{x^2 dx}{x^4-1} = \frac{1}{2}\int \left\{\frac{1}{x^2+1} + \frac{1}{x^2-1}\right\} dx.$

$$= \frac{1}{2}\int \frac{dx}{x^2+1} + \frac{1}{4}\int \left\{\frac{1}{x-1} - \frac{1}{x+1}\right\} dx$$

$$= \frac{1}{2}\tan^{-1}x + \frac{1}{4}\log\frac{x-1}{x+1}.$$

Ex. 6. $\int \frac{x^3 dx}{1+x^2} = \int \left\{x - \frac{x}{1+x^2}\right\} dx$

$$= \frac{x^2}{2} - \log(1+x^2)^{\frac{1}{2}}.$$

27.] The following integrals might be determined by one or other of the preceding methods: but the process of integration by parts leads to a result more convenient, and better suited in most cases for finding the definite integral.

Integration of $\frac{dx}{(x^2+a^2)^n}$.

In the formula, $\int u\,dv = uv - \int v\,du$,

$$\text{let } u = (x^2+a^2)^{-n}, \qquad dv = dx;$$

$$\therefore \quad du = -\frac{2nx\,dx}{(x^2+a^2)^{n+1}}, \qquad v = x.$$

$$\therefore \int \frac{dx}{(x^2+a^2)^n} = \frac{x}{(x^2+a^2)^n} + 2n\int \frac{x^2\,dx}{(x^2+a^2)^{n+1}}$$

$$= \text{- - - -} + 2n\int \frac{x^2+a^2-a^2}{(x^2+a^2)^{n+1}}\,dx$$

$$= \text{- - - -} + 2n\int \frac{dx}{(x^2+a^2)^n} - 2na^2\int \frac{dx}{(x^2+a^2)^{n+1}}.$$

$$\therefore \quad 2na^2\int \frac{dx}{(x^2+a^2)^{n+1}} = \frac{x}{(x^2+a^2)^n} + (2n-1)\int \frac{dx}{(x^2+a^2)^n};$$

for n write $n-1$,

$$2(n-1)a^2\int \frac{dx}{(x^2+a^2)^n} = \frac{x}{(x^2+a^2)^{n-1}} + (2n-3)\int \frac{dx}{(x^2+a^2)^{n-1}},$$

$$\int \frac{dx}{(x^2+a^2)^n} = \frac{x}{2(n-1)a^2(x^2+a^2)^{n-1}} + \frac{2n-3}{a^2(2n-2)}\int \frac{dx}{(x^2+a^2)^{n-1}}.$$

Now, as the integral in the last term of the right-hand member of the equation is of the same form as the original integral, but has the index of its denominator less by unity, we may, the same process be repeated successively until finally $n = 1$, in which case the formula fails to give a determinate result, but the integral becomes $\int \frac{dx}{x^2+a^2}$, and we have, see (Art. 15),

$$\int \frac{dx}{x^2+a^2} = \frac{1}{a}\tan^{-1}\frac{x}{a}.$$

The method is known by the name of Integration by Successive Reduction.

Ex. 1. Integration of $\frac{dx}{(x^2+a^2)^3}$.

$$\int \frac{dx}{(x^2+a^2)^3} = \frac{x}{4a^2(x^2+a^2)^2} + \frac{3}{4a^2}\int \frac{dx}{(x^2+a^2)^2}$$

$$= \text{- - - - - -} + \frac{3}{4a^2}\left\{\frac{x}{2a^2(x^2+a^2)} + \frac{1}{2a^2}\int \frac{dx}{x^2+a^2}\right\}$$

$$= \frac{x}{4a^2(x^2+a^2)^2} + \frac{3x}{8a^4(x^2+a^2)} + \frac{3}{8a^5}\tan^{-1}\frac{x}{a}.$$

28.] Integration of $\dfrac{dx}{(a^2-x^2)^n}$.

A process exactly parallel to that of the last article gives the following formula,

$$\int\frac{dx}{(a^2-x^2)^n} = \frac{x}{2(n-1)a^2(a^2-x^2)^{n-1}} + \frac{1}{a^2}\frac{2n-3}{2n-2}\int\frac{dx}{(a^2-x^2)^{n-1}}; \quad (57)$$

and by successive reduction the last integral becomes

$$\int\frac{dx}{a^2-x^2} = \frac{1}{2a}\log\frac{a+x}{a-x}; \quad (58)$$

the formula failing to give a definite result, when $n = 1$.

Ex. 1. $$\int\frac{dx}{(a^2-x^2)^2} = \frac{x}{2a^2(a^2-x^2)} + \frac{1}{2a^2}\int\frac{dx}{a^2-x^2}$$
$$= \frac{x}{2a^2(a^2-x^2)} + \frac{1}{4a^3}\log\frac{a+x}{a-x}.$$

29.] Integration of $\dfrac{x^m\,dx}{(x^2+a^2)^n}$.

$$\int\frac{x^m\,dx}{(x^2+a^2)^n} = \int\frac{x\,dx}{(x^2+a^2)^n}x^{m-1}.$$

And in the formula $\int u\,dv = uv - \int v\,du$,

$$\text{let}\quad dv = \frac{x\,dx}{(x^2+a^2)^n}, \qquad u = x^{m-1};$$

$$\therefore\quad v = -\frac{1}{2(n-1)(x^2+a^2)^{n-1}}, \qquad du = (m-1)x^{m-2}dx;$$

$$\therefore \int\frac{x^m\,dx}{(x^2+a^2)^n} = -\frac{x^{m-1}}{2(n-1)(x^2+a^2)^{n-1}} + \frac{m-1}{2n-2}\int\frac{x^{m-2}dx}{(x^2+a^2)^{n-1}}. \quad (59)$$

By which means the integration of the original element is made to depend on that of another element of precisely the same form, but whose numerator and denominator are of lower dimensions; and thus, by successive and similar reductions, the integral will be brought either into a fundamental form or to that of Art. 27; the formula, it will be observed, failing when $n = 1$.

Ex. 1. $\int\dfrac{x^2\,dx}{(x^2+a^2)^4}$: here $m = 2$, $n = 4$.

$$\therefore\quad \int\frac{x^2\,dx}{(x^2+a^2)^4} = -\frac{x}{6(x^2+a^2)^3} + \frac{1}{6}\int\frac{dx}{(x^2+a^2)^3};$$

and the latter integral has been determined in Art. 27, so that it is unnecessary to repeat it.

Ex. 2. $\int \frac{x^3\,dx}{(x^2+a^2)^2}$: here $m = 3$, $n = 2$.

$$\therefore \int \frac{x^3\,dx}{(x^2+a^2)^2} = -\frac{x^2}{2(x^2+a^2)} + \int \frac{x\,dx}{x^2+a^2}$$

$$= -\frac{x^2}{2(x^2+a^2)} + \frac{1}{2}\log(x^2+a^2).$$

30.] Integration of $\frac{x^m\,dx}{(a^2-x^2)^n}$.

A process exactly parallel to that of the last article gives

$$\int \frac{x^m\,dx}{(a^2-x^2)^n} = \frac{x^{m-1}}{2(n-1)(a^2-x^2)^{n-1}} - \frac{m-1}{2n-2}\int \frac{x^{m-2}\,dx}{(a^2-x^2)^{n-1}}. \quad (60)$$

By which means the indices in both numerator and denominator are diminished, though the form is unchanged: and by a similar reduction we shall arrive at an integral either of a fundamental form, or of that of Art. 28; the formula fails, when $n = 1$.

Ex. $\int \frac{x^5\,dx}{(a^2-x^2)^2}$: here $m = 5$, $n = 2$; therefore

$$\int \frac{x^5\,dx}{(a^2-x^2)^2} = \frac{x^4}{2(a^2-x^2)} - 2\int \frac{x^3\,dx}{a^2-x^2}$$

$$= \cdots\cdots + 2\int \left\{x - \frac{a^2 x}{a^2-x^2}\right\} dx$$

$$= \frac{x^4}{2(a^2-x^2)} + x^2 + a^2\log(a^2-x^2).$$

SECTION 3.—*Integration of Irrational Algebraical Functions.*

31.] Integration of $\frac{dx}{(a^2-x^2)^{\frac{1}{2}}}$.

Since $d.\sin^{-1}\frac{x}{a} = \frac{dx}{(a^2-x^2)^{\frac{1}{2}}}$;

$$\therefore \int \frac{dx}{(a^2-x^2)^{\frac{1}{2}}} = \sin^{-1}\frac{x}{a}. \quad (61)$$

And again, since $d.\cos^{-1}\frac{x}{a} = \frac{-dx}{(a^2-x^2)^{\frac{1}{2}}}$;

$$\therefore \int \frac{-dx}{(a^2-x^2)^{\frac{1}{2}}} = \cos^{-1}\frac{x}{a}. \quad (62)$$

Now here is an apparent inconsistency; for it would appear from (61) and (62) that

$$\sin^{-1}\frac{x}{a} + \cos^{-1}\frac{x}{a} = \int\frac{dx}{(a^2-x^2)^{\frac{1}{2}}} - \int\frac{dx}{(a^2-x^2)^{\frac{1}{2}}} = 0; \quad (63)$$

which result is not correct; because $\sin^{-1}\frac{x}{a} + \cos^{-1}\frac{x}{a} = \frac{\pi}{2}$: we must therefore have recourse to the *accurate* process of definite integration; and let us take both integrals between the same limits; say, between the limits 0 and x; then by (11), Art. 5,

$$\int_0^x \frac{dx}{(a^2-x^2)^{\frac{1}{2}}} = \left[\sin^{-1}\frac{x}{a}\right]_0^x = \sin^{-1}\frac{x}{a}; \quad (64)$$

$$-\int_0^x \frac{dx}{(a^2-x^2)^{\frac{1}{2}}} = \left[\cos^{-1}\frac{x}{a}\right]_0^x = \cos^{-1}\frac{x}{a} - \frac{\pi}{2}; \quad (65)$$

therefore by addition

$$\sin^{-1}\frac{x}{a} + \cos^{-1}\frac{x}{a} = \frac{\pi}{2}, \quad (66)$$

which is a correct result; (61) and (62) therefore are not simultaneously true in the forms of indefinite integrals.

32.] Integration of $\frac{dx}{(2ax-x^2)^{\frac{1}{2}}}$.

Since $d.\text{versin}^{-1}\frac{x}{a} = \frac{dx}{(2ax-x^2)^{\frac{1}{2}}}$;

$$\therefore \int\frac{dx}{(2ax-x^2)^{\frac{1}{2}}} = \text{versin}^{-1}\frac{x}{a}. \quad (67)$$

Or thus,

$$\int\frac{dx}{(2ax-x^2)^{\frac{1}{2}}} = \int\frac{-d(a-x)}{\{a^2-(a-x)^2\}^{\frac{1}{2}}} = \cos^{-1}\frac{a-x}{a} = \text{versin}^{-1}\frac{x}{a}.$$

33.] The integration of $\frac{dx}{x(x^2-a^2)^{\frac{1}{2}}}$.

$$\int\frac{dx}{x(x^2-a^2)^{\frac{1}{2}}} = \int\frac{dx}{x^2\left(1-\frac{a^2}{x^2}\right)^{\frac{1}{2}}} = -\frac{1}{a}\int\frac{d\left(\frac{a}{x}\right)}{\left(1-\frac{a^2}{x^2}\right)^{\frac{1}{2}}}$$

$$= \frac{1}{a}\cos^{-1}\frac{a}{x}, \text{ by reason of equation (62)},$$

$$= \frac{1}{a}\sec^{-1}\frac{x}{a}. \quad (68)$$

34.] The integration of $\dfrac{dx}{(a^2+x^2)^{\frac{1}{2}}}$.

Let $a^2+x^2 = z^2$; $\qquad \therefore\ x\,dx = z\,dz$;

$$\therefore\ \frac{dx}{z} = \frac{dz}{x} = \frac{dx+dz}{x+z};$$

$$\begin{aligned}\therefore\ \int\frac{dx}{(a^2+x^2)^{\frac{1}{2}}} &= \int\frac{dx}{z} = \int\frac{dx+dz}{x+z}\\ &= \log(x+z)\\ &= \log\{x+(a^2+x^2)^{\frac{1}{2}}\}. \qquad (69)\end{aligned}$$

Similarly $\displaystyle\int\frac{dx}{(x^2\pm a^2)^{\frac{1}{2}}} = \log\{x+(x^2\pm a^2)^{\frac{1}{2}}\}. \qquad (70)$

35.] Integration of $\dfrac{dx}{(2ax+x^2)^{\frac{1}{2}}}$, and of $\dfrac{dx}{x(a^2-x^2)^{\frac{1}{2}}}$.

$$\int\frac{dx}{(2ax+x^2)^{\frac{1}{2}}} = \int\frac{d(x+a)}{\{(x+a)^2-a^2\}^{\frac{1}{2}}} = \log\{x+a+(2ax+x^2)^{\frac{1}{2}}\}, \quad (71)$$

by reason of (70).

Again,

$$\begin{aligned}\int\frac{dx}{x(a^2-x^2)^{\frac{1}{2}}} &= \int\frac{dx}{x^2\left(\frac{a^2}{x^2}-1\right)^{\frac{1}{2}}} = -\frac{1}{a}\int\frac{d\left(\frac{a}{x}\right)}{\left(\frac{a^2}{x^2}-1\right)^{\frac{1}{2}}}\\ &= -\frac{1}{a}\log\left\{\frac{a}{x}+\left(\frac{a^2}{x^2}-1\right)^{\frac{1}{2}}\right\}\\ &= -\frac{1}{a}\log\frac{a+(a^2-x^2)^{\frac{1}{2}}}{x}\\ &= \frac{1}{a}\log\frac{x}{a+(a^2-x^2)^{\frac{1}{2}}};\\ &= \frac{1}{a}\log\frac{a-(a^2-x^2)^{\frac{1}{2}}}{x}. \qquad (72)\end{aligned}$$

36.] The relations existing between some of these last integrals deserve consideration; for taking the definite integral of (69) between the limits x and 0, we have

$$\int_0^x\frac{dx}{(a^2+x^2)^{\frac{1}{2}}} = \log\left\{\frac{x+(a^2+x^2)^{\frac{1}{2}}}{a}\right\}; \qquad (73)$$

but the left-hand member may be put under the following form, whereby its integral is determined by equation (61): viz.

$$\frac{1}{\sqrt{-1}}\int_0^x\frac{d(x\sqrt{-1})}{\{a^2-(x\sqrt{-1})^2\}^{\frac{1}{2}}} = \frac{1}{\sqrt{-1}}\sin^{-1}\frac{x\sqrt{-1}}{a}; \qquad (74)$$

and the identity of (73) and (74) may thus be proved:

$$\text{Let} \quad \frac{1}{\sqrt{-1}}\sin^{-1}\frac{x\sqrt{-1}}{a} = z,$$

$$\therefore \quad \frac{x}{a} = \frac{1}{\sqrt{-1}}\sin(z\sqrt{-1})$$

$$= \frac{e^{-z}-e^{z}}{-2},$$

replacing $\sin(z\sqrt{-1})$ in terms of its exponential value as given by equation (30), Art. 61, of Vol. I;

$$\therefore \quad e^{z} = \frac{x+(a^2+x^2)^{\frac{1}{2}}}{a};$$

$$\therefore \quad z = \frac{1}{\sqrt{-1}}\sin^{-1}\frac{x\sqrt{-1}}{a} = \log\frac{x+(a^2+x^2)^{\frac{1}{2}}}{a}: \tag{75}$$

which shews that the two results though different in form are identical in value. Similarly might $\int_0^x \frac{dx}{a^2+x^2}$ be put in the form $\frac{1}{\sqrt{-1}}\int_0^x \frac{d(x\sqrt{-1})}{a^2-(x\sqrt{-1})^2}$, and be integrated according to Article 15, and the result shewn to be equal to $\frac{1}{a}\tan^{-1}\frac{x}{a}$.

37.] Integration of $\frac{dx}{(a+bx\pm cx^2)^{\frac{1}{2}}}$.

First let c be positive; then,

$$\int\frac{dx}{(a+bx+cx^2)^{\frac{1}{2}}} = \frac{1}{\sqrt{c}}\int\frac{d\left(x+\frac{b}{2c}\right)}{\left\{\left(x+\frac{b}{2c}\right)^2-\frac{b^2-4ac}{4c^2}\right\}^{\frac{1}{2}}}$$

$$= \frac{1}{\sqrt{c}}\log\left\{x+\frac{b}{2c}+\left(x^2+\frac{b}{c}x+\frac{a}{c}\right)^{\frac{1}{2}}\right\}, \tag{76}$$

by reason of equation (70).

Again, let c be negative; then

$$\int\frac{dx}{(a+bx-cx^2)^{\frac{1}{2}}} = \frac{1}{\sqrt{c}}\int\frac{d\left(x-\frac{b}{2c}\right)}{\left\{\frac{4ac+b^2}{4c^2}-\left(x-\frac{b}{2c}\right)^2\right\}^{\frac{1}{2}}}$$

$$= \frac{1}{\sqrt{c}}\sin^{-1}\left\{\frac{2cx-b}{(4ac+b^2)^{\frac{1}{2}}}\right\}, \tag{77}$$

by reason of equation (61).

Ex. 1. $$\int \frac{dx}{(x^2+2x+2)^{\frac{1}{2}}} = \int \frac{d(x+1)}{\{(x+1)^2+1\}^{\frac{1}{2}}}$$
$$= \log\{x+1+(x^2+2x+2)^{\frac{1}{2}}\}.$$

Ex. 2. $$\int \frac{dx}{(2-2x-x^2)^{\frac{1}{2}}} = \int \frac{d(x+1)}{\{3-(x+1)^2\}^{\frac{1}{2}}} = \sin^{-1}\frac{x+1}{\sqrt{3}}.$$

38.] Integration of $\dfrac{dx}{x(a+bx+cx^2)^{\frac{1}{2}}}$.

$$\int \frac{dx}{x(a+bx+cx^2)^{\frac{1}{2}}} = \int \frac{dx}{x^2\left(\frac{a}{x^2}+\frac{b}{x}+c\right)^{\frac{1}{2}}}$$
$$= -\int \frac{dz}{(az^2+bz+c)^{\frac{1}{2}}}, \text{ if } x = \frac{1}{z};$$

whereby the integral is reduced to one or other of the forms (76) or (77).

Ex. 1. $$\int \frac{dx}{x(1-2x+3x^2)^{\frac{1}{2}}} = \int \frac{dx}{x^2\left(\frac{1}{x^2}-\frac{2}{x}+3\right)^{\frac{1}{2}}}$$
$$= -\int \frac{dz}{(z^2-2z+3)^{\frac{1}{2}}}, \text{ if } x = \frac{1}{z},$$
$$= -\int \frac{d(z-1)}{\{(z-1)^2+2\}^{\frac{1}{2}}}$$
$$= -\log\{z-1+(z^2-2z+3)^{\frac{1}{2}}\}$$
$$= -\log\frac{1-x+(1-2x+3x^2)^{\frac{1}{2}}}{x}.$$

39.] The integration of $(a^2-x^2)^{\frac{1}{2}}dx$, and of $(a^2+x^2)^{\frac{1}{2}}dx$.

In these cases it is convenient to employ the method of integration by parts given by the formula,

$$\int u\,dv = uv - \int v\,du. \tag{78}$$

To determine $\int (a^2-x^2)^{\frac{1}{2}}dx$.

Let $u = (a^2-x^2)^{\frac{1}{2}}$, $\quad dv = dx$;

$\therefore\ du = \dfrac{-x\,dx}{(a^2-x^2)^{\frac{1}{2}}}$, $\quad v = x$.

Substituting which in the formula (78),

$$\int (a^2-x^2)^{\frac{1}{2}} dx = x(a^2-x^2)^{\frac{1}{2}} + \int \frac{x^2 dx}{(a^2-x^2)^{\frac{1}{2}}}.$$

Then, adding and subtracting a^2 in the numerator of the last quantity, and writing the fraction in two parts, and cancelling in the second part $(a^2-x^2)^{\frac{1}{2}}$, which occurs in both numerator and denominator, we have

$$\begin{aligned}\int (a^2-x^2)^{\frac{1}{2}} dx &= x(a^2-x^2)^{\frac{1}{2}} + \int \frac{a^2-(a^2-x^2)}{(a^2-x^2)^{\frac{1}{2}}} dx \\ &= x(a^2-x^2)^{\frac{1}{2}} + a^2 \int \frac{dx}{(a^2-x^2)^{\frac{1}{2}}} - \int (a^2-x^2)^{\frac{1}{2}} dx \\ &= x(a^2-x^2)^{\frac{1}{2}} + a^2 \sin^{-1}\frac{x}{a} - \int (a^2-x^2)^{\frac{1}{2}} dx;\end{aligned}$$

$$\therefore \quad \int (a^2-x^2)^{\frac{1}{2}} dx = \frac{x(a^2-x^2)^{\frac{1}{2}}}{2} + \frac{a^2}{2} \sin^{-1}\frac{x}{a}. \qquad (79)$$

To determine $\int (a^2+x^2)^{\frac{1}{2}} dx$.

$$\text{Let} \quad u = (a^2+x^2)^{\frac{1}{2}}, \qquad dv = dx;$$

$$du = \frac{x\,dx}{(a^2+x^2)^{\frac{1}{2}}}, \qquad v = x;$$

Then, following a process similar to that of the last integral,

$$\begin{aligned}\int (a^2+x^2)^{\frac{1}{2}} dx &= x(a^2+x^2)^{\frac{1}{2}} - \int \frac{x^2 dx}{(a^2+x^2)^{\frac{1}{2}}} \\ &= x(a^2+x^2)^{\frac{1}{2}} - \int \frac{a^2+x^2-a^2}{(a^2+x^2)^{\frac{1}{2}}} dx \\ &= x(a^2+x^2)^{\frac{1}{2}} + a^2 \int \frac{dx}{(a^2+x^2)^{\frac{1}{2}}} - \int (a^2+x^2)^{\frac{1}{2}} dx.\end{aligned}$$

$$\therefore \quad 2\int (a^2+x^2)^{\frac{1}{2}} dx = x(a^2+x^2)^{\frac{1}{2}} + a^2 \log\{x+(a^2+x^2)^{\frac{1}{2}}\},$$

by reason of equation (69);

$$\therefore \quad \int (a^2+x^2)^{\frac{1}{2}} dx = \frac{x(a^2+x^2)^{\frac{1}{2}}}{2} + \frac{a^2}{2} \log\{x+(a^2+x^2)^{\frac{1}{2}}\}. \quad (80)$$

Similarly it may be shewn that

$$\int (x^2-a^2)^{\frac{1}{2}} dx = \frac{x(x^2-a^2)^{\frac{1}{2}}}{2} - \frac{a^2}{2} \log\{x+(x^2-a^2)^{\frac{1}{2}}\}. \quad (81)$$

40.] Integration of $(2ax-x^2)^{\frac{1}{2}}dx$, and of $(2ax+x^2)^{\frac{1}{2}}dx$.

$$\int(2ax-x^2)^{\frac{1}{2}}dx = \int\{a^2-(x-a)^2\}^{\frac{1}{2}}d(x-a);$$

which latter integral, if $x-a=z$, becomes $\int\{a^2-z^2\}^{\frac{1}{2}}dz$, and is therefore of the form (79); and we have

$$\int(2ax-x^2)^{\frac{1}{2}}dx = \frac{(x-a)(2ax-x^2)^{\frac{1}{2}}}{2}+\frac{a^2}{2}\sin^{-1}\frac{x-a}{a}. \quad (82)$$

Again,

$$\int(2ax+x^2)^{\frac{1}{2}}dx = \int\{(x+a)^2-a^2\}^{\frac{1}{2}}d(x+a)$$

$$= \int(z^2-a^2)^{\frac{1}{2}}dz, \text{ if } z = x+a;$$

$$= \frac{z(z^2-a^2)^{\frac{1}{2}}}{2}-\frac{a^2}{2}\log\{z+(z^2-a^2)^{\frac{1}{2}}\}, \text{ by equation (81)},$$

$$= \frac{(x+a)(2ax+x^2)^{\frac{1}{2}}}{2}-\frac{a^2}{2}\log\{x+a+(2ax+x^2)^{\frac{1}{2}}\}. \quad (83)$$

41.] Integration of $\frac{dx}{(a^2\pm x^2)^{\frac{3}{2}}}$, and of $\frac{dx}{(a+bx+cx^2)^{\frac{3}{2}}}$.

$$\int\frac{dx}{(a^2\pm x^2)^{\frac{3}{2}}} = \int\frac{dx}{x^3\left(\frac{a^2}{x^2}\pm 1\right)^{\frac{3}{2}}}$$

$$= -\frac{1}{2a^2}\int\frac{d\left(\frac{a^2}{x^2}\pm 1\right)}{\left(\frac{a^2}{x^2}\pm 1\right)^{\frac{3}{2}}}$$

$$= -\frac{1}{2a^2}\int\left(\frac{a^2}{x^2}\pm 1\right)^{-\frac{3}{2}}d\left(\frac{a^2}{x^2}\pm 1\right)$$

$$= \frac{1}{a^2}\left(\frac{a^2}{x^2}\pm 1\right)^{-\frac{1}{2}}$$

$$= \frac{x}{a^2(a^2\pm x^2)^{\frac{1}{2}}}. \quad (84)$$

Hence $\int\frac{dx}{(a+bx+cx^2)^{\frac{3}{2}}} = \frac{1}{c^{\frac{3}{2}}}\int\frac{dx}{\left\{\left(x+\frac{b}{2c}\right)^2+\frac{4ac-b^2}{4c^2}\right\}^{\frac{3}{2}}}$.

Let therefore $x+\frac{b}{2c} = z$; $\quad\therefore\; dx = dz$;

$\frac{4ac-b^2}{4c^2} = k^2$; $\quad$ whence we have

$$\int \frac{dx}{(a+bx+cx^2)^{\frac{3}{2}}} = \frac{1}{c^{\frac{3}{2}}} \int \frac{dz}{(z^2+k^2)^{\frac{3}{2}}}; \text{ and by equation (84),}$$

$$= \frac{1}{c^{\frac{3}{2}}} \frac{1}{k^2} \frac{z}{(z^2+k^2)^{\frac{1}{2}}}$$

$$= \frac{2(2cx+b)}{(4ac-b^2)(a+bx+cx^2)^{\frac{1}{2}}}. \qquad (85)$$

42.] Examples illustrative of the preceding methods.

Ex. 1. $\int \frac{(m+nx)\,dx}{(a+bx+cx^2)^{\frac{1}{2}}}$

$$= \frac{n}{2c} \int \frac{(2cx+b)\,dx}{(a+bx+cx^2)^{\frac{1}{2}}} + \left(m - \frac{nb}{2c}\right) \int \frac{dx}{(a+bx+cx^2)^{\frac{1}{2}}}$$

$$= \frac{n}{c}(a+bx+cx^2)^{\frac{1}{2}} + \frac{2cm-nb}{2c} \int \frac{dx}{(a+bx+cx^2)^{\frac{1}{2}}};$$

the latter term of which has been integrated in Art. 37.

Ex. 2. $\int \frac{x\,dx}{(a^4-x^4)^{\frac{1}{2}}} = \frac{1}{2} \int \frac{d.x^2}{\{(a^2)^2-(x^2)^2\}^{\frac{1}{2}}}$

$$= \frac{1}{2} \int \frac{dz}{\{(a^2)^2-z^2\}^{\frac{1}{2}}}, \text{ if } x^2 = z,$$

$$= \frac{1}{2} \sin^{-1} \frac{z}{a^2}$$

$$= \frac{1}{2} \sin^{-1} \frac{x^2}{a^2}.$$

Ex. 3. $\int \frac{x\,dx}{(2ax-x^2)^{\frac{1}{2}}} = \int \frac{a-(a-x)}{(2ax-x^2)^{\frac{1}{2}}}\,dx$

$$= \int \frac{a\,dx}{(2ax-x^2)^{\frac{1}{2}}} - \int \frac{(a-x)\,dx}{(2ax-x^2)^{\frac{1}{2}}}$$

$$= a \operatorname{versin}^{-1} \frac{x}{a} - (2ax-x^2)^{\frac{1}{2}}.$$

Ex. 4. $\int \left(\frac{a+x}{a-x}\right)^{\frac{1}{2}} dx = \int \frac{(a+x)\,dx}{(a^2-x^2)^{\frac{1}{2}}}$

$$= \int \frac{a\,dx}{(a^2-x^2)^{\frac{1}{2}}} + \int \frac{x\,dx}{(a^2-x^2)^{\frac{1}{2}}}$$

$$= a \sin^{-1} \frac{x}{a} - (a^2-x^2)^{\frac{1}{2}}.$$

Ex. 5. $$\int\frac{dx}{x^3(a^2+x^2)^{\frac{1}{2}}}=\int\frac{dx}{x^3\left(\frac{a^2}{x^2}+1\right)^{\frac{1}{2}}}$$
$$=-\frac{1}{2a^2}\int\left(\frac{a^2}{x^2}+1\right)^{-\frac{1}{2}}d\left(\frac{a^2}{x^2}+1\right)$$
$$=-\frac{1}{a^2}\left(\frac{a^2}{x^2}+1\right)^{\frac{1}{2}}=-\frac{(a^2+x^2)^{\frac{1}{2}}}{a^2x}.$$

Section 4.—*Integration of Irrational Functions by Rationalization.*

43.] Many infinitesimal elements involving irrational quantities may by a judicious substitution be transformed into equivalent integral and rational functions, and consequently integrated by the methods which have been investigated in the first two sections of the present chapter; the process of such transformation is called Rationalization, and we proceed to inquire into the conditions requisite for its application in certain cases.

To find the integral of $x^m(a+bx^n)^{\frac{p}{q}}\,dx$, where m, n, p, q are constants, integral or fractional, positive or negative.

Let $a+bx^n=z^q$; $\quad\therefore\quad x=\left(\frac{z^q-a}{b}\right)^{\frac{1}{n}}$;

$$dx=\frac{qz^{q-1}}{nb^{\frac{1}{n}}}(z^q-a)^{\frac{1-n}{n}}dz;$$

$$\therefore\quad\int x^m(a+bx^n)^{\frac{p}{q}}dx=\frac{q}{nb^{\frac{m+1}{n}}}\int z^{p+q-1}(z^q-a)^{\frac{m+1}{n}-1}dz. \quad (86)$$

If therefore $\frac{m+1}{n}$ is an integer, $(z^q-a)^{\frac{m+1}{n}-1}$ is of a rational form: and if $\frac{m+1}{n}-1$ is positive, $(z^q-a)^{\frac{m+1}{n}-1}$ may be expanded by the Binomial theorem, and each term of it having been multiplied by z^{p+q-1} may be integrated by means of Art. 11. And if $\frac{m+1}{n}-1$ is negative, the integration may be accomplished by means of Section 2, and chiefly by the Reduction-formulæ of Arts. 27–30.

44.] Again, as the element-function $x^m(a+bx^n)^{\frac{p}{q}}dx$ may be written in the form

$$x^{m+\frac{np}{q}}(b+ax^{-n})^{\frac{p}{q}}dx; \tag{87}$$

and as this is the same as that of equation (86), it follows that by substituting $b+ax^{-n} = z^q$,

the result will be rational if $\frac{m+1}{n}+\frac{p}{q}$ is an integer; and we shall be able to integrate by known methods.

Hence we may by means of rationalization determine

$$\int x^m(a+bx^n)^{\frac{p}{q}}dx,$$

(1) when $\frac{m+1}{n}$ is an integer, by substituting $a+bx^n = z^q$;

(2) when $\frac{m+1}{n}+\frac{p}{q}$ is an integer, by substituting $b+ax^{-n}=z^q$.*

45.] Examples of the two preceding articles.

Ex. 1. $\int x(a+bx)^{\frac{3}{2}}dx$.

In this case $m=1$, $n=1$, $\frac{p}{q}=\frac{3}{2}$; $\therefore$ $\frac{m+1}{n}=2$, and is integral.

Let $a+bx = z^2$; $\therefore$ $x = \frac{z^2-a}{b}$; $dx = \frac{2z\,dz}{b}$;

$$\begin{aligned}\therefore \int x(a+bx)^{\frac{3}{2}}dx &= \frac{2}{b^2}\int (z^2-a)z^4 dz \\ &= \frac{2}{b^2}\int (z^6-az^4)dz \\ &= \frac{2}{b^2}\left\{\frac{z^7}{7}-\frac{az^5}{5}\right\} \\ &= \frac{2(a+bx)^{\frac{5}{2}}}{b^2}\left\{\frac{a+bx}{7}-\frac{a}{5}\right\}.\end{aligned}$$

Ex. 2. $\int \frac{x^3dx}{(a^2+x^2)^{\frac{1}{2}}}$.

In this case $m=3$, $n=2$, $\frac{p}{q}=-\frac{1}{2}$; $\frac{m+1}{n}=2$, and is integral.

* For other methods of Rationalization, and indeed for a complete collection of integrals of all kinds, the reader is referred to "Sammlung von Integraltafeln," von Ferdinand Minding; Berlin, 1849.

Let $a^2+x^2 = z^2$; $\therefore\ x = (z^2-a^2)^{\frac{1}{2}}$; $dx = \dfrac{z\,dz}{(z^2-a^2)^{\frac{1}{2}}}$;

$$\therefore \int \frac{x^3\,dx}{(a^2+x^2)^{\frac{1}{2}}} = \int (z^2-a^2)\,dz$$
$$= \frac{z^3}{3} - a^2 z$$
$$= (a^2+x^2)^{\frac{1}{2}}\,\frac{x^2-2a^2}{3}.$$

Ex. 3. $\displaystyle\int \frac{x^2\,dx}{(a^2+x^2)^{\frac{5}{2}}}$.

Here $m=2$, $n=2$, $\dfrac{p}{q} = -\dfrac{5}{2}$; $\dfrac{m+1}{n}+\dfrac{p}{q} = -1$, and is integral.

$$\therefore \int \frac{x^2\,dx}{(a^2+x^2)^{\frac{5}{2}}} = \int \frac{x^2\,dx}{x^5\left(\frac{a^2}{x^2}+1\right)^{\frac{5}{2}}}$$
$$= \int \frac{dx}{x^3\left(\frac{a^2}{x^2}+1\right)^{\frac{5}{2}}}.$$

Let $\dfrac{a^2}{x^2}+1 = z^2$; $\therefore\ x = \dfrac{a}{(z^2-1)^{\frac{1}{2}}}$, $dx = \dfrac{-az\,dz}{(z^2-1)^{\frac{3}{2}}}$;

$$\therefore \int \frac{x^2\,dx}{(a^2+x^2)^{\frac{5}{2}}} = -\frac{1}{a^2}\int \frac{dz}{z^4}$$
$$= \frac{1}{3a^2}\,\frac{1}{z^3}$$
$$= \frac{x^3}{3a^2(a^2+x^2)^{\frac{3}{2}}}.$$

46.] Integration of $\mathrm{R}\left\{x^n,\ \left(\dfrac{a+bx}{c+ex}\right)^{\frac{p}{q}},\ \left(\dfrac{a+bx}{c+ex}\right)^{\frac{r}{s}},\ \ldots\ldots\right\} dx$, where R is the symbol of a rational function.

Let l be the least common multiple of the denominators of the fractional indices; and let us assume

$$\frac{a+bx}{c+ex} = z^l; \quad \therefore\ x = \frac{cz^l-a}{b-ez^l}, \quad dx = \frac{l(bc-ae)z^{l-1}}{(b-ez^l)^2}\,dz.$$

Also $$\left(\frac{a+bx}{c+ex}\right)^{\frac{p}{q}} = z^{\frac{pl}{q}}, \quad \left(\frac{a+bx}{c+ex}\right)^{\frac{r}{s}} = z^{\frac{rl}{s}}, \ldots\ldots;$$

all of which are rational; and therefore

$$\int \mathrm{R}\left\{x^n,\ \left(\frac{a+bx}{c+ex}\right)^{\frac{p}{q}},\ \left(\frac{a+bx}{c+ex}\right)^{\frac{r}{s}},\ldots\ldots\right\} dx = \int \mathrm{R}_1(z)\,dz; \quad (88)$$

where $\mathrm{R}_1(z)$ denotes a rational function of z.

47.] If $a = e = 0$, the integral takes the form

$$\int \mathrm{R}\left\{x^m,\ x^{\frac{p}{q}},\ x^{\frac{r}{s}}, \ldots\ldots\right\} dx; \tag{89}$$

in which case we must assume as above; viz., if l is the least common multiple of the denominators $q, s, \ldots\ldots$,

$$x = z^l;$$

and then integrate.

Ex. 1. $\displaystyle\int \frac{1-x^{\frac{1}{2}}}{1-x^{\frac{1}{3}}}\,dx.$

Let $x = z^6$; $\quad\therefore\ dx = 6z^5 dz$;

$$\int \frac{1-x^{\frac{1}{2}}}{1-x^{\frac{1}{3}}}\,dx = 6\int \frac{1-z^3}{1-z^2} z^5 dz$$

$$= 6\int \left\{z^6+z^4-z^3+z^2-z+1-\frac{1}{z+1}\right\} dz$$

$$= 6\left\{\frac{z^7}{7}+\frac{z^5}{5}-\frac{z^4}{4}+\frac{z^3}{3}-\frac{z^2}{2}+z-\log(z+1)\right\}$$

$$= 6\left\{\frac{x^{\frac{7}{6}}}{7}+\frac{x^{\frac{5}{6}}}{5}-\frac{x^{\frac{2}{3}}}{4}+\frac{x^{\frac{1}{2}}}{3}-\frac{x^{\frac{1}{3}}}{2}+x^{\frac{1}{6}}-\log(x^{\frac{1}{6}}+1)\right\}.$$

Other methods of rationalizing irrational functions by means of substitution must be left to the ingenuity of the student.

In many cases however when the irrational term is of one or other of the following forms we may rationalize the expression by the substitutions indicated; if there is involved

$$(a+bx)^{\frac{1}{2}}; \qquad \text{let } x = \frac{y^2-a}{b};$$

$$(a^2+b^2x^2)^{\frac{1}{2}}; \qquad x = \frac{a}{b}\frac{1-y^2}{2y};$$

$$(a^2-b^2x^2)^{\frac{1}{2}}; \qquad x = \frac{a}{b}\frac{1-y^2}{1+y^2}.$$

SECTION 5.—*Integration of Irrational Functions by successive Reduction.*

48.] Our object in the present Section is, by means of integration by parts, to make the integrals of certain irrational functions depend on those of similar forms and lower indices, and thereby to reduce them to forms whose integrals are fundamental or have been already determined.

A careful examination of the process of reduction of the rational forms which were integrated by this method in Articles 27–30, and of one or two of the following formulæ, will give a clearer insight into the method than any general remarks and rules, and therefore I proceed at once to examples.

49.] Integration of $\dfrac{x^n\,dx}{(a^2-x^2)^{\frac{1}{2}}}$.

In all cases we use the same formula, viz. $\int u\,dv = uv - \int v\,du$.

Now $$\int \frac{x^n\,dx}{(a^2-x^2)^{\frac{1}{2}}} = \int \frac{x\,dx}{(a^2-x^2)^{\frac{1}{2}}}\,x^{n-1}.$$

Let $u = x^{n-1}$, $\qquad dv = \dfrac{x\,dx}{(a^2-x^2)^{\frac{1}{2}}}$;

$du = (n-1)x^{n-2}dx$, $\qquad v = -(a^2-x^2)^{\frac{1}{2}}$.

$$\therefore \int \frac{x^n\,dx}{(a^2-x^2)^{\frac{1}{2}}} = -x^{n-1}(a^2-x^2)^{\frac{1}{2}} + (n-1)\int x^{n-2}(a^2-x^2)^{\frac{1}{2}}dx$$

$$= -x^{n-1}(a^2-x^2)^{\frac{1}{2}} + (n-1)\int \frac{(a^2-x^2)x^{n-2}\,dx}{(a^2-x^2)^{\frac{1}{2}}}$$

$$= \cdots + (n-1)a^2\int \frac{x^{n-2}\,dx}{(a^2-x^2)^{\frac{1}{2}}} - (n-1)\int \frac{x^n\,dx}{(a^2-x^2)^{\frac{1}{2}}};$$

$$\therefore\ n\int \frac{x^n\,dx}{(a^2-x^2)^{\frac{1}{2}}} = -x^{n-1}(a^2-x^2)^{\frac{1}{2}} + (n-1)a^2\int \frac{x^{n-2}\,dx}{(a^2-x^2)^{\frac{1}{2}}},$$

$$\int \frac{x^n\,dx}{(a^2-x^2)^{\frac{1}{2}}} = -\frac{x^{n-1}(a^2-x^2)^{\frac{1}{2}}}{n} + \frac{n-1}{n}a^2\int \frac{x^{n-2}\,dx}{(a^2-x^2)^{\frac{1}{2}}}. \qquad (90)$$

By means of which the integral will at last depend on,

if n is odd, $\displaystyle\int \frac{x\,dx}{(a^2-x^2)^{\frac{1}{2}}} = -(a^2-x^2)^{\frac{1}{2}}$;

if n is even, $\displaystyle\int \frac{dx}{(a^2-x^2)^{\frac{1}{2}}} = \sin^{-1}\frac{x}{a}$.

The formula, it will be observed, is always applicable when n is odd; but if n is even, ultimately, when $n = 0$, it becomes infinite, and fails to give a determinate result.

Ex. 1. $$\int \frac{x^3\,dx}{(a^2-x^2)^{\frac{1}{2}}} = -\frac{x^2(a^2-x^2)^{\frac{1}{2}}}{3} + \frac{2a^2}{3}\int \frac{x\,dx}{(a^2-x^2)^{\frac{1}{2}}}$$

$$= -\frac{x^2(a^2-x^2)^{\frac{1}{2}}}{3} - \frac{2a^2}{3}(a^2-x^2)^{\frac{1}{2}}.$$

Ex. 2. $$\int \frac{x^2\,dx}{(a^2-x^2)^{\frac{1}{2}}} = -\frac{x}{2}(a^2-x^2)^{\frac{1}{2}} + \frac{a^2}{2}\sin^{-1}\frac{x}{a}.$$

50.] Similarly may it be shewn that

$$\int \frac{x^n\,dx}{(a^2+x^2)^{\frac{1}{2}}} = \frac{x^{n-1}(a^2+x^2)^{\frac{1}{2}}}{n} - \frac{n-1}{n}a^2\int \frac{x^{n-2}\,dx}{(a^2+x^2)^{\frac{1}{2}}}; \qquad (91)$$

and the integral will at last depend on,

if n is odd, $$\int \frac{x\,dx}{(a^2+x^2)^{\frac{1}{2}}} = (a^2+x^2)^{\frac{1}{2}};$$

if n is even, $$\int \frac{dx}{(a^2+x^2)^{\frac{1}{2}}} = \log\{x+(a^2+x^2)^{\frac{1}{2}}\}.$$

51.] Integration of $\dfrac{x^n\,dx}{(2ax-x^2)^{\frac{1}{2}}}$.

$$\int \frac{x^n\,dx}{(2ax-x^2)^{\frac{1}{2}}} = -\int \frac{(a-x)\,dx}{(2ax-x^2)^{\frac{1}{2}}}x^{n-1} + a\int \frac{x^{n-1}\,dx}{(2ax-x^2)^{\frac{1}{2}}}.$$

To integrate the first term; let

$$u = x^{n-1}, \qquad dv = \frac{(a-x)\,dx}{(2ax-x^2)^{\frac{1}{2}}};$$

$$du = (n-1)x^{n-2}dx, \qquad v = (2ax-x^2)^{\frac{1}{2}};$$

$$\begin{aligned}\int \frac{x^n\,dx}{(2ax-x^2)^{\frac{1}{2}}} &= -x^{n-1}(2ax-x^2)^{\frac{1}{2}} + (n-1)\int x^{n-2}(2ax-x^2)^{\frac{1}{2}}dx \\ &\qquad + a\int \frac{x^{n-1}\,dx}{(2ax-x^2)^{\frac{1}{2}}} \\ &= -x^{n-1}(2ax-x^2)^{\frac{1}{2}} + (n-1)\int \frac{(2ax-x^2)\,x^{n-2}\,dx}{(2ax-x^2)^{\frac{1}{2}}} \\ &\qquad + a\int \frac{x^{n-1}\,dx}{(2ax-x^2)^{\frac{1}{2}}} \\ &= -x^{n-1}(2ax-x^2)^{\frac{1}{2}} + 2a(n-1)\int \frac{x^{n-1}\,dx}{(2ax-x^2)^{\frac{1}{2}}} \\ &\qquad -(n-1)\int \frac{x^n\,dx}{(2ax-x^2)^{\frac{1}{2}}} + a\int \frac{x^{n-1}\,dx}{(2ax-x^2)^{\frac{1}{2}}};\end{aligned}$$

$$\therefore\ n\int \frac{x^n\,dx}{(2ax-x^2)^{\frac{1}{2}}} = -x^{n-1}(2ax-x^2)^{\frac{1}{2}} + a(2n-1)\int \frac{x^{n-1}\,dx}{(2ax-x^2)^{\frac{1}{2}}};$$

$$\int \frac{x^n\,dx}{(2ax-x^2)^{\frac{1}{2}}} = -\frac{x^{n-1}(2ax-x^2)^{\frac{1}{2}}}{n} + \frac{2n-1}{n}a\int \frac{x^{n-1}\,dx}{(2ax-x^2)^{\frac{1}{2}}}. \qquad (92)$$

By which process the integral will at last depend on

$$\int \frac{dx}{(2ax-x^2)^{\frac{1}{2}}} = \text{versin}^{-1}\frac{x}{a}.$$

Ex. 1. $\displaystyle\int \frac{x^2\,dx}{(2ax-x^2)^{\frac{1}{2}}}$

$$\therefore \quad = \frac{-x(2ax-x^2)^{\frac{1}{2}}}{2} + \frac{3}{2}a\int\frac{x\,dx}{(2ax-x^2)^{\frac{1}{2}}}$$

$$= \;\cdot\;\cdot\;\cdot\;\cdot\;\cdot\;\cdot\;\cdot\; + \frac{3a}{2}\left\{-(2ax-x^2)^{\frac{1}{2}} + a\,\text{versin}^{-1}\frac{x}{a}\right\}$$

$$= -\frac{x+3a}{2}(2ax-x^2)^{\frac{1}{2}} + \frac{3a^2}{2}\text{versin}^{-1}\frac{x}{a}.$$

52.] Integration of $\displaystyle\frac{x^n\,dx}{(2ax+x^2)^{\frac{1}{2}}}$.

By a process exactly similar to that of the last article it may be proved that

$$\int\frac{x^n\,dx}{(2ax+x^2)^{\frac{1}{2}}} = \frac{x^{n-1}(2ax+x^2)^{\frac{1}{2}}}{n} - \frac{2n-1}{n}a\int\frac{x^{n-1}\,dx}{(2ax+x^2)^{\frac{1}{2}}}; \quad (93)$$

so that ultimately the integral depends on

$$\int\frac{dx}{(2ax+x^2)^{\frac{1}{2}}} = \log\{x+a+(2ax+x^2)^{\frac{1}{2}}\}.$$

53.] Integration of $(a^2-x^2)^{\frac{n}{2}}dx$, where n is odd.

On comparing $\int (a^2-x^2)^{\frac{n}{2}}dx$ with $\int u\,dv$, let

$$u = (a^2-x^2)^{\frac{n}{2}}, \qquad dv = dx;$$

$$du = -nx(a^2-x^2)^{\frac{n}{2}-1}dx, \qquad v = x;$$

$$\therefore \int (a^2-x^2)^{\frac{n}{2}}dx = x(a^2+x^2)^{\frac{n}{2}} + n\int (a^2-x^2)^{\frac{n}{2}-1}x^2\,dx$$

$$= x(a^2-x^2)^{\frac{n}{2}} + n\int (a^2-x^2)^{\frac{n}{2}-1}\{a^2-(a^2-x^2)\}\,dx$$

$$= x(a^2-x^2)^{n} + na^2\int (a^2-x^2)^{\frac{n}{2}-1}dx - n\int (a^2-x^2)^{\frac{n}{2}}dx;$$

$$\therefore \quad (n+1)\int (a^2-x^2)^{\frac{n}{2}}dx = x(a^2-x^2)^{\frac{n}{2}} + na^2\int (a^2-x^2)^{\frac{n}{2}-1}dx;$$

$$\therefore \int (a^2-x^2)^{\frac{n}{2}}dx = \frac{x(a^2-x^2)^{\frac{n}{2}}}{n+1} + \frac{n}{n+1}a^2\int (a^2-x^2)^{\frac{n}{2}-1}dx. \quad (94)$$

By which means the process of reduction may be continued until

$n = -1$, when the formula becomes infinite, and therefore fails to give a determinate result; but in that case we have

$$\int \frac{dx}{(a^2-x^2)^{\frac{1}{2}}} = \sin^{-1}\frac{x}{a}.$$

Ex. 1. $\displaystyle\int (a^2-x^2)^{\frac{1}{2}}\,dx = \frac{x(a^2-x^2)^{\frac{1}{2}}}{2} + \frac{a^2}{2}\sin^{-1}\frac{x}{a}.$

Ex. 2. $\displaystyle\int (a^2-x^2)^{\frac{3}{2}}\,dx = \frac{x(a^2-x^2)^{\frac{3}{2}}}{4} + \frac{3a^2}{4}\int (a^2-x^2)^{\frac{1}{2}}\,dx$

$$= \frac{x(a^2-x^2)^{\frac{3}{2}}}{4} + \frac{3a^2}{4}\left\{\frac{x(a^2-x^2)^{\frac{1}{2}}}{2} + \frac{a^2}{2}\sin^{-1}\frac{x}{a}\right\}$$

$$= \frac{x}{4}(a^2-x^2)^{\frac{3}{2}} + \frac{3a^2x}{8}(a^2-x^2)^{\frac{1}{2}} + \frac{3a^4}{8}\sin^{-1}\frac{x}{a}.$$

54.] Similarly it may be shewn that

$$\int (a^2+x^2)^{\frac{n}{2}}\,dx = \frac{x(a^2+x^2)^{\frac{n}{2}}}{n+1} - \frac{n}{n+1}a^2\int (a^2+x^2)^{\frac{n}{2}-1}\,dx; \qquad (95)$$

and the integral finally depends on

$$\int \frac{dx}{(a^2+x^2)^{\frac{1}{2}}} = \log\{x+(a^2+x^2)^{\frac{1}{2}}\}.$$

55.] Integration of $\dfrac{dx}{(a^2-x^2)^{\frac{n}{2}}}$, where n is odd.

On comparing $\displaystyle\int \frac{dx}{(a^2-x^2)^{\frac{n}{2}}}$ with the typical form $\displaystyle\int u\,dv$,

$$\text{let } u = (a^2-x^2)^{-\frac{n}{2}}, \qquad dv = dx;$$

$$du = \frac{n\,x\,dx}{(a^2-x^2)^{\frac{n}{2}+1}}, \qquad v = x;$$

$$\therefore \int \frac{dx}{(a^2-x^2)^{\frac{n}{2}}} = \frac{x}{(a^2-x^2)^{\frac{n}{2}}} - n\int \frac{x^2\,dx}{(a^2-x^2)^{\frac{n}{2}+1}}$$

$$= \frac{x}{(a^2-x^2)^{\frac{n}{2}}} + n\int \frac{a^2-x^2-a^2}{(a^2-x^2)^{\frac{n}{2}+1}}\,dx$$

$$= \frac{x}{(a^2-x^2)^{\frac{n}{2}}} + n\int \frac{dx}{(a^2-x^2)^{\frac{n}{2}}} - na^2\int \frac{dx}{(a^2-x^2)^{\frac{n}{2}+1}};$$

$$\therefore \int \frac{dx}{(a^2-x^2)^{\frac{n}{2}+1}} = \frac{1}{na^2}\frac{x}{(a^2-x^2)^{\frac{n}{2}}} + \frac{n-1}{na^2}\int \frac{dx}{(a^2-x^2)^{\frac{n}{2}}};$$

for n substitute $n-2$; so that

$$\int \frac{dx}{(a^2-x^2)^{\frac{n}{2}}} = \frac{1}{(n-2)a^2}\frac{x}{(a^2-x^2)^{\frac{n}{2}-1}} + \frac{n-3}{n-2}\frac{1}{a^2}\int \frac{dx}{(a^2-x^2)^{\frac{n}{2}-1}}; \qquad (96)$$

so that finally, when $n = 3$,

$$\int \frac{dx}{(a^2-x^2)^{\frac{3}{2}}} = \frac{x}{a^2(a^2-x^2)^{\frac{1}{2}}}.$$

Ex. 1. $\displaystyle\int \frac{dx}{(a^2-x^2)^{\frac{5}{2}}} = \frac{1}{3a^2}\,\frac{x}{(a^2-x^2)^{\frac{3}{2}}} + \frac{2}{3a^2}\,\frac{x}{a^2(a^2-x^2)^{\frac{1}{2}}}$

56.] Integration of $\dfrac{dx}{x^n(a^2-x^2)^{\frac{1}{2}}}$.

$$\int \frac{dx}{x^n(a^2-x^2)^{\frac{1}{2}}} = \int \frac{1}{x^{n+1}}\,\frac{x\,dx}{(a^2-x^2)^{\frac{1}{2}}}.$$

Let $\displaystyle dv = \frac{x\,dx}{(a^2-x^2)^{\frac{1}{2}}}, \qquad u = x^{-(n+1)};$

$$v = -(a^2-x^2)^{\frac{1}{2}}, \qquad du = -(n+1)x^{-(n+2)}\,dx;$$

$$\therefore \int \frac{dx}{x^n(a^2-x^2)^{\frac{1}{2}}} = -\frac{(a^2-x^2)^{\frac{1}{2}}}{x^{n+1}} - (n+1)\int \frac{(a^2-x^2)^{\frac{1}{2}}}{x^{n+2}}\,dx$$

$$= -\frac{(a^2-x^2)^{\frac{1}{2}}}{x^{n+1}} - (n+1)\int \frac{(a^2-x^2)}{x^{n+2}(a^2-x^2)^{\frac{1}{2}}}\,dx$$

$$= -\frac{(a^2-x^2)^{\frac{1}{2}}}{x^{n+1}} - (n+1)a^2\int \frac{dx}{x^{n+2}(a^2-x^2)^{\frac{1}{2}}} + (n+1)\int \frac{dx}{x^n(a^2-x^2)^{\frac{1}{2}}};$$

$$\therefore \quad (n+1)a^2\int \frac{dx}{x^{n+2}(a^2-x^2)^{\frac{1}{2}}} = -\frac{(a^2-x^2)^{\frac{1}{2}}}{x^{n+1}} + n\int \frac{dx}{x^n(a^2-x^2)^{\frac{1}{2}}};$$

for n substitute $n-2$, and divide by $(n-1)a^2$;

$$\int \frac{dx}{x^n(a^2-x^2)^{\frac{1}{2}}} = -\frac{(a^2-x^2)^{\frac{1}{2}}}{(n-1)a^2x^{n-1}} + \frac{n-2}{n-1}\,\frac{1}{a^2}\int \frac{dx}{x^{n-2}(a^2-x^2)^{\frac{1}{2}}}. \quad (97)$$

Finally, when n is even, the formula is applicable, and we have

$$\int \frac{dx}{x^2(a^2-x^2)^{\frac{1}{2}}} = -\frac{(a^2-x^2)^{\frac{1}{2}}}{a^2x};$$

and when n is odd, the last integral is,

$$\int \frac{dx}{x(a^2-x^2)^{\frac{1}{2}}} = \frac{1}{a}\log\frac{a-(a^2-x^2)^{\frac{1}{2}}}{x}.$$

57.] Integration of $\dfrac{x^n\,dx}{(a+bx)^{\frac{1}{2}}}$.

On comparing $\displaystyle\int \frac{x^n\,dx}{(a+bx)^{\frac{1}{2}}}$ with $\int u\,dv$, let

$$u = x^n, \qquad dv = (a+bx)^{-\frac{1}{3}}dx;$$

$$du = nx^{n-1}dx, \qquad v = \frac{3}{2b}(a+bx)^{\frac{2}{3}};$$

$$\therefore \int\frac{x^n dx}{(a+bx)^{\frac{1}{3}}} = \frac{3x^n(a+bx)^{\frac{2}{3}}}{2b} - \frac{3n}{2b}\int x^{n-1}(a+bx)^{\frac{2}{3}}dx$$

$$= \frac{3x^n(a+bx)^{\frac{2}{3}}}{2b} - \frac{3n}{2b}\int\frac{x^{n-1}(a+bx)dx}{(a+bx)^{\frac{1}{3}}}$$

$$= \frac{3x^n(a+bx)^{\frac{2}{3}}}{2b} - \frac{3na}{2b}\int\frac{x^{n-1}dx}{(a+bx)^{\frac{1}{3}}} - \frac{3n}{2}\int\frac{x^n dx}{(a+bx)^{\frac{1}{3}}};$$

$$\therefore \left(1+\frac{3n}{2}\right)\int\frac{x^n dx}{(a+bx)^{\frac{1}{3}}} = \frac{3x^n(a+bx)^{\frac{2}{3}}}{2b} - \frac{3na}{2b}\int\frac{x^{n-1}dx}{(a+bx)^{\frac{1}{3}}};$$

$$\therefore \int\frac{x^n dx}{(a+bx)^{\frac{1}{3}}} = \frac{3x^n(a+bx)^{\frac{2}{3}}}{(3n+2)b} - \frac{3n}{3n+2}\frac{a}{b}\int\frac{x^{n-1}dx}{(a+bx)^{\frac{1}{3}}}. \qquad (98)$$

Which formula is applicable for all values of n; and at last, when $n = 0$,

$$\int\frac{dx}{(a+bx)^{\frac{1}{3}}} = \frac{3}{2b}(a+bx)^{\frac{2}{3}}.$$

Ex. 1. $$\int\frac{x^2 dx}{(a+bx)^{\frac{1}{3}}} = \frac{3x^2(a+bx)^{\frac{2}{3}}}{8b} - \frac{6}{8}\frac{a}{b}\int\frac{x\,dx}{(a+bx)^{\frac{1}{3}}}$$

$$= \frac{3x^2(a+bx)^{\frac{2}{3}}}{8b} - \frac{3a}{4b}\left\{\frac{3x(a+bx)^{\frac{2}{3}}}{5b} - \frac{3}{5}\frac{a}{b}\frac{3}{2b}(a+bx)^{\frac{2}{3}}\right\}$$

$$= (a+bx)^{\frac{2}{3}}\left\{\frac{3x^2}{8b} - \frac{9ax}{20b^2} + \frac{27a^2}{40b^3}\right\}.$$

58.] Integration of $\dfrac{x^n dx}{(a+bx+cx^2)^{\frac{1}{2}}}$.

$$\int\frac{x^n dx}{(a+bx+cx^2)^{\frac{1}{2}}} = \frac{1}{2c}\int\frac{(2cx+b-b)dx}{(a+bx+cx^2)^{\frac{1}{2}}}x^{n-1}$$

$$= \frac{1}{2c}\int\frac{(2cx+b)dx}{(a+bx+cx^2)^{\frac{1}{2}}}x^{n-1} - \frac{b}{2c}\int\frac{x^{n-1}dx}{(a+bx+cx^2)^{\frac{1}{2}}}; \qquad (99)$$

On comparing $\int\frac{(2cx+b)dx}{(a+bx+cx^2)^{\frac{1}{2}}}x^{n-1}$ with $\int u\,dv$,

let $dv = \dfrac{(2cx+b)dx}{(a+bx+cx^2)^{\frac{1}{2}}}, \qquad u = x^{n-1};$

$\therefore \quad v = 2(a+bx+cx^2)^{\frac{1}{2}}, \qquad du = (n-1)x^{n-2}dx;$

$$\int \frac{(2cx+b)\,dx}{(a+bx+cx^2)^{\frac{1}{2}}}x^{n-1}$$

$$= 2x^{n-1}(a+bx+cx^2)^{\frac{1}{2}} - 2(n-1)\int x^{n-2}(a+bx+cx^2)^{\frac{1}{2}}\,dx$$

$$= 2x^{n-1}(a+bx+cx^2)^{\frac{1}{2}} - 2(n-1)\int \frac{(a+bx+cx^2)x^{n-2}\,dx}{(a+bx+cx^2)^{\frac{1}{2}}}$$

$$= 2x^{n-1}(a+bx+cx^2)^{\frac{1}{2}} - 2(n-1)\,a\int \frac{x^{n-2}\,dx}{(a+bx+cx^2)^{\frac{1}{2}}}$$

$$-2(n-1)b\int \frac{x^{n-1}\,dx}{(a+bx+cx^2)^{\frac{1}{2}}} - 2(n-1)c\int \frac{x^n\,dx}{(a+bx+cx^2)^{\frac{1}{2}}}.$$

Substituting which in (99), and adding and reducing,

$$\int \frac{x^n\,dx}{(a+bx+cx^2)^{\frac{1}{2}}} = \frac{x^{n-1}(a+bx+cx^2)^{\frac{1}{2}}}{nc} - \frac{n-1}{n}\frac{a}{c}\int \frac{x^{n-2}\,dx}{(a+bx+cx^2)^{\frac{1}{2}}}$$

$$-\frac{2n-1}{2n}\frac{b}{c}\int \frac{x^{n-1}\,dx}{(a+bx+cx^2)^{\frac{1}{2}}}; \quad (100)$$

so that the last integrals become

$$\int \frac{x\,dx}{(a+bx+cx^2)^{\frac{1}{2}}} = \frac{(a+bx+cx^2)^{\frac{1}{2}}}{c} - \frac{b}{2c}\int \frac{dx}{(a+bx+cx^2)^{\frac{1}{2}}}; \text{ and}$$

$\int \frac{dx}{(a+bx+cx^2)^{\frac{1}{2}}}$, which has been integrated in Art. 87.

Ex. 1. $\int \frac{x^2\,dx}{(2-2x+x^2)^{\frac{1}{2}}}$; here $n = 2$, $a = 2$, $b = -2$, $c = 1$.

$$\therefore \int \frac{x^2\,dx}{(2-2x+x^2)^{\frac{1}{2}}} = \frac{x}{2}(2-2x+x^2)^{\frac{1}{2}} - \int \frac{dx}{(2-2x+x^2)^{\frac{1}{2}}}$$

$$+\frac{3}{2}\int \frac{x\,dx}{(2-2x+x^2)^{\frac{1}{2}}};$$

but $\int \frac{x\,dx}{(2-2x+x^2)^{\frac{1}{2}}} = (2-2x+x^2)^{\frac{1}{2}} + \int \frac{dx}{(2-2x+x^2)^{\frac{1}{2}}}$;

and $\int \frac{dx}{(2-2x+x^2)^{\frac{1}{2}}} = \int \frac{d(x-1)}{\{(x-1)^2+1\}^{\frac{1}{2}}}$

$$= \log\{x-1+(2-2x+x^2)^{\frac{1}{2}}\}.$$

$$\therefore \int \frac{x^2\,dx}{(2-2x+x^2)^{\frac{1}{2}}}$$

$$= \frac{x+3}{2}(2-2x+x^2)^{\frac{1}{2}} + \frac{1}{2}\log\{x-1+(2-2x+x^2)^{\frac{1}{2}}\}.$$

CHAPTER III.

INTEGRATION OF LOGARITHMIC AND CIRCULAR FUNCTIONS.

SECTION 1.—*Integration of Exponential and Logarithmic Functions.*

59.] Integration of $a^x dx$; and of $e^{mx} dx$.

Since $\quad d.a^x = \log a . a^x dx;$

$$\int \log a . a^x dx = a^x;$$

$$\therefore \int a^x dx = \frac{1}{\log a} a^x; \tag{1}$$

$$\therefore \int e^x dx = e^x. \tag{2}$$

Also since $\quad d.e^{mx} = m e^{mx} dx;$

$$\int e^{mx} dx = \frac{e^{mx}}{m}. \tag{3}$$

60.] Integration of $x^n e^{ax} dx$, where n is a positive integer.

On comparing $\int x^n e^{ax} dx$ with $\int u\, dv$;

$$\text{let} \quad u = x^n, \qquad dv = e^{ax} dx;$$

$$du = n x^{n-1} dx, \qquad v = \frac{e^{ax}}{a};$$

$$\therefore \int x^n e^{ax} dx = \frac{x^n e^{ax}}{a} - \frac{n}{a} \int x^{n-1} e^{ax} dx; \tag{4}$$

By which formula the integral ultimately becomes

$$\int e^{ax} dx = \frac{e^{ax}}{a}.$$

Ex. 1. $\displaystyle\int x^3 e^{ax} dx = \frac{x^3 e^{ax}}{a} - \frac{3}{a} \int x^2 e^{ax} dx$

$$= \frac{x^3 e^{ax}}{a} - \frac{3}{a} \left\{ \frac{x^2 e^{ax}}{a} - \frac{2}{a} \int x e^{ax} dx \right\}$$

$$= \frac{x^3 e^{ax}}{a} - \frac{3x^2 e^{ax}}{a^2} + \frac{6}{a^2} \left\{ \frac{x e^{ax}}{a} - \frac{1}{a^2} e^{ax} \right\}$$

$$= e^{ax} \left\{ \frac{x^3}{a} - \frac{3x^2}{a^2} + \frac{6x}{a^3} - \frac{6}{a^4} \right\}.$$

61.] Integration of $\frac{e^{ax}\,dx}{x^n}$.

On comparing $\int \frac{e^{ax}\,dx}{x^n}$ with $\int u\,dv$,

$$\text{let } \quad u = e^{ax}, \qquad dv = \frac{dx}{x^n};$$

$$du = a e^{ax}\,dx, \qquad v = -\frac{1}{(n-1)\,x^{n-1}};$$

$$\therefore \quad \int \frac{e^{ax}\,dx}{x^n} = -\frac{e^{ax}}{(n-1)\,x^{n-1}} + \frac{a}{n-1}\int \frac{e^{ax}\,dx}{x^{n-1}}. \qquad (5)$$

By which formula the integral finally becomes

$$\int \frac{e^{ax}\,dx}{x},$$

which does not admit of integration in finite terms, but may be expressed in a series: for since

$$e^{ax} = 1 + \frac{ax}{1} + \frac{a^2x^2}{1.2} + \frac{a^3x^3}{1.2.3} + \ldots\ldots,$$

$$\frac{e^{ax}}{x} = \frac{1}{x} + a + \frac{a^2x}{1.2} + \frac{a^3x^2}{1.2.3} + \ldots\ldots;$$

$$\therefore \quad \int \frac{e^{ax}\,dx}{x} = \log x + ax + \frac{a^2x^2}{1.2^2} + \frac{a^3x^3}{1.2.3^2} + \ldots\ldots. \qquad (6)$$

Ex. 1. $\int \frac{e^x\,dx}{x^2} = -\frac{e^x}{x} + \int \frac{e^x\,dx}{x}$

$$= -\frac{e^x}{x} + \log x + \frac{x}{1} + \frac{x^2}{1.2^2} + \frac{x^3}{1.2.3^2} + \ldots.$$

62.] The preceding are the integrals of the simpler exponential functions; other combinations however often admit of reduction to algebraic forms by means of substitution, and thereby of integration by the methods of the last Chapter; of these some examples are subjoined.

Ex. 1. $\int \frac{e^{2x}-1}{e^{2x}+1}\,dx = \int \frac{e^x - e^{-x}}{e^x + e^{-x}}\,dx$

$$= \log(e^x + e^{-x}).$$

Ex. 2. $\int e^{e^x} e^x\,dx = e^{e^x}$.

Ex. 3. $\int \frac{e^x\,x\,dx}{(1+x)^2} = \int \left\{ \frac{e^x\,dx}{1+x} - \frac{e^x\,dx}{(1+x)^2} \right\}$

$$= \int \left\{ \frac{1}{1+x}\,d.e^x + e^x\,d.\frac{1}{1+x} \right\}$$

$$= \frac{e^x}{1+x}.$$

63.] Integration of $x^m (\log x)^n dx$.

On comparing $\int (\log x)^n x^m dx$ with $\int u\, dv$,

$$\text{let} \quad u = (\log x)^n, \qquad dv = x^m dx;$$

$$du = n(\log x)^{n-1}\frac{dx}{x}, \qquad v = \frac{x^{m+1}}{m+1};$$

$$\therefore \int (\log x)^n x^m dx = \frac{(\log x)^n x^{m+1}}{m+1} - \frac{n}{m+1}\int (\log x)^{n-1} x^m dx; \quad (7)$$

by means of which the last integral becomes

$$\int x^m dx = \frac{x^{m+1}}{m+1}.$$

Ex. 1. $$\int x^4 (\log x)^2 dx = \frac{(\log x)^2 x^5}{5} - \frac{2}{5}\int \log x . x^4 dx$$

$$= \frac{x^5 (\log x)^2}{5} - \frac{2}{5}\left\{\frac{x^5 \log x}{5} - \frac{x^5}{25}\right\}$$

$$= \frac{x^5}{5}\left\{(\log x)^2 - \frac{2}{5}\log x + \frac{2}{25}\right\}.$$

Ex. 2. $$\int \log x\, dx = x \log x - x.$$

64.] Examples of integration of various logarithmic functions.

Ex. 1. $$\int \frac{dx}{x}\log x = \int \log x . d(\log x)$$

$$= \frac{(\log x)^2}{2}.$$

Ex. 2. $$\int \frac{dx}{x \log x} = \int \frac{d.\log x}{\log x} = \log \log x$$

$$= \log^2 x.$$

Ex. 3. $$\int \frac{x \log x\, dx}{(a^2+x^2)^{\frac{1}{2}}} = \int \frac{x\, dx}{(a^2+x^2)^{\frac{1}{2}}} \log x$$

$$= (a^2+x^2)^{\frac{1}{2}} \log x - \int \frac{(a^2+x^2)^{\frac{1}{2}}}{x} dx$$

$$= (a^2+x^2)^{\frac{1}{2}} \log x - \int \frac{a^2 dx}{x(a^2+x^2)^{\frac{1}{2}}} - \int \frac{x\, dx}{(a^2+x^2)^{\frac{1}{2}}}$$

$$= (a^2+x^2)^{\frac{1}{2}} \log x + a \log \frac{a+(a^2+x^2)^{\frac{1}{2}}}{x} - (a^2+x^2)^{\frac{1}{2}}.$$

SECTION 2.—*Integration of Circular Functions.*

65.] Integration of the fundamental circular functions.

Since $\quad d.\cos mx = -m \sin mx \, dx;$

$$\int \sin mx \, dx = -\frac{\cos mx}{m}; \tag{8}$$

$$\therefore \int \sin x \, dx = -\cos x. \tag{9}$$

Again, since $\quad d.\sin mx = m \cos mx \, dx;$

$$\int \cos mx . dx = \frac{\sin mx}{m}; \tag{10}$$

$$\therefore \int \cos x \, dx = \sin x. \tag{11}$$

Again, since $\quad d.\tan mx = m (\sec mx)^2 dx;$

$$\int (\sec mx)^2 dx = \frac{1}{m} \tan mx; \tag{12}$$

$$\therefore \int (\sec x)^2 dx = \int \{1 + (\tan x)^2\} dx = \tan x. \tag{13}$$

Again, since $\quad d.\cot mx = -m (\operatorname{cosec} mx)^2 dx;$

$$\int (\operatorname{cosec} mx)^2 dx = \int \frac{dx}{(\sin mx)^2} = -\frac{\cot mx}{m}; \tag{14}$$

$$\therefore \int (\operatorname{cosec} x)^2 dx = \int \frac{dx}{(\sin x)^2} = -\cot x. \tag{15}$$

Similarly, $$\int \sec mx \tan mx \, dx = \frac{1}{m} \sec mx; \tag{16}$$

$$\int \operatorname{cosec} mx \cot mx \, dx = -\frac{1}{m} \operatorname{cosec} mx. \tag{17}$$

66.] And all the preceding formulæ are of course true when for x any function of x, say $f(x)$, is substituted, provided that dx is replaced by $f'(x)\,dx$. Thus

$$\int \sin (mx+n)\, dx = \frac{1}{m} \int \sin (mx+n)\, d(mx+n)$$

$$= -\frac{1}{m} \cos (mx+n).$$

$$\int \sin (x^3)\, x^2 dx = \frac{1}{3} \int \sin (x^3)\, d.x^3 = -\frac{1}{3} \cos (x^3).$$

$$\int \cos (mx^2 + nx + p)\,(2mx+n)\, dx = \sin (mx^2 + nx + p).$$

67.] Integration of other circular functions by transformation into the fundamental forms.

$$\int \tan x\, dx = \int \frac{\sin x\, dx}{\cos x} = -\int \frac{d.\cos x}{\cos x}$$
$$= -\log \cos x = \log \sec x. \qquad (18)$$

$$\int \cot x\, dx = \int \frac{\cos x\, dx}{\sin x} = \int \frac{d.\sin x}{\sin x}$$
$$= \log \sin x. \qquad (19)$$

$$\int \frac{dx}{\sin x} = \int \frac{\sin x\, dx}{(\sin x)^2} = -\int \frac{d.\cos x}{1-(\cos x)^2}$$
$$= -\frac{1}{2}\log \frac{1+\cos x}{1-\cos x}, \text{ by (16), Art. 15,}$$
$$= \log\left(\frac{1-\cos x}{1+\cos x}\right)^{\frac{1}{2}} = \log \tan \frac{x}{2}. \qquad (20)$$

$$\int \frac{dx}{\cos x} = \int \frac{\cos x\, dx}{(\cos x)^2} = \int \frac{d.\sin x}{1-(\sin x)^2}$$
$$= \frac{1}{2}\log \frac{1+\sin x}{1-\sin x} = \log \frac{\cos\frac{x}{2}+\sin\frac{x}{2}}{\cos\frac{x}{2}-\sin\frac{x}{2}}$$
$$= \log \tan\left(\frac{\pi}{4}+\frac{x}{2}\right). \qquad (21)$$

$$\int \frac{dx}{\sin x \cos x} = \int \frac{(\sec x)^2 dx}{\tan x}$$
$$= \int \frac{d.\tan x}{\tan x}$$
$$= \log \tan x. \qquad (22)$$

$$\int \frac{dx}{(\sin x)^2(\cos x)^2} = \int \frac{(\sin x)^2+(\cos x)^2}{(\sin x)^2(\cos x)^2}dx$$
$$= \int \{(\sec x)^2 + (\operatorname{cosec} x)^2\}\, dx$$
$$= \tan x - \cot x. \qquad (23)$$

$$\int (\tan x)^2 dx = \int \{(\sec x)^2 - 1\}\, dx$$
$$= \tan x - x. \qquad (24)$$

$$\int (\cot x)^2 dx = \int \{(\operatorname{cosec} x)^2 - 1\}\, dx$$
$$= -\cot x - x. \qquad (25)$$

$$\int \frac{dx}{a+b\cos x} = \int \frac{dx}{a\left\{\left(\cos\frac{x}{2}\right)^2 + \left(\sin\frac{x}{2}\right)^2\right\} + b\left\{\left(\cos\frac{x}{2}\right)^2 - \left(\sin\frac{x}{2}\right)^2\right\}}$$

$$= \int \frac{\left(\sec\frac{x}{2}\right)^2 dx}{a+b+(a-b)\left(\tan\frac{x}{2}\right)^2}; \tag{26}$$

(α) let a be greater than b; then the last expression becomes

$$\int \frac{dx}{a+b\cos x} = \frac{2}{a-b}\int \frac{d.\tan\frac{x}{2}}{\frac{a+b}{a-b} + \left(\tan\frac{x}{2}\right)^2}$$

$$= \frac{2}{(a^2-b^2)^{\frac{1}{2}}}\tan^{-1}\left\{\left(\frac{a-b}{a+b}\right)^{\frac{1}{2}}\tan\frac{x}{2}\right\}. \tag{27}$$

(β) let a be less than b, then (26) becomes

$$\int \frac{dx}{a+b\cos x} = \frac{2}{b-a}\int \frac{d.\tan\frac{x}{2}}{\frac{b+a}{b-a} - \left(\tan\frac{x}{2}\right)^2},$$

$$= \frac{1}{(b^2-a^2)^{\frac{1}{2}}}\log\frac{(b+a)^{\frac{1}{2}} + (b-a)^{\frac{1}{2}}\tan\frac{x}{2}}{(b+a)^{\frac{1}{2}} - (b-a)^{\frac{1}{2}}\tan\frac{x}{2}} \tag{28}$$

$$= \frac{1}{(b^2-a^2)^{\frac{1}{2}}}\log\frac{(b+a)^{\frac{1}{2}}\cos\frac{x}{2} + (b-a)^{\frac{1}{2}}\sin\frac{x}{2}}{(b+a)^{\frac{1}{2}}\cos\frac{x}{2} - (b-a)^{\frac{1}{2}}\sin\frac{x}{2}}. \tag{29}$$

$$\int \frac{dx}{a+b\sin x} = \int \frac{dx}{a\left\{\left(\sin\frac{x}{2}\right)^2 + \left(\cos\frac{x}{2}\right)^2\right\} + 2b\sin\frac{x}{2}\cos\frac{x}{2}}$$

$$= \int \frac{\left(\sec\frac{x}{2}\right)^2 dx}{a + a\left(\tan\frac{x}{2}\right)^2 + 2b\tan\frac{x}{2}}$$

$$= \frac{2}{a}\int \frac{d\left(\tan\frac{x}{2} + \frac{b}{a}\right)}{\left(\tan\frac{x}{2} + \frac{b}{a}\right)^2 + \frac{a^2-b^2}{a^2}}$$

$$\int \frac{dx}{a+b\sin x} = \frac{2}{(a^2-b^2)^{\frac{1}{2}}} \tan^{-1}\left\{\frac{a}{(a^2-b^2)^{\frac{1}{2}}}\left(\tan\frac{x}{2}+\frac{b}{a}\right)\right\}, \text{ if } b<a,$$

$$= \frac{2}{(a^2-b^2)^{\frac{1}{2}}} \tan^{-1} \frac{a\sin\frac{x}{2}+b\cos\frac{x}{2}}{(a^2-b^2)^{\frac{1}{2}}\cos\frac{x}{2}}. \qquad (30)$$

$$= \frac{1}{(b^2-a^2)^{\frac{1}{2}}} \log \frac{\tan\frac{x}{2}+\frac{b}{a}-\frac{(b^2-a^2)^{\frac{1}{2}}}{a}}{\tan\frac{x}{2}+\frac{b}{a}+\frac{(b^2-a^2)^{\frac{1}{2}}}{a}}, \text{ if } b>a,$$

$$= \frac{1}{(b^2-a^2)^{\frac{1}{2}}} \log \frac{a\tan\frac{x}{2}+b-(b^2-a^2)^{\frac{1}{2}}}{a\tan\frac{x}{2}+b+(b^2-a^2)^{\frac{1}{2}}}. \qquad (31)$$

$$\int \frac{dx}{a+b\tan x} = \int \frac{\cos x\, dx}{a\cos x+b\sin x}$$

$$= \frac{1}{b}\int \frac{(b\cos x-a\sin x)\,dx}{b\sin x+a\cos x} + \frac{a}{b}\int \frac{\sin x\, dx}{b\sin x+a\cos x}$$

$$= \frac{\log(b\sin x+a\cos x)}{b} + \frac{a}{b^2}\int \frac{b\sin x+a\cos x}{b\sin x+a\cos x}dx - \frac{a^2}{b^2}\int \frac{\cos x\, dx}{b\sin x+a\cos x}$$

$$= \frac{\log(b\sin x+a\cos x)}{b} + \frac{a}{b^2}x - \frac{a^2}{b^2}\int \frac{dx}{a+b\tan x};$$

$$\therefore \quad \left(1+\frac{a^2}{b^2}\right)\int \frac{dx}{a+b\tan x} = \frac{\log(b\sin x+a\cos x)}{b} + \frac{a}{b^2}x;$$

$$\int \frac{dx}{a+b\tan x} = \frac{b}{a^2+b^2}\log(b\sin x+a\cos x) + \frac{ax}{a^2+b^2}. \qquad (32)$$

$$\int \sin mx \cos nx\, dx = \frac{1}{2}\int \{\sin(m+n)x+\sin(m-n)x\}\, dx$$

$$= -\frac{1}{2}\left\{\frac{\cos(m+n)x}{m+n}+\frac{\cos(m-n)x}{m-n}\right\}. \qquad (33)$$

$$\int \cos mx \cos nx\, dx = \frac{1}{2}\int \{\cos(m+n)x+\cos(m-n)x\}\, dx$$

$$= \frac{1}{2}\left\{\frac{\sin(m+n)x}{m+n}+\frac{\sin(m-n)x}{m-n}\right\}. \qquad (34)$$

Similarly may the integrals of the product of three or more sines and cosines be determined.

Ex. 1. $$\int \frac{dx}{2+\cos x} = \int \frac{dx}{2+\left(\cos\frac{x}{2}\right)^2 - \left(\sin\frac{x}{2}\right)^2}$$

$$= \int \frac{\left(\sec\frac{x}{2}\right)^2 dx}{2+2\left(\tan\frac{x}{2}\right)^2 + 1 - \left(\tan\frac{x}{2}\right)^2}$$

$$= 2\int \frac{d.\tan\frac{x}{2}}{3+\left(\tan\frac{x}{2}\right)^2}$$

$$= \frac{2}{3^{\frac{1}{2}}} \tan^{-1}\left\{\frac{1}{3^{\frac{1}{2}}}\tan\frac{x}{2}\right\}.$$

Ex. 2. $$\int \sin\frac{2x}{3}\cos\frac{4x}{3}\,dx = \frac{1}{2}\int\left(\sin 2x - \sin\frac{2x}{3}\right)dx$$

$$= -\frac{1}{2}\left\{\frac{\cos 2x}{2} - \frac{3}{2}\cos\frac{2x}{3}\right\}$$

$$= \frac{3}{4}\cos\frac{2x}{3} - \frac{1}{4}\cos 2x.$$

Ex. 3. $$\int \sin(mx+\alpha)\cos(nx+\beta)\,dx$$

$$= \frac{1}{2}\int\Big(\sin\{(m+n)x+\alpha+\beta\} + \sin\{(m-n)x+\alpha-\beta\}\Big)dx$$

$$= -\frac{1}{2}\left\{\frac{\cos\{(m+n)x+\alpha+\beta\}}{m+n} + \frac{\cos\{(m-n)x+\alpha-\beta\}}{m-n}\right\}.$$

Ex. 4. $$\int \frac{dx}{a(\cos x)^2 + b(\sin x)^2} = \int \frac{dx\,(\sec x)^2}{a+b(\tan x)^2}$$

$$= \int \frac{d.\tan x}{a+b(\tan x)^2}$$

$$= \frac{1}{(ab)^{\frac{1}{2}}}\tan^{-1}\left\{\frac{b^{\frac{1}{2}}\tan x}{a^{\frac{1}{2}}}\right\}.$$

68.] Integration of $(\sin x)^n dx$, and of $(\cos x)^n dx$.

These integrals may be determined by integration by parts; for since

$$\int (\sin x)^n dx = \int (\sin x)^{n-1} \sin x\, dx,$$

on comparing the above with $\int u\,dv$,

let $u = (\sin x)^{n-1}$, $\qquad dv = \sin x\,dx$;

$du = (n-1)(\sin x)^{n-2}\cos x\,dx$, $\qquad v = -\cos x$;

$$\therefore \int (\sin x)^n dx$$

$$= -\cos x (\sin x)^{n-1} + (n-1) \int (\sin x)^{n-2} (\cos x)^2 dx$$

$$= -\cos x (\sin x)^{n-1} + (n-1) \int (\sin x)^{n-2} \{1-(\sin x)^2\} dx$$

$$= -\cos x (\sin x)^{n-1} + (n-1) \int (\sin x)^{n-2} dx - (n-1) \int (\sin x)^n dx;$$

$$\therefore \int (\sin x)^n dx = \frac{-\cos x (\sin x)^{n-1}}{n} + \frac{n-1}{n} \int (\sin x)^{n-2} dx. \quad (35)$$

By means of which the integral is finally reduced to,

if n is even, $\int dx = x$;

if n is odd, $\int \sin x\, dx = -\cos x$.

Again, in a similar manner

$$\int (\cos x)^n dx = \frac{\sin x (\cos x)^{n-1}}{n} + \frac{n-1}{n} \int (\cos x)^{n-2} dx; \quad (36)$$

and the last integral becomes,

if n is even, $\int dx = x$;

if n is odd, $\int \cos x\, dx = \sin x$.

If n is odd, the following method is more convenient:

Let $n = 2m+1$;

$$\therefore \int (\sin x)^{2m+1} dx = \int \{1-(\cos x)^2\}^m \sin x\, dx$$

$$= -\int \left\{1 - m(\cos x)^2 + \frac{m(m-1)}{1.2} (\cos x)^4 - \dots \right\} d.\cos x$$

$$= -\left\{\cos x - \frac{m(\cos x)^3}{3} + \frac{m(m-1)}{1.2} \frac{(\cos x)^5}{5} - \dots \right\}. \quad (37)$$

Also in a similar manner

$$\int (\cos x)^{2m+1} dx = \int (\cos x)^{2m} \cos x\, dx$$

$$= \int \{1-(\sin x)^2\}^m d.\sin x$$

$$= \int \left\{1 - m(\sin x)^2 + \frac{m(m-1)}{1.2} (\sin x)^4 - \dots \right\} d.\sin x$$

$$= \sin x - \frac{m(\sin x)^3}{3} + \frac{m(m-1)}{1.2} \frac{(\sin x)^5}{5} - \dots \quad (38)$$

Ex. 1. $$\int (\sin x)^4 dx = \frac{-\cos x (\sin x)^3}{4} + \frac{3}{4}\int (\sin x)^2 dx$$

$$= -\frac{\cos x (\sin x)^3}{4} + \frac{3}{4}\left\{\frac{-\cos x \sin x}{2} + \frac{x}{2}\right\}$$

$$= \frac{-\cos x (\sin x)^3}{4} - \frac{3 \sin x \cos x}{8} + \frac{3x}{8}.$$

Ex. 2. $$\int (\cos x)^5 dx = \int \{1-(\sin x)^2\}^2 \cos x\, dx$$

$$= \int \{1-2(\sin x)^2 + (\sin x)^4\}\, d.\sin x$$

$$= \sin x - \frac{2}{3}(\sin x)^3 + \frac{(\sin x)^5}{5}.$$

69.] Integration of $(\sin x)^n dx$, and of $(\cos x)^n dx$, in terms of sines and cosines of multiple arcs.

As $(\sin x)^n$ and $(\cos x)^n$ may be expressed in terms of sines and cosines of multiple arcs by the method of Arts. 62, 63, Vol. I, and as these latter are easily integrated; this process is convenient, especially for the purpose of definite integration; but since the general term admits of various forms according to the form of n, the application of the method will be better exhibited by means of examples.

Ex. 1. To integrate $(\sin x)^6 dx$.

Employing the same abbreviation as in Art. 63, Vol. I, let

$$2\sqrt{-1} \sin x = z - \frac{1}{z};$$

$$\therefore \quad -2^6 (\sin x)^6 = z^6 - 6z^4 + 15z^2 - 20 + \frac{15}{z^2} - \frac{6}{z^4} + \frac{1}{z^6}$$

$$= 2\cos 6x - 12\cos 4x + 30\cos 2x - 20;$$

$$(\sin x)^6 = -\frac{1}{2^5}\{\cos 6x - 6\cos 4x + 15\cos 2x - 10\};$$

$$\therefore \quad \int (\sin x)^6 dx = -\frac{1}{2^5}\left\{\frac{\sin 6x}{6} - \frac{6\sin 4x}{4} + \frac{15\sin 2x}{2} - 10x\right\}.$$

Ex. 2. $\int (\cos x)^3 dx$.

$$\text{let} \quad 2\cos x = z + \frac{1}{z};$$

$$\therefore \quad 2^3 (\cos x)^3 = z^3 + 3z + \frac{3}{z} + \frac{1}{z^3}$$

$$= 2\cos 3x + 6\cos x;$$

$$\therefore \quad \int (\cos x)^3 \, dx = \frac{1}{2^2} \int \Big\{ \cos 3x + 3 \cos x \Big\} \, dx$$

$$= \frac{1}{2^2} \Big\{ \frac{\sin 3x}{3} + 3 \sin x \Big\}.$$

70.] Integration of $\frac{dx}{(\sin x)^n}$, and of $\frac{dx}{(\cos x)^n}$.

$$\int \frac{dx}{(\sin x)^n} = \int \frac{(\cos x)^2 + (\sin x)^2}{(\sin x)^n} \, dx$$

$$= \int \frac{\cos x \, dx}{(\sin x)^n} \cos x + \int \frac{dx}{(\sin x)^{n-2}};$$

and, integrating by parts,

$$\int \frac{\cos x \, dx}{(\sin x)^n} \cos x = -\frac{\cos x}{(n-1)(\sin x)^{n-1}} - \frac{1}{n-1} \int \frac{dx}{(\sin x)^{n-2}};$$

$$\therefore \quad \int \frac{dx}{(\sin x)^n} = -\frac{\cos x}{(n-1)(\sin x)^{n-1}} + \frac{n-2}{n-1} \int \frac{dx}{(\sin x)^{n-2}}; \qquad (39)$$

by means of which the last integral becomes,

if n is even, $\int \frac{dx}{(\sin x)^2} = -\frac{\cos x}{\sin x} = -\cot x$;

if n is odd, $\int \frac{dx}{\sin x} = \log \tan \frac{x}{2}$, by (20), Art. 67.

Again, by a similar process,

$$\int \frac{dx}{(\cos x)^n} = \frac{\sin x}{(n-1)(\cos x)^{n-1}} + \frac{n-2}{n-1} \int \frac{dx}{(\cos x)^{n-2}}; \qquad (40)$$

by means of which the last integral becomes,

if n is even, $\int \frac{dx}{(\cos x)^2} = \frac{\cos x}{\sin x} = \tan x$;

if n is odd, $\int \frac{dx}{\cos x} = \log \tan \left(\frac{\pi}{4} + \frac{x}{2} \right)$, by (21), Art. 67.

In cases wherein n is even, the integrals are more conveniently found in terms of cotangents and tangents: thus, if $n = 2m$,

$$\int \frac{dx}{(\sin x)^n} = \int (\operatorname{cosec} x)^{2m} \, dx$$

$$= \int \{1 + (\cot x)^2\}^{m-1} (\operatorname{cosec} x)^2 \, dx$$

$$= -\int \Big\{ 1 + (m-1)(\cot x)^2 + \frac{(m-1)(m-2)}{1.2} (\cot x)^4 + \dots \Big\} \, d.\cot x$$

$$= -\Big\{ \cot x + (m-1) \frac{(\cot x)^3}{3} + \frac{(m-1)(m-2)}{1.2} \frac{(\cot x)^5}{5} + \dots \Big\}. \quad (41)$$

Similarly

$$\int \frac{dx}{(\cos x)^{2m}} = \int (\sec x)^{2m-2} (\sec x)^2 dx$$

$$= \int \{1+(\tan x)^2\}^{m-1} d.\tan x$$

$$= \tan x + (m-1)\frac{(\tan x)^3}{3} + \dots \quad (42)$$

Ex. 1. $\int \frac{dx}{(\sin x)^3} = -\frac{\cos x}{2(\sin x)^2} + \frac{1}{2}\log\tan\frac{x}{2}.$

Ex. 2. $\int \frac{dx}{(\cos x)^4} = \int (\sec x)^2 (\sec x)^2 dx$

$$= \int \{1+(\tan x)^2\}\, d.\tan x$$

$$= \tan x + \frac{(\tan x)^3}{3}.$$

71.] Integration of $(\sin x)^m (\cos x)^n dx$.

The value of this integral can easily be found when either m or n or both are uneven positive integers; and when $m+n$ is an even negative integer.

(α) Let $m = 2r+1$;

$$\int (\sin x)^m (\cos x)^n dx = \int (\sin x)^{2r+1} (\cos x)^n dx$$

$$= \int \{1-(\cos x)^2\}^r (\cos x)^n \sin x\, dx$$

$$= -\int \{1-(\cos x)^2\}^r (\cos x)^n d.\cos x; \quad (43)$$

of which expression each term, after expanding $\{1-(\cos x)^2\}^r$, may be integrated immediately.

(β) Similarly may $(\sin x)^m (\cos x)^n dx$ be integrated, when n is of the form $2r+1$.

(γ) Let $m+n = -2r$;

$$\therefore \int (\sin x)^m (\cos x)^n dx = \int (\tan x)^m (\cos x)^{n+m} dx$$

$$= \int (\tan x)^m (\sec x)^{2r} dx$$

$$= \int (\tan x)^m \{1+(\tan x)^2\}^{r-1} d.\tan x; \quad (44)$$

each term of which after expansion is immediately integrable.

Ex. 1. $\int (\sin x)^2 (\cos x)^3 dx = \int (\sin x)^2 \{1-(\sin x)^2\}\, d.\sin x$

$$= \frac{(\sin x)^3}{3} - \frac{(\sin x)^5}{5}.$$

Ex. 2. $$\int (\sin x)^3 (\cos x)^4 dx = \int \{1-(\cos x)^2\} (\cos x)^4 \sin x\, dx$$
$$= -\int \{(\cos x)^4 - (\cos x)^6\}\, d.\cos x$$
$$= -\frac{(\cos x)^5}{5} + \frac{(\cos x)^7}{7}.$$

Ex. 3. $$\int \frac{(\sin x)^2}{(\cos x)^4} dx = \int (\tan x)^2 (\sec x)^2 dx$$
$$= \frac{(\tan x)^3}{3}.$$

72.] When neither of the three above-mentioned conditions as to m and n is fulfilled, we must have recourse to integration by parts, and proceed as follows:

$$\int (\sin x)^m (\cos x)^n dx = \int (\sin x)^m \cos x\, dx\, (\cos x)^{n-1};$$

on comparing which with the typical form $\int u\, dv$,

let $dv = (\sin x)^m \cos x\, dx,\quad u = (\cos x)^{n-1};$

$$v = \frac{(\sin x)^{m+1}}{m+1}, \qquad du = -(n-1)(\cos x)^{n-2} \sin x\, dx;$$

$$\therefore \int (\sin x)^m (\cos x)^n dx$$
$$= \frac{(\sin x)^{m+1}(\cos x)^{n-1}}{m+1} + \frac{n-1}{m+1}\int (\sin x)^{m+2}(\cos x)^{n-2} dx; \quad (45)$$

which is an useful form when m is negative and n is positive.

Also similarly

$$\int (\sin x)^m (\cos x)^n dx$$
$$= \frac{-(\cos x)^{n+1}(\sin x)^{m-1}}{n+1} + \frac{m-1}{n+1}\int (\cos x)^{n+2}(\sin x)^{m-2} dx; \quad (46)$$

which is useful when n is negative and m is positive.

Also the last term of the right-hand member of (46) may be written in the form,

$$\int (\cos x)^n (\sin x)^{m-2} (\cos x)^2 dx$$
$$= \int (\cos x)^n (\sin x)^{m-2} \{1-(\sin x)^2\}\, dx$$
$$= \int (\cos x)^n (\sin x)^{m-2} dx - \int (\cos x)^n (\sin x)^m dx;$$

substituting which in (46) and reducing, we have

$$\int (\sin x)^m (\cos x)^n dx$$

$$= -\frac{(\cos x)^{n+1} (\sin x)^{m-1}}{n+m} + \frac{m-1}{n+m} \int (\cos x)^n (\sin x)^{m-2} dx. \quad (47)$$

Similarly may other formulæ be constructed; but the form of the element to be integrated will usually suggest various modifications by which it may be transformed into some known integral.

Ex. 1. $\int \frac{(\cos x)^4}{(\sin x)^2} dx = -\frac{(\cos x)^3}{\sin x} - 3\int (\cos x)^2 dx$

$$= -\frac{(\cos x)^3}{\sin x} - 3\left\{\frac{\sin x \cos x}{2} + \frac{x}{2}\right\}.$$

Ex. 2. $\int \frac{dx}{(\sin x)^5 \cos x} = \int \frac{(\sec x)^6 dx}{(\tan x)^5}$

$$= \int \frac{\{1+(\tan x)^2\}^2}{(\tan x)^5} d.\tan x$$

$$= \int \{(\tan x)^{-5} + 2(\tan x)^{-3} + (\tan x)^{-1}\} d.\tan x$$

$$= -\frac{1}{4(\tan x)^4} - \frac{1}{(\tan x)^2} + \log \tan x.$$

73.] Integration of $(\tan x)^n dx$, and of $(\cot x)^n dx$.

$$\int (\tan x)^n dx = \int (\tan x)^{n-2} (\tan x)^2 dx$$

$$= \int (\tan x)^{n-2} \{(\sec x)^2 - 1\} dx$$

$$= \int (\tan x)^{n-2} d.\tan x - \int (\tan x)^{n-2} dx$$

$$= \frac{(\tan x)^{n-1}}{n-1} - \int (\tan x)^{n-2} dx. \quad (48)$$

Similarly

$$\int (\cot x)^n dx = -\frac{(\cot x)^{n-1}}{n-1} - \int (\cot x)^{n-2} dx. \quad (49)$$

These formulæ give definite results for even values of n, but ultimately fail when n is odd; in which cases however, by (18) and (19), Art. 67,

$$\int \tan x\, dx = \log \sec x,$$

$$\int \cot x\, dx = \log \sin x.$$

74.] Integration of $x^n \cos x\, dx$.

$$\int x^n \cos x\, dx = x^n \sin x - n\int x^{n-1} \sin x\, dx$$

$$= x^n \sin x - n\left\{-x^{n-1}\cos x + (n-1)\int x^{n-2}\cos x\, dx\right\}$$

$$= x^n \sin x + nx^{n-1}\cos x - n(n-1)\int x^{n-2}\cos x\, dx. \quad (50)$$

Similarly it may be shewn that

$$\int x^n \sin x\, dx = -x^n\cos x + nx^{n-1}\sin x - n(n-1)\int x^{n-2}\sin x\, dx. \quad (51)$$

By means of which the last integral becomes

$$\int \cos x\, dx = \sin x, \text{ or } \int \sin x\, dx = -\cos x.$$

Ex. 1. $\int x^3 \cos x\, dx = x^3 \sin x + 3x^2\cos x - 6x\sin x - 6\cos x.$

Similarly may formulæ be constructed for determining

$$\int x^n \sin kx\, dx, \quad \text{and} \quad \int x^n \cos kx\, dx. \quad (52)$$

And hence we may integrate infinitesimal elements of the forms

$$x^n(\sin x)^m\, dx, \qquad x^n(\cos x)^m\, dx; \quad (53)$$

for if $(\sin x)^m$ and $(\cos x)^m$ are expressed in terms of the sines and cosines of the multiple arcs, by means of Arts. 62, 63, Vol. I, then each term of the integral will be of one of the forms (52), and may be integrated accordingly.

75.] Integration of $e^{ax}(\cos x)^n dx$, and of $e^{ax}(\sin x)^n dx$.

$$\int (\cos x)^n e^{ax}\, dx = \frac{e^{ax}(\cos x)^n}{a} + \frac{n}{a}\int (\cos x)^{n-1}\sin x\, e^{ax}\, dx$$

$$= \frac{e^{ax}(\cos x)^n}{a} + \frac{n}{a}\left\{\frac{(\cos x)^{n-1}\sin x\, e^{ax}}{a}\right.$$

$$\left. - \frac{1}{a}\int \{(\cos x)^n - (n-1)(\cos x)^{n-2}(\sin x)^2\}\, e^{ax}\, dx\right\}$$

$$= \frac{e^{ax}(\cos x)^n}{a} + \frac{ne^{ax}(\cos x)^{n-1}\sin x}{a^2}$$

$$- \frac{n}{a^2}\int \{n(\cos x)^n - (n-1)(\cos x)^{n-2}\}\, e^{ax}\, dx;$$

$$\therefore \left(1 + \frac{n^2}{a^2}\right)\int (\cos x)^n e^{ax}\, dx$$

$$= \frac{e^{ax}(\cos x)^{n-1}\{a\cos x + n\sin x\}}{a^2} + \frac{n(n-1)}{a^2}\int (\cos x)^{n-2} e^{ax}\, dx;$$

$$\int (\cos x)^n e^{ax}\,dx$$
$$= \frac{e^{ax}(\cos x)^{n-1}\{a\cos x + n\sin x\}}{n^2+a^2} + \frac{n(n-1)}{n^2+a^2}\int (\cos x)^{n-2} e^{ax}\,dx. \quad (54)$$

Similarly may a formula be found for $\int (\sin x)^n e^{ax}\,dx$.

Ex. 1. $\int e^{ax}(\cos x)^2\,dx = \dfrac{e^{ax}\cos x\,(a\cos x + 2\sin x)}{4+a^2} + \dfrac{2}{4+a^2}\,\dfrac{e^{ax}}{a}$.

76.] Integration of $e^{ax}\cos nx\,dx$, and of $e^{ax}\sin nx\,dx$.

$$\int \cos nx\, e^{ax}\,dx = \frac{\cos nx\, e^{ax}}{a} + \frac{n}{a}\int \sin nx\, e^{ax}\,dx$$
$$= \frac{\cos nx\, e^{ax}}{a} + \frac{n}{a}\left\{\frac{\sin nx\, e^{ax}}{a} - \frac{n}{a}\int \cos nx\, e^{ax}\,dx\right\};$$
$$\therefore\quad \int \cos nx\, e^{ax}\,dx = \frac{e^{ax}(a\cos nx + n\sin nx)}{a^2+n^2}. \quad (55)$$

Similarly
$$\int \sin nx\, e^{ax}\,dx = \frac{e^{ax}(a\sin nx - n\cos nx)}{a^2+n^2}. \quad (56)$$

These results may also be obtained as follows, by expressing $\sin nx$ and $\cos nx$ in terms of their exponential values:

$$\int e^{ax}\cos nx\,dx = \frac{1}{2}\int e^{ax}\{e^{nx\sqrt{-1}} + e^{-nx\sqrt{-1}}\}\,dx$$
$$= \frac{1}{2}\int \{e^{(a+n\sqrt{-1})x} + e^{(a-n\sqrt{-1})x}\}\,dx$$
$$= \frac{1}{2}\left\{\frac{e^{(a+n\sqrt{-1})x}}{a+n\sqrt{-1}} + \frac{e^{(a-n\sqrt{-1})x}}{a-n\sqrt{-1}}\right\}$$
$$= \frac{e^{ax}}{2(a^2+n^2)}\{a(e^{n\sqrt{-1}x} + e^{-n\sqrt{-1}x}) - n\sqrt{-1}(e^{n\sqrt{-1}x} - e^{-n\sqrt{-1}x})\}$$
$$= \frac{e^{ax}}{a^2+n^2}\{a\cos nx + n\sin nx\}.$$

Or thus: Let $S_1 = \int e^{ax}\cos nx\,dx,\quad S_2 = \int e^{ax}\sin nx\,dx$;

$$\therefore\quad S_1 + S_2\sqrt{-1} = \int e^{ax}\{\cos nx + \sqrt{-1}\sin nx\}\,dx$$
$$= \int e^{(a+n\sqrt{-1})x}\,dx$$
$$= \frac{e^{(a+n\sqrt{-1})x}}{a+n\sqrt{-1}}$$
$$= \frac{e^{ax}}{a^2+n^2}(\cos nx + \sqrt{-1}\sin nx)(a - n\sqrt{-1});$$

and therefore, equating possible and impossible parts,

$$S_1 = \frac{e^{ax}}{a^2+n^2}\{a\cos nx + n\sin nx\},$$

$$S_2 = \frac{e^{ax}}{a^2+n^2}\{a\sin nx - n\cos nx\}.$$

77.] Integration of $f(x)\sin^{-1}x\,dx$, $f(x)\tan^{-1}x\,dx$, &c.

Integrals of these forms must be determined by integration by parts; the method is best exhibited by examples such as follow:

Ex. 1. $$\int \sin^{-1}x\,dx = x\sin^{-1}x - \int \frac{x\,dx}{(1-x^2)^{\frac{1}{2}}}$$
$$= x\sin^{-1}x + (1-x^2)^{\frac{1}{2}}.$$

Ex. 2. $$\int \sin^{-1}x\frac{dx}{(1-x^2)^{\frac{1}{2}}} = \int \sin^{-1}x\,d.\sin^{-1}x = \frac{1}{2}(\sin^{-1}x)^2.$$

Ex. 3. $$\int \frac{x^2\,dx}{1+x^2}\tan^{-1}x = \int \frac{1+x^2-1}{1+x^2}\tan^{-1}x\,dx$$
$$= \int \tan^{-1}x\,dx - \int \tan^{-1}x\frac{dx}{1+x^2}$$
$$= x\tan^{-1}x - \frac{1}{2}\log(1+x^2) - \frac{1}{2}(\tan^{-1}x)^2.$$

Ex. 4. $$\int \frac{dx\,e^{a\tan^{-1}x}}{(1+x^2)^{\frac{3}{2}}} = \int \frac{dx}{(1+x^2)^{\frac{3}{2}}}e^{a\tan^{-1}x}$$
$$= \frac{x}{(1+x^2)^{\frac{1}{2}}}e^{a\tan^{-1}x} - a\int \frac{x\,dx}{(1+x^2)^{\frac{3}{2}}}e^{a\tan^{-1}x}$$
$$= \frac{x e^{a\tan^{-1}x}}{(1+x^2)^{\frac{1}{2}}} - a\left\{-\frac{e^{a\tan^{-1}x}}{(1+x^2)^{\frac{1}{2}}} + a\int \frac{dx}{(1+x^2)^{\frac{3}{2}}}e^{a\tan^{-1}x}\right\};$$
$$\therefore\ (1+a^2)\int \frac{dx}{(1+x^2)^{\frac{3}{2}}}e^{a\tan^{-1}x} = \frac{(x+a)e^{a\tan^{-1}x}}{(1+x^2)^{\frac{1}{2}}};$$
$$\int \frac{e^{a\tan^{-1}x}\,dx}{(1+x^2)^{\frac{3}{2}}} = \frac{(x+a)e^{a\tan^{-1}x}}{(1+a^2)(1+x^2)^{\frac{1}{2}}}.$$

78.] Integration of $\dfrac{dx}{(a+b\cos x)^n}$.

$$\int \frac{dx}{(a+b\cos x)^n} = \int \frac{(a+b\cos x)\,dx}{(a+b\cos x)^{n+1}}$$
$$= \int \frac{a\,dx}{(a+b\cos x)^{n+1}} + b\int \cos x\,dx\,\frac{1}{(a+b\cos x)^{n+1}}$$
$$= \int \frac{a\,dx}{(a+b\cos x)^{n+1}} + b\left\{\frac{\sin x}{(a+b\cos x)^{n+1}} - (n+1)\int \frac{b(\sin x)^2\,dx}{(a+b\cos x)^{n+2}}\right\}$$
$$= \frac{b\sin x}{(a+b\cos x)^{n+1}} + a\int \frac{dx}{(a+b\cos x)^{n+1}} - (n+1)\int \frac{b^2 - b^2(\cos x)^2}{(a+b\cos x)^{n+2}}dx;$$

but

$$\int \frac{b^2 - b^2 (\cos x)^2}{(a + b\cos x)^{n+2}} dx = \int \frac{b^2 - a^2 + 2a(a + b\cos x) - (a + b\cos x)^2}{(a + b\cos x)^{n+2}} dx$$

$$= (b^2 - a^2) \int \frac{dx}{(a + b\cos x)^{n+2}} + 2a \int \frac{dx}{(a + b\cos x)^{n+1}} - \int \frac{dx}{(a + b\cos x)^n};$$

substituting which, we have

$$(n+1)(b^2 - a^2) \int \frac{dx}{(a + b\cos x)^{n+2}}$$

$$= \frac{b\sin x}{(a + b\cos x)^{n+1}} - (2n+1)a \int \frac{dx}{(a + b\cos x)^{n+1}} + n \int \frac{dx}{(a + b\cos x)^n};$$

for n write $n-2$; therefore

$$(n-1)(b^2 - a^2) \int \frac{dx}{(a + b\cos x)^n}$$

$$= \frac{b\sin x}{(a + b\cos x)^{n-1}} - (2n-3)a \int \frac{dx}{(a + b\cos x)^{n-1}} + (n-2) \int \frac{dx}{(a + b\cos x)^{n-2}};$$

$$\therefore \int \frac{dx}{(a + b\cos x)^n}$$

$$= \frac{b\sin x}{(n-1)(b^2 - a^2)(a + b\cos x)^{n-1}} - \frac{(2n-3)a}{(n-1)(b^2 - a^2)} \int \frac{dx}{(a + b\cos x)^{n-1}}$$

$$+ \frac{n-2}{(n-1)(b^2 - a^2)} \int \frac{dx}{(a + b\cos x)^{n-2}}. \quad (57)$$

By which means the integral becomes ultimately reduced to

$$\int \frac{dx}{a + b\cos x},$$

the value of which has been determined in Art. 67.

79.] Many of the algebraical functions which have been integrated in Chapter II may by substitution be transformed into circular functions, and in some cases the integrals may be determined with greater facility; and by a reverse process many of the circular functions which have been integrated in the present Chapter may be transformed into algebraical functions. The method is exhibited by the following examples:

Ex. 1. $\int \frac{dx}{a^2 + x^2}$.

Let $x = a\tan\theta$; $\therefore dx = a(\sec\theta)^2 d\theta$;

$$\therefore \int \frac{dx}{a^2 + x^2} = \int \frac{a(\sec\theta)^2 d\theta}{a^2 (\sec\theta)^2}$$

$$= \frac{1}{a} \int d\theta = \frac{\theta}{a}$$

$$= \frac{1}{a} \tan^{-1} \frac{x}{a}.$$

Ex. 2. $\int \frac{dx}{(a^2+x^2)^n}$.

Let $x = a\tan\theta$; $\therefore$ $dx = a(\sec\theta)^2 d\theta$;

$$\therefore \int \frac{dx}{(a^2+x^2)^n} = \int \frac{a(\sec\theta)^2 d\theta}{a^{2n}(\sec\theta)^{2n}} = \frac{1}{a^{2n-1}}\int (\cos\theta)^{2n-2} d\theta;$$

which last integral is of the form (36), Art. 68, and may be integrated by the reduction-formula therein given.

Ex. 3. $\int \frac{dx}{(a^2-x^2)^{\frac{1}{2}}}$.

Let $x = a\sin\theta$; $\therefore$ $dx = a\cos\theta\, d\theta$;

$$\therefore \int \frac{dx}{(a^2-x^2)^{\frac{1}{2}}} = \int \frac{a\cos\theta\, d\theta}{a\cos\theta}$$

$$= \int d\theta$$

$$= \theta = \sin^{-1}\frac{x}{a}.$$

Ex. 4. $\int \frac{dx}{(a^2+x^2)^{\frac{1}{2}}}$.

Let $x = a\tan\theta$; $\therefore$ $dx = a(\sec\theta)^2 d\theta$;

$$\int \frac{dx}{(a^2+x^2)^{\frac{1}{2}}} = \int \sec\theta\, d\theta$$

$$= \int \frac{\sec\theta+\tan\theta}{\sec\theta+\tan\theta}\sec\theta\, d\theta$$

$$= \int \frac{(\sec\theta)^2+\sec\theta\tan\theta}{\tan\theta+\sec\theta}\, d\theta$$

$$= \log\{\tan\theta+\sec\theta\}$$

$$= \log\left\{\frac{x}{a}+\left(1+\frac{x^2}{a^2}\right)^{\frac{1}{2}}\right\}$$

$$= \log\frac{x+(a^2+x^2)^{\frac{1}{2}}}{a}.$$

Ex. 5. $\int (a^2-x^2)^{\frac{1}{2}}\, dx$.

Let $x = a\sin\theta$; $\therefore$ $dx = a\cos\theta\, d\theta$;

$$\int (a^2-x^2)^{\frac{1}{2}} = a^2\int (\cos\theta)^2 d\theta$$

$$= a^2\left\{\frac{\sin\theta\cos\theta}{2}+\frac{\theta}{2}\right\}$$

$$= \frac{x(a^2-x^2)^{\frac{1}{2}}}{2}+\frac{a^2}{2}\sin^{-1}\frac{x}{a}.$$

Ex. 6. $\int \frac{x^n dx}{(a^2 - x^2)^{\frac{1}{2}}}$.

Let $x = a \sin\theta$; $\therefore\ dx = a\cos\theta\, d\theta$;

$$\int \frac{x^n dx}{(a^2 - x^2)^{\frac{1}{2}}} = a^n \int (\sin\theta)^n d\theta,$$

which, according as n is odd or even, is by Art. 68, equation (35), equal to

$$a^n \left\{ \frac{-\cos\theta(\sin\theta)^{n-1}}{n} - \frac{n-1}{n(n-2)}\cos\theta(\sin\theta)^{n-3} - \frac{(n-1)(n-3)}{n(n-2)(n-4)}\cos\theta(\sin\theta)^{n-5} - \ldots\ldots \ldots - \frac{(n-1)(n-3)\ldots 4.2}{n(n-2)(n-4)\ldots 5.3}\cos\theta \right\}; \quad (58)$$

or to

$$a^n \left\{ \frac{-\cos\theta(\sin\theta)^{n-1}}{n} - \frac{n-1}{n(n-2)}\cos\theta(\sin\theta)^{n-3} - \frac{(n-1)(n-3)}{n(n-2)(n-4)}\cos\theta(\sin\theta)^{n-5} - \ldots\ldots \ldots + \frac{(n-1)(n-3)\ldots 3.1}{n(n-2)(n-4)\ldots 4.2}\theta \right\}; \quad (59)$$

and replacing θ in terms of x, the results are identical with (90) in Art. 49.

80.] This and the preceding Chapters contain an account of almost all the known methods for finding indefinite integrals. Very few indeed they are, and they may be reduced to two or three general heads; so that most of the labour consists in transforming given element-functions into other and equivalent forms of which the integrals are known. Should any one ask why the number of known indefinite integrals is so small, the reply is easy: we have no means of expressing them; our materials fail: it is not because the Calculus as a system of rules for integrating and disintegrating (or differentiating) fails; but it is because the materials, on which it has to operate, fail. Other functions and other combinations of variables are required beside those which we now have. A word or two will shew how this is. On an examination of the several forms of functions to which differentiation leads, it will be seen that certain forms do not occur. Thus $(a_0 + a_1 x + a_2 x^2 + a_3 x^3 + a_4 x^4)^{\frac{1}{2}} dx$, $\frac{dx}{(a^2 - x^2)^{\frac{1}{2}}(a^2 - e^2 x^2)^{\frac{1}{2}}}$, $e^{-x^2} dx$, $\{1 - e^2(\sin\phi)^2\}^{\frac{1}{2}} d\phi$, are forms of functions, to which differentiation does not lead; that is, we do not in the Differential Calculus meet with the functions of which these are the derived functions. Also again, when differentiation is performed on a given function, in most cases it changes the nature of it, and reduces it from a more complex and transcendental to a more simple form: thus $\log x$ is by differentiation changed into $(x)^{-1}dx$, that is, into an algebraical form; $\sin^{-1} x$, $\tan^{-1} x$, ... similarly give rise to algebraical expressions: in the reverse process therefore of integration the simple functions are changed into more complex ones; algebraical functions may not produce other algebraical functions, but may become logarithmic or circular. In order then that logarithmic and circular functions should *generally* be integrated, there must be other transcendents higher than they are, and of which they are the typical infinitesimal-elements: but such functions do not as yet *generally* exist; and until they have been discovered, studied, and had their values calculated and tabulated for given values of their variable subjects in the same way as logarithmic and circular functions have been treated, it is vain to seek for indefinite integrals of the (at present) highest transcendents.

Whenever therefore in the sequel we meet with the expression "cannot be integrated," let the exact force of it be borne in mind; it is not meant that the infinitesimal element-function to which the expression is applied is not the element of *some* finite function, for doubtless such a primary function exists, and it *may* be a question of *time* only when functions will have been examined with accuracy sufficient to have their values tabulated and their properties understood: but it is meant that such an infinitesimal function is not the element of any circular, logarithmic, or algebraical function which has already been the subject of complete analysis and examination; and thus that the integral cannot be expressed in terms of the ordinary functions or symbols with which we are familiar. Many instances of this incomplete state of the science will occur hereafter.

CHAPTER IV.

ON DEFINITE INTEGRATION AND ON DEFINITE INTEGRALS.

SECTION 1.—*Definite Integrals determined by means of Indefinite Integration.*

81.] In most future cases of the application of the Calculus, and indeed in all problems into which integration enters, the solution depends on one or more definite integrals; for most of our problems depending on a continuous law, which can be expressed only by an element-function, the results of that law can be definitely found only when the definite integral of that element-function can be determined. Hence arises the importance of definite integration, and of the evaluation of definite integrals. In Chapter I. the general notion of a definite integral has been explained, and some general theorems of such integrals have been demonstrated. I propose in the present Chapter to investigate other theorems, and chiefly with the view to the evaluation of the integrals. Great care will be needed in the inquiry, as the subject is of a very delicate nature; for a definite integral is the sum of a series, and the usual analytical difficulties as to convergency and divergency are inherent in it. The terms of the series are also infinitesimal, and the number of them is infinite, and these quantities, various as to their orders, cannot be subjects of combination and of calculation without considerable risk of error. Neither does any general method exist for the evaluation of definite integrals; a method universally applicable is a desideratum: it will appear in the sequel that certain methods are adapted to particular forms of integrals, but no general principle has as yet been discovered which includes all these several processes. The difficulty of the inquiry is hereby increased, because there is no general rule to which all the cases must conform. I have however had the benefit of consulting the great work on

Definite Integrals, by D. Bierens de Haan*, and have found it of use for verification as well as for guidance; and many of the examples given in the following pages have been selected from it, and from its companion volume of Tables of Integrals. The latter indeed exhausts that part of the subject in its present state.

82.] The method for the evaluation of a definite integral which first presents itself, is that of deriving it from the indefinite integral when the latter can be found. The method depends on the equation (11), Art. 5; viz.,

$$\int_{x_0}^{x_n} \text{F}'(x)\, dx = \Big[\text{F}(x)\Big]_{x_0}^{x_n}; \tag{1}$$

$$= \text{F}(x_n) - \text{F}(x_0); \tag{2}$$

that is, the definite integral is the excess of the value of the indefinite integral when the superior limit is substituted for the variable over that when the inferior limit is substituted. The following are examples of this method.

Ex. 1. $\displaystyle\int_0^1 x^n\, dx = \left[\frac{x^{n+1}}{n+1}\right]_0^1 = \frac{1}{n+1}.$

Ex. 2. $\displaystyle\int_0^\infty e^{-x}\, dx = \Big[-e^{-x}\Big]_0^\infty = 1.$

$$\int_0^1 e^{ax}\, dx = \left[\frac{e^{ax}}{a}\right]_0^1 = \frac{e^a - 1}{a}.$$

Ex. 3. $$\int_0^\infty \frac{dx}{a^2+x^2} = \frac{1}{a}\left[\tan^{-1}\frac{x}{a}\right]_0^\infty = \frac{\pi}{2a}. \tag{3}$$

$$\int_0^a \frac{dx}{a^2+x^2} = \frac{\pi}{4a} = \int_a^\infty \frac{dx}{a^2+x^2}. \tag{4}$$

$$\int_{-\infty}^\infty \frac{dx}{a^2+x^2} = \frac{\pi}{a}. \tag{5}$$

Ex. 4. $\displaystyle\int_0^1 \frac{1-x^m}{1-x}\, dx = \left[x + \frac{x^2}{2} + \frac{x^3}{3} + \dots + \frac{x^m}{m}\right]_0^1$

$$= 1 + \frac{1}{2} + \frac{1}{3} + \dots + \frac{1}{m}.$$

This is a series of terms in harmonical progression: consequently

* Exposé de la Théorie des Propriétés, des Formules de Transformation, et des Méthodes d'Évaluation des Intégrales Définies; par D. Bierens de Haan. Amsterdam, C. G. Van der Post, 1860, 1862: publiée par l'Académie Royale des Sciences à Amsterdam.

the sum of such a series may be expressed as a definite integral, although it cannot be expressed in an algebraical form.

Ex. 5. $\int_0^a \frac{dx}{(a^2-x^2)^{\frac{1}{2}}} = \left[\sin^{-1}\frac{x}{a}\right]_0^a = \frac{\pi}{2}.$

$$\int_{-a}^a \frac{dx}{(a^2-x^2)^{\frac{1}{2}}} = \pi.$$

Ex. 6. $\int_0^a (a^2-x^2)^{\frac{1}{2}} dx = \left[\frac{x}{2}(a^2-x^2)^{\frac{1}{2}} + \frac{a^2}{2}\sin^{-1}\frac{x}{a}\right]_0^a = \frac{\pi a^2}{4}.$

Ex. 7. $\int_{-\infty}^{+\infty} \frac{dx}{(x-a)^2+b^2} = \frac{1}{b}\left[\tan^{-1}\frac{x-a}{b}\right]_{-\infty}^{\infty} = \frac{\pi}{b}.$

Ex. 8. $\int_0^\infty e^{-ax}\cos bx\, dx = \left[\frac{e^{-ax}(b\sin bx - a\cos bx)}{a^2+b^2}\right]_0^\infty$

$$= \frac{a}{a^2+b^2}. \tag{6}$$

Hence if $a = 0$, and $b = 1$, $\int_0^\infty \cos x\, dx = 0$; but

$$\int_0^\infty \cos x\, dx = \left[\sin x\right]_0^\infty = \sin\infty;$$

$$\therefore \quad \sin\infty = 0.$$

Ex. 9. $\int_0^\infty e^{-ax}\sin bx\, dx = \left[\frac{e^{-ax}(-a\sin bx - b\cos bx)}{a^2+b^2}\right]_0^\infty$

$$= \frac{b}{a^2+b^2}. \tag{7}$$

Hence if $a = 0$, $b = 1$, $\int_0^\infty \sin x\, dx = 1$; but

$$\int_0^\infty \sin x\, dx = \left[-\cos x\right]_0^\infty = -\cos\infty + 1;$$

$$\therefore \quad -\cos\infty + 1 = 1;$$

$$\therefore \quad \cos\infty = 0.$$

These are two remarkable instances of the mode in which definite integration may be applied to the determination of quantities apparently indeterminate. More will be said on the subject when the integrals are determined by a different process hereafter. See Art. 86.

Ex. 10. $\int_0^\infty \frac{dx}{(1+x^2)^n}$

$$= \left[\frac{x}{2(n-1)(1+x^2)^{n-1}} + \frac{2n-3}{2n-2}\int \frac{dx}{(1+x^2)^{n-1}}\right]_0^\infty$$

$$\int_0^\infty \frac{dx}{(1+x^2)^n} = \frac{2n-3}{2n-2}\int_0^\infty \frac{dx}{(1+x^2)^{n-1}}$$

$$= \frac{(2n-3)(2n-5)\ldots 5.3.1}{(2n-2)(2n-4)\ldots 6.4.2}\int_0^\infty \frac{dx}{1+x^2}$$

$$= \text{- - - - - - - -} \Big[\tan^{-1}x\Big]_0^\infty$$

$$= \frac{(2n-3)(2n-5)\ldots 5.3.1}{(2n-2)(2n-4)\ldots 6.4.2}\,\frac{\pi}{2}.$$

Ex. 11. $\displaystyle\int_0^{\frac{\pi}{4}} (\tan x)^n\,dx = \left[\frac{(\tan x)^{n-1}}{n-1} - \int (\tan x)^{n-2}\,dx\right]_0^{\frac{\pi}{4}}$

$$= \frac{1}{n-1} - \int_0^{\frac{\pi}{4}} (\tan x)^{n-2}\,dx;$$

so that if n is even,

$$\int_0^{\frac{\pi}{4}} (\tan x)^n\,dx = \frac{1}{n-1} - \frac{1}{n-3} + \ldots \pm \frac{1}{3} \mp 1 \pm \frac{\pi}{4};$$

and if n is odd,

$$\int_0^{\frac{\pi}{4}} (\tan x)^n\,dx = \frac{1}{n-1} - \frac{1}{n-3} + \ldots \pm \frac{1}{2} \mp \frac{1}{2}\log\left(\frac{1}{2}\right).$$

Ex. 12. $\displaystyle\int_0^\pi \cos mx \cos nx\,dx = \frac{1}{2}\left[\frac{\sin(m+n)x}{m+n} + \frac{\sin(m-n)x}{m-n}\right]_0^\pi$

$$= 0, \text{ if } m \text{ is not equal to } n. \qquad (8)$$

If $m = n$, the integral becomes,

$$\int_0^\pi (\cos mx)^2\,dx = \left[\frac{x}{2} + \frac{\sin 2mx}{4m}\right]_0^\pi$$

$$= \frac{\pi}{2}. \qquad (9)$$

Similarly $\displaystyle\int_0^\pi \sin mx \sin nx\,dx = 0$, if m is not equal to n; (10)

$$= \frac{\pi}{2}, \text{ if } m = n. \qquad (11)$$

Ex. 13. $\displaystyle\int_0^\infty e^{-x}x^n\,dx = \left[-e^{-x}x^n + n\int e^{-x}x^{n-1}\,dx\right]_0^\infty$ *

$$= n\int_0^\infty e^{-x}x^{n-1}\,dx$$

$$= n(n-1)(n-2)\ldots 3.2.1\int_0^\infty e^{-x}\,dx$$

$$= n(n-1)(n-2)\ldots 3.2.1\Big[-e^{-x}\Big]_0^\infty$$

$$= n(n-1)(n-2)\ldots 3.2.1. \qquad (12)$$

* See Example 3, Art. 125, Vol. I.

Ex. 14. $$\int_0^1 \frac{x^n dx}{(1-x^2)^{\frac{1}{2}}} = \Big[-\frac{x^{n-1}(1-x^2)^{\frac{1}{2}}}{n} + \frac{n-1}{n}\int \frac{x^{n-2}dx}{(1-x^2)^{\frac{1}{2}}}\Big]_0^1$$
$$= \frac{n-1}{n}\int_0^1 \frac{x^{n-2}dx}{(1-x^2)^{\frac{1}{2}}};$$

If therefore n is even,

$$\int_0^1 \frac{x^n dx}{(1-x^2)^{\frac{1}{2}}} = \frac{(n-1)(n-3)\dots 3.1}{n(n-2)\dots 4.2}\int_0^1 \frac{dx}{(1-x^2)^{\frac{1}{2}}}$$
$$= \frac{(n-1)(n-3)\dots 3.1}{n(n-2)\dots 4.2}\Big[\sin^{-1} x\Big]_0^1$$
$$= \frac{(n-1)(n-3)\dots 3.1}{n(n-2)\dots 4.2}\,\frac{\pi}{2}. \qquad (13)$$

And if n is odd,

$$\int_0^1 \frac{x^n dx}{(1-x^2)^{\frac{1}{2}}} = \frac{(n-1)(n-3)\dots 4.2}{n(n-2)\dots 5.3}\int_0^1 \frac{x\,dx}{(1-x^2)^{\frac{1}{2}}}$$
$$= \frac{(n-1)(n-3)\dots 4.2}{n(n-2)\dots 5.3}\Big[-(1-x^2)^{\frac{1}{2}}\Big]_0^1$$
$$= \frac{(n-1)(n-3)\dots 4.2}{n(n-2)\dots 5.3}. \qquad (14)$$

The remark made at the end of Art. 6 is of great importance in reference to examples such as this and Ex. 5, viz. that the value of the infinitesimal element corresponding to the superior limit is excluded, while that corresponding to the inferior limit is included in the definite integral; for were this not the case, as $\frac{x^n}{(1-x^2)^{\frac{1}{2}}}$ becomes equal to ∞, when $x = 1$, the integrals would not satisfy the conditions, which the theory of such summation requires: but as the limit unity, being the superior limit in the above examples and that which renders infinite the infinitesimal element, is not included, the definite integrals are correct.

Again, since $\frac{x^n}{(1-x^2)^{\frac{1}{2}}}$ is, for all values of x between 0 and 1, intermediate to $\frac{x^{n-1}}{(1-x^2)^{\frac{1}{2}}}$ and $\frac{x^{n+1}}{(1-x^2)^{\frac{1}{2}}}$, therefore $\int_0^1 \frac{x^n dx}{(1-x^2)^{\frac{1}{2}}}$ is intermediate to $\int_0^1 \frac{x^{n-1}dx}{(1-x^2)^{\frac{1}{2}}}$ and $\int_0^1 \frac{x^{n+1}dx}{(1-x^2)^{\frac{1}{2}}}$; hence, if n is even, $\frac{\pi}{2}\,\frac{1.3\dots(n-3)(n-1)}{2.4\dots(n-2)n}$ is intermediate to $\frac{2.4\dots(n-4)(n-2)}{3.5\dots(n-3)(n-1)}$ and $\frac{2.4\dots(n-2)n}{3.5\dots(n-1)(n+1)}$;

$$\therefore \quad \frac{\pi}{2}\frac{1.3\ldots(n-3)(n-1)}{2.4\ldots(n-4)(n-2)}$$

$$= \frac{2.4\ldots(n-4)(n-2)}{3.5\ldots(n-3)(n-1)} \times \left\{\text{a quantity} > \frac{n}{n+1}, < 1\right\};$$

and therefore if n is a very large number, we have the following approximate value of π,

$$\pi = 2\,\frac{2.2.4.4.6.6\ldots\ldots}{1.3.3.5.5.7\ldots\ldots}; \tag{15}$$

an equivalent for π which was first discovered by Dr. Wallis, and will be of considerable use in the sequel. These results are also deducible from $\int_0^{\frac{\pi}{2}} (\sin x)^n dx$ and $\int_0^{\frac{\pi}{2}} (\cos x)^n dx$, the indefinite integrals of which are given in Art. 68.

Ex. 15. $\int_{-\infty}^{\infty} \frac{x^{2m}\,dx}{x^{2n}+1}$, where m is less than n by unity at least.

Decompose $\frac{x^{2m}}{x^{2n}+1}$ into partial fractions by the process of Chapter II, Section 2. Hence, since

$$\frac{F(x)}{f'(x)} = \frac{x^{2m}}{2nx^{2n-1}} = -\frac{x^{2m+1}}{2n},$$

the coefficients of

$$\frac{1}{x-\cos\frac{\pi}{2n} - \sqrt{-1}\sin\frac{\pi}{2n}} \text{ and of } \frac{1}{x-\cos\frac{\pi}{2n} + \sqrt{-1}\sin\frac{\pi}{2n}}$$

are respectively $-\frac{1}{2n}\left\{\cos\frac{2m+1}{2n}\pi + \sqrt{-1}\sin\frac{2m+1}{2n}\pi\right\}$ and $-\frac{1}{2n}\left\{\cos\frac{2m+1}{2n}\pi - \sqrt{-1}\sin\frac{2m+1}{2n}\pi\right\}$; and this pair of fractions may be combined into the single fraction,

$$-\frac{1}{n}\,\frac{\left(x-\cos\frac{\pi}{2n}\right)\cos\frac{2m+1}{2n}\pi - \sin\frac{\pi}{2n}\sin\frac{2m+1}{2n}\pi}{x^2-2x\cos\frac{\pi}{2n}+1};$$

and all the other pairs of conjugate factors of $x^{2n}+1$ will produce similar fractions, the several denominators being

$$x^2-2x\cos\frac{3\pi}{2n}+1,\ x^2-2x\cos\frac{5\pi}{2n}+1, \ldots x^2-2x\cos\frac{2n-1}{2n}\pi+1.$$

Now each of these fractions gives an integral of the following form; viz.:

$$\int_{-\infty}^{\infty} \frac{A(x-a)+Bb}{(x-a)^2+b^2}\,dx = \left[\frac{A}{2}\log\{(x-a)^2+b^2\} + B\tan^{-1}\frac{x-a}{b}\right]_{-\infty}^{\infty}$$

$$= B\pi.$$

Consequently

$$\int_{-\infty}^{\infty} \frac{x^{2m}\,dx}{x^{2n}+1} = \frac{\pi}{n}\left\{\sin\frac{2m+1}{2n}\pi + \sin 3\frac{2m+1}{2n}\pi + \dots \right.$$

$$\left. \dots + \sin(2n-1)\frac{2m+1}{2n}\pi\right\} \quad (16)$$

$$= \frac{\pi}{n}\operatorname{cosec}\frac{2m+1}{2n}\pi. \quad (17)$$

By a similar process it may be shewn that

$$\int_{-\infty}^{\infty} \frac{x^{2m}\,dx}{x^{2n}-1} = -\frac{\pi}{n}\left\{\sin\frac{2m+1}{n}\pi + \sin 2\frac{2m+1}{n}\pi + \dots \right.$$

$$\left. \dots + \sin(n-1)\frac{2m+1}{n}\pi\right\} \quad (18)$$

$$= -\frac{\pi}{n}\cot\frac{2m+1}{2n}\pi. \quad (19)$$

As very many instances of the evaluation of a definite integral by means of the corresponding indefinite integral will occur in the sequel, it is unnecessary to add others here; and I may observe, that many of the preceding have been investigated, because they will become hereafter the subject-matter for illustration of general theorems.

SECTION 2.—*The Change of Limits in Definite Integrals.—The Resolution of a Definite Integral into two or more connected by addition or subtraction.—Cauchy's Principal Value.*

83.] We proceed now to farther researches into the general theory of definite integrals. That our notion of a definite integral may be clear and precise, it must be borne in mind that the symbol $\int_{x_0}^{x_n} F'(x)\,dx$, as well as its equivalent $F(x_n) - F(x_0)$, is only a concise expression for the sum of the series given in Art. 4: so that we have

$$\int_{x_0}^{x_n} F'(x)\,dx = F(x_n) - F(x_0) \quad (20)$$

$$= (x_1 - x_0)F'(x_0) + (x_2 - x_1)F'(x_1) + \dots$$

$$\dots + (x_n - x_{n-1})F'(x_{n-1}), \quad (21)$$

wherein $x_1, x_2, x_3, \dots x_{n-1}$ are the values of x corresponding to the $(n-1)$ points of partition of $x_n - x_0$, each part being infinitesimal, and n being infinite; and wherein $F'(x)$ is finite and continuous for all employed values of its subject-variable.

Now whenever a definite integral or its properties are the subjects of inquiry, the definite integral must be considered as the sum of the series given in (21), and its theorems are true because they are true of (21). This is the point of view from which definite integrals have been considered in Art. 8, and from which they are always to be considered. Hence it is evident that if $F'(x)$ does not change sign between x_n and x_0, but is always either positive or negative, the definite integral of $F'(x)\,dx$ taken between these limits is likewise positive or negative. If however $F'(x)$ changes sign, the definite integral of $F'(x)\,dx$ will be positive or negative, according as the positive or negative part of the series is the greater.

84.] Subject to the condition that $F'(x)$ does not become infinite or discontinuous between the limits of integration, the limits may be altered, and the value of the integral will generally be changed thereby. Thus the integral becomes a function of the limits, and may be treated as such. The case of a continuous variation of the limits, and of the consequent variation of the integral, will be considered in Section 3 of the present Chapter. But it is necessary at once to investigate certain simple cases of a change of limits; and in the first place the effect of a reversal of the limits.

By the definition of a definite integral,

$$\begin{aligned}\int_{x_n}^{x_0} F'(x)\,dx &= (x_{n-1}-x_n)\,F'(x_n) + (x_{n-2}-x_{n-1})\,F'(x_{n-1}) + \dots \\ &\qquad \dots + (x_1-x_2)\,F'(x_2) + (x_0-x_1)\,F'(x_1) \qquad (22)\\ &= -\{(x_1-x_0)F'(x_0+x_1-x_0) + (x_2-x_1)F'(x_1+x_2-x_1) + \dots \\ &\qquad \dots + (x_n-x_{n-1})F'(x_{n-1}+x_n-x_{n-1})\}. \qquad (23)\end{aligned}$$

But as $x_1-x_0,\ x_2-x_1, \dots x_n-x_{n-1}$ are infinitesimals; by (6), Art. 18, Vol. I,

$$\left.\begin{aligned}F'(x_0+x_1-x_0) &= F'(x_0) + (x_1-x_0)\,F''(x_0),\\ F'(x_1+x_2-x_1) &= F'(x_1) + (x_2-x_1)\,F''(x_1),\\ &\dots\dots\dots\dots\dots\dots\\ F'(x_{n-1}+x_n-x_{n-1}) &= F'(x_{n-1}) + (x_n-x_{n-1})\,F''(x_{n-1}).\end{aligned}\right\} \quad (24)$$

Consequently, substituting these values in (23),

$$\int_{x_n}^{x_0} F'(x)\,dx = -\{(x_1-x_0)\,F'(x_0)+(x_2-x_1)\,F'(x_1)+\ldots$$
$$\ldots+(x_n-x_{n-1})\,F'(x_{n-1})$$
$$+(x_1-x_0)^2F''(x_0)+(x_2-x_1)^2F''(x_1)+\ldots+(x_n-x_{n-1})^2F''(x_{n-1})\}. \quad (25)$$

Now the last row of terms in the right-hand member of this equation involves infinitesimals of a higher order than those in the upper rows; and the number of terms is the same in both: these latter must therefore be neglected in the sum; and we have

$$\int_{x_n}^{x_0} F'x)\,dx = -\{(x_1-x_0)\,F'(x_0)+(x_2-x_1)\,F'(x_1)+\ldots$$
$$\ldots+(x_n-x_{n-1})\,F'(x_{n-1})\}$$
$$= -\int_{x_0}^{x_n} F'(x)\,dx; \quad (26)$$

and thus the effect of a reversal of the limits of a definite integral is the change of sign of the integral.

It is however to be observed that the two members of (26) are not absolutely identical; but that they differ by a quantity, Δ say, which is an infinitesimal of an order which must be neglected; so that we have

$$\int_{x_n}^{x_0} F'(x)\,dx = -\int_{x_0}^{x_n} F'(x)\,dx - \Delta; \quad (27)$$

where

$$\Delta = (x_1-x_0)^2F''(x_0)+(x_2-x_1)^2F''(x_1)+\ldots+(x_n-x_{n-1})^2F''(x_{n-1}). \quad (28)$$

So many applications of this Theorem will occur in the sequel, that it is unnecessary to insert examples here.

85.] A definite integral of which x_n and x_0 are the limits is equal to the sum of a series of definite integrals of the same element-function, provided that the extreme limits are the same, and the several intermediate limits are continuously additive.

Let $\int_{x_0}^{x_n} F'(x)\,dx$ be the integral under consideration; and let us suppose $x_n - x_0$, which is the range through which the integration is to be effected, to be divided into n finite parts; and to the several points of partition let $x_1, x_2, \ldots x_{n-1}$ refer; so that we have the identity,

$$x_n - x_0 = (x_1-x_0)+(x_2-x_1)+\ldots+(x_n-x_{n-1}). \quad (29)$$

Also let the finite intervals x_1-x_0, x_2-x_1, ... x_n-x_{n-1} be divided each into infinitesimal parts; of which let x_1-x_0 contain a parts, to the points of partition of which let $a_1, a_2, \ldots a_{a-1}$ correspond; let x_2-x_1 contain b parts, to the points of partition of which let

$\beta_1, \beta_2, \ldots \beta_{b-1}$ correspond; and so on: and let $x_n - x_{n-1}$ contain k parts, to the points of partition of which let $\kappa_1, \kappa_2, \ldots \kappa_{k-1}$ correspond; so that $a, b, \ldots k$ are infinities; then, by the definition of a definite integral,

$$\begin{aligned}
&\int_{x_0}^{x_n} F'(x)\,dx \\
&= (a_1 - x_0)F'(x_0) + (a_2 - a_1)F'(a_1) + \ldots + (x_1 - a_{a-1})F'(a_{a-1}) \\
&\quad + (\beta_1 - x_1)F'(x_1) + (\beta_2 - \beta_1)F'(\beta_1) + \ldots + (x_2 - \beta_{b-1})F'(\beta_{b-1}) \\
&\quad + \ldots\ldots\ldots\ldots\ldots\ldots \\
&\quad + (\kappa_1 - x_{n-1})F'(x_{n-1}) + (\kappa_2 - \kappa_1)F'(\kappa_1) + \ldots + (x_n - \kappa_{k-1})F'(\kappa_{k-1}) \quad (30) \\
&= \int_{x_0}^{x_1} F'(x)\,dx + \int_{x_1}^{x_2} F'(x)\,dx + \ldots + \int_{x_{n-1}}^{x_n} F'(x)\,dx; \quad (31)
\end{aligned}$$

consequently the sum of the latter integrals is absolutely equal to the given integral; and thus the theorem enunciated at the beginning of the Article is proved.

Thus, to take the special case of the partition of the range into two parts; if ξ is a value of x intermediate to x_n and x_0,

$$\int_{x_0}^{x_n} F'(x)\,dx = \int_{x_0}^{\xi} F'(x)\,dx + \int_{\xi}^{x_n} F'(x)\,dx. \quad (32)$$

86.] The following are examples in illustration of this Theorem.

Ex. 1.

$$\int_0^{\infty} \cos x\,dx = \int_0^{\frac{\pi}{2}} \cos x\,dx + \int_{\frac{\pi}{2}}^{\pi} \cos x\,dx + \int_{\pi}^{\frac{3\pi}{2}} \cos x\,dx + \int_{\frac{3\pi}{2}}^{2\pi} \cos x\,dx + \ldots$$

$$\Big[\sin x\Big]_0^{\infty} = \Big[\sin x\Big]_0^{\frac{\pi}{2}} + \Big[\sin x\Big]_{\frac{\pi}{2}}^{\pi} + \Big[\sin x\Big]_{\pi}^{\frac{3\pi}{2}} + \Big[\sin x\Big]_{\frac{3\pi}{2}}^{2\pi} + \ldots$$

$$\begin{aligned}
\sin\infty &= 1 - 1 - 1 + 1 + 1 - 1 - 1 \ldots \\
&= 1 - 2 + 2 - 2 + \ldots \\
&= 1 - \frac{2}{1+1} \\
&= 0.
\end{aligned}$$

Ex. 2.

$$\int_0^{\infty} \sin x\,dx = \int_0^{\frac{\pi}{2}} \sin x\,dx + \int_{\frac{\pi}{2}}^{\pi} \sin x\,dx + \int_{\pi}^{\frac{3\pi}{2}} \sin x\,dx + \int_{\frac{3\pi}{2}}^{2\pi} \sin x\,dx + \ldots$$

$$\begin{aligned}
-\cos\infty + 1 &= 1 + 1 - 1 - 1 + 1 + 1 - \ldots \\
&= 2 - 2 + 2 - \ldots \\
&= \frac{2}{1+1} = 1; \\
\therefore\ \cos\infty &= 0.
\end{aligned}$$

Which results are in accordance with those demonstrated by another process in Ex. 8 and 9 of Art. 82.

Various opinions have been expressed by mathematicians as to the correctness of these values of $\sin \infty$ and of $\cos \infty$; they are quantities apparently indeterminate, because as the arc is increased by 2π, $\sin x$ and $\cos x$ pass through all values between $+1$ and -1, $+1$ being the maximum and -1 being the minimum; and because there is no reason, *à priori*, why any one value between these limits should be taken rather than any other. Moral expectation might lead to the choice of the average value, which is zero; but moral expectation is not mathematical demonstration. The indeterminateness too of the value is also inherent in the preceding series; so that this process of evaluation would hardly be considered rigorous; for zero is not the value of either of the series, unless the number of the terms of the series is of the form $2m$ in the former series and of the form $4m$ in the latter. When however the number of terms of a series is infinite, what is the form of that number? Is it par or impar? Is it par par, or par impar? On the answer to these questions must an opinion as to the rigorousness of proof of the preceding process depend. And the answer depends on the view taken of infinity. If infinity, which is the superior limit of the integral, is capable of discontinuous increase by units, the preceding process will probably be considered to be wanting in rigorousness; but if infinity admits of only continuous increase, so that 2π may be considered an infinitesimal increment of an infinite arc, then the preceding process will probably be considered sufficient. On the view taken of infinities and infinitesimals, in our theory of definite integration, the process of evaluation given in Ex. 8 and 9, Art. 82, is apparently free from objection. Indeed for every positive value of a, however small, these equations are arithmetically true; and the results may be shewn to be true by actual summation.*

87.] Again, if ξ is a value of x not included between x_n and x_0, but lying beyond the range of integration, say beyond x_n; then, if $F'(x)$ is finite and continuous for all values of x between x_0 and ξ, by (31) we have

$$\int_{x_0}^{\xi} F'(x)\,dx = \int_{x_0}^{x_n} F'(x)\,dx + \int_{x_n}^{\xi} F'(x)\,dx;$$

* See De Morgan: Differential and Integral Calculus. London. 1842. Page 571.

$$\therefore \int_{x_0}^{x_n} F'(x)\,dx = \int_{x_0}^{\xi} F'(x)\,dx - \int_{x_n}^{\xi} F'(x)\,dx$$

$$= \int_{x_0}^{\xi} F'(x)\,dx + \int_{\xi}^{x_n} F'(x)\,dx; \tag{33}$$

the sign of the last integral having been changed by the reversal of the limits.

As this process admits of extension to any one or more values of x beyond $x_n - x_0$, provided that $F'(x)$ does not become infinite or discontinuous within the range of integration, it follows that the theorem enunciated at the beginning of Art. 85 may be enlarged so as to include all values of x, which by algebraical addition give $x_n - x_0$, over which range the sum of the element-functions is to be taken.

88.] Hence also may be deduced two theorems of great use in the evaluation of definite integrals.

Firstly, if ξ is the arithmetic mean between x_n and x_0 the limits of integration, and if $F'(x)$ has the same value and the same sign at equal distances from ξ on either side of it, that is, if $F'(\xi - x) = F'(\xi + x)$; then the series, the sums of which are denoted by $\int_{x_0}^{\xi} F'(x)\,dx$ and $\int_{\xi}^{x_n} F'(x)\,dx$, will consist of terms which are, term by term, equal to each other; and consequently

$$\int_{x_0}^{\xi} F'(x)\,dx = \int_{\xi}^{x_n} F'(x)\,dx. \quad \text{Hence}$$

$$\int_{x_0}^{x_n} F'(x)\,dx = \int_{x_0}^{\xi} F'(x)\,dx + \int_{\xi}^{x_n} F'(x)\,dx$$

$$= 2\int_{x_0}^{\xi} F'(x)\,dx. \tag{34}$$

If ξ, the arithmetic mean of the two limits, is zero, so that $x_0 = -x_n$, and $F'(-x) = -F'(x)$, then

$$\int_{-x_n}^{x_n} F'(x)\,dx = 2\int_{0}^{x_n} F'(x)\,dx. \tag{35}$$

The following are examples of these theorems.

Ex. 1. $\int_0^{\pi} \sin x\,dx = 2\int_0^{\frac{\pi}{2}} \sin x\,dx = 2.$

Ex. 2. $\int_{-\frac{\pi}{2}}^{\frac{\pi}{2}} \cos x\,dx = 2\int_0^{\frac{\pi}{2}} \cos x\,dx = 2.$

Ex. 3. $\int_{-a}^{a}(a^2-x^2)^{\frac{1}{2}}\,dx = 2\int_0^a (a^2-x^2)^{\frac{1}{2}}\,dx = \frac{\pi a^2}{2}.$

Ex. 4. $\int_{-\infty}^{\infty} x^{2n}e^{-a^2x^2}\,dx = 2\int_0^{\infty} x^{2n}e^{-a^2x^2}\,dx.$

Also generally, if R denotes a rational function,

Ex. 5. $\int_0^{\pi} \mathrm{R}(\sin x)\,dx = 2\int_0^{\frac{\pi}{2}} \mathrm{R}(\sin x))\,dx.$ (36)

Ex. 6. $\int_{-a}^{a} \mathrm{R}(x^2)\,dx = 2\int_0^{a} \mathrm{R}(x^2)\,dx.$ (37)

Secondly, if ξ is an arithmetic mean between x_n and x_0, the limits of integration, and if $\mathrm{F}'(x)$ has the same value, but of different signs, at equal distances from ξ on either side of it, that is, if $\mathrm{F}'(\xi+x) = -\mathrm{F}'(\xi-x)$; then the series, the sums of which are denoted by $\int_{x_0}^{\xi}\mathrm{F}'(x)\,dx$ and $\int_{\xi}^{x_n}\mathrm{F}'(x)\,dx$, will consist of terms which are, term by term, equal to each other, and of contrary signs; and consequently

$$\int_{x_0}^{\xi}\mathrm{F}'(x)\,dx = -\int_{\xi}^{x_n}\mathrm{F}'(x)\,dx;$$

so that the sum of the two definite integrals is zero. Hence

$$\int_{x_0}^{x_n}\mathrm{F}'(x)\,dx = \int_{x_0}^{\xi}\mathrm{F}'(x)\,dx + \int_{\xi}^{x_n}\mathrm{F}'(x)\,dx$$
$$= 0. \qquad (38)$$

If $\xi = 0$, so that $x_0 = -x_n$, and $\mathrm{F}'(x) = -\mathrm{F}'(-x)$, then

$$\int_{-x_n}^{x_n}\mathrm{F}'(x)\,dx = 0. \qquad (39)$$

The following are examples of these theorems.

Ex. 1. $\int_0^{\pi}\cos x\,dx = \int_0^{\frac{\pi}{2}}\cos x\,dx + \int_{\frac{\pi}{2}}^{\pi}\cos x\,dx$
$= 0;$

since $\cos\left(\frac{\pi}{2}+x\right) = -\cos\left(\frac{\pi}{2}-x\right)$.

Ex. 2. $\int_{-\frac{\pi}{2}}^{\frac{\pi}{2}}\sin x\,dx = 0.$

Ex. 3. $\int_{-\infty}^{\infty} x^{2n+1}e^{-a^2x^2}\,dx = 0.$

Also, generally, if R denotes a rational function,

Ex. 4. $$\int_0^{\pi} \mathrm{R}\,\{\sin x,\ (\cos x)^2\} \cos x\, dx = 0. \tag{40}$$

89.] Hitherto in evaluating $\int_{x_0}^{x_n} \mathrm{F}'(x)\,dx$, $\mathrm{F}'(x)$ has been assumed to be finite and continuous for all employed values of its subject-variable; and our inquiry has been restricted to cases wherein this condition is fulfilled. Suppose however ξ to be a value of x, within the range of integration, for which $\mathrm{F}'(x)$ becomes either infinite or discontinuous, the theory of a definite integral which is given in Art. 85 enables us to treat of this case. Let us divide the integral into two parts, as follows;

$$\int_{x_0}^{x_n} \mathrm{F}'(x)\,dx = \int_{x_0}^{\xi} \mathrm{F}'(x)\,dx + \int_{\xi}^{x_n} \mathrm{F}'(x)\,dx. \tag{41}$$

Now the integrals in the right-hand member do not admit of treatment, because they have no determinate value according to the principles of the preceding Articles, inasmuch as the limits include values for which the element-function is infinite. Suppose however i to be a general symbol for an infinitesimal, and μ and ν to be two arbitrary and undetermined constants; and let ξ, which is the superior limit in the first integral of the right-hand member of (41), be replaced by $\xi - \mu i$, and let ξ which is the inferior limit in the second integral be replaced by $\xi + \nu i$, so that in neither integral does the range include that value of the subject-variable, for which the element-function is infinite or discontinuous. Then if the definite integrals in the right-hand member of (41) are determined for these limits, μ, ν, and i will generally enter into them; and if after integration 0 is substituted for i, the required definite integral will be found. Thus we have

$$\int_{x_0}^{x_n} \mathrm{F}'(x)\,dx = \int_{x_0}^{\xi-\mu i} \mathrm{F}'(x)\,dx + \int_{\xi+\nu i}^{x_n} \mathrm{F}'(x)\,dx, \tag{42}$$

i being replaced by 0 in the result.

If in the determination of a definite integral it is convenient to extend the range of integration beyond the original limits of integration, say to ξ beyond the superior limit x_n, and if ξ is a value for which the element-function is infinite or discontinuous; then the limits must be taken as follows;

$$\int_{x_0}^{x_n} \mathrm{F}'(x)\,dx = \int_{x_0}^{\xi-\mu i} \mathrm{F}'(x)\,dx + \int_{\xi-\kappa i}^{x_n} \mathrm{F}'(x)\,dx, \tag{43}$$

and zero must be substituted for i in the result. A similar formula will be required if ξ is a value below the inferior limit.

Similarly if $F'(x)$ is infinite or discontinuous for many values of x, say for $x_1, x_2, \dots x_{n-1}$, lying within the range of integration; if i is the general symbol of an infinitesimal, and $\mu_1, \nu_1, \mu_2, \nu_2 \dots$ are positive and arbitrary constants, then

$$\int_{x_0}^{x_n} F'(x)\,dx = \int_{x_0}^{x_1-\mu_1 i} F'(x)\,dx + \int_{x_1+\nu_1 i}^{x_2-\mu_2 i} F'(x) + \dots\dots + \int_{x_{n-1}+\nu_{n-1} i}^{x_n} F'(x)\,dx, \quad (44)$$

i being in the result replaced by zero.

The following are instances of the process.

Ex. 1. In $\int_{-1}^{1} \frac{dx}{x}$, the limits include zero, for which the element-function, x^{-1}, $= \infty$; we must consequently divide the integral into two parts, of which the limits will be $-\mu i$ and -1, and 1 and νi: thus,

$$\int_{-1}^{1} \frac{dx}{x} = \int_{-1}^{-\mu i} \frac{dx}{x} + \int_{\nu i}^{1} \frac{dx}{x}$$

$$= \log \mu i - \log \nu i = \log \frac{\mu}{\nu}; \quad (45)$$

which quantity is definite as to form, but indefinite as to value, because μ and ν are thus far undetermined quantities.

Ex. 2. In the integral $\int_0^{\pi} \frac{dx}{a+b\cos x}$, if a is greater than b, the denominator is always positive, and, see (27), Art. 67,

$$\int_0^{\pi} \frac{dx}{a+b\cos x} = \frac{2}{(a^2-b^2)^{\frac{1}{2}}} \left[\tan^{-1}\left\{\left(\frac{a-b}{a+b}\right)^{\frac{1}{2}} \tan\frac{x}{2}\right\}\right]_0^{\pi}$$

$$= \frac{\pi}{(a^2-b^2)^{\frac{1}{2}}}. \quad (46)$$

But if a is less than b, the denominator $= 0$, and changes sign from $+$ to $-$, when $x = \cos^{-1}\frac{-a}{b}$; that is, when

$$x = 2\tan^{-1}\left(\frac{b+a}{b-a}\right)^{\frac{1}{2}}.$$

Let this value of $x = \alpha$; so that the element-function $= \infty$, when $x = \alpha$; consequently,

$$\int_0^{\pi} \frac{dx}{a+b\cos x} = \int_0^{\alpha-\mu i} \frac{dx}{a+b\cos x} + \int_{\alpha+\nu i}^{\pi} \frac{dx}{a+b\cos x}; \quad (47)$$

now, since $\tan\frac{\alpha}{2} = \left(\frac{b+a}{b-a}\right)^{\frac{1}{2}}$, $\frac{\sin\frac{\alpha}{2}}{(b+a)^{\frac{1}{2}}} = \frac{\cos\frac{\alpha}{2}}{(b-a)^{\frac{1}{2}}} = \frac{1}{(2b)^{\frac{1}{2}}}$;

consequently, by (28), Art. 67, substituting in (47),

$$\int_0^{\pi} \frac{dx}{a+b\cos x}$$

$$= \frac{1}{(b^2-a^2)^{\frac{1}{2}}} \left\{ \left[\log \frac{\sin \frac{a+x}{2}}{\sin \frac{a-x}{2}} \right]_0^{a-\mu i} + \left[\log \frac{\sin \frac{x+a}{2}}{\sin \frac{x-a}{2}} \right]_{a+\nu i}^{\pi} \right\}$$

$$= \frac{1}{(b^2-a^2)^{\frac{1}{2}}} \log \frac{\sin\left(a-\frac{\mu i}{2}\right) \sin \frac{\nu i}{2}}{\sin\left(a+\frac{\nu i}{2}\right) \sin \frac{\mu i}{2}}; \tag{48}$$

which $= \frac{0}{0}$, when $i = 0$; but, on evaluation, we have

$$\int_0^{\pi} \frac{dx}{a+b\cos x} = \frac{1}{(b^2-a^2)^{\frac{1}{2}}} \log \frac{\nu}{\mu}; \tag{49}$$

which is determinate in form, but undetermined in value, because μ and ν are constants thus far arbitrary.

If $a = b$, $\quad \displaystyle\int_0^{\pi} \frac{dx}{a(1+\cos x)} = \frac{1}{2a} \int_0^{\pi} \left(\sec \frac{x}{2}\right)^2 dx$

$$= \infty. \tag{50}$$

Hence we have this remarkable result;

$$\int_0^{\pi} \frac{dx}{a+b\cos x} = \frac{\pi}{(a^2-b^2)^{\frac{1}{2}}}, \; = \infty, \; = \frac{1}{(b^2-a^2)^{\frac{1}{2}}} \log \frac{\nu}{\mu},$$

according as a is greater than, equal to, or less than b.

90.] The case in which one of the limits of a definite integral is ∞, either positive or negative, and the element-function corresponding to that limit $= \infty$, may be treated by a process similar to that of the preceding Article: for we may replace $\pm \infty$ by $\pm \frac{1}{\mu i}$, and in the evaluated definite integral substitute 0 for i. Thus if $F'(x) = \pm \infty$ when $x = \pm \infty$, either sign being taken,

$$\int_{-\infty}^{\infty} F'(x)\, dx = \int_{-\frac{1}{\mu i}}^{\frac{1}{\nu i}} F'(x)\, dx, \tag{51}$$

i being replaced by 0 in the result. Thus,

$$\int_{-\infty}^{\infty} \frac{(x-a)\, dx}{(x-a)^2-b^2} = \int_{-\frac{1}{\mu i}}^{\frac{1}{\nu i}} \frac{(x-a)\, dx}{(x-a)^2-b^2}$$

$$= \left[\log\{(x-a)^2-b^2\}^{\frac{1}{2}}\right]_{-\frac{1}{\mu i}}^{\frac{1}{\nu i}}$$

$$= \log \frac{\mu}{\nu}, \tag{52}$$

when $i = 0$; which quantity is determinate in form, but undetermined in value, because μ and ν are arbitrary constants.

91.] The values of the definite integrals given in the two preceding Articles are indeterminate because they involve the arbitrary constants μ and ν; and they are consequently called *general definite integrals*: there is however no sufficient reason why μ and ν should not be equal, and why indeed μ and ν should not each be equal to unity. If $\mu = \nu = 1$, then the integral assumes a determinate form, and has been called by Cauchy the ***principal value of the definite integral.*** Thus for the principal value of the definite integral given in (42), we have

$$\int_{x_0}^{x_n} F'(x)\,dx = \int_{x_0}^{\xi-i} F'(x)\,dx + \int_{\xi+i}^{x_n} F'(x)\,dx. \tag{53}$$

And as the general value of $\int_{-1}^{1} \frac{dx}{x} = \log \frac{\mu}{\nu}$; so the principal value is 0; since if $\mu = \nu = 1$, $\log \frac{\mu}{\nu} = 0$.

Also when a is less than b, for the general value of $\int_0^{\pi} \frac{dx}{a+b\cos x}$ we have, by (49),

$$\int_0^{\pi} \frac{dx}{a+b\cos x} = \frac{1}{(b^2-a^2)^{\frac{1}{2}}} \log \frac{\nu}{\mu};$$

and as the right-hand member $= 0$, when $\mu = \nu = 1$, the principal value of the integral $= 0$. Thus then,

$$\int_0^{\pi} \frac{dx}{a+b\cos x} = \frac{\pi}{(a^2-b^2)^{\frac{1}{2}}}, \quad = \infty, \quad = 0,$$

according as a is greater than, equal to, or less than b.

Hence it appears that $\frac{(a^2-b^2)^{\frac{1}{2}}}{\pi} \int_0^{\pi} \frac{dx}{a+b\cos x}$ is a discontinuous function of a and b; it is equal to 1 for all values of a greater than b; it is equal to ∞, when $a = b$, and it is equal to 0 for all values of a less than b; and thus it abruptly changes its value from a constant quantity to zero, by passing through ∞, when a becomes equal to b. Suppose this definite integral to represent the ordinate of a locus of which a is the abscissa; the locus will be a straight line, parallel to the x-axis, at a distance, $= 1$, from it, for all values of the abscissa greater than b; and for all values of the abscissa less than b, it will be a straight line coincident with the axis of x; and the ordinate is infinite when the abscissa $= b$.

Thus a definite integral may be a new species of transcendent, and may express a discontinuous function; indeed it often does so; in the sequel many cases will occur, and we shall have occasion to exhibit this particular characteristic in some very striking and important forms.

92.] Indefinite integrals often take a form which is incompatible with the remarks made at the end of Art. 83: although the element-function has the same sign throughout the range of integration, yet the indefinite integral has a contrary sign. Thus for instance,

$$\int (\cot x)^2 dx = \int \{(\operatorname{cosec} x)^2 - 1\} dx$$
$$= -\cot x - x; \qquad (54)$$

that is, the sum of a series of positive quantities is a negative quantity: which it of course cannot be. The preceding theory removes the apparent contradiction. Let the limits, whatever they may be, be introduced, and let the integrals be definite; then the result is correct. Thus,

$$\int_{\frac{\pi}{4}}^{\frac{\pi}{2}} (\cot x)^2 dx = \Big[-\cot x - x\Big]_{\frac{\pi}{4}}^{\frac{\pi}{2}}$$
$$= 1 - \frac{\pi}{2} + \frac{\pi}{4} = \frac{4-\pi}{4}; \qquad (55)$$

which is a positive result.

Also again,

$$\int_{-x_0}^{x_n} \frac{dx}{x^{2m}} = -\frac{1}{2m-1}\left\{\frac{1}{x_n^{2m-1}} + \frac{1}{x_0^{2m-1}}\right\}; \qquad (56)$$

which is an absurd result; because the right-hand member is negative, although all the element-functions of which it is the sum are positive. But x^{-2m} is infinite, when $x = 0$, 0 being a value included within the range of integration. We must therefore divide the integral into two parts, and we will take the principal value of each: thus,

$$\int_{-x_0}^{x_n} \frac{dx}{x^{2m}} = \int_{-x_0}^{-\iota} \frac{dx}{x^{2m}} + \int_{\iota}^{x_n} \frac{dx}{x^{2m}}$$
$$= -\frac{1}{2m-1}\left\{\left[\frac{1}{x^{2m-1}}\right]_{-x_0}^{-\iota} + \left[\frac{1}{x^{2m-1}}\right]_{\iota}^{x_n}\right\}$$
$$= \frac{1}{2m-1}\left\{\frac{2}{\iota^{2m-1}} - \frac{1}{x_n^{2m-1}} - \frac{1}{x_0^{2m-1}}\right\}; \qquad (57)$$

which is equal to ∞, when $\iota = 0$; and this is the correct value of the integral.

SECTION 3.—*The Transformation of Definite Integrals by a Change of Variable.*

93.] The general theory of the transformation of a single definite integral by means of a change of variable, and the effects of the substitution of a new variable, have been explained in Theorem IV, Art 8: and various examples of substitution have already been made in Chapters II and III, wherein indefinite integrals have been investigated. Further applications of the theory are now required, and with respect to a change of the limits of integration as well as to that of the element-function; the latter part of the theory alone having been required in the preceding Chapters.

The equivalence of the following expressions is evident on inspection.

$$\int_{x_0}^{x_n} \mathrm{F}'(x)\,dx = \int_{x_0}^{x_n} \mathrm{F}'(z)\,dz. \tag{58}$$

$$\int_{x_0}^{x_n} \mathrm{F}'(x+a)\,dx = \int_{x_0+a}^{x_n+a} \mathrm{F}'(x)\,dx. \tag{59}$$

$$\int_{x_0}^{x_n} \mathrm{F}'(ax+b)\,dx = \frac{1}{a}\int_{ax_0+b}^{ax_n+b} \mathrm{F}'(x)\,dx. \tag{60}$$

If it is required to transform $\int_{x_0}^{x_n} \mathrm{F}'(x)\,dx$ into an equivalent definite integral of which the limits are z_n and z_0, both z_n and z_0 being finite quantities, the following substitution may be made. Let

$$x = x_n \frac{z-z_0}{z_n-z_0} + x_0 \frac{z_n-z}{z_n-z_0}; \tag{61}$$

in which case, $z = z_0$ when $x = x_0$; and $z = z_n$ when $x = x_n$; and

$$dx = \frac{x_n - x_0}{z_n-z_0}\,dz; \tag{62}$$

so that

$$\int_{x_0}^{x_n} \mathrm{F}'(x)\,dx = \frac{x_n-x_0}{z_n-z_0}\int_{z_0}^{z_n} \mathrm{F}'\left(x_n\frac{z-z_0}{z_n-z_0} + x_0\frac{z_n-z}{z_n-z_0}\right)dz. \tag{63}$$

The following are instances of this formula.

Suppose that the limits of the transformed integral are 1 and 0: then, $z_n = 1$, $z_0 = 0$; and

$$\int_{x_0}^{x_n} \mathrm{F}'(x)\,dx = (x_n - x_0)\int_0^1 \mathrm{F}'\{x_0 + (x_n - x_0)x\}\,dx; \tag{64}$$

the variable z having been replaced by x in the right-hand member by reason of (58). This is the formula of transformation from the old limits x_n and x_0 to new limits 1 and 0.

Other cases of the preceding formula are the following: let $z_n = x_n - x_0$, and $z_0 = 0$: then

$$\int_{x_0}^{x_n} \text{F}'(x)\,dx = \int_0^{x_n - x_0} \text{F}'(x + x_0)\,dx. \tag{65}$$

Let $z_n = x_0$, $z_0 = x_n$, so that the limits may be reversed; then

$$\int_{x_0}^{x_n} \text{F}'(x)\,dx = -\int_{x_n}^{x_0} \text{F}'(x_n + x_0 - x)\,dx$$

$$= \int_{x_0}^{x_n} \text{F}'(x_n + x_0 - x)\,dx; \tag{66}$$

and, if $x_0 = 0$,

$$\int_0^{x_n} \text{F}'(x)\,dx = \int_0^{x_n} \text{F}'(x_n - x)\,dx. \tag{67}$$

The formula of transformation given in (61) is inapplicable when either z_n or $z_0 = \infty$. If however a transformation is required so that the new limits may be ∞ and 0; then let

$$x = x_0 + \frac{x_n - x_0}{1 + z}; \tag{68}$$

in which case $z = \infty$ when $x = x_0$, and $z = 0$ when $x = x_n$; and

Ex. 1. By (17), Art. 82, $\int_{-\infty}^{\infty} \frac{x^{2m}\,dx}{x^{2n}+1} = \frac{\pi}{n} \operatorname{cosec} \frac{2m+1}{2n}\pi.$

Let $x^{2n} = z$; then $2nx^{2n-1}dx = dz$.

$$\therefore \int_{-\infty}^{\infty} \frac{x^{2m}\,dx}{x^{2n}+1} = 2\int_0^{\infty} \frac{x^{2m}\,dx}{x^{2n}+1}$$

$$= \frac{1}{n}\int_0^{\infty} \frac{z^{\frac{2m+1}{2n}-1}\,dz}{z+1}.$$

Let $\frac{2m+1}{2n} = p$; and for z substitute x in the last integral;

then $$\int_0^{\infty} \frac{x^{p-1}\,dx}{x+1} = \pi \operatorname{cosec} p\pi. \tag{72}$$

Again in this last integral, let $x = (\tan\theta)^2$; then

$$\int_0^{\frac{\pi}{2}} (\tan\theta)^{2p-1}\,d\theta = \frac{\pi}{2}\operatorname{cosec} p\pi.$$

Ex. 2. Similarly $$\int_0^{\infty} \frac{x^{p-1}\,dx}{x-1} = -\pi \cot p\pi. \tag{73}$$

Ex. 3. Let $u = \int_0^{\frac{\pi}{2}} (\sin x)^2\,dx$; and for x substitute $\frac{\pi}{2} - x$;

then $$u = -\int_{\frac{\pi}{2}}^{0} (\cos x)^2\,dx = \int_0^{\frac{\pi}{2}} (\cos x)^2\,dx;$$

consequently adding this to the former value of u,

$$2u = \int_0^{\frac{\pi}{2}} \{(\sin x)^2 + (\cos x)^2\}\,dx$$

$$= \frac{\pi}{2};$$

$$\therefore \quad u = \int_0^{\frac{\pi}{2}} (\sin x)^2\,dx = \int_0^{\frac{\pi}{2}} (\cos x)^2\,dx = \frac{\pi}{4}. \tag{74}$$

Thus the definite integral is evaluated for the given limits without a previous knowledge of the indefinite integral. Many similar instances will occur hereafter; and indeed the investigation of these values is a capital part of our Treatise.

Ex. 4. $u = \int_0^{\frac{\pi}{2}} \log \sin x\,dx.$

For x substitute $\frac{\pi}{2} - x$; therefore $u = \int_0^{\frac{\pi}{2}} \log\cos x\,dx.$

$$\therefore \quad 2u = \int_0^{\frac{\pi}{2}} \{\log \sin x + \log \cos x\}\, dx$$

$$= \int_0^{\frac{\pi}{2}} \log \frac{\sin 2x}{2}\, dx.$$

In this last integral for $2x$ substitute x; then

$$2u = \frac{1}{2}\int_0^{\pi} \{\log \sin x - \log 2\}\, dx$$

$$= \int_0^{\frac{\pi}{2}} \log \sin x\, dx - \frac{\pi}{2}\log 2$$

$$= u - \frac{\pi}{2}\log 2;$$

$$\therefore \quad u = \int_0^{\frac{\pi}{2}} \log \sin x\, dx = \int_0^{\frac{\pi}{2}} \log \cos x\, dx = -\frac{\pi}{2}\log 2. \quad (75)$$

Ex. 5.

$$\int_0^{\infty} \frac{\sin x}{x}\, dx = \int_0^{\frac{\pi}{2}} \frac{\sin x}{x}\, dx + \int_{\frac{\pi}{2}}^{\pi} \frac{\sin x}{x}\, dx + \int_{\pi}^{\frac{3\pi}{2}} \frac{\sin x}{x}\, dx + \int_{\frac{3\pi}{2}}^{2\pi} \frac{\sin x}{x}\, dx + \ldots$$

Let all the integrals in the right-hand member be brought to the same limits, $\frac{\pi}{2}$ and 0; for this purpose in the second integral, let x be replaced by $\pi - x$; then

$$\int_{\frac{\pi}{2}}^{\pi} \frac{\sin x}{x}\, dx = \int_0^{\frac{\pi}{2}} \frac{\sin x}{\pi - x}\, dx.$$

In the third integral, let x be replaced by $\pi + x$; then

$$\int_{\pi}^{\frac{3\pi}{2}} \frac{\sin x}{x}\, dx = -\int_0^{\frac{\pi}{2}} \frac{\sin x}{\pi + x}\, dx.$$

In the fourth integral, let x be replaced by $2\pi - x$; then

$$\int_{\frac{3\pi}{2}}^{2\pi} \frac{\sin x}{x}\, dx = -\int_0^{\frac{\pi}{2}} \frac{\sin x}{2\pi - x}\, dx;$$

and so on; therefore

$$\int_0^{\infty} \frac{\sin x}{x}\, dx$$

$$= \int_0^{\frac{\pi}{2}} \sin x \left\{\frac{1}{x} + \frac{1}{\pi - x} - \frac{1}{\pi + x} - \frac{1}{2\pi - x} + \frac{1}{2\pi + x} + \ldots\right\} dx;$$

but $\dfrac{1}{x} + \dfrac{1}{\pi - x} - \dfrac{1}{\pi + x} - \dfrac{1}{2\pi - x} + \dfrac{1}{2\pi + x} + \ldots = \operatorname{cosec} x;$

$$\therefore \quad \int_0^\infty \frac{\sin x}{x}\,dx = \int_0^{\frac{\pi}{2}} \sin x \operatorname{cosec} x\,dx$$

$$= \frac{\pi}{2}. \tag{76}$$

Hence, if x is replaced by kx,

$$\int_0^\infty \frac{\sin kx}{x}\,dx = \frac{\pi}{2}. \tag{77}$$

Ex. 6. $\displaystyle\int_0^\pi \frac{x\,dx}{a^2-b^2(\cos x)^2} = \frac{1}{2a}\int_0^\pi \left\{\frac{x}{a+b\cos x} + \frac{x}{a-b\cos x}\right\} dx.$

In the second integral in the right-hand member, for x sub stitute $\pi - x$; then

$$\int_0^\pi \frac{x\,dx}{a-b\cos x} = \int_0^\pi \frac{(\pi-x)\,dx}{a+b\cos x};$$

$$\therefore \quad \int_0^\pi \frac{x\,dx}{a^2-b^2(\cos x)^2} = \frac{1}{2a}\int_0^\pi \left\{\frac{x}{a+b\cos x} + \frac{\pi-x}{a+b\cos x}\right\} dx$$

$$= \frac{\pi}{2a}\int_0^\pi \frac{dx}{a+b\cos x}; \tag{78}$$

the value of which integral has been already determined in Art. 91.

95.] The following are cases in which the element-function contains a general function into which the subject-variable enters in a special combination, and in which the definite integral is simplified by transformation. The function denoted by f in the following examples is finite and continuous for all employed values of its subject-variable.

Ex. 1. $\displaystyle\int_0^\infty f(x^n + x^{-n}) \log x \frac{dx}{x}$

$$= \int_0^1 f(x^n+x^{-n}) \log x \frac{dx}{x} + \int_1^\infty f(x^n+x^{-n}) \log x \frac{dx}{x}.$$

In the second integral of the right-hand member for x substitute $\frac{1}{x}$; then

$$\int_1^\infty f(x^n+x^{-n}) \log x \frac{dx}{x} = -\int_0^1 f(x^n+x^{-n}) \log x \frac{dx}{x};$$

$$\therefore \quad \int_0^\infty f(x^n+x^{-n}) \log x \frac{dx}{x} = 0. \tag{79}$$

Ex. 2. By the same substitution it may be shewn that

$$\int_0^\infty f(x^n+x^{-n}) \log x \frac{dx}{1+x^2} = 0.$$

Ex. 3. $\int_0^\infty f(x^n+x^{-n})\tan^{-1}x\frac{dx}{x}$

$$=\int_0^1 f(x^n+x^{-n})\tan^{-1}x\frac{dx}{x}+\int_1^\infty f(x^n+x^{-n})\tan^{-1}x\frac{dx}{x}.$$

In the second integral of the right-hand member let x be replaced by $\frac{1}{x}$; then

$$\int_1^\infty f(x^n+x^{-n})\tan^{-1}x\frac{dx}{x} = -\int_1^0 f(x^n+x^{-n})\tan^{-1}\frac{1}{x}\frac{dx}{x}$$

$$=\int_0^1 f(x^n+x^{-n})\left(\frac{\pi}{2}-\tan^{-1}x\right)\frac{dx}{x};$$

$$\therefore \int_0^\infty f(x^n+x^{-n})\tan^{-1}x\frac{dx}{x} = \frac{\pi}{2}\int_0^1 f(x^n+x^{-n})\frac{dx}{x}. \quad (80)$$

Ex. 4. By a similar substitution it may be shewn that

$$\int_0^\infty f(x^n+x^{-n})\frac{dx}{x} = 2\int_0^1 f(x^n+x^{-n})\frac{dx}{x}. \quad (81)$$

Ex. 5. Consider the integral $\int_0^\pi x f\{\sin x, (\cos x)^2\}\,dx$; then since $\sin x$ and $(\cos x)^2$ are unaltered when x is replaced by $\pi-x$;

$$\therefore \int_0^\pi x f\{\sin x, (\cos x)^2\}\,dx = -\int_\pi^0 (\pi-x) f\{\sin x, (\cos x)^2\}\,dx$$

$$= \int_0^\pi (\pi-x) f\{\sin x, (\cos x)^2\}\,dx;$$

$$\therefore \int_0^\pi x f\{\sin x, (\cos x)^2\}\,dx = \frac{\pi}{2}\int_0^\pi f\{\sin x, (\cos x)^2\}\,dx. \quad (82)$$

The following is an example of this theorem:

$$\int_0^\pi x\log\sin x\,dx = \frac{\pi}{2}\int_0^\pi \log\sin x\,dx$$

$$= -\frac{\pi^2}{2}\log 2. \quad (83)$$

Many cases of transformation and of a consequent simplification of definite integrals will occur in the sequel: and the preceding are sufficient for illustrations of the process.

SECTION 4.—*On the Differentiation and Integration of a Definite Integral with respect to a Variable Parameter.*

96.] As a definite integral in its most general form involves the limits of integration, and any constants which may have been contained in the element-function, it may be considered as a function of these quantities, and treated accordingly. Thus, if u is a definite integral, of which x_n and x_0 are the limits, and a is a parameter involved in the element-function, so that

$$u = \int_{x_0}^{x_n} \mathrm{F}'(a, x)\, dx, \tag{84}$$

u may be treated as a function of x_n, x_0, a. And subject to the condition, which is always necessary, that the element-function is finite and continuous for all employed values of its subject-variable, u may be treated as a continuous function of these three quantities, and differentiated and integrated accordingly. These processes we propose now to develop; and we shall have many applications of them; and from definite integrals, which have already been determined, others will be derived. In the most general case x_n, x_0, and a may be considered as three independent variables; so that the total differential of u will consist of three partial differentials. There will also be particular cases where one, or two, of these three quantities will vary, the others being constant; these however may be treated as special forms of the general case.

Let the right-hand member of (84) be written at length; then

$$u = (x_1 - x_0)\,\mathrm{F}'(a, x_0) + (x_2 - x_1)\,\mathrm{F}'(a, x_1) + \dots$$
$$\dots + (x_n - x_{n-1})\,\mathrm{F}'(a, x_{n-1}). \tag{85}$$

(1) Let x_n be increased by an increment $x_{n+1} - x_n$, which is dx_n; then the corresponding increment of u is $(x_{n+1} - x_n)\,\mathrm{F}'(a, x_n)$; so that for the partial differential of u due to the variation of x_n, we have

$$\left(\frac{du}{dx_n}\right) = \mathrm{F}'(a, x_n); \tag{86}$$

(2) Let x_0 be increased by its increment $x_1 - x_0$, which is dx_0; so that the range of integration commences at x_1; and u is diminished by $(x_1 - x_0)\,\mathrm{F}'(a, x_0)$; consequently

$$\left(\frac{du}{dx_0}\right) = -\mathrm{F}'(a, x_0); \tag{87}$$

Hereby we have the two partial differentials of the definite integral with respect to the limits.

(3) Let a vary; then, from (85),

$$\frac{du}{da} = (x_1-x_0)\frac{d}{da}\text{F}'(a,x_0)+(x_2-x_1)\frac{d}{da}\text{F}'(a,x_1)+\dots$$

$$\dots+(x_n-x_{n-1})\frac{d}{da}\text{F}'(a,x_{n-1}) \quad (88)$$

$$=\int_{x_0}^{x_n}\frac{d}{da}\text{F}'(a,x)\,dx\,; \quad (89)$$

which gives the partial differential of u with respect to a. Consequently for the total differential of u we have

$$\text{D}\int_{x_0}^{x_n}\text{F}'(a,x)\,dx$$

$$=\text{F}'(a,x_n)\,dx_n-\text{F}'(a,x_0)\,dx_0+\int_{x_0}^{x_n}\frac{d.\text{F}'(a,x)}{da}\,da\,dx. \quad (90)$$

Hence, if x_n, x_0, a are all functions of an independent variable t, when t varies,

$$\frac{\text{D}}{dt}\int_{x_0}^{x_n}\text{F}'(a,x)\,dx$$

$$=\text{F}'(a,x_n)\frac{dx_n}{dt}-\text{F}'(a,x_0)\frac{dx_0}{dt}+\int_{x_0}^{x_n}\frac{d.\text{F}'(a,x)}{da}\frac{da}{dt}\,dx. \quad (91)$$

The process by which this equation has been found is commonly called differentiation under the sign of integration with respect to a variable parameter. Leibnitz has called it *Differentiatio de Curvâ in Curvam.* The meaning of this remark will be plain from the following geometrical interpretation of (91).

Let $\text{P}_0\text{P}\text{P}_n$, see Fig. 46, be the curve whose equation is $y=\text{F}'(a,x)$; let $\text{OM}_0=x_0$, $\text{OM}_n=x_n$, then, as explained in Art. 3, the area $\text{M}_0\text{M}_n\text{P}_n\text{P}_0=\int_{x_0}^{x_n}\text{F}'(a,x)\,dx.$

Let the parameter a vary, and first let the element-function alone vary, and let the new position of the curve which is due to the variation of a be $\text{LP}'_0\text{NP}'_n$, so that the area becomes increased by the quadrilateral $\text{LP}_0\text{P}_n\text{N}$: therefore

$$\text{LP}_0\text{P}_n\text{N}=\int_{x_0}^{x_n}\frac{d.\text{F}'(a,x)}{da}\,da\,dx\,;$$

next let the limits vary, so that by the change of a, OM_0 becomes OM'_0, and OM_n becomes OM'_n; and therefore the area is increased by $\text{P}_n\text{M}_n\text{M}'_n\text{N}'$ and diminished by $\text{P}_0\text{M}_0\text{M}'_0\text{L}'$, which are respectively represented by $\text{F}'(a,x_n)\,dx_n$ and $\text{F}'(a,x_0)\,dx_0$. But when all these variations are simultaneous, the definite integral expresses the area $\text{P}'_0\text{M}'_0\text{M}'_n\text{P}'_n$ instead of $\text{P}_0\text{M}_0\text{M}_n\text{P}_n$, the two quadrilaterals LL', NN' being omitted, because they are infinitesimals of a higher order; being, in fact, quadrilaterals, each of whose sides is an infinitesimal, and which are therefore infinitesimals of the second order.

97.] Of the theorems of the preceding Article, that contained in equation (89) is the most useful for our present inquiry, although the others are important in some subsequent physical investigations. Now (89) may be expressed in the following form.

$$\frac{d}{da}\int_{x_0}^{x_n} \mathrm{F}'(a,x)\,dx = \int_{x_0}^{x_n} \frac{d.\mathrm{F}'(a,x)}{da}\,dx. \tag{92}$$

Hence it appears that the differential of a definite integral with respect to a variable parameter involved in the element-function is the definite integral of the differential of the element-function with respect to that variable parameter. The two operations therefore of differentiation and of integration, effected as they are with respect to different variables, and thus independent of each other, may be interchanged without any alteration of the result.

The process of differentiating a definite integral with respect to a variable parameter involved in its element-function having been thus established, the operation may be repeated on the function thus found: and repeated as often as the circumstances require; and as the order of these operations is indifferent, the final result will be expressed in the following form;

$$\frac{d^r}{da^r}\int_{x_0}^{x_n} \mathrm{F}'(a,x)\,dx = \int_{x_0}^{x_n} \frac{d^r.\mathrm{F}'(a,x)}{da^r}\,dx. \tag{93}$$

The complete expression given in (91) may of course be subjected to repeated differentiations in the same manner as the part of it given in (92). But as we shall not require the general result, it is not necessary to insert it here.

98.] The following are examples of (92) and (93), in which from a known definite integral by means of differentiation with respect to a variable contained in its element-function a new definite integral is evaluated.

Ex. 1. Let the known definite integral be

$$\int_0^x \frac{dx}{a^2+x^2} = \frac{1}{a}\tan^{-1}\frac{x}{a}:$$

then, taking the a-differential,

$$\int_0^x \frac{-2a\,dx}{(a^2+x^2)^2} = -\frac{1}{a^2}\tan^{-1}\frac{x}{a} - \frac{x}{a(a^2+x^2)};$$

$$\therefore \int_0^x \frac{dx}{(a^2+x^2)^2} = \frac{1}{2a^3}\tan^{-1}\frac{x}{a} + \frac{x}{2a^2(a^2+x^2)}.$$

Ex. 2. Also, since $\int_0^\infty \frac{dx}{a^2+x^2} = \frac{\pi}{2a}$,

$$\int_0^\infty \frac{-2a\,dx}{(a^2+x^2)^2} = -\frac{\pi}{2a^2};$$

$$\therefore \int_0^\infty \frac{dx}{(a^2+x^2)^2} = \frac{\pi}{4a^3};$$

and after n differentiations,

$$\int_0^\infty \frac{dx}{(a^2+x^2)^n} = \frac{(2n-3)(2n-5)\dots 3.1}{(2n-2)(2n-4)\dots 4.2}\frac{\pi}{2}a^{-(2n-1)}. \qquad (94)$$

Ex. 3. Since, if a is greater than b, by (27), Art. 67,

$$\int_0^x \frac{dx}{a+b\cos x} = \frac{2}{(a^2-b^2)^{\frac{1}{2}}}\tan^{-1}\left\{\left(\frac{a-b}{a+b}\right)^{\frac{1}{2}}\tan\frac{x}{2}\right\};$$

therefore differentiating with respect to a, and reducing,

$$\int_0^x \frac{dx}{(a+b\cos x)^2}$$

$$= \frac{-b}{a^2-b^2}\frac{\sin x}{a+b\cos x} + \frac{2a}{(a^2-b^2)^{\frac{3}{2}}}\tan^{-1}\left\{\left(\frac{a-b}{a+b}\right)^{\frac{1}{2}}\tan\frac{x}{2}\right\};$$

$$\therefore \int_0^\pi \frac{dx}{(a+b\cos x)^2} = \frac{\pi a}{(a^2-b^2)^{\frac{3}{2}}}. \qquad (95)$$

the latter result is also the a-differential of (46).

Ex. 4. Since $\int_0^\infty e^{-ax}\,dx = a^{-1}$;

therefore differentiating n times with respect to a,

$$\int_0^\infty x^n e^{-ax}\,dx = 1.2.3\dots n a^{-n}.$$

Ex. 5. Since by (6), Art. 82,

$$\int_0^\infty e^{-ax}\cos bx\,dx = \frac{a}{a^2+b^2}$$

$$= \frac{1}{2}\left\{\frac{1}{a+b\sqrt{-1}} + \frac{1}{a-b\sqrt{-1}}\right\};$$

therefore differentiating both members $n-1$ times with respect to a, we have

$$\int_0^\infty e^{-ax}x^{n-1}\cos bx\,dx$$

$$= \frac{1.2.3\dots(n-1)}{2}\{(a+b\sqrt{-1})^{-n}+(a-b\sqrt{-1})^{-n}\};$$

so that if $b = a \tan \alpha$,

$$\int_0^{\infty} e^{-ax} x^{n-1} \cos bx\, dx = \frac{1.2.3\ldots(n-1)}{(a^2+b^2)^{\frac{n}{2}}} \cos\left(n \tan^{-1}\frac{b}{a}\right). \quad (96)$$

Similarly

$$\int_0^{\infty} e^{-ax} x^{n-1} \sin bx\, dx = \frac{1.2.3\ldots(n-1)}{(a^2+b^2)^{\frac{n}{2}}} \sin\left(n \tan^{-1}\frac{b}{a}\right). \quad (97)$$

Ex. 6. Since $\int_0^{\frac{\pi}{2}} \frac{dx}{a^2(\cos x)^2 + b^2(\sin x)^2} = \frac{\pi}{2ab}$; therefore taking successively the a- and the b-differential, we have

$$\left.\begin{aligned} \int_0^{\frac{\pi}{2}} \frac{(\cos x)^2\, dx}{\{a^2(\cos x)^2+b^2(\sin x)^2\}^2} &= \frac{\pi}{4a^3 b}, \\ \int_0^{\frac{\pi}{2}} \frac{(\sin x)^2\, dx}{\{a^2(\cos x)^2+b^2(\sin x)^2\}^2} &= \frac{\pi}{4ab^3}; \end{aligned}\right\} \quad (98)$$

and by addition of these,

$$\int_0^{\frac{\pi}{2}} \frac{dx}{\{a^2(\cos x)^2+b^2(\sin x)^2\}^2} = \frac{\pi(a^2+b^2)}{4a^3b^3}. \quad (99)$$

If we differentiate this integral again with respect to a and b successively, and add the results, we have

$$\int_0^{\frac{\pi}{2}} \frac{dx}{\{a^2(\cos x)^2+b^2(\sin x)^2\}^3} = \frac{\pi}{16}\,\frac{3a^4+2a^2b^2+3b^4}{a^5b^5}; \quad (100)$$

the same process may be repeated, and other integrals will be evaluated.

99.] As a definite integral may be considered a continuous function of a variable parameter involved in the element-function, and thus be the subject of differentiation with respect to this parameter; so may it also be the element-function of a definite integral, when the arbitrary parameter continuously varies: the element-function being finite and continuous for all employed values of the parameter as well as for those of its original subject-variable. Let u be the given definite integral; then

$$u = \int_{x_0}^{x_n} \mathrm{F}'(a,x)\,dx = (x_1-x_0)\,\mathrm{F}'(a,x_0)+(x_2-x_1)\,\mathrm{F}'(a,x_1)+\ldots$$
$$\ldots+(x_n-x_{n-1})\,\mathrm{F}'(a,x_{n-1}).$$

Now let us suppose the parameter a to vary continuously from

a_0 to a_n, and let the distance $a_n - a_0$ be divided into n infinitesimal parts, to the points of partition of which let $a_1, a_2, \dots a_{n-1}$ correspond; then

$$\int_{a_0}^{a_n} u\,da = \int_{a_0}^{a_n} da \int_{x_0}^{x_n} \mathrm{F}'(a,x)\,dx$$

$$\begin{aligned}
&= (x_1-x_0)\{(a_1-a_0)\mathrm{F}'(a_0,x_0)+(a_2-a_1)\mathrm{F}'(a_1,x_0)+\dots \\
&\qquad\qquad \dots+(a_n-a_{n-1})\mathrm{F}'(a_{n-1},x_0)\} \\
&\quad +(x_2-x_1)\{(a_1-a_0)\mathrm{F}'(a_0,x_1)+(a_2-a_1)\mathrm{F}'(a_1,x_1)+\dots \\
&\qquad\qquad \dots+(a_n-a_{n-1})\mathrm{F}'(a_{n-1},x_1)\} \\
&\quad + \dots\dots\dots\dots\dots\dots \\
&\quad +(x_n-x_{n-1})\{(a_1-a_0)\mathrm{F}'(a_0,x_{n-1})+(a_2-a_1)\mathrm{F}'(a_1,x_{n-1})+\dots \\
&\qquad\qquad \dots+(a_n-a_{n-1})\mathrm{F}'(a_{n-1},x_{n-1})\} \\
&= (a_1-a_0)\{(x_1-x_0)\mathrm{F}'(a_0,x_0)+(x_2-x_1)\mathrm{F}'(a_0,x_1)+\dots \\
&\qquad\qquad \dots+(x_n-x_{n-1})\mathrm{F}'(a_0,x_{n-1})\} \\
&\quad +(a_2-a_1)\{(x_1-x_0)\mathrm{F}'(a_1,x_0)+(x_2-x_1)\mathrm{F}'(a_1,x_1)+\dots \\
&\qquad\qquad \dots+(x_n-x_{n-1})\mathrm{F}'(a_1,x_{n-1})\} \\
&\quad + \dots\dots\dots\dots\dots\dots \\
&\quad +(a_n-a_{n-1})\{(x_1-x_0)\mathrm{F}'(a_{n-1},x_0)+(x_2-x_1)\mathrm{F}'(a_{n-1},x_1)+\dots \\
&\qquad\qquad \dots+(x_n-x_{n-1})\mathrm{F}'(a_{n-1},x_{n-1})\} \qquad (101)
\end{aligned}$$

$$\begin{aligned}
&= (a_1-a_0)\int_{x_0}^{x_n}\mathrm{F}'(a_0,x)\,dx+(a_2-a_1)\int_{x_0}^{x_n}\mathrm{F}'(a_1,x)\,dx+\dots \\
&\qquad\qquad \dots+(a_n-a_{n-1})\int_{x_0}^{x_n}\mathrm{F}'(a_{n-1},x)\,dx \\
&= \int_{x_0}^{x_n}\{(a_1-a_0)\mathrm{F}'(a_0,x)+(a_2-a_1)\mathrm{F}'(a_1,x)+\dots \\
&\qquad\qquad \dots+(a_n-a_{n-1})\mathrm{F}'(a_{n-1},x)\}\,dx \\
&= \int_{x_0}^{x_n} dx\int_{a_0}^{a_n}\mathrm{F}'(a,x)\,da. \qquad (102)
\end{aligned}$$

Thus the a-integral of the x-integral of the element-function is equal to the x-integral of the a-integral; and thus the order of these two processes, effected as they are with respect to variables independent of each other, may be interchanged without any alteration of the result. This theorem is called the inversion of the order of integrations; and the members of (102) are called double integrals.

In the preceding investigation it is assumed that a and all its values are independent of x and all its values. The inquiry into the properties of double integrals will be extended to the cases where this restriction is not made; but the preceding is sufficient for our present purpose.

The principle of single integration of a definite integral with respect to a parameter contained in its element-function having been thus established, and the order of the operations having been shewn to be indifferent, the process may be repeated, subject of course to the same conditions. So that we have to r successive integrations,

$$\int_{a_0}^{a_n} da \int_{a_0}^{a_n} da \ldots \int_{a_0}^{a_n} da \int_{x_0}^{x_n} \mathrm{F}'(a, x)\, dx$$
$$= \int_{x_0}^{x_n} dx \int_{a_0}^{a_n} da \int_{a_0}^{a_n} da \ldots \int_{a_0}^{a_n} \mathrm{F}'(a, x)\, da. \quad (103)$$

Integrals of this kind are called multiple integrals, of which double, triple integrals are particular forms, according as two, three, ... integration-processes are involved. The general theory is of great importance in subsequent investigations, and will be a subject of inquiry hereafter: but the preceding is sufficient at present.

100.] The following are examples of the process.

Ex. 1. Since $\int_0^1 x^{n-1} dx = \frac{1}{n}$;

$$\therefore \int_a^b dn \int_0^1 x^{n-1} dx = \int_a^b \frac{dn}{n};$$

but $\int_a^b dn \int_0^1 x^{n-1} dx = \int_0^1 dx \int_a^b x^{n-1} dn = \int_0^1 \frac{x^{b-1} - x^{a-1}}{\log x} dx$;

and $\int_a^b \frac{dn}{n} = \log \frac{b}{a}$;

$$\therefore \int_0^1 \frac{x^{b-1} - x^{a-1}}{\log x} dx = \log \frac{b}{a}. \quad (104)$$

Ex. 2. Since $\int_0^\infty e^{-ax} dx = \frac{1}{a}$;

$$\therefore \int_c^a da \int_0^\infty e^{-ax} dx = \int_c^a \frac{da}{a};$$

$$\therefore \int_0^\infty dx \int_c^a e^{-ax} da = \int_0^\infty \frac{e^{-cx} - e^{-ax}}{x} dx = \log \frac{a}{c}. \quad (105)$$

Ex. 3. Since $\int_0^\infty e^{-ax} \sin bx\, dx = \frac{b}{a^2 + b^2}$;

$$\therefore \int_a^\infty da \int_0^\infty e^{-ax} \sin bx\, dx = \int_a^\infty \frac{b\,da}{a^2+b^2};$$

$$\therefore \int_0^\infty dx \int_a^\infty e^{-ax} \sin bx\, da = \int_0^\infty \frac{e^{-ax} \sin bx\, dx}{x} = \tan^{-1}\frac{b}{a}. \quad (106)$$

Let $a = 0$; then $\int_0^\infty \frac{\sin bx}{x}\, dx = \frac{\pi}{2}$; (107)

which is the same result as (77), Art. 94.

Ex. 4. By means of the preceding integral may another integral be determined which expresses a remarkable discontinuous function, of which considerable use will be made in the sequel.

Since $2 \sin mx \cos nx = \sin (m+n)x + \sin (m-n)x$,

$$\int_0^\infty \frac{e^{-ax} \sin mx \cos nx}{x}\, dx = \int_0^\infty e^{-ax} \frac{\sin (m+n)x + \sin (m-n)x}{2x}\, dx$$

$$= \frac{1}{2}\left\{\tan^{-1}\frac{m+n}{a} + \tan^{-1}\frac{m-n}{a}\right\}, \text{ if } n \text{ is between } +m \text{ and } -m;$$

$$= \frac{1}{2}\left\{\tan^{-1}\frac{n+m}{a} - \tan^{-1}\frac{n-m}{a}\right\}, \text{ if } n \text{ is beyond } +m \text{ or } -m.$$

Let $a = 0$; then

$$\int_0^\infty \frac{\sin mx \cos nx}{x}\, dx = \frac{1}{2}\left\{\frac{\pi}{2} + \frac{\pi}{2}\right\} = \frac{\pi}{2}, \text{ if } n \text{ is between } +m \text{ and } -m;$$

$$= \frac{1}{2}\left\{\frac{\pi}{2} - \frac{\pi}{2}\right\} = 0, \text{ if } n \text{ is beyond } +m \text{ or } -m;$$

so that

$$\frac{2}{\pi}\int_0^\infty \frac{\sin mx \cos nx}{x}\, dx = 1, \text{ if } n \text{ is between } +m \text{ and } -m; \quad (108)$$

$$= 0, \text{ if } n \text{ is beyond } +m \text{ or } -m. \quad (109)$$

If $n = m$,

$$\int_0^\infty \frac{e^{-ax} \sin mx \cos mx}{x}\, dx = \frac{1}{2}\int_0^\infty \frac{e^{-ax} \sin 2mx}{x}\, dx = \frac{1}{2}\tan^{-1}\frac{2m}{a};$$

$$\therefore \text{ if } a = 0, \quad \int_0^\infty \frac{\sin mx \cos mx}{x}\, dx = \frac{\pi}{4}. \quad (110)$$

Hence, if positive values only of m and n are taken,

$\frac{2}{\pi}\int_0^\infty \frac{\sin mx \cos nx}{x}\, dx = 1$, for all values of n less than m; $= \frac{1}{2}$, when $n = m$; and $= 0$, for all values of n greater than m.

Ex. 5. Since $\int_0^{\infty} \frac{dx}{(a^2+x^2)(b^2+x^2)} = \frac{\pi}{2ab(a+b)}$;

$$\therefore \int_0^a da \int_0^{\infty} \frac{2a\,dx}{(a^2+x^2)(b^2+x^2)} = \int_0^a \frac{\pi\,da}{b(a+b)};$$

$$\therefore \int_0^{\infty} \frac{dx}{b^2+x^2} \int_0^a \frac{2a\,da}{a^2+x^2} = \frac{\pi}{b} \int_0^a \frac{da}{a+b};$$

$$\therefore \int_0^{\infty} \log\left(\frac{a^2+x^2}{x^2}\right) \frac{dx}{b^2+x^2} = \frac{\pi}{b} \log \frac{a+b}{b}. \qquad (111)$$

Ex. 6. It is required to evaluate $\int_0^{\infty} e^{-ax^2} dx$.

Substitute y for ax^2; then

$$\int_0^{\infty} e^{-ax^2} dx = \int_0^{\infty} e^{-y} \frac{dy}{2(ay)^{\frac{1}{2}}};$$

$$\therefore \int_0^{\infty} e^{-a} da \int_0^{\infty} e^{-ax^2} dx = \int_0^{\infty} e^{-a} da \int_0^{\infty} e^{-y} \frac{dy}{2(ay)^{\frac{1}{2}}};$$

$$\therefore \int_0^{\infty} dx \int_0^{\infty} e^{-a(1+x^2)} da = \frac{1}{2} \int_0^{\infty} \frac{e^{-a} da}{a^{\frac{1}{2}}} \int_0^{\infty} \frac{e^{-y} dy}{y^{\frac{1}{2}}}.$$

Now, since a and y are two quantities independent of each other,

$$\int_0^{\infty} \frac{e^{-a} da}{a^{\frac{1}{2}}} = \int_0^{\infty} \frac{e^{-y} dy}{y^{\frac{1}{2}}};$$

and as each, being a definite integral, is a constant, the repetition is equivalent to multiplication; and we have

$$\int_0^{\infty} \frac{dx}{1+x^2} = \frac{1}{2} \left\{ \int_0^{\infty} \frac{e^{-y} dy}{y^{\frac{1}{2}}} \right\}^2;$$

in the right-hand member replace y by ax^2; then, since the left-hand member $= \frac{\pi}{2}$,

$$\int_0^{\infty} e^{-ax^2} dx = \frac{1}{2} \left(\frac{\pi}{a}\right)^{\frac{1}{2}}; \qquad (112)$$

an integral which we shall hereafter find of considerable importance. If $a = 1$,

$$\int_0^{\infty} e^{-x^2} dx = \frac{\pi^{\frac{1}{2}}}{2}; \qquad (113)$$

$$\therefore \int_{-\infty}^{\infty} e^{-x^2} dx = \pi^{\frac{1}{2}}. \qquad (114)$$

101.] When a definite integral, which is to be evaluated, is differentiated or integrated with respect to a variable parameter contained in the element-function, another definite integral is produced, which can frequently be evaluated by means of its indefinite integral; in other cases the new integral can often be integrated either by parts or by some other method: and sometimes the original definite integral will arise in the process. When this is the case, a differential equation is formed, the solution of which will give the value of the definite integral. Sometimes again the definite integral will arise, when two or more differentiations or integrations have been effected on the original integral. The following examples illustrate the method, and the development of it will be best understood by them.

Ex. 1. Evaluate $u = \int_0^\infty \frac{e^{-ax} - e^{-bx} \cos rx}{x} dx.$

Take the r-differential: then

$$\frac{du}{dr} = \int_0^\infty e^{-bx} \sin rx \, dx$$

$$= \frac{r}{b^2+r^2}; \text{ by (7), Art. 82.}$$

$$\therefore \quad du = \frac{r\,dr}{b^2+r^2}.$$

As the solution is definite, the integrals of both sides of this equation must be definite. Let the limits of r be r and 0; and when $r = 0$, the value of $u = \log \frac{b}{a}$, by (105);

$$\therefore \quad u - \log \frac{b}{a} = \log \frac{(r^2+b^2)^{\frac{1}{2}}}{b};$$

$$\therefore \quad u = \int_0^\infty \frac{e^{-ax} - e^{-bx} \cos rx}{x} dx = \log \frac{(r^2+b^2)^{\frac{1}{2}}}{a}. \qquad (115)$$

Ex. 2. Evaluate $u = \int_0^\infty \frac{\tan^{-1} ax}{b^2+x^2} \frac{dx}{x};$

$$\therefore \quad \frac{du}{da} = \int_0^\infty \frac{dx}{(1+a^2x^2)(b^2+x^2)};$$

$$= \frac{\pi}{2b(1+ab)};$$

$$\therefore \quad u = \int_0^\infty \frac{\tan^{-1} ax}{b^2+x^2} \frac{dx}{x} = \frac{\pi}{2b^2} \log(1+ab); \qquad (116)$$

since $u = 0$, when $a = 0$.

Ex. 3. Evaluate $u = \int_0^\infty e^{-x^2 - \frac{a^2}{x^2}} dx$.

$$\therefore \quad \frac{du}{da} = -2a \int_0^\infty e^{-x^2 - \frac{a^2}{x^2}} \frac{dx}{x^2}.$$

For x substitute $\frac{a}{x}$; $\therefore \quad \frac{du}{da} = -2 \int_0^\infty e^{-x^2 - \frac{a^2}{x^2}} dx.$

$$= -2u;$$

$$\therefore \quad \frac{du}{u} = -2\, da;$$

$$\therefore \quad u = \int_0^\infty e^{-x^2 - \frac{a^2}{x^2}} dx = \frac{\pi^{\frac{1}{2}}}{2} e^{-2a}; \tag{117}$$

since by (113) $u = \frac{\pi^{\frac{1}{2}}}{2}$, when $a = 0$.

Ex. 4. Evaluate $u = \int_0^\infty e^{-a^2x^2} \cos bx\, dx$.

$$\therefore \quad \frac{du}{db} = -\int_0^\infty e^{-a^2x^2} x \sin bx\, dx$$

$$= \left[\frac{e^{-a^2x^2} \sin bx}{2a^2}\right]_0^\infty - \frac{b}{2a^2} \int_0^\infty e^{-a^2x^2} \cos bx\, dx$$

$$= -\frac{b}{2a^2} u;$$

$$\therefore \quad \frac{du}{u} = -\frac{b}{2a^2} db;$$

$$\therefore \quad u = \int_0^\infty e^{-a^2x^2} \cos bx\, dx = \frac{\pi^{\frac{1}{2}}}{2a} e^{-\frac{b^2}{4a^2}}; \tag{118}$$

since by (112) $u = \frac{\pi^{\frac{1}{2}}}{2a}$, when $b = 0$.

Ex. 5. Evaluate $u = \int_0^\infty \frac{e^{-ax} (\sin bx)^2}{x^2} dx$.

$$\therefore \quad \frac{du}{db} = 2 \int_0^\infty \frac{e^{-ax} \sin bx \cos bx}{x} dx;$$

$$\frac{d^2u}{db^2} = 2 \int_0^\infty e^{-ax} \cos 2bx\, dx$$

$$= 2 \left[\frac{e^{-ax}(2b \sin 2bx - a \cos 2bx)}{a^2 + 4b^2}\right]_0^\infty$$

$$= \frac{2a}{a^2 + 4b^2};$$

$$\therefore \quad \frac{du}{db} = \tan^{-1}\frac{2b}{a}; \text{ since } \frac{du}{db} = 0, \text{ when } b = 0.$$

$$\therefore u = \int_0^\infty \frac{e^{-ax}(\sin bx)^2}{x^2} dx = b\tan^{-1}\frac{2b}{a} - \frac{a}{4}\log\frac{a^2+4b^2}{a^2}; \quad (119)$$

since $u = 0$, when $b = 0$.

Ex. 6. Evaluate $u = \int_0^\infty \frac{\sin bx}{a^2+x^2}\frac{dx}{x}$.

$$\frac{du}{db} = \int_0^\infty \frac{\cos bx\,dx}{a^2+x^2}$$

$$\frac{d^2u}{db^2} = -\int_0^\infty \frac{x\sin bx\,dx}{a^2+x^2}$$

$$= -\int_0^\infty \frac{\sin bx}{x}dx + a^2\int_0^\infty \frac{\sin bx}{a^2+x^2}\frac{dx}{x}$$

$$= -\frac{\pi}{2} + a^2 u, \text{ by (107), Art. 100.}$$

$$\therefore \quad \frac{2\,du\,d^2u}{db^2} = 2a^2u\,du - \pi\,du;$$

$$\left(\frac{du}{db}\right)^2 = a^2u^2 - \pi u + \frac{\pi^2}{4a^2};$$

since $u = 0$, and $\frac{du}{db} = \frac{\pi}{2a}$, when $b = 0$;

$$\therefore \quad \frac{du}{\frac{\pi}{2a} - au} = db;$$

$$\therefore \quad u = \int_0^\infty \frac{\sin bx}{a^2+x^2}\frac{dx}{x} = \frac{\pi}{2a^2}\{1 - e^{-ab}\}. \quad (120)$$

Hence also equating the values of $\frac{du}{db}$ and of $\frac{d^2u}{db^2}$ derived from this equation with the values given in the preceding process,

$$\frac{du}{db} = \int_0^\infty \frac{\cos bx\,dx}{a^2+x^2} = \frac{\pi}{2a}e^{-ab}; \quad (121)$$

$$-\frac{d^2u}{db^2} = \int_0^\infty \frac{x\sin bx\,dx}{a^2+x^2} = \frac{\pi}{2}e^{-ab}. \quad (122)$$

Ex. 7. Evaluate $u = \int_0^1 \frac{x^{m-1} - x^{-m}}{(1+x)\log x}dx$;

$$\therefore \quad \frac{du}{dm} = \int_0^1 \frac{x^{m-1} + x^{-m}}{1+x}dx;$$

in the right-hand member replace x by $\frac{1}{x}$; then

$$\frac{du}{dm} = \int_1^\infty \frac{x^{m-1}+x^{-m}}{1+x}\,dx;$$

$$\therefore \text{ by addition } 2\frac{du}{dm} = \int_0^\infty \frac{x^{m-1}+x^{-m}}{1+x}\,dx;$$

$$= \int_0^\infty \frac{x^{m-1}dx}{1+x} + \int_0^\infty \frac{x^{-m}dx}{1+x};$$

in the second integral of the right-hand member replace x by $\frac{1}{x}$;

then
$$\int_0^\infty \frac{x^{-m}dx}{1+x} = \int_0^\infty \frac{x^{m-1}dx}{1+x};$$

$$\therefore \quad \frac{du}{dm} = \int_0^\infty \frac{x^{m-1}dx}{1+x};$$

$$= \frac{\pi}{\sin m\pi}, \text{ by (72), Art. 94.}$$

$$\therefore \quad u = \int_0^1 \frac{x^{m-1}-x^{-m}}{(1+x)\log x}\,dx = \log\tan\frac{m\pi}{2}. \qquad (123)$$

SECTION 5.—*Definite Integrals involving Impossible Quantities.—Cauchy's Method of Evaluating Definite Integrals.*

102.] In our researches into the properties of definite integrals, contained in the preceding sections of the present Chapter, both the element-function and the limits of integration have been taken to be possible quantities. These quantities have been thus restricted, that there might be less risk of error in a subject of so delicate a kind, and because it was expedient to confine within the narrowest bounds the circumstances under which the element-function might become infinite or discontinuous; this latter being an event which is in all cases to be excluded, or at least to be separately considered. Impossible quantities however as well as possible are continuous; and we propose now to extend the theory, and to evaluate definite integrals into which impossible quantities enter, either in the limits or in the element-functions. We shall hereby be led also to a process of evaluation devised by

M. Cauchy*; and which is of so great generality, that it includes almost all the known values of these functions.

As a definite integral is the sum of a series, of which in this case some of the terms may be possible, and others may be impossible; so for the determinateness of the sum of such a series it is requisite that it should be convergent; and thus that none of the terms should be infinite, in either the possible or the impossible part of the series. And consequently for no value of the variable within the range of integration must the element-function become infinite. When this is the case and the sum of the series is determined, that sum will consist of two parts, one of which is possible and the other impossible; and these will severally be equal to the corresponding parts of the definite integral. Thus, if

$$\int_{x_0}^{x_n} \{f'(x) + \sqrt{-1}\,\phi'(x)\}\,dx = \text{A} + \sqrt{-1}\,\text{B},$$

then
$$\int_{x_0}^{x_n} f'(x)\,dx = \text{A}, \qquad \int_{x_0}^{x_n} \phi'(x)\,dx = \text{B}. \tag{124}$$

If however within the range of integration a value of the variable occurs, either possible or impossible, for which the element-function becomes infinite, then we may have recourse to the method explained in Art. 89: we may divide the integral into two parts, and take the range of one part from the lower limit to within an infinitesimal of the value of the variable for which the element-function is infinite; and take the range of the other from within an infinitesimal of that value to the superior limit; if these two infinitesimal parts are equal, so that in the symbols of Art. 89, $\mu = \nu = 1$, the resulting sum will be the principal value of the integral; (see Art. 91). In this process however a part of the original definite integral is excluded or lost; viz., that within an infinitesimal range above and below the value of the variable for which the element-function is infinite. This part may be either finite or even zero; consequently it must be determined, and added to the other two integrals. It is called the correction for infinity or for discontinuity. Instances of it will occur in the sequel.

103.] The method of M. Cauchy depends on the application of

* The original memoir of Cauchy, in which the method was explained, was read to the French Institute on Aug. 22, 1814; and is contained in the first volume of "Mémoires des Savans Étrangers, Paris, 1827.

the theory of double integration, which has been explained in Art. 99, to the following identical equation.

Let $u = f(z)$, where $z = \phi(x, y)$; then

$$\left(\frac{du}{dx}\right) = f'(z)\,\frac{dz}{dx}; \qquad \left(\frac{du}{dy}\right) = f'(z)\,\frac{dz}{dy};$$

$$\therefore \quad \left(\frac{d^2u}{dy\,dx}\right) = \frac{d}{dy}\left\{f'(z)\,\frac{dz}{dx}\right\}; \qquad \left(\frac{d^2u}{dx\,dy}\right) = \frac{d}{dx}\left\{f'(z)\,\frac{dz}{dy}\right\};$$

$$\therefore \quad \frac{d}{dy}\left\{f'(z)\,\frac{dz}{dx}\right\} = \frac{d}{dx}\left\{f'(z)\,\frac{dz}{dy}\right\}; \tag{125}$$

which is an identical equation, as both members express the same quantity, whether x and y are possible or impossible, and whatever is the form of f. Let both sides be multiplied by $dx\,dy$; and then let them be the element-functions of a double integration with respect to x and y; the limits of x being x_n and x_0, and those of y being y_n and y_0; these four limits being independent of x and y, so that the order of integration may be inverted, if it is necessary. Also let us in the first place assume that the element-function does not become infinite or discontinuous for any values of the variables within the ranges of integration. Then we have

$$\int_{x_0}^{x_n} dx \int_{y_0}^{y_n} \frac{d}{dy}\left\{f'(z)\,\frac{dz}{dx}\right\} dy = \int_{y_0}^{y_n} dy \int_{x_0}^{x_n} \frac{d}{dx}\left\{f'(z)\,\frac{dz}{dy}\right\} dx; \tag{126}$$

which is an absolute identity, and may be called M. Cauchy's equation, as it is the basis of his method of evaluating definite integrals of it.

For the first general application of this equation, let

$$z = x + y\sqrt{-1};$$

then (126) becomes

$$\int_{x_0}^{x_n} dx \int_{y_0}^{y_n} \frac{d}{dy} f'(x+y\sqrt{-1})\, dy$$

$$= \sqrt{-1}\int_{y_0}^{y_n} dy \int_{x_0}^{x_n} \frac{d}{dx} f'(x+y\sqrt{-1})\, dx;$$

and effecting in each member the integration which stands first, we have

$$\int_{x_0}^{x_n} \{f'(x+y_n\sqrt{-1}) - f'(x+y_0\sqrt{-1})\}\, dx$$

$$= \sqrt{-1}\int_{y_0}^{y_n} \{f'(x_n+y\sqrt{-1}) - f'(x_0+y\sqrt{-1})\}\, dy. \tag{127}$$

Let us apply this equation to some examples; and for a first application let $f'(z) = e^{-az^2}$;

$$\therefore f'(x+y\sqrt{-1}) = e^{-a(x^2-y^2)}(\cos 2axy - \sqrt{-1}\sin 2axy); \quad (128)$$

then from (127), we have

$$e^{ay_n^2}\int_{x_0}^{x_n} e^{-ax^2}\{\cos 2axy_n - \sqrt{-1}\sin 2axy_n\}\,dx$$

$$-e^{ay_0^2}\int_{x_0}^{x_n} e^{-ax^2}\{\cos 2axy_0 - \sqrt{-1}\sin 2axy_0\}\,dx$$

$$= \sqrt{-1}\left\{e^{-ax_n^2}\int_{y_0}^{y_n} e^{ay^2}\{\cos 2ax_n y - \sqrt{-1}\sin 2ax_n y\}\,dy\right.$$

$$\left.-e^{-ax_0^2}\int_{y_0}^{y_n} e^{ay^2}\{\cos 2ax_0 y - \sqrt{-1}\sin 2ax_0 y\}\,dy\right\}; \quad (129)$$

from which two equations may be formed by taking separately the possible and the impossible parts.

Let $x_n = \infty$; so that $e^{-ax_n^2} = 0$; therefore

$$e^{ay_n^2}\int_{x_0}^{\infty} e^{-ax^2}\cos 2axy_n\,dx - e^{ay_0^2}\int_{x_0}^{\infty} e^{-ax^2}\cos 2axy_0\,dx$$

$$= -e^{-ax_0^2}\int_{y_0}^{y_n} e^{ay^2}\sin 2ax_0 y\,dy; \quad (130)$$

$$e^{ay_n^2}\int_{x_0}^{\infty} e^{-ax^2}\sin 2axy_n\,dx - e^{ay_n^2}\int_{x_0}^{\infty} e^{-ax^2}\sin 2axy_0\,dx$$

$$= e^{-ax_0^2}\int_{y_0}^{y_n} e^{ay^2}\cos 2ax_0 y\,dy. \quad (131)$$

Moreover, to simplify further, let $x_0 = y_0 = 0$; then

$$e^{ay_n^2}\int_0^{\infty} e^{-ax^2}\cos 2ay_n x\,dx - \int_0^{\infty} e^{-ax^2}\,dx = 0;$$

$$e^{ay_n^2}\int_0^{\infty} e^{-ax^2}\sin 2ay_n x\,dx = \int_0^{y_n} e^{ay^2}\,dy;$$

so that, as $\int_0^{\infty} e^{-ax^2}\,dx = \frac{1}{2}\left(\frac{\pi}{a}\right)^{\frac{1}{2}}$, see (112), Art. 100, we have

$$\int_0^{\infty} e^{-ax^2}\cos 2ay_n x\,dx = \frac{1}{2}\left(\frac{\pi}{a}\right)^{\frac{1}{2}} e^{-ay_n^2} \quad (132)$$

$$\int_0^{\infty} e^{-ax^2}\sin 2ay_n x\,dx = e^{-ay_n^2}\int_0^{y_n} e^{ay^2}\,dy. \quad (133)$$

(132) is the same result as (118), Art. 101.

104.] For another application of (126), let $z = x(a+y\sqrt{-1})$; then

$$\frac{dz}{dx} = a+y\sqrt{-1}, \quad \frac{dz}{dy} = x\sqrt{-1};$$

so that (126) becomes

$$\int_{x_0}^{x_n} dx \int_{y_0}^{y_n} \frac{d}{dy}\{(a+y\sqrt{-1})f'(z)\}\, dy$$

$$= \int_{y_0}^{y_n} dy \int_{x_0}^{x_n} \frac{d}{dx}\{x\sqrt{-1}f'(z)\}\, dx; \quad (134)$$

and effecting in each member the integration which stands first, we have

$$\int_{x_0}^{x_n} \{(a+y_n\sqrt{-1})f'(ax+xy_n\sqrt{-1})$$

$$-(a+y_0\sqrt{-1})f'(ax+xy_0\sqrt{-1})\}\, dx$$

$$= \sqrt{-1}\int_{y_0}^{y_n} \{x_n f'(ax_n+x_n y\sqrt{-1})$$

$$-x_0 f'(ax_0+x_0 y\sqrt{-1})\}\, dy. \quad (135)$$

In application of this equation let us suppose

$$f'(z) = z^{m-1}e^{-z}; \quad x_n = \infty, \quad y_n = b, \quad x_0 = y_0 = 0;$$

then $f'(z)$ does not become infinite for any value of its subject-variable within the range of integration, and $f'(\infty) = 0 = f'(0)$. Consequently from (135),

$$(a+b\sqrt{-1})^m \int_0^\infty x^{m-1}e^{-(a+b\sqrt{-1})x}\, dx = \int_0^\infty (ax)^{m-1}e^{-ax}\, d.ax;$$

$$\therefore \quad (a+b\sqrt{-1})^m \int_0^\infty x^{m-1}e^{-(a+b\sqrt{-1})x}\, dx = \int_0^\infty x^{m-1}e^{-x}\, dx. \quad (136)$$

The integral in the right-hand member is evidently a function of m only; let us abbreviate the notation, and anticipating the notation of the seventh section, let us denote it by $\Gamma(m)$; so that

$$\int_0^\infty x^{m-1}e^{-x}\, dx = \Gamma(m); \quad (137)$$

Also let $a = k\cos\alpha$, $b = k\sin\alpha$; then (136) becomes

$$\int_0^\infty x^{m-1}e^{-ax}(\cos bx - \sqrt{-1}\sin bx)\, dx$$

$$= \frac{\cos m\alpha - \sqrt{-1}\sin m\alpha}{(a^2+b^2)^{\frac{m}{2}}}\Gamma(m); \quad (138)$$

and equating possible and impossible parts

$$\int_0^\infty x^{m-1}e^{-ax}\cos bx\, dx = \frac{\Gamma(m)}{(a^2+b^2)^{\frac{m}{2}}}\cos\left(m\tan^{-1}\frac{b}{a}\right); \quad (139)$$

$$\int_0^\infty x^{m-1}e^{-ax}\sin bx\, dx = \frac{\Gamma(m)}{(a^2+b^2)^{\frac{m}{2}}}\sin\left(m\tan^{-1}\frac{b}{a}\right); \quad (140)$$

which integrals have already been evaluated in (97) and (98), Art. 98.

Again multiplying (140) by $\sqrt{-1}$, and adding (139); and replacing the circular functions by their exponential equivalents, we have

$$\int_0^\infty x^{m-1} e^{-(a-b\sqrt{-1})x} dx = \frac{e^{m\alpha\sqrt{-1}}}{(a^2+b^2)^{\frac{m}{2}}} \Gamma(m). \tag{141}$$

Let $a = 0$; in which case $\alpha = \frac{\pi}{2}$: then we have

$$\int_0^\infty x^{m-1} e^{bx\sqrt{-1}} dx = \frac{e^{\frac{m\pi}{2}\sqrt{-1}}}{b^m} \Gamma(m); \tag{142}$$

This definite integral is of a form which will hereafter be considered at length.

For another application of (135) let $f'(z) = \frac{z^{m-1}}{1+z}$, where $2 > m > 1$; also let $x_n = \infty$, $y_n = b$; $x_0 = y_0 = 0$; so that $f'(z)$ does not become infinite for any value of its subject-variable within the range of integration, and $f'(\infty) = f'(0) = 0$; consequently from (135),

$$(a+b\sqrt{-1})^m \int_0^\infty \frac{x^{m-1} dx}{1+ax+bx\sqrt{-1}} = \int_0^\infty \frac{(ax)^{m-1}}{1+ax} d(ax)$$

$$= \frac{\pi}{\sin m\pi}, \text{ by (72), Art. 94.}$$

Then, if $b = a \tan \alpha$,

$$\int_0^\infty \frac{(1+ax-bx\sqrt{-1})\, x^{m-1} dx}{1+2ax+(a^2+b^2)x^2} = \frac{\pi}{\sin m\pi} \, \frac{\cos m\alpha - \sqrt{-1}\sin m\alpha}{(a^2+b^2)^{\frac{m}{2}}};$$

and separating possible and impossible parts,

$$\int_0^\infty \frac{(1+ax)x^{m-1} dx}{1+2ax+(a^2+b^2)x^2} = \frac{\pi}{(a^2+b^2)^{\frac{m}{2}} \sin m\pi} \cos\left(m \tan^{-1}\frac{b}{a}\right); \tag{143}$$

$$\int_0^\infty \frac{bx^m dx}{1+2ax+(a^2+b^2)x^2} = \frac{\pi}{(a^2+b^2)^{\frac{m}{2}} \sin m\pi} \sin\left(m \tan^{-1}\frac{b}{a}\right). \tag{144}$$

Hence also, if $b = \sin \alpha$, $a = \cos \alpha$,

$$\int_0^\infty \frac{(1+x\cos\alpha)\, x^{m-1} dx}{1+2x\cos\alpha+x^2} = \frac{\pi \cos m\alpha}{\sin m\pi}; \tag{145}$$

$$\int_0^\infty \frac{x^m dx}{1+2x\cos\alpha+x^2} = \frac{\pi \sin m\alpha}{\sin\alpha \sin m\pi}. \tag{146}$$

105.] The preceding process can always be safely employed, so long as $f'(z)$ does not become infinite or discontinuous within the limits of integration. If however $f'(z)$ does become infinite for particular values of x and y within the range of integration, say when $x = \xi$, and $y = \eta$; then the circumstance requires close examination; and to simplify the inquiry I will take that particular form of z given in Art. 103; viz.,

$$z = x+y\sqrt{-1};$$

in which case $f'(\xi+\eta\sqrt{-1}) = \infty$.

Let the definite integral be divided into two parts with the following limits; for the former part let the range of the x-integration extend from x_0 to $\xi-i$; and for the latter part from $\xi+i$ to x_n, i being an infinitesimal according to the theory of Art. 89; the range of the y-integration being in both parts from y_n to y_0. Thus the part of the integral corresponding to the infinite value of the element-function will be excluded; but ultimately if $i = 0$, the whole value will be included. Under these circumstances the integral given in (127) consists of the two following integrals:

$$\int_{x_0}^{\xi-i} \{f'(x+y_n\sqrt{-1})-f'(x+y_0\sqrt{-1})\}\,dx$$
$$= \sqrt{-1}\int_{y_0}^{y_n} \{f'(\xi-i+y\sqrt{-1})-f'(x_0+y\sqrt{-1})\}\,dy;$$

$$\int_{\xi+i}^{x_n} \{f'(x+y_n\sqrt{-1})-f'(x+y_0\sqrt{-1})\}\,dx$$
$$= \sqrt{-1}\int_{y_0}^{y_n} \{f'(x_n+y\sqrt{-1})-f'(\xi+i+y\sqrt{-1})\}\,dy;$$

Now let these two integrals be added; then if $i = 0$, the sum of the left-hand members is

$$\int_{x_0}^{x_n} \{f'(x+y_n\sqrt{-1})-f'(x+y_0\sqrt{-1})\}\,dx,$$

and is determinate, because the range of integration does not include the values of the variables for which the element-function is infinite. Also the sum of the right-hand members is

$$\sqrt{-1}\int_{y_0}^{y_n} \{f'(x_n+y\sqrt{-1})-f'(x_0+y\sqrt{-1})\}\,dy$$
$$-\sqrt{-1}\int_{y_0}^{y_n} \{f'(\xi+i+y\sqrt{-1})-f'(\xi-i+y\sqrt{-1})\}\,dy,$$

of which the former part is determinate; the latter part is not so; it would be equal to 0, if $i = 0$, were it not that within the range of the y-integration y takes the value η, when x has that of ξ; and consequently the variables have those values for which the element-function is infinite. This latter part may therefore be finite, and must be determined in each case. Thus the definite integral found in the ordinary way is to be diminished by this quantity; let us denote it by Δ; so that

$$\Delta = \sqrt{-1}\int_{y_0}^{y_n}\{f'(\xi+i+y\sqrt{-1})-f'(\xi-i+y\sqrt{-1})\}\,dy. \quad (147)$$

This quantity is called the correction for infinity or for discontinuity; and its value may be determined as follows.

(1) When η, the value of y for which the element-function becomes infinite, is between y_n and y_0, so that the preceding definite integral includes values of the element-function on both sides of that value; then as for all other values of y when $i = 0$, the whole function under the sign of integration in (147) is zero, the limits may be extended from y_n and y_0 to $+\infty$ and $-\infty$ without any change of value of the integral, so that

$$\Delta = \sqrt{-1}\int_{-\infty}^{\infty}\{f'(\xi+i+y\sqrt{-1})-f'(\xi-i+y\sqrt{-1})\}\,dy. \quad (148)$$

(2) If $\eta = y_0$, the inferior limit, only those values of the element-function which lie on the positive side of $y = \eta = y_0$ are included in the definite integral; so that all those lying beyond that limit, and up to infinity, may be included without change of value of the integral: in this case

$$\Delta = \sqrt{-1}\int_{\eta}^{\infty}\{f'(\xi+i+y\sqrt{-1})-f'(\xi-i+y\sqrt{-1})\}\,dy. \quad (149)$$

(3) If $\eta = y_n$, the superior limit, only those values of the element-function which lie on the negative side of $y = \eta = y_n$ are included in the definite integral: in this case the lower limit of integration may be extended to $-\infty$ without any change of value of the integral; and we have

$$\Delta = \sqrt{-1}\int_{-\infty}^{\eta}\{f'(\xi+i+y\sqrt{-1})-f'(\xi-i+y\sqrt{-1})\}\,dy. \quad (150)$$

I may observe in passing that the correctness of this change of limits in the value of Δ given in (147), without any change of value in the integral, may also be demonstrated by a transformation of variable in (147), by replacing y by $\eta+iu$, where i is an infinitesimal, and u is the variable. I have preferred however the general reasoning given above.

These values of Δ may be further simplified. Since

$$f'(x+y\sqrt{-1}) = \infty,$$

when $x = \xi$ and $y = \eta$, $f'(z) = \infty$ when $z = \xi+\eta\sqrt{-1} = \zeta$ say. Now the factor $z-\zeta$ may enter into $f'(z)$ in any power: I shall assume that it enters in only the first power, leaving the student to refer to the original memoirs of Cauchy for the more complicated case; and accordingly I shall suppose $(z-\zeta)f'(z)$ to be finite when $z = \zeta$. Let

$$(z-\zeta)f'(z) = \text{F}(z). \tag{151}$$

Then, omitting the limits of Δ, which will be hereafter supplied according to (148) or (149) or (150),

$$\Delta = \sqrt{-1}\int\left\{\frac{\text{F}(\xi+i+y\sqrt{-1})}{i+(y-\eta)\sqrt{-1}} - \frac{\text{F}(\xi-i+y\sqrt{-1})}{-i+(y-\eta)\sqrt{-1}}\right\}dy. \tag{152}$$

As this integral vanishes, when $i=0$, for all values of y except those near to η,

$$\text{F}(\xi+i+\eta\sqrt{-1}) = \text{F}(\xi-i+\eta\sqrt{-1}) = \text{F}(\xi+\eta\sqrt{-1})$$
$$= \text{B (say)}; \tag{153}$$

and we may place this quantity outside the sign of integration: so that

$$\begin{aligned}\Delta &= \text{B}\sqrt{-1}\int\left\{\frac{1}{i+(y-\eta)\sqrt{-1}} - \frac{1}{-i+(y-\eta)\sqrt{-1}}\right\}dy \\ &= 2\text{B}\sqrt{-1}\int\frac{i\,dy}{i^2+(y-\eta)^2} \\ &= 2\text{B}\sqrt{-1}\left[\tan^{-1}\frac{y-\eta}{i}\right].\end{aligned} \tag{154}$$

Hence if the value of Δ is that given in (148), the limits of integration are $+\infty$ and $-\infty$; and

$$\Delta = 2\pi\text{B}\sqrt{-1}; \tag{155}$$

and if Δ has the value given in (149) or in (150), so that the limits are ∞ and η, or η and $-\infty$, then

$$\Delta = \pi\text{B}\sqrt{-1}; \tag{156}$$

thus if the value of y, which makes the element-function equal to infinity, is a limit of the y-integration, the correction for infinity or for discontinuity is only one half of its value when this is not the case.

In the preceding inquiry we have investigated the circumstances and the necessary correction when the element-function becomes infinite at given values of its subject-variables. This may occur however many times within the range of integration; and in this case the correction must be introduced for each set of variables. Thus, suppose $f'(z) = \infty$, when $x = \xi_1$, $y = \eta_1$; $x = \xi_2$, $y = \eta_2$; ... $x = \xi_n$, $y = \eta_n$; and the corrections corresponding to these values and determined as above to be $\Delta_1, \Delta_2, \dots \Delta_n$: then the whole correction for infinity or discontinuity is the sum of these several corrections; and if $\text{E}_1, \text{E}_2, \dots \text{E}_n$ correspond to them, the whole correction

$$\Delta = 2\pi\sqrt{-1}\,\{\text{E}_1 + \text{E}_2 + \dots + \text{E}_n\}$$

$$= 2\pi\sqrt{-1}\,\Sigma.\text{E}; \tag{157}$$

remembering always to diminish any one of these partial corrections by one half, if the variable corresponding to it is a limit of the y-integration.

Hence, finally, we have

$$\int_{x_0}^{x_n} \{f(x+y_n\sqrt{-1}) - f(x+y_0\sqrt{-1})\}\,dx$$

$$= \sqrt{-1}\int_{y_0}^{y_n} \{f(x_n+y\sqrt{-1}) - f(x_0+y\sqrt{-1})\}\,dy$$

$$-2\pi\sqrt{-1}\,\Sigma.\text{E}. \tag{158}$$

106.] The preceding equation is too general for our present purpose; and I propose to take certain particular cases of it, from which definite integrals may be determined.

Let $x_n = +\infty$, $x_0 = -\infty$; $y_n = +\infty$, $y_0 = 0$; so that the range of the x-integration includes all positive and negative values of x, and that of the y-integration includes all positive but no negative values of y: and thus in seeking the values of x and y, viz., ξ and η, for which $f'(x+y\sqrt{-1}) = \infty$, ξ may be negative as well as positive, but η must always be positive.

Let us moreover suppose that $f'(x+y\sqrt{-1}) = 0$, when $x = \pm\infty$, whatever is the value of y; and $= 0$, when $y = +\infty$, whatever is the value of x. Then under all these circumstances, we have from (158)

$$\int_{-\infty}^{\infty} f'(x)\,dx = 2\pi\sqrt{-1}\,\Sigma.\text{E}. \tag{159}$$

The following are examples of this formula.

Ex. 1. $\int_{-\infty}^{\infty} \frac{\phi(x)}{1+x^2}\,dx$. Here, when x is replaced by $x+y\sqrt{-1}$,

$\frac{\phi(x)}{1+x^2} = 0$, when $x = \pm\infty$, and when $y = \infty$; also the element-function $= \infty$, when $x = \pm\sqrt{-1}$; of which values only the upper one is to be taken, because y, which is the coefficient of $\sqrt{-1}$, is capable of only positive values. In this case by (158),

$$\text{R} = \frac{\phi(x)}{x+\sqrt{-1}}, \text{ when } x = \sqrt{-1}, \ = \frac{\phi(\sqrt{-1})}{2\sqrt{-1}};$$

$$\therefore \int_{-\infty}^{\infty} \frac{\phi(x)}{1+x^2} dx = \pi\phi(\sqrt{-1}) \tag{160}$$

In illustration of this formula take the following examples;

(1) Let $\phi(x) = e^{ax\sqrt{-1}} = \cos ax + \sqrt{-1}\sin ax$; in which case it is to be observed that

$$f'(x+y\sqrt{-1}) = \frac{e^{-ay}(\cos ax+\sqrt{-1}\sin ax)}{1+x^2-y^2+2xy\sqrt{-1}};$$

which vanishes when $x = \pm\infty$, whatever is the value of y; and, when $y = +\infty$, whatever is the value of x; thus,

$$\phi(\sqrt{-1}) = e^{-a};$$

and consequently

$$\int_{-\infty}^{\infty} \frac{\cos ax+\sqrt{-1}\sin ax}{1+x^2} dx = \pi e^{-a};$$

and separating possible and impossible parts,

$$\int_{-\infty}^{\infty} \frac{\cos ax}{1+x^2} dx = \pi e^{-a}; \tag{161}$$

$$\int_{-\infty}^{\infty} \frac{\sin ax}{1+x^2} dx = 0. \tag{162}$$

(2) Let $\phi(x) = x^{m-1}$, where $2 > m > 0$, so that when x is replaced by $x+y\sqrt{-1}$, $\frac{x^{m-1}}{1+x^2} = 0$, when $x = \pm\infty$, and when $y = \infty$;

$$\int_{-\infty}^{\infty} \frac{x^{m-1}}{1+x^2} dx = \pi(\sqrt{-1})^{m-1} = \pi(-1)^{\frac{m-1}{2}}.$$

Now
$$\int_{-\infty}^{0} \frac{x^{m-1}}{1+x^2} dx = (-1)^{m-1}\int_{0}^{\infty} \frac{x^{m-1}}{1+x^2} dx;$$

$$\therefore \int_{-\infty}^{\infty} \frac{x^{m-1}}{1+x^2} dx = \{1+(-1)^{m-1}\}\int_{0}^{\infty} \frac{x^{m-1}}{1+x^2} dx;$$

$$\therefore \int_{0}^{\infty} \frac{x^{m-1}}{1+x^2} dx = \frac{\pi(-1)^{\frac{m-1}{2}}}{1+(-1)^{m-1}} = \frac{\pi}{(-1)^{-\frac{m-1}{2}}+(-1)^{\frac{m-1}{2}}}.$$

But $(-1)^{\frac{m-1}{2}} = (\cos\pi + \sqrt{-1}\sin\pi)^{\frac{m-1}{2}}$

$$= \cos\frac{(m-1)\pi}{2} + \sqrt{-1}\sin\frac{(m-1)\pi}{2};$$

and $(-1)^{-\frac{m-1}{2}} = (\cos\pi + \sqrt{-1}(\sin\pi)^{-\frac{m-1}{2}}$

$$= \cos\frac{(m-1)\pi}{2} - \sqrt{-1}\sin\frac{(m-1)\pi}{2}.$$

$$\therefore \quad \int_0^\infty \frac{x^{m-1}dx}{1+x^2} = \frac{\pi}{2\cos\frac{(m-1)\pi}{2}} = \frac{\pi}{2\sin\frac{m\pi}{2}}. \tag{163}$$

Ex. 2. $\int_{-\infty}^{\infty}\frac{\phi(x)}{1-x^2}dx$. Here, when x is replaced by $x+y\sqrt{-1}$, $\frac{\phi(x)}{1-x^2}=0$, when $x=\pm\infty$, and when $y=\infty$. Now the element-function $=\infty$, when $x=\pm 1$, both of which values lie within the limits of x; but as their values are possible, $\eta=0$ for both; and as 0 is the inferior limit of y, only one half of the general correction for infinity is to be taken in each case; so that

$$\Delta = \pi\sqrt{-1}\left\{\frac{-\phi(x)}{x+1}\right\} \text{ (when } x=1),$$

$$+\pi\sqrt{-1}\left\{\frac{-\phi(x)}{x-1}\right\} \text{ (when } x=-1),$$

$$= \frac{\pi}{2}\sqrt{-1}\{\phi(-1)-\phi(1)\};$$

$$\therefore \quad \int_{-\infty}^{\infty}\frac{\phi(x)}{1-x^2}dx = \frac{\pi}{2}\sqrt{-1}\{\phi(-1)-\phi(1)\}. \tag{164}$$

Thus, if $\phi(x)=x^{m-1}$, where $2>m>0$; then if x is replaced by $x+y\sqrt{-1}$, $\frac{x^{m-1}}{1-x^2}=0$, when $x=\pm\infty$, whatever is the value of y, and $=\infty$ when $y=\infty$, whatever is the value of x; and

$$\int_{-\infty}^{\infty}\frac{x^{m-1}dx}{1-x^2} = \frac{\pi\sqrt{-1}}{2}\{(-1)^{m-1}-(1)^{m-1}\};$$

and consequently, as in the second case of the preceding example,

$$\int_0^\infty \frac{x^{m-1}dx}{-\;{}^2} = \frac{\pi\sqrt{-1}}{2}\,\frac{(-1)^{m-1}-(1)^{m-1}}{1+(-1)^{m-1}}$$

$$= \frac{\pi}{2}\cot\frac{m\pi}{2}. \tag{165}$$

If in (163) and (165) $m=2p$, and x is replaced by $\sqrt{x}$, then

$$\int_0^{\infty} \frac{x^{p-1}\,dx}{1+x} = \pi \operatorname{cosec} p\pi ; \tag{166}$$

$$\int_0^{\infty} \frac{x^{p-1}\,dx}{1-x} = \pi \cot p\pi ; \tag{167}$$

which are results of the same form as (72) and (73), Art. 94. In this case however p may be any positive proper fraction.

107.] If the form of $f'(x)$ is such that all the corrections for infinity and discontinuity vanish; then $\Delta = 0$; and from (159) we have

$$\int_{-\infty}^{\infty} f'(x)\,dx = 0. \tag{168}$$

All these conditions are fulfilled when

$$f'(x) = \frac{e^{ax\sqrt{-1}} - e^{-a}}{1+x^2} = \frac{\cos ax + \sqrt{-1}\sin ax - e^{-a}}{1+x^2};$$

for (1) the element-function $= 0$, when $x = \pm\infty$, whatever is the value of y, and $= 0$ when $y = \infty$, whatever is the value of x. (2) $\Delta = \pi\phi(\sqrt{-1}) = 0$; since $e^{ax\sqrt{-1}} - e^{-a} = 0$, when $x = \sqrt{-1}$.

$$\therefore \int_{-\infty}^{\infty} \frac{\cos ax + \sqrt{-1}\sin ax - e^{-a}}{1+x^2}\,dx = 0;$$

so that equating possible and impossible parts,

$$\int_{-\infty}^{\infty} \frac{\cos ax\,dx}{1+x^2} = e^{-a}\int_{-\infty}^{\infty} \frac{dx}{1+x^2}$$

$$= \pi e^{-a}; \tag{169}$$

$$\int_{-\infty}^{\infty} \frac{\sin ax\,dx}{1+x^2} = 0; \tag{170}$$

the same results as (161) and (162).

108.] The limits of the x-integration in the preceding Articles are ∞ and $-\infty$; they may however be changed to ∞ and 0 by the following process. Since

$$\int_{-\infty}^{\infty} f'(x)\,dx = \int_0^{\infty} f'(x)\,dx + \int_{-\infty}^{0} f'(x)\,dx$$

$$= \int_0^{\infty} f'(x)\,dx + \int_0^{\infty} f'(-x)\,dx$$

$$\therefore \int_0^{\infty} \{f'(x) + f'(-x)\}\,dx = \int_{-\infty}^{\infty} f'(x)\,dx$$

$$= \Delta. \tag{171}$$

The following are examples of this equation.

Ex. 1. $$\int_0^\infty \frac{\phi(x)-\phi(-x)}{x}\,dx = \int_0^\infty \left\{\frac{\phi(x)}{x}+\frac{\phi(-x)}{-x}\right\}dx$$
$$= \int_{-\infty}^\infty \frac{\phi(x)}{x}\,dx$$
$$= \pi\sqrt{-1}\,\phi(0). \qquad (172)$$

Ex. 2. $$\int_0^\infty \left\{\frac{\phi(x)}{a+x}+\frac{\phi(-x)}{a-x}\right\}dx = \int_{-\infty}^\infty \frac{\phi(x)}{a+x}\,dx$$
$$= 2\pi\sqrt{-1}\,\phi(-a). \qquad (173)$$

Ex. 3. $$\int_0^\infty \frac{\phi(x)+\phi(-x)}{x^2+a^2}\,dx = \int_{-\infty}^\infty \frac{\phi(x)}{x^2+a^2}\,dx$$
$$= \frac{\pi}{a}\,\phi(a\sqrt{-1}). \qquad (174)$$

Thus, if $\phi(x) = e^{bx\sqrt{-1}}$,
$$\int_0^\infty \frac{\cos bx\,dx}{x^2+a^2} = \frac{\pi}{2a}e^{-ab}. \qquad (175)$$
which is the same as (121), Art. 101.

Ex. 4. $$\int_0^\infty \frac{\phi(x)+\phi(-x)}{x^2-a^2}\,dx = \int_{-\infty}^\infty \frac{\phi(x)}{x^2-a^2}\,dx$$
$$= \frac{\pi}{2a}\sqrt{-1}\,\{\phi(a)-\phi(-a)\}. \qquad (176)$$

Thus, if $\phi(x) = e^{bx\sqrt{-1}}$,
$$\int_0^\infty \frac{\cos bx\,dx}{x^2-a^2} = -\frac{\pi}{2a}\sin ab. \qquad (177)$$

Ex. 5. $$\int_0^\infty \{\phi(x)-\phi(-x)\}\frac{x\,dx}{x^2-a^2} = \int_{-\infty}^\infty \frac{x\,\phi(x)}{x^2-a^2}\,dx$$
$$= \frac{\pi\sqrt{-1}}{2}\{\phi(a)+\phi(-a)\}. \qquad (178)$$

Thus, if $\phi(x) = e^{bx\sqrt{-1}}$,
$$\int_0^\infty \frac{x\sin bx}{x^2-a^2}\,dx = \frac{\pi}{2}\cos ab. \qquad (179)$$
$$\therefore \text{ if } a = 0, \quad \int_0^\infty \frac{\sin bx}{x}\,dx = \frac{\pi}{2}; \qquad (180)$$
the last integral being the same as (77), Art. 94, and as (107), Art. 100.

Ex. 6. $$\int_0^\infty \{\phi(x)-\phi(-x)\}\frac{x\,dx}{x^2+a^2} = \int_{-\infty}^\infty \frac{x\,\phi(x)}{x^2+a^2}\,dx$$
$$= \pi\sqrt{-1}\,\phi(a\sqrt{-1}). \qquad (181)$$

Thus, if $\phi(x) = e^{bx\sqrt{-1}}$,

$$\int_0^{\infty} \frac{x \sin bx\, dx}{x^2+a^2} = \frac{\pi}{2} e^{-ab}; \tag{182}$$

which is the same result as (122), Art. 101.

Ex. 7. $$\int_0^{\infty} \frac{\phi(x)-\phi(-x)}{x(x^2+a^2)}\, dx = \frac{\pi\sqrt{-1}}{a^2} \{\phi(0)-\phi(a\sqrt{-1})\}. \tag{183}$$

Thus, if $\phi(x) = e^{bx\sqrt{-1}}$,

$$\int_0^{\infty} \frac{\sin bx\, dx}{x(x^2+a^2)} = \frac{\pi}{2a^2} \{1-e^{-ab}\}; \tag{184}$$

the same result as (120), Art. 101.

109.] M. Cauchy has also made another application of the general principle of the inversion of order of integration in a double integral which it is expedient to insert, as it exhibits the applicability of the principle to another form of function.

Let $f'(z)$ be the element-function of the required definite integral, wherein z is a variable whose modulus is r and whose argument is θ; so that

$$\begin{aligned} z &= r\{\cos\theta + \sqrt{-1}\sin\theta\} \\ &= re^{\theta\sqrt{-1}}; \end{aligned} \tag{185}$$

and consequently the element-function is $f'(re^{\theta\sqrt{-1}})$; and let us suppose this element-function, as also its first derived function, to be finite and continuous for all values of its modulus less than R. Now, since

$$\frac{d}{dr} f'(re^{\theta\sqrt{-1}}) = e^{\theta\sqrt{-1}} f''(re^{\theta\sqrt{-1}}),$$

and $$\frac{d}{d\theta} f'(re^{\theta\sqrt{-1}}) = r\sqrt{-1}\, e^{\theta\sqrt{-1}} f''(re^{\theta\sqrt{-1}});$$

$$\therefore\ \frac{d}{dr} f'(re^{\theta\sqrt{-1}}) = \frac{1}{r\sqrt{-1}} \frac{d}{d\theta} f'(re^{\theta\sqrt{-1}}). \tag{186}$$

Let the two members of this identity be multiplied by $dr\, d\theta$; and let the limits of r be r and 0; and of θ, $2\pi+a$ and a; then

$$\int_a^{2\pi+a} d\theta \int_0^r \frac{d}{dr} f'(re^{\theta\sqrt{-1}}) dr = \frac{1}{\sqrt{-1}} \int_0^r \frac{dr}{r} \int_a^{2\pi+a} \frac{d}{d\theta} f'(re^{\theta\sqrt{-1}}) d\theta. \tag{187}$$

Let us moreover suppose $f'(re^{\theta\sqrt{-1}})$ to be such that, when $2\pi+\theta$ is substituted for θ, its value is unaltered; then the right-hand member of (187) is manifestly zero; and since

$$\int_0^r \frac{d}{dr} f'(re^{\theta\sqrt{-1}})\, dr = f'(re^{\theta\sqrt{-1}}) - f'(0);$$

$$\therefore \int_a^{2\pi+a} f'(re^{\theta\sqrt{-1}})\, d\theta = f'(0) \int_a^{2\pi+a} d\theta$$

$$= 2\pi f'(0). \qquad (188)$$

If $a = 0$, then

$$\int_0^{2\pi} f'(re^{\theta\sqrt{-1}})\, d\theta = 2\pi f'(0). \qquad (189)$$

In (188) replace θ by $-\theta$; and a by -2π, then

$$\int_0^{2\pi} f'(re^{-\theta\sqrt{-1}})\, d\theta = 2\pi f'(0). \qquad (190)$$

Thus (188) is true when θ is replaced by $-\theta$.

If the subject-variable z, given in (185), is replaced throughout by $x+z$ where x is a variable independent of z, so that the element-function is $f'(x+re^{\theta\sqrt{-1}})$; then $f'(0) = f'(x)$, and we have

$$f'(x) = \frac{1}{2\pi} \int_a^{2\pi+a} f'(x+re^{\theta\sqrt{-1}})\, d\theta; \qquad (191)$$

so that any function of x, $f'(x)$, where x is possible or impossible, may be expressed as a definite integral, of which the element-function is the original function, provided that $f'(x+re^{\theta\sqrt{-1}})$, is finite and continuous for all values of the modulus, and that the value of $f'(x+re^{\theta\sqrt{-1}})$ is not altered, when $2\pi+\theta$ is substituted for θ.

Ex. 1. Let $f'(z) = \dfrac{1}{1-z}$; thus $f'(z)$ is finite and continuous for all values of z less than 1: in this case then the modulus r must be less than 1; therefore by (189),

$$\int_0^{2\pi} \frac{d\theta}{1-re^{\theta\sqrt{-1}}} = 2\pi;$$

but $$\int_0^{2\pi} \frac{d\theta}{1-re^{\theta\sqrt{-1}}} = \int_0^{2\pi} \frac{(1-re^{-\theta\sqrt{-1}})\, d\theta}{1-r(e^{\theta\sqrt{-1}}+e^{-\theta\sqrt{-1}})+r^2}$$

$$= \int_0^{2\pi} \frac{1-r\cos\theta + r\sqrt{-1}\sin\theta}{1-2r\cos\theta+r^2}\, d\theta;$$

so that equating possible and impossible parts, we have for all values of r less than 1,

$$\int_0^{2\pi} \frac{(1-r\cos\theta)\,d\theta}{1-2r\cos\theta+r^2} = 2\pi, \tag{192}$$

$$\int_0^{2\pi} \frac{\sin\theta\,d\theta}{1-2r\cos\theta+r^2} = 0. \tag{193}$$

Ex. 2. Let $f'(z) = e^{az}$. As this is finite and continuous for all values of z, the modulus r may have any value; so that by (189),

$$\int_0^{2\pi} e^{ar(\cos\theta+\sqrt{-1}\sin\theta)}\,d\theta = 2\pi;$$

$$\therefore \int_0^{2\pi} e^{ar\cos\theta}\cos(ar\sin\theta)\,d\theta = 2\pi; \tag{194}$$

$$\int_0^{2\pi} e^{ar\cos\theta}\sin(ar\sin\theta)\,d\theta = 0. \tag{195}$$

Ex. 3. Let $f'(z) = \log(1-z)$; which is finite and continuous for all values of z, and of the modulus r, less than 1. Also if

$$1-z = 1-r\cos\theta - r\sqrt{-1}\sin\theta = \rho(\cos\phi+\sqrt{-1}\sin\phi);$$

$$\therefore\quad 1-r\cos\theta = \rho\cos\phi, \qquad -r\sin\theta = \rho\sin\phi;$$

$$\therefore\quad \rho^2 = 1-2r\cos\theta+r^2,$$

$$\cos\phi = \frac{1-r\cos\theta}{(1-2r\cos\theta+r^2)^{\frac{1}{2}}}, \quad \sin\phi = \frac{-r\sin\theta}{(1-2r\cos\theta+r^2)^{\frac{1}{2}}}.$$

Now by these values it is evident that ρ, $\cos\phi$ and $\sin\phi$, and consequently $1-z$ have the same values when $\theta = 2\pi+a$, as when $\theta = a$; hence $\log(1-z)$ satisfies the necessary conditions;

$$\therefore \int_0^{2\pi} \log(1-re^{\theta\sqrt{-1}})\,d\theta = 0 \quad .$$

but $\log(1-re^{\theta\sqrt{-1}}) = \log(\rho e^{\phi\sqrt{-1}}) = \log\rho+\phi\sqrt{-1}$; therefore separating the possible and the impossible parts,

$$\int_0^{2\pi} \log(1-2r\cos\theta+r^2)\,d\theta = 0. \tag{196}$$

$$\int_0^{2\pi} \tan^{-1}\frac{-r\sin\theta}{1-r\cos\theta}\,d\theta = 0. \tag{197}$$

If r is greater than 1, then $\frac{1}{r}$ is less than 1, and

$$\int_0^{2\pi} \log(1-2r\cos\theta+r^2)\,d\theta$$

$$= \int_0^{2\pi} \log r^2\,d\theta + \int_0^{2\pi} \log\left(1-\frac{2}{r}\cos\theta+\frac{1}{r^2}\right)d\theta$$

$$= 4\pi\log r. \tag{198}$$

If $r = 1$,

$$\int_0^{2\pi} \log 4\left(\sin\frac{\theta}{2}\right)^2 d\theta = 4\pi\log 2 + 2\int_0^{2\pi}\log\sin\frac{\theta}{2}\,d\theta$$

$$= 4\pi\log 2 + 4\int_0^{\pi}\log\sin\theta\,d\theta$$

$$= 4\pi\log 2 - 4\pi\log 2, \text{ by (75), Art. 94;}$$

$$= 0. \qquad (199)$$

110.] This is all the account that I am able to give, consistently with the scope of the present treatise, of the method of evaluating definite integrals devised by M. Cauchy. Much more might be said both on the simple cases which have been investigated, and on the numerous applications of the resulting formulæ, as well as on definite integrals, the direct forms of the element-functions of which are more complicated. But for all these I must refer the student to the original memoirs of M. Cauchy.

In the course of his studies the reader will find that when the factor, for which $f(z) = \infty$, is of the form $(z-\zeta)^m$, the correction for infinity takes a more complicated form than that given in (155); and M. Cauchy has devised for the determination of it a new process involving new symbols and a new algorithm, which he calls Calcul des Résidus. This however is one of the higher parts of the subject which I have not attempted to develope. It will be observed that almost all the definite integrals to which the method has been applied have also been deduced by other processes; indeed Cauchy has drawn hardly any results which had not been demonstrated by other methods. The method requires very great caution: but theoretically there are scarcely any limits to the extent of application of the equation (126), which is the fundamental theorem of the method. It is the nearest approach to a general method of evaluating definite integrals that has, as yet, been discovered. It will also be observed that in the preceding Articles the subject-variable z of the definite integral has been restricted to the very particular form $z = x + y\sqrt{-1}$; although it is theoretically very general, being any function of x and y. The application of the general theorem to another form, viz. $z = x(a + y\sqrt{-1})$ has been briefly and imperfectly made in Art. 104: and the correction for infinity, which is one of the most useful parts of the method, has not been made in that case. I may in conclusion remark that the process requires the determination of the roots of $f'(z) = \infty$; so that it can be employed only when the roots can be found.

SECTION 6.—*Methods of Approximating to the Value of a Definite Integral.*

111.] If the correct value of a definite integral cannot be determined by any of the methods explained in the preceding sections, yet there are many methods by which an approximate value of it can frequently be found; and the limits also can be determined within which the error of approximation lies. These methods will be investigated in this section.

We must recur to the precise definition of a definite integral given in equations (20) and (21), Art. 83; and we have

$$\int_{x_0}^{x_n} \mathrm{F}'(x)\,dx = \mathrm{F}(x_n)-\mathrm{F}(x_0)$$

$$=(x_1-x_0)\,\mathrm{F}'(x_0)+(x_2-x_1)\,\mathrm{F}'(x_1)+\ldots+(x_n-x_{n-1})\,\mathrm{F}'(x_{n-1}); \quad (200)$$

subject to the condition that $\mathrm{F}'(x)$ does not become infinite for any value of x within the range of integration.

Now (200) is a series of terms each of which is the product of two factors. One of the factors is an element of the range of integration, and the other is a given function of the variable, the variable having a given value; thus the latter factor is a quantity completely defined; but the former factor is an element of the range, and is arbitrary, provided that it is infinitesimal, because the mode of partition of the range is arbitrary. Different modes of partition are suitable to different element-functions; but doubtless that which is most generally applicable, and which is also the most simple, is the partition of the range into equal elements. Let us adopt this mode; and accordingly let us suppose x_n-x_0 to be divided into n equal parts, each of which $= i$; so that $x_n-x_0 = ni$, and

$$x_1-x_0 = x_2-x_1 = \ldots\ldots = x_n-x_{n-1} = i. \quad (201)$$

then (200) becomes

$$\int_{x_0}^{x_n} \mathrm{F}'(x)\,dx$$

$$= i\left\{\mathrm{F}'(x_0)+\mathrm{F}'(x_0+i)+\mathrm{F}'(x_0+2i)+\ldots\mathrm{F}'\{x_0+(n-1)i\}\right\}; \quad (202)$$

and for the evaluation of the definite integral it is necessary to find the sum of the series of functions contained in the right-hand member. This sum will generally be a function of n and i; and consequently of n only, since $x_n-x_0 = ni$; and if in it ∞ is substituted for n, so that i is an infinitesimal, the resulting value will be the value of the given definite integral. Although some examples of this mode of evaluating a definite integral have been given in Art. 9, it is desirable to add others; for the notion of a definite integral, as the sum of a series, cannot be too frequently impressed on the student, whether for the sake of an exact idea, or for the purposes to which this Calculus will be applied in both the present and the subsequent volumes of our course.

Ex. 1. $$\int_0^\infty \frac{\sin x}{x}\,dx = i\left\{\frac{\sin 0}{0} + \frac{\sin i}{i} + \frac{\sin 2i}{2i} + \ldots + \frac{\sin\infty}{\infty}\right\}$$
$$= i + \frac{\sin i}{1} + \frac{\sin 2i}{2} + \frac{\sin 3i}{3} + \ldots \text{ ad infinitum}$$
$$= i + \frac{\pi}{2} - \frac{i}{2}$$
$$= \frac{\pi}{2}, \text{ when } i \text{ is infinitesimal.} \qquad (203)$$

the same result as (76), Art. 94.

Ex. 2. $\int_0^\pi \log(a^2 - 2a\cos\theta + 1)\,d\theta$, when a is greater than 1.

Let the range be divided into n equal parts, each of which $= \frac{\pi}{n}$; then the definite integral

$$= \frac{\pi}{n}\left\{\log(a-1)^2 + \log\left(a^2 - 2a\cos\frac{\pi}{n} + 1\right)\right.$$
$$\left. + \log\left(a^2 - 2a\cos\frac{2\pi}{n} + 1\right) + \ldots + \log\left(a^2 - 2a\cos\frac{n-1}{n}\pi + 1\right)\right\}$$
$$= \frac{\pi}{n}\log\left\{(a-1)^2\left(a^2 - 2a\cos\frac{\pi}{n} + 1\right)\left(a^2 - 2a\cos\frac{2\pi}{n} + 1\right)\ldots\right.$$
$$\left.\ldots\left(a^2 - 2a\cos\frac{n-1}{n}\pi + 1\right)\right\}$$

Now by Art 64, Vol. I,

$$a^{2n} - 1 = (a-1)\left(a^2 - 2a\cos\frac{\pi}{n} + 1\right)\left(a^2 - 2a\cos\frac{2\pi}{n} + 1\right)\ldots$$
$$\ldots\left(a^2 - 2a\cos\frac{n-1}{n}\pi + 1\right)(a+1).$$

$$\therefore \int_0^\pi \log(a^2 - 2a\cos\theta + 1)\,d\theta = \frac{\pi}{n}\log\frac{a-1}{a+1}(a^{2n} - 1)$$
$$= 2\pi\log a. \qquad (204)$$

which is the same result as (198), Art. 109.

112.] If however the sum of the series in the right-hand member of (202) cannot be expressed in general terms of n, yet an approximation may be made to it, and may be carried to any extent. For this purpose let the range be divided into a finite number of equal parts; say, into n equal parts; and let the value of the element-functions corresponding to the commencement of each part of the range be calculated; then the product of the sum of all these element-functions and of the element of the range will be the approximate value of the definite integral; and the larger n is, the nearer will the value thus determined be to the true value of the integral. This process may of course involve long and intricate calculations; yet the process is theoretically perfect, and may always be applied. A small value of n will be sufficient for a first approximation, and this is frequently of considerable practical use.

For an example of the method let us take the definite integral $\int_0^1 \frac{dx}{1+x^2}$. I have chosen one, the value of which is known, that the approximate results may be compared with the true result.

Let the range, which $= 1$, be divided into four equal parts, each of which $= \frac{1}{4}$: then

$$\begin{aligned}\int_0^1 \frac{dx}{1+x^2} &= \frac{1}{4}\left\{1+\frac{16}{17}+\frac{4}{5}+\frac{16}{25}\right\}\\ &= .25+.234+.2+.16\\ &= .844.\end{aligned}$$

Again, let the range be divided into ten equal parts, each of which $= \frac{1}{10}$; then

$$\begin{aligned}\int_0^1 \frac{dx}{1+x^2} &= \frac{1}{10}\left\{1+\frac{100}{101}+\frac{100}{104}+\frac{100}{109}+\frac{100}{116}+\frac{100}{125}+\frac{100}{136}\right.\\ &\qquad\left.+\frac{100}{149}+\frac{100}{164}+\frac{100}{181}\right\}\\ &= \frac{1}{10}\{1+.99+.9615+.9174+.862+.8+.7352\\ &\qquad +.6711+.6097+.5524\}\\ &= .80993.\end{aligned}$$

Now the true value of $\int_0^1 \frac{dx}{1+x^2} = \frac{\pi}{4} = .78539$; so that the errors in the two preceding calculations are respectively + .059 and + .02454.

113.] Now let the reader refer to the geometrical interpretation of integration given in Art. 3, and compare with it the process of the preceding Article. Let $y = F'(x)$ be the equation to the plane curve represented in fig. 16; let $OM_0 = x_0$, $OM = x$, $OM_n = x_n$; $MN = dx$, $MP = y = F'(x)$: so that $F'(x)\,dx$ expresses the area of MPQN, when $MN = dx =$ an infinitesimal. Evidently therefore will $\int_{x_0}^{x_n} F'(x)\,dx$ express the area contained between the curve, the extreme ordinates M_0P_0 and M_nP_n, and the x-axis. Under these circumstances the range $M_n M_0$ is divided into an infinite number of elements. If however the number of parts into which the range is divided is finite, and the sum of the element-functions corresponding to these is calculated, that sum will be only an approximate expression for the value of the area. For suppose MN, in fig. 16, to be a part of the range, $= i$, say; then $i\,F'(x) = MN \times MP =$ the rectangle MPRN; which is short of the corresponding part of the curvilinear area by the area of PRQ: and as a similar result is true of each part of the range, the sum of all the rectangles will be less than the required area by the sum of all the similar triangular pieces. If the bounding curve approaches the axis of x, as x increases, the sum of the areas corresponding to the finite partition of the range of the x-integration will be greater than the true result by the sum of similar triangles. The differences however between the true results and the approximate results thus determined will be less, according as the number of parts into which the range is divided is greater.

The process of thus approximating to the value of a curvilinear area by the geometrical expression of the mode of approximate integration explained in the preceding Article is so exact, that the latter has been called Approximate Integration by summation of ordinates at equal finite intervals.

This geometrical illustration suggests a more exact process of approximation. It is plain that the product of MN into the semi-sum of MP and NQ is nearer to the true value of the area MNQP than $MN \times MP$. So that if I represents the required area, or definite integral, and if $y_0, y_1, y_2, \ldots y_n$ denote the ordinates, or the element-functions, corresponding to $x_0, x_0+i, \ldots x_n$, then

$$I = i\left\{\frac{y_0+y_1}{2} + \frac{y_1+y_2}{2} + \ldots + \frac{y_{n-1}+y_n}{2}\right\};$$

$$\therefore \int_{x_0}^{x_n} \mathrm{F}'(x)\,dx = i\left\{\frac{\mathrm{F}'(x_0)}{2} + \mathrm{F}'(x_0+i) + \mathrm{F}'(x_0+2i) + \dots \right.$$
$$\left. \dots + \mathrm{F}'(x_n - i) + \frac{\mathrm{F}'(x_n)}{2}\right\}. \quad (205)$$

Of this Theorem we have the following interesting application. Let $\mathrm{F}'(x) = \log x$, and let the limits of integration be m and $m+n$; and let us suppose $i = 1$; then

$$\int_m^{m+n} \log x\,dx = \frac{1}{2}\log m + \log(m+1) + \log(m+2) + \dots$$
$$\dots + \log(m+n-1) + \frac{1}{2}\log(m+n)$$
$$= \log m + \log(m+1) + \dots + \log(m+n-1)$$
$$+ \frac{1}{2}\{\log(m+n) - \log m\}.$$

But $\displaystyle\int_m^{m+n} \log x\,dx = \Big[x\log x - x\Big]_m^{m+n}$
$$= (m+n)\log(m+n) - m\log m - n.$$

$$\therefore \quad \log m + \log(m+1) + \log(m+2) + \dots + \log(m+n-1)$$
$$= (m+n)\log(m+n) - m\log m - n - \frac{1}{2}\{\log(m+n) - \log m\}$$
$$= \left(m+n-\frac{1}{2}\right)\log(m+n) - \left(m - \frac{1}{2}\right)\log m - n;$$

and taking numbers instead of logarithms, we have

$$m(m+1)(m+2)\dots(m+n-1) = (m+n)^{m+n-\frac{1}{2}}\, m^{-m+\frac{1}{2}} e^{-n}. \quad (206)$$

Thus the right-hand member is an approximate value of the product of n numbers integral or fractional, in arithmetical progression, of which the common difference is unity. If $m = 1$,

$$1.2.3\dots n = (1+n)^{n+\frac{1}{2}} e^{-n}. \quad (207)$$

Now the difference between the true value and the approximate value given in (206) of $m(m+1)\dots(m+n-1)$ will become less, the larger n is, because the greater the range of integration is, the smaller proportionally becomes the difference between the successive elements, which we have assumed to be unity. Let us suppose $n = \infty$; then since, when $n = \infty$,

$$(m+n)^{m+n-\frac{1}{2}} = n^{m+n-\frac{1}{2}}\left(1+\frac{m}{n}\right)^n \left(1+\frac{m}{n}\right)^{m-\frac{1}{2}}$$
$$= n^{m+n-\frac{1}{2}} e^m;$$

therefore from (206),

$$m(m+1)(m+2)\dots(m+n-1) = m^{-m+\frac{1}{2}} e^{m} n^{m+n-\frac{1}{2}} e^{-n}, \quad (208)$$

when $n = \infty$.

This is indeed only an approximate value for the product of the factorials, but we shall hereafter be able to correct it for the particular case given in (208), viz. when $n = \infty$.

Again, if (205) is applied to the Example in Art. 112, and the range is divided into ten equal parts, the result $= 7.8493$, which is less than the true value by only .0046, and is a much more exact approximation than those found in the preceding Article.

114.] The partition of the range of integration into equal finite parts and the calculation of the element-function corresponding to each part, produces other formulæ for the determination of approximate values of definite integrals; and as these are also of considerable practical use in Mensuration it is necessary to demonstrate them.

Let the element-function be $F(x)$; and let the range $x_n - x_0$ be divided into n equal parts, each of which $= i$; so that $x_n - x_0 = ni$.

Also let the difference between $F(x+i)$ and $F(x)$, which is generally finite, since i is finite, be denoted by $\Delta F(x)$; so that we have

$$F(x+i) = F(x) + \Delta F(x); \quad (209)$$

in reference to which it may be observed that Δ becomes d, when i becomes dx. Let the variable x receive another finite increment i; then from (209),

$$\begin{aligned} F(x+2i) &= F(x+i) + \Delta F(x+i) \\ &= F(x) + \Delta F(x) + \Delta \{F(x) + \Delta F(x)\} \\ &= F(x) + 2\Delta F(x) + \Delta\Delta F(x). \end{aligned}$$

Let $\Delta\Delta$ be denoted by Δ^2; this substitution being merely a matter of notation, and independent of any supposition as to whether Δ denotes an operation subject to the index law. Similarly let $\Delta\Delta^2$ be denoted by Δ^3; and so on; then

$$F(x+2i) = F(x) + 2\Delta F(x) + \Delta^2 F(x).$$

Again, let x receive another increment i; then

$$F(x+3i) = F(x) + 3\Delta F(x) + 3\Delta^2 F(x) + \Delta^3 F(x);$$

the law of the coefficients being evidently that of the Binomial Theorem. Consequently, if x is increased m times successively by i, we shall have

$$F(x+mi) = F(x) + \frac{m}{1}\Delta F(x) + \frac{m(m-1)}{1.2}\Delta^2 F(x)$$
$$+ \frac{m(m-1)(m-2)}{1.2.3}\Delta^3 F(x) + \dots \quad (210)$$

Now this theorem is generally true for all positive and integral values of m. To accommodate it to the present problem, let x be replaced by x_0, and let $x_0 + mi = x$; then

$$F(x) = F(x_0) + \Delta F(x_0)\frac{x-x_0}{i} + \Delta^2 F(x_0)\frac{(x-x_0)(x-x_0-i)}{1.2.i^2}$$
$$+ \Delta^3 F(x_0)\frac{(x-x_0)(x-x_0-i)(x-x_0-2i)}{1.2.3.i^3} + \dots \quad (211)$$

Hence by integration through the range $x_n - x_0$,

$$\int_{x_0}^{x_n} F(x)\,dx = F(x_0)(x_n-x_0)$$
$$+ \frac{\Delta F(x_0)}{i}\frac{(x_n-x_0)^2}{2} + \frac{1}{1.2}\frac{\Delta^2 F(x_0)}{i^2}\left\{\frac{(x_n-x_0)^3}{3} - \frac{i(x_n-x_0)^2}{2}\right\}$$
$$+ \frac{1}{1.2.3}\frac{\Delta^3 F(x_0)}{i^3}\left\{\frac{(x_n-x_0)^4}{4} - i(x_n-x_0)^3 + i^2(x_n-x_0)^2\right\}$$
$$+ \frac{1}{1.2.3.4}\frac{\Delta^4 F(x_0)}{i^4}\left\{\frac{(x_n-x_0)^5}{5} - i\frac{3(x_n-x_0)^4}{2}\right.$$
$$\left. + i^2\frac{11(x_n-x_0)^3}{3} - 3i^3(x_n-x_0)^2\right\}$$
$$+ \dots\dots\dots\dots\dots\dots ; \quad (212)$$

replacing $x_n - x_0$ by ni, this becomes

$$\int_{x_0}^{x_n} F(x)\,dx = i\left\{nF(x_0) + \frac{n^2}{2}\Delta F(x_0) + \left(\frac{n^3}{3} - \frac{n^2}{2}\right)\frac{\Delta^2 F(x_0)}{1.2}\right.$$
$$+ \left(\frac{n^4}{4} - n^3 + n^2\right)\frac{\Delta^3 F(x_0)}{1.2.3}$$
$$\left. + \left(\frac{n^5}{5} - \frac{3n^4}{2} + \frac{11n^3}{3} - 3n^2\right)\frac{\Delta^4 F(x_0)}{1.2.3.4} + \dots\right\}; \quad (213)$$

the subsequent terms of which are easily calculated. In this series let $\Delta F(x_0)$, $\Delta^2 F(x_0)$, $\Delta^3 F(x_0)$ be replaced by the quantities which they represent: for this purpose let $y_0, y_1, y_2, y_3 \dots$ denote $F(x_0)$, $F(x_0+i)$, $F(x_0+2i)$, $F(x_0+3i)$, …; then, from (207),

$$\left.\begin{aligned} \Delta F(x_0) &= \Delta y_0 = y_1 - y_0; \\ \Delta^2 F(x_0) &= \Delta y_1 - \Delta y_0, \\ &= y_2 - 2y_1 + y_0; \\ \text{similarly} \quad \Delta^3 F(x_0) &= y_3 - 3y_2 + 3y_1 - y_0; \\ \dots & \dots\dots\dots ; \end{aligned}\right\} \quad (214)$$

so that from a geometrical point of view $y_0, y_1, y_2, y_3, \ldots$ are equidistant ordinates of the curve $y = F(x)$.

Now these quantities $\Delta F(x_0)$, $\Delta^2 F(x_0), \ldots$ are called the successive differences of $F(x_0)$; and in all cases to which the preceding formulæ of approximation can be successfully applied, it is necessary that the values of these differences should rapidly decrease; this circumstance may easily be verified, and it is found to be true in all cases of tabulated functions. Thus, if $F(x)$ is a rational algebraical function of n dimensions, $\Delta^n F(x) = 0$. On this hypothesis then we shall deduce from (213) successive approximate values to the definite integral.

Thus, if we omit all terms after $\Delta^2 F(x_0)$, we have

$$\int_{x_0}^{x_n} F(x)\,dx = i\left\{nF(x_0) + \frac{n^2}{2}\Delta F(x_0) + \left(\frac{n^3}{3} - \frac{n^2}{2}\right)\frac{\Delta^2 F(x_0)}{1.2.}\right\}$$

$$= i\left\{y_0\left(n - \frac{3n^2}{4} + \frac{n^3}{6}\right) + y_1\left(n^2 - \frac{n^3}{3}\right) + y_2\left(\frac{n^3}{6} - \frac{n^2}{4}\right)\right\}. \quad (215)$$

Let $n = 2$; $$\int_{x_0}^{x_n} F(x)\,dx = \frac{i}{3}\{y_0 + 4y_1 + y_2\}; \quad (216)$$

which is a theorem frequently applied in mensuration.

Let $n = 4$; $$\int_{x_0}^{x_n} F(x)\,dx = \frac{4i}{3}\{2y_0 - 4y_1 + 5y_2\}. \quad (217)$$

From (216) may be deduced a rule, originally discovered by Thomas Simpson, and frequently employed in mensuration.

Let the range $x_n - x_0$ be divided into n equal parts, where n is an even number, each of which $= i$; then as (216) gives the value of the integral for any two consecutive parts, taking the sum of all, we shall have,

$$\int_{x_0}^{x_n} F'(x)\,dx = \frac{i}{3}\{y_0 + 4y_1 + y_2 + y_2 + 4y_3 + y_4 + \cdots$$
$$\cdots + y_{n-2} + 4y_{n-1} + y_n\}$$
$$= \frac{i}{3}\{y_0 + y_n + 4(y_1 + y_3 + \cdots + y_{n-1})$$
$$+ 2(y_2 + y_4 + \cdots + y_{n-2})\}. \quad (218)$$

Again, if all terms in (213) after $\Delta^3 F(x_0)$ are omitted, we have

$$\int_{x_0}^{x_n} F(x)\,dx = i\left\{y_0\left(n - \frac{11n^2}{12} + \frac{n^3}{3} - \frac{n^4}{24}\right)\right.$$
$$+ y_1\left(\frac{3n^2}{2} - \frac{5n^3}{6} + \frac{n^4}{8}\right) + y_2\left(-\frac{3n^2}{4} + \frac{2n^3}{3} - \frac{n^4}{8}\right)$$
$$\left. + y_3\left(\frac{n^2}{6} - \frac{n^3}{6} + \frac{n^4}{24}\right)\right\}. \quad (219)$$

Let $n = 3$; then

$$\int_{x_0}^{x_n} \text{F}(x)\,dx = \frac{3i}{8}\{y_0+3y_1+3y_2+y_3\}. \tag{220}$$

Let $n = 4$; then

$$\int_{x_0}^{x_n} \text{F}(x)\,dx = \frac{4i}{3}\{2y_1-y_2+2y_3\}. \tag{221}$$

If $\Delta^3 \text{F}(x_0) = 0$, as is the case when $\text{F}(x) = \frac{a}{b^2}x^2$, (220) and (221) give results strictly accurate.

115.] An approximation may also be made by the following process, which assigns certain limits within which the true value of the definite integral lies. Let $\text{F}'(x)$ be the element-function of the definite integral; and let $f'(x)$ and $\phi'(x)$ be two other functions of x, the values of which are respectively greater and less than $\text{F}'(x)$, for every value of x within the range of integration; so that for every such value

$$f'(x) > \text{F}'(x) > \phi'(x);$$

Moreover let us take $f'(x)$ and $\phi'(x)$, such that the definite integrals, of which they are the element-functions, may be determined: then evidently, by the definition of a definite integral,

$$\int_{x_0}^{x_n} f'(x)\,dx > \int_{x_0}^{x_n} \text{F}'(x)\,dx > \int_{x_0}^{x_n} \phi'(x)\,dx. \tag{222}$$

Let us investigate by this method an approximate value of the definite integral $\int_0^{\frac{1}{2}} \frac{dx}{(1-x^3)^{\frac{1}{2}}}$: since for all values of x included within the range of integration,

$$\frac{1}{(1-x^2)^{\frac{1}{2}}} > \frac{1}{(1-x^3)^{\frac{1}{2}}} > 1;$$

$$\therefore \quad \int_0^{\frac{1}{2}} \frac{dx}{(1-x^2)^{\frac{1}{2}}} > \int_0^{\frac{1}{2}} \frac{dx}{(1-x^3)^{\frac{1}{2}}} > \int_0^{\frac{1}{2}} dx;$$

$$\frac{\pi}{6} > \int_0^{\frac{1}{2}} \frac{dx}{(1-x^3)^{\frac{1}{2}}} > \frac{1}{2};$$

$$\therefore \quad .52358 > \int_0^{\frac{1}{2}} \frac{dx}{(1-x^3)^{\frac{1}{2}}} > .5; \tag{223}$$

so that limits are assigned very near together within which the value of the given definite integral lies.

116.] The definition given in (200) suggests other means of

obtaining approximate values of definite integrals, and of ascertaining certain properties relating to those values.

Let $F'(x)$ be the element-function: then if x_n is greater than x_0, and $F'(x)$ does not change sign within the range of integration, $\int_{x_0}^{x_n} F'(x)\,dx$ has the same sign as $F'(x)$. If however $F'(x)$ changes sign, the integral will be positive or negative, according as the positive or negative part of the series is the greater.

Since, in the series on the right hand member of (200), x_1-x_0, x_2-x_1, ... x_n-x_{n-1} are all quantities of the same sign, by preliminary Theorem III of Vol. I. the sum of the series is equal to the product of the sum of all these quantities and some mean value of the other factor: consequently

$$\int_{x_0}^{x_n} F'(x)\,dx = (x_n-x_0)\,F'\{x_0+\theta(x_n-x_0)\}, \tag{224}$$

wherein θ denotes an undetermined proper positive fraction. Then, the limits of the value of the definite integral, given by substituting 0 and 1 severally for θ, are $(x_n-x_0)\,F'(x_0)$ and $(x_n-x_0)\,F'(x_n)$, between which the value of the definite integral lies.

The geometrical interpretation of (224) deserves a passing notice. Let $y = F'(x)$ be the equation to the curve P_0PQP_n in fig. 16; Then as the definite integral expresses the area $P_0P_nM_nM_0$, and as $M_0M_n = x_n-x_0$, the equation shews that the area is equal to that of a rectangle one of whose sides is M_0M_n, and the other is an ordinate to the curve corresponding to an abscissa intermediate to x_n and x_0.

If the left-hand member of (200) is determined by means of its indefinite integral; then

$$F(x_n)-F(x_0) = (x_n-x_0)\,F'\{x_0+\theta(x_n-x_0)\}, \tag{225}$$

which equation has been already found in Art. 111, Vol. I.

If the range x_n-x_0 is infinitesimal,

$$\int_{x_0}^{x_n} F'(x)\,dx = (x_n-x_0)\,F'(x_0); \tag{226}$$

which is the first term of the series, of which the definite integral is the sum.

If the element-function $F'(x)$ is the product of two functions, $f(x)$ and $\phi'(x)$, of which $\phi'(x)$ has the same sign for all values of x within the range of integration; then

$$\int_{x_0}^{x_n} f(x)\phi'(x)\,dx = (x_1-x_0)f(x_0)\phi'(x_0)+(x_2-x_1)f(x_1)\phi'(x_1)+\dots$$
$$\dots+(x_n-x_{n-1})\phi'(x_{n-1}). \quad (227)$$

Now the right-hand member of this equation consists of a series of terms, into each of which a factor enters, which is of the form $(x_j-x_i)\,\phi'(x_i)$, and is always of the same sign; therefore by Preliminary Theorem III, Vol. I.,

$$\int_{x_0}^{x_n} f(x)\,\phi'(x)\,dx$$
$$= f\{x_0+\theta\,(x_n-x_0)\}\,\{(x_1-x_0)\,\phi'(x_0)+(x_2-x_1)\,\phi'(x_1)+\dots$$
$$\dots+(x_n-x_{n-1})\,\phi'(x_{n-1})\}$$
$$= f\{x_0+\theta\,(x_n-x_0)\}\int_{x_0}^{x_n}\phi'(x)\,dx. \quad (228)$$

Equation (224) is evidently a particular case of this theorem.

If the indefinite integral of the right-hand member can be determined, this equation becomes

$$\int_{x_0}^{x_n} f(x)\phi'(x)\,dx = f\{x_0+\theta\,(x_n-x_0)\}\,\{\phi(x_n)-\phi(x_0)\}. \quad (229)$$

The following is an example of this theorem;

$$\int_{x_0}^{x_n} \text{F}'(x)\,dx = \int_{x_0}^{x_n} x\,\text{F}'(x)\,\frac{dx}{x}$$
$$= \{x_0+\theta(x_n-x_0)\}\,\text{F}'\{x_0+\theta(x_n-x_0)\}\log\frac{x_n}{x_0}. \quad (230)$$

Hence it follows that if G and L are respectively the greatest and the least values of $f(x)$ within the range of integration,

$$\text{G}\int_{x_0}^{x_n}\phi'(x)\,dx > \int_{x_0}^{x_n} f(x)\,\phi'(x)dx > \text{L}\int_{x_0}^{x_n}\phi'(x)\,dx. \quad (231)$$

These theorems are frequently of considerable use in determining the limits of value of a definite integral; and also in shewing the value which a definite integral takes for a particular value of a constant contained in its element-function. Thus we may determine the value of $\int_0^{\infty}\frac{e^{-ax}}{e^x+e^{-x}}\frac{dx}{(x)^{\frac{1}{2}}}$, when $a=\infty$. On comparing this with the formula given in (228), let $f(x) = \frac{1}{e^x+e^{-x}}$; which $=0$, when $x=\infty$; and $=\frac{1}{2}$, when $x=0$, so that the mean value of this factor is mean between these quantities. Also

$$\int_0^{\infty}\frac{e^{-ax}\,dx}{x^{\frac{1}{2}}} = 2\int_0^{\infty} e^{-ax^2}dx = \left(\frac{\pi}{a}\right)^{\frac{1}{2}}; \text{ see (112), Art. 100};$$

so that $$\int_0^{\infty} \frac{e^{-ax}\, dx}{x^{\frac{1}{2}}} = 0, \text{ if } a = \infty;$$

$$\therefore \int_0^{\infty} \frac{e^{-ax}}{e^x + e^{-x}} \frac{dx}{x^{\frac{1}{2}}} = 0, \text{ if } a = \infty. \tag{232}$$

117.] Of the various processes which have been devised for the purpose of approximating to the value of definite integrals, some consist in the development of the element-function into a series, and in the subsequent integration of each term separately; in others, series are formed, either by integration by parts, or by some other method, in general terms, and these are applied to the particular element-function.

The following series was devised by John Bernoulli.

Let $\int_{x_0}^{x_n} \text{F}'(x)\, dx$ be the definite integral whose value is required: then, integrating by parts, we have

$$\int \text{F}'(x)\, dx = x\, \text{F}'(x) - \int x\, \text{F}''(x)\, dx;$$

$$\int x\, \text{F}''(x)\, dx = \frac{x^2}{1.2} \text{F}''(x) - \int \frac{x^2}{1.2} \text{F}'''(x)\, dx;$$

$$\int \frac{x^2}{1.2} \text{F}'''(x)\, dx = \frac{x^3}{1.2.3} \text{F}'''(x) - \int \frac{x^3}{1.2.3} \text{F}^{\text{iv}}(x)\, dx;$$

. . . .

$$\therefore \int_{x_0}^{x_n} \text{F}'(x)\, dx$$

$$= \Big[\frac{x}{1} \text{F}'(x) - \frac{x^2}{1.2} \text{F}''(x) + \frac{x^3}{1.2.3} \text{F}'''(x) - \dots \Big]_{x_0}^{x_n}. \tag{233}$$

This theorem may be proved, and also its general term may be found, by means of Taylor's series; and hereby the limits of the value will be determined.

By Taylor's series we have

$$\text{F}(x+h) = \text{F}(x) + \text{F}'(x)\frac{h}{1} + \text{F}''(x)\frac{h^2}{1.2} + \dots + \text{F}^r(x+\theta h)\frac{h^r}{1.2\dots r}.$$

Let h be replaced by $-x$,

$$\therefore \quad \text{F}(x) = \text{F}(0) + \text{F}'(x)\frac{x}{1} - \text{F}''(x)\frac{x^2}{1.2.} + \text{F}'''(x)\frac{x^3}{1.2.3} - \dots\dots$$

$$\dots (-)^{r-1} \text{F}^r(x - \theta x)\frac{x^r}{1.2.3\dots r};$$

so that replacing $\text{F}(x)$ by $\int \text{F}'(x)\, dx$, and taking the definite integral of both members, we have

$$\int_{x_0}^{x_n} \mathrm{F}'(x)\,dx = \Big[\mathrm{F}'(x)\frac{x}{1} - \mathrm{F}''(x)\frac{x^2}{1.2} + \mathrm{F}'''(x)\frac{x^3}{1.2.3} - \dots$$
$$\dots (-)^{r-1}\mathrm{F}^r(x-\theta x)\frac{x^r}{1.2.3\dots r}\Big]_{x_0}^{x_n}; \qquad (234)$$

omitting $\mathrm{F}(0)$, because it disappears in the definite integral.

118.] Other series useful for the present purpose may also be derived from Taylor's Theorem.

From (84), Art. 74, Vol. I, we have

$$\mathrm{F}(x_n) = \mathrm{F}(x_0) + \mathrm{F}'(x_0)\frac{x_n - x_0}{1} + \mathrm{F}''(x_0)\frac{(x_n-x_0)^2}{1.2} + \dots$$
$$\dots + \mathrm{F}^r\{x_0 + \theta(x_n - x_0)\}\frac{(x_n-x_0)^r}{1.2.3\dots r};$$

$$\therefore \int_{x_0}^{x_n} \mathrm{F}'(x)\,dx = \mathrm{F}(x_n) - \mathrm{F}(x_0)$$
$$= \mathrm{F}'(x_0)\frac{x_n - x_0}{1} + \mathrm{F}''(x_0)\frac{(x_n-x_0)^2}{1.2} + \dots$$
$$\dots + \mathrm{F}^r\{x_0 + \theta(x_n - x_0)\}\frac{(x_n-x_0)^r}{1.2.3\dots r}; \qquad (235)$$

the right-hand member of which rapidly converges, if $x_n - x_0$ is a small quantity; and if $x_n - x_0$ is infinitesimal, taking two terms, we have

$$\int_{x_0}^{x_n} \mathrm{F}'(x)\,dx = \mathrm{F}'(x_0)\frac{x_n - x_0}{1} + \mathrm{F}''(x_0)\frac{(x_n-x_0)^2}{1.2}; \qquad (236)$$

which is in fact the same result as (225); and if only the first term is taken, the result is the same as (226).

Again, suppose the range $x_n - x_0$ to be finite, and to be divided into n parts, each of which is equal to i, so that $x_n - x_0 = ni$; then

$$\int_{x_0}^{x_n} \mathrm{F}'(x)\,dx = \int_{x_0}^{x_0+i} \mathrm{F}'(x)\,dx + \int_{x_0+i}^{x_0+2i} \mathrm{F}'(x)\,dx + \dots + \int_{x_0+(n-1)i}^{x_n} \mathrm{F}'(x)\,dx;$$

and replacing the definite integrals by their values in (236), we have

$$\int_{x_0}^{x_n} \mathrm{F}'(x)\,dx = i\{\mathrm{F}'(x_0) + \mathrm{F}'(x_0+i) + \dots + \mathrm{F}'(x_0+(n-1)i)\}$$
$$+ \frac{i^2}{1.2}\{\mathrm{F}''(x_0) + \mathrm{F}''(x_0+i) + \dots + \mathrm{F}''(x_0+(n-1)i)\}; \qquad (237)$$

a nearer approximation may of course be made by including more terms of (235).

119.] We come now to a method of approximate integration, which may be very largely, and indeed almost universally applied.

It is called the method of integration by series. And although it is generally inexpedient to expand a function into series, yet as this may be a case where all other methods fail, we must take the expansion as the best possible solution, and examine it as closely as we can.

The process suggests itself at once. Let the element-function be expanded into a series of terms, such that each term may be integrated separately. Let these integrations be effected; hereby a new series is formed, which is the value of the given integral. If the sum of this series can be expressed in general and finite terms, that sum is the value of the definite integral; but when this cannot be found, an equality will exist between the definite integral and the series, which is often of considerable use, as to the definite integrals and the series. It must be borne in mind, that if a series adequately expresses a function, the series must be convergent: it is necessary therefore that the series to which the definite integral is equal should be convergent: this will always be the case if the series, into which the element-function is expanded, is convergent for all values of the variable within the range of integration; and may be so, even when that series is not convergent for all these values.

Let $F'(x)$ be the element-function; and let us suppose $F'(x)$ to be capable of expansion into a series of the form

$$u_0+u_1+u_2+\ldots+u_m+\ldots,$$

which is convergent for all values of x within the range of integration; and let us suppose R_m to be the sum of all the terms after the $(m+1)$th, so that, the series being convergent, R_m becomes smaller the greater m is, and ultimately is infinitesimal when $m=\infty$; then we have

$$F'(x) = u_0+u_1+u_2+\ldots\ldots+u_m+R_m, \tag{238}$$

$$\int_{x_0}^{x_n} F'(x)\,dx = \int_{x_0}^{x_n} u_0\,dx + \int_{x_0}^{x_n} u_1\,dx + \ldots$$
$$\ldots + \int_{x_0}^{x_n} u_m\,dx + \int_{x_0}^{x_n} R_m\,dx. \tag{239}$$

Now, by (224), $\int_{x_0}^{x_n} R_m\,dx = (x_n-x_0)\times$ some value of R_m intermediate to those corresponding to x_n and x_0; but since R_m becomes infinitesimal, when m becomes infinite, so will also its mean value; and therefore $\int_{x_0}^{x_n} R_m\,dx$ becomes infinitesimal and must be neglected. Hence

$$\int_{x_0}^{x_n} \mathrm{F}'(x)\,dx = \int_{x_0}^{x_n} u_0\,dx + \int_{x_0}^{x_n} u_1\,dx + \int_{x_0}^{x_n} u_2\,dx + \dots \qquad (240)$$

The following are examples of this method of integration.

Ex. 1. $$\int_{x_0}^{x_n} \frac{dx}{(1+x^4)^{\frac{1}{2}}} = \int_{x_0}^{x_n} \Big(1 - \frac{1}{2}x^4 + \frac{1.3}{2.4}x^8 - \frac{1.3.5}{2.4.6}x^{12} + \dots\Big)\,dx$$

$$= \Big[x - \frac{1}{2}\frac{x^5}{5} + \frac{1.3}{2.4}\frac{x^9}{9} - \frac{1.3.5}{2.4.6}\frac{x^{13}}{13} + \dots\Big]_{x_0}^{x_n}.$$

Ex. 2. $$\int_0^x e^{-x^2}\,dx = \int_0^x \Big(1 - \frac{x^2}{1} + \frac{x^4}{1.2} - \frac{x^6}{1.2.3} + \dots\Big)\,dx$$

$$= x - \frac{x^3}{3} + \frac{1}{1.2}\frac{x^5}{5} - \frac{1}{1.2.3}\frac{x^7}{7} + \dots$$

120.] If the element-function is capable of expansion by Maclaurin's Theorem, then the theorem given in (239) takes the following form. Since

$$\mathrm{F}'(x) = \mathrm{F}'(0) + \mathrm{F}''(0)\frac{x}{1} + \mathrm{F}'''(0)\frac{x^2}{1.2} + \dots$$

$$\dots + \mathrm{F}^r(0)\frac{x^{r-1}}{1.2.3\dots(r-1)} + \mathrm{F}^{r+1}(\theta x)\frac{x^r}{1.2.3\dots r}; \qquad (241)$$

$$\int_{x_0}^{x_n} \mathrm{F}'(x)\,dx = \Big[\mathrm{F}'(0)x + \mathrm{F}''(0)\frac{x^2}{1.2} + \mathrm{F}'''(0)\frac{x^3}{1.2.3} + \dots$$

$$\dots + \mathrm{F}^r(0)\frac{x^r}{1.2.3\dots r}\Big]_{x_0}^{x_n} + \int_{x_0}^{x_n} \mathrm{F}^{r+1}(\theta x)\frac{x^r}{1.2.3\dots r}\,dx. \qquad (242)$$

When the series (241) is convergent, the last term in (242) is infinitesimal, and must be neglected; in this case

$$\int_{x_0}^{x_n} \mathrm{F}'(x)\,dx = \Big[\mathrm{F}'(0)x + \mathrm{F}''(0)\frac{x^2}{1.2} + \mathrm{F}'''(0)\frac{x^3}{1.2.3} + \dots\Big]_{x_0}^{x_n}$$

$$= \mathrm{F}'(0)(x_n - x_0) + \mathrm{F}''(0)\frac{x_n^2 - x_0^2}{1.2} + \mathrm{F}'''(0)\frac{x_n^3 - x_0^3}{1.2.3} + \dots \qquad (243)$$

121.] A similar process may be conveniently adopted when the element-function is the product of two functions, one of which may be expanded into a convergent series by Maclaurin's or some other equivalent Theorem. Thus suppose $\mathrm{F}'(x) = f(x)\,\phi'(x)$; and that

$$f(x) = f(0) + f'(0)\frac{x}{1} + f''(0)\frac{x^2}{1.2} + \dots$$

$$\dots + f^{r-1}(0)\frac{x^{r-1}}{1.2.3\dots(r-1)} + f^r(\theta x)\frac{x^r}{1.2.3\dots r};$$

then

$$\int_{x_0}^{x_n} f(x)\,\phi'(x)\,dx = f(0)\int_{x_0}^{x_n}\phi'(x)\,dx + f'(0)\int_{x_0}^{x_n}\phi'(x)\,x\,dx$$
$$+\frac{f''(0)}{1.2}\int_{x_0}^{x_n}\phi'(x)\,x^2\,dx+\ldots+\frac{f^{r-1}(0)}{1.2.3\ldots(r-1)}\int_{x_0}^{x_n}\phi'(x)\,x^{r-1}\,dx$$
$$+\frac{1}{1.2.3\ldots r}\int_{x_0}^{x_n}\phi'(x)\,f^r(\theta x)\,x^r\,dx. \quad (244)$$

If the series for $f(x)$ is convergent, the last term of this equation must be omitted; and we have

$$\int_{x_0}^{x_n} f(x)\,\phi'(x)\,dx = f(0)\int_{x_0}^{x_n}\phi'(x)\,dx + f'(0)\int_{x_0}^{x_n}\phi'(x)\,x\,dx$$
$$+\frac{f''(0)}{1.2}\int_{x_0}^{x_n}\phi'(x)\,x^2\,dx+\ldots \quad (245)$$

in which the original definite integral is expressed in a series of other definite integrals of a more simple form. The success of the method indeed depends on the possibility of integrating the several terms in the right-hand member. The following is an example of the process.

Ex. 1. $\displaystyle\int_0^1 \frac{dx}{\{(1-x^2)(1-e^2x^2)\}^{\frac{1}{2}}}$

$$=\int_0^1 \frac{dx}{(1-x^2)^{\frac{1}{2}}}\left\{1+\frac{1}{2}e^2x^2+\frac{1.3}{2.4}e^4x^4+\frac{1.3.5}{2.4.6}e^6x^6+\ldots\right\}.$$

Now the general term of this series is $\dfrac{c_m x^m}{(1-x^2)^{\frac{1}{2}}}$, where m is even: and by (13), Art. 82, if m is even,

$$\int_0^1 \frac{x^m\,dx}{(1-x^2)^{\frac{1}{2}}} = \frac{(m-1)(m-3)\ldots 3.1}{m(m-2)\ldots 4.2}\,\frac{\pi}{2};$$

so that substituting these in the right-hand member we have

$$\int_0^1 \frac{dx}{\{(1-x^2)(1-e^2x^2)\}^{\frac{1}{2}}}$$
$$=\frac{\pi}{2}\left\{1+\left(\frac{1}{2}\right)^2e^2+\left(\frac{1.3}{2.4}\right)^2e^4+\left(\frac{1.3.5}{2.4.6}\right)^2e^6+\ldots\right\}; \quad (246)$$

which is a rapidly converging series, if e is a small quantity.

As many examples of this process will occur in the sequel, it is unnecessary to add others here.

SECTION 7.—*On the Gamma-Function, and allied Definite Integrals.*

122.] One of the most important definite integrals both in itself on account of the many peculiar properties which it exhibits, and in its application and the definite integrals allied to it, is $\int_0^\infty e^{-x}x^{n-1}\,dx$. As the limits of this integral are ∞ and 0, it is evidently a function of n only; and the symbol $\Gamma(n)$, devised by Legendre, has been of late ordinarily employed to denote it; so that we have

$$\Gamma(n) = \int_0^\infty e^{-x}x^{n-1}\,dx. \tag{247}$$

For this reason and for the sake of a distinctive name, the definite integral has been called the Gamma-function. It has also been called by Legendre the second Eulerian Integral, because the properties of it were originally investigated by Euler. I propose in the present section to develope these properties in an analysis of the function, so far as they fall within the scope of a general elementary treatise on the Integral Calculus.

The definition of the Gamma-function is the definite integral given in (247), and the properties of it will, at first at least, be deduced from this integral. n, which is the subject-variable, is called the argument of the function.

The following are other and equivalent forms of the Gamma-function, and are derived from (247) by transformation.

(1) For e^{-x} substitute x; then

$$\Gamma(n) = \int_0^1 \left(\log\frac{1}{x}\right)^{n-1} dx. \tag{248}$$

(2) For x^n substitute x; then

$$\Gamma(n) = \frac{1}{n}\int_0^\infty e^{-x^{\frac{1}{n}}}\,dx; \tag{249}$$

so that the definite integrals given in (248) and (249) are definitions of the Gamma-function, as well as (247). Poisson investigated the properties of it in the form (247), and Legendre in the form (248).

The following is also evident, and is a theorem of great importance;

$$\int_0^\infty e^{-ax}x^{n-1}\,dx = \frac{\Gamma(n)}{a^n}; \tag{250}$$

a particular case of this equation, viz. when n is positive and integral, has been proved in Ex. 4, Art. 98.

123.] In the first place I propose to shew that $\Gamma(n)$, as defined by (247), is finite and determinate for all positive values of n. For this purpose let the definite integral be divided into the three following integrals, the sum of which is equal to it;

$$\int_0^\infty e^{-x}x^{n-1}dx = \int_0^i e^{-x}x^{n-1}dx + \int_i^{x_1} e^{-x}x^{n-1}dx + \int_{x_1}^\infty e^{-x}x^{n-1}dx, \quad (251)$$

where i is an infinitesimal, and x_1 is a finite quantity, to which a convenient value will be given. As to the first of these three integrals, by (228) we have

$$\int_0^i e^{-x}x^{n-1}dx = e^{-\theta i}\int_0^i x^{n-1}dx = e^{-\theta i}\left[\frac{x^n}{n}\right]_0^i = \frac{e^{\theta i}i^n}{n},$$

which is an infinitesimal. As to the second integral, the limits of integration are finite, and the element-function does not become infinite within the range; consequently the definite integral is finite and determinate. As to the third integral, let x_1 be that value of x for which, and for all quantities greater than which, e^{-x} is less than $x^{-(n+1)}$; so that

$$\int_{x_1}^\infty e^{-x}x^{n-1}dx < \int_{x_1}^\infty \frac{x^{n-1}}{x^{n+1}}dx$$

$$< \frac{1}{x_1};$$

which is a finite quantity; and consequently the whole definite integral denoted by $\Gamma(n)$ is finite and determinate.

As to x_1; we must shew that our hypothesis of its value is possible; we have assumed e^{-x_1} to be less than $x_1^{-(n+1)}$; and therefore, taking logarithms, $\frac{x_1}{\log x_1}$ is greater than $n+1$; and this is possible because $\frac{x}{\log x} = \infty$, when $x = 1$; $= e$, when $x = e$, and this is the minimum value; $= \infty$, when $x = \infty$.

$\Gamma(n)$ is also a positive quantity because all the values of the element-function within the range of integration are positive.

It is evidently a continuous function of n; because as n continuously varies, the values of the definite integral will also continuously vary. The continuity of the function may also be demonstrated by means of the n-differential of it.

Thus, differentiating (247) with respect to n,

$$\frac{d.\Gamma(n)}{dn} = \int_0^\infty e^{-x} \log x \, x^{n-1} \, dx ; \tag{252}$$

which is evidently determinate for all values of x within the range of integration; and consequently $\Gamma(n)$ varies continuously as n varies continuously.

124.] The following are some values of $\Gamma(n)$ corresponding to particular values of the argument.

(1) Let n be negative; then, if x_1 is a positive finite quantity,

$$\Gamma(-n) = \int_0^\infty e^{-x} x^{-n-1} \, dx = \int_0^{x_1} \frac{e^{-x} \, dx}{x^{n+1}} + \int_{x_1}^\infty \frac{e^{-x} \, dx}{x^{n+1}} . \tag{253}$$

Now, giving approximate values to these two latter integrals by means of the theorem contained in (228), if θ is a positive proper fraction, and x_2 is a value of x intermediate to x_1 and ∞, we have,

$$\begin{aligned} \Gamma(-n) &= e^{-\theta x_1} \Big[-\frac{1}{n x^n} \Big]_0^{x_1} + \frac{e^{-x_1}}{x_2^{\,n+1}} \\ &= \infty . \end{aligned} \tag{254}$$

(2) Let $n = 0$; then, employing the same symbols and theorems,

$$\begin{aligned} \Gamma(0) &= \int_0^{x_1} \frac{e^{-x} \, dx}{x} + \int_{x_1}^\infty \frac{e^{-x} \, dx}{x} \\ &= e^{-\theta x_1} \log \left(\frac{x_1}{0} \right) + \frac{e^{-x_1}}{x_2} \\ &= \infty. \end{aligned} \tag{255}$$

Thus it appears that the definition of $\Gamma(n)$ given in (247) is applicable only when n is a positive quantity. It will appear hereafter that another definition may be given of the function which will place it on a wider basis, and will not exclude all except positive values of n.

(3) Let $n = 1$; then

$$\Gamma(1) = \int_0^\infty e^{-x} \, dx = 1. \tag{256}$$

(4) Let $n = 2$; then

$$\begin{aligned} \Gamma(2) &= \int_0^\infty e^{-x} x \, dx \\ &= \Big[-x e^{-x} - e^{-x} \Big]_0^\infty \\ &= 1. \end{aligned} \tag{257}$$

Taking these values in connection with (252), we can determine the general course of the value of the Gamma-function; or in other words we can trace the curve $y = \Gamma(x)$.

Expressing the equation (252) in the following manner,

$$\frac{d.\Gamma(n)}{dn} = \int_1^\infty e^{-x} \log x \, x^{n-1} \, dx - \int_0^1 e^{-x} \log\left(\frac{1}{x}\right) x^{n-1} \, dx, \quad (258)$$

it appears that $\frac{d.\Gamma(n)}{dn}$ is equal to the difference of two definite integrals, which are necessarily positive; the former of which increases, and the latter decreases, as n increases; therefore $\frac{d^2\Gamma(n)}{dn^2}$ is positive, and $\Gamma(n)$ has a minimum value corresponding to that value of n for which $\frac{d.\Gamma(n)}{dn} = 0$. It is clear then that $\Gamma(n)$ has one minimum value; and since $\Gamma(0) = \infty$, $\Gamma(1) = 1$, $\Gamma(2) = 1$, that minimum must correspond to a value of n greater than 1 and less than 2; and the minimum value of $\Gamma(n)$ is less than 1. Also, beyond that value, $\Gamma(n)$ increases as n increases; and $\Gamma(n) = \infty$, when $n = \infty$.

125.] In (250) let n be replaced by $m+n$, and let a be replaced by $1+z$, where z is a new variable independent of x; then

$$\int_0^\infty e^{-(1+z)x} x^{m+n-1} \, dx = \frac{\Gamma(m+n)}{(1+z)^{m+n}}. \quad (259)$$

Let both members of this equation be multiplied by z^{n-1}, and let the z-integral be taken between the limits ∞ and 0; then

$$\int_0^\infty dz \int_0^\infty e^{-(1+z)x} z^{n-1} x^{m+n-1} \, dx = \Gamma(m+n) \int_0^\infty \frac{z^{n-1} \, dz}{(1+z)^{m+n}}. \quad (260)$$

Now, as x and z are independent variables, the order of the integrations may be changed; and consequently we have from the left-hand member of (260),

$$\int_0^\infty dz \int_0^\infty e^{-(1+z)x} z^{n-1} x^{m+n-1} \, dx$$

$$= \int_0^\infty e^{-x} x^{m-1} dx \int_0^\infty e^{-xz} (xz)^{n-1} \, d.xz$$

$$= \Gamma(m) \Gamma(n);$$

substituting which in (260), and replacing z by x, we have,

$$\int_0^\infty \frac{x^{n-1} \, dx}{(1+x)^{m+n}} = \frac{\Gamma(m)\Gamma(n)}{\Gamma(m+n)}. \quad (261)$$

In this process no restriction has been put on the values of m and n, except that they are positive quantities.

The integral in the left-hand member of this equation has been called by Legendre the first Eulerian Integral and is of considerable importance in its relation to the Gamma-function. It is evidently a function of two parameters m and n, and has been denoted by the symbol $\mathrm{B}(m,n)$, being called the Beta-function. So that for the definition of this function we have

$$\mathrm{B}(m,n) = \int_0^\infty \frac{x^{n-1}\,dx}{(1+x)^{m+n}} = \frac{\Gamma(m)\Gamma(n)}{\Gamma(m+n)}. \tag{262}$$

If in the definite integral x is replaced by $\frac{1}{x}$, then

$$\int_0^\infty \frac{x^{n-1}\,dx}{(1+x)^{m+n}} = \int_0^\infty \frac{x^{m-1}\,dx}{(1+x)^{m+n}};$$

$$\therefore\ \mathrm{B}(m,n) = \mathrm{B}(n,m) = \int_0^\infty \frac{x^{n-1}\,dx}{(1+x)^{m+n}} = \int_0^\infty \frac{x^{m-1}\,dx}{(1+x)^{m+n}}; \tag{263}$$

so that the value of the Beta-function is unaltered by the interchange of m and n. This theorem might also have been inferred from the symmetry with respect to m and n of the last member of (262).

As the Beta-function is a function of two variables m and n, it evidently represents a surface; and if x, y, z are the coordinates to any point on it,

$$z = \mathrm{B}(x,y) = \frac{\Gamma(x)\Gamma(y)}{\Gamma(x+y)};$$

and the general course of the surface may be traced from the previously known values of the Gamma-function.

126]. The following are other and equivalent forms of the Beta-function, being derived from (262) by transformation.

$$(1)\quad \mathrm{B}(m,n) = \int_0^\infty \frac{x^{n-1}\,dx}{(1+x)^{m+n}}$$

$$= \int_0^1 \frac{x^{n-1}\,dx}{(1+x)^{m+n}} + \int_1^\infty \frac{x^{n-1}\,dx}{(1+x)^{m+n}}.$$

In the second of these latter integrals let x be replaced by $\frac{1}{x}$;

$$\text{then}\quad \int_1^\infty \frac{x^{n-1}\,dx}{(1+x)^{m+n}} = \int_0^1 \frac{x^{m-1}\,dx}{(1+x)^{m+n}};$$

$$\therefore\ \mathrm{B}(m,n) = \int_0^1 \frac{(x^{m-1}+x^{n-1})\,dx}{(1+x)^{m+n}}; \tag{264}$$

in which m and n enter symmetrically, and consequently we have another proof of (263).

(2) Let x, in (263), be replaced by e^x-1; then

$$\text{B}(m,n) = \int_0^{\infty} (e^x-1)^{n-1} e^{-(m+n-1)x} dx \tag{265}$$

$$= \int_0^{\infty} (e^x-1)^{m-1} e^{-(m+n-1)x} dx. \tag{266}$$

(3) In (263) let x be replaced by $\frac{x}{1-x}$; then

$$\text{B}(m,n) = \int_0^1 x^{n-1}(1-x)^{m-1} dx = \int_0^1 x^{m-1}(1-x)^{n-1} dx; \tag{267}$$

which is a definite integral of great importance in the theory of Probabilities.

(4) In (267) let x be replaced by $\frac{x}{a}$; then

$$\text{B}(m,n) = \int_0^a \frac{x^{m-1}(a-x)^{n-1} dx}{a^{m+n-1}}. \tag{268}$$

$$\therefore \quad \int_0^a x^{m-1}(a-x)^{n-1} dx = a^{m+n-1}\,\text{B}(m,n) \tag{269}$$

$$= a^{m+n-1} \frac{\Gamma(m)\,\Gamma(n)}{\Gamma(m+n)}. \tag{270}$$

(5) In (267) let x be replaced by $(\sin\theta)^2$; then

$$\text{B}(m,n) = 2\int_0^{\frac{\pi}{2}} (\sin\theta)^{2n-1}(\cos\theta)^{2m-1} d\theta. \tag{271}$$

All these values of $\text{B}(m,n)$ are of course equivalents of $\frac{\Gamma(m)\Gamma(n)}{\Gamma(m+n)}$, since this is the relation which exists between the Beta- and the Gamma-functions; and since by it the Beta-function may be expressed in terms of the Gamma-function, it is unnecessary to consider separately the properties of both, so that we shall henceforth investigate the properties of only the Gamma-function.

127.] The first fundamental theorem of the Gamma-function.

By (261) and (263),

$$\frac{\Gamma(m)\Gamma(n)}{\Gamma(m+n)} = \int_0^{\infty} \frac{x^{m-1} dx}{(1+x)^{m+n}};$$

Therefore if $m = 1$, in which case $\Gamma(m) = \Gamma(1) = 1$,

$$\frac{\Gamma(n)}{\Gamma(n+1)} = \int_0^{\infty} \frac{dx}{(1+x)^{n+1}}$$

$$= \frac{1}{n}; \tag{272}$$

$$\therefore \quad \Gamma(n+1) = n\,\Gamma(n). \tag{273}$$

This is the first fundamental theorem of the Gamma-function.

The following theorems are deductions from it.

(1) Let $n-1, n-2, n-3, \ldots n-m$, be successively substituted for n in (273); then

$$\begin{aligned} \Gamma(n) &= (n-1)\Gamma(n-1) \\ &= (n-1)(n-2)\Gamma(n-2) \\ &= (n-1)(n-2)\ldots(n-m)\Gamma(n-m). \end{aligned} \tag{274}$$

And replacing n by $n+m$

$$\Gamma(n+m) = (n+m-1)(n+m-2)\ldots(n+1)\,n\,\Gamma(n): \tag{275}$$

so that $\Gamma(n+m)$ depends on $\Gamma(n)$. Consequently if the value of $\Gamma(n)$ is known for all values of n comprised between 0 and 1, or between 1 and 2, or generally between any two numbers the difference between which $= 1$, the value of the function will be known for other real values of the argument.

(2) If in (274) n is an integral as well as positive number,

$$\begin{aligned} \Gamma(n) &= (n-1)(n-2)\ldots 3.2\,\Gamma(2) \\ &= (n-1)(n-2)\ldots 3.2.1\,\Gamma(1) \\ &= (n-1)(n-2)\ldots 3.2.1, \end{aligned} \tag{276}$$

since by (256), $\Gamma(1) = 1$. So that if n is a positive integral number, $\Gamma(n)$ is the product of all integral numbers from 1 to $n-1$, both inclusive. For this reason $\Gamma(n)$ has been called the factorial function. Thus $\Gamma(n)$ is known for all positive integral numbers.

(3) If $n = 0$, then from (272),

$$\begin{aligned} \Gamma(0) &= \frac{\Gamma(1)}{0} \\ &= \infty\text{, since } \Gamma(1) = 1; \end{aligned}$$

which confirms the theorem given in (255).

Hence also we have a particular form of $B(m,n)$ which deserves notice. Let $m = n =$ a positive integer; then

$$\begin{aligned} B(n,n) &= \frac{\{\Gamma(n)\}^2}{\Gamma(2n)} \\ &= \frac{1^2.2^2.3^2.\ldots(n-1)^2}{1.2.3.\ldots(2n-1)} \\ &= \frac{1}{2^{n-1}}\,\frac{1.2.3\ldots(n-1)}{1.3.5\ldots(2n-1)}. \end{aligned}$$

The particular form of the Gamma-function given in (276) leads to another definition of it which we shall hereafter find of considerable use. From (248) we have

$$\Gamma(n) = \int_0^1 \left(\log\frac{1}{x}\right)^{n-1} dx;$$

and since $\log\left(\frac{1}{x}\right) = m(1-x^{\frac{1}{m}})$, when $m = \infty$;

$$\Gamma(n) = \int_0^1 m^{n-1}(1-x^{\frac{1}{m}})^{n-1} dx, \text{ when } m = \infty.$$

Let x be replaced by x^m: then

$$\begin{aligned}\Gamma(n) &= m^n \int_0^1 (1-x)^{n-1} x^{m-1} dx, \text{ when } m = \infty; \\ &= m^n \mathrm{B}(m, n), \\ &= m^n \frac{\Gamma(m)\Gamma(n)}{\Gamma(m+n)}, \text{ when } m = \infty, \\ &= \frac{1.2.3\ldots(m-1)}{n(n+1)\ldots(n+m-1)} m^n, \text{ when } m = \infty, \qquad (277)\end{aligned}$$

if m and n are integers. This equivalent has been taken by Gauss as the definition of the Gamma-function; and from it he has derived in his celebrated memoir* all the properties of the function.

128.] I may in passing observe that (273) may be deduced by integration by parts from the definite integral which defines $\Gamma(n)$;

thus
$$\begin{aligned}\Gamma(n+1) &= \int_0^\infty e^{-x} x^n dx \\ &= \Big[-e^{-x} x^n\Big]_0^\infty + n\int_0^\infty e^{-x} x^{n-1} dx \\ &= n\Gamma(n);\end{aligned}$$

because n being positive, the integrated part vanishes at both limits. And if n is also an integer, we shall derive by successive integration the theorem given in (276); viz.,

$$\Gamma(n) = (n-1)(n-2)\ldots 3.2.1.$$

And I may also observe that this being the case, if m and n are both integers, (262) may be proved as follows by indefinite integration;

$$\begin{aligned}&\int_0^\infty \frac{x^{n-1} dx}{(1+x)^{m+n}} \\ &= \left[-\frac{1}{m+n-1}\frac{x^{n-1}}{(1+x)^{m+n-1}}\right]_0^\infty + \frac{n-1}{m+n-1}\int_0^\infty \frac{x^{n-2} dx}{(1+x)^{m+n-1}}\end{aligned}$$

* Commentationes recentiores Societatis Scientiarum Gottingensis; Vol. I, Gottingen, 1812.

$$\int_0^\infty \frac{x^{n-1}\,dx}{(1+x)^{m+n}} = \frac{(n-1)}{m+n-1}\int_0^\infty \frac{x^{n-2}\,dx}{(1+x)^{m+n-1}}$$

$$= \frac{(n-1)(n-2)\dots 3.2.1}{(m+n-1)(m+n-2)\dots(m+1)}\int_0^\infty \frac{dx}{(1+x)^{m+1}}$$

$$= \frac{(n-1)(n-2)\dots 3.2.1}{(m+n-1)(m+n-2)\dots(m+1)m}$$

$$= \frac{(n-1)(n-2)\dots 2.1.(m-1)(m-2)\dots 2.1}{(m+n-1)(m+n-2)\dots\dots\dots\dots 2.1}$$

$$= \frac{\Gamma(m)\,\Gamma(n)}{\Gamma(m+n)}.$$

As the risk of error is great in a subject of so delicate a nature as the evaluation of definite integrals, it is expedient to verify the theorems, whenever verification is possible. Thus, although in the general theorem given in (262), m and n are not necessarily integral numbers, yet the preceding process proves the truth of the theorem when they are integers.

129.] Second fundamental theorem of the Gamma-function.

In (262), let $m+n = 1$; then

$$\Gamma(n)\,\Gamma(1-n) = \int_0^\infty \frac{x^{n-1}\,dx}{1+x}$$

$$= \frac{\pi}{\sin n\pi}, \text{ by reason of (72) Art. 94;} \qquad (278)$$

which is the second fundamental theorem of the Gamma-function. It is subject to no other condition than that n and $1-n$ are both positive numbers: so that n is a positive proper fraction.

From this theorem it follows that if the value of $\Gamma(n)$ is known for all values of n from 0 to $\frac{1}{2}$, it is also known for all values of n from $\frac{1}{2}$ to 1. And consequently from this theorem taken in connection with the first general theorem we learn that if the value of $\Gamma(n)$ is known for all values of n between two numbers whose difference is $\frac{1}{2}$, it is also known for all other values of n.

The following are deductions from the preceding.

(1) Since $n\,\Gamma(n) = \Gamma(n+1)$,

$$\Gamma(1+n)\,\Gamma(1-n) = \frac{n\pi}{\sin n\pi}. \qquad (279)$$

(2) Let $n = \frac{1}{2}$; then from (278),

$$\Gamma\left(\frac{1}{2}\right) = \int_0^\infty e^{-x}\frac{dx}{x^{\frac{1}{2}}} = \pi^{\frac{1}{2}}.$$

In this equation let x be replaced by x^2; then

$$\Gamma\left(\frac{1}{2}\right) = 2\int_0^\infty e^{-x^2}dx = \int_{-\infty}^\infty e^{-x^2}dx = \pi^{\frac{1}{2}}; \qquad (280)$$

which value has been already found in (114), Art. 100.

This may be determined from (262) without the intervention of (278). Thus in (262) let $m = n = \frac{1}{2}$; then

$$\left\{\Gamma\left(\frac{1}{2}\right)\right\}^2 = \int_0^\infty \frac{dx}{x^{\frac{1}{2}}(1+x)}$$

$$= 2\int_0^\infty \frac{dx}{1+x^2}, \text{ if } x \text{ is replaced by } x^2,$$

$$= \pi;$$

$$\therefore \quad \Gamma\left(\frac{1}{2}\right) = \pi^{\frac{1}{2}}.$$

(3) In (278) let n be successively replaced by

$$\frac{1}{n}, \frac{2}{n}, \frac{3}{n}, \ldots \frac{n-2}{n}, \frac{n-1}{n};$$

then we have

$$\Gamma\left(\frac{1}{n}\right)\Gamma\left(\frac{n-1}{n}\right) = \frac{\pi}{\sin\frac{\pi}{n}},$$

$$\Gamma\left(\frac{2}{n}\right)\Gamma\left(\frac{n-2}{n}\right) = \frac{\pi}{\sin\frac{2\pi}{n}},$$

$$\cdot \quad \cdot \quad \cdot \quad \cdot \quad \cdot \quad \cdot \quad \cdot \quad \cdot \quad \cdot$$

$$\Gamma\left(\frac{n-1}{n}\right)\Gamma\left(\frac{1}{n}\right) = \frac{\pi}{\sin\frac{(n-1)\pi}{n}};$$

then taking the products of all the right-hand members and of all the left-hand members separately, we have

$$\left\{\Gamma\left(\frac{1}{n}\right)\Gamma\left(\frac{2}{n}\right)\ldots\Gamma\left(\frac{n-1}{n}\right)\right\}^2 = \frac{\pi^{n-1}}{\sin\frac{\pi}{n}\sin\frac{2\pi}{n}\ldots\sin\frac{(n-1)\pi}{n}}. \qquad (281)$$

Now, by (52), Art. 64, Vol. I., substituting $2n$ for n, we have

$$\frac{x^{2n}-1}{x^2-1} = \left(x^2-2x\cos\frac{\pi}{n}+1\right)\left(x^2-2x\cos\frac{2\pi}{n}+1\right)\dots$$
$$\dots\left(x^2-2x\cos\frac{(n-1)\pi}{n}+1\right). \quad (282)$$

In this equation let $+1$ and -1 be successively substituted for x; then

$$n = 2^{n-1}\left(1-\cos\frac{\pi}{n}\right)\left(1-\cos\frac{2\pi}{n}\right)\left(1-\cos\frac{3\pi}{n}\right)\dots$$
$$\dots\left(1-\cos\frac{(n-1)\pi}{n}\right); \quad (283)$$

$$n = 2^{n-1}\left(1+\cos\frac{\pi}{n}\right)\left(1+\cos\frac{2\pi}{n}\right)\left(1+\cos\frac{3\pi}{n}\right)\dots$$
$$\dots\left(1+\cos\frac{(n-1)\pi}{n}\right); \quad (284)$$

therefore, taking the product of these two and extracting the square root,

$$n = 2^{n-1}\sin\frac{\pi}{n}\sin\frac{2\pi}{n}\sin\frac{3\pi}{n}\dots\sin\frac{(n-1)\pi}{n}; \quad (285)$$

so that substituting in the denominator of (281), and extracting the square root,

$$\Gamma\left(\frac{1}{n}\right)\Gamma\left(\frac{2}{n}\right)\dots\Gamma\left(\frac{n-1}{n}\right) = \frac{(2\pi)^{\frac{n-1}{2}}}{n^{\frac{1}{2}}}. \quad (286)$$

130.] The form of the preceding equation suggests the means of determining the value of $\int_n^{n+1}\log\Gamma(x)\,dx$.

Let us first consider $\int_0^1\log\Gamma(x)\,dx$, and suppose the range of integration to be $1-i$, where $i = \frac{1}{n}$ and is an infinitesimal; so that the value of $\log\Gamma(x)$ at the inferior limit, being ∞, may be excluded; then, if $n = \infty$,

$$\int_i^1\log\Gamma(x)\,dx = \left\{\log\Gamma\left(\frac{1}{n}\right)+\log\Gamma\left(\frac{2}{n}\right)+\log\Gamma\left(\frac{3}{n}\right)+\dots\right.$$
$$\left.\dots+\log\Gamma\left(\frac{n-1}{n}\right)\right\}\frac{1}{n}$$
$$= \left\{\frac{n-1}{2}\log 2\pi - \frac{1}{2}\log n\right\}\frac{1}{n}$$
$$= \frac{1}{2}\log 2\pi, \text{ when } n = \infty.$$
$$\therefore \int_0^1\log\Gamma(x)\,dx = \frac{1}{2}\log 2\pi. \quad (287)$$

Now $\int_0^{n+1} \log \Gamma(x)\,dx = \int_0^1 \log \Gamma(x)\,dx + \int_1^{n+1} \log \Gamma(x)\,dx$

$$= \frac{1}{2}\log 2\pi + \int_0^n \log \Gamma(x+1)\,dx$$

$$= \frac{1}{2}\log 2\pi + \int_0^n \log x\,dx + \int_0^n \log \Gamma(x)\,dx.$$

$$\therefore \quad \int_n^{n+1} \log \Gamma(x)\,dx = \frac{1}{2}\log 2\pi + n(\log n - 1). \tag{288}$$

131.] As a primary object of this research however is the determination of the value of $\Gamma(n)$ for all positive values of n, we must investigate other general properties of the function, so as to supply the deficiencies of the two former general theorems. They are sufficient for the determination of values of $\Gamma(n)$, when the values are known for all numbers lying between two numbers, the difference between which is $\frac{1}{2}$; but they fail to give the latter values. In the course of the inquiry too we shall be led to some new and important theorems of the Gamma-function, and to another proof of Gauss's definition of it.

Taking the n-differential of the Gamma-function, as in (252), we have

$$\frac{d.\Gamma(n)}{dn} = \int_0^\infty e^{-x} x^{n-1} \log x\,dx. \tag{289}$$

Now by an artifice due, I believe, to M. Cauchy*, which is of great use in the evaluation of definite integrals, we may replace $\log x$ by any definite integral which is equivalent to it; and if the limits of this equivalent are constant, the order of the integrations is arbitrary; and they may consequently be taken in that which is most convenient to the problem.

Thus, since

$$\frac{1}{x} = \int_0^\infty e^{-xz}\,dz;$$

$$\int_1^x \frac{dx}{x} = \int_0^\infty dz \int_1^x e^{-xz}\,dx,$$

$$\therefore \quad \log x = \int_0^\infty \frac{e^{-z} - e^{-xz}}{z}\,dz; \tag{290}$$

replacing $\log x$ in (289) by this value, we have

$$\frac{d.\Gamma(n)}{dn} = \int_0^\infty e^{-x} x^{n-1}\,dx \int_0^\infty \frac{e^{-z} - e^{-xz}}{z}\,dz$$

$$= \int_0^\infty \frac{dz}{z}\left\{e^{-z}\int_0^\infty e^{-x} x^{n-1}\,dx - \int_0^\infty e^{-(1+z)x} x^{n-1}\,dx\right\}. \tag{291}$$

* Exercises d'Analyse, tome II, p. 379. Paris, 1841.

But $\int_0^\infty e^{-x} x^{n-1} dx = \Gamma(n)$, by the definition of the Gamma-function; and by (250), $\int_0^\infty e^{-(1+z)x} x^{n-1} dx = \frac{\Gamma(n)}{(1+z)^n}$; so that substituting these in (291), and placing $\Gamma(n)$ outside the sign of integration, we have

$$\frac{d.\Gamma(n)}{dn} = \Gamma(n) \int_0^\infty \left\{ e^{-z} - \frac{1}{(1+z)^n} \right\} \frac{dz}{z}; \tag{292}$$

and dividing both sides by $\Gamma(n)$,

$$\frac{d.\log\Gamma(n)}{dn} = \int_0^\infty \left\{ e^{-z} - \frac{1}{(1+z)^n} \right\} \frac{dz}{z}. \tag{293}$$

Let the n-integral of this equation be taken for the limits n and 1; then, bearing in mind that $\Gamma(1)=1$, so that $\log\Gamma(1)=0$,

$$\begin{aligned} \log\Gamma(n) &= \int_1^n dn \int_0^\infty \left\{ e^{-z} - \frac{1}{(1+z)^n} \right\} \frac{dz}{z} \\ &= \int_0^\infty \left\{ (n-1)e^{-z} + \frac{(1+z)^{-n} - (1+z)^{-1}}{\log(1+z)} \right\} \frac{dz}{z}. \end{aligned} \tag{294}$$

This equation may be simplified. For, if $n = 2$, $\Gamma(2) = 1$, and $\log\Gamma(2) = 0$: accordingly

$$0 = \int_0^\infty \left\{ e^{-z} + \frac{(1+z)^{-2} - (1+z)^{-1}}{\log(1+z)} \right\} \frac{dz}{z};$$

$\therefore \quad \log\Gamma(n)$

$$= \int_0^\infty \{(1+z)^{-n} - (1+z)^{-1} + (n-1)z(1+z)^{-2}\} \frac{dz}{z\log(1+z)}. \tag{295}$$

To simplify this; let $1+z = e^y$; then

$$\log\Gamma(n) = \int_0^\infty \left\{ (n-1)e^{-y} - \frac{e^{-y} - e^{-ny}}{1-e^{-y}} \right\} \frac{dy}{y}. \tag{296}$$

Hence, taking the n-differential,

$$\frac{d.\log\Gamma(n)}{dn} = \int_0^\infty \left\{ \frac{e^{-y}}{y} - \frac{e^{-ny}}{1-e^{-y}} \right\} dy. \tag{297}$$

132.] These equations lead to another fundamental theorem of the Gamma-function, which was discovered by Gauss, and is given by him in his previously cited memoir. The theorem is more commonly called the third fundamental theorem of the Gamma-function; and is useful in reducing the number of particular values of the function which must be determined by direct calculation.

Let the value of (293), when $n = 1$, be $-\text{E}$; so that

$$\frac{d.\log \Gamma(n)}{dn} \text{ (when } n=1) = \int_0^\infty \left\{ e^{-z} - \frac{1}{1+z} \right\} \frac{dz}{z} = -\text{E}; \qquad (298)$$

the quantity being negative, because, when $n = 1$, $\Gamma(n)$ decreases as n increases; then

$$\frac{d.\log \Gamma(n)}{dn} = -\text{E} + \int_0^\infty \left\{ \frac{1}{1+z} - \frac{1}{(1+z)^n} \right\} \frac{dz}{z}. \qquad (299)$$

Let $1+z$ be replaced by $\frac{1}{z}$; then

$$\frac{d.\log \Gamma(n)}{dn} = -\text{E} + \int_0^1 \frac{1-z^{n-1}}{1-z} dz. \qquad (300)$$

Let n be successively replaced by $n+\frac{1}{r}$, $n+\frac{2}{r}$, ... $n+\frac{r-1}{r}$; and let the sum of (300) and of all these several terms be taken; then

$$\frac{d}{dn} \log \left\{ \Gamma(n) \Gamma\left(n+\frac{1}{r}\right) \Gamma\left(n+\frac{2}{r}\right) \dots \Gamma\left(n+\frac{r-1}{r}\right) \right\}$$

$$= -r\text{E} + \int_0^1 \left\{ r - z^{n-1} \left(1 + z^{\frac{1}{r}} + z^{\frac{2}{r}} + \dots + z^{\frac{r-1}{r}} \right) \right\} \frac{dz}{1-z}$$

$$= -r\text{E} + \int_0^1 \left\{ \frac{r}{1-z} - \frac{z^{n-1}}{1-z^{\frac{1}{r}}} \right\} dz$$

$$= -r\text{E} + r \int_0^1 \left\{ \frac{r z^{r-1}}{1-z^r} - \frac{z^{rn-1}}{1-z} \right\} dz, \qquad (301)$$

if z is replaced by z^r.

Also, let n be replaced by rn in (300); then

$$\frac{d.\log \Gamma(rn)}{dn} = -r\text{E} + r \int_0^1 \frac{1-z^{rn-1}}{1-z} dz; \qquad (302)$$

and subtracting this from (301), we have

$$\frac{d}{dn} \log \frac{\Gamma(n) \Gamma\left(n+\frac{1}{r}\right) \dots \Gamma\left(n+\frac{r-1}{r}\right)}{\Gamma(nr)} = r \int_0^1 \left\{ \frac{r z^{r-1}}{1-z^r} - \frac{1}{1-z} \right\} dz$$

$$= r \left[\log \frac{1-z}{1-z^r} \right]_0^1$$

$$= -r \log r. \qquad (303)$$

Let the definite integral of this equation be taken for the limits n and $\frac{1}{r}$; then, since by (286),

$$\Gamma\left(\frac{1}{r}\right) \Gamma\left(\frac{2}{r}\right) \Gamma\left(\frac{3}{r}\right) \dots \Gamma\left(\frac{r-1}{r}\right) = \frac{(2\pi)^{\frac{r-1}{2}}}{r^{\frac{1}{2}}},$$

$$\Gamma(n)\Gamma\left(n+\frac{1}{r}\right)\Gamma\left(n+\frac{2}{r}\right)\dots\Gamma\left(n+\frac{r-1}{r}\right)$$
$$= (2\pi)^{\frac{r-1}{2}} r^{\frac{1}{2}-rn}\,\Gamma(rn); \quad (304)$$

which is the third fundamental theorem of the Gamma-function.

The following are particular forms of this general theorem.

Let $r = 2$; $$\Gamma(n)\Gamma\left(n+\frac{1}{2}\right) = \frac{\pi^{\frac{1}{2}}}{2^{2n-1}}\Gamma(2n); \quad (305)$$

let $r=3$; $$\Gamma(n)\Gamma\left(n+\frac{1}{3}\right)\Gamma\left(n+\frac{2}{3}\right) = \frac{2\pi}{3^{3n-\frac{1}{2}}}\Gamma(3n); \quad (306)$$

consequently, if $n = \frac{1}{2}$, from (305) we have $\Gamma\left(\frac{1}{2}\right) = \pi^{\frac{1}{2}}$, as we have determined several times heretofore.

We have said that the theorem is useful for the reduction of the number of values of $\Gamma(n)$, which must be found by direct calculation. In illustration of this, suppose that the values of $\Gamma(n)$ are required for all arguments at an interval of $\frac{1}{6}$th between two successive integers. By the first general theorem these values depend on those of $\Gamma(n)$ at an interval of $\frac{1}{6}$th between 0 and 1; that is, on $\Gamma\left(\frac{1}{6}\right)$, $\Gamma\left(\frac{2}{6}\right)$, $\Gamma\left(\frac{3}{6}\right)$, $\Gamma\left(\frac{4}{6}\right)$, $\Gamma\left(\frac{5}{6}\right)$. Of these the third, viz. $\Gamma\left(\frac{3}{6}\right)$, $= \Gamma\left(\frac{1}{2}\right) = \pi^{\frac{1}{2}}$, and is known. Also, by the second general theorem, we have

$$\Gamma\left(\frac{1}{6}\right)\Gamma\left(\frac{5}{6}\right) = \frac{\pi}{\sin\frac{\pi}{6}} = 2\pi, \quad (307)$$

$$\Gamma\left(\frac{2}{6}\right)\Gamma\left(\frac{4}{6}\right) = \frac{\pi}{\sin\frac{\pi}{3}} = \frac{2\pi}{3^{\frac{1}{2}}}. \quad (308)$$

And by (305), if $n = \frac{1}{6}$, $$\Gamma\left(\frac{1}{6}\right)\Gamma\left(\frac{4}{6}\right) = 2^{\frac{2}{3}}\pi^{\frac{1}{2}}\Gamma\left(\frac{2}{6}\right). \quad (309)$$

Consequently if $\Gamma\left(\frac{1}{6}\right)$ is known, all the others are known. In other similar cases the number of the values of the Gamma-function, which must be determined by direct arithmetical summation, may be reduced by means of the theorem contained in (304).

133.] We now come to the investigation of series for the direct calculation of $\log \Gamma(n)$. Returning to (297) we have,

$$\frac{d.\log \Gamma(n)}{dn} = \int_0^\infty \left\{ \frac{e^{-y}}{y} - \frac{e^{-ny}}{1-e^{-y}} \right\} dy. \quad (310)$$

$$\therefore \frac{d^2 \log \Gamma(n)}{dn^2} = \int_0^\infty \frac{y e^{-ny}}{1-e^{-y}} dy$$

$$= \int_0^\infty y e^{-ny} \{1 + e^{-y} + e^{-2y} + e^{-3y} + \ldots\} dy$$

$$= \int_0^\infty y \{e^{-ny} + e^{-(n+1)y} + e^{-(n+2)y} + \ldots\} dy$$

$$= \frac{1}{n^2} + \frac{1}{(n+1)^2} + \frac{1}{(n+2)^2} + \frac{1}{(n+3)^2} + \ldots; \quad (311)$$

and this is evidently a convergent series for all values of n, except negative integers.

Let us take the n-integral of (311), between the limits n and 1; then as from (297) and (298) we have

$$\frac{d.\log \Gamma(n)}{dn} (\text{when } n = 1) = \int_0^\infty \left\{ \frac{e^{-y}}{y} - \frac{e^{-y}}{1-e^{-y}} \right\} dy = -\mathrm{E}, \quad (312)$$

therefore

$$\frac{d.\log \Gamma(n)}{dn} + \mathrm{E} = \left(1 - \frac{1}{n}\right) + \left(\frac{1}{2} - \frac{1}{n+1}\right) + \left(\frac{1}{3} - \frac{1}{n+2}\right) + \ldots; \quad (313)$$

the right-hand member being a convergent series. E is called Euler's Constant, and is of greater importance in analytical investigations than any other constant except those denoted by π and e; the analytical value of it is given in (312); the numerical value will be determined hereafter.

Let us again take the n-integral of (313) between the limits n and 1; then, as $\Gamma(n) = 1$, when $n = 1$, and $\log \Gamma(n) = 0$,

$$\log \Gamma(n) = -\mathrm{E}(n-1) + \left(\frac{n-1}{1} - \log \frac{n}{1}\right) + \left(\frac{n-1}{2} - \log \frac{n+1}{2}\right)$$
$$+ \left(\frac{n-1}{3} - \log \frac{n+2}{3}\right) + \ldots. \quad (314)$$

In this series let $n = 2$; then, since $\log \Gamma(2) = \log 1 = 0$,

$$\mathrm{E} = \left(\frac{1}{1} - \log \frac{2}{1}\right) + \left(\frac{1}{2} - \log \frac{3}{2}\right) + \left(\frac{1}{3} - \log \frac{4}{3}\right) + \ldots; \quad (315)$$

multiplying which by $(n-1)$, and subtracting from (314), we have

$$\log\Gamma(n) = (n-1)\log\frac{2}{1} - \log\frac{n}{1} + (n-1)\log\frac{3}{2} - \log\frac{n+1}{2} + \ldots\ldots$$
$$+ (n-1)\log\frac{m+1}{m} - \log\frac{n+m-1}{m} + \ldots. \quad (316)$$

Let the sum of all the terms after the mth in this series $= \log(1+i)$; where $i = 0$, when $m = \infty$; then

$$\log\Gamma(n) = (n-1)\log\frac{2}{1} - \log\frac{n}{1} + (n-1)\log\frac{3}{2} - \log\frac{n+1}{2} + \ldots$$
$$+ (n-1)\log\frac{m+1}{m} - \log\frac{n+m-1}{m} + \log(1+i); \quad (317)$$

$$\Gamma(n) = \frac{1.2.3\ldots m\,(m+1)^{n-1}}{n(n+1)(n+2)\ldots(n+m-1)}(1+i), \text{ when } m = \infty, i = 0;$$
$$= \frac{1.2.3\ldots(m-1)}{n(n+1)(n+2)\ldots(n+m-1)}m^n, \text{ when } m = \infty. \quad (318)$$

This is Gauss' definition of $\Gamma(n)$, and is the equivalent which has already been proved in Art. 127.

From (315) we have the following value of E;

$$E = \frac{1}{1} + \frac{1}{2} + \frac{1}{3} + \ldots + \frac{1}{m} - \log m, \text{ when } m = \infty. \quad (319)$$

The numerical value of E may be determined by this series; but the calculation is very long and laborious.

134.] The calculation of $\Gamma(n)$.

In (311) let $n+1$ be substituted for n; then

$$\frac{d^2\log\Gamma(n+1)}{dn^2} = \frac{1}{(n+1)^2} + \frac{1}{(n+2)^2} + \frac{1}{(n+3)^2} + \ldots;$$

$$\therefore \quad \frac{d^r\log\Gamma(n+1)}{dn^r}$$
$$= (-)^r\,1.2.3\ldots(r-1)\left\{\frac{1}{(n+1)^r} + \frac{1}{(n+2)^r} + \frac{1}{(n+3)^r} + \ldots\right\}; \quad (320)$$

Also for the sake of conciseness let

$$S_r = \frac{1}{1^r} + \frac{1}{2^r} + \frac{1}{3^r} + \ldots. \quad (321)$$

Now applying Maclaurin's Theorem to $\log\Gamma(1+n) = F(n)$, say; $F(0) = 0$, since $\Gamma(1) = 1$; $F'(n) = \dfrac{d.\log\Gamma(1+n)}{dn}$; therefore $F'(0) = -E$, by (312); and $F''(0), \ldots$ are given by (320), so that

$$\log\Gamma(1+n) = -En + S_2\frac{n^2}{2} - S_3\frac{n^3}{3} + S_4\frac{n^4}{4} - \ldots; \quad (322)$$

this series however is not sufficiently convergent for purposes of calculation.

Since however $\log(1+n) = n - \frac{n^2}{2} + \frac{n^3}{3} - \frac{n^4}{4} + \ldots;$ therefore by addition,

$$\log \Gamma(1+n)$$
$$= -\log(1+n) + (1-E)\frac{n}{1} + (S_2-1)\frac{n^2}{2} - (S_3-1)\frac{n^3}{3} + \ldots, \quad (323)$$

which is a series sufficiently convergent. Also another, more convergent, may be deduced from it.

Replace n by $-n$; then $\log \Gamma(1-n)$

$$= -\log(1-n) - (1-E)\frac{n}{1} + (S_2-1)\frac{n^2}{2} + (S_3-1)\frac{n^3}{3} + \ldots; \quad (324)$$

and subtracting this from (323), we have

$$\log \Gamma(1+n) - \log \Gamma(1-n)$$
$$= \log\frac{1-n}{1+n} + 2\left\{(1-E)\frac{n}{1} - (S_3-1)\frac{n^3}{3} - (S_5-1)\frac{n^5}{5} - \ldots\right\}. \quad (325)$$

Now, by (279), $\log \Gamma(1+n) + \log \Gamma(1-n) = \log\frac{n\pi}{\sin n\pi}$; therefore, adding this to (325), and dividing the result by 2, we have,

$$\log \Gamma(1+n) = \frac{1}{2}\log\frac{n\pi}{\sin n\pi} + \frac{1}{2}\log\frac{1-n}{1+n} + (1-E)\frac{n}{1}$$
$$- (S_3-1)\frac{n^3}{3} - (S_5-1)\frac{n^5}{5} - \ldots. \quad (326)$$

Since now it is only necessary, as before observed, to calculate $\Gamma(n)$ for all values of n between two numbers whose difference is $\frac{1}{2}$, it will be sufficient to apply the preceding formula for all values of n between 0 and $\frac{1}{2}$; or by reason of the third fundamental theorem, between 0 and a quantity less than $\frac{1}{2}$; thus, as n is a small quantity, (326) will give the required value with great facility. These are also convenient numbers, because they give those values of $\Gamma(n)$ near to which it has its minimum value.

It is necessary however to determine the value of E.

Let $n = 1$ in (326); then $\log \Gamma(1+n) = \log \Gamma(2) = 0$; and $\frac{1}{2}\log\frac{n\pi(1-n)}{(1+n)\sin n\pi} = \frac{1}{2}\log\frac{1}{2}$, when $n = 1$; so that

$$1-\text{E} = \frac{1}{2}\log 2 + \frac{1}{3}(\text{S}_3-1) + \frac{1}{5}(\text{S}_5-1) + \frac{1}{7}(\text{S}_7-1) + \dots; \quad (327)$$

whence it is found that $\text{E} = .57721566 \dots\dots$; (328)
which is the arithmetical value of Euler's Constant*.

135.] We may also hereby determine the value of n for which $\Gamma(1+n)$ is a minimum, and thus complete the remarks as to the course of the function made in Art. 124.

When $\Gamma(1+n)$ is a minimum, $\frac{d.\Gamma(1+n)}{dn} = 0$; and consequently $\frac{d.\log\Gamma(1+n)}{dn} = 0$. Now taking the n-differential of (323), we have

$$0 = -\frac{1}{1+n} + 1 - \text{E} + (\text{S}_2-1)n - (\text{S}_3-1)n^2 + (\text{S}_4-1)n^3 \dots;$$

and if the coefficients are calculated, it will be found by a process of approximation that n lies between .4 and .5; and that

$$1+n = 1.461632\dots\dots. \quad (329)$$

We can also hence deduce an expression for the sum of a series of terms in harmonic progression: for, from (300) we have,

$$\frac{d.\log\Gamma(n)}{dn} + \text{E} = \int_0^1 \frac{1-z^{n-1}}{1-z}\,dz$$

$$= \int_0^1 \{1+z+z^2+\dots+z^{n-2}\}\,dz, \text{ if } n \text{ is an integer;}$$

$$= \frac{1}{1} + \frac{1}{2} + \frac{1}{3} + \dots + \frac{1}{n-1}; \quad (330)$$

which gives an analytical expression for the sum of the harmonic series. $\frac{d.\log\Gamma(n)}{dn}$ may of course be replaced by either of its values given in Art. 131.

136.] If we accept Gauss' definition of the Gamma-function, given in (277), the fundamental theorems may be deduced from it; thus we have

$$\Gamma(n) = \frac{1.2.3\dots(m-1)}{n(n+1)(n+2)\dots(n+m-1)}m^n, \text{ when } m = \infty;$$

$$\therefore \Gamma(n+1) = \frac{1.2.3\dots(m-1)}{(n+1)(n+2)\dots(n+m)}m^{n+1}, \text{ when } m=\infty; \quad (331)$$

* On certain discrepancies existing in the calculated values of E, see an Article by Oettinger, Crelle's Journal, Vol. LX. p. 375.

$$\therefore \quad \frac{\Gamma(n+1)}{\Gamma(n)} = \frac{nm}{n+m} = n, \text{ when } m = \infty;$$

$$\therefore \quad \Gamma(n+1) = n\Gamma(n); \tag{332}$$

which is the first fundamental theorem of the Gamma-function.

Again, replacing n by $-n$ in (331),

$$\Gamma(1-n) = \frac{1.2.3\ldots(m-1)}{(1-n)(2-n)\ldots(m-n)} m^{1-n};$$

therefore

$$\Gamma(1+n)\Gamma(1-n) = \frac{(1.2.3\ldots m)^2}{(1^2-n^2)(2^2-n^2)\ldots(m^2-n^2)}, \text{ when } m = \infty,$$

$$= \frac{1}{\left(1-\frac{n^2}{1^2}\right)\left(1-\frac{n^2}{2^2}\right)\left(1-\frac{n^2}{3^2}\right)\ldots\left(1-\frac{n^2}{m^2}\right)};$$

$$= \frac{\pi n}{\sin \pi n}, \text{ by (135), Art. 89, Vol. I}; \tag{333}$$

which is the second fundamental theorem of the Gamma-function. The third theorem may be proved by a similar process.

137.] A remarkable value of $\Gamma(n+1)$, when $n = \infty$, may also be deduced from Wallis' value of π, which is given in (15), Art. 82.

Since $\Gamma(n+1) = 1.2.3\ldots n$,

$$\frac{\Gamma(n+1)}{n^n} = \frac{1}{n}\frac{2}{n}\frac{3}{n}\ldots\frac{n-1}{n}\frac{n}{n}, \tag{334}$$

which is a small fraction of 1, and evidently is infinitesimal when n is infinite; let this quantity $= f(n)$; so that

$$1.2.3\ldots n = n^n f(n); \tag{335}$$

where $f(n)$ is an infinitesimal, when $n = \infty$. Now by Wallis' Theorem, when n is an integer,

$$\frac{\pi}{2} = \frac{2^2.4^2.6^2\ldots(2n-2)^2 2n}{1.3^2.5^2.\ldots(2n-1)^2}, \text{ when } n = \infty;$$

$$\therefore \quad (n\pi)^{\frac{1}{2}} = \frac{2.4.6\ldots(2n-2)2n}{1.3.5\ldots(2n-1)},$$

$$= \frac{2^2.4^2.6^2\ldots(2n-2)^2(2n)^2}{1.2.3.4\ldots(2n-1)2n},$$

$$= \frac{\{1.2.3\ldots(n-1)n\}^2 2^{2n}}{1.2.3\ldots(2n-1)2n},$$

$$= \frac{n^{2n}\{f(n)\}^2 2^{2n}}{(2n)^{2n} f(2n)}.$$

$$\therefore \quad \frac{f(2n)}{(2\pi \times 2n)^{\frac{1}{2}}} = \frac{\{f(n)\}^2}{2\pi n};$$

which is a functional equation of the form, $\phi(2n) = \{\phi(n)\}^2$; and of which evidently a solution is $\phi(n) = e^{an}$, where a is an undetermined constant; so that $\frac{f(n)}{(2\pi n)^{\frac{1}{2}}} = e^{an}$, and consequently (335) becomes

$$1.2.3 \ldots n = n^n (2\pi n)^{\frac{1}{2}} e^{an}, \text{ when } n = \infty.$$

Now to determine a; replace n by $n+1$; then

$$1.2.3 \ldots n(n+1) = (n+1)^{n+1} \{2\pi(n+1)\}^{\frac{1}{2}} e^{a(n+1)}, \text{ when } n = \infty;$$

and dividing the latter by the former

$$(n+1) = (n+1)\left(\frac{n+1}{n}\right)^{n+\frac{1}{2}} e^a;$$

$$\therefore \quad n^{n+\frac{1}{2}} = (n+1)^{n+\frac{1}{2}} e^a;$$

and, taking logarithms, $\left(n+\frac{1}{2}\right)\log\left(1+\frac{1}{n}\right) = -a$; whence, by evaluation, when $n = \infty$, $a = -1$; and thus,

$$1.2.3 \ldots n = (2\pi)^{\frac{1}{2}} n^{n+\frac{1}{2}} e^{-n}, \text{ when } n = \infty. \qquad (336)$$

This result enables us to correct the error of the approximate formula (208), Art. 113, and to give a more general theorem.

Let us suppose c to be the ratio of the first to the second member of that formula, so that c nearly $= 1$; then

$$m(m+1)(m+2)\ldots(m+n-1) = \mathrm{c}\, m^{-m+\frac{1}{2}} e^m n^{m+n-\frac{1}{2}} e^{-n}, \qquad (337)$$

when $n = \infty$.

Let $m = 1$; then $1.2.3 \ldots n = \mathrm{c}\, e^{1-n} n^{n+\frac{1}{2}}$, when $n = \infty$; and equating this value to that given in (336), $\mathrm{c} = \frac{(2\pi)^{\frac{1}{2}}}{e}$; hence from (337),

$$m(m+1)(m+2)\ldots(m+n-1) = (2\pi)^{\frac{1}{2}}\left(\frac{n}{m}\right)^{m-\frac{1}{2}} n^n e^{m-n-1}, \qquad (338)$$

when $n = \infty$.

Also (338) may be further simplified; since

$$\Gamma(m) = \frac{1.2.3\ldots n}{m(m+1)\ldots(m+n-1)} n^{m-1}, \text{ when } n = \infty;$$

$$m(m+1)(m+2)\ldots(m+n-1) = \frac{1.2.3\ldots n}{\Gamma(m)} n^{m-1}, \text{ - - - - };$$

$$= \frac{(2\pi)^{\frac{1}{2}}}{\Gamma(m)} n^{m+n-\frac{1}{2}} e^{-n}, \text{ - - - - }; \qquad (339)$$

which is a general formula, of which (336) is the particular case, when $m = 1$. Also in (339), let $m = \frac{1}{2}$: then, since $\Gamma\left(\frac{1}{2}\right) = \pi^{\frac{1}{2}}$,

$$1.3.5\dots(2n-1) = 2^{n+\frac{1}{2}} n^n e^{-n}, \text{ when } n = \infty. \qquad (340)$$

138.] Many other curious and important properties of the Gamma-function may be deduced from the preceding equations, and will be found in monographs on Eulerian Integrals, and in treatises where these properties are specially investigated. They scarcely fall within the narrower scope to which this treatise is limited: and I must refer the student to other works for them. I purpose to conclude this section with the application of the Gamma-function to the evaluation of some important definite integrals, and to the theory of series.

Let us first consider the more general form of the Gamma-function; viz. $\int_0^\infty e^{-ax^m} x^{n-1}\, dx$.

In this integral let $-ax^m$ be replaced by $-x$; then

$$\int_0^\infty e^{-ax^m} x^{n-1}\, dx = \frac{1}{ma^{\frac{n}{m}}} \int_0^\infty e^{-x} x^{\frac{n}{m}-1} = \frac{\Gamma\left(\frac{n}{m}\right)}{ma^{\frac{n}{m}}}. \qquad (341)$$

The following are particular cases of this result

$$\int_0^\infty e^{-a^2x^2} x^{n-1}\, dx = \frac{1}{2a^n}\Gamma\left(\frac{n}{2}\right).$$

$$\int_0^\infty e^{-a^2x^2}\, dx = \frac{1}{2a}\Gamma\left(\frac{1}{2}\right) = \frac{\pi^{\frac{1}{2}}}{2a}.$$

and as the value of this last integral is not changed when x is replaced by $-x$,

$$\int_{-\infty}^\infty e^{-a^2x^2}\, dx = 2\int_0^\infty e^{-a^2x^2}\, dx = \frac{\pi^{\frac{1}{2}}}{a}. \qquad (342)$$

Hence we have the following integrals;

$$\int_0^\infty e^{-a^2x^2}(e^{2bx}+e^{-2bx})\, dx = \frac{1}{2}\int_{-\infty}^\infty e^{-a^2x^2}(e^{2bx}+e^{-2bx})\, dx;$$

but
$$\int_{-\infty}^\infty e^{-a^2x^2+2bx}\, dx = e^{\frac{b^2}{a^2}}\int_{-\infty}^\infty e^{-\left(ax-\frac{b}{a}\right)^2} dx$$
$$= e^{\frac{b^2}{a^2}} \frac{\pi^{\frac{1}{2}}}{a};$$

similarly,
$$\int_{-\infty}^\infty e^{-a^2x^2-2bx} dx = e^{\frac{b^2}{a^2}} \frac{\pi^{\frac{1}{2}}}{a};$$

therefore, by addition,

$$\int_{-\infty}^{\infty} e^{-a^2x^2}(e^{2bx}+e^{-2bx})\,dx = \frac{2\pi^{\frac{1}{2}}}{a} e^{\frac{b^2}{a^2}};$$

$$\therefore \int_0^{\infty} e^{-a^2x^2}(e^{2bx}+e^{-2bx})\,dx = \frac{\pi^{\frac{1}{2}}}{a} e^{\frac{b^2}{a^2}}. \qquad (343)$$

Also by subtraction,

$$\int_{-\infty}^{\infty} e^{-a^2x^2}(e^{2bx}-e^{-2bx})\,dx = 0. \qquad (344)$$

If in these integrals b is replaced by $b\sqrt{-1}$, we have

$$\left.\begin{aligned} \int_0^{\infty} e^{-a^2x^2}\cos 2bx\,dx &= \frac{\pi^{\frac{1}{2}}}{2a} e^{-\frac{b^2}{a^2}}; \\ \int_0^{\infty} e^{-a^2x^2}\sin 2bx\,dx &= 0. \end{aligned}\right\} \qquad (345)$$

Next let us take the more general form of the Beta-function viz. $\int_0^{\infty} \frac{x^{n-1}\,dx}{(1+ax^r)^{m+n}}$.

Let ax^r be replaced by x; then

$$\int_0^{\infty} \frac{x^{n-1}\,dx}{(1+ax^r)^{m+n}} = \frac{1}{ra^{\frac{n}{r}}} \int_0^{\infty} \frac{x^{\frac{n}{r}-1}\,dx}{(1+x)^{m+n}}$$

$$= \frac{1}{ra^{\frac{n}{r}}} \mathrm{B}\left(\frac{n}{r}, m+n-\frac{n}{r}\right)$$

$$= \frac{1}{ra^{\frac{n}{r}}} \frac{\Gamma\left(\frac{n}{r}\right)\Gamma\left(m+n-\frac{n}{r}\right)}{\Gamma(m+n)}. \qquad (346)$$

If in this last expression $m+n = 1$,

$$\int_0^{\infty} \frac{x^{n-1}\,dx}{1+ax^r} = \frac{1}{ra^{\frac{n}{r}}} \Gamma\left(\frac{n}{r}\right)\Gamma\left(1-\frac{n}{r}\right)$$

$$= \frac{\pi}{ra^{\frac{n}{r}}} \operatorname{cosec} \frac{n\pi}{r}, \text{ by (278), Art. 129.} \qquad (347)$$

which is indeed the same result as (17), Art. 82.

Hence also, if $r = 2n$,

$$\int_0^{\infty} \frac{x^{n-1}\,dx}{1+ax^{2n}} = \frac{\pi}{2na^{\frac{1}{2}}};$$

which result may be proved by common integration.

139.] The values of the two integrals $\int_0^{\infty} e^{-ax} x^{n-1} \cos bx\,dx$, and $\int_0^\infty e^{-ax}x^{n-1}\sin bx\,dx$, when n is a positive quantity greater than unity, also depend immediately on the Gamma-function.

$$\text{Let } c = \int_0^\infty e^{-ax}x^{n-1}\cos bx\,dx;$$

$$s = \int_0^\infty e^{-ax}x^{n-1}\sin bx\,dx;$$

$$\therefore\quad c+s\sqrt{-1} = \int_0^\infty e^{-(a-b\sqrt{-1})x}x^{n-1}dx;$$

$$= \frac{\Gamma(n)}{(a-b\sqrt{-1})^n}, \text{ by (250) Art. 122.}$$

Let $a = k\cos\alpha$, $b = k\sin\alpha$, so that $k^2 = a^2+b^2$; then

$$c+s\sqrt{-1} = \frac{\Gamma(n)}{(a^2+b^2)^{\frac{n}{2}}}(\cos n\alpha+\sqrt{-1}\sin n\alpha);$$

so that equating possible and impossible parts, and replacing c and s by their values,

$$\int_0^\infty e^{-ax}x^{n-1}\cos bx\,dx = \frac{\Gamma(n)\cos n\alpha}{(a^2+b^2)^{\frac{n}{2}}}; \tag{348}$$

$$\int_0^\infty e^{-ax}x^{n-1}\sin bx\,dx = \frac{\Gamma(n)\sin n\alpha}{(a^2+b^2)^{\frac{n}{2}}}; \tag{349}$$

which results have already been found by Cauchy's method in Art. 104.

It will be observed that in the preceding process of evaluation the exponent of e contains an impossible as well as a possible part, and it may consequently be supposed that the method is not rigorously exact. Much might be said on the subject; let it however suffice to say that as the possible part of the exponent is negative, and as only positive values of x are included within the range of integration, the results are doubtless correct. They may also be verified in the following way.

Expanding $\cos bx$ we have

$$\int_0^\infty e^{-ax}x^{n-1}\cos bx\,dx = \int_0^\infty e^{-ax}\left\{x^{n-1}-b^2\frac{x^{n+1}}{1.2}+b^4\frac{x^{n+3}}{1.2.3.4}-\dots\right\}dx.$$

$$= \frac{\Gamma(n)}{a^n}-\frac{b^2}{1.2}\frac{\Gamma(n+2)}{a^{n+2}}+\frac{b^4}{1.2.3.4}\frac{\Gamma(n+4)}{a^{n+4}}-\dots,$$

$$= \frac{\Gamma(n)}{a^n}\left\{1-\frac{n(n+1)}{1.2}\frac{b^2}{a^2}+\frac{n(n+1)(n+2)(n+3)}{1.2.3.4}\frac{b^4}{a^4}-\dots\right\};$$

$$\int_0^\infty e^{-ax}x^{n-1}\cos bx\,dx = \frac{\Gamma(n)}{2a^n}\left\{\left(1+\sqrt{-1}\frac{b}{a}\right)^{-n}+\left(1-\sqrt{-1}\frac{b}{a}\right)^{-n}\right\}$$

$$= \frac{\Gamma(n)\cos n\alpha}{(a^2+b^2)^{\frac{n}{2}}}, \tag{350}$$

if $b = a\tan\alpha$. Similarly may (349) be verified.

140.] If in the preceding expressions $a = 0$, then $\alpha = \frac{\pi}{2}$; and (348) and (349) become, when n is a positive quantity,

$$\int_0^\infty x^{n-1}\cos bx\,dx = \frac{\Gamma(n)}{b^n}\cos\frac{n\pi}{2}, \tag{351}$$

$$\int_0^\infty x^{n-1}\sin bx\,dx = \frac{\Gamma(n)}{b^n}\sin\frac{n\pi}{2}; \tag{352}$$

Hence also, replacing $\Gamma(n)$ by its value given in (278),

$$\int_0^\infty x^{n-1}\cos bx\,dx = \frac{\pi}{b^n\Gamma(1-n)\sin n\pi}\cos\frac{n\pi}{2},$$

$$= \frac{\pi}{2b^n\Gamma(1-n)}\operatorname{cosec}\frac{n\pi}{2}; \tag{353}$$

$$\int_0^\infty x^{n-1}\sin bx\,dx = \frac{\pi}{2b^n\Gamma(1-n)}\sec\frac{n\pi}{2}; \tag{354}$$

$1-n$ being a positive proper fraction.

Let $1-n$ be replaced by m, when m is a positive proper fraction;

$$\int_0^\infty \frac{\cos bx\,dx}{x^m} = \frac{\pi b^{m-1}}{2\Gamma(m)}\sec\frac{m\pi}{2}; \tag{355}$$

$$\int_0^\infty \frac{\sin bx\,dx}{x^m} = \frac{\pi b^{m-1}}{2\Gamma(m)}\operatorname{cosec}\frac{m\pi}{2}. \tag{356}$$

Hence if $m = 1$, which is its superior value,

$$\int_0^\infty \frac{\sin bx}{x}\,dx = \frac{\pi}{2}; \tag{357}$$

which is the value already determined in (76), Art. 94.

If $m = \frac{1}{2}$,

$$\int_0^\infty \frac{\cos bx\,dx}{x^{\frac{1}{2}}} = \int_0^\infty \frac{\sin bx\,dx}{x^{\frac{1}{2}}} = \left(\frac{\pi}{2b}\right)^{\frac{1}{2}}; \tag{358}$$

since $\Gamma\left(\frac{1}{2}\right) = \pi^{\frac{1}{2}}$.

Hence, if x is replaced by x^2,

$$\int_0^\infty \cos bx^2\, dx = \int_0^\infty \sin bx^2\, dx = \frac{\pi^{\frac{1}{2}}}{2^{\frac{3}{2}} b^{\frac{1}{2}}}. \qquad (359)$$

Hence also, $\displaystyle\int_0^\infty e^{bx^2\sqrt{-1}}\, dx = \int_0^\infty \{\cos bx^2 + \sqrt{-1}\sin bx^2\}\, dx.$

$$= \frac{(1+\sqrt{-1})\pi^{\frac{1}{2}}}{2^{\frac{3}{2}} b^{\frac{1}{2}}},$$

$$= \frac{\pi^{\frac{1}{2}}}{2b^{\frac{1}{2}}} e^{\frac{\pi}{4}\sqrt{-1}}. \qquad (360)$$

Hence also, $\displaystyle\int_{-\infty}^\infty e^{bx^2\sqrt{-1}}\, dx = \frac{\pi^{\frac{1}{2}}}{b^{\frac{1}{2}}} e^{\frac{\pi}{4}\sqrt{-1}}. \qquad (361)$

And if in this last integral, x is replaced by $x + \frac{c}{b}$,

$$\int_{-\infty}^\infty e^{(bx^2+2cx)\sqrt{-1}}\, dx = \frac{\pi^{\frac{1}{2}}}{b^{\frac{1}{2}}} e^{\left(\frac{\pi}{4}-\frac{c^2}{b}\right)\sqrt{-1}}. \qquad (362)$$

Also, from (355) and (356), we have

$$\int_0^\infty \frac{e^{bx\sqrt{-1}}\, dx}{x^m} = \frac{\pi b^{m-1}}{2\Gamma(m)}\left(\sec\frac{m\pi}{2} + \sqrt{-1}\operatorname{cosec}\frac{m\pi}{2}\right). \qquad (363)$$

141.] Also from these formulæ two other integrals of considerable importance may be deduced.

By reason of (250), Art. 122,

$$\int_0^\infty e^{-(a-x\sqrt{-1})z} z^{n-1}\, dz = \frac{\Gamma(n)}{(a-x\sqrt{-1})^n},$$

$$= \frac{\Gamma(n)}{(a^2+x^2)^n}(a+x\sqrt{-1})^n;$$

and dividing both members by x^m, and then making them the element-functions of an x-integral between the limits ∞ and 0, we have

$$\Gamma(n)\int_0^\infty \frac{(a+x\sqrt{-1})^n\, dx}{(a^2+x^2)^n\; x^m} = \int_0^\infty \frac{dx}{x^m}\int_0^\infty e^{-(a-x\sqrt{-1})z} z^{n-1}\, dz$$

$$= \int_0^\infty e^{-az} z^{n-1}\, dz \int_0^\infty e^{xz\sqrt{-1}} \frac{dx}{x^m}$$

$$= \int_0^\infty e^{-az} z^{n-1}\, dz\, \frac{\pi z^{m-1}}{2\Gamma(m)}\left(\sec\frac{m\pi}{2} + \sqrt{-1}\operatorname{cosec}\frac{m\pi}{2}\right)$$

$$= \frac{\pi}{2\Gamma(m)}\left(\sec\frac{m\pi}{2} + \sqrt{-1}\operatorname{cosec}\frac{m\pi}{2}\right)\int_0^\infty z^{m+n-2} e^{-az}\, dz$$

$$= \frac{\pi\Gamma(m+n-1)}{2a^{m+n-1}\Gamma(m)}\left(\sec\frac{m\pi}{2} + \sqrt{-1}\operatorname{cosec}\frac{m\pi}{2}\right). \qquad (364)$$

In the left-hand member, let $a = k\cos\alpha$, $x = k\sin\alpha$, so that $x = a\tan\alpha$; then

$$\int_0^{\infty} \frac{\cos n\alpha + \sqrt{-1}\sin n\alpha}{(a^2+x^2)^{\frac{n}{2}}} \frac{dx}{x^m}$$

$$= \frac{\pi\Gamma(m+n-1)}{2a^{m+n-1}\Gamma(m)\Gamma(n)}\left(\sec\frac{m\pi}{2} + \sqrt{-1}\operatorname{cosec}\frac{m\pi}{2}\right); \quad (365)$$

and equating possible and impossible parts,

$$\int_0^{\infty} \frac{\cos\left(n\tan^{-1}\frac{x}{a}\right)dx}{(a^2+x^2)^{\frac{n}{2}}x^m} = \frac{\pi}{2a^{m+n-1}}\frac{\Gamma(m+n-1)}{\Gamma(m)\Gamma(n)}\sec\frac{m\pi}{2}; \quad (366)$$

$$\int_0^{\infty} \frac{\sin\left(n\tan^{-1}\frac{x}{a}\right)dx}{(a^2+x^2)^{\frac{n}{2}}x^m} = \frac{\pi}{2a^{m+n-1}}\frac{\Gamma(m+n-1)}{\Gamma(m)\Gamma(n)}\operatorname{cosec}\frac{m\pi}{2}; \quad (367)$$

where a is a positive quantity, $m+n-1$ is a positive quantity, and m is a positive proper fraction.

If in these last formulæ x is replaced by $a\tan\theta$, then

$$\int_0^{\frac{\pi}{2}} \cos n\theta(\cos\theta)^{n-2}(\cot\theta)^m d\theta = \frac{\pi}{2}\frac{\Gamma(m+n-1)}{\Gamma(m)\Gamma(n)}\sec\frac{m\pi}{2}; \quad (368)$$

$$\int_0^{\frac{\pi}{2}} \sin n\theta(\cos\theta)^{n-2}(\cot\theta)^m d\theta = \frac{\pi}{2}\frac{\Gamma(m+n-1)}{\Gamma(m)\Gamma(n)}\operatorname{cosec}\frac{m\pi}{2}. \quad (369)$$

If in this last equation $m = 1$,

$$\int_0^{\frac{\pi}{2}} \frac{\sin n\theta}{\sin\theta}(\cos\theta)^{n-1}d\theta = \frac{\pi}{2}. \quad (370)$$

142.] Another important use of the Gamma-function, and of its allied integrals, is the evaluation of a definite integral in terms of a series, and conversely the summation of a series in terms of a definite integral. This method, however, is only an application of the general process of development explained in the preceding section; and especially of that particular form given in Art. 121; but it is useful, because the same definite integral may be made to enter into every term of the development by reason of the properties of the Gamma-function, and consequently will be a common factor. Moreover we shall hereby arrive at certain series which have been called hypergeometrical, the series being of the type of geometrical series, but the coefficients of the several terms being more complex.

The following is the fundamental theorem on which the application depends.

Let $F(x)$ be a function of x, expanded into a convergent series of the form,

$$F(x)=A_0 f(x,0)+A_1 f(x,1)+A_2 f(x,2)+\ldots+A_k f(x,k)+\ldots; \quad (371)$$

and let both sides of this equation be multiplied by $\phi(x)\,dx$, and let the x-integral be taken between the limits x_n and x_0, it being supposed that none of the element-functions is infinite for any value of its subject-variable between these limits; and let us moreover suppose that

$$\int_{x_0}^{x_n} \phi(x) f(x,k)\,dx = B_k \int_{x_0}^{x_n} \phi(x) f(x,0)\,dx; \quad (372)$$

then

$$\int_{x_0}^{x_n} \phi(x)\,F(x)\,dx$$

$$= \{A_0 B_0 + A_1 B_1 + \ldots + A_k B_k + \ldots\} \int_{x_0}^{x_n} \phi(x) f(x,0)\,dx. \quad (373)$$

Thus, if $\int_{x_0}^{x} \phi(x) f(x,0) dx$ is known, the definite integral in the left-hand member of (373) will be given in terms of a series; and conversely the series will be given in terms of a definite integral.

If the value of $\int_{x_0}^{x_n} \phi(x)\,F(x)\,dx$ is known, the sum of the series will be known also. The following are examples of the process.

143.] Let us in the first place employ the Gamma-function for the purpose of forming the definite integrals; and let us suppose $F(x)$ to be a convergent series in ascending powers of x; so that

$$F(x) = a_0 + a_1 x + a_2 x^2 + a_3 x^3 + \ldots; \quad (374)$$

then $\int_0^{\infty} e^{-x} x^{n-1} F(x)\,dx$

$$= a_0 \int_0^{\infty} e^{-x} x^{n-1} dx + a_1 \int_0^{\infty} e^{-x} x^{n} dx + a_2 \int_0^{\infty} e^{-x} x^{n+1} dx + \ldots$$

$$= a_0 \Gamma(n) + a_1 \Gamma(n+1) + a_2 \Gamma(n+2) + \ldots \quad (375)$$

$$= \Gamma(n)\{a_0 + a_1 n + a_2 n(n+1) + a_3 n(n+1)(n+2) + \ldots\}. \quad (376)$$

$$\therefore \quad a_0 + a_1 n + a_2 n(n+1) + a_3 n(n+1)(n+2) + \ldots$$

$$= \frac{1}{\Gamma(n)} \int_0^{\infty} e^{-x} x^{n-1} \{a_0 + a_1 x + a_2 x^2 + \ldots\}\,dx. \quad (377)$$

The following is an example of the process;

$$\int_0^{\infty} e^{-x} x^{n-1} \cos(2bx^{\frac{1}{2}})\,dx$$

$$= \Gamma(n)\left\{1 - \frac{2n}{1}\frac{b^2}{1} + \frac{2^2 n(n+1)}{1.3}\frac{b^4}{1.2} - \frac{2^3 n(n+1)(n+2)}{1.3.5}\frac{b^6}{1.2.3} + \ldots\right\};$$

and if $n = \dfrac{1}{2}$,

$$\int_0^\infty e^{-x} x^{-\frac{1}{2}} (\cos 2bx^{\frac{1}{2}})\, dx = \Gamma\left(\frac{1}{2}\right)\left\{1 - \frac{b^2}{1} + \frac{b^4}{1.2} - \frac{b^6}{1.2.3} + \dots\right\}$$
$$= \pi^{\frac{1}{2}} e^{-b^2}. \quad (378)$$

Again, let us suppose $F(x)$ to be a convergent series in ascending powers of e^{-x}; so that

$$F(x) = a_0 + a_1 e^{-x} + a_2 e^{-2x} + a_3 e^{-3x} + \dots; \quad (379)$$

then

$$\int_0^\infty e^{-bx} x^{n-1} F(x)\, dx = a_0 \int_0^\infty e^{-bx} x^{n-1}\, dx + a_1 \int_0^\infty e^{-(b+1)x} x^{n-1}\, dx$$
$$+ a_2 \int_0^\infty e^{-(b+2)x} x^{n-1}\, dx + \dots \quad (380)$$
$$= a_0 \frac{\Gamma(n)}{b^n} + a_1 \frac{\Gamma(n)}{(b+1)^n} + a_2 \frac{\Gamma(n)}{(b+2)^n} + \dots,$$
$$= \Gamma(n) \left\{ \frac{a_0}{b^n} + \frac{a_1}{(b+1)^n} + \frac{a_2}{(b+2)^n} + \dots \right\}; \quad (381)$$

and consequently, if $b = 1$,

$$\int_0^\infty e^{-x} x^{n-1} \{a_0 + a_1 e^{-x} + a_2 e^{-2x} + \dots\}\, dx$$
$$= \Gamma(n) \left\{ \frac{a_0}{1^n} + \frac{a_1}{2^n} + \frac{a_2}{3^n} + \dots \right\}; \quad (382)$$

and if $a_0 = a_1 = a_2 = \dots = 1$,

$$\int_0^\infty \frac{e^{-x}}{1-e^{-x}} x^{n-1}\, dx = \Gamma(n) \left\{ \frac{1}{1^n} + \frac{1}{2^n} + \frac{1}{3^n} + \dots \right\}. \quad (383)$$

144.] In the next place let us employ the Beta-function, and its equivalent in terms of the Gamma-function, for the purpose of forming the definite integrals; and let us suppose $F(x)$ to be a series in ascending powers of x; so that

$$F(x) = a_0 + a_1 x + a_2 x^2 + a_3 x^3 + \dots; \quad (384)$$

the series being convergent for all values of x between 1 and 0; then

$$\int_0^1 x^{m-1} (1-x)^{n-1} F(x)\, dx = a_0 \int_0^1 x^{m-1} (1-x)^{n-1}\, dx$$
$$+ a_1 \int_0^1 x^m (1-x)^{n-1}\, dx + a_2 \int_0^1 x^{m+1} (1-x)^{n-1}\, dx + \dots$$
$$= a_0 \frac{\Gamma(m)\Gamma(n)}{\Gamma(m+n)} + a_1 \frac{\Gamma(m+1)\Gamma(n)}{\Gamma(m+n+1)} + a_2 \frac{\Gamma(m+2)\Gamma(n)}{\Gamma(m+n+2)} + \dots$$
$$= \frac{\Gamma(m)\Gamma(n)}{\Gamma(m+n)} \left\{ a_0 + a_1 \frac{m}{m+n} + a_2 \frac{m(m+1)}{(m+n)(m+n+1)} + \dots \right\}. \quad (385)$$

Hence

$$\int_0^1 x^{m-1}(1-x)^{n-1}(1-ax)^{-r}dx$$

$$= \int_0^1 x^{m-1}(1-x)^{n-1}\left(1+\frac{r}{1}ax+\frac{r(r+1)}{1.2}a^2x^2+\dots\right)dx, \quad (386)$$

$$= \frac{\Gamma(m)\Gamma(n)}{\Gamma(m+n)}\left\{1+\frac{r}{1}\frac{m}{m+n}a+\frac{r(r+1)}{1.2}\frac{m(m+1)}{(m+n)(m+n+1)}a^2+\dots\right\}.$$

If in this equation $r = m+n$,

$$\int_0^1 x^{m-1}(1-x)^{n-1}(1-ax)^{-(m+n)}dx$$

$$= \frac{\Gamma(m)\Gamma(n)}{\Gamma(m+n)}\left\{1+\frac{m}{1}a+\frac{m(m+1)}{1.2}a^2+\frac{m(m+1)(m+2)}{1.2.3}a^3+\dots\right\}$$

$$= \frac{\Gamma(m)\Gamma(n)}{\Gamma(m+n)}(1-a)^{-m}. \quad (387)$$

Hence also, if in (386) $a = 1$,

$$\int_0^1 x^{m-1}(1-x)^{n-r-1}dx$$

$$= \frac{\Gamma(m)\Gamma(n)}{\Gamma(m+n)}\left\{1+\frac{r}{1}\frac{m}{m+n}\frac{r(r+1)}{1.2}\frac{m(m+1)}{(m+n)(m+n+1)}+\dots\right\};$$

but $\int_0^1 x^{m-1}(1-x)^{n-r-1} = \frac{\Gamma(m)\Gamma(n-r)}{\Gamma(m+n-r)}$;

$$\therefore \quad 1+\frac{r}{1}\frac{m}{m+n}+\frac{r(r+1)}{1.2}\frac{m(m+1)}{(m+n)(m+n+1)}+\dots$$

$$= \frac{\Gamma(m+n)\Gamma(n-r)}{\Gamma(n)\Gamma(m+n-r)}. \quad (388)$$

If in this last series $r = 1$, then

$$1+\frac{m}{m+n}+\frac{m(m+1)}{(m+n)(m+n+1)}+\dots = \frac{\Gamma(m+n)\Gamma(n-1)}{\Gamma(n)\Gamma(m+n-1)},$$

$$= \frac{m+n-1}{n-1}. \quad (389)$$

The value of the following integral may also be determined by means of the properties of the Beta-function.

$$\int_0^\infty \frac{x^n dx}{1+2x\cos a+x^2} = \int_0^\infty \frac{x^n}{(1+x)^2}\left\{1-\frac{4x}{(1+x)^2}\left(\sin\frac{a}{2}\right)^2\right\}^{-1}dx$$

$$= \int_0^\infty \frac{x^n dx}{(1+x)^2}+\left(2\sin\frac{a}{2}\right)^2\int_0^\infty \frac{x^{n+1}dx}{(1+x)^4}+\left(2\sin\frac{a}{2}\right)^4\int_0^\infty \frac{x^{n+2}dx}{(1+x)^6}+\dots$$

$$\int_0^\infty \frac{x^n\,dx}{1+2x\cos a+x^2}$$

$$= \frac{\Gamma(1+n)\,\Gamma(1-n)}{\Gamma(2)} + \left(2\sin\frac{a}{2}\right)^2 \frac{\Gamma(2+n)\,\Gamma(2-n)}{\Gamma(4)}$$

$$+ \left(2\sin\frac{a}{2}\right)^4 \frac{\Gamma(3+n)\,\Gamma(3-n)}{\Gamma(6)} + \dots$$

$$= \frac{\Gamma(1+n)\,\Gamma(1-n)}{\Gamma(2)} \left\{ 1 + \frac{1^2-n^2}{1.2.3}\left(2\sin\frac{a}{2}\right)^2 \right.$$

$$\left. + \frac{(1^2-n^2)(2^2-n^2)}{1.2.3.4.5}\left(2\sin\frac{a}{2}\right)^4 + \dots \right\}. \quad (390)$$

This however is only a particular case of a much more general theorem which is of use in the higher applied mathematics.

145.] In (386) let n be replaced by $n-m$, then

$$1+\frac{r}{1}\frac{m}{n}a+\frac{r(r+1)}{1.2}\frac{m(m+1)}{n(n+1)}a^2+\frac{r(r+1)(r+2)}{1.2.3}\frac{m(m+1)(m+2)}{n(n+1)(n+2)}a^3+\dots$$

$$= \frac{\Gamma(n)}{\Gamma(m)\,\Gamma(n-m)}\int_0^1 x^{m-1}(1-x)^{n-m-1}(1-ax)^{-r}\,dx. \quad (391)$$

The series in the left-hand member of this equation is that in pursuit of the properties of which Gauss has been led into his inquiries respecting the Gamma-function. His original memoir was read to the Academy of Sciences at Gottingen on Jan. 30, 1812, and is contained in the first volume of the Commentationes Recentiores of that Society. The series is evidently a function of four variables, r, m, n, a, and is the general type to which very many known series conform. This is the reason why he has investigated its properties. Let it be denoted by the symbol $F(r, m, n, a)$, so that

$$1+\frac{r}{1}\frac{m}{n}a+\frac{r(r+1)}{1.2}\frac{m(m+1)}{n(n+1)}a^2+\frac{r(r+1)(r+2)}{1.2.3}\frac{m(m+1)(m+2)}{n(n+1)(n+2)}a^3+\dots$$

$$= F(r, m, n, a) \quad (392).$$

This series is evidently convergent for all possible values of a, less than 1; and for all impossible values of a, of which the modulus is less than 1. Hence, under these circumstances, the right-hand member of (391) may be used as the equivalent value of it. The function $F(r, m, n, a)$, is called the Gaussian Function, and the equivalent series is called the Gaussian series.

The following are particular cases of the series.

$$(1+a)^m = 1+\frac{m}{1}a+\frac{m(m-1)}{1.2}a^2+\frac{m(m-1)(m-2)}{1.2.3}a^3+\dots$$

$$= F(n,-m,n,-a). \qquad (393)$$

$$e^a = 1+\frac{a}{1}+\frac{a^2}{1.2}+\frac{a^3}{1.2.3}+\dots$$

$$= F\left(1,\ m,\ 1,\frac{a}{m}\right), \text{ when } m = \infty. \qquad (394)$$

$$\log(1+a) = a-\frac{a^2}{2}+\frac{a^3}{3}-\frac{a^4}{4}+\dots$$

$$= a\,F(1,1,2,-a). \qquad (395)$$

$$\log\left(\frac{1+a}{1-a}\right) = 2a\,F\left(\frac{1}{2},\ 1,\ \frac{3}{2},\ a^2\right). \qquad (396)$$

$$\sin a = a\,F\left(r,m,\frac{3}{2},\ \frac{-a^2}{4rm}\right), \text{ when } r = m = \infty. \qquad (397)$$

$$\cos a = F\left(r,\ m,\frac{1}{2},\frac{-a^2}{4rm}\right), \text{ when } r = m = \infty. \qquad (398)$$

$$\sin ma = m\sin a\,F\left(\frac{m+1}{2},\ \frac{-m+1}{2},\ \frac{3}{2},\ (\sin a)^2\right);$$

so that the properties of all these functions, as well as of many others which are developable into series of the form (392), are the same as those of the right-hand member of (391).

In all these cases, when $a = 1$, we have from (391) and (392),

$$F(r,m,n,1) = \frac{\Gamma(n)}{\Gamma(m)\,\Gamma(n-m)}\int_0^1 x^{m-1}(1-x)^{n-m-r-1}\,dx$$

$$= \frac{\Gamma(n)\,\Gamma(n-m-r)}{\Gamma(n-m)\,\Gamma(n-r)}. \qquad (399)$$

146.] In these latter applications of the Gamma-function, the argument will frequently be negative; and we have shewn in Art. 124, that the function, as defined by $\int_0^\infty e^{-x}x^{n-1}\,dx$, is always infinite when n is negative. But these latter applications are legitimate deductions from the fundamental theorems of the Gamma-function; so that it is clear that they express properties of a function of wider extent than that of the definite integral. Consequently a wider definition must be given to the function.

That due to Gauss and given in (318) is sufficient for the purpose; as it holds good for negative as well as for positive values of the argument. The following theorems are easily deduced from it.

$$\Gamma(1-n) = -n\Gamma(-n); \tag{400}$$

$$\Gamma(1+n)\Gamma(-n) = -\frac{\pi}{\sin n\pi}; \tag{401}$$

$$\Gamma(-n)\Gamma\left(-n+\frac{1}{r}\right)\Gamma\left(-n+\frac{2}{r}\right)\dots\Gamma\left(-n+\frac{r-1}{r}\right)$$
$$= (2\pi)^{\frac{r-1}{2}} r^{\frac{1}{2}+rn}\Gamma(-nr). \tag{402}$$

From (401) it is evident that, if n is an integer, $\Gamma(-n) = \infty$, since $\sin n\pi = 0$. So that the Gamma-function is infinite for all negative integral values of the argument; and will be finite when n is not an integer.

If $n = \frac{1}{2}$, from (401) we have

$$\Gamma\left(-\frac{1}{2}\right) = -2\pi^{\frac{1}{2}}; \tag{403}$$

$$\therefore \quad \Gamma\left(-\frac{3}{2}\right) = \frac{4}{3}\pi^{\frac{1}{2}}. \tag{404}$$

Section 8.—*The Logarithm-Integral.*

147.] The definite integral $\int_0^{x_n} \frac{dx}{\log x}$ is closely allied to the Gamma-function, being indeed the form taken by that function, as expressed in (248), when $n = 0$; except that in this case the superior limit is the general value x_n. This integral moreover deserves notice, because it is one of the few which have been tabulated, the necessary calculations having been made by Soldner of Munich as long ago as the year 1809. The integral is also instrumental in the determination of many other definite integrals, and occurs in the solution of certain physical problems.

Soldner devised the symbol $li.x_n$ to denote it: *li* being the initial letters of logarithm-integral, by which name he called it. Thus we have for the definition of the logarithm-integral

$$li.x_n = \int_0^{x_n} \frac{dx}{\log x}. \tag{405}$$

Let x in the right-hand member be replaced by e^{-x}; then

$$li.x_n = -\int_{-\log x_n}^{\infty} \frac{e^{-x}\,dx}{x}; \qquad (406)$$

which may also be taken as the definition of the logarithm-integral.

If x_n is greater than 1, the range of integration in (406) includes a value, viz. $x = 0$, for which the element-function $= \infty$; we must therefore divide the integral into two parts and take the principal value of each, making the range of each approach infinitesimally near to 1; so that if i is an infinitesimal, we may express (406) in the following form,

$$li.x_n = -\int_{i}^{\infty} \frac{e^{-x}dx}{x} - \int_{-\log x_n}^{-i} \frac{e^{-x}dx}{x}; \qquad (407)$$

and the value of this is to be determined, when $i = 0$.

Now by the definition of E, Euler's Constant, given in (298), when $i = 0$,

$$\int_{i}^{\infty} \frac{e^{-x}dx}{x} = -\text{E} + \int_{i}^{\infty} \frac{dx}{x(1+x)};$$

$$= -\text{E} + \left[\log\frac{x}{1+x}\right]_{i}^{\infty};$$

$$= -\text{E} - \log\frac{i}{1+i}. \qquad (408)$$

Also, taking the second integral in the right-hand member of (407), expanding e^{-x}, and integrating, we have

$$\int_{-\log x_n}^{-i} \frac{e^{-x}\,dx}{x} = \left[\log x - x + \frac{x^2}{1.2^2} - \frac{x^3}{1.2.3^2} + \dots\right]_{-\log x_n}^{-i}$$

$$= \log\frac{i}{\log x_n} - \log x_n - \frac{(\log x_n)^2}{1.2^2} - \frac{(\log x_n)^3}{1.2.3^2} - \dots, \qquad (409)$$

omitting terms involving i and its powers, which must be neglected. Consequently, substituting in (407), we have

$$li.x_n = \text{E} + \log\log x_n + \log x_n + \frac{(\log x_n)^2}{1.2^2} + \frac{(\log x_n)^3}{1.2.3^2} + \dots \qquad (410)$$

If x_n is less than 1, although in (406) the element-function never becomes infinite within the range of integration, yet the preceding process is not applicable, because the result contains the term $\log\log x_n$; and this is impossible when x_n is less than 1, for $\log x_n$ is in that case negative, and $\log\log x_n$ is the logarithm of a negative quantity. In this case let us suppose $(x_n)^{-1}$ to be the subject-variable of *li*; then from (406) we have

$$li.\frac{1}{x_n} = -\int_{\log x_n}^{\infty} \frac{e^{-x}\,dx}{x};$$

and if we expand the element-function, and integrate as in the preceding case,

$$li.\frac{1}{x_n} = \text{E} + \log\log x_n - \log x_n + \frac{(\log x_n)^2}{1.2^2} - \frac{(\log x_n)^3}{1.2.3^2} + \dots; \quad (411)$$

in which every term is real, since x_n is greater than 1. Thus by (410) or (411) the logarithm-integral may be calculated for all values of the subject-variable; and they have been employed by Soldner for that purpose; but it is beyond the scope of this work to enter into the details of the calculation.

148.] As another instance of the logarithm-integral, let the subject-variable be $(1+x_n)$; so that by (405),

$$li.(1+x_n) = \int_{-1}^{x_n} \frac{dx}{\log(1+x)}.$$

Now
$$\frac{1}{\log(1+x)} = \frac{1}{x}\left\{1 - \frac{x}{2} + \frac{x^2}{3} - \frac{x^3}{4} + \dots\right\}^{-1}$$
$$= \frac{1}{x}\left\{\text{A}_0 + \text{A}_1\frac{x}{1} + \text{A}_2\frac{x^2}{1.2} + \text{A}_3\frac{x^3}{1.2.3} + \dots\right\},$$

where $\text{A}_0, \text{A}_1, \text{A}_2, \dots$ are coefficients of the successive powers of x, which may be calculated by the process of Derivation, as explained in Art. 95 of Vol. I. The values thereby found are

$$\text{A}_0 = 1,\ \text{A}_1 = \frac{1}{2},\ \text{A}_2 = -\frac{1}{6},\ \text{A}_3 = \frac{1}{4},\ \text{A}_4 = -\frac{19}{120}, \dots.$$

Now as $x = 0$ occurs within the range of integration, and is that value for which the element-function becomes infinite, the definite integral must be divided into two parts, and we have

$$li.(1+x_n) = \int_{\iota}^{x_n} \frac{1}{x}\left\{\text{A}_0 + \text{A}_1\frac{x}{1} + \text{A}_2\frac{x^2}{1.2} + \dots\right\}dx$$
$$+ \int_{-1}^{-\iota} \frac{1}{x}\left\{\text{A}_0 + \text{A}_1\frac{x}{1} + \text{A}_2\frac{x^2}{1.2} + \dots\right\}dx,$$
$$= \text{E} + \text{A}_0 \log x_n + \text{A}_1\frac{x_n}{1} + \text{A}_2\frac{x_n^{\,2}}{1.2^2} + \text{A}_3\frac{x_n^{\,3}}{1.2.3^2} + \dots; \quad (412)$$

the sum of all the quantities in the definite integral, which are independent of the powers of x_n in the series, being equal to E when x_n is an infinitesimal, because in that case

$$li.(1+x_n)-\log x_n = \text{E}.$$

The following are indefinite integrals which are expressed in terms of the logarithm-integral.

(1) $\int \frac{dx}{\log x} = li.x.$ (2) $\int \frac{x^m dx}{\log x} = li.x^{m+1}.$

(3) $\int \frac{e^x dx}{x} = li.e^x.$ (4) $\int e^{e^x} dx = li.e^{e^x}.$

(5) $\int \frac{e^x dx}{a+x} = e^{-a} li.e^{a+x}.$ (6) $\int \frac{dx}{x \log \log x} = li \log x.$

(7) $\int li.x\, dx = x\, li.x - li.x^2.$

CHAPTER V.

ON SUCCESSIVE INTEGRATION OF AN EXPLICIT FUNCTION OF ONE VARIABLE.

149.] In the preceding Chapters, so far as integration with respect to any one variable has been considered, the infinitesimal element which is the subject of integration has been an infinitesimal of the first order: and consequently the integral of that infinitesimal element has been a finite quantity. In the general theory of integration however, no restriction is put on the order of the element; the element may be an infinitesimal of any order; and the effect of integration on it will be the reduction of that order by unity. Thus if an element is of the nth order, the first integral of that element will be an element of the $(n-1)$th order; and the integral of that element will be of the $(n-2)$th order, and so on; until ultimately, if the integration-process is carried on so far, the nth integral will be a finite quantity.

Now we will suppose the superior limit in all these successive integrations, to be the same general variable x, the inferior limit being constant although arbitrary in each case. It is evident then that each integration will introduce an additional term, which will be a function of the inferior limit. It is convenient to represent this additional term by a constant; and thus, in the entire process, n additional terms, or n arbitrary constants will be introduced; and the final integral is not considered complete unless it contains them.

150.] Suppose $F(x)$ to be a function of x finite and continuous for all values of its subject-variable within the range for which we employ it; and suppose its derived functions to be $F'(x)$, $F''(x), \dots F^n(x)$, and to be subject to like conditions: then,

$$\int F'(x)\,dx = F(x) + C_n, \qquad (1)$$

$$\int F''(x)\,dx = F'(x) + C_{n-1}, \qquad (2)$$

. . . .

$$\int F^n(x)\,dx = F^{n-1}(x) + C_1; \qquad (3)$$

and therefore from (2)

$$\iint \mathrm{F}''(x)\,dx\,dx = \int \mathrm{F}'(x)\,dx + \mathrm{C}_{n-1}x$$

$$= \mathrm{F}(x) + \mathrm{C}_{n-1}x + \mathrm{C}_n. \tag{4}$$

Similarly

$$\iiint \mathrm{F}'''(x)\,dx\,dx\,dx = \mathrm{F}(x) + \mathrm{C}_{n-2}\frac{x^2}{1.2} + \mathrm{C}_{n-1}x + \mathrm{C}_n; \tag{5}$$

and therefore, abbreviating the notation by writing $\int^n dx^n$ for $\iint \ldots dx\,dx$, when the latter series involves n symbols of integration,

$$\int^n \mathrm{F}^n(x)\,dx^n = \mathrm{F}(x) + \mathrm{C}_1\frac{x^{n-1}}{1.2\ldots(n-1)} + \mathrm{C}_2\frac{x^{n-2}}{1.2\ldots(n-2)} + \ldots$$
$$\ldots + \mathrm{C}_{n-1}x + \mathrm{C}_n, \tag{6}$$

$\mathrm{C}_1, \mathrm{C}_2, \ldots \mathrm{C}_n$ being n arbitrary constants. This is the general expression for the nth integral of an element of the nth order.

The following are particular cases of the preceding general form.

$$\int^n dx^n = \frac{x^n}{1.2.3\ldots n} + \mathrm{C}_1\frac{x^{n-1}}{1.2\ldots(n-1)} + \ldots + \mathrm{C}_{n-1}x + \mathrm{C}_n.$$

$$\int^n e^x\,dx^n = e^x + \mathrm{C}_1\frac{x^{n-1}}{1.2\ldots(n-1)} + \ldots + \mathrm{C}_{n-1}x + \mathrm{C}_n.$$

$$\int^n \log x\,dx^n = \frac{x^n}{1.2.3\ldots n}\left\{\log x - \left(\frac{1}{1} + \frac{1}{2} + \frac{1}{3} + \ldots + \frac{1}{n}\right)\right\}$$
$$+ \mathrm{C}_1\frac{x^{n-1}}{1.2.3\ldots(n-1)} + \ldots + \mathrm{C}_{n-1}x + \mathrm{C}_n.$$

151.] Many useful theorems of successive integration may be deduced by means of the Calculus of Operations, of which a concise account is given in Chapter XIX. of Vol. I. In the first Chapter of the present volume the reciprocal relation of the differential calculus and the integral calculus has been explained, and the processes of the latter have been proved to be inverse of those of the former. Also the inverse character of the symbols of the latter calculus relatively to those of the former, has been exhibited in Art. 7, Chapter I. Now, as the operations of differentiation and derivation are subject to the commutative and distributive laws, and as the symbols of the operations are subject to the index law, so will the symbols of the inverse operations be also subject to the same laws; and the transition from the symbols of one calculus to those of the other will consist in only a change of sign of the index. Thus, as d and $\int$ are the symbols of differentiation and integration respectively, we have as in Art. 7,

$$\int = d^{-1}; \tag{7}$$

and consequently $\int dx = \left(\frac{d}{dx}\right)^{-1}$; (8)

and as these symbols are accordant with the index law,

$$\int\int\int \ldots \text{ to } n \text{ symbols} = \int^n = d^{-n}; \tag{9}$$

$$\int^n (dx)^n = \left(\frac{d}{dx}\right)^{-n}. \tag{10}$$

These laws are applied in the following examples; care being of course taken to select those cases which satisfy the required laws.

Ex. 1. Since $\frac{d^n \sin(mx+a)}{dx^n} = m^n \sin\left(mx+a+n\frac{\pi}{2}\right)$;

therefore replacing n by $-n$,

$$\int^n \sin(mx+a)\, dx^n = \frac{1}{m^n}\sin\left(mx+a-n\frac{\pi}{2}\right);$$

and if $n = 1$,

$$\int \sin(mx+a)\, dx = \frac{1}{m}\sin\left(mx+a-\frac{\pi}{2}\right)$$

$$= -\frac{1}{m}\cos(mx+a).$$

Similarly $\int^n \cos mx\, dx^n = \frac{1}{m^n}\cos\left(mx-n\frac{\pi}{2}\right)$;

and if $n = 1$, $\int \cos mx\, dx = \frac{1}{m}\cos\left(mx-\frac{\pi}{2}\right)$

$$= \frac{1}{m}\sin mx.$$

Ex. 2. By Ex. 6, Art. 54, Vol. I, if $m = a\tan\phi$,

$$\frac{d^n(e^{ax}\sin mx)}{dx^n} = (a^2+m^2)^{\frac{n}{2}} e^{ax}\sin(mx+n\phi);$$

$$\therefore \int^n e^{ax}\sin mx\, dx^n = \frac{e^{ax}\sin(mx-n\phi)}{(a^2+m^2)^{\frac{n}{2}}};$$

and if $n = 1$,

$$\int e^{ax}\sin mx\, dx = \frac{e^{ax}\sin(mx-\phi)}{(a^2+m^2)^{\frac{1}{2}}}.$$

152.] Again, taking Leibnitz's Series for $\frac{d^n . uv}{dx^n}$, where u and v are explicit functions of x, and making n negative, we have, as in Art. 425, Vol. I,

$$\int^n uv\, dx^n = u\int^n v\, dx^n - n\frac{du}{dx}\int^{n+1} v\, dx^{n+1} + \frac{n(n+1)}{1.2}\frac{d^2u}{dx^2}\int^{n+2} v\, dx^{n+2} - \dots \quad (11)$$

Let $n = 1$; then

$$\int uv\, dx = u\int v\, dx - \frac{du}{dx}\int^2 v\, dx^2 + \frac{d^2u}{dx^2}\int^3 v\, dx^3 - \dots; \quad (12)$$

and if $v = 1$,

$$\int u\, dx = u\int dx - \frac{du}{dx}\int^2 dx^2 + \frac{d^2u}{dx^2}\int^3 dx^3 - \dots$$

$$= u\frac{x}{1} - \frac{du}{dx}\frac{x^2}{1.2} + \frac{d^2u}{dx^2}\frac{x^3}{1.2.3} - \dots; \quad (13)$$

which is Bernoulli's Series for the determination of an integral; see equation (233), Art. 117.

Equation (11) will apparently not give a finite result for the integral, unless the series in the right-hand member is continued to an infinite number of terms; or unless the derived functions of u should vanish after a certain term. The limit of the sum may however be determined by the following process.

Considering $\frac{d}{dx}$ to affect u only, and $\int dx$ to affect v only, and in the right-hand member of the equation separating symbols of operation from their subjects, we have

$$\int^n uv\, dx^n = \left\{\int^n dx^n - n\frac{d}{dx}\int^{n+1} dx^{n+1} + \frac{n(n+1)}{1.2}\frac{d^2}{dx^2}\int^{n+2} dx^{n+2} - \dots\right\} uv$$

$$= \int^n dx^n \left\{1 - n\frac{d}{dx}\int dx + \frac{n(n+1)}{1.2}\frac{d^2}{dx^2}\int^2 dx^2 - \dots\right\} uv$$

$$= \frac{\int^n dx^n}{\left\{1 + \frac{d}{dx}\int dx\right\}^n} uv; \quad (14)$$

and taking symbols of operation only,

$$\int^n dx^n = \frac{\int^n dx^n}{\left\{1 + \frac{d}{dx}\int dx\right\}^n}; \quad (15)$$

remembering that the symbol in the left-hand member refers to the integral of uv, whereas those on the right-hand side refer to either u or v according to the preceding hypothesis. In the symbolical form the limit can easily be expressed by means of the general expression for the limit of Maclaurin's Series.

In (14), let $n = 1$; then

$$\int uv\,dx = \frac{\int dx}{1+\frac{d}{dx}\int dx}\,uv; \tag{16}$$

$$\text{or,}\quad \int dx = \int dx\left\{1+\frac{d}{dx}\int dx\right\}^{-1}. \tag{17}$$

The preceding theorems are deduced from laws of expansion, and from developments which the integration-symbol is subject to, and are thus far independent of any condition as to their subjects. Certain conditions however are requisite, and are the same as those which have already been explained. That is, if the final results are determinate, the integrals must be definite, and none of the elements must be infinite within the range of integration.

153.] The following cases, in which certain subject element-functions are supplied, will be useful in the sequel.

In (14) let $v = e^{ax}$; so that $\int v\,dx = \frac{e^{ax}}{a}$; then

$$\begin{aligned}\int^n u e^{ax} dx^n &= \frac{\frac{1}{a^n}}{\left(1+\frac{1}{a}\frac{d}{dx}\right)^n} e^{ax} u \\ &= \left(a+\frac{d}{dx}\right)^{-n} e^{ax} u \\ &= e^{ax}\left(a+\frac{d}{dx}\right)^{-n} u,\end{aligned} \tag{18}$$

since $\frac{d}{dx}$ affects u only: and consequently,

$$\left(\frac{d}{dx}+a\right)^{-n} u = e^{-ax}\int^n u e^{ax}\,dx^n. \tag{19}$$

and if $n = 1$,

$$\left(\frac{d}{dx}+a\right)^{-1} u = e^{-ax}\int u e^{ax}\,dx: \tag{20}$$

both of these formulæ will be of great use in the sequel.

Again, in (16), let $v = e^{\frac{x}{a}}$, $u = x^n$; so that $\int v\,dx = ae^{\frac{x}{a}}$;

$$\therefore \int e^{\frac{x}{a}} x^n\,dx = ae^{\frac{x}{a}}\left\{1 + a\frac{d}{dx}\right\}^{-1} x^n$$

$$= ae^{\frac{x}{a}}\{x^n - anx^{n-1} + a^2 n(n-1)x^{n-2} - \ldots\}. \quad (21)$$

Again, in (12) let $u = e^{ax}$; so that $\frac{du}{dx} = ae^{ax}$, $\frac{d^2u}{dx^2} = a^2e^{ax}\ldots$;

$$\int ve^{ax}\,dx = e^{ax}\left\{\int v\,dx - a\int^2 v\,dx^2 + a^2\int^3 v\,dx^3 - \ldots\right\} \quad (22)$$

$$= e^{ax}\frac{\int dx}{1 + a\int dx}v.$$

CHAPTER VI.

THE APPLICATION OF SINGLE INTEGRATION TO QUESTIONS OF GEOMETRY.

SECTION 1.—*Rectification of Plane Curves referred to Rectangular Coordinates.*

154.] IN the present Chapter I propose to consider some of the most simple applications of single integration to questions of geometry, reserving the more complex problems of space until the higher parts of the Integral Calculus, which are required for their solution, have been more fully developed. And, in the first place, I will shew how this calculus enables us to determine, either exactly or approximately, the length of a plane curve between two given points in terms of the coordinates of these points. Its length will thus be compared with that of a straight line; and hence the process has been called the Rectification of Curves.

Let $y = f(x)$, or $F(x, y) = c$, be the equation of a plane curve referred to rectangular coordinates; and let it be required to determine the length of the curve between the points (x_0, y_0) and (x_n, y_n); that is, to determine the length of a straight line, along which, if the curve is made to roll (not to slide), the extremities of it will coincide with those of the curve. Now, adopting the notation of Vol. 1, Art. 218, let ds be an infinitesimal length-element of the curve, then the required length is the integral of ds between the specified limits; but

$$ds = \{dx^2 + dy^2\}^{\frac{1}{2}}; \qquad (1)$$

and therefore the required length $= \int \{dx^2 + dy^2\}^{\frac{1}{2}}$; (2)

the integral being taken between the given limits.

Let s represent the length of the curve; then if the equation is $y = f(x)$, so that $dy = f'(x)\,dx$,

$$ds = \{1 + (f'(x))^2\}^{\frac{1}{2}}\,dx;$$

$$\therefore \quad s = \int_{x_0}^{x_n} \{1 + (f'(x))^2\}^{\frac{1}{2}}\,dx. \qquad (3)$$

And if the equation to the curve is of the form $x = f(y)$, so that $dx = f'(y)\,dy$, then

$$s = \int_{y_0}^{y_n} \{1+(f'(y))^2\}^{\frac{1}{2}} dy. \qquad (4)$$

As the radical expression in (1) involves an ambiguity of sign which is continued in (3) and (4), and as s is an absolute length, we must choose that sign which the circumstances of the problem require; that is, ds and dx or dy must be taken with the same or different signs according as x and y increase or decrease when s increases.

Although I have given the general formulæ (3) and (4) which express the length of a curve as a definite integral, yet in the sequel, as the following examples shew, it will be more convenient to deduce the expression for s directly from the equation to the curve.

155.] Examples of rectification of plane curves.

Ex. 1. The circle; see fig. 3.

Let the centre be the origin; and let the arc AP_n, whose length is required, begin at A, and be measured from A towards B; so that, if $AP = s$, $OM = x$, x decreases as s increases; let $OM_n = x_n$; then since $x^2+y^2 = a^2$,

$$-\frac{dx}{y} = \frac{dy}{x} = \frac{ds}{a}; \qquad \therefore \quad ds = -\frac{a\,dx}{(a^2-x^2)^{\frac{1}{2}}}.$$

$$\therefore \text{ the length of the arc } AP_n = \int_a^{x_n} \frac{-a\,dx}{(a^2-x^2)^{\frac{1}{2}}}$$

$$= \Big[a\cos^{-1}\frac{x}{a}\Big]_a^{x_n} = a\cos^{-1}\frac{x_n}{a}. \qquad (5)$$

$$\text{Hence the length of the quadrant } AB = \int_a^0 \frac{-a\,dx}{(a^2-x^2)^{\frac{1}{2}}}$$

$$= \Big[a\cos^{-1}\frac{x}{a}\Big]_a^0 = \frac{\pi a}{2}.$$

$\therefore$ The perimeter of the circle $= 2\pi a$. (6)

Hence also if $OM_0 = x_0$, then

$$\text{the arc } P_n P_0 = \int_{x_0}^{x_n} \frac{-a\,dx}{(a^2-x^2)^{\frac{1}{2}}}$$

$$= \Big[a\cos^{-1}\frac{x}{a}\Big]_{x_0}^{x_n}$$

$$= a\left\{\cos^{-1}\frac{x_n}{a} - \cos^{-1}\frac{x_0}{a}\right\}. \qquad (7)$$

Ex. 2. The Parabola; fig. 4.

Let the arc OP_n, whose length is required, begin at the vertex; let $\text{OM}_n = x_n$, $\text{M}_n\text{P}_n = y_n$; and let the equation to the parabola be $y^2 = 4ax$. Then

$$\frac{dy}{2a} = \frac{dx}{y} = \frac{ds}{(y^2+4a^2)^{\frac{1}{2}}}; \qquad \therefore \quad ds = \frac{(y^2+4a^2)^{\frac{1}{2}}\,dy}{2a};$$

$$\therefore \text{ the length of OP}_n = \int_0^{y_n} \frac{(y^2+4a^2)^{\frac{1}{2}}}{2a}\,dy$$

$$= \frac{1}{2a}\Big[\frac{y}{2}(y^2+4a^2)^{\frac{1}{2}} + 2a^2 \log\{y+(y^2+4a^2)^{\frac{1}{2}}\}\Big]_0^{y_n}$$

$$= \frac{y_n(y_n^2+4a^2)^{\frac{1}{2}}}{4a} + a\log\frac{y_n+(y_n^2+4a^2)^{\frac{1}{2}}}{2a}; \qquad (8)$$

as appears by equation (80), Art. 39; and as y_n may be the ordinate to any point on the parabola, we may replace it by the general value y, so that the length of the arc of the parabola beginning at the vertex is equal to

$$\frac{y}{4a}(y^2+4a^2)^{\frac{1}{2}} + a\log\frac{y+(y^2+4a^2)^{\frac{1}{2}}}{2a}.$$

And if for y we write $2(ax)^{\frac{1}{2}}$,

$$\text{the length} = (ax+x^2)^{\frac{1}{2}} + a\log\frac{x^{\frac{1}{2}}+(a+x)^{\frac{1}{2}}}{a^{\frac{1}{2}}}. \qquad (9)$$

Thus if $x = a$, the length of the arc between the vertex and the extremity of the latus rectum is equal to $a\{2^{\frac{1}{2}}+\log(1+2^{\frac{1}{2}})\}$.

Ex. 3. The Cycloid.

(1) Let the highest point be origin, see fig. 5; and let the arc begin at the vertex; then since

$$y = (2ax-x^2)^{\frac{1}{2}} + a\,\text{versin}^{-1}\frac{x}{a},$$

$$\frac{dy}{(2a-x)^{\frac{1}{2}}} = \frac{dx}{x^{\frac{1}{2}}} = \frac{ds}{(2a)^{\frac{1}{2}}}; \qquad \therefore \quad ds = \Big(\frac{2a}{x}\Big)^{\frac{1}{2}} dx.$$

$$\therefore \text{ the length of OP}_n = \int_0^{x_n}\Big(\frac{2a}{x}\Big)^{\frac{1}{2}} dx = 2(2ax_n)^{\frac{1}{2}}; \qquad (10)$$

and as x_n may be the abscissa to any point on the curve, we may replace it by the general value x; and then

$$s = 2(2ax)^{\frac{1}{2}}; \qquad \therefore \quad s^2 = 8ax. \qquad (11)$$

$$\text{Hence the length of OB} = \int_0^{2a}\Big(\frac{2a}{x}\Big)^{\frac{1}{2}} dx$$

$$= \Big[2(2ax)^{\frac{1}{2}}\Big]_0^{2a} = 4a; \qquad (12)$$

consequently the whole length of the cycloid is $8a$; that is, four times the diameter of the generating circle.

Since (11) expresses a general relation between the length of an arc of the cycloid measured from the vertex and the abscissa to the extremity of that arc, it may be and frequently will be employed hereafter as the equation to the cycloid. If OQA, fig. 5, is the generating circle of the cycloid, the equation shews that the arc OP = twice the chord OQ.

(2) Let the starting point be origin; see fig. 6, and let

OM $= x$, MP $= y$; OM$_n = x_n$, M$_n$P$_n = y_n$; so that

$$x = a \operatorname{versin}^{-1}\frac{y}{a} - (2ay-y^2)^{\frac{1}{2}}.$$

$$\therefore \quad \frac{dx}{y} = \frac{dy}{(2ay-y^2)^{\frac{1}{2}}} = \frac{ds}{(2ay)^{\frac{1}{2}}};$$

$$\therefore \quad \text{OP}_n = \int_0^{y_n} \left(\frac{2a}{2a-y}\right)^{\frac{1}{2}} dy$$

$$= -\left[2(2a)^{\frac{1}{2}}(2a-y)^{\frac{1}{2}}\right]_0^{y_n}$$

$$= 2(2a)^{\frac{1}{2}}\{(2a)^{\frac{1}{2}}-(2a-y_n)^{\frac{1}{2}}\}; \qquad (13)$$

$$\therefore \quad \text{OB} = \int_0^{2a}\left(\frac{2a}{2a-y}\right)^{\frac{1}{2}} dy$$

$$= 4a;$$

which result is the same as that found in the former case.

Ex. 4. The Tractory; see fig. 2.

Let the required arc begin at A; and let the ordinate at its extremity be y_n; then since

$$\frac{-dy}{y} = \frac{dx}{(a^2-y^2)^{\frac{1}{2}}} = \frac{ds}{a};$$

$$\text{the required arc} = \int_a^{y_n} \frac{-a\,dy}{y}$$

$$= \left[-a\log y\right]_a^{y_n}$$

$$= a\log\frac{a}{y_n}; \qquad (14)$$

and writing for y_n the general value y, we have

$$s = a\log\frac{a}{y}. \qquad (15)$$

Ex. 5. To determine the values of m and n, for which the curves expressed by the equation $a^m y^n = x^{m+n}$ are rectifiable.

Since
$$a^{\frac{m}{n}} y = x^{\frac{m+n}{n}},$$
$$a^{\frac{m}{n}} dy = \frac{m+n}{n} x^{\frac{m}{n}} dx;$$
$$\therefore \quad ds = \left\{1 + \left(\frac{m+n}{n} a^{-\frac{m}{n}}\right)^2 x^{\frac{2m}{n}}\right\}^{\frac{1}{2}} dx;$$
$$s = \int \left\{1 + \left(\frac{m+n}{n} a^{-\frac{m}{n}}\right)^2 x^{\frac{2m}{n}}\right\}^{\frac{1}{2}} dx; \qquad (16)$$

on comparing which with equation (86), Art. 43, the conditions requisite for integration by rationalization are, that either $\frac{n}{2m}$ or $\frac{n}{2m} + \frac{1}{2}$ should be an integer.

From the former of these conditions we have
$$\frac{m+n}{n} = \frac{3}{2}, \text{ or } = \frac{5}{4}, \text{ or } = \frac{7}{6}, \ldots; \qquad (17)$$

and from the latter
$$\frac{m+n}{n} = \frac{2}{1}, \text{ or } = \frac{4}{3}, \text{ or } = \frac{6}{5}, \ldots. \qquad (18)$$

Ex. 6. In the curve whose equation is $ay^2 = x^3$, shew that the length of the arc from the origin to the point (x, y)
$$= \frac{1}{27 a^{\frac{1}{2}}} \{(4a+9x)^{\frac{3}{2}} - (4a)^{\frac{3}{2}}\}.$$

Ex. 7. To determine the length of the arc of the catenary, measured from its lowest point to any point on the curve; see fig. 7.
$$y = \frac{a}{2}\left\{e^{\frac{x}{a}} + e^{-\frac{x}{a}}\right\};$$
$$\therefore \quad dy = \frac{1}{2}\left\{e^{\frac{x}{a}} - e^{-\frac{x}{a}}\right\} dx,$$
$$ds = \frac{1}{2}\left\{e^{\frac{x}{a}} + e^{-\frac{x}{a}}\right\} dx;$$

and taking a *general* value x, which will refer to any point on the curve, for the superior limit,
$$s = \int_0^x \frac{1}{2}\left\{e^{\frac{x}{a}} + e^{-\frac{x}{a}}\right\} dx$$

$$\therefore \quad s = \frac{a}{2}\left[e^{\frac{x}{a}} - e^{-\frac{x}{a}}\right]_0^x$$

$$= \frac{a}{2}\left\{e^{\frac{x}{a}} - e^{-\frac{x}{a}}\right\}; \tag{19}$$

which is the same result as that found in Ex. 4, Art. 293, Vol. I. Hence

$$s^2 = y^2 - a^2; \tag{20}$$

consequently the arc measured from the lowest point of the catenary is equal to the side of a right-angled triangle, of which $y\,(=\text{MP})$ is the hypothenuse, and $a\,(=\text{MN})$ is the other side; that is, AP = PN. This result is also evident by reason of the relation which exists between the equitangential curve and its evolute, which is the catenary.

Ex. 8. Prove that the whole length of the hypocycloid whose equation is $x^{\frac{2}{3}} + y^{\frac{2}{3}} = a^{\frac{2}{3}}$, see fig. 10, is equal to $6a$.

Ex. 9. The equation to a curve being $e^y = \frac{e^x + 1}{e^x - 1}$, prove that the length of the arc between (x_0, y_0) and (x_n, y_n)

$$= \log \frac{e^{x_n} - e^{-x_n}}{e^{x_0} - e^{-x_0}}.$$

156.] The process of rectification is frequently simplified by the use of a subsidiary angle, the form of which is suggested by the equation to the curve. The following are examples of this method.

Ex. 1. Let the circle referred to the centre as origin be expressed by the two equations $x = a\cos\theta$, $y = a\sin\theta$; then

$$dx = -a\sin\theta\, d\theta, \quad dy = a\cos\theta\, d\theta;$$

$$\therefore \quad ds = a\, d\theta;$$

$$s = a(\theta_n - \theta_0), \tag{21}$$

if s is the length of the arc between the points to which θ_n and θ_0 correspond. Thus

$$\text{the length of the quadrant} = a\int_0^{\frac{\pi}{2}} d\theta$$

$$= \frac{\pi a}{2}.$$

Ex. 2. It is required to find the length of the evolute of the ellipse defined by the equation $\left(\frac{x}{\alpha}\right)^{\frac{2}{3}} + \left(\frac{y}{\beta}\right)^{\frac{2}{3}} = 1$.

Let the curve be expressed by means of the two simultaneous equations $x = a(\cos\theta)^3$, $y = \beta(\sin\theta)^3$;

$$\therefore \quad dx = -3a(\cos\theta)^2 \sin\theta\, d\theta, \quad dy = 3\beta(\sin\theta)^2\cos\theta\, d\theta,$$

$$ds = 3\{a^2(\cos\theta)^2 + \beta^2(\sin\theta)^2\}^{\frac{1}{2}} \sin\theta\cos\theta\, d\theta$$

$$= -\frac{3}{4}\left\{\frac{a^2+\beta^2}{2} + \frac{a^2-\beta^2}{2}\cos 2\theta\right\}^{\frac{1}{2}} d.\cos 2\theta.$$

therefore the whole length

$$= -3\int_0^{\frac{\pi}{2}} \left\{\frac{a^2+\beta^2}{2} + \frac{a^2-\beta^2}{2}\cos 2\theta\right\}^{\frac{1}{2}} d.\cos 2\theta$$

$$= 4\frac{a^3-\beta^3}{a^2-\beta^2}. \tag{22}$$

If $a = \beta$, the result is the same as that given in Ex. 8 of the preceding Article.

Ex. 3. Determine the length of the arc of the parabola, $x^{\frac{1}{2}} + y^{\frac{1}{2}} = a^{\frac{1}{2}}$, contained between the coordinate axes.

Let $\qquad x = a(\cos\theta)^4, \quad y = a(\sin\theta)^4$;

$$\therefore \quad dx = -4a(\cos\theta)^3\sin\theta\, d\theta, \quad dy = 4a(\sin\theta)^3\cos\theta\, d\theta;$$

$$\therefore \quad ds = 4a\{(\cos\theta)^4 + (\sin\theta)^4\}^{\frac{1}{2}}\sin\theta\cos\theta\, d\theta$$

$$= -\frac{a}{2^{\frac{1}{2}}}\{1+(\cos 2\theta)^2\}^{\frac{1}{2}} d.\cos 2\theta;$$

$$\therefore \quad \text{the length of the arc} = -\frac{a}{2^{\frac{1}{2}}}\int_0^{\frac{\pi}{2}}\{1+(\cos 2\theta)^2\}^{\frac{1}{2}} d.\cos 2\theta$$

$$= a + \frac{a}{2^{\frac{1}{2}}}\log(2^{\frac{1}{2}}+1). \tag{23}$$

Ex. 4. The cycloid being defined by the equations

$$x = a(1-\cos\theta), \quad y = a(\theta+\sin\theta),$$

prove that the length of the arc beginning at the vertex

$$= 4a\sin\frac{\theta}{2}.$$

157.] In all the preceding examples of rectification, the value of the definite integral which expresses the length of the curve has been determined by means of the indefinite integral: this however is plainly not possible in all cases, and we are obliged to have recourse to those methods of evaluating definite integrals which have been explained in Chapter IV. The number of such cases is infinite; and consequently I shall consider only those which have some special interest. The most important perhaps is that of elliptic arcs; on account of the forms of the definite integrals, the geometrical properties derivable from them, the large number of problems whose solutions depend on these integrals, the history of these functions, and the treatises which have been written on them; and especially on account of the large generalizations and developments which double periodic functions, more general than elliptic integrals, have received at the hands of Abel and Jacobi, and which are now being reduced into systematic treatises. As however it is beyond the scope of the present work to give a systematic account of these discoveries and of the properties of these high transcendents, I may refer the student to (1) Théorie des fonctions doublement périodiques, par M M. Briot et Bouquet; Paris, Mallet-Bachelier, 1859: (2) an Appendix by M. Hermite to the new edition of Lacroix's Differential and Integral Calculus; edited by M M. Hermite and J. A. Serret, Paris, Mallet-Bachelier, 1862: (3) Theorie der Elliptischen Functionen, von Dr. H. Durège; Leipsig, Teubner, 1861: (4) Die Lehre von den Elliptischen Integralen und den Theta-Functionen von K. H. Schellbach; Reimer, Berlin, 1864.

I propose however to take the simple problem of the rectification of the arc of an ellipse. Let the equation to the ellipse be

$$\frac{x^2}{a^2} + \frac{y^2}{b^2} = 1; \tag{24}$$

and let us suppose s to be measured in such a manner that s increases as x increases; then if e is the eccentricity of the ellipse, and x and x_0 refer to two points P and P_0, see fig. 9, on the curve, x being greater than x_0, and s being measured from the point nearer the minor axis towards the major axis,

$$s = \int_{x_0}^{x} \left(\frac{a^2 - e^2 x^2}{a^2 - x^2}\right)^{\frac{1}{2}} dx. \tag{25}$$

Now the indefinite integral of the element in the right-hand member of this equation cannot be determined; (see Art. 80). As however e is less than unity, and x is less than a, we may employ the method of Art. 121, and expand one of the factors of the element-function into a converging series, in ascending powers of $\frac{ex}{a}$, and thereby obtain an approximate value of the integral. Let us assume the arc, whose length is required, to begin at B, the extremity of the minor axis, so that in (25) $x_0 = 0$; then since

$$\frac{(a^2-e^2x^2)^{\frac{1}{2}}}{(a^2-x^2)^{\frac{1}{2}}} = \frac{a}{(a^2-x^2)^{\frac{1}{2}}}\left(1-\frac{e^2x^2}{a^2}\right)^{\frac{1}{2}}$$

$$= \frac{a}{(a^2-x^2)^{\frac{1}{2}}}\left\{1-\frac{e^2x^2}{2a^2}-\frac{e^4x^4}{2.4.a^4}-\frac{1.3.e^6x^6}{2.4.6a^6}-\dots\right\};$$

the arc BP, see fig. 8,

$$= a\int_0^x \frac{dx}{(a^2-x^2)^{\frac{1}{2}}}\left\{1-\frac{e^2x^2}{2a^2}-\frac{e^4x^4}{2.4.a^4}-\frac{1.3.e^6x^6}{2.4.6a^6}-\dots\right\}; \quad (26)$$

and the length of the quadrant of the ellipse

$$= a\int_0^a \frac{dx}{(a^2-x^2)^{\frac{1}{2}}}\left\{1-\frac{e^2x^2}{2a^2}-\frac{e^4x^4}{2.4.a^4}-\frac{1.3.e^6x^6}{2.4.6a^6}-\dots\right\}; \quad (27)$$

but by equation (13), Art. 82, if n is even,

$$\int_0^a \frac{x^n\,dx}{a^n(a^2-x^2)^{\frac{1}{2}}} = \frac{(n-1)(n-3)\dots 3.1}{n(n-2)\dots 4.2}\,\frac{\pi}{2};$$

$$\therefore \int_0^a \frac{a\,dx}{(a^2-x^2)^{\frac{1}{2}}}\left\{1-\frac{e^2x^2}{2a^2}-\frac{e^4x^4}{2.4.a^4}-\frac{1.3.e^6x^6}{2.4.6.a^6}-\dots\right\}$$

$$= \frac{\pi a}{2}\left\{1-\left(\frac{1}{2}\right)^2e^2-\frac{1}{3}\left(\frac{1.3}{2.4}\right)^2e^4-\frac{1}{5}\left(\frac{1.3.5}{2.4.6}\right)^2e^6-\dots\right\}; \quad (28)$$

and therefore the perimeter of the ellipse

$$= 2\pi a\left\{1-\left(\frac{1}{2}\right)^2e^2-\frac{1}{3}\left(\frac{1.3}{2.4}\right)^2e^4-\frac{1}{5}\left(\frac{1.3.5}{2.4.6}\right)^2e^6-\dots\right\}. \quad (29)$$

158.] Although the length of an elliptic arc cannot be expressed generally in finite terms of the rectangular coordinates of its extremities, yet some properties of the arcs may be deduced from the differential of the arc given in (25), and from other equivalent expressions, which deserve a passing notice.

Let the ellipse be defined by the two equations

$$x = a\cos\phi, \qquad y = b\sin\phi; \quad (30)$$

then, supposing, as heretofore, s and x simultaneously to increase,

$$ds = -a\{1-e^2(\cos\phi)^2\}^{\frac{1}{2}}\,d\phi, \quad (31)$$

$$s = -a\int_{\phi_0}^{\phi}\{1-e^2(\cos\phi)^2\}^{\frac{1}{2}}\,d\phi. \quad (32)$$

Hence the perimeter of the ellipse

$$= 4a\int_{\frac{\pi}{2}}^{0}\{1-e^2(\cos\phi)^2\}^{\frac{1}{2}}\,d\phi. \quad (33)$$

Again, taking the equation to the ellipse given by (24), if τ is the acute angle between the x-axis and the tangent at (x, y), ds may be expressed in terms of τ and $d\tau$, as follows;

$$\tan\tau = -\frac{dy}{dx} = \frac{b^2x}{a^2y};$$

$$\therefore x = \frac{a^2\sin\tau}{\{a^2(\sin\tau)^2+b^2(\cos\tau)^2\}^{\frac{1}{2}}}, \quad y = \frac{b^2\cos\tau}{\{a^2(\sin\tau)^2+b^2(\cos\tau)^2\}^{\frac{1}{2}}}; \quad (34)$$

and if $\rho =$ the radius of curvature of the ellipse at (x, y),

$$\rho = \frac{\{a^4y^2+b^4x^2\}^{\frac{3}{2}}}{a^4b^4} = \frac{a^2b^2}{\{a^2(\sin\tau)^2+b^2(\cos\tau)^2\}^{\frac{3}{2}}};$$

Now $ds = \rho\, d\tau$; consequently replacing ρ by its equivalent

$$ds = \frac{a^2b^2\,d\tau}{\{a^2(\sin\tau)^2+b^2(\cos\tau)^2\}^{\frac{3}{2}}} \quad (35)$$

$$= \frac{a(1-e^2)\,d\tau}{\{1-e^2(\cos\tau)^2\}^{\frac{3}{2}}}, \quad (36)$$

the positive sign being taken, so that s increases as τ increases: consequently

$$s = a(1-e^2)\int_{\tau_0}^{\tau}\frac{d\tau}{\{1-e^2(\cos\tau)^2\}^{\frac{3}{2}}}; \quad (37)$$

which is another expression for the length of an elliptic arc.

159.] The definite integrals given in (32) and (37) bear a remarkable relation to each other depending on the equation connecting ϕ and τ. Equating the two values of x given in (30) and (34), we have

$$\cos\phi = \frac{\sin\tau}{\{1-e^2(\cos\tau)^2\}^{\frac{1}{2}}}; \quad (38)$$

$$\therefore \quad \cos\phi\cos\tau = \frac{\sin\tau\cos\tau}{\{1-e^2(\cos\tau)^2\}^{\frac{1}{2}}};$$

$$d.\cos\phi\cos\tau = -\frac{1}{e^2}\{1-e^2(\cos\tau)^2\}^{\frac{1}{2}}d\tau + \frac{1-e^2}{e^2}\frac{d\tau}{\{1-e^2(\cos\tau)^2\}^{\frac{3}{2}}} \quad (39)$$

Now, multiplying this equation by ae^2, and taking the integral of it for the limits τ and τ_0, and taking ϕ and ϕ_0 to be the values of ϕ, at the points to which τ and τ_0 correspond, we have

$$ae^2(\cos\phi\cos\tau-\cos\phi_0\cos\tau_0)$$

$$= -a\int_{\tau_0}^{\tau}\{1-e^2(\cos\tau)^2\}^{\frac{1}{2}}d\tau + a(1-e^2)\int_{\tau_0}^{\tau}\frac{d\tau}{\{1-e^2(\cos\tau)^2\}^{\frac{3}{2}}}. \quad (40)$$

Of the two definite integrals in the right-hand member, the latter, by reason of (37), represents the length of the arc contained between two points to which τ and τ_0 correspond, and which arc is measured from the minor axis towards the major axis; and by reason of (32) the former with a positive sign represents the length of the arc contained between two points to which ϕ and ϕ_0, when they are equal to τ and τ_0 respectively, correspond, and which arc is measured from the major axis towards the minor axis, because s increases as ϕ increases; so that if σ and s denote respectively the former and the latter definite integrals in (40),

$$s-\sigma = ae^2\{\cos\phi\cos\tau-\cos\phi_0\cos\tau_0\}. \tag{41}$$

Let x and x_0 be the abscissæ to the extremities of s, and ξ and ξ_0 the abscissæ to those of σ; then $x = a\cos\phi$, $x_0 = a\cos\phi_0$, $\xi = a\cos\tau$, $\xi_0 = a\cos\tau_0$; so that (41) becomes

$$s-\sigma = \frac{e^2}{a}\left\{x\,\xi-x_0\,\xi_0\right\}. \tag{42}$$

that is, the difference between the length of two elliptic arcs, determined under the preceding conditions, is expressed as a function of the abscissæ of their extremities. The discovery of this theorem is due to Fagnani.

This theorem is exhibited geometrically in fig 9; let PP_0 be the arc whose length is s, P and P_0 being the points to which τ and τ_0 correspond: viz. $\text{P}_0\text{T}_0\text{O} = \tau_0$, $\text{PTO} = \tau$; and let $\text{RON} = \phi$, $\text{R}_0\text{ON}_0 = \phi_0$, where $\phi = \tau$, $\phi_0 = \tau_0$; and from R and R_0 let the ordinates RN, R_0N_0 be drawn, cutting the ellipse in Q and Q_0; then the arc $\text{QQ}_0 = \sigma$; and $\text{OM} = x$, $\text{OM}_0 = x_0$; $\text{ON} = \xi$, $\text{ON}_0 = \xi_0$; and therefore from (42)

$$\text{PP}_0-\text{QQ}_0 = \frac{e^2}{a}\{\text{OM}\times\text{ON}-\text{OM}_0\times\text{ON}_0\}. \tag{43}$$

If P_0 is at B, that is, if the arc is measured from the extremity of the minor axis, then $\tau_0 = 0$, $x_0 = 0$, $\phi_0 = 0$, $\xi_0 = a$, and Q_0 is at A; in which case,

$$\text{BP}-\text{AQ} = \frac{e^2}{a}\,\text{OM}\times\text{ON}; \tag{44}$$

the abscissæ of the points P and Q being connected by the equation (38), which in terms of x and ξ is

$$a^4-a^2x^2-a^2\xi^2+e^2x^2\xi^2 = 0. \tag{45}$$

From O draw a perpendicular OZ on the tangent to the ellipse at the point P which is (x, y); then

$$\text{PZ}^2 = \text{OP}^2 - \text{OZ}^2$$
$$= \frac{a^2-x^2}{a^2-e^2x^2}e^4x^2$$
$$= \frac{e^4}{a^2}x^2\xi^2.$$

so that (44) takes the geometrical form,

$$\text{BP}-\text{AQ} = \text{PZ}.$$

If the points P and Q coincide, then $\xi = x$, and from (45)

$$x = \frac{1}{e}\{a(a-b)\}^{\frac{1}{2}};$$

$$\therefore \quad \text{BP}-\text{AP} = a-b;$$

that is, the difference of the arcs into which the elliptic quadrant is divided is equal to the difference of the semi-axes.

160.] Before we leave the subject which is treated of in the present section, it is desirable to say a few words on certain forms of a general character of those curves which being expressed in terms of x and y, are susceptible of rectification; that is, to consider under what circumstances we can find a general integral of $ds^2 = dx^2+dy^2$.

Now a general form which evidently satisfies this equation is

$$ds^2 = (dx\cos a - dy\sin a)^2 + (dx\sin a + dy\cos a)^2, \qquad (46)$$

where a represents an arbitrary angle: and it is satisfied by

$$ds = dx\cos a - dy\sin a,$$
$$0 = dx\sin a + dy\cos a;$$

whence integrating

$$s = x\cos a - y\sin a + f(a),$$
$$0 = x\sin a + y\cos a + \phi(a);$$

where $f(a)$ and $\phi(a)$ are two arbitrary constants of integration. Now to combine these so that they may form an envelope and thus a curve, let us take the a-differential of each; then

$$0 = -x\sin a - y\cos a + f'(a),$$
$$0 = x\cos a - y\sin a + \phi'(a);$$
$$\therefore \quad \phi(a) = -f'(a),$$
$$\phi'(a) = -f''(a);$$

and consequently we have

$$\left.\begin{aligned} s &= x\cos a - y\sin a + f(a) \\ 0 &= x\sin a + y\cos a - f'(a) \\ 0 &= x\cos a - y\sin a - f''(a); \end{aligned}\right\} \qquad (47)$$

where f represents an arbitrary function; and hence we have

$$\left.\begin{aligned} x &= \sin a f'(a)+\cos a f''(a) \\ y &= \cos a f'(a)-\sin a f''(a) \\ s &= f(a)+f''(a); \end{aligned}\right\} \qquad (48)$$

the values of which manifestly satisfy (1).

There are also other general forms which satisfy (1): such as

$$x = s\,\phi(a)+a,$$
$$y = s\{1-(\phi(a))^2\}^{\frac{1}{2}}+\psi(a),$$

where ϕ and ψ denote arbitrary functions: and taking the a-differentials we have

$$0 = s\phi'(a)+1,$$
$$0 = \frac{-s\,\phi(a)\,\phi'(a)}{\{1-(\phi(a))^2\}^{\frac{1}{2}}}+\psi'(a);$$

hence we have the system of equations

$$s^2 = \{x-a\}^2+\{y-\psi(a)\}^2$$
$$0 = (x-a)+\{y-\psi(a)\}\,\psi'(a),$$
$$0 = -1-\{\psi'(a)\}^2+\{y-\psi(a)\}\,\psi''(a),$$

which are plainly equivalent to those by means of which the equation to an evolute is determined from that to the involute: and which is accordant with the fact that all evolutes are rectifiable.

It is worth observing, that if the equations to a plane curve in terms of x and y are of the form of the second and third in (47), or of the first two in (48), the length of the curve is given by the remaining one of each group.

SECTION 2.—*Rectification of Plane Curves referred to Polar Coordinates.*

161.] Let $F(r,\theta) = 0$ be the equation to a plane curve in terms of polar coordinates; and let it be required to determine the length of the curve between the points (r_n, θ_n), and (r_0, θ_0): Now if ds is an infinitesimal length-element, by Vol. I, Art. 269,

$$ds = \{dr^2+r^2\,d\theta^2\}^{\frac{1}{2}};$$

$$\therefore \quad s = \int\{dr^2+r^2\,d\theta^2\}^{\frac{1}{2}}, \qquad (49)$$

the integral being taken between the limits assigned by the conditions of the problem.

If the equation to the curve admits of being put into the form $r = f(\theta)$, then

$$dr = f'(\theta)\,d\theta;$$

$$s = \int_{\theta_0}^{\theta_n} \{(f(\theta))^2 + (f'(\theta))^2\}^{\frac{1}{2}}\,d\theta. \tag{50}$$

And if the equation to the curve is put into the form $\theta = f(r)$, so that $d\theta = f'(r)\,dr$,

$$s = \int_{r_0}^{r_n} \{1 + r^2(f'(r))^2\}^{\frac{1}{2}}\,dr. \tag{51}$$

162.] Examples of rectification of plane curves referred to polar coordinates.

Ex. 1. The spiral of Archimedes measured from the pole.

Let the equation to the curve be $r = a\theta$; so that $dr = a\,d\theta$;

$$\begin{aligned} \therefore \quad s &= \int_0^{\theta_n} a(1+\theta^2)^{\frac{1}{2}}\,d\theta \\ &= \frac{a}{2}\Big[\theta(1+\theta^2)^{\frac{1}{2}} + \log\{\theta + (1+\theta^2)^{\frac{1}{2}}\}\Big]_0^{\theta_n} \\ &= \frac{a}{2}\{\theta_n(1+\theta_n^2)^{\frac{1}{2}} + \log(\theta_n + (1+\theta_n^2)^{\frac{1}{2}})\}; \end{aligned} \tag{52}$$

and if s is expressed in terms of r_n,

$$s = \frac{r_n(a^2+r_n^2)^{\frac{1}{2}}}{2a} + \frac{a}{2}\log\frac{r_n + (a^2+r_n^2)^{\frac{1}{2}}}{a}. \tag{53}$$

Ex. 2. The Logarithmic Spiral.

Let the equation to the curve be $r = a^{\theta}$; then since

$$dr = \log a . a^{\theta}\,d\theta = \log a . r\,d\theta,$$

$$\begin{aligned} s &= \int_{r_0}^{r_n} \frac{\{1+(\log a)^2\}^{\frac{1}{2}}}{\log a}\,dr \\ &= \frac{\{1+(\log a)^2\}^{\frac{1}{2}}}{\log a}(r_n - r_0); \end{aligned} \tag{54}$$

a result which immediately follows from the fact that the curve cuts all its radii vectores at a constant angle, and therefore that the difference between any two radii vectores is equal to the projection of the length of the curve between the corresponding points on a line to which it is inclined at the constant angle.

Ex. 3. The Circle, the extremity of a diameter being the pole. In this case the equation is $r = 2a\cos\theta$; consequently

$$ds = 2a\,d\theta;$$

$$\therefore \quad s = 2a\,(\theta_n - \theta_0). \tag{55}$$

If the arc begins at the extremity of the diameter, $\theta_0 = 0$; so that

$$s = 2a\,\theta_n; \tag{56}$$

and if $\theta_n = \frac{\pi}{2}$, the semi-perimeter $= \pi a$.

Ex. 4. The whole length of the cardioid, whose equation is $r = a(1+\cos\theta)$, is $8a$.

Ex. 5. If the equation to the lemniscata is $r^2 = a^2\cos 2\theta$, and $s =$ the length of a loop,

$$s = 2a^2\int_a^0 \frac{dr}{(a^4-r^4)^{\frac{1}{2}}} = a\int_{-\frac{\pi}{4}}^{\frac{\pi}{4}} \frac{d\theta}{(\cos 2\theta)^{\frac{1}{2}}}.$$

This last value may be expressed in terms of the Gamma-function. For let 2θ be replaced by $\frac{\pi}{2} - 2\theta$; then

$$s = a\int_0^{\frac{\pi}{2}} \frac{d\theta}{(\sin 2\theta)^{\frac{1}{2}}}.$$

In (271), Art. 126, let $m = n = \frac{1}{4}$; then

$$2\int_0^{\frac{\pi}{2}} \frac{d\theta}{(\sin\theta\cos\theta)^{\frac{1}{2}}} = 2^{\frac{3}{2}}\int_0^{\frac{\pi}{2}} \frac{d\theta}{(\sin 2\theta)^{\frac{1}{2}}} = \mathrm{B}\left(\frac{1}{4}, \frac{1}{4}\right)$$

$$= \frac{\left\{\Gamma\left(\frac{1}{4}\right)\right\}^2}{\Gamma\left(\frac{1}{2}\right)} = \frac{\left\{\Gamma\left(\frac{1}{4}\right)\right\}^2}{\pi^{\frac{1}{2}}};$$

$$\therefore \quad s = a\,2^{-\frac{3}{2}}\pi^{-\frac{1}{2}}\left\{\Gamma\left(\frac{1}{4}\right)\right\}^2; \quad \text{and} \quad \Gamma\left(\frac{1}{4}\right) = \left(\frac{s}{a}\right)^{\frac{1}{2}} 2^{\frac{3}{4}}\pi^{\frac{1}{4}}; \tag{57}$$

which gives a geometrical interpretation of $\Gamma\left(\frac{1}{4}\right)$.

163.] If the equation to the curve is given in terms of r and p, then, by Vol. I, Art. 271, (23),

$$ds = \frac{r\,dr}{(r^2-p^2)^{\frac{1}{2}}}; \tag{58}$$

$$\therefore \quad s = \int \frac{r\,dr}{(r^2-p^2)^{\frac{1}{2}}}; \tag{59}$$

the integral being taken between limits assigned by the problem.

Ex. 1. To find the length of the involute of the circle, whose equation is $r^2 = a^2+p^2$, between any two given points on it.

$$\therefore \quad s = \int_{r_0}^{r_n} \frac{r\,dr}{a} = \frac{1}{2a}(r_n^2-r_0^2). \tag{60}$$

and if s begins at the point where the involute leaves the circle, $r_0 = a$; and

$$s = \frac{r^2-a^2}{2a} = \frac{p^2}{2a}. \tag{61}$$

Ex. 2.] It is required to find the whole length of the hypocycloid whose equation is $x^{\frac{2}{3}}+y^{\frac{2}{3}} = a^{\frac{2}{3}}$; see fig. 10.

The equation in terms of r and p is $3p^2 = a^2-r^2$;

$$\therefore \quad \text{the whole length} = -8\int_a^{\frac{a}{2}} \frac{3^{\frac{1}{2}}\,r\,dr}{(4r^2-a^2)^{\frac{1}{2}}}$$

$$= -8\,\frac{3^{\frac{1}{2}}}{4}\Big[(4r^2-a^2)^{\frac{1}{2}}\Big]_a^{\frac{a}{2}} = 6a. \tag{62}$$

SECTION 3.—*The Rectification of Curves in Space.*

164.] The infinitesimal length-element of a curve in space, whether plane or non-plane, has been determined in Art. 341, Vol. I, and is given by the equation

$$ds = (dx^2+dy^2+dz^2)^{\frac{1}{2}}; \tag{63}$$

so that, by integration between the given limits, the length of the arc of the curve may be found.

If the equations to the curve are given in the form $x = f(z)$, $y = \phi(z)$, then

$$dx = f'(z)\,dz, \qquad dy = \phi'(z)\,dz;$$

$$\therefore \quad ds = \{(f'(z))^2+(\phi'(z))^2+1\}^{\frac{1}{2}}\,dz; \tag{64}$$

$$s = \int_{z_0}^{z_n} \{(f'(z))^2+(\phi'(z))^2+1\}^{\frac{1}{2}}\,dz. \tag{65}$$

If the equations are given in terms of another variable, say ϕ, and are of the forms

$$x = f(\phi), \qquad y = F(\phi), \qquad z = \psi(\phi);$$

then $\quad dx = f'(\phi)\,d\phi, \quad dy = F'(\phi)\,d\phi, \quad dz = \psi'(\phi)\,d\phi;$

$$s = \int_{\phi_0}^{\phi_n} \{(f'(\phi))^2+(F'(\phi))^2+(\psi'(\phi))^2\}^{\frac{1}{2}}\,d\phi. \tag{66}$$

And if the equations are given in the form

$$F(x, y, z), = 0, \quad f(x, y, z), = 0, \tag{67}$$

ds may be found by means of the total differentials of these two functions.

Ex. 1. To determine the length of the helix between two given points.

Taking the equations to the curve as found in Vol. 1, Art. 347, (32),

$$x = a\cos\phi, \qquad y = a\sin\phi, \qquad z = ka\phi;$$

$$\therefore \quad dx = -a\sin\phi\, d\phi, \qquad dy = a\cos\phi\, d\phi, \qquad dz = ka\, d\phi;$$

$$ds^2 = dx^2 + dy^2 + dz^2 = a^2(1+k^2)\, d\phi^2;$$

$$\begin{aligned} \therefore \quad s &= a(1+k^2)^{\frac{1}{2}} \int_{\phi_0}^{\phi_n} d\phi \\ &= a(1+k^2)^{\frac{1}{2}} (\phi_n - \phi_0) \\ &= \frac{(1+k^2)^{\frac{1}{2}}}{k} (z_n - z_0). \end{aligned}$$

If the arc begins at the point where $z_0 = 0$, then

$$s = \frac{(1+k^2)^{\frac{1}{2}}}{k} z_n; \tag{68}$$

a result which also follows immediately from the geometrical generation of the curve.

Ex. 2. To determine the length of the curve formed by the intersection of two right cylinders, of which one is parabolic and perpendicular to the plane of (y, x), and the other is cycloidal and perpendicular to the plane of (x, z).

Let the equations to the director-curves of the cylinders be

$$y^2 = 4cx; \quad \text{and} \quad z = a\,\text{versin}^{-1}\frac{x}{a} + (2ax - x^2)^{\frac{1}{2}};$$

$$\therefore \quad dy = \left(\frac{c}{x}\right)^{\frac{1}{2}} dx; \qquad dz = \left(\frac{2a-x}{x}\right)^{\frac{1}{2}} dx;$$

$$\therefore \quad ds^2 = dx^2 + \frac{c}{x} dx^2 + \frac{2a-x}{x} dx^2;$$

$$\begin{aligned} \text{and} \quad s &= \int_0^{x_n} (c+2a)^{\frac{1}{2}} \frac{dx}{x^{\frac{1}{2}}} \\ &= 2(c+2a)^{\frac{1}{2}} x_n^{\frac{1}{2}}. \end{aligned} \tag{69}$$

Ex. 3. Prove that the length of the curve of intersection of the elliptic cylinder, $a^2y^2 + b^2x^2 = a^2b^2$, with the sphere $x^2 + y^2 + z^2 = a^2$, is equal to $2\pi a$.

In Vol. XIII of Liouville's Journal is a memoir by M. J. A. Serret which contains a solution, by a process somewhat similar to that given in Art. 160, of the equation

$$ds^2 = dx^2 + dy^2 + dz^2;$$

and observations on the mode of solution will be found in Art. 8 of Note I, appended to Liouville's Edition of Monge's Application d'Analyse.

165.]. It is frequently more convenient for purposes of rectification, as well as for other problems which will be subsequently investigated, to refer the position of a point in space to polar coordinates, the system of which is as follows.

Let E, fig. 42, be the point in space whose position is to be determined. Take a fixed point O for origin, and through it draw a plane which I will call the fundamental plane, and, to fix our thoughts, will assume to be horizontal. Through O draw Oz perpendicular to, and also any line Ox in, the fundamental plane. Join OE; OE is the radius vector of E and is denoted by r, and the angle EOz, at which OE is inclined to Oz, $= \theta$. Let a plane be drawn through Oz and OE, and let ON be the line in which this plane cuts the fundamental plane; then as these two planes are perpendicular to each other, ON is the projection of OE on the fundamental plane. ON is called the curtate radius vector, and is denoted by ρ; so that ON $= \rho = r \sin \theta$. Also let another plane be drawn through Oz and Ox, and let ϕ be the angle between it and the plane zOE; or, what is the same thing, let ϕ be the angle xON; then these quantities r, θ, ϕ completely determine the position of E, and are consequently called the polar coordinates of E; so that when they are given, the place of E is also given. This system is equivalent to a spherical system of coordinates, and is that by which position on the surface of the earth is usually determined. O is the centre of the earth, Oz is the polar axis, the fundamental plane is the plane of the equator, the plane zOx is that from which longitude is measured, and in the English system passes through Greenwich Observatory; thus ϕ is the longitude and θ is the co-latitude, and when they are given, position on the surface of a sphere is given.

If r is of given length, and θ and ϕ vary, the extremity of r describes a sphere. If θ is constant and ϕ varies, r describes a right conical surface, which intersects a spherical surface along a parallel of latitude. If ϕ is constant and θ varies, r describes a meridian plane. It is evident that the sphere, the cone, and the meridian plane intersect orthogonally at every common point.

To compare this system with that of rectangular coordinates, let the fundamental plane be that of (x, y); and in it let Oy be drawn perpendicular to Ox: let E be (x,y,z); then, since OM $= x$, MN $= y$, NP $= z$; OE $= r$, ON $= \rho$, zOE $= \theta$, xON $= \phi$, therefore

$$\left.\begin{array}{l} z = r\cos\theta, \\ \rho = r\sin\theta; \end{array}\right\} \quad \left.\begin{array}{l} x = \rho\cos\phi = r\sin\theta\cos\phi, \\ y = \rho\sin\phi = r\sin\theta\sin\phi; \end{array}\right\} \tag{70}$$

and by means of these equivalents, an equation may be transformed from one system to another.

In reference to the problem of rectification, let the preceding equivalents for x, y, z be differentiated; then we have

$$dx = \sin\theta\cos\phi\, dr + r\cos\theta\cos\phi\, d\theta - r\sin\theta\sin\phi\, d\phi,$$

$$dy = \sin\theta\sin\phi\, dr + y\cos\theta\sin\phi\, d\theta + r\sin\theta\cos\phi\, d\phi,$$

$$dz = \cos\theta\, dr \qquad - r\sin\theta\, d\theta;$$

$$\therefore \quad ds = (dx^2 + dy^2 + dz^2)^{\frac{1}{2}}$$

$$= \{dr^2 + r^2 d\theta^2 + r^2(\sin\theta)^2 d\phi^2\}^{\frac{1}{2}}. \tag{71}$$

This last expression however may be determined as follows.

As ds is the distance between two points infinitesimally near to one another in space, let E and H be the two points, see fig. 42, where E is (r, θ, ϕ), and H is $(r + dr, \theta + d\theta, \phi + d\phi)$. Let EJ $= dr$; and, ϕ being constant, let the radius vector OE revolve through an infinitesimal angle $d\theta$ in the meridian plane thus determined, so that E describes the small circular arc EF which $= r\,d\theta$; and finally let the meridian plane revolve through an infinitesimal angle $d\phi$, whereby, θ being constant, E describes the small circular arc EI which $=$ NQ $= \rho\, d\phi = r\sin\theta\, d\phi$. Now these three lines EF, EI, EJ, which meet at E, form such a system that each is at right angles to the other two; they may consequently be considered a system of rectangular coordinates, which assigns the coordinates of the point H; and consequently, as EH $= ds$,

$$ds = \{dr^2 + r^2\, d\theta^2 + r^2(\sin\theta)^2\, d\phi^2)^{\frac{1}{2}}.$$

Ex. 1. Determine the length of the curve of intersection of the right cone and the sphere whose equations are respectively

$$\theta = a, \text{ and } r = a.$$

Here $\qquad ds = a\sin a\, d\phi;$

$$\therefore \quad \text{the whole length} = 2\pi a \sin a. \tag{72}$$

Ex. 2. A sphere of radius a is intersected by a right circular cylinder whose diameter is a and which passes through the centre of the sphere; find the length of the curve of intersection.

The equations of the sphere and the cylinder are respectively $r = a$, and $\rho = a\cos\phi$; so that $\sin\theta = \cos\phi$; and consequently

$$s = 4a\int_0^{\frac{\pi}{2}} \{1+(\cos\phi)^2\}^{\frac{1}{2}}\, d\phi$$

$$= 4a2^{\frac{1}{2}}\int_0^{\frac{\pi}{2}} \left\{1-\frac{(\sin\phi)^2}{2}\right\}^{\frac{1}{2}} d\phi;$$

on comparing which with the expression for the perimeter of an ellipse given in (33), Art. 158, it appears that the length of the curve is equal to that of an ellipse of which the major axis $= 2^{\frac{3}{2}}a$, and the eccentricity $= \dfrac{1}{2^{\frac{1}{2}}}$.

Ex. 3. Determine the length of the curve of intersection of the right cone, whose equation is $\theta = a$, with the skew helicoid whose equation is $r\cos\theta = k\phi$.

$$s = k\tan a\int_0^{\phi} \{(\operatorname{cosec} a)^2+\phi^2\}^{\frac{1}{2}}\, d\phi$$

$$= \frac{k\tan a}{2}\left\{\phi\{(\operatorname{cosec} a)^2+\phi^2\}^{\frac{1}{2}}+(\operatorname{cosec} a)^2\log\frac{\phi+\{(\operatorname{cosec} a)^2+\phi^2\}^{\frac{1}{2}}}{\operatorname{cosec} a}\right\}.$$

This problem may also be stated in the following form. Find the length of the path described by a point which moving on the surface of a right circular cone uniformly recedes from the vertex while its meridian plane revolves uniformly.

In this aspect the equations are, $r = at$, $\theta = a$, $\phi = bt$, where t is a variable, and a and b are constants; and the expression for s is of the same form as the preceding.

Ex. 4. Find the equation to that curve on the surface of a sphere which cuts all the meridian lines at the same angle a.

Let the radius of the sphere $= a$; then as $a\,d\theta$ is the projection of ds on the meridian line, the differential equation of the curve is $a\,d\theta = \cos a\, ds$; whence we have

$$\frac{d\theta}{\sin\theta} = \cot a\, d\phi.$$

$$\therefore\quad \tan\frac{\theta}{2} = e^{\phi\cot a},$$

the limits being such that $\theta = \frac{\pi}{2}$, when $\phi = 0$. Hence from the differential equation of the curve, if s is measured from the point where $\theta = \frac{\pi}{2}$, $\phi = 0$,

$$s = a \sec \alpha \left(\theta - \frac{\pi}{2}\right);$$

so that when the curve reaches the pole and $\theta = \pi$, $s = \frac{\pi a}{2} \sec \alpha$; in which case $\phi = \infty$; whence it appears that the number of revolutions made by the curve on the surface of the sphere is infinite. The curve is called the Loxodrome.

SECTION 4.—*Investigation of various properties of Curves depending on the length of the Arc and other quantities in terms of which the equation is expressed. The Intrinsic Equation of a Curve.*

166.] In the preceding sections we have expressed the length of a curve between two given points, in terms of the coordinates of those points or at least of one of the coordinates. I propose now to investigate other theorems of curves depending on the length of the arc between two given points, and certain properties of the curve at these points; and in the first place I will take the inverse problem, and find the equation of a curve in terms of x and y, when an equation is given in terms of s and of one or both of the coordinates x and y.

Let us suppose the relation between s, x, and y to be given in the explicit form

$$s = f(x, y); \tag{73}$$

then

$$ds = \left(\frac{df}{dx}\right) dx + \left(\frac{df}{dy}\right) dy; \tag{74}$$

$$\therefore \quad (dx^2 + dy^2)^{\frac{1}{2}} = \left(\frac{df}{dx}\right) dx + \left(\frac{df}{dy}\right) dy; \tag{75}$$

which will give an equation in terms of x, y, dx, dy; and from which in many cases the integral equation may be deduced by the processes already explained; in other cases processes will be required, which belong to a more advanced part of our treatise.

Ex. 1. Let $\quad s = x \cos \alpha + y \sin \alpha + c$;

$$\therefore \quad ds = dx \cos \alpha + dy \sin \alpha;$$

$$dx^2 + dy^2 = (\cos \alpha)^2 dx^2 + 2 \sin \alpha \cos \alpha \, dx \, dy + (\sin \alpha)^2 dy^2.$$

$$(\sin \alpha)^2 dx^2 - 2 \sin \alpha \cos \alpha \, dx \, dy + (\cos \alpha)^2 dy^2 = 0,$$

$$\sin \alpha \, dx - \cos \alpha \, dy = 0;$$

and taking the integral between the limits x, y and x_0, y_0, we have

$$\frac{x-x_0}{\cos a} = \frac{y-y_0}{\sin a}; \qquad (76)$$

which expresses a straight line perpendicular to the line

$$x\cos a + y\sin a + c = 0.$$

Ex. 2. $$s^2 = x^2+y^2.$$

$$\therefore \quad s\,ds = x\,dx + y\,dy;$$

$$(x^2+y^2)(dx^2+dy^2) = (x\,dx+y\,dy)^2,$$

$$y^2\,dx^2 - 2xy\,dx\,dy + x^2\,dy^2 = 0;$$

$$\frac{dx}{x} - \frac{dy}{y} = 0;$$

$$\frac{x}{x_0} = \frac{y}{y_0}; \qquad (77)$$

which is the equation to a straight line passing through the origin; and which evidently satisfies the given relation.

Ex. 3. $s^2 = 8ax. \quad \therefore \quad ds = \left(\frac{2a}{x}\right)^{\frac{1}{2}} dx;$

$$x(dx^2+dy^2) = 2a\,dx^2;$$

$$\therefore \quad dy = \left(\frac{2a-x}{x}\right)^{\frac{1}{2}} dx = \frac{2a-x}{(2ax-x^2)^{\frac{1}{2}}}\,dx.$$

$$\therefore \quad y = a\,\text{versin}^{-1}\frac{x}{a} + (2ax-x^2)^{\frac{1}{2}}; \qquad (78)$$

which is the equation to a cycloid, whose vertex is at the origin, and the radius of whose generating circle is a. The problem is evidently the inverse one to that of Ex. 3, Art. 155.

Ex. 4. If $s^3 = ax^2$, the equation to the curve in terms of x and y is $x^{\frac{2}{3}}+y^{\frac{2}{3}} = k^{\frac{2}{3}}$, where $4a^{\frac{2}{3}} = 9k^{\frac{2}{3}}$.

This example is a particular case of a more general problem. Determine the values of m and n, when the equation to a curve can be expressed in finite terms of x and y, the defining property of the curve being $s^{m+n} = a^m x^n$.

Ex. 5. If $s^2 = y^2-a^2$, the curve is the catenary whose equation is

$$y = \frac{a}{2}\left\{e^{\frac{x}{a}} + e^{-\frac{x}{a}}\right\}.$$

Ex. 6. What plane curves satisfy the equation $s = a\theta$?

$$ds = a\,d\theta;$$

$$\therefore \quad dr^2 + r^2\,d\theta^2 = a^2\,d\theta^2;$$

$$\frac{\pm\, dr}{(a^2-r^2)^{\frac{1}{2}}} = d\theta;$$

$$\therefore \quad r = a\cos\theta; \quad \text{or } r = a\sin\theta.$$

which is the equation to a circle, whose diameter is a, and whose pole is at a point on the circumference.

Ex. 7. If $s = a\dfrac{dy}{dx}$, the equation to the curve is

$$y = \frac{a}{2}\left\{e^{\frac{x}{a}} + e^{-\frac{x}{a}}\right\}.$$

Ex. 8. If $s = a\tan^{-1}\dfrac{dy}{dx}$, then $y^2+x^2 = a^2$. ?

Ex. 9. If $s = a\tan^{-1}\dfrac{r\,d\theta}{dr}$, then $r = a\cos\theta$.

Other examples involving a relation between the length of a curve, and quantities contained in the equation to the curve will be given hereafter, when the integrals of more complicated functions have been investigated.

167.] The fundamental relations which exist between the length-element of a curve, its projections on the x- or on the y-axis, and the angle at which either the tangent or the normal at the point (x, y) is inclined to the x-axis, suggest another mode of expressing a curve which is a sufficient defining property of it.

Two quantities are of course sufficient to define the plane curve; and I propose to take the length or the length-element, and the angle which the normal makes with the x-axis, for the purpose; we shall hereby obtain a relation which will express the length of the arc as a function, implicit or explicit, of the angle at which the normals at its extremities are inclined to each other. As this equation is evidently independent of any origin or of any system of coordinates to which the curve may be referred, it has been named by Dr. Whewell* the intrinsic equation to the curve.

Let dx and dy be the projections of ds, the length-element, on the axes of x and y respectively, and let ψ be the angle which the normal at (x, y), makes with the x-axis, and let ρ be the radius of curvature at the point (x, y): then

$$\frac{dx}{\sin\psi} = \frac{dy}{\cos\psi} = ds; \tag{79}$$

$$\text{also} \quad ds = \rho\, d\psi. \tag{80}$$

* See two Memoirs by Dr. Whewell in vol. VIII, p. 659, and vol. IX. p. 150, of the Cambridge Philosophical Transactions.

Now, if by means of these equations, and that of a curve given in terms of x and y, x and y are eliminated, the resulting equation is of the form

$$F(s, \psi) = 0, \tag{81}$$

and this equation is the intrinsic equation to the curve.

If the equation to the curve, of which the intrinsic equation is to be found, is given in polar coordinates, then if $\phi = \tan^{-1}\frac{r\,d\theta}{dr}$,

$$d\psi = d\theta + d\phi; \tag{82}$$

and by means of this equation and that to the curve, a relation may be determined in terms of s and ψ, of the form (81).

The process of finding the intrinsic equation by means of the ordinary equation in terms of x and y, or of r and θ, may of course be inverted, and the general equation may be found from the intrinsic equation, if the requisite integrations are possible.

168.] Examples of intrinsic equations.

Ex. 1. The intrinsic equation of the circle.

If in (80) we replace ρ by a, which we will take to be the radius of the circle,

$$s = a\psi, \tag{83}$$

which is the intrinsic equation to the circle.

Ex. 2. The intrinsic equation of the parabola, $y^2 = 4ax$.

$$\tan\psi = \frac{dx}{dy} = \left(\frac{x}{a}\right)^{\frac{1}{2}};$$

$$\therefore \quad x = a(\tan\psi)^2.$$

$$dx = 2a\tan\psi(\sec\psi)^2 d\psi,$$

$$ds = \frac{dx}{\sin\phi} = 2a(\sec\psi)^3 d\psi.$$

Consequently if s begins at the vertex of the parabola, at which point $\psi = 0$,

$$s = a\left\{\log\tan\left(\frac{\pi}{4}+\frac{\psi}{2}\right) + \tan\psi\sec\psi\right\}; \tag{84}$$

$$\text{also} \quad \rho = \frac{ds}{d\psi} = 2a(\sec\psi)^3.$$

Ex. 3. The intrinsic equation of the ellipse.

$$s = \int_{\psi_0}^{\psi} \frac{a(1-e^2)\,d\psi}{\{1-e^2(\sin\psi)^2\}^{\frac{3}{2}}}; \tag{85}$$

which result is identical with that given in (37), Art. 158.

Ex. 4. The intrinsic equation of the cycloid.

$$y = (2ax - x^2)^{\frac{1}{2}} + a \operatorname{versin}^{-1}\frac{x}{a}.$$

$$\therefore \quad \frac{dx}{dy} = \tan\psi = \left(\frac{x}{2a-x}\right)^{\frac{1}{2}};$$

$$x = 2a(\sin\psi)^2. \tag{86}$$

Therefore $$ds = \frac{dx}{\sin\psi} = 4a\cos\psi\, d\psi;$$

$$s = 4a(\sin\psi - \sin\psi_0). \tag{87}$$

If s begins at the highest point (the vertex) of the cycloid, $\psi_0 = 0$, so that in this case $s = 4a\sin\psi$.

Ex. 5. The intrinsic equation of the catenary whose equation is

$$y = \frac{a}{2}\left\{e^{\frac{x}{a}} + e^{-\frac{x}{a}}\right\}, \text{ is } \quad s = a\{\cot\psi - \cot\psi_0\};$$

and if s begins at the lowest point of the curve, $s = a\cot\psi$.

In this case $\rho = -a(\operatorname{cosec}\psi)^2$.

Ex. 6. In the curve whose equation is $x^{\frac{2}{3}} + y^{\frac{2}{3}} = a^{\frac{2}{3}}$, if s begins when $\psi = 0$,

$$s = \frac{3a}{4}(1 - \cos 2\psi) = \frac{3a}{2}(\sin\psi)^2. \tag{88}$$

$$\because \quad \rho = \frac{3a}{2}\sin 2\psi.$$

Ex. 7. If $y = e^{\frac{x}{a}}$,

$$s = a\left\{\operatorname{cosec}\psi - \operatorname{cosec}\psi_0 - \frac{1}{2}\log\frac{(1+\sin\psi)(1-\sin\psi_0)}{(1-\sin\psi)(1+\sin\psi_0)}\right\}. \tag{89}$$

Ex. 8. If $r = a^{\theta}$; then in the formula (82), $d\phi = 0$; and $d\psi = d\theta$; so that $\theta = \psi - \psi_0$, where ψ_0 is the value of ψ when $\theta = 0$; then $$ds = \{1 + (\log a)^2\}^{\frac{1}{2}} a^{\theta}\, d\theta;$$

$$\therefore \quad s = \frac{\{1+(\log a)^2\}^{\frac{1}{2}}}{\log a}(a^{\theta} - 1) \tag{90}$$

$$= \frac{\{1+(\log a)^2\}^{\frac{1}{2}}}{\log a} a^{-\psi_0}\{a^{\psi} - a^{\psi_0}\} \tag{91}$$

if $s = 0$, when $\theta = 0$; that is, when $\psi = \psi_0$.

Hence also $\rho = \{1 + (\log a)^2\}^{\frac{1}{2}} a^{\psi - \psi_0}$.

Ex. 9. Find the intrinsic equation to the Epicycloid whose equations are

$$\left.\begin{aligned} x &= (a+b)\cos\theta - b\cos\frac{a+b}{b}\theta, \\ y &= (a+b)\sin\theta - b\sin\frac{a+b}{b}\theta. \end{aligned}\right\} \qquad (92)$$

From the preceding equations and (79) we have

$$\frac{\sin\frac{a+b}{b}\theta - \sin\theta}{\sin\psi}\,d\theta = \frac{\cos\theta - \cos\frac{a+b}{b}\theta}{\cos\psi}\,d\theta = \frac{ds}{a+b};$$

$$\therefore \quad \sin\left(\frac{a+b}{b}\theta+\psi\right) = \sin(\theta+\psi);$$

$$\therefore \quad \theta = \frac{b}{a+2b}(\pi-2\psi).$$

Now let us suppose s to begin at the cusp, where $\theta = 0$, and $\psi = \frac{\pi}{2}$; so that ψ decreases as s increases; then we have

$$\frac{ds}{a+b} = -\frac{4b}{a+2b}\sin\frac{a}{a+2b}\left(\frac{\pi}{2}-\psi\right)d\psi;$$

$$\therefore \quad s = \frac{4b(a+b)}{a}\left\{1-\cos\frac{a}{a+2b}\left(\frac{\pi}{2}-\psi\right)\right\} \qquad (93)$$

$$= \frac{8b(a+b)}{a}\left\{\sin\frac{a}{a+2b}\left(\frac{\pi}{4}-\frac{\psi}{2}\right)\right\}^2; \qquad (94)$$

which is the required intrinsic equation.

If $b=a$, the epicycloid becomes the cardioide, and (94) becomes

$$s = 16a\left(\sin\frac{\pi-2\psi}{12}\right)^2. \qquad (95)$$

And in this case if $\psi = -\pi$, $s = 8a$, which is half the perimeter of the cardioide.

Ex. 10. The intrinsic equation to the involute of the circle is

$$s = \frac{a}{2}\left(\psi-\frac{\pi}{2}\right)^2. \qquad (96)$$

SECTION 5.—*Involutes of Plane Curves.*

169.] Another geometrical problem of considerable interest, which requires single integration in its solution, is that of the discovery of an involute of a curve, when the curve is given. Our investigation will comprise the determination of certain general formulæ which express reciprocal properties of evolutes and involutes, and the application of these formulæ to certain examples.

Let $A\Pi$, fig. 11, be a part of the curve whose involute is to be found. Let $ON = \xi$, $N\Pi = \eta$, and let the equation to $A\Pi$ be

$$\eta = f(\xi); \tag{97}$$

let the length-element of this curve be $d\sigma$, and let $P\Pi$ be the tangent at Π, whose extremity P is the generating point of the involute; so that $P\Pi$ is the radius of curvature of the involute. Let $P\Pi = \rho$, then, by Vol. I, Art. 292, equation (40),

$$d\sigma = d\rho;$$

$$\therefore \quad \rho = \sigma + c; \tag{98}$$

c being a constant, the value of which depends on the position of the point on $A\Pi$, at which σ begins. Thus in fig. 11, if $A\Pi = \sigma$, and $\Pi P = A\Pi$, then $\sigma = \rho$, and $c = 0$; and if $P\Pi$ be longer than $A\Pi$, then c is the excess of length; that is, if a string of the length $P\Pi$ is wrapped round $A\Pi$, and ultimately becomes a tangent at A, c is the length of the string that remains when the wrapping is complete.

Let $OM = x$, $MP = y$; then, since $\frac{d\eta}{d\xi} = \tan \Pi TN$,

$$\frac{\sin \Pi TN}{d\eta} = \frac{\cos \Pi TN}{d\xi} = \frac{1}{d\sigma}; \tag{99}$$

and from the geometry of the figure,

$$\left.\begin{aligned} x &= ON - NM = \xi - \rho\frac{d\xi}{d\sigma}, \\ y &= N\Pi - \Pi R = \eta - \rho\frac{d\eta}{d\sigma}; \end{aligned}\right\} \tag{100}$$

in which equations ρ must be expressed in terms of ξ and η; and ξ and η having been eliminated from them and (98), the resulting equation will contain x and y only, and be that to the required involute.

170.] From (100) by differentiation we have

$$\left.\begin{aligned} dx &= d\xi - d\xi - \rho d.\frac{d\xi}{d\sigma} = -\rho d.\frac{d\xi}{d\sigma}, \\ dy &= d\eta - d\eta - \rho d.\frac{d\eta}{d\sigma} = -\rho d.\frac{d\eta}{d\sigma}; \end{aligned}\right\} \tag{101}$$

$$\therefore \quad \frac{dy}{dx} = \frac{d \cdot \frac{d\eta}{d\sigma}}{d \cdot \frac{d\xi}{d\sigma}}. \tag{102}$$

But by Art. 285, Vol. I, (21) and (22), the numerator and denominator of the right-hand side of (102) are proportional to the direction-cosines of the tangent of the evolute; and therefore we conclude that it is perpendicular to the tangent of the involute.

Again, squaring and adding the two equations (101), we have

$$dx^2 + dy^2 = \rho^2 \left\{ \left(d \cdot \frac{d\xi}{d\sigma}\right)^2 + \left(d \cdot \frac{d\eta}{d\sigma}\right)^2 \right\}.$$

But if ρ' is the radius of curvature of the evolute at the point (ξ, η), then, by Vol. I, Art. 285, (19),

$$\frac{d\sigma^2}{\rho'^2} = \left(d \cdot \frac{d\xi}{d\sigma}\right)^2 + \left(d \cdot \frac{d\eta}{d\sigma}\right)^2;$$

$$\therefore \quad \frac{ds}{\rho} = \pm \frac{d\sigma}{\rho'};$$

$$\text{but} \quad d\sigma = d\rho;$$

and therefore by means of (34), Art. 311, Vol. I,

$$\frac{\rho'}{\rho} = \pm \frac{3\,ds\,d^2s\,(dx\,d^2y - dy\,d^2x) - ds^2\,(dx\,d^3y - dy\,d^3x)}{(dx\,d^2y - dy\,d^2x)^2}. \tag{103}$$

For a problem in illustration of these equations, I will suppose $\rho' = k\rho$, so that

$$ds = \frac{\rho}{\rho'}\,d\sigma = \frac{d\rho}{k};$$

but if s and ψ are intrinsic coordinates, $ds = \rho\,d\psi$;

$$\therefore \quad \frac{d\rho}{\rho} = k\,d\psi;$$

$$\therefore \quad \rho = ae^{k\psi}, \quad \text{and} \quad s = \frac{a}{k}e^{k\psi}; \tag{104}$$

which is the intrinsic equation to the curve.

171.] Examples of involutes.

Ex. 1. To find the involute to the catenary, the generating point being in contact with it at its lowest point.

$$\eta = \frac{a}{2}\left\{ e^{\frac{\xi}{a}} + e^{-\frac{\xi}{a}} \right\}.$$

By equations (19) and (20), Art. 155,

$$\sigma = \frac{a}{2}\left\{e^{\frac{\xi}{a}} - e^{-\frac{\xi}{a}}\right\},$$

$$\sigma^2 = \eta^2 - a^2 = \rho^2;$$

$$\therefore \quad \frac{d\sigma}{d\eta} = \frac{\eta}{\sigma}, \quad \frac{d\sigma}{d\xi} = \frac{\eta}{a},$$

$$y = \eta - \frac{\sigma^2}{\eta}, \qquad x = \xi - \sigma\frac{a}{\eta},$$

$$= \frac{a^2}{\eta}; \qquad = \xi - (a^2-y^2)^{\frac{1}{2}};$$

$$\therefore \quad \eta = \frac{a^2}{y}, \qquad \xi = x + (a^2-y^2)^{\frac{1}{2}};$$

$$\therefore \quad \frac{a^2}{y} = \frac{a}{2}\left\{e^{\frac{x}{a}+\left(1-\frac{y^2}{a^2}\right)^{\frac{1}{2}}} + e^{-\frac{x}{a}-\left(1-\frac{y^2}{a^2}\right)^{\frac{1}{2}}}\right\}.$$

$$\therefore \quad x = a\log\frac{a+(a^2-y^2)^{\frac{1}{2}}}{y} - (a^2-y^2)^{\frac{1}{2}};$$

the equation to the tractory, the form of which is evident from fig. 7.

Ex. 2. To find the equation to the involute of the cycloid, the generating point being in contact with it at its vertex.

Let the cycloid be placed as in fig. 12; and let ON $=\xi$, NП $=\eta$; OM $= x$, MP $= y$; then the equation to the cycloid is

$$\xi = a \operatorname{versin}^{-1}\frac{\eta}{a} + (2a\eta - \eta^2)^{\frac{1}{2}};$$

$$\frac{d\xi}{(2a-\eta)^{\frac{1}{2}}} = \frac{d\eta}{\eta^{\frac{1}{2}}} = \frac{d\sigma}{(2a)^{\frac{1}{2}}};$$

$$\therefore \quad \sigma = 2(2a\eta)^{\frac{1}{2}} = \rho;$$

$$y = \eta - 2\eta, \qquad x = \xi - 2(2a\eta - \eta^2)^{\frac{1}{2}},$$

$$= -\eta; \qquad = \xi - 2(-2ay - y^2)^{\frac{1}{2}};$$

$$\therefore \quad \eta = -y; \qquad \xi = x + 2(-2ay - y^2)^{\frac{1}{2}};$$

therefore by substitution

$$x = a \operatorname{versin}^{-1}\frac{-y}{a} - (-2ay - y^2)^{\frac{1}{2}};$$

the equation to a cycloid in an inverted position, as OPD in the figure, and lying below the axis of x.

Ex. 3. To find the involute of a point.

Let the coordinates of the point be $\xi = a$, $\eta = b$; and let

$c =$ the length of the string which is attached to the point, and whose extremity generates the involute; then

$$x = a - c\frac{d\xi}{d\sigma}, \qquad y = b - c\frac{d\eta}{d\sigma};$$

$$\therefore \quad (x-a)^2 + (y-b)^2 = c^2\frac{d\xi^2 + d\eta^2}{d\sigma^2}$$
$$= c^2;$$

which is the equation to a circle whose centre is at the point, and whose radius is equal to c.

As the involute of a point is a circle, so conversely the evolute of a circle is a point.

Ex. 4. To find the equation of the involute of the semicubical parabola whose equation is $27a\eta^2 = 4\xi^3$, the length of ΠP being longer by $2a$ than the arc AΠ. Fig. 13.

$$\frac{d\eta}{2\xi^2} = \frac{d\xi}{9a\eta} = \frac{d\sigma}{(4\xi^4 + 81a^2\eta^2)^{\frac{1}{2}}} = \frac{d\sigma}{2\xi^{\frac{3}{2}}(\xi+3a)^{\frac{1}{2}}};$$

$$\therefore \quad d\sigma = \left(\frac{\xi+3a}{3a}\right)^{\frac{1}{2}} d\xi;$$

$$\sigma = \frac{2}{3(3a)^{\frac{1}{2}}}\{(\xi+3a)^{\frac{3}{2}} - (3a)^{\frac{3}{2}}\}$$
$$= \frac{2(\xi+3a)^{\frac{3}{2}}}{3(3a)^{\frac{1}{2}}} - 2a;$$

Now $\quad \rho = \sigma + 2a$, by the conditions of the problem;

$$\therefore \quad \rho = \frac{2(\xi+3a)^{\frac{3}{2}}}{3(3a)^{\frac{1}{2}}};$$

therefore by equations (100),

$$\xi = 3x + 6a, \text{ and } \xi = \frac{3y^2}{4a};$$

$$\therefore \quad y^2 = 4a(x+2a);$$

the equation to a parabola, situated as in the figure.

172.] On involutes of curves referred to polar coordinates.

Let AP be the curve, fig. 14, whose involute is to be determined; and let its equation be

$$r = f(p). \qquad (105)$$

Let PP′ be the tangent at P, P′ being the generating point of the involute. Draw from the pole S, SY perpendicular to PP′, and SY′

perpendicular to P′Y′, which is the tangent to the involute at P′. Then SP $= r$, SY $= p$; SP′ $= r'$, SY′ $= p'$; and our object is to find the relation between r' and p'. Let ds represent a length-element of the original curve; and, since PP′ is the radius of curvature of the involute at P′, let PP′ $= \rho'$; then, by (40), Art. 292, Vol. I, $d\rho' = ds$.

$$\therefore \quad \rho' = s \pm c;$$

and from the geometry,

$$r'^2 = p^2 + p'^2, \tag{106}$$

$$r^2 = r'^2 + \rho'^2 - 2\rho' p'; \tag{107}$$

and after the elimination of r, p, s from these equations there will remain an expression in terms of r' and p', which is the equation to the involute.

Ex. 1. To find the equation to the involute of a circle.

Let the centre of the circle be the pole: then, if $a =$ the radius, its equation in terms of r and p is

$$r = p = a;$$

whence (106) gives $\quad r'^2 - p'^2 = a^2;$

which is the equation to the involute of the circle.

This equation however may be found in the following way.

In fig. 31, let QSA $= \theta$, so that the arc AQ $= a\theta$; then, if S is the origin, and P is (x, y), PQ $= a\theta$, and

$$\left.\begin{aligned} y &= a \sin\theta - a\theta \cos\theta \\ x &= a\cos\theta + a\theta \sin\theta \end{aligned}\right\}; \tag{108}$$

which simultaneous equations are those to the involute of the circle. From them we have

$$x^2 + y^2 = a^2 + a^2\theta^2;$$

$$x\cos\theta + y\sin\theta = a;$$

$$\therefore \quad x \cos\frac{(x^2+y^2-a^2)^{\frac{1}{2}}}{a} + y \sin\frac{(x^2+y^2-a^2)^{\frac{1}{2}}}{a} = a;$$

which is the equation to the involute of the circle in terms of x and y.

Also, since from (108) $dy = a\theta \sin\theta \, d\theta$, $dx = a\theta\cos\theta \, d\theta$,

$$\begin{aligned} \therefore \quad p &= \frac{y\,dx - x\,dy}{ds} \\ &= -a\theta; \\ \therefore \quad p^2 &= a^2\theta^2 \\ &= r^2 - a^2. \end{aligned}$$

Ex. 2. To find the equation to the involute of the logarithmic spiral.

Let a be the constant angle at which the curve cuts all the radii vectores; then its equation is

$$p = r \sin a. \tag{109}$$

Therefore, see fig. 15, if PP' is equal to the length of the curve from the pole to the point P, and if PP' $= \rho'$, by (54), Art. 162,

$$\rho' = \int_0^r \frac{dr}{\cos a} = r \sec a. \tag{110}$$

From (107), completing the square,

$$\begin{aligned}(\rho'-p')^2 &= r^2 - r'^2 + p'^2 \\ &= r^2 - p^2 \\ &= r^2 (\cos a)^2; \\ \therefore \quad \rho' &= p' + r \cos a;\end{aligned}$$

and substituting for ρ' from (110),

$$\begin{aligned}p' &= r \sec a - r \cos a \\ &= r \sin a \tan a;\end{aligned}$$

$$\begin{aligned}\therefore \quad \text{from (106)}, \quad r'^2 &= p'^2 + p^2 \\ &= p'^2 + r^2 (\sin a)^2 \\ &= p'^2 + p'^2 (\cot a)^2 \\ &= p'^2 (\operatorname{cosec} a)^2; \\ \therefore \quad p' &= r' \sin a;\end{aligned}$$

the equation to a logarithmic spiral, similar to the original one; that is, which cuts all its radii vectores at a constant angle, the same as that of the original spiral. From (110) it is evident that PSP' is a right-angle, and therefore SP'Y' $=$ SPY $= a$: the involute therefore is also the locus of the extremity of the polar subtangent.

SECTION 6.—*Various problems of Plane Geometry solved by means of Single Definite Integration.*

173.] Curves and classes of curves are generally defined by means of some salient geometrical property, and the equation of the curve is the expression of this property by means of mathematical symbols. Now, as many of these principal properties are only capable of expression in terms of differentials, it is evident that the equation which defines a curve or a class of curves may be given in terms of these differentials in combination with, or independent of, finite coordinates, whether rectangular or polar. When the equation is given in this form, it is called the differential equation of the curve; and it is frequently necessary to obtain the finite equation from the differential equation. For this purpose integration is required.

The problem in its most general form is entirely beyond the present power of mathematical analysis. Many special forms of it however are capable of solution; and although in a subsequent part of this Treatise other methods will be investigated, yet it is expedient at once to apply to such problems the process of single definite integration so far as it is applicable, both for the sake of the geometrical problems, and for the illustration of the process of integration, which it so well exhibits. In some of the following examples, integration will be performed more than once; but only a series of successive integrations will be required, and these will be of the nature explained in Chapter V, Art. 149, 150. In all cases the limits will be given, and the integrations will be definite.

174.] I will in the first place take that class of problems in which $\frac{dy}{dx}$ is given in terms of x and y in a simple form, capable of integration. This equation, geometrically interpreted, assigns the law of the ratio of the simultaneous increments of the coordinates at any point of a plane curve in terms of those coordinates; that is, gives the value of the geometrical tangent of the angle between the tangent to the curve at any point and either of the rectangular axes, in terms of the coordinates of the point.

Ex. 1. Let $\frac{dy}{dx} = \frac{2a}{y}$; and let us assume that the curve passes through the origin, so that when $x = 0$, $y = 0$.

$$\therefore \quad \int_0^y y\,dy = \int_0^x 2a\,dx;$$

$$\therefore \quad y^2 = 4ax.$$

Ex. 2. $\frac{dy}{dx} = -\frac{x}{y}$; and let us suppose $x = a$, when $y = 0$.

$$\therefore \int_0^y y\,dy + \int_a^x x\,dx = 0;$$

$$\therefore \quad y^2 + x^2 - a^2 = 0;$$

$$x^2 + y^2 = a^2.$$

Ex. 3. $\frac{dy}{dx} = \left(\frac{2a-x}{x}\right)^{\frac{1}{2}}$; and let us suppose that the origin is on the curve.

$$\int_0^y dy = \int_0^x \left(\frac{2a-x}{x}\right)^{\frac{1}{2}} dx;$$

$$y = \int_0^x \frac{(2a-x)\,dx}{(2ax-x^2)^{\frac{1}{2}}}$$

$$= (2ax-x^2)^{\frac{1}{2}} + a \operatorname{versin}^{-1}\frac{x}{a};$$

which is the equation to a cycloid, of which the vertex is the origin.

175.] The process is similar when $\frac{d^2y}{dx^2}$ is given in terms of x and y. In this case however, as the first integration will produce $\frac{dy}{dx}$ it is necessary that the value of this quantity at the limits, or at least at one limit, should be given. This condition requires that the direction of the axes should be given. Also, as the second integration will produce a finite equation in terms of x and y, of which the values at one limit at least must be given, this condition assigns the place of the origin.

Ex. 1. $\frac{d^2y}{dx^2} = \frac{2a^2}{x^3}$; and let us suppose $\frac{dy}{dx} = 0$, when $x = \infty$; $y = 0$, when $x = \infty$.

$$\therefore \int_0^{\frac{dy}{dx}} d.\frac{dy}{dx} = \int_\infty^x \frac{2a^2dx}{x^3},$$

$$\frac{dy}{dx} = -\frac{a^2}{x^2};$$

$$\int_0^y dy = -\int_\infty^x \frac{a^2}{x^2}dx,$$

$$y = \frac{a^2}{x};$$

$$\therefore \quad xy = a^2.$$

If the values at the inferior limits were such that when $x = a$,

$\frac{dy}{dx} = 0$, $y = 0$, so that the curve should touch the x-axis at a distance $= a$, from the origin; then

$$\frac{dy}{dx} = -\frac{a^2}{x^2} + 1;$$

$$\therefore \quad xy = (x-a)^2.$$

Ex. 2. $\frac{d^2 y}{dx^2} = -\frac{a^3}{y^3}$; and let us suppose when $y = a$, $\frac{dy}{dx} = 0$, $x = 0$.

$$\therefore \int_0^{\frac{dy}{dx}} \frac{2\,dy\,d^2y}{dx^2} = \int_a^y \frac{-2a^2\,dy}{y^3},$$

$$\left(\frac{dy}{dx}\right)^2 = \frac{a^2}{y^2} - 1;$$

$$\int_a^y \frac{-y\,dy}{(a^2-y^2)^{\frac{1}{2}}} = \int_0^x dx,$$

$$(a^2-y^2)^{\frac{1}{2}} = x;$$

$$x^2 + y^2 = a^2.$$

Ex. 3. If $\frac{d^2 y}{dx^2} = y$, and $x = 0$, and $\frac{dy}{dx} = 0$, when $y = a$, then

$$y = \frac{a}{2}\left\{e^{\frac{x}{a}} + e^{-\frac{x}{a}}\right\}.$$

176.] Geometrical problems depending on the integration of differential quantities.

Ex. 1. Find the curve of which the subnormal is constant.

Taking the value of the subnormal given in (41), Art. 219, Vol. I, we have

$$\frac{y\,dy}{dx} = a;$$

and taking the origin to be on the curve, we have

$$y^2 = 2ax.$$

Ex. 2. Find the curve, of which the subtangent is constant.

$$y = e^{\frac{x}{a}}.$$

Ex. 3. Find the equation to the curve, of which the tangent is constant.

Taking the value of the tangent given in (42), Art. 219, Vol. I, we have

$$y^2\left(1 + \frac{dx^2}{dy^2}\right) = a^2;$$

$$\therefore \quad dx = -\frac{(a^2-y^2)^{\frac{1}{2}}}{y}\,dy,$$

if we suppose y to decrease, as x increases. Let us also assume $y = a$, when $x = 0$; then

$$\int_0^x dx = -\int_a^y \frac{(a^2-y^2)^{\frac{1}{2}}}{y}\,dy$$

$$= -\int_a^y \left\{ \frac{a^2}{y\,(a^2-y^2)^{\frac{1}{2}}} - \frac{y}{(a^2-y^2)^{\frac{1}{2}}} \right\} dy;$$

$$\therefore \quad x = a \log \frac{a+(a^2-y^2)^{\frac{1}{2}}}{y} - (a^2-y^2)^{\frac{1}{2}}; \qquad (111)$$

which is the equation to the equitangential curve; see Vol. I, Art. 200, equation (27).

Ex. 4. Find the equation to the curve whose normal is constant.

In this case, by (42), Art. 219, Vol. I,

$$y^2\left(1+\frac{dy^2}{dx^2}\right) = a^2; \qquad \therefore \quad dx = \frac{\pm\, y\,dy}{(a^2-y^2)^{\frac{1}{2}}};$$

and assuming $x = 0$, when $y = a$, we have $x^2+y^2 = a^2$.

Ex. 5. Find the equation to the curve all the normals to which pass through the same point (a, b).

This property is expressed by the equation

$$(x-a)\,dx+(y-b)\,dy = 0;$$

$$\therefore \quad (x-a)^2+(y-b)^2 = c^2;$$

which is the equation to a circle, whose radius is c.

Ex. 6. Find the curves in which the sum of (1) the abscissa and the subnormal, (2) the ordinate and the subnormal, is constant.

$$(1) \quad (x-a)^2+y^2 = c^2;$$

$$(2) \quad x+y+a\log(a-y) = c.$$

Ex. 7. Find the curve in which the tangent is equal to the radius vector of the point of contact.

$$y^2\left(1+\frac{dx^2}{dy^2}\right) = y^2+x^2.$$

$$\therefore \quad \frac{dy}{y} \pm \frac{dx}{x} = 0; \qquad \therefore \quad \log\frac{y}{b} \pm \log\frac{x}{a} = 0.$$

$$\therefore \quad (1) \quad \frac{y}{b} = \frac{x}{a}; \qquad (2) \quad xy = ab;$$

so that the required line is either straight, or an equilateral hyperbola, according as the lower or the upper sign is taken.

Ex. 8. Find the curve which cuts all its radii vectores at the same angle.

Let this angle be $\tan^{-1} c$; then $r\dfrac{d\theta}{dr} = c$;

$$\therefore \quad \frac{dr}{r} = \frac{d\theta}{c};$$

and let us suppose $r = a$, when $\theta = 0$; then

$$\log\frac{r}{a} = \frac{\theta}{c}; \qquad \therefore \quad r = ae^{\frac{\theta}{c}}.$$

Ex. 9. Find the curve in which the perpendicular on the tangent is equal to the part of the tangent intercepted between the foot of the perpendicular and the point of contact.

This is a particular case of the preceding example; that, viz., in which $c = 1$; so that the equation of the curve is $r = ae^{\theta}$.

It may also be solved in the following way; by the condition we have $p^2 = r^2 - p^2$; so that $r^2 = 2p^2$;

$$\frac{1}{p^2} = 2u^2 = u^2 + \frac{du^2}{d\theta^2};$$

$$\therefore \quad -\frac{du}{u} = \frac{dr}{r} = d\theta;$$

$$\therefore \quad r = ae^{\theta}.$$

Ex. 10. The curve in which the polar subtangent is constant is the reciprocal spiral.

Ex. 11. The curve in which the polar subnormal is constant is the spiral of Archimedes.

Ex. 12. Find the curve in which the perpendicular from the origin on the tangent is equal to the abscissa.

$$p = r\cos\theta;$$

$$\therefore \quad u^2(\sec\theta)^2 = u^2 + \frac{du^2}{d\theta^2};$$

$$\therefore \quad r = 2a\cos\theta.$$

Ex. 13. Find the curve of which the radius of curvature is equal to a.

$$\frac{r\,dr}{dp} = a;$$

$$\therefore \quad r^2 = 2ap;$$

$$r = 2a\cos\theta.$$

Ex. 14. If $p^2 = ar$, the equation to the curve is

$$r = \frac{2a}{1+\cos\theta}.$$

Ex. 15. If the angle between the radius vector and the curve $= \theta$, the curve is a circle.

Ex. 16. If the subnormal is equal to the abscissa, the curve is a hyperbola.

Ex. 17. If the radius of curvature $= p$, the curve is the involute of the circle.

CHAPTER VII.

THE APPLICATION OF SINGLE DEFINITE INTEGRATION TO THE THEORY OF SERIES.

SECTION 1.—*On the Convergence and Divergence of Series.*

177.] The subject of series has come under our notice many times in the preceding parts of our treatise, and in various forms. At one time processes have been investigated for the development of functions into series; as Maclaurin's and Taylor's theorems, and others subordinate to and cognate to them; at other times, as in the Integral Calculus, the sums of series have been determined: that is, functions have been found of which the given series are the developments. These two processes are evidently inverse to each other. Now, when a function is developed into a series, that series taken to a finite number of terms, or even to an infinite number of terms, cannot always be assumed to be, and used as, the equivalent of the function; it is necessary to calculate what has been called "the remainder" in the theorems of Maclaurin and Taylor: and it is only when this remainder fulfils certain conditions, that the series can be used as an adequate equivalent of the function. Similarly there are certain conditions which a series must satisfy when it is adequately represented by a given function, or when in other words, the series can be summed, and that function is the sum; and there are also certain limits of the variable within which the summation is possible, and outside of which it is impossible. These thus far have not been discussed; and I have assumed only that knowledge of the subject which will be found in ordinary treatises on algebra.

The subject however requires more complete and precise investigation, and arises now in the regular course of our Treatise on the Integral Calculus; because single definite integration supplies a theorem on convergence and divergence of series, which determines whether a given series is capable of summation or not; and which is of wider application than any heretofore suggested. It is indeed one of the most important applications of the calculus in its relation to algebra and algebraical development. Definite integration also yields a new, and in some respects a more convenient form of the remainders of the theorems of Taylor and Maclaurin; and gives development of certain functions in periodic terms, which are not only curious, but are also important in the application of pure mathematics to questions of physics. The investigation of these theorems will occupy the present chapter; but previously to that inquiry it is desirable to give a succinct account of series.

178.] A series is a succession of terms, of which the number is infinite, placed one after another in a given order, and formed according to an uniform and determinate law, so that each term is a function of the place which it holds in the series. The function which expresses each in terms of its place gives the law of the series; and it is evident that the place of a term in a series cannot be changed without a change of law of the series.

The following are the forms of series which we shall take;

$$u_1+u_2+u_3+\ldots\ldots+u_n+u_{n+1}+\ldots; \tag{1}$$

$$f(a)+f(a+1)+f(a+2)+\ldots+f(a+n-1)+f(a+n)+\ldots; \tag{2}$$

$$\frac{1}{f(a)}+\frac{1}{f(a+1)}+\frac{1}{f(a+2)}+\ldots+\frac{1}{f(a+n-1)}+\frac{1}{f(a+n)}+\ldots; \tag{3}$$

the nth or the general term being respectively $u_n, f(a+n-1)$, $\frac{1}{f(a+n-1)}$: and consequently the law of the series is given, when these terms are expressed as functions of n.

For conciseness of notation we shall express these series severally by the symbols $\sum_1^\infty u_x$; $\sum_a^\infty f(x)$; $\sum_a^\infty \frac{1}{f(x)}$; which indicate the sum of a series of terms formed by assigning to x all integer values from the inferior limit to infinity. This notation is analogous to that of a definite integral.

A series is said to be convergent, when the sum of an infinite number of terms of it is a finite quantity; and this finite quantity is called the sum of the series.

A series is said to be divergent, when the sum of an infinite number of terms of it is infinite; in this case it is said that the series cannot be summed.

Thus if s_n is the sum of the first n terms of the series given above, if the series is convergent, s_n is finite, when $n = \infty$. If the series is divergent, $S_n = \infty$, when $n = \infty$. When $n = \infty$, the sum of the series is denoted by S.

The sum of all the terms of a series after the nth is called the remainder of the series. Let it be denoted by R; then taking the form of the series given in (1),

$$R = u_{n+1} + u_{n+2} + \dots \qquad (4)$$

$$= S - S_n. \qquad (5)$$

It is to be observed that although the sum of a convergent series is a finite quantity, yet frequently that quantity cannot be determined in the form of a definite function. It may be capable of expression as a definite integral, which does not admit of evaluation; or it may be expressed as a function of certain letters contained in the series, as the notation of Art. 145 signifies: but in many cases it does not admit of more definite determination.

The remainder of a convergent series is infinitesimal, when $n = \infty$; so that S_n and S_{n+r} differ by an infinitesimal, which must be neglected when $n = \infty$. Hence it follows that, when a series is convergent, not only must u_n be infinitesimal, and consequently be neglected, when $n = \infty$, but also R, which is the sum of all these infinitesimals, must also be an infinitesimal, and be neglected when $n = \infty$. Thus although the vanishing of each term, when $n = \infty$, is true of all convergent series, yet this circumstance does not afford a sufficient test of convergence. It is a necessary characteristic, but is evidently not a sufficient criterion.

It will be observed that these definitions have respect to the terminal terms of the series; that is, those for which $n = \infty$; thus although a series may begin divergently and continue divergently for a given number of terms, yet if it ultimately converges, it will satisfy the condition of convergence, and will be classed as a convergent series.

179.] Before we enter on the investigation of a strict criterion of convergence, it is worth while to examine two series of typical forms: viz., series in geometrical and in harmonical progression: for we shall hereby exemplify certain phenomena of series, the number of the terms of which is infinity; and we shall shew the necessity of either determining the convergence of a series when it and the function of which it is supposed to be the development are used as equivalents: or of ascertaining the remainder so that an exact equality may exist between the function and the series

180.] We now come to the investigation of general theorems for the determination of convergence and divergence of series. I shall arrange them as far as possible in order of simplicity, for that is the order in which they are most conveniently applied to any particular series.

If the terms of a series decrease in absolute magnitude, and are, or ultimately become, alternately positive and negative, the series is convergent.

Let the series be $u_1 - u_2 + u_3 - u_4 + \dots;$ (12)
the terms of which become less and less. Then the series may be expressed in the two following forms

$$(u_1 - u_2) + (u_3 - u_4) + (u_5 - u_6) + \dots;$$

and

$$u_1 - (u_2 - u_3) - (u_4 - u_5) - \dots;$$

every quantity within brackets being positive. Whence it is evident that the sum of the series is greater than $u_1 - u_2$, and is less than u_1; consequently the series is convergent.

Ex. 1. Thus the series $\sum_1^\infty \frac{1}{(2x-1)2x}$ is convergent.

The series is $\frac{1}{1.2} + \frac{1}{3.4} + \frac{1}{5.6} + \dots;$ which may be expressed in the form $1 - \frac{1}{2} + \frac{1}{3} - \frac{1}{4} + \dots;$ which is by the preceding theorem greater than 1, and less than $\frac{1}{2}$.

Ex. 2. Thus also $\sum_1^\infty \frac{1}{(2x-1)2x(2x+1)}$ is convergent.

181.] The series $u_1 + u_2 + u_3 + \dots$, all of whose terms are positive, is convergent or divergent according as the ratio of u_{x+1} to u_x, when $x = \infty$, is less or greater than unity.

Let r be the value of $\frac{u_{x+1}}{u_x}$, when $x = \infty$; and firstly let r be less than 1; let ρ be greater than r and less than 1; then for some determinate value of x, and for all other greater values, the ratio of u_{x+1} to u_x is less than ρ; so that if x is that value,

$$u_{x+1} < \rho u_x, \qquad u_{x+2} < \rho u_{x+1}, \qquad u_{x+3} < \rho u_{x+2}, \dots\dots;$$

consequently

$$u_x + u_{x+1} + u_{x+2} + \dots < u_x \{1 + \rho + \rho^2 + \dots\}$$
$$< \frac{u_x}{1-\rho},$$

since ρ is less than 1. And this quantity is finite; consequently the given series is convergent. Its terms evidently become less and less, because they are less than the corresponding terms of a decreasing geometrical series.

If r is greater than 1, let ρ be greater than 1 and less than r; then for some determinate value of x, and for all other greater values,

$$u_{x+1} > \rho u_x, \quad u_{x+2} > \rho u_{x+1}, \quad u_{x+3} > \rho u_{x+2}, \quad \ldots;$$

consequently,

$$u_x + u_{x+1} + u_{x+2} + \ldots > u_x(1+\rho+\rho^2+\ldots);$$

the right-hand member of which $= \infty$, when the number of the terms is infinite; thus the sum of the series is infinite, and the series is accordingly divergent. If $r = 1$, we can make no inference; and in this case the series may be either convergent or divergent.

If the series is arranged in ascending powers of a variable t, say, and is such that $u_x = A_x t^x$; then $\frac{u_{x+1}}{u_x} = \frac{A_{x+1}}{A_x} t$; and if $\frac{A_{x+1}}{A_x} = r$, when $x = \infty$, the discriminating quantity $= rt$; so that the series is convergent or divergent accordingly as t is less than or greater than r^{-1}. If $t = r^{-1}$, we can affirm nothing as to convergence or divergence.

Ex. 1. Let the series be $1 + \frac{t}{1} + \frac{t^2}{1.2} + \frac{t^3}{1.2.3} + \ldots$; then

$$\frac{u_{x+1}}{u_x} = \frac{t}{x} = 0, \text{ when } x = \infty;$$

thus, whatever is the value of t, the series is convergent. I may however observe, that if t is greater than 1, the series begins divergently, and begins to converge at that term for which $\frac{t}{x}$ is less than 1; that is, when x is greater than t.

The test of convergence and divergence given in this Article is of all perhaps the most easy of application, and is accordingly that which should be at once employed. And it is only when the ratio of u_{x+1} to u_x is equal to 1, that the test fails, and we are obliged to have recourse to other criteria.

182.] If $f(x)$ is a function of x, positive in sign, and decreasing in value as x increases from $x = a$ to $x = \infty$, and ultimately equal to 0, when $x = \infty$, then the series $\sum_a^\infty f(x)$, that is,

$$f(a)+f(a+1)+f(a+2)+\ldots \tag{13}$$

will be convergent or divergent according as $\int_a^\infty f(x)\,dx$ is finite or infinite.

By the theory of definite integration

$$\int_a^\infty f(x)\,dx = \int_a^{a+1} f(x)\,dx + \int_{a+1}^{a+2} f(x)\,dx + \int_{a+2}^{a+3} f(x)\,dx + \ldots; \quad (14)$$

and applying the theorem given in (228), Art. 116, where θ is the general symbol of a positive proper fraction,

$$\begin{aligned}\int_a^\infty f(x)\,dx &= f(a+\theta)\int_a^{a+1} dx + f(a+1+\theta)\int_{a+1}^{a+2} dx + \ldots, \\ &= f(a+\theta) + f(a+1+\theta) + f(a+2+\theta) + \ldots. \quad (15)\end{aligned}$$

Now as θ is a positive proper fraction, 1 and 0 are its limits; we may consequently deduce from this equation two inequalities corresponding to these values of θ; and as $f(x)$ is positive, and decreases as x increases, when $\theta = 0$, the second member of (15) is greater than the first; and when $\theta = 1$, the second member is less than the first; hence

$$f(a) + f(a+1) + f(a+2) + \ldots\ldots > \int_a^\infty f(x)\,dx; \quad (16)$$

$$f(a+1) + f(a+2) + \ldots\ldots < \int_a^\infty f(x)\,dx. \quad (17)$$

Then as the sum of the given series is greater than $\int_a^\infty f(x)\,dx$, and is less than $\int_a^\infty f(x)\,dx + f(a)$, where $f(a)$ is a finite quantity, it is evident that the sum is infinite or finite, that is, the series is divergent or convergent, according as $\int_0^\infty f(x)\,dx$ is infinite or finite. This quantity I shall call the discriminating quantity.

This is the most general criterion of the convergence and divergence of series which has as yet been discovered. The difficulty of its application consists in the integration. By a farther inquiry however into the general theory of series, we shall be led to a classification of them as to order and degree, and the preceding theorem will supply certain derivative tests convenient for the purpose.

The investigation also gives narrow limits within which the sum of a series of the form (13) is contained.

In (17) let a be replaced by $a-1$; then

$$f(a) + f(a+1) + f(a+2) + \ldots < \int_{a-1}^\infty f(x)\,dx. \quad (18)$$

And if the sum of the series = s, from this and (16) it follows that

$$\int_{a-1}^{\infty} f(x)\,dx > s > \int_{a}^{\infty} f(x)\,dx; \tag{19}$$

so that s is contained within these two definite integrals.

183.] The following are examples in which the preceding test is applied.

Ex. 1. The series in geometrical progression $a, ar, ar^2, \ldots,$ where r is a positive quantity.

Taking the general term of the series to be ar^x, the successive terms of the series will be formed by putting $x = 0, 1, 2, 3, \ldots$; so that the discriminating quantity is $\int_0^{\infty} ar^x dx$. Now

$$\int_0^{\infty} ar^x dx = \left[\frac{ar^x}{\log r}\right]_0^{\infty}$$

$$= -\frac{a}{\log r}, \text{ if } r \text{ is less than } 1;$$

$$= \infty, \text{ if } r \text{ is not less than } 1.$$

Hence a series in geometrical progression is convergent or divergent according as r is less than, or not less than, 1.

Ex. 2. Let the series be

$$\frac{1}{a^m} + \frac{1}{(a+1)^m} + \frac{1}{(a+2)^m} + \ldots; \text{ which is } \sum_a^{\infty} x^{-m}. \tag{20}$$

In this example the discriminating quantity is $\int_a^{\infty} x^{-m} dx$. Now

$$\int_a^{\infty} x^{-m} dx = \frac{\infty^{-m+1} - a^{-m+1}}{-m+1}$$

$$= \frac{a^{1-m}}{m-1}, \text{ if } m \text{ is greater than } 1;$$

$$= \infty, \text{ if } m \text{ is not greater than } 1.$$

Hence the series is convergent or divergent according as m is greater, or is not greater, than 1. If $m = 1$, the series is harmonic, and we have hereby another proof that such a series is divergent.

Ex. 3. $\sum_a^{\infty} x^{-1}(\log x)^{-m}$ is convergent or divergent according as m is greater than, or not greater than, 1.

Here the discriminating condition is $\int_a^{\infty} \frac{dx}{x(\log x)^m}$, and

$$\int_a^\infty \frac{dx}{x(\log x)^m} = \frac{(\log \infty)^{-m+1}-(\log a)^{-m+1}}{-m+1}$$

$$= \frac{(\log a)^{-m+1}}{m-1}, \text{ if } m \text{ is greater than } 1;$$

$$= \infty, \text{ if } m \text{ is not greater than } 1.$$

Ex. 4. $\sum_a^\infty \{x \log x (\log \log x)^m\}^{-1}$ is convergent or divergent, according as m is greater, or is not greater, than 1.

Ex. 5. Generally if $\log\log\log \dots$ (to n symbols) $\dots x$ is denoted by $\log^n x$, then $\sum_a^\infty \{x \log x \log^2 x \dots \log^{n-1} x (\log^n x)^m\}^{-1}$ is convergent or divergent, according as m is greater than, or not greater than, 1.

Ex. 6. $\sum_0^\infty \frac{1}{a^2+x^2}$ is convergent.

Ex. 7. $\sum_2^\infty \frac{\log x}{x}$ is divergent. Hence $2^{\frac{1}{2}} 3^{\frac{1}{3}} 4^{\frac{1}{4}} \dots n^{\frac{1}{n}} = \infty$, when $n = \infty$.

Ex. 8. $\sum_2^\infty \left(\frac{\log x}{x}\right)^2$ is convergent.

184.] Although the criterion of the preceding article is theoretically always sufficient, yet it is in many cases inapplicable, because the definite integral which gives the discriminating quantity cannot be found. We can however derive from it certain other criteria, which are frequently easy of application; this is effected by a comparison of the given series, with other series which are known to be convergent or divergent.

Let there be two series

$$\text{U} = u_1+u_2+u_3+\dots \tag{21}$$

$$\text{V} = v_1+v_2+v_3+\dots; \tag{22}$$

the ratio of the nth terms of which is equal to a finite quantity k, when $n = \infty$: so that $u_n = k\,v_n$, when $n = \infty$; then these two series are either both convergent or both divergent.

Let us commence with the nth terms respectively of the two series; and let $u_n = k_n v_n$, $u_{n+1} = k_{n+1} v_{n+1}$, $u_{n+2} = k_{n+2} v_{n+2}, \dots$; then since $k_n = k_{n+1} = k_{n+2} = \dots = k$, when $n = \infty$; so when n is very great, the difference between k and each of these quantities is an infinitesimal; and we shall have generally $k_n = k \pm i$, where i is an infinitesimal; and ultimately when $n = \infty$, $i = 0$, and

$$u_n+u_{n+1}+u_{n+2}+\dots = k\{v_n+v_{n+1}+v_{n+2}+\dots\}. \tag{23}$$

Hence it appears that if either of these sums is a finite quantity, the other is also finite; and consequently the series are either both convergent or both divergent.

If series of the form (21) and (22) are so related to each other, that when $n = \infty$, $u_n = kv_n$, k being a finite quantity, they are called comparable. If however $k = \infty$ or $= 0$, they are said to be incomparable. If $k = \infty$, (21) is incomparably higher than (22). If $k = 0$, (22) is incomparably higher than (21).

Hence it is evident that if a series is convergent, all comparable series and all incomparably lower series are also convergent. If a series is divergent, all comparable series and also incomparably higher series are also divergent.

185.] I proceed now to apply this theory; and I shall take for the series of comparison those of which the character as to convergence or divergence has been demonstrated in Art. 183.

Let the series, the character of which is to be determined, be $\sum_a^\infty \frac{1}{f(x)}$, in which $f(x) = \infty$, when $x = \infty$: and let us in the first place compare it with $\sum_a^\infty x^{-m}$, of which the character is determined in Ex. 2 of the preceding Article. Let us take m to be such that these series are comparable; then if k is the ratio of the corresponding terms when $x = \infty$,

$$k = \frac{f(x)}{x^m} = \frac{\infty}{\infty}, \text{ when } x = \infty;$$

$$= \frac{f'(x)}{m x^{m-1}}$$

$=$ a finite quantity, suppose, when $x = \infty$: so that $f'(x)$ and x^{m-1} are infinities of the same order;

$$\therefore \quad k = \frac{x f'(x)}{m x^m} = k \frac{x f'(x)}{m f(x)};$$

$$\therefore \quad m = \frac{x f'(x)}{f(x)}, \text{ when } x = \infty. \tag{24}$$

and as $\sum_a^\infty x^{-m}$ is convergent when m is greater than 1, and divergent when m is less than 1; so will the given series $\sum_a^\infty \frac{1}{f(x)}$ be convergent if $\frac{x f'(x)}{f(x)}$, when $x = \infty$, is greater than 1; and divergent if $\frac{x f'(x)}{f(x)}$, when $x = \infty$, is less than 1. Hence the discriminating quantity of

$$\sum_a^\infty \frac{1}{f(x)} \text{ is } \frac{xf'(x)}{f(x)}, \text{ when } x = \infty. \tag{25}$$

I may in passing observe that this discriminating quantity may be expressed in the form

$$\frac{d.\log f(x)}{d.\log x}. \tag{26}$$

In this form it was first given by Cauchy.

If the series, whose character is to be determined, is $\sum_a^\infty f(x)$, then, if we replace $f(x)$ by $\frac{1}{f(x)}$, the preceding quantity becomes $-\frac{xf'(x)}{f(x)}$; and consequently the discriminating quantity of

$$\sum_a^\infty f(x) \text{ is } -\frac{xf'(x)}{f(x)}, \text{ when } x = \infty; \tag{27}$$

and the series is convergent when this quantity is greater than 1, and is divergent when it is less than 1.

The quantity (27) may also be put into the form

$$-\frac{d.\log f(x)}{d.\log x}. \tag{28}$$

Ex. 1. Shew that the series $\sum_1^\infty \frac{1}{x(x+1)}$ is convergent.

If $f(x) = x(x+1)$; then $\frac{xf'(x)}{f(x)} = \frac{2x^2+x}{x^2+x} = 2$, when $x = \infty$. Consequently the series is convergent.

Ex. 2. Determine the character as to convergence or divergence of the series $\sum_1^\infty \frac{(x+1)^a}{x^{a+b}}$.

Here from (25), if $f(x) = \frac{x^{a+b}}{(x+1)^a}$,

$$\frac{xf'(x)}{f(x)} = \frac{bx^2+(a+b)x}{x^2+x} = b, \text{ if } x = \infty.$$

Consequently the series is convergent if b is greater than 1, and divergent if b is less than 1.

Ex. 3. $\sum_2^\infty \frac{1}{x^a(\log x)^b}$ is convergent when a is greater than 1, and is divergent when a is less than 1.

Ex. 4. $\sum_1^\infty \frac{1}{x^{\frac{1}{2}}}$ is divergent.

186.] If the discriminating quantity determined in the last article $= 1$, when $x = \infty$; the test fails, and the series may thus far be either convergent or divergent. In this case let us compare the given series with $\sum_a^\infty \{x(\log x)^m\}^{-1}$, the character of which is determined in Ex. 3, Art. 183, and which is therein shewn to be convergent or divergent according as m is greater than or less than 1. In this case

$$k = \frac{f(x)}{x(\log x)^m} = \frac{\infty}{\infty}, \text{ when } x = \infty;$$

$$= \frac{f'(x)}{(\log x)^m + m(\log x)^{m-1}}, \text{ when } x = \infty;$$

= a finite quantity, suppose, when $x = \infty$; so that

$$k = \frac{f'(x)\log x}{(m+\log x)f(x)}kx;$$

$$\therefore \quad m = \log x\left\{\frac{xf'(x)}{f(x)} - 1\right\}, \text{ when } x = \infty. \qquad (29)$$

which is the required discriminating quantity; and the series $\sum_a^\infty \frac{1}{f(x)}$ is convergent or divergent according as this quantity is greater than or less than 1.

Ex. 1. Determine the character as to convergence and divergence of the series $\sum_1^\infty (a^{\frac{1}{x}} - 1)$.

In this case the discriminating quantity

$$= \log x\left\{\frac{a^{\frac{1}{x}}\log a}{x(a^{\frac{1}{x}}-1)} - 1\right\}$$

$$= 0, \text{ when } x = \infty:$$

consequently the series is divergent.

Ex. 2. The series $\sum_1^\infty \frac{1}{x^{\frac{x+1}{x}}}$ is divergent.

187.] If however the discriminating quantity given in (29) is equal to 1, the test fails, and the series may thus far be either convergent or divergent. In this case we may compare the series with $\sum_a^\infty \{x\log x(\log^2 x)^m\}^{-1}$, which has been proved to be convergent or divergent according as m is greater than or less than 1; and comparing $\sum_a^\infty \frac{1}{f(x)}$ with this series, and making it comparable with it, we find as in the last two articles,

$$m = \log^2 x\left\{\log x\left(\frac{xf'(x)}{f(x)} - 1\right) - 1\right\}, \text{ when } x = \infty. \qquad (30)$$

and the given series is convergent or divergent according as m is greater than or less than 1.

If the value of m given in (30) = 1, we must proceed in a similar manner to another of the series given in Ex. 5, Art 183: and so on. Hereby we obtain the following series of discriminating quantities, viz.

$$\frac{xf'(x)}{f(x)},\ \log x\left\{\frac{xf'(x)}{f(x)}-1\right\},\ \log^2 x\left\{\log x\left(\frac{xf'(x)}{f(x)}-1\right)-1\right\},\ldots$$

$$\ldots \text{ when } x=\infty; \qquad (31)$$

of which the law of formation is evidently

$$Q_n = \log^{n-1} x\,\{Q_{n-1}-1\}, \qquad (32)$$

where

$$Q_1 = \frac{xf'(x)}{f(x)}. \qquad (33)$$

188.] Another series of discriminating quantities has been derived by Raabe from the preceding; and in certain forms of functions these are applied with greater facility.

Take (1) to be the form of the series; then since $u_x = \frac{1}{f(x)}$;

$$\begin{aligned}\frac{u_x}{u_{x+1}}-1 &= \frac{f(x+1)-f(x)}{f(x)}\\ &= \frac{f'(x+\theta)}{f(x)}\\ &= \frac{f'(x)}{f(x)}, \text{ when } x=\infty;\end{aligned}$$

and thus $\frac{f'(x)}{f(x)}$ may be replaced by $\frac{u_x}{u_{x+1}}-1$.

Hence Q_1, which is given in (33), $= x\left\{\frac{u_x}{u_{x+1}}-1\right\}$; (34)

and the discriminating quantities given in (31) and (32) become changed accordingly.

This latter form of the discriminating quantity is convenient, when a series of factorials enters into the general term of the series, as the following examples will shew.

Ex. 1. Determine the character as to convergence and divergence of $\sum_1^{\infty} \frac{1.3.5\ldots(2x-1)}{2.4.6\ldots 2x}\,\frac{1}{2x+1}$.

$$x\left\{\frac{u_x}{u_{x+1}}-1\right\} = \frac{6x^2+5x}{(2x+1)^2} = \frac{3}{2}, \text{ when } x=\infty;$$

hence the series is convergent.

Ex. 2. The series $\sum_1^{\infty} \frac{1.3.5\ldots(2x-1)}{2.4.6\ldots 2x}$ is divergent.

Another form has been given to the preceding theorems by

M. Paucker of St. Petersbourg*: but as they are exactly equivalent, it is unnecessary to insert them.

189.] All the preceding investigations, as to the convergence and divergence of series, may be summed up in the following practical directions to the student. Examine first the ratio of u_{x+1} to u_x when $x = \infty$; according as that ratio is greater than or less than 1, the series is divergent or convergent. If that ratio = 1, then examine in order the series of discriminating quantities given in (31), or in the equivalent forms as modified by (34); then according as that discriminating quantity, which is the first not to be equal to 1 when $x = \infty$, is greater or less than 1, so will the series be convergent or divergent.

I cannot conclude this section without acknowledging my obligations to Mr. De Morgan, M. Duhamel, and H. Raabe; but above all to the critical Memoir of M. J. Bertrand on the convergence of series, in Vol. VII, of Liouville's Journal, in which he compares the several criteria of the three mentioned writers, and demonstrates the identity of them, and consequently shews that the extent of applicability is the same in all.

I would also refer the reader to two Memoirs on the Theory of Series by Ossian Bonnet. The former on Convergence and Divergence, given in Liouville, Vol. VIII, p. 73; the latter in Mém. Cour. de l'Acad. Roy. de Belgique, Tome XXIII, in which he treats especially of Periodic Series.

SECTION 2.—*The Series of Taylor and Maclaurin.*

190.] In the first article of the preceding section, it has been remarked that there are two cases in which an exact equality exists between a function and the series into which it is developed; and consequently in which these quantities may be employed interchangeably with each other. These cases are (1) that in which the series is convergent, and the number of terms is infinite; (2) that in which the remainder, that is the sum of all the terms after the nth, can be determined, and added to the preceding n terms. It has also been observed that a series may be convergent for certain values of its variable and divergent for other values. Now in the proofs of the theorems of Taylor and Maclaurin, which are given in Chapter VI. of Vol. I, it is shewn that

* See Crelle's Journal. Vol. XLII, p. 138.

the functions which are to be expanded must in the first place be such, that neither they nor their derived functions up to the $(n-1)$th inclusively must be infinite within certain limits, when the development is effected as far as the nth term. This condition is previous to any test which determines the convergence or divergence of the series; this latter character must be determined by the criteria of the preceding section.

If Taylor's Theorem is given in the form

$$F(a+h) = F(a)+F'(a)\frac{h}{1} + \dots + F^{x-1}(a)\frac{h^{x-1}}{1.2.3\dots(x-1)} + \dots, \quad (35)$$

then taking the discriminating quantity in the form (34), we have

$$x\left\{\frac{u_x}{u_{x+1}} - 1\right\} = x\left\{\frac{F^{x-1}(a)}{F^x(a)}\frac{x}{h} - 1\right\}; \quad (36)$$

and consequently, the series (35) is convergent or divergent, according as (36), when $x = \infty$, is greater than or less than 1.

Again, taking Maclaurin's series in the form

$$F(h) = F(0)+F'(0)\frac{h}{1} + \dots + F^{x-1}(0)\frac{h^{x-1}}{1.2.3\dots(x-1)} + \dots, \quad (37)$$

this series will be convergent or divergent according as

$$x\left\{\frac{F^{x-1}(0)}{F^x(0)}\frac{x}{h} - 1\right\}, \text{ when } x = \infty, \quad (38)$$

is greater than or less than 1.

191.] In the preceding volume the remainders of Taylor's and Maclaurin's series have also been investigated; but the forms are indeterminate, because they depend on θ which is an undetermined positive proper fraction. But the remainders can be found in more precise forms as definite integrals. These we proceed to determine; and the inquiry is also otherwise important, because it yields new proofs of these theorems, subject however to the same conditions, as to continuity and finiteness of the function and the derived functions, as those which are required in the previous proofs.

Let $F'(x+h-z)$ be a function, which is finite and continuous for all employed values of its subject-variable; let it be the element-function of a z-integration, and let h and 0 be the limits of z; then

$$\int_0^h F'(x+h-z)\,dz = \Big[-F(x+h-z)\Big]_0^h \quad (39)$$

$$= F(x+h) - F(x). \quad (40)$$

Let us suppose the several derived functions of $F'(x+h-z)$, up to the nth inclusive, to be finite and continuous for all employed values of their subject-variables; then, by a series of successive integrations by parts, we have

$$\int_0^h F'(x+h-z)\,dz = \Big[z F'(x+h-z)\Big]_0^h + \int_0^h F''(x+h-z)\,z\,dz$$

$$= \frac{h}{1} F'(x) + \Big[\frac{z^2}{1.2} F''(x+h-z)\Big]_0^h + \int_0^h F'''(x+h-z)\frac{z^2}{1.2}\,dz$$

$$= \frac{h}{1} F'(x) + \frac{h^2}{1.2} F''(x) + \Big[\frac{z^3}{1.2.3} F'''(x+h-z)\Big]_0^h$$
$$+ \int_0^h F''''(x+h-z)\frac{z^3}{1.2.3}\,dz$$

$$= \frac{h}{1} F'(x) + \frac{h^2}{1.2} F''(x) + \frac{h^3}{1.2.3} F'''(x)$$
$$+ \int_0^h F''''(x+h-z)\frac{z^3}{1.2.3}\,dz;$$

and so on for n integrations, until

$$\int_0^h F'(x+h-z)\,dz = F'(x)\frac{h}{1} + F''(x)\frac{h^2}{1.2} + F'''(x)\frac{h^3}{1.2.3} + \dots$$
$$\dots + F^{n-1}(x)\frac{h^{n-1}}{1.2.3\dots(n-1)} + \int_0^h F^n(x+h-z)\frac{z^{n-1}}{1.2\dots(n-1)}\,dz; \quad (41)$$

and replacing the left-hand member in terms of its equivalent from (40), we have

$$F(x+h) = F(x) + F'(x)\frac{h}{1} + F''(x)\frac{h^2}{1.2} + F'''(x)\frac{h^3}{1.2.3} + \dots$$
$$\dots + F^{n-1}(x)\frac{h^{n-1}}{1.2.3\dots(n-1)} + \int_0^h F^n(x+h-z)\frac{z^{n-1}dz}{1.2\dots(n-1)}, \quad (42)$$

which is Taylor's Theorem; and in which the equivalence of the two members of the equation is perfect, and without any indeterminateness. The remainder is given by the definite integral $\int_0^h F^n(x+h-z)\frac{z^{n-1}dz}{1.2.3\dots(n-1)}$; if this can be determined in a finite form, the expansion of $F(x+h)$ is completely exhibited; and if the remainder is zero, when $n = \infty$, it must be neglected when the number of the terms of the series is infinite. This will of course always be the case, when the series is convergent. In Vol. I, Art. 134, (16), the remainder of Taylor's series is expressed in the form $\frac{h^n}{1.2.3\dots n} F^n(x+\theta h)$, where θ is a positive proper fraction. The equivalence of these two forms of remainders is easily demonstrated. For by the theorem contained in (228), Art. 116, we have

$$\int_0^h F^n(x+h-z)\frac{z^{n-1}\,dz}{1.2.3\ldots(n-1)} = F^n(x+\theta h)\int_0^h \frac{z^{n-1}\,dz}{1.2.3\ldots(n-1)}$$

$$= F^n(x+\theta h)\frac{h^n}{1.2.3\ldots n}, \qquad (43)$$

where θ denotes a positive proper fraction; but as we have no means generally for determining its value, the definite integral is the more exact expression of the remainder.

192.] Maclaurin's Theorem for the development of a function of x may also be demonstrated in a similar manner; and its remainder may be expressed as a definite integral.

Let $F'(x-z)$ be a function of z which is finite and continuous for all employed values of its subject-variables, the value of z ranging from 0 to x; then

$$\int_0^x F'(x-z)\,dz = \Big[-F(x-z)\Big]_{z=0}^{z=x}$$

$$= F(x)-F(0). \qquad (44)$$

And suppose also that all the derived functions of $F'(x-z)$ up to the nth are finite and continuous for all employed values of its subject-variables, and for all values of z between 0 and x; then by integration by parts we have

$$\int_0^x F'(x-z)\,dz = \Big[z F'(x-z)\Big]_0^x + \int_0^x F''(x-z)\frac{z}{1}\,dz$$

$$= \frac{x}{1}F'(0) + \Big[F''(x-z)\frac{z^2}{1.2}\Big]_0^x + \int_0^x F'''(x-z)\frac{z^2}{1.2}\,dz$$

$$= \frac{x}{1}F'(0) + \frac{x^2}{1.2}F''(0) + \ldots$$

$$\ldots + \frac{x^{n-1}}{1.2.3\ldots(n-1)}F^{n-1}(0) + \int_0^x F^n(x-z)\frac{z^{n-1}\,dz}{1.2\ldots(n-1)}; \quad (45)$$

and replacing the left-hand member from (44), we have

$$F(x) = F(0)+F'(0)\frac{x}{1}+F''(0)\frac{x^2}{1.2}+F'''(0)\frac{x^3}{1.2.3}+\ldots$$

$$\ldots + F^{n-1}(0)\frac{x^{n-1}}{1.2.3\ldots(n-1)} + \int_0^x F^n(x-z)\frac{z^{n-1}dz}{1.2\ldots(n-1)}, \quad (46)$$

which is Maclaurin's series; and in which the equivalence of the two members is perfect, and without any indeterminateness.

The remainder of the series is given by the definite integral $\int_0^x F^n(x-z)\frac{z^{n-1}dz}{1.2\ldots(n-1)}$; and if this definite integral can be evaluated, the expansion of $F(x)$ is completely exhibited; and if it $=0$, when $n=\infty$, it must be omitted when the number of terms of the series is infinite. This is the case when the series is convergent.

The remainder of Maclaurin's series, as it is given in (18), Art. 134, Vol. I. is $F^n(\theta x)\frac{x^n}{1.2.3\ldots n}$; and this is equivalent to the preceding definite integral; for by (228), Art. 116,

$$\int_0^x F^n(x-z)\frac{z^{n-1}dz}{1.2.3\ldots(n-1)} = F^n(\theta x)\int_0^x \frac{z^{n-1}dz}{1.2.3\ldots(n-1)}$$

$$= F^n(\theta x)\frac{x^n}{1.2.3\ldots n}, \qquad (47)$$

where θ denotes a positive proper fraction; but as we have generally no means of determining it, the definite integral is the more exact expression of the remainder.

SECTION 3.—*The Development of Series by means of Single Definite Integration.*

193.] In discussing the several methods which have been employed for the approximate determination of the value of a definite integral, when that value cannot be found directly and in finite terms, we shewed how in many cases the element-function or part of the element-function might be expanded into a converging series; whereby it would consist of a series of terms, of each of which the definite integral could be determined; and thus the value of the whole definite integral could be, at least approximately, found. This process we explained and illustrated in Arts. 119–121; and have applied in Art. 157 to the rectification of an ellipse of small eccentricity.

Also in exhibiting the uses of the Gamma-function and of its allied integrals in Arts. 143–145 we were incidentally led to the sum of certain series of a complex character under the form of definite integrals.

Now the correctness of these processes is based on the following theorem. If an element-function can be expressed in a convergent series, so that the series is equivalent to the element-function and can be used instead of it, then the definite integral of the element is equal to that of the series; the limits being the same in both cases, and neither the element-function nor any term of the series being infinite or discontinuous for any value of the variable within the range of integration. Consequently, if the definite integral of all the terms of the series can be found, the sum of them is the value of the original definite integral.

In the preceding cases this method has been employed for the purpose of approximating to the value of a definite integral, when that value cannot otherwise be found. Here however we propose to apply the method to integrals, the values of which can otherwise be determined; so that the definite integral will be expressed as a series; and thus either the definite integral may be considered as the sum of a series, or the series may be considered as the expansion of the definite integral. Consequently the process is that of summation or of expansion according to the point of view whence it is considered.

194.] The following are examples of the process:

Ex. 1. Since the following series found by the binomial theorem is convergent for all values of x greater than -1,

$$\frac{1}{1+x} = 1-x+x^2-x^3+\dots; \qquad (48)$$

$$\therefore \quad \int_0^x \frac{dx}{1+x} = \int_0^x \{1-x+x^2-x^3+\dots\}\,dx;$$

$$\log(1+x) = x-\frac{x^2}{2}+\frac{x^3}{3}-\frac{x^4}{4}+\dots. \qquad (49)$$

Ex. 2. By the binomial theorem,

$$\frac{1}{1+x^2} = 1-x^2+x^4-x^6+\dots; \qquad (50)$$

$$\therefore \quad \int_0^x \frac{dx}{1+x^2} = \int_0^x \{1-x^2+x^4-x^6+\dots\}\,dx;$$

$$\tan^{-1}x = x-\frac{x^3}{3}+\frac{x^5}{5}-\frac{x^7}{7}+\dots; \qquad (51)$$

which series is convergent for all values of x less than 1, and is useful for the calculation of π. Also if $x = 1$,

$$\frac{\pi}{4} = 1-\frac{1}{3}+\frac{1}{5}-\frac{1}{7}+\frac{1}{9}; \qquad (52)$$

$$\therefore \quad \frac{\pi}{8} = \frac{1}{1.3}+\frac{1}{5.7}+\frac{1}{9.11}+\dots. \qquad (53)$$

In the case where x is greater than 1, that is, when $\tan^{-1}x$ is intermediate to $\frac{\pi}{4}$ and $\frac{\pi}{2}$, the following series is more convergent; since

$$\frac{1}{1+x^2} = \frac{1}{x^2}\left(1+\frac{1}{x^2}\right)^{-1}$$

$$= \frac{1}{x^2} - \frac{1}{x^4} + \frac{1}{x^6} - \dots;$$

$$\therefore \int_x^\infty \frac{dx}{1+x^2} = \int_x^\infty \left\{\frac{1}{x^2} - \frac{1}{x^4} + \frac{1}{x^6} - \dots\right\} dx$$

$$\therefore \frac{\pi}{2} - \tan^{-1}x = \frac{1}{x} - \frac{1}{3x^3} + \frac{1}{5x^5} - \frac{1}{7x^7} + \dots;$$

$$\therefore \tan^{-1}x = \frac{\pi}{2} - \frac{1}{x} + \frac{1}{3x^3} - \frac{1}{5x^5} + \dots; \tag{54}$$

which series is however the same as (51), when x is replaced by $\frac{1}{x}$.

Ex. 3. Again, since

$$\frac{1}{(1-x^2)^{\frac{1}{2}}} = 1 + \frac{1}{2}x^2 + \frac{1.3}{2.4}x^4 + \frac{1.3.5}{2.4.6}x^6 + \dots;$$

$$\therefore \int_0^x \frac{dx}{(1-x^2)^{\frac{1}{2}}} = \int_0^x \left\{1 + \frac{1}{2}x^2 + \frac{1.3}{2.4}x^4 + \frac{1.3.5}{2.4.6}x^6 + \dots\right\} dx;$$

$$\sin^{-1}x = x + \frac{1}{2}\frac{x^3}{3} + \frac{1.3}{2.4}\frac{x^5}{5} + \frac{1.3.5}{2.4.6}\frac{x^7}{7} + \dots. \tag{55}$$

Hence, if $x = 1$,

$$\frac{\pi}{2} = 1 + \frac{1}{2}\frac{1}{3} + \frac{1.3}{2.4}\frac{1}{5} + \frac{1.3.5}{2.4.6}\frac{1}{7} + \dots. \tag{56}$$

This series however converges too slowly to be of use for the calculation of π. It is to be observed too that although we have taken the superior limit to be 1, which makes the element-function infinite, yet the series is correct; because the value of the element-function corresponding to the superior limit is not included in the definite integral.

Again, in (55), let $x = \frac{1}{2}$; then

$$\frac{\pi}{6} = \frac{1}{2} + \frac{1}{2}\frac{1}{3(2)^3} + \frac{1.3}{2.4}\frac{1}{5(2)^5} + \frac{1.3.5}{2.4.6}\frac{1}{7(2)^7} + \dots. \tag{57}$$

Also, since $\sin^{-1}x+\cos^{-1}x = \frac{\pi}{2}$, therefore

$$\cos^{-1}x = \frac{\pi}{2} - \sin^{-1}x$$

$$= \frac{\pi}{2} - x - \frac{1}{2}\frac{x^3}{3} - \frac{1.3}{2.4}\frac{x^5}{5} - \dots \qquad (58)$$

And if in this equation x is replaced by $\frac{1}{x}$,

$$\sec^{-1}x = \frac{\pi}{2} - \frac{1}{x} - \frac{1}{2}\frac{1}{3x^3} - \frac{1.3}{2.4}\frac{1}{5x^5} - \dots \qquad (59)$$

195.] The preceding examples are sufficient to illustrate the process which is the subject of the present section. The equivalences given in them may again be the subjects of definite integration, and thereby will other series be found, and their sums will be found; these latter however will be in the form of definite integrals; and consequently the sums of the series will be expressed as determinate quantities only when these integrals can be evaluated. The following are examples of this process.

Ex. 1. Let the two members of (55) be the element-functions of a definite integral whose limits are x and 0; so that

$$\int_0^x \sin^{-1}x\,dx = \int_0^x \left\{ x + \frac{1}{2}\frac{x^3}{3} + \frac{1.3}{2.4}\frac{x^5}{5} + \dots\dots \right\} dx. \qquad (60)$$

$$\therefore \quad \Big[x\sin^{-1}x + (1-x^2)^{\frac{1}{2}} \Big]_0^x = \Big[\frac{x^2}{2} + \frac{1}{2}\frac{x^4}{3.4} + \frac{1.3}{2.4}\frac{x^6}{5.6} + \dots \Big]_0^x$$

$$x\sin^{-1}x + (1-x^2)^{\frac{1}{2}} - 1 = \frac{x^2}{2} + \frac{1}{2}\frac{x^4}{3.4} + \frac{1.3}{2.4}\frac{x^6}{5.6} + \dots \qquad (61)$$

Hence, if $x = 1$,

$$\frac{\pi}{2} = 1 + \frac{1}{2} + \frac{1}{2}\frac{1}{3.4} + \frac{1.3}{2.4}\frac{1}{5.6} + \dots \qquad (62)$$

Similarly, if $x = \frac{1}{2}$,

$$\frac{\pi}{12} = 1 - \frac{3^{\frac{1}{2}}}{2} + \frac{1}{4}\left\{ \frac{1}{2} + \frac{1}{2}\frac{1}{3.4}\frac{1}{2^2} + \frac{1.3}{2.4}\frac{1}{5.6}\frac{1}{2^4} + \dots \right\}; \qquad (63)$$

which is a series sufficiently converging for the calculation of π. The terms which are contained in (63) give the true value to four places of decimals.

Ex. 2. Let both members of (55) be multiplied by x, and

then made the subjects of definite integration with the limits x and 0, we have

$$\int_0^x x\sin^{-1}x\,dx = \int_0^x \left\{x^3 + \frac{1}{2}\frac{x^4}{3} + \frac{1.3}{2.4}\frac{x^6}{5} + \frac{1.3.5}{2.4.6}\frac{x^8}{7} + \dots\right\} dx.$$

$$\frac{2x^2-1}{4}\sin^{-1}x + \frac{x}{4}(1-x^2)^{\frac{1}{2}}$$

$$= \frac{x^3}{3} + \frac{1}{2}\frac{x^5}{3.5} + \frac{1.3}{2.4}\frac{x^7}{5.7} + \frac{1.3.5}{2.4.6}\frac{x^9}{7.9} + \dots. \quad (64)$$

Ex. 3. Again, let both members of (55) be multiplied by $\frac{dx}{(1-x^2)^{\frac{1}{2}}}$; and let the definite integral be taken of both members for the limits 1 and 0; then

$$\int_0^1 \sin^{-1}x \frac{dx}{(1-x^2)^{\frac{1}{2}}} = \int_0^1 \left\{x + \frac{1}{2}\frac{x^3}{3} + \frac{1.3}{2.4}\frac{x^5}{5} + \dots\right\} \frac{dx}{(1-x^2)^{\frac{1}{2}}}.$$

And applying to each term of the right-hand member the theorem contained in (14), Art. 82, we have

$$\left[\frac{1}{2}(\sin^{-1}x)^2\right]_0^1 = \left[-(1-x^2)^{\frac{1}{2}}\left\{\frac{1}{1^2} + \frac{1}{3^2} + \frac{1}{5^2} + \frac{1}{7^2} + \dots\right\}\right]_0^1;$$

$$\therefore \quad \frac{\pi^2}{8} = \frac{1}{1^2} + \frac{1}{3^2} + \frac{1}{5^2} + \dots. \quad (65)$$

Ex. 4. If in (49) x is replaced by $-x$,

$$-\log(1-x) = x + \frac{x^2}{2} + \frac{x^3}{3} + \frac{x^4}{4} + \dots;$$

$$\therefore \quad -\frac{\log(1-x)}{x} = 1 + \frac{x}{2} + \frac{x^2}{3} + \frac{x^3}{4} + \dots;$$

$$-\int_0^1 \frac{\log(1-x)}{x}dx = \left[x + \frac{x^2}{2^2} + \frac{x^3}{3^2} + \frac{x^4}{4^2} + \dots\right]_0^1$$

$$= \frac{1}{1^2} + \frac{1}{2^2} + \frac{1}{3^2} + \frac{1}{4^2} + \dots; \quad (66)$$

and thus the sum of the series in the right-hand member is expressed as the definite integral given in the left-hand member. But the sum of the series in the right-hand member $= \frac{\pi^2}{6}$, by (139), Art. 89, Vol. I; consequently

$$-\int_0^1 \frac{\log(1-x)}{x}dx = -\int_0^1 \log(1-x)\,d.\log x = \frac{\pi^2}{6}. \quad (67)$$

Ex. 5. $$\int_0^1 \frac{\tan^{-1}x}{x}dx = \frac{1}{1^2} - \frac{1}{3^2} + \frac{1}{5^2} - \frac{1}{7^2} + \dots. \quad (68)$$

Ex. 6. Since $\int \tan^{-1} x \, dx = x \tan^{-1} x - \log (1+x^2)^{\frac{1}{2}}$,

$$\frac{\pi}{4} - \log 2^{\frac{1}{2}} = \frac{1}{1.2} - \frac{1}{3.4} + \frac{1}{5.6} - \frac{1}{7.8} + \dots \qquad (69)$$

Section 4.—*On Periodic Series, and on Fourier's Integral.*

196.] The series, which, as the equivalents of certain functions, have been made the subjects of definite integration in the preceding section, have been derived from these functions primarily by the binomial theorem; and thus are perhaps the most simple of their kind. Series however formed by other processes of development may be the subjects of integration; and we intend in this section to consider one of the most important of these; viz., periodic series. The subject is important; for it exhibits new and extraordinary results which are given by no other process; and in the application of mathematical analysis to physical problems affords the solution of many intricate and curious questions, and gives expression to certain very peculiar laws. An instance of these has been adduced in Art. 15, Vol. I. in illustration of discontinuous functions; for the sum of the series therein quoted vanishes for all values of the subject-variable a, except when some multiple of π is substituted for a, when the sum takes an indeterminate form; thus the sum of the series varies discontinuously, although each term varies continuously. In this section these and similar properties of periodic functions and series will be investigated and traced to their origin.

A periodic series is that whose terms contain sines, or cosines, of the subject-variable, and of multiples of that variable. Thus

$$A_1 \cos x + A_2 \cos 2x + \dots + A_n \cos nx + \dots$$

is a periodic series. Whatever the sum of it may be, that sum goes through a succession of values as x increases from 0 to 2π; because every term of the series has at the end of that period the same value which it had at the beginning; and this succession is repeated as x increases from 2π to 4π; from 4π to 6π; and so on; so that whatever n is, 2π is the period of the function.

Now the primary problem is to determine the conditions for which a given function, say $f(x)$, is capable of expression in the form of a periodic series, and to express it in that form. Thus, if

$$f(x) = A_0 + A_1 \cos x + A_2 \cos 2x + \ldots + A_n \cos nx + \ldots$$
$$+ B_1 \sin x + B_2 \sin 2x + \ldots + B_n \sin nx + \ldots, \tag{70}$$

the problem is the determination of the unknown constants, the A's and the B's.

It will be observed that in the right-hand member of the equivalence a non-periodic term A_0 is introduced; the most general form has hereby been given to the assumed equivalent of $f(x)$; so that if $f(x)$ is capable of expansion in periodic terms only, $A_0 = 0$; the result will shew whether this is the case or not.

To determine A_0, let both members of (70) be the element-functions of a definite integral whose limits are $a+2\pi$ and a, when a is undetermined; then since

$$\int_a^{a+2\pi} \sin ix\, dx = 0, \qquad \int_a^{a+2\pi} \cos ix\, dx = 0;$$

$$\therefore \int_a^{a+2\pi} f(x)\, dx = A_0 \int_a^{a+2\pi} dx$$
$$= 2\pi A_0;$$

$$\therefore A_0 = \frac{1}{2\pi} \int_a^{a+2\pi} f(x)\, dx. \tag{71}$$

Thus $A_0 = 0$, or the series equivalent to $f(x)$ consists of periodic terms only, when $\int_a^{a+2\pi} f(x) dx = 0$.

Again, let both the members of (70) be multiplied firstly by $\cos ix\, dx$, and secondly by $\sin ix\, dx$; and in each case let the definite integrals of both members be taken for the same limits as before; then since

$$\int_a^{a+2\pi} \cos ix \cos jx\, dx = \int_a^{a+2\pi} \cos ix \sin jx\, dx = \int_a^{a+2\pi} \sin ix \sin jx\, dx = 0; \tag{72}$$

$$\int_a^{a+2\pi} (\cos ix)^2\, dx = \pi, \qquad \int_a^{a+2\pi} (\sin ix)^2\, dx = \pi; \tag{73}$$

$$\therefore \int_a^{a+2\pi} f(x) \cos ix\, dx = \pi A_i, \qquad \int_a^{a+2\pi} f(x) \sin ix\, dx = \pi B_i; \tag{74}$$

$$A_i = \frac{1}{\pi} \int_a^{a+2\pi} f(x) \cos ix\, dx, \qquad B_i = \frac{1}{\pi} \int_a^{a+2\pi} f(x) \sin ix\, dx; \tag{75}$$

and consequently, if in these formulæ i is successively replaced by $1, 2, 3, \ldots n, \ldots$, the A's and the B's will be determined. Let these be substituted in (70); and, to give greater distinctness to the expression, let x, the subject-variable under the symbol of integration be replaced by z; then

$$
\begin{aligned}
f(x) &= \frac{1}{2\pi}\int_{a}^{a+2\pi} f(z)\,dz \\
&\quad + \frac{\cos x}{\pi}\int_{a}^{a+2\pi} f(z)\cos z\,dz + \frac{\cos 2x}{\pi}\int_{a}^{a+2\pi} f(z)\cos 2z\,dz + \dots \\
&\quad + \frac{\sin x}{\pi}\int_{a}^{a+2\pi} f(z)\sin z\,dz + \frac{\sin 2x}{\pi}\int_{a}^{a+2\pi} f(z)\sin 2z\,dz + \dots \quad (76)\\
&= \frac{1}{2\pi}\int_{a}^{a+2\pi} f(z)\,dz \\
&\quad + \frac{1}{\pi}\left\{\int_{a}^{a+2\pi} f(z)\cos(x-z)\,dz + \int_{a}^{a+2\pi} f(z)\cos 2(x-z)\,dz + \dots\right\} \quad (77)\\
&= \frac{1}{2\pi}\int_{a}^{a+2\pi} f(z)\,dz + \frac{1}{\pi}\sum_{n=1}^{n=\infty}\int_{a}^{a+2\pi} f(z)\cos n(x-z)\,dz; \quad (78)
\end{aligned}
$$

where the symbol $\sum_{n=1}^{n=\infty}$ denotes the sum of all the definite integrals in the quantity following it, which are obtained by replacing n successively by 1, 2, 3, ... ∞. Thus the form and the value of the coefficients in the equivalent for $f(x)$ which is given in (70) are determined.

In the preceding process it has been assumed that $f(x)$ is capable of development in the given form; the complete solution however of the problem requires that the conditions requisite for the possibility of this development should be ascertained. It is also assumed that the right-hand member of (70) is equivalent to $f(x)$; but when this is the case, the series must be convergent; and accordingly it is necessary to shew that the series is convergent in form, and to assign the limits of value of the several terms within which it is convergent. Now the theory of convergence and the tests of convergence, which have been investigated in the first section of the present Chapter, are sufficient for the purpose; and if they are applied it will be at once seen that the required characteristics are satisfied, and that consequently an equivalence is established between the two members of (78).

197.] We need not however pursue the subject in this direction; for another course of investigation, leading to a more general result, is open to us; and I shall assume the preceding series to be only a suggestion of the form of the required development; and acting on this presumption, investigate the series by an exact and precise process.

Now whatever is the angle denoted by θ, we have the following equivalence,

$$\frac{1}{2}+\cos\theta+\cos 2\theta+\ldots+\cos n\theta = \frac{\sin\left(n+\frac{1}{2}\right)\theta}{2\sin\frac{\theta}{2}}. \qquad (79)$$

Let θ be replaced by $x-z$; and this being done, let every term be multiplied by $f(z)dz$; and let the definite integral of each term thus found be taken for the limits b and a, $f(z)$ being finite and continuous throughout this range of integration; then

$$\frac{1}{2}\int_a^b f(z)\,dz+\int_a^b f(z)\cos(x-z)\,dz+\int_a^b f(z)\cos 2(x-z)\,dz+\ldots$$
$$+\int_a^b f(z)\cos n(x-z)\,dz = \int_a^b f(z)\frac{\sin\left(n+\frac{1}{2}\right)(x-z)}{2\sin\frac{x-z}{2}}\,dz. \qquad (80)$$

It is evident that every term in the left-hand member is unaltered when x is increased by 2π. Consequently both members denote periodic functions of x, whose period is 2π, whatever is the value of n; and if we trace the value of either member of (80) through this range, that value will be repeated when b and a are replaced by $b+2k\pi$ and $a+2k\pi$ respectively; we may therefore without loss of generality confine our attention within limits of integration, such that $b-a$ is not greater than 2π.

When $n=\infty$, the number of terms in the first member of (80) is infinite, and the sum of them is denoted by the value which the second member takes when $n=\infty$. As this is the case of the series which we have to investigate, this latter value must be determined.

Its value evidently depends on the relative values of x and the several values of z, and we shall have three cases; (1), when x falls within the range of integration; (2), when x coincides with either limit of the integration; (3), when x falls beyond the range.

(1). Let x fall within the range of integration; that is, let x be greater than a and less than b; then, when $z=x$, the right-hand member of (80) takes an indeterminate form, and must be evaluated. Now the element-function vanishes for all other values of z; for since $\left(n+\frac{1}{2}\right)(x-z)$ varies by 2π, when z varies by

$\frac{4\pi}{2n+1}$, which $= 0$, when $n = \infty$; it is evident that over that range of z, which is infinitesimal, $\frac{f(z)}{\sin\frac{x-z}{2}}$ may be considered constant, and $\int \sin\left(n+\frac{1}{2}\right)(x-z)\,dz = 0$.

Hence the right-hand member of (80) $= 0$, except when $z = x$. To determine the value when $z = x$, we may consider it only for values of z infinitesimally near to x; that is, if ι is an infinitesimal, we may take the limits of the z-integration to be $x-\iota$ and $x+\iota$; and if we replace $z-x$ by ξ, ξ is the variable of integration, and is always infinitesimal, its limits of integration being $-\iota$ and $+\iota$; so that

$$\int_a^b f(z)\frac{\sin\left(n+\frac{1}{2}\right)(x-z)}{\sin\frac{x-z}{2}}dz = \int_{x-\iota}^{x+\iota} f(z)\frac{\sin\left(n+\frac{1}{2}\right)(x-z)}{\sin\frac{x-z}{2}}dz \quad (81)$$

$$= \int_{-\iota}^{\iota} f(x+\xi)\frac{\sin\left(n+\frac{1}{2}\right)\xi}{\sin\frac{\xi}{2}}d\xi;$$

and as ξ is always infinitesimal $f(x+\xi)$ may be replaced by $f(x)+\xi f'(x)$, and $\sin\frac{\xi}{2}$ by $\frac{\xi}{2}$; so that

$$\int_a^b f(z)\frac{\sin\left(n+\frac{1}{2}\right)(x-z)}{\sin\frac{x-z}{2}}dz$$

$$= 2f(x)\int_{-\iota}^{\iota}\frac{\sin\left(n+\frac{1}{2}\right)\xi}{\xi}d\xi + 2f'(x)\int_{-\iota}^{\iota}\sin\left(n+\frac{1}{2}\right)\xi\,d\xi.$$

Of the integrals in the right-hand member of this equation, the latter vanishes, because for all values of ξ, $\sin\left(n+\frac{1}{2}\right)\xi = 0$, when $n = \infty$. And to determine the value of the former, we may enlarge the range to $-\infty$ and $+\infty$, because all the elements of that definite integral vanish when $n = \infty$, except those corresponding to small values of ξ. Hence

$$\int_{-i}^{i} \frac{\sin\left(n+\frac{1}{2}\right)\xi}{\xi} d\xi = \int_{-\infty}^{\infty} \frac{\sin\left(n+\frac{1}{2}\right)\xi}{\xi} d\xi \tag{82}$$

$$= \pi; \quad \text{by (77), Art. 94;} \tag{83}$$

$$\therefore \quad \int_a^b f(z) \frac{\sin\left(n+\frac{1}{2}\right)(x-z)}{\sin\frac{x-z}{2}} dz = 2\pi f(x); \tag{84}$$

and substituting in (80),

$$f(x) = \frac{1}{2\pi}\int_a^b f(z)\,dz + \frac{1}{\pi}\int_a^b f(z)\cos(x-z)\,dz$$
$$+ \frac{1}{\pi}\int_a^b f(z)\cos 2(x-z)\,dz + \dots$$
$$= \frac{1}{2\pi}\int_a^b f(z)\,dz + \frac{1}{\pi}\sum_{n=1}^{n=\infty}\int_a^b f(z)\cos n(x-z)\,dz; \tag{85}$$

which gives the periodic series into which $f(x)$ may be developed, when x falls within the limits of the z-integration.

As the general term of this series is $\frac{1}{\pi}\int_a^b f(z)\cos n\,(x-z)\,dz$, and this is equal to

$$\frac{\cos nx}{\pi}\int_a^b f(z)\cos nz\,dz + \frac{\sin nx}{\pi}\int_a^b f(z)\sin nz\,dz,$$

it is of the form $\text{A}_n \cos nx + \text{B}_n \sin nx$; and consequently by means of (85) $f(x)$ is developed into a series of the form given in (70), Art. 196. This series is however more general, because the limits of integration in it are b and a, whereas those in (70) Art. 196 are $a+2\pi$ and a.

(2). Let $x =$ one of the limits of the z-integration; $= b$, say, the superior limit; then the limits of integration in the right-hand member of (81) must be b and $b-i$, otherwise values of z would be included which are not within the range of integration; and consequently the limits of ξ in the ξ-integration are 0 and $-i$; and the value of the definite integral in the right-hand member of (82) is $\frac{\pi}{2}$, the limits being 0 and $-\infty$; and the series is consequently equal to $\frac{\pi}{2}f(b)$. Similarly if $x = a$, the inferior limit, the limits in the right-hand member of (81) are $a+i$ and a; so that the limits of ξ in the ξ-integration are i and 0; and the value of the definite integral in the right-hand member of

(82) is $\frac{\pi}{2}$, the limits being ∞ and 0; and the series is consequently equal to $\frac{\pi}{2}f(a)$. Hence if x is equal to the limit b or a, we have respectively

$$\frac{\pi}{2}f(b) = \frac{1}{2}\int_a^b f(z)\,dz + \sum_{n=1}^{n=\infty}\int_a^b f(z)\cos n(b-z)\,dz; \qquad (86)$$

$$\frac{\pi}{2}f(a) = \frac{1}{2}\int_a^b f(z)\,dz + \sum_{n=1}^{n=\infty}\int_a^b f(z)\cos n(a-z)\,dz. \qquad (87)$$

(3). Let x fall beyond the range of integration; that is, let x be less than a, or be greater than b; then, as we have shewn in case (1), the element-function vanishes for all values of z, and the right-hand member of (80) always $= 0$. Thus

$$\frac{1}{2}\int_a^b f(z)\,dz + \sum_{n=1}^{n=\infty}\int_a^b f(z)\cos n(x-z)\,dz = 0. \qquad (88)$$

Thus in recapitulation, $b-a$ being not greater than 2π,

$$\frac{1}{2}\int_a^b f(z)\,dz + \sum_{n=1}^{n=\infty}\int_a^b f(z)\cos n(x-z)\,dz \qquad (89)$$

represents a function of x which is periodic, and of which the period is 2π; and which

$$\left.\begin{array}{llll} =\frac{\pi}{2}f(a), & =\pi f(x), & =\frac{\pi}{2}f(b), & =0, \\ \text{according as} & & & \\ x=a, & a<x<b, & x=b, & b<x<2\pi+a; \end{array}\right\} \qquad (90)$$

and these are the values of (89) through a complete period, and are repeated in both directions. This theorem contains the doctrine of periodic series in its most general form.

198.] The preceding results may be geometrically interpreted, and the interpretation is interesting and curious, inasmuch as it exhibits the discontinuity of the function in a striking manner. Take the right-hand member of (85) to be the ordinate of a plane curve, whose abscissa is x; so that if the left-hand member is denoted by y, $y=f(x)$ is the equation to the curve. Then for all values of x between b and a exclusively, the locus is represented by the curve $y=f(x)$; when $x=a$, $y=\frac{f(a)}{2}$; and when $x=b$, $y=\frac{f(b)}{2}$; so that the locus then becomes two points which are evidently discontinuous points. Also for values of x greater than b, and for values less than a, $y = 0$, so that the locus becomes the x-axis. This last part of the locus however is subject to a certain restriction; for since $\sin\frac{x-z}{2} = 0$, not only when $z = x$, but also when $z = x+2\pi$, $= x+4\pi$, $= \ldots = x+2k\pi$; therefore the right-hand member of (85) $= f(x)$, not only when x has values between b and a, but also when x has values between $b+2\pi$ and $a+2\pi$, between $b+4\pi$ and $a+4\pi$, ...; accordingly the curve, which is the locus between $x = b$ and $x = a$, is repeated at intervals, such that the distance between two similar points $= 2\pi$. Also as $y = \frac{f(a)}{2}$, when $x = a$, so $y = \frac{f(a)}{2}$ when $x = a+2\pi$, $= a+4\pi$, $= \ldots$; and $y = \frac{f(b)}{2}$ when $x = b$, $= b+2\pi$, $= b+4\pi$, $= \ldots$; and consequently we have a repetition of the points of discontinuity. Also the portions of the x-axis which the series expresses are of a finite length, and these finite portions are repeated: for as $y = 0$ for values of x greater than b, so this zero-value continues until $x = a+2\pi$, when y abruptly takes the value $\frac{f(a)}{2}$; and then as x increases, $y = f(x)$; and this locus continues until $x = b+2\pi$, when y abruptly takes the value $\frac{f(b)}{2}$; and again as x increases, $y = 0$; and this zero-value continues until $x = a+4\pi$, when another point of discontinuity occurs. And so on to infinity, in both the positive and negative directions. Thus $2\pi-(b-a)$ is the length of each portion of the x-axis which the series represents, and these are repeated at intervals, each of which $= b-a$.

If $b-a = 2\pi$, so that the limits of the z-integration are a and $a+2\pi$, then these portions of the x-axis disappear, and the locus consists of discontinuous branches of the curve $y = f(x)$; and at those values of x at which the discontinuity takes place,

$$y = \frac{1}{2}\{f(a)+f(\pi+a)\}. \tag{91}$$

199.] This last remark leads to a very important result. It shews that $f(x)$ need not be a continuous function, and that it may have points of discontinuity between the limits of the z-integration. When however this is the case, the integrals must be taken separately over the ranges through which the continuity of the functions extends; and the values at the limits must be given by means of (86) and (87). Thus, if $f(x)$ becomes discontinuous when $x=a$, and the value of $f(x)$ changes abruptly from A_1 to A_2, then, when $x=a$, $y=\frac{A_1+A_2}{2}$; similarly, if there is a point of discontinuity when $x=\beta$, and the value of $f(x)$ changes abruptly from B_1 to B_2, then at that point $y=\frac{B_1+B_2}{2}$; and so on.

200.] We now return to the general theorems contained in (90), and consider certain forms which they assume for particular values of the limits, for particular forms of the variables, and in some special cases.

As the period of the function is 2π, if the difference between the limits of the z-integration is 2π, the values of the functions for the variables contained between these limits recur without intervals; and if we take either 0 and 2π, or $-\pi$ and π, to be limits of integration, there is no loss of generality. In these cases we have the following theorems.

$$\left.\begin{array}{l}\frac{1}{2}\int_0^{2\pi} f(z)\,dz+\sum_{n=1}^{n=\infty}\int_0^{2\pi} f(z)\cos n(x-z)\,dz \\ \quad =\frac{\pi}{2}f(0), \quad =\pi f(x), \quad =\frac{\pi}{2}f(2\pi), \text{ when} \\ x=0, \quad 0<x<2\pi, \quad x=2\pi. \end{array}\right\} \tag{92}$$

$$\left.\begin{array}{l}\frac{1}{2}\int_{-\pi}^{\pi} f(z)\,dz+\sum_{n=1}^{n=\infty}\int_{-\pi}^{\pi} f(z)\cos n(x-z)\,dz \\ \quad =\frac{\pi}{2}f(-\pi), \quad =\pi f(x), \quad =\frac{\pi}{2}f(\pi), \text{ when} \\ x=-\pi, \quad -\pi<x<\pi, \quad x=\pi. \end{array}\right\} \tag{93}$$

If the limits of the z-integration are π and 0; then

$$\left.\begin{array}{l}\frac{1}{2}\int_0^{\pi} f(z)\,dz+\sum_{n=1}^{n=\infty}\int_0^{\pi} f(z)\cos n(x-z)\,dz \\ \quad =0, \quad =\frac{\pi}{2}f(0) \quad =\pi f(x), \quad =\frac{\pi}{2}f(\pi), \text{ when} \\ -\pi<x<0, \; x=0, \quad 0<x<\pi, \quad x=\pi; \end{array}\right\} \tag{94}$$

and these assign the values of the integrals through a complete period.

Now if in (94) x is replaced by $-x$; then

$$\left.\begin{array}{l}\frac{1}{2}\int_0^{\pi} f(z)\,dz + \sum_{n=1}^{n=\infty}\int_0^{\pi} f(z)\cos n(x+z)\,dz \\ = \pi f(-x), \quad = \frac{\pi}{2}f(0), \quad = 0, \qquad = \frac{\pi}{2}f(\pi), \text{ when} \\ -\pi < x < 0, \quad x = 0, \quad 0 < x < \pi, \quad x = \pi.\end{array}\right\} \quad (95)$$

As the limits of integration are the same in (94) and (95), we may add these equations, and we have

$$\left.\begin{array}{l}\int_0^{\pi} f(z)\,dz + 2\sum_{n=1}^{n=\infty}\cos nx\int_0^{\pi} f(z)\cos nz\,dz \\ = \pi f(-x), \quad = \pi f(0), \quad = \pi f(x), \quad = \pi f(\pi), \text{ when} \\ -\pi < x < 0, \quad x = 0, \quad 0 < x < \pi, \quad x = \pi;\end{array}\right\} \quad (96)$$

thus the definite integral in the left-hand member of (96) is equal to $\pi f(x)$ for all values contained between 0 and π inclusively; and to $\pi f(-x)$ for all values between $-\pi$ and 0 inclusively.

Again, subtracting (95) from (94),

$$\left.\begin{array}{l}2\sum_{n=1}^{n=\infty}\sin nx\int_0^{\pi} f(z)\sin nz\,dz \\ = -\pi f(-x), \quad = 0, \quad = \pi f(x), \quad = 0, \text{ when} \\ -\pi < x < 0, \quad x = 0, \quad 0 < x < \pi, \quad x = \pi;\end{array}\right\} \quad (97)$$

so that the definite integral in the left-hand member of (97) $= -\pi f(-x)$ for values of x between $-\pi$ and 0 exclusively, $= 0$ at the limits of the z-integration, and $= \pi f(x)$ for values of x between 0 and π exclusively.

If $f(x)$ is such a function of x that $f(-x) = f(x)$; then the function has the same sign for positive and negative values of x, and it will consist of the series of cosines given in (96). And if $f(x)$ is such a function of x that $f(-x) = -f(x)$, then $f(x)$ will have opposite signs for positive and negative values of x, and will consist of the series of sines given in (97).

201.] Another form, which is frequently required, may be given to these theorems by means of a change of form of the variables x and z.

Let x and z be respectively replaced by $\frac{\pi x}{c}$ and $\frac{\pi z}{c}$, so that the period of the function becomes $2c$ instead of 2π. Also let the functions $f\left(\frac{\pi x}{c}\right)$ and $f\left(\frac{\pi z}{c}\right)$ be replaced respectively by $f(x)$ and $f(z)$. Then (92) takes the following form,

$$\left.\begin{aligned}&\frac{1}{2}\int_0^{2c} f(z)\,dz + \sum_{n=1}^{n=\infty}\int_0^{2c} f(z)\cos\frac{n\pi(x-z)}{c}\,dz\\ &\quad = \frac{c}{2}f(0), \quad = cf(x), \qquad = \frac{c}{2}f(2c), \text{ when}\\ &x=0, \qquad 0<x<2c, \qquad x=2c.\end{aligned}\right\} \tag{98}$$

From (93), we have

$$\left.\begin{aligned}&\frac{1}{2}\int_{-c}^{c} f(z)\,dz + \sum_{n=1}^{n=\infty}\int_{-c}^{c} f(z)\cos\frac{n\pi(x-z)}{c}\,dz\\ &\quad = \frac{c}{2}f(-c), \quad = cf(x), \qquad = \frac{c}{2}f(c), \text{ when}\\ &x=-c, \qquad -c<x<c, \qquad x=c.\end{aligned}\right\} \tag{99}$$

From (96), we have

$$\left.\begin{aligned}&\int_0^{c} f(z)\,dz + 2\sum_{n=1}^{n=\infty}\cos\frac{n\pi x}{c}\int_0^{c} f(z)\cos\frac{n\pi z}{c}\,dz\\ &\quad = cf(-x), \quad = cf(0), \quad = cf(x), \quad = cf(c), \text{ when}\\ &-c<x<0, \qquad x=0, \quad 0<x<c, \quad x=c.\end{aligned}\right\} \tag{100}$$

And from (97), we have

$$\left.\begin{aligned}&2\sum_{n=1}^{n=\infty}\sin\frac{n\pi x}{c}\int_0^{c} f(z)\sin\frac{n\pi z}{c}\,dz\\ &\quad = -cf(-x), \quad = 0, \qquad = cf(x), \quad = 0, \text{ when}\\ &-c<x<0, \qquad x=0, \quad 0<x<c, \quad x=c.\end{aligned}\right\} \tag{101}$$

The remarks made in the preceding article as to the values at the limits, and as to the forms of functions, are, mutatis mutandis, also applicable to the present forms of functions.

202.] The following are examples of these several theorems.

Ex. 1. Let $f(x)=x$; then applying the general theorem in (90),

$$\frac{1}{2}\int_a^b z\,dz + \sum_{n=1}^{n=\infty}\int_a^b z\cos n(x-z)\,dz = \frac{b^2-a^2}{4}$$
$$+\sum_{n=1}^{n=\infty}\left\{\frac{b\sin n(b-x)-a\sin n(a-x)}{n}+\frac{\cos n(b-x)-\cos n(a-x)}{n^2}\right\}$$
$= \frac{\pi a}{2}$, $= \pi x$, $= \frac{\pi b}{2}$, $=0$, according as $x=a$, $a<x<b$, $x=b$, $b<x<2\pi+a$. And these values give the value of the function through a complete period.

If the limits of the z-integration are 2π and 0, then by (92),

$$\frac{1}{2}\int^{2\pi} z\,dz + \sum_{n=1}^{n=\infty}\int_0^{2\pi} z\cos n(x-z)\,dz = \pi^2 - 2\pi\sum_{n=1}^{n=\infty}\frac{\sin nx}{n}$$

$= 0$, $= \pi x$, $= \pi^2$, according as $x = 0$, $0 < x < 2\pi$, $x = 2\pi$. Consequently, for all values of x between 2π and 0 exclusively,

$$x = \pi - 2\left\{\frac{\sin x}{1} + \frac{\sin 2x}{2} + \frac{\sin 3x}{3} + \dots\right\}; \qquad (102)$$

and thus x is expressed in a series of sines of multiples of x. The geometrical interpretation of this result deserves notice. If the right-hand member of (102), which is a function of x, represents the ordinate of a locus, say, y, the equation to that locus is $y = x$. That is, the locus is a straight line passing through the origin and making an angle of 45° with the x-axis. That line however is limited by $x = 0$ and $x = 2\pi$, and (102) expresses no part of that line beyond these limits. If $x = 2\pi$, $y = \pi$; and as the point expressed by these equations is not on the line, it is a point of discontinuity. As this geometrical interpretation of (102) recurs at every interval of 2π, so the equation represents a series of broken lines and of points of discontinuity exactly similar to those described above.

If in (102) x is replaced by $\pi - x$, then for all values of x between $-\pi$ and π exclusively,

$$x = 2\left\{\frac{\sin x}{1} - \frac{\sin 2x}{2} + \frac{\sin 3x}{3} - \dots\right\}. \qquad (103)$$

This series may also be deduced from (93).

Again, applying (96) we have the following result,

$$\begin{aligned}\int_0^\pi z\,dz + 2\sum_{n=1}^{n=\infty} \cos nx \int_0^\pi z \cos nz\,dz &= \frac{\pi^2}{2} + 2\sum_{n=1}^{n=\infty} \cos nx\,\frac{\cos n\pi - 1}{n^2} \\ &= \frac{\pi^2}{2} - 4\left\{\frac{\cos x}{1^2} + \frac{\cos 3x}{3^2} + \frac{\cos 5x}{5^2} + \dots\right\}\end{aligned}$$

$= -\pi x$, $= 0$, $= \pi x$, $= \pi^2$, according as $-\pi < x < 0$, $x = 0$, $0 < x < \pi$, $x = \pi$. Hence, as (96) is true at the limits, we have for all values of x contained between 0 and π inclusive,

$$x = \frac{\pi}{2} - \frac{4}{\pi}\left\{\frac{\cos x}{1^2} + \frac{\cos 3x}{3^2} + \frac{\cos 5x}{5^2} + \dots\right\}; \qquad (104)$$

and for all values of x between $-\pi$ and 0 inclusive,

$$-x = \frac{\pi}{2} - \frac{4}{\pi}\left\{\frac{\cos x}{1^2} + \frac{\cos 3x}{3^2} + \frac{\cos 5x}{5^2} + \dots\right\}. \qquad (105)$$

Thus x is expressed in a series of cosines of uneven multiples of x.

If x is replaced by $x+\frac{\pi}{2}$, then by (104) for all values of x between $-\frac{\pi}{2}$ and $\frac{\pi}{2}$ inclusive,

$$x = \frac{4}{\pi}\left\{\frac{\sin x}{1^2} - \frac{\sin 3x}{3^2} + \frac{\sin 5x}{5^2} - \frac{\sin 7x}{7^2} + \dots\right\}. \qquad (106)$$

Also applying (97), we have

$$2\sum_{n=1}^{n=\infty} \sin nx \int_0^{\pi} z \sin nz\, dz = -2\pi \sum_{n=1}^{n=\infty} \frac{\sin nx \cos n\pi}{n}$$

$$= 2\pi\left\{\frac{\sin x}{1} - \frac{\sin 2x}{2} + \frac{\sin 3x}{3} - \dots\right\}$$

$= \pi x, = 0, = \pi x, = 0$, according as $-\pi < x < 0$, $x = 0$, $0 < x < \pi$, $x = \pi$; and consequently for all values of x between $-\pi$ and π exclusively

$$x = 2\left\{\frac{\sin x}{1} - \frac{\sin 2x}{2} + \frac{\sin 3x}{3} - \dots\right\}; \qquad (107)$$

which is the same as (103).

From these series others may be deduced by the substitution of particular values for the variables, provided that such values are within the limits of application. Thus if $x = 0$ in (104),

$$\frac{1}{1^2} + \frac{1}{3^2} + \frac{1}{5^2} + \dots = \frac{\pi^2}{8}. \qquad (108)$$

Also these series may be the subjects of differentiation. Thus from (106) by differentiation we have for all values of x between $-\frac{\pi}{2}$ and $\frac{\pi}{2}$ inclusively,

$$\frac{\pi}{4} = \frac{\cos x}{1} - \frac{\cos 3x}{3} + \frac{\cos 5x}{5} - \dots; \qquad (109)$$

and from (107), for all values between $-\pi$ and π exclusively,

$$\frac{1}{2} = \cos x - \cos 2x + \cos 3x - \dots. \qquad (110)$$

Also these series may be the element-functions of definite integration according to the method of the preceding section. Thus if we take the two members of (107) to be the element-functions of a definite integral for the limits x and 0, then, when x is between $-\pi$ and π,

$$\frac{x^2}{4} = c - \frac{\cos x}{1^2} + \frac{\cos 2x}{2^2} - \frac{\cos 3x}{3^2} + \dots;$$

where $$c = \frac{1}{1^2} - \frac{1}{2^2} + \frac{1}{3^2} - \frac{1}{4^2} + \dots \tag{111}$$

$$= \frac{1}{1^2} + \frac{1}{3^2} + \frac{1}{5^2} + \dots - \frac{1}{4}\left\{\frac{1}{1^2} + \frac{1}{2^2} + \frac{1}{3^2} + \dots\right\}$$

$$= \frac{\pi^2}{8} - \frac{1}{4}\left\{\frac{1}{1^2} + \frac{1}{2^2} + \frac{1}{3^2} + \dots\right\};$$

$$\therefore \quad \frac{\pi^2}{2} - 4c = \frac{1}{1^2} + \frac{1}{2^2} + \frac{1}{3^2} + \dots : \tag{112}$$

adding which to (111), we have

$$\frac{\pi^2}{2} - 3c = 2\left\{\frac{1}{1^2} + \frac{1}{3^2} + \frac{1}{5^2} + \dots\right\}$$

$$= \frac{\pi^2}{4}, \quad \text{by (108)};$$

$$\therefore \quad c = \frac{\pi^2}{12} = \frac{1}{1^2} - \frac{1}{2^2} + \frac{1}{3^2} - \frac{1}{4^2} + \dots;$$

$$\therefore \quad \frac{x^2}{4} = \frac{\pi^2}{12} - \frac{\cos x}{1^2} + \frac{\cos 2x}{2^2} - \frac{\cos 3x}{3^2} + \dots; \tag{113}$$

and if in (112) c is replaced by its value,

$$\frac{1}{1^2} + \frac{1}{2^2} + \frac{1}{3^2} + \dots = \frac{\pi^2}{6}. \tag{114}$$

Also taking the members of (104) to be the element-functions of a definite integral for the limits x and 0, where x is not greater than π,

$$\frac{x^2}{2} = \frac{\pi x}{2} - \frac{4}{\pi}\left\{\frac{\sin x}{1^3} + \frac{\sin 3x}{3^3} + \frac{\sin 5x}{5^3} + \dots\right\}; \tag{115}$$

and if $x = \frac{\pi}{2}$,

$$\frac{\pi^3}{32} = \frac{1}{1^3} - \frac{1}{3^3} + \frac{1}{5^3} - \frac{1}{7^3} + \dots. \tag{116}$$

Ex. 2. Let $f(x) = 1$; then the theorems contained in (92) and (93) are evidently identities. But taking (97),

$$\frac{\sin x}{1} + \frac{\sin 3x}{3} + \frac{\sin 5x}{5} + \dots = \frac{\pi}{4}, \tag{117}$$

for all values of x between 0 and π exclusively; and $= -\frac{\pi}{4}$, for all values of x between $-\pi$ and 0 exclusively; and $= 0$, when $x = -\pi$, $= 0$, and $= \pi$.

Ex. 3. Let $f(x) = \sin x$. Then by the general theorem in (89) we have

$$\sin x = \left[\frac{-\cos x}{2\pi} + \frac{z \sin x}{2\pi} - \frac{\cos(2z-x)}{4\pi}\right]_{z=a}^{z=b}$$

$$+ \frac{1}{\pi} \sum_{n=2}^{n=\infty} \left[\frac{n \sin n(z-x) \sin z + \cos n(z-x) \cos z}{n^2-1}\right]_{z=a}^{z=b}. \quad (118)$$

The applications of (92) and (93) lead to an indeterminate quantity, and ultimately, on evaluation, to an identity.

On applying (96), we have

$$\pi \sin x = 2 - 4\left\{\frac{\cos 2x}{2^2-1} + \frac{\cos 4x}{4^2-1} + \frac{\cos 6x}{6^2-1} + \dots\right\}, \quad (119)$$

for all values of x between 0 and π inclusively.

Hence if $x = 0$, and if $x = \pi$,

$$\frac{1}{2^2-1} + \frac{1}{4^2-1} + \frac{1}{6^2-1} + \dots = \frac{1}{2}; \quad (120)$$

and if $x = \frac{\pi}{2}$,

$$\frac{1}{2^2-1} - \frac{1}{4^2-1} + \frac{1}{6^2-1} - \frac{1}{8^2-1} + \dots = \frac{\pi-2}{4}; \quad (121)$$

and by addition of (120) and (121), and by subtraction of (121) from 120,

$$\frac{1}{2^2-1} + \frac{1}{6^2-1} + \frac{1}{10^2-1} + \dots = \frac{\pi}{8}; \quad (122)$$

$$\frac{1}{4^2-1} + \frac{1}{8^2-1} + \frac{1}{12^2-1} + \dots = \frac{4-\pi}{8}. \quad (123)$$

Also differentiating (119), for all values of x between 0 and π exclusively,

$$\cos x = \frac{8}{\pi}\left\{\frac{\sin 2x}{2^2-1} + \frac{2 \sin 4x}{4^2-1} + \frac{3 \sin 6x}{6^2-1} + \dots\right\}. \quad (124)$$

Ex. 4.] Let $f(x) = \cos mx$, where m is a proper fraction; then from (96),

$$\pi \cos mx = \int_0^\pi \cos mz\, dz + 2 \sum_{n=1}^{n=\infty} \cos nx \int_0^\pi \cos mz \cos nz\, dz$$

$$= \frac{\sin m\pi}{m} + 2m \sin m\pi \sum_{n=1}^{n=\infty} \frac{\cos nx \cos n\pi}{m^2-n^2}; \quad (125)$$

therefore

$$\frac{\pi}{2} \frac{\cos mx}{\sin m\pi} = \frac{1}{2m} + m\left\{\frac{\cos x}{1^2-m^2} - \frac{\cos 2x}{2^2-m^2} + \frac{\cos 3x}{3^2-m^2} - \dots\right\}; \quad (126)$$

and this is true for all values of x from $-\pi$ to π inclusively.

If $x = \pi$,

$$\frac{\pi}{2} \cot m\pi = \frac{1}{2m} - \frac{m}{1^2-m^2} - \frac{m}{2^2-m^2} - \frac{m}{3^2-m^2} - \dots; \quad (127)$$

also differentiating (126),

$$\frac{\pi}{2}\frac{\sin mx}{\sin m\pi} = \frac{\sin x}{1^2-m^2} - \frac{2\sin 2x}{2^2-m^2} + \frac{3\sin 3x}{3^2-m^2} - \dots; \tag{128}$$

which is true for all values of x from $-\pi$ to π exclusively.

Ex. 5. Let $f(x) = e^{mx}+e^{-mx}$. Then from (96),

$$\frac{e^{mx}+e^{-mx}}{e^{m\pi}-e^{-m\pi}} = \frac{1}{m\pi} - \frac{2m}{\pi}\left\{\frac{\cos x}{m^2+1} - \frac{\cos 2x}{m^2+2^2} + \frac{\cos 3x}{m^2+3^2} - \dots\right\}; \tag{129}$$

which is true for all values of x from $-\pi$ to π inclusively.

If $x = \pi$, (129) gives

$$\frac{\pi}{2m}\frac{e^{m\pi}+e^{-m\pi}}{e^{m\pi}-e^{-m\pi}} = \frac{1}{2m^2} + \frac{1}{m^2+1^2} + \frac{1}{m^2+2^2} + \frac{1}{m^2+3^2} + \dots. \tag{130}$$

Hence, if $m = 0$,

$$\frac{1}{1^2} + \frac{1}{2^2} + \frac{1}{3^2} + \dots = \frac{\pi^2}{6}. \tag{131}$$

Also differentiating (129),

$$\frac{\pi}{2}\frac{e^{mx}-e^{-mx}}{e^{m\pi}-e^{-m\pi}} = \frac{\sin x}{m^2+1} - \frac{2\sin 2x}{m^2+2^2} + \frac{3\sin 3x}{m^2+3^2} - \dots; \tag{132}$$

which is true for all values of x from $-\pi$ to π exclusively.

And if in (132) x is replaced by $\pi - x$, then

$$\frac{\pi}{2}\frac{e^{m(\pi-x)}-e^{-m(\pi-x)}}{e^{m\pi}-e^{-m\pi}} = \frac{\sin x}{m^2+1} + \frac{2\sin 2x}{m^2+2^2} + \frac{3\sin 3x}{m^2+3^2} + \dots; \tag{133}$$

which is true for all values of x from 0 to 2π exclusively.

203.] The preceding examples are sufficient to illustrate the general theorems of periodic series. Many others will be found in a memoir entitled "Die Lagrangesche Formel und die Reihensummirung durch dieselbe," by J. Dienger, Crelle's Journal, Vol. XXXIV, p. 75. The theory however admits of consideration from another point of view, and thereby has another important application; viz. that to discontinuous functions. The fact has been explained in Art. 199, and it is now only requisite to exhibit some cases of its application. Herein I shall take the theorems given in Art. 201, of which the period is $2c$ instead of 2π as in the preceding examples.

Ex. 1. Find a periodic function of x, whose period is $2c$, which is equal to 1 for all values of x between 0 and c, and is equal to -1 for all values of x between $-c$ and 0.

As the values of $f(x)$ in this case are to be equal, but of opposite signs for the positive and negative values of x, we must take the theorem given in (101); and we have

$$f(x) = \frac{2}{c}\sum_{n=1}^{n=\infty} \sin\frac{n\pi x}{c}\int_0^c \sin\frac{n\pi z}{c}\,dz,$$

$$= \frac{4}{\pi}\left\{\sin\frac{\pi x}{c} + \frac{1}{3}\sin\frac{3\pi x}{c} + \frac{1}{5}\sin\frac{5\pi x}{c} + \ldots\right\}. \quad (134)$$

Also the value of this series, at the limits 0 and c, $= 0$. Thus we have the following geometrical interpretation. Let the right-hand member of (134) express the ordinate of a locus; then that locus between $x = -c$ and $x = 0$, both exclusively, is a straight line at a distance $= -1$ from the x-axis; when $x = 0$, it is a point on the x-axis; between $x = 0$ and $x = c$, both exclusively, it is a straight line at a distance $= 1$ from the x-axis; and when $x = c$, it is a point on the x-axis. And as we have traced the locus through a complete period, these lines and points are repeated ad infinitum in both directions.

Ex. 2. Find a periodic function of x, whose period is $2c$, which is equal to $-mx$, for all values of x between $-c$ and 0; and $= mx$ for all values of x between 0 and c.

As the values of $f(x)$ in this case are to be equal, and of the same sign for equal, positive, and negative values of x, we must apply the series of cosines given in (100); whence we have

$$f(x) = \frac{cm}{2} - \frac{4cm}{\pi^2}\left\{\cos\frac{\pi x}{c} + \frac{1}{3^2}\cos\frac{3\pi x}{c} + \frac{1}{5^2}\cos\frac{5\pi x}{c} + \ldots\right\}; (135)$$

and this is true at the limits as well as for other values of x.

This admits of a remarkable geometrical interpretation; let the right-hand member of (135) express the ordinate of a locus; then that locus consists (1) of the portion of the straight line, whose equation is $y = -mx$, contained between $x = -c$ and $x = 0$, the line terminating abruptly at these values of x; (2) of the portion of the line whose equation is $y = mx$, contained between $x = 0$, and $x = c$, this line also terminating abruptly at these values of x. Also these lines are repeated ad infinitum in both directions. So that the locus consists of a series of straight lines forming the sides of isosceles triangles, the lengths of whose bases are $2c$ placed continuously along the x-axis on its positive side, and whose base angle $= \tan^{-1} m$.

Ex. 3. Find a periodic function of x whose period is $2c$, which is equal to mx for all values of x between $2c$ and 0.

From (98) we have

$$f(x) = \frac{m}{2c}\int_0^{2c} z\,dz + \frac{m}{c}\sum_{n=1}^{n=\infty}\int_0^{2c} z\cos\frac{n\pi(x-z)}{c}\,dz$$

$$= cm - \frac{2cm}{\pi}\left\{\sin\frac{\pi x}{c} + \frac{1}{2}\sin\frac{2\pi x}{c} + \frac{1}{3}\sin\frac{3\pi x}{c} + \dots\right\}; \quad (136)$$

which is true for all values of x between 0 and $2c$ exclusively; and $= 0$, when $x = 0$; and $= mc$, when $x = 2c$.

Thus, if (136) is taken to express the ordinate of a locus, that locus consists of (1) the length of a straight line, whose equation is $y = mx$, contained within the limits $x = 0$ and $x = 2c$; (2) of a point $x = 2c$, $y = mc$, which is a discontinuous point; (3) of similar lines and points repeated ad infinitum in both directions.

Ex. 4. Find a periodic function of x, whose period is $2c$, which $= a$ for all values of x between $x = 0$ and $x = k$; and $= b$ for all values of x between $x = k$ and $x = 2c$.

$$f(x) = \frac{1}{2c}\int_0^k a\,dz + \frac{1}{c}\sum_{n=1}^{n=\infty}\int_0^k a\cos\frac{n\pi(x-z)}{c}\,dz$$

$$+ \frac{1}{2c}\int_k^{2c} b\,dz + \frac{1}{c}\sum_{n=1}^{n=\infty}\int_k^{2c} b\cos\frac{n\pi(x-z)}{c}\,dz$$

$$= b + \frac{k(a-b)}{2c} + \frac{2(a-b)}{\pi}\sum_{n=1}^{n=\infty}\frac{1}{n}\sin\frac{nk\pi}{2c}\cos\frac{n\pi(k-2x)}{2c}. \quad (137)$$

Also at the limits, $f(0) = f(k) = f(2c) = \dfrac{a+b}{2}$.

If $k = c$, so that $f(x) = a$ for all values of x between 0 and c, and $= b$ for all values of x between c and $2c$,

$$f(x) = \frac{a+b}{2} + \frac{2(a-b)}{\pi}\left\{\sin\frac{\pi x}{c} + \frac{1}{3}\sin\frac{3\pi x}{c} + \frac{1}{5}\sin\frac{5\pi x}{5} + \dots\right\}; \quad (138)$$

and at the limits, $f(0) = f(c) = f(2c) = \dfrac{a+b}{2}$.

If $b = -a$,

$$f(x) = \frac{4a}{\pi}\left\{\sin\frac{\pi x}{c} + \frac{1}{3}\sin\frac{3\pi x}{c} + \dots\right\}; \quad (139)$$

and at the limits, $f(0) = f(c) = f(2c) = 0$. This is the result already found in example 1. The geometrical interpretation is similar to that of that example.

Ex. 5. Determine $f(x)$, a periodic function of x, whose period is 2π, which $= x$ from $x = 0$ to $x = a$; $= a$ from $x = a$ to $x = \pi - a$; and $= \pi - x$ from $x = \pi - a$ to $x = \pi$; and is also such that $f(-x) = -f(x)$.

In this case we may take (97), and expand in a series of sines; and we have

$$f(x) = \frac{2}{\pi}\sum_{n=1}^{n=\infty} \sin nx \left\{\int_0^a z\sin nz\,dz + \int_a^{\pi-a} a\sin nz\,dz + \int_{\pi-a}^{\pi}(\pi-z)\sin nz\,dz\right\}$$

$$= \frac{4}{\pi}\left\{\sin a\sin x + \frac{1}{3^2}\sin 3a\sin 3x + \frac{1}{5^2}\sin 5a\sin 5x + \dots\right\}. \quad (140)$$

The geometrical interpretation is as follows. The locus consists of three straight lines, which, with the x-axis, bound a trapezium, the first and third being inclined to the x-axis at 45° and 135° respectively; and the second line being parallel to that axis at a distance a from it; the base of the trapezium on the x-axis $= \pi$, and the length of the opposite side $= \pi - 2a$. On the negative side of the origin is an equal trapezium, which is below the axis of x. And these two trapeziums are repeated continually along the x-axis.

If $a = \frac{\pi}{2}$, the trapeziums become isosceles triangles, and

$$f(x) = \frac{4}{\pi}\left\{\sin x - \frac{1}{3^2}\sin 3x + \frac{1}{5^2}\sin 5x - \dots\right\}. \quad (141)$$

204.] Thus far we have assumed the limits of the definite integral (99) to be finite quantities; let us now consider the form which that equation takes when $c = \infty$. And to avoid all difficulty let us assume that $f(x)$ never $= \infty$, and $= 0$ when $x = -\infty$ and when $x = \infty$. From (99) we have

$$f(x) = \frac{1}{2c}\int_{-c}^{c} f(z)\,dz + \frac{1}{c}\sum_{n=1}^{n=\infty}\int_{-c}^{c} f(z)\cos\frac{n\pi(x-z)}{c}\,dz. \quad (142)$$

Now $\frac{1}{2c}\int_{-c}^{c} f(z)\,dz = 0$, when $c = \infty$, under the conditions to which z is subject; so that the value of $f(x)$ is given by the second term of (142). In this let $\frac{n\pi}{c} = u$; then as n is increased by 1, u is increased by $\frac{\pi}{c}$; which is infinitesimal, when $c = \infty$, and so must be denoted by du; and the summation becomes integration, the range of integration being the same as that of the summation; and thus the range of u extends from 0 to ∞; hence we have

$$f(x) = \frac{1}{\pi}\int_0^{\infty}\int_{-\infty}^{\infty} f(z)\cos u(x-z)\,dz\,du. \quad (143)$$

This theorem is called Fourier's Theorem*; and is of great importance in higher analytical investigations.

If the limits of the u-integration are $-\infty$ and ∞, then

$$f(x) = \frac{1}{2\pi}\int_{-\infty}^{\infty}\int_{-\infty}^{\infty} f(z)\cos u(x-z)\,dz\,du. \tag{144}$$

If $f(x) = f(-x)$, then

$$\int_{-\infty}^{\infty} f(z)\cos uz\,dz = 2\int_{0}^{\infty} f(z)\cos uz\,dz, \quad \int_{-\infty}^{\infty} f(z)\sin uz\,dz = 0;$$

therefore from (143),

$$f(x) = \frac{2}{\pi}\int_{0}^{\infty}\int_{0}^{\infty} f(z)\cos ux\cos uz\,dz\,du. \tag{145}$$

This theorem, as (100) shews, is true, when x is equal to either of the limits of the z-integration.

If $f(x) = -f(-x)$, then by a similar process,

$$f(x) = \frac{2}{\pi}\int_{0}^{\infty}\int_{0}^{\infty} f(z)\sin ux\sin uz\,dz\,du. \tag{146}$$

This theorem, as (101) shews, is not true, when x is equal to either of the limits of the z-integration.

205.] The applications of these theorems are very numerous, and the values of many definite integrals may be determined by them. For it is evident that if either one or the other of the integrations can be effected, the value of the remaining definite integral will also be determined. The theorem may also be applied to problems similar to those of Arts. 202, 203.

Ex. 1. Find a function of x which $= 1$ for all values of x between -1 and 1, and $= 0$ for all other values.

Here $f(x) = f(-x)$, and we have evidently from (100),

$$\begin{aligned} f(x) &= \frac{2}{\pi}\int_{0}^{\infty}\int_{0}^{1} \cos ux\cos uz\,dz\,du \\ &= \frac{2}{\pi}\int_{0}^{\infty} \frac{\cos ux \sin u}{u}\,du. \end{aligned} \tag{147}$$

Ex. 2. Let $f(x) = e^{-ax}$ from $x = 0$ to $x = \infty$; and $f(x) = e^{ax}$ from $x = -\infty$ to $x = 0$; then $f(x)$ satisfies the conditions required for the theorem, and also those in (145); and we have

$$\begin{aligned} e^{-ax} &= \frac{2}{\pi}\int_{0}^{\infty}\int_{0}^{\infty} e^{-az}\cos uz\cos ux\,dz\,du \\ &= \frac{2}{\pi}\int_{0}^{\infty} \frac{a\cos ux}{a^2+u^2}\,du; \end{aligned}$$

* See Théorie Analytique de la Chaleur; par M. Fourier, pp. 431, 445.

$$\therefore \int_0^\infty \frac{\cos ux\, du}{a^2+u^2} = \frac{\pi}{2a} e^{-ax}; \tag{148}$$

which result is the same as (121), Art. 101.

Ex. 3. Let $f(x) = \sin ax$, from $x = 0$ to $x = \infty$; and $f(x) = \sin(-ax)$, from $x = -\infty$ to $x = 0$; then $f(x)$ satisfies the conditions required in (145); and we have

$$\begin{aligned}\sin ax &= \frac{2}{\pi}\int_0^\infty\int_0^\infty \sin az \cos uz \cos ux\, dz\, du \\ &= \frac{2a}{\pi}\int_0^\infty \frac{\cos ux\, du}{a^2-u^2};\end{aligned}$$

$$\therefore \int_0^\infty \frac{\cos ux\, du}{a^2-u^2} = \frac{\pi}{2a}\sin ax; \tag{149}$$

which result is the same as (177), Art. 108.

Ex. 4. Let $f(x) = \int_a^b \phi(y)e^{-xy}dy$; and let us suppose $f(x) = f(-x)$; then from (145),

$$\int_a^b \phi(y)e^{-xy}dy = \frac{2}{\pi}\int_0^\infty\int_0^\infty\int_a^b \phi(y)e^{-zy}\cos ux \cos uz\, dy\, dz\, du. \tag{150}$$

The right-hand member of this equation is a triple integral, in which all the limits of integration are constant. Consequently, as we have demonstrated in Art. 99, the truth of the result is independent of the order in which the integrations are effected; and since by (6), Art. 82,

$$\int_0^\infty e^{-zy}\cos uz\, dz = \frac{y}{u^2+y^2};$$

$$\therefore \int_a^b \phi(y)e^{-xy}dy = \frac{2}{\pi}\int_0^\infty\int_a^b \phi(y)\frac{y\cos ux}{u^2+y^2}dy\, du; \tag{151}$$

in which a double definite integral is expressed in terms of a single definite integral.

Let $x = 0$, as this value is included in the theorem; then

$$\int_a^b \phi(y)dy = \frac{2}{\pi}\int_0^\infty\int_a^b \phi(y)\frac{y}{u^2+y^2}dy\, du. \tag{152}$$

Let $\phi(y) = (1-y^2)^{-\frac{1}{2}}$; and let $b = 1$, $a = 0$; then

$$\frac{\pi^2}{4} = \int_0^\infty\int_0^1 \frac{y\, dy\, du}{(u^2+y^2)(1-y^2)^{\frac{1}{2}}} \tag{153}$$

$$= \frac{1}{2}\int_0^\infty \log\frac{(1+u^2)^{\frac{1}{2}}+1}{(1+u^2)^{\frac{1}{2}}-1}\,\frac{du}{(1+u^2)^{\frac{1}{2}}}. \tag{154}$$

CHAPTER VIII.

ON MULTIPLE INTEGRATION, AND THE TRANSFORMATION OF MULTIPLE INTEGRALS.

SECTION 1.—*On Double, Triple, and Multiple Integration.*

206.] Although hitherto in this treatise the element, whose integral is to be found, has been primarily an infinitesimal of the first order, yet it is evident, and it has been so stated in the first article, that such need not always be the case. Infinitesimals of higher orders may be subjects of integration equally as much as those of the first order. Several times indeed, but only incidentally, have such arisen in previous investigations; and subject to a particular condition whereby we have avoided a difficulty which is inherent in those of a more general character. Element-functions of an order higher than the first have been investigated in Art. 99, and it has therein been shewn how they may be the subjects of successive integration; and in the inquiry into definite integrals that theory has been frequently applied; as in Cauchy's method of evaluation in Art. 103; in the investigation of the properties of the Gamma-function, and in the application of those properties, and in Fourier's Theorem. Also again in Chapter V, an element of the nth order in terms of dx, involving an element-function of a single variable, has been the subject of n successive integrations.

Now in all three cases the limits of integration have been constants; and thus, as we have demonstrated in Art. 99, the order of integration has been indifferent. These circumstances have indeed removed the difficulties of the subject, but they have also restricted the problem within very narrow limits. It is necessary however to consider it in its utmost generality. Very many subsequent applications of the Calculus depend on multiple integration; and perspicuity is required in this subject more perhaps than in any other. I propose therefore to consider the theory generally in this chapter, and to apply it to geometrical and other problems in subsequent chapters.

207.] In the general problem of multiple integration the infinitesimal element is a product of factors one of which is a function of n variables independent of each other, and the others are differentials of these variables, each of whch enters in the first degree only. Thus if $x_1, x_2, x_3, \ldots x_n$, are n independent variables, the infinitesimal element is of the form,

$$\text{F}(x_1, x_2, \ldots x_n)\, dx_1\, dx_2\, dx_3 \ldots dx_n. \qquad (1)$$

This is evidently an infinitesimal of the nth order; and consequently n successive integrations must be effected on it as their subject-function, before we arrive at the finite quantity, which is the object of search. And if the problem is definite, these several integrations must be definite; and it is also necessary that the element-function should not become infinite or discontinuous for any value of a variable within the several ranges of integration. In the preceding parts of our treatise, wherever an element of the form (1) has been the subject of integration, the limits of integration have been constant; in the general problem however this will not be the case; for the limits will be, or may be, functions of all the variables which enter into the element-function of that integration. Now these are new circumstances requiring an extension of language and an extension of symbols.

Integration, as heretofore, means the summation of a series, and the limits of integration assign the first and last terms of the series. When an element of the form (1) is the subject of integration, and the finite function is sought of which (1) is the infinitesimal element, n summations must be effected, and in a prescribed order; let us take the order of integrations to be that in which the differentials stand in (1); so that the integration is first to be effected with respect to x_1; let X_1, x_1 be the limits of this integration; let $\text{X}_1 - \text{x}_1$ be divided into n infinitesimal parts; and let $\xi_1, \xi_2, \ldots \xi_{n-1}$ be the values of x_1 corresponding to the points of partition. As x_1 is independent of the other variables, viz., $x_2, x_3, \ldots x_n$, which enter into the element (1), so will $\xi_1, \xi_2, \ldots \xi_{n-1}$, which are particular values of x_1, be also independent of these variables; and consequently in the summation of the series found by giving x_1 successive values, $x_2, x_3, \ldots x_n$, and also $dx_2, dx_3, \ldots dx_n$, are to be considered constant. Thus the series which is first to be summed is of the form

$$\{ \text{F}(\text{x}_1, x_2, \ldots x_n)(\xi_1 - \text{x}_1) + \text{F}(\xi_1, x_2, \ldots x_n)(\xi_2 - \xi_1) + \ldots$$
$$\ldots + \text{F}(\xi_{n-1}, x_2, \ldots x_n)(\text{X}_1 - \xi_{n-1})\}\, dx_2\, dx_3 \ldots dx_n; \quad (2)$$

and the sum of this series, if $F(x_1, x_2, \ldots x_n)\,dx_1$ is the x_1-partial-differential of $f(x_1, x_2, \ldots x_n)$, is

$$\{f(X_1, x_2, x_3, \ldots x_n) - f(x_1, x_2, x_3, \ldots x_n)\}\, dx_2\, dx_3 \ldots dx_n. \quad (3)$$

In the general case X_1 and x_1 will be functions of $x_2, x_3, \ldots x_n$, that is, of all the variables in the original function except x_1.

This process is called the integration of the element with respect to x_1, or the x_1-integration of the given element; X_1 and x_1 being limits of the x_1-integration; and by it the element takes the form,

$$F(x_2, x_3, \ldots x_n)\, dx_2\, dx_3 \ldots dx_n. \quad (4)$$

This is again to be treated with regard to x_2, as the element given in (1) has been treated with regard to x_1; if X_2, x_2 are the limits of this second integration, they will generally be functions of all the variables except x_1 and x_2, and the result will be an element of the form,

$$F(x_3, x_4, \ldots x_n)\, dx_3\, dx_4 \ldots dx_n. \quad (5)$$

This process is called the x_2-integration of the element given in (4), and X_2, x_2, are called the limits of the x_2-integration.

The process is to be repeated for all the successive variables in the order in which the differentials are placed in (1); and if X_n and x_n are the limits of the x_n-integration, which is the last, these will be constants, and the final integral will be of the form,

$$F(X_n) - F(x_n); \quad (6)$$

and will of course be independent of all the variables involved in the original element. This is the quantity which is the object of search.

208.] The following is the system of symbols which I shall adopt. Let the integral be generally expressed by

$$\int\int\int \ldots\ldots \int F(x_1, x_2, \ldots x_n)\, dx_1\, dx_2 \ldots dx_{n-1}\, dx_n; \quad (7)$$

the order in which the differentials are placed being that in which the successive integrations are to be effected. Thus the x_1-integration is the first, and the x_n-integration is the last. If the limits of integration are given, so that the integral is definite, then it will be expressed in the following form

$$\int_{x_n}^{X_n}\int_{x_{n-1}}^{X_{n-1}} \ldots\ldots \int_{x_2}^{X_2}\int_{x_1}^{X_1} F(x_1, x_2, \ldots x_n)\, dx_1\, dx_2 \ldots dx_n: \quad (8)$$

where X_1 and x_1 are, or may be, functions of all the variables except x_1; X_2 and x_2 are, or may be, functions of all the variables except x_1 and x_2: and so on; and lastly X_n, x_n are constants.

It will be observed that the order of the symbols of integration in (8) is the reverse of that of the differentials. The reason is, that the first definite integral becomes the subject of, and is thus included within the symbols of, the second integration; and so on. Thus $\int_{x_2}^{X_2} dx_2$ denotes an operation which is performed subsequently to that denoted by $\int_{x_1}^{X_1} dx_1$; and the quantity which is the result of the latter becomes the subject of the former, and is consequently properly included within its symbol.

Another arrangement of the symbols is the following, which plainly expresses the order of the successive integrations;

$$\int_{x_n}^{X_n} dx_n \int_{x_{n-1}}^{X_{n-1}} dx_{n-1} \ldots\ldots \int_{x_1}^{X_1} \mathrm{F}(x_1, x_2, \ldots x_n)\, dx_1; \qquad (9)$$

and as this system is sometimes more convenient than (8), it will be employed whenever that is the case.

Integrals of the form (7) are called ***multiple integrals***; and those of the form (8) in which the limits are given are called ***definite multiple integrals***. When n integrations are to be effected, they are called multiple integrals of the nth order. Particular forms of these are ***double integrals, triple integrals***, ... according as two, three, or more integrations are to be effected.

209.] For the truth of these theorems on multiple integrals it is necessary, as it has been before observed, that the element-function should not become infinite or discontinuous for any value of the variable within the ranges of integration. Subject to this condition the integrations may be effected in that order which is most convenient. In a definite multiple integral, when the limits are constant, the order of integration is indifferent, as we have demonstrated from first principles in Art. 99; but if the limits are functions of the variables, the order cannot be changed without a change of the limits; and this is frequently a work of considerable difficulty; because when a multiple integral is *definite*, it is required to find the sum of the values which the element has for all values of the variables within a certain district; now this district is assigned by the limits; and although it may theoretically be possible to determine the boundary of this district in many ways, yet practically one particular mode may be more convenient than another, or may be the only possible mode.

Thus for example, suppose that it is required to find the value of $\iint dy\,dx$ for all points of the plane area contained within a cycloid and its base, the highest point being the origin, and the perpendicular to the base being the x-axis; then if we denote $a \operatorname{versin}^{-1} \frac{x}{a} + (2ax - x^2)^{\frac{1}{2}}$ by Y;

$$\iint dy\,dx = \int_0^{2a}\int_{-\text{Y}}^{\text{Y}} dy\,dx$$

$$= 2\int_0^{2a}\left\{a \operatorname{versin}^{-1}\frac{x}{a} + (2ax-x^2)^{\frac{1}{2}}\right\} dx$$

$$= 3\pi a^2;$$

which is the required result. But the order of integration cannot be changed, because the limits of the x-integration cannot be determined from the equation of the cycloid; it is a transcendental equation, and x cannot be expressed explicitly in terms of y by means of it.

Theoretically however the order of integration is arbitrary, although the equation which fix the boundaries of the district through which the integration extends, may constrain us to take a particular order. Many illustrations of this remark will occur in the following chapters; wherein also will be shewn the convenience of one order in preference to another.

210.] The following are examples of definite multiple integrals.

Ex. 1. $$\int_0^{\frac{\pi}{2}}\int_0^x \frac{x\,dy\,dx}{x^2+y^2} = \int_0^{\frac{\pi}{2}} \frac{\pi}{4}\,dx = \frac{\pi^2}{16}.$$

Ex. 2. $$\int_0^a\int_0^x\int_0^y xyz\,dz\,dy\,dx = \int_0^a\int_0^x \frac{xy^3}{2}\,dy\,dx = \int_0^a \frac{x^5}{8}\,dx = \frac{a^6}{48}.$$

Ex. 3. $$\int_0^a\int_0^{a-x}\int_0^{a-x-y} dz\,dy\,dx = \int_0^a\int_0^{a-} (a-x-y)\,dy\,dx = \int_0^a \frac{(a-x)^2}{2}\,dx = \frac{a^3}{6}.$$

Ex. 4. Find the value of $r^3\,dr\,d\theta$ through the area bounded by the circle whose equation is (1) $r = a$; (2) $r = 2a\cos\theta$.

(1) $$\int_0^{2\pi}\int_0^a r^3\,dr\,d\theta = \frac{\pi a^4}{2}.$$

(2) $$\int_{-\frac{\pi}{2}}^{\frac{\pi}{2}}\int_0^{2a\cos\theta} r^3\,dr\,d\theta = 4a^4\int_{-\frac{\pi}{2}}^{\frac{\pi}{2}}(\cos\theta)^4\,d\theta$$
$$= \frac{a^4}{2}\int_{-\frac{\pi}{2}}^{\frac{\pi}{2}}(\cos 4\theta + 4\cos 2\theta + 3)\,d\theta$$
$$= \frac{3\pi a^4}{2}.$$

As the sequel will contain many examples of multiple integration, it is unnecessary to give others in this place.

Section 2.—*On the Transformation of Multiple Integrals.*

211.] The evaluation of multiple integrals may frequently be greatly facilitated by a change of variables. The process is called Transformation of Multiple Integrals; and I propose in the present section to consider it in its most general form, and also to give various examples of it.

Let us take the integrals given in (7) and (8) to be the types of multiple integrals of the nth order; and, for the present omitting the limits, let

$$\text{I} = \int\int\int\ldots\ldots\int \text{F}(x_1, x_2, x_3, \ldots x_n)\,dx_1\,dx_2\ldots dx_n. \quad (10)$$

Let the new variables, in terms of which the original variables are to be expressed, be $\xi_1, \xi_2, \ldots \xi_n$; and let us in the first place suppose the equations of relation to be given in the following explicit forms; viz.

$$\left.\begin{aligned} x_1 &= f_1(\xi_1, \xi_2, \ldots\ldots \xi_n), \\ x_2 &= f_2(\xi_1, \xi_2, \ldots\ldots \xi_n), \\ &\cdots\cdots\cdots\cdots \\ x_n &= f_n(\xi_1, \xi_2, \ldots\ldots \xi_n); \end{aligned}\right\} \quad (11)$$

Let $\alpha_1, \beta_1, \ldots \rho_1$ be the partial derived functions of x_1 with regard to $\xi_1, \xi_2, \ldots \xi_n$ respectively; and let $\alpha_2, \beta_2, \ldots \rho_2$ be the similar derived functions of x_2; and so on; then

$$\left.\begin{aligned} dx_1 &= \alpha_1\,d\xi_1 + \beta_1\,d\xi_2 + \ldots\ldots + \rho_1\,d\xi_n, \\ dx_2 &= \alpha_2\,d\xi_1 + \beta_2\,d\xi_2 + \ldots\ldots + \rho_2\,d\xi_n, \\ &\cdots\cdots\cdots\cdots \\ dx_n &= \alpha_n\,d\xi_1 + \beta_n\,d\xi_2 + \ldots\ldots + \rho_n\,d\xi_n. \end{aligned}\right\} \quad (12)$$

In article (207) I have explained how $x_2, x_3, \ldots x_n$ remain constant during the variation of x_1, that is, during the x_1-integration; consequently in the calculation of the quantity which is to be substituted for dx_1, it must be consistent with the conditions, $dx_2 = dx_3 = \ldots = dx_n = 0$. Hence we have

$$\left.\begin{array}{rl} dx_1 &= a_1 d\xi_1 + \beta_1 d\xi_2 + \ldots\ldots + \rho_1 d\xi_n, \\ 0 &= a_2 d\xi_1 + \beta_2 d\xi_2 + \ldots\ldots + \rho_2 d\xi_n, \\ \ldots & \ldots\ldots\ldots\ldots\ldots \\ 0 &= a_n d\xi_1 + \beta_n d\xi_2 + \ldots\ldots + \rho_n d\xi_n; \end{array}\right\} \tag{13}$$

whence, employing the symbols of determinants,

$$dx_1 \Sigma . \pm \beta_2 \gamma_3 \ldots \rho_n = d\xi_1 \Sigma . \pm a_1 \beta_2 \gamma_3 \ldots \rho_n;$$

$$\therefore \quad dx_1 = \frac{\Sigma . \pm a_1 \beta_2 \gamma_3 \ldots\ldots \rho_n}{\Sigma . \pm \beta_2 \gamma_3 \ldots\ldots \rho_n} d\xi_1; \tag{14}$$

the numerator of the right-hand member of which is the determinant of the system in the right-hand member of (12); and the denominator is the first minor determinant of the same system.

Again, in calculating dx_2, $dx_1 = dx_3 = dx_4 = \ldots = dx_n = 0$; and by reason of (14), $d\xi_1 = 0$ when $dx_1 = 0$; so that we have

$$\left.\begin{array}{rl} dx_2 &= \beta_2 d\xi_2 + \gamma_2 d\xi_3 + \ldots\ldots + \rho_2 d\xi_n, \\ 0 &= \beta_3 d\xi_2 + \gamma_3 d\xi_3 + \ldots\ldots + \rho_3 d\xi_n, \\ \ldots & \ldots\ldots\ldots\ldots\ldots \\ 0 &= \beta_n d\xi_2 + \gamma_n d\xi_3 + \ldots\ldots + \rho_n d\xi_n; \end{array}\right\} \tag{15}$$

whence, employing the symbols of determinants,

$$dx_2 \Sigma . \pm \gamma_3 \delta_4 \ldots \rho_n = d\xi_2 \Sigma . \pm \beta_2 \gamma_3 \ldots \rho_n;$$

$$\therefore \quad dx_2 = \frac{\Sigma . \pm \beta_2 \gamma_3 \ldots \rho_n}{\Sigma . \pm \gamma_3 \delta_4 \ldots \rho_n} d\xi_2. \tag{16}$$

And proceeding in the same manner we shall ultimately have, ν being the letter preceding ρ,

$$dx_{n-1} = \frac{\Sigma . \pm \nu_{n-1} \rho_n}{\rho_n} d\xi_{n-1}; \tag{17}$$

$$dx_n = \rho_n d\xi_n; \tag{18}$$

so that, multiplying together the left-hand and the right-hand members of (14), (16), ..., we have

$$dx_1 \, dx_2 \, dx_3 \ldots dx_n = \Sigma . \pm a_1 \beta_2 \gamma_3 \ldots \rho_n . d\xi_1 \, d\xi_2 \, d\xi_3 \ldots d\xi_n; \tag{19}$$

and in $F(x_1, x_2, \ldots x_n)$, replacing $x_1, x_2, \ldots x_n$ by their values in terms of $\xi_1, \xi_2, \ldots \xi_n$, and denoting the function thereby obtained by $\Phi(\xi_1, \xi_2, \ldots \xi_n)$, we have

$$\int\int\ldots\ldots\int F(x_1, x_2, \ldots x_n)\, dx_1\, dx_2 \ldots dx_n$$

$$= \int\int\ldots\ldots\int \Phi(\xi_1, \xi_2, \ldots \xi_n)\, \Sigma . \pm a_1 \beta_2 \ldots \rho_n . d\xi_1\, d\xi_2 \ldots d\xi_n. \quad (20)$$

212.] Again, let us take the integral (10) to be the subject of transformation, and suppose the n equations connecting $x_1, x_2, \ldots x_n, \xi_1, \xi_2, \ldots \xi_n$ to be implicit, and to be of the forms

$$f_1 = 0, \quad f_2 = 0, \quad \ldots f_n = 0; \quad (21)$$

also let the several partial derived functions of these be denoted by letters according to the following scheme; viz.,

$$\left.\begin{array}{l} a_1\, dx_1 + b_1\, dx_2 + \ldots + r_1\, dx_n + a_1\, d\xi_1 + \ldots + \rho_1\, d\xi_n = 0, \\ a_2\, dx_1 + b_2\, dx_2 + \ldots + r_2\, dx_n + a_2\, d\xi_1 + \ldots + \rho_2\, d\xi_n = 0, \\ \cdot\quad\cdot\quad\cdot\quad\cdot\quad\cdot\quad\cdot\quad\cdot\quad\cdot\quad\cdot\quad\cdot\quad\cdot\quad\cdot \\ a_n\, dx_1 + b_n\, dx_2 + \ldots + r_n\, dx_n + a_n\, d\xi_1 + \ldots + \rho_n\, d\xi_n = 0. \end{array}\right\} \quad (22)$$

Now when x_1 varies, $dx_2 = dx_3 = \ldots = dx_n = 0$; so that in calculation the quantity which is to replace dx_1, we have

$$\left.\begin{array}{l} a_1\, d\xi_1 + \beta_1\, d\xi_2 + \ldots + \rho_1\, d\xi_n = -a_1\, dx_1, \\ a_2\, d\xi_1 + \beta_2\, d\xi_2 + \ldots + \rho_2\, d\xi_n = -a_2\, dx_1, \\ \cdot\quad\cdot\quad\cdot\quad\cdot\quad\cdot\quad\cdot\quad\cdot\quad\cdot\quad\cdot\quad\cdot \\ a_n\, d\xi_1 + \beta_n\, d\xi_2 + \ldots + \rho_n\, d\xi_n = -a_n\, dx_1; \end{array}\right\} \quad (23)$$

and consequently by elimination,

$$d\xi_1\, \Sigma . \pm a_1 \beta_2 \ldots \rho_n = -dx_1 . \pm a_1 \beta_2 \ldots \rho_n;$$

$$\therefore \quad dx_1 = -\frac{\Sigma . \pm a_1 \beta_2 \ldots \rho_n}{\Sigma . \pm a_1 \beta_2 \ldots \rho_n}\, d\xi_1; \quad (24)$$

now substituting this value for dx_1 in the given element of integration, and eliminating $x_1, \xi_2, \xi_3, \ldots \xi_n$ from that element by means of (21), it becomes a function of $\xi_1, x_2, x_3, \ldots x_n$; and consequently $\xi_1, x_3, \ldots x_n$ are all constant, when x_2 varies; so that in calculating the quantity which is to replace dx_2,

$$d\xi_1 = dx_3 = \ldots = dx_n = 0;$$

and therefore from (22),

$$\left.\begin{array}{l} a_1\, dx_1 + \beta_1\, d\xi_2 + \ldots + \rho_1\, d\xi_n = -b_1\, dx_2, \\ a_2\, dx_1 + \beta_2\, d\xi_2 + \ldots + \rho_2\, d\xi_n = -b_2\, dx_2, \\ \cdot\quad\cdot\quad\cdot\quad\cdot\quad\cdot\quad\cdot\quad\cdot\quad\cdot\quad\cdot\quad\cdot \\ a_n dx_1 + \beta_n\, d\xi_2 + \ldots + \rho_n\, d\xi_n = -b_n\, dx_2; \end{array}\right\} \quad (25)$$

and therefore

$$dx_2 = -\frac{\Sigma . \pm a_1 \beta_2 \ldots \rho_n}{\Sigma . \pm b_1 a_2 \ldots \rho_n}\, d\xi_2. \quad (26)$$

And proceeding by processes similar to the above,

$$dx_n = -\frac{\Sigma . \pm a_1 b_2 \dots \rho_n}{\Sigma . \pm a_1 b_2 \dots r_n} d\xi_n ; \tag{27}$$

and therefore by substitution

$$dx_1\, dx_2 \dots dx_n = (-)^n \frac{\Sigma . \pm a_1 \beta_2 \dots \rho_n}{\Sigma . \pm a_1 b_2 \dots r_n} d\xi_1\, d\xi_2 \dots d\xi_n . \tag{28}$$

Equation (19) is evidently only a particular case of (28); that viz., in which $a_1 = b_2 = c_3 = \dots = r_n = 1$, and all the other partial derived functions vanish.

Hence we have the following theorem;

In transforming the multiple integral

$$\iint \dots\dots \int \mathrm{F}(x_1, x_2, \dots x_n)\, dx_1\, dx_2 \dots dx_n$$

into its equivalent in terms of $\xi_1, \xi_2, \dots \xi_n$, where these variables are related by the n equations $f_1 = 0$, $f_2 = 0$, $f_3 = 0, \dots f_n = 0$; $dx_1\, dx_2 \dots dx_n$ is to be replaced by $\pm \frac{\Delta_1}{\Delta} d\xi_1\, d\xi_2 \dots d\xi_n$; where

$$\Delta_1 = \begin{vmatrix} \left(\frac{df_1}{d\xi_1}\right), & \left(\frac{df_1}{d\xi_2}\right), & \dots\dots & \left(\frac{df_1}{d\xi_n}\right) \\ \left(\frac{df_2}{d\xi_1}\right), & \left(\frac{df_2}{d\xi_2}\right), & \dots\dots & \left(\frac{df_2}{d\xi_n}\right) \\ \cdot & \cdot & \cdot & \cdot \\ \left(\frac{df_n}{d\xi_1}\right), & \left(\frac{df_n}{d\xi_2}\right), & \dots\dots & \left(\frac{df_n}{d\xi_n}\right) \end{vmatrix} ; \tag{29}$$

$$\Delta = \begin{vmatrix} \left(\frac{df_1}{dx_1}\right), & \left(\frac{df_1}{dx_2}\right), & \dots\dots & \left(\frac{df_1}{dx_n}\right) \\ \left(\frac{df_2}{dx_1}\right), & \left(\frac{df_2}{dx_2}\right), & \dots\dots & \left(\frac{df_2}{dx_n}\right) \\ \cdot & \cdot & \cdot & \cdot \\ \left(\frac{df_n}{dx_1}\right), & \left(\frac{df_n}{dx_2}\right), & \dots\dots & \left(\frac{df_n}{dx_n}\right) \end{vmatrix} . \tag{30}$$

The sign of the result being ambiguous, because it depends on the order of integration in the transformed integral, and that order is obviously arbitrary*.

213.] The following are examples illustrative of the preceding theorem.

* A full discussion of the theory of the transformation of multiple integrals will be found in a paper on the subject by M. Catalan, in the Mémoires couronnés par l'Académie de Bruxelles, Tome XIV, 1839, 1840.

Ex. 1. Take the double integral $\int\int \mathrm{F}(x,y)\,dy\,dx$; and (1) let the equations of transformation be

$$x = \xi\cos a - \eta\sin a, \quad y = \xi\sin a + \eta\cos a.$$

$$\therefore \quad dx = \cos a\,d\xi - \sin a\,d\eta,$$
$$dy = \sin a\,d\xi + \cos a\,d\eta;$$

$$\therefore \quad \Delta = 1, \quad \Delta_1 = (\cos a)^2 + (\sin a)^2 = 1.$$

$$\therefore \quad dy\,dx = d\eta\,d\xi. \tag{31}$$

(2) Let the equations of transformation be

$$x = r\cos\theta, \quad y = r\sin\theta.$$

$$\therefore \quad \left.\begin{aligned} dx &= \cos\theta\,dr - r\sin\theta\,d\theta, \\ dy &= \sin\theta\,dr + r\cos\theta\,d\theta; \end{aligned}\right\}$$

$$\therefore \quad \Delta = 1, \quad \Delta_1 = r(\cos\theta)^2 + r(\sin\theta)^2 = r;$$

$$\therefore \quad dy\,dx = r\,dr\,d\theta. \tag{32}$$

These cases are obviously those of transformation from a rectangular system of axes, to another rectangular system and to a polar system respectively.

(3) More generally, if the equations of transformation are given in the form,

$$x = f_1(\xi,\eta), \quad y = f_2(\xi,\eta); \quad \text{then}$$

$$\left.\begin{aligned} dx &= \left(\frac{df_1}{d\xi}\right)d\xi + \left(\frac{df_1}{d\eta}\right)d\eta, \\ dy &= \left(\frac{df_2}{d\xi}\right)d\xi + \left(\frac{df_2}{d\eta}\right)d\eta; \end{aligned}\right\} \tag{33}$$

$$\therefore \quad dx\,dy = \pm\left\{\left(\frac{df_1}{d\xi}\right)\left(\frac{df_2}{d\eta}\right) - \left(\frac{df_1}{d\eta}\right)\left(\frac{df_2}{d\xi}\right)\right\}d\xi\,d\eta. \tag{34}$$

Ex. 2. Let the integral be the triple integral

$$\int\int\int \mathrm{F}(x,y,z)\,dz\,dy\,dx:$$

and (1) let the equations of transformation be those from a system of rectangular axes to another system of rectangular axes; that is, let

$$\left.\begin{aligned} x &= a_1\xi + b_1\eta + c_1\zeta, \\ y &= a_2\xi + b_2\eta + c_2\zeta, \\ z &= a_3\xi + b_3\eta + c_3\zeta; \end{aligned}\right\} \tag{35}$$

then, since $\dfrac{b_2c_3 - c_2b_3}{a_1} = \dfrac{b_3c_1 - c_3b_1}{a_2} = \dfrac{b_1c_2 - c_1b_2}{a_3} = 1,$

$$\text{and} \qquad a_1^2 + a_2^2 + a_3^2 = 1;$$

$$\therefore \quad \Delta = 1, \qquad \Delta_1 = 1;$$

$$dz\, dy\, dx = d\zeta\, d\eta\, d\xi. \tag{36}$$

(2) Let $x = r\sin\theta\cos\phi, \quad y = r\sin\theta\sin\phi, \quad z = r\cos\theta$;

$$\left.\begin{aligned} \therefore\quad dx &= \sin\theta\cos\phi\, dr + r\cos\theta\cos\phi\, d\theta - r\sin\theta\sin\phi\, d\phi, \\ dy &= \sin\theta\sin\phi\, dr + r\cos\theta\sin\phi\, d\theta + r\sin\theta\cos\phi\, d\phi, \\ dz &= \cos\theta\, dr \qquad - r\sin\theta\, d\theta; \\ \therefore\quad \Delta &= 1; \qquad \Delta_1 = r^2\sin\theta; \end{aligned}\right\} \tag{37}$$

$$\therefore \quad dz\, dy\, dx = r^2 \sin\theta\, dr\, d\theta\, d\phi. \tag{38}$$

These latter equations are those of transformation from a rectangular to a polar system in geometry of three dimensions.

Ex. 3. Let

$$\left.\begin{aligned} x_1 &= r\cos\theta_1, \\ x_2 &= r\sin\theta_1\cos\theta_2, \\ x_3 &= r\sin\theta_1\sin\theta_2\cos\theta_3, \\ &\cdots\cdots\cdots \\ x_n &= r\sin\theta_1\sin\theta_2\ldots\sin\theta_{n-1}; \end{aligned}\right\} \tag{39}$$

whence, squaring and adding,

$$x_1^2 + x_2^2 + \ldots + x_n^2 = r^2; \tag{40}$$

which equation we shall find it convenient to use instead of the last of the group (39); differentiating these with respect to $x_1, x_2, \ldots x_n$, we have

$$\left.\begin{aligned} dx_1 + \quad 0 \quad + \ldots + \quad 0 \quad &= 0, \\ 0 \quad + \quad dx_2 + \ldots + \quad 0 \quad &= 0, \\ \cdots\cdots\cdots\cdots\cdots \\ 2x_1\, dx_1 + 2x_2\, dx_2 + \ldots + 2x_n\, dx_n &= 0; \end{aligned}\right\} \tag{41}$$

whence $\Delta = 2x_n$.

Differentiating (40) and (39) with respect to $r, \theta_1, \ldots \theta_{n-1}$,

$$\left.\begin{aligned} 2r\, dr \qquad\qquad\qquad\qquad &= 0, \\ \cos\theta_1\, dr - r\sin\theta_1\, d\theta_1 \qquad\qquad &= 0, \\ - - - - - - - r\sin\theta_1\sin\theta_2\, d\theta_2 \qquad &= 0, \\ \cdots\cdots\cdots\cdots\cdots \\ - - - - - - - - - r\sin\theta_1\sin\theta_2\ldots\sin\theta_{n-1}\, d\theta_{n-1} &= 0; \end{aligned}\right\}$$

whence $\Delta_1 = (-)^{n-1} 2r^n (\sin\theta_1)^{n-1} (\sin\theta_2)^{n-2} \ldots \sin\theta_{n-1}$; so that

$$\frac{\Delta_1}{\Delta} = \pm r^{n-1} (\sin\theta_1)^{n-2} (\sin\theta_2)^{n-3} \ldots \sin\theta_{n-2}\, dr\, d\theta_1\, d\theta_2 \ldots d\theta_{n-1};$$

and consequently,

$$dx_1\, dx_2 \ldots dx_n$$
$$= \pm r^{n-1} (\sin\theta_1)^{n-2} (\sin\theta_2)^{n-3} \ldots \sin\theta_{n-2}\, dr\, d\theta_1\, d\theta_2 \ldots d\theta_{n-1}. \quad (42)$$

If $n = 3$, we have

$$dx_1\, dx_2\, dx_3 = \pm r^2 \sin\theta_1\, dr\, d\theta_1\, d\theta_2;$$

which is the transformation given in the latter part of the preceding example.

Ex. 4. Let the equations of transformation be

$$\left.\begin{aligned} x_1 &= 1-\xi_1, \\ x_2 &= \xi_1(1-\xi_2), \\ x_3 &= \xi_1 \xi_2 (1-\xi_3), \\ &\cdots\cdots \\ x_n &= \xi_1 \xi_2 \ldots \xi_{n-1}(1-\xi_n); \end{aligned}\right\} \quad (43)$$

$$\begin{aligned} \therefore\quad dx_1 &= -d\xi_1, \\ dx_2 &= (1-\xi_2)\, d\xi_1 - \xi_1\, d\xi_2, \\ dx_3 &= \xi_2(1-\xi_3)\, d\xi_1 + \xi_1(1-\xi_3)\, d\xi_2 - \xi_1 \xi_2\, d\xi_3, \\ &\cdots\cdots \\ dx_n &= \xi_2 \xi_3 \ldots \xi_{n-1}(1-\xi_n)\, d\xi_1 + \ldots\ldots - \xi_1 \xi_2 \ldots \xi_{n-1}\, d\xi_n; \end{aligned}$$

$$\therefore\quad \Delta = 1; \qquad \Delta_1 = (-)^n \xi_1^{n-1} \xi_2^{n-2} \ldots \xi_{n-1};$$

$$\therefore\quad dx_1\, dx_2 \ldots dx_n = \pm \xi_1^{n-1} \xi_2^{n-2} \ldots \xi_{n-1}\, d\xi_1\, d\xi_2 \ldots d\xi_n. \quad (44)$$

Hence we have the following transformation, when $n = 2$;

$$x_1 = 1-\xi_1,$$
$$x_2 = \xi_1(1-\xi_2);$$
$$\therefore\quad dx_1\, dx_2 = \xi_1\, d\xi_1\, d\xi_2;$$
$$\iint x_1^m x_2^n\, dx_1\, dx_2 = \iint \xi_1^{n+1}(1-\xi_1)^m (1-\xi_2)^n\, d\xi_1\, d\xi_2.$$

Ex. 5. To transform into its equivalent $dx\, dy\, dz$, when $x = lr$, $y = mr$, $z = nr$; l, m, and n being subject to the condition

$$l^2 + m^2 + n^2 = 1.$$

By a process similar to those above, we have the following results:

$$dx\, dy\, dz = \frac{r^2\, dr\, dm\, dn}{l} = \frac{r^2\, dr\, dn\, dl}{m} = \frac{r^2\, dr\, dl\, dm}{n}. \quad (45)$$

214.] If the original integral is definite, the transformed one will also be definite; for the latter is to be equivalent to the former in all respects, and consequently the values of the variables in both integrals must extend over the same district of values. Thus the principle on which the limits of the new integral are to be assigned is the same in all cases, although the particular values will depend on the peculiar circumstances of each case. The limits of the original integral, and the equations of relation, will be found sufficient for the determination of the limits of the transformed integral. The following case frequently occurs, and illustrates the principle.

Let the definite integral, which is the subject of transformation, be of the form (8), viz.,

$$\mathrm{I} = \int_{x_n}^{X_n} \int_{x_{n-1}}^{X_{n-1}} \dots\dots \int_{x_2}^{X_2} \int_{x_1}^{X_1} \mathrm{F}(x_1, x_2, \dots x_n)\, dx_1\, dx_2 \dots dx_n; \quad (46)$$

and let us suppose the integral to include all values of the variables for which the function $f(x_1, x_2, \dots x_n)$ is negative; so that X_1 and x_1 are functions of all the variables except x_1; X_2 and x_2 are functions of all the variables except x_2 and x_1; and so on: X_{n-1} and x_{n-1} being functions of x_n only; and X_n and x_n being constants. Now suppose $\xi_1, \xi_2, \dots \xi_n$ to be new variables, connected with the former variables by equations of relation; and let the new integral be

$$\mathrm{I} = \int_{\text{Ξ}_n}^{\Xi_n} \int_{\text{Ξ}_{n-1}}^{\Xi_{n-1}} \dots\dots \int_{\text{Ξ}_1}^{\Xi_1} \Phi(\xi_1, \xi_2, \dots \xi_n)\, d\xi_1\, d\xi_2 \dots d\xi_n; \quad (47)$$

also suppose the equation, which determines the limits, to become $\phi(\xi_1, \xi_2, \dots \xi_n)$; so that the integral includes all values of $\xi_1, \xi_2, \dots \xi_n$ for which $\phi(\xi_1, \xi, \dots \xi_n)$ is negative; then the limits must be determined as follows;

Observing the order of the successive integrations as indicated by the order in which the differentials are arranged, Ξ_1 and Ξ_1 are functions of all the variables except ξ_1, and are determined by putting $\phi = 0$, for thus will be included all values of ξ_1 for which ϕ is negative. By this process the new element-function will involve $\xi_2, \xi_3, \dots \xi_n$; and the limits of integration must include all values of the variables for which ξ_1 is not impossible. To determine this condition, let it be remembered that when, by the variation of certain quantities, the roots of an equation having been real become impossible, in the course of their change they must have been equal; the limits therefore of the new integral will be given by the conditions for which two at least of the values of ξ_1, as found from $\phi = 0$, are equal; that is, by the simultaneous equations $\phi = 0$, and $\left(\frac{d\phi}{d\xi_1}\right) = 0$. From these two equations therefore if we eliminate ξ_1, we shall have a resulting equation of the form $\phi_1(\xi_2, \xi_3, \ldots \xi_n) = 0$, from which Ξ_2, ɪ_2 are to be determined, because they are the values of ξ_2 which satisfy $\phi_1 = 0$; and thus we shall include all values of ξ_2 for which the expression ϕ_1 is negative. Similarly by eliminating ξ_2 between the equations $\phi_1 = 0$, and $\left(\frac{d\phi_1}{d\xi_2}\right) = 0$, the limiting values of ξ_3 will be determined. And the limits of all the other variables will be determined by similar processes.

Hence in recapitulation; the limits of ξ_1 are determined by solving for ξ_1 the equation $\phi = 0$; those of ξ_2 by solving for ξ_2 the equation which results from the elimination of ξ_1 between $\phi = 0$, and $\left(\frac{d\phi}{d\xi_1}\right) = 0$; those of ξ_3 by solving for ξ_3 the equation which results from the elimination of ξ_1 and ξ_2 between $\phi = 0$, $\left(\frac{d\phi}{d\xi_1}\right) = 0$, and $\left(\frac{d\phi}{d\xi_2}\right) = 0$; and so for the others.

If the equations $\phi = 0$, $\phi_1 = 0$, ..., give only two values of $\xi_1, \xi_2, \ldots$ for the limits, then the case is free from difficulty, and the new definite integral assumes the form in which it is written above: but if any of these equations has more than two roots, and if there is nothing in the conditions of the problem which excludes them, we must resolve I into a series of integrals according to the roots, and the limits of the several integrals will always be given by the equations found as above *.

As an example, let us suppose a triple definite integral to include all variables which refer to points lying within the surface of a given ellipsoid, and to exclude all others. Then if the triple integral is

$$\text{I} = \iiint \text{F}(x, y, z)\, dz\, dy\, dx,$$

all values of the variables are to be included for which

$$\frac{x^2}{a^2} + \frac{^2}{b^2} + \frac{z^2}{c^2} - 1 = f(x, y, z)$$

is negative. Now if the integrations are to be effected in the order in which the differentials are placed, the z-integration

* See a Memoir by M. Ostrogradsky on the Calculus of Variations in the Memoirs of the Imperial Academy of St. Petersbourg, Vol. I, 1838, p. 46. The Memoir has been reprinted in Crelle's Mathematical Journal, Vol. XV. p. 332.

comes first, and its limits are given by $f(x, y, z) = 0$; whence we have

$$z = \pm c\left\{1 - \frac{x^2}{a^2} - \frac{y^2}{b^2}\right\}^{\frac{1}{2}}$$
$$= \pm \text{Z}, \text{ say}.$$

The y-integration comes next; and its limits are given by the elimination of z by means of $f(x, y, z) = 0$, and $\left(\frac{df}{dz}\right) = 0$; whence we have

$$y = \pm b\left\{1 - \frac{x^2}{a^2}\right\}^{\frac{1}{2}}$$
$$= \pm \text{Y}, \text{ say};$$

and the x-integration comes last; and its limits are given by the elimination of z and y by means of $f(x, y, z) = 0$, $\left(\frac{df}{dz}\right) = 0$, $\left(\frac{df}{dy}\right) = 0$; whence we have

$$x = \pm a.$$

so that

$$\text{I} = \int_{-a}^{a}\int_{-\text{Y}}^{\text{Y}}\int_{-\text{Z}}^{\text{Z}} \text{F}(x, y, z)\, dz\, dy\, dx.$$

215.] Examples of transformation of definite multiple integrals.

Ex. 1. The double integral $\iint (x^2+y^2)\, dy\, dx$ extends to all points within the area of the circle $x^2+y^2 = a^2$; it is required to express it in terms of polar coordinates.

If $\text{Y} = (a^2-x^2)^{\frac{1}{2}}$,

$$\text{I} = \int_{-a}^{a}\int_{-\text{Y}}^{\text{Y}} (x^2+y^2)\, dy\, dx = \int_{0}^{2\pi}\int_{0}^{a} r^3\, dr\, d\theta. \qquad (48)$$

Ex. 2. Transform into its equivalent in terms of polar coordinates the double integral

$$\int_{-a}^{a}\int_{-\text{Y}}^{\text{Y}} dy\, dx, \text{ where } \text{Y} = \frac{b}{a}(a^2-x^2)^{\frac{1}{2}}.$$

Since $x = r\cos\theta$, $y = r\sin\theta$, therefore by (32), $dy\, dx = r\, dr\, d\theta$; also the given limits shew that the district of the integral is bounded by the equation $a^2y^2+b^2x^2-a^2b^2 = 0$; and this expressed in polar coordinates is $r^2\{a^2(\sin\theta)^2+b^2(\cos\theta)^2\}-a^2b^2=0$; so that if $\text{r} = \frac{ab}{\{a^2(\sin\theta)^2+b^2(\cos\theta)^2\}^{\frac{1}{2}}}$,

$$\int_{-a}^{a}\int_{-\text{Y}}^{\text{Y}} dy\, dx = \int_{0}^{2\pi}\int_{0}^{\text{r}} r\, dr\, d\theta. \qquad (49)$$

Ex. 3. If $z^2 = a^2 - x^2 - y^2$, $Y^2 = a^2 - x^2$; then in terms of polar coordinates, see Ex. 2, Art. 213,

$$\int_{-a}^{a}\int_{-Y}^{Y}\int_{-Z}^{Z}\frac{dz\,dy\,dx}{(x^2+y^2+z^2)^{\frac{1}{2}}} = \int_0^{2\pi}\int_0^{\pi}\int_0^{a} r\sin\theta\,dr\,d\theta\,d\phi.$$

Ex. 4. In the definition of the Gamma-function given in (247), Art. 122, let x be replaced by x^2; then

$$\Gamma(m) = 2\int_0^{\infty} e^{-x^2} x^{2m-1} dx; \tag{50}$$

similarly $\quad \Gamma(n) = 2\int_0^{\infty} e^{-y^2} y^{2n-1} dy.$

$$\therefore \quad \Gamma(m)\Gamma(n) = 4\int_0^{\infty}\int_0^{\infty} e^{-(x^2+y^2)} x^{2m-1} y^{2n-1}\,dy\,dx. \tag{51}$$

In this definite integral let $x = r\cos\theta$, $y = r\sin\theta$; then as the limits of x and y are such that all positive values of x and y, that is, the values of x and y corresponding to all points in the first quadrant of the plane (x, y), are included in the integral, the limits of r and θ must be co-extensive; hence

$$\Gamma(m)\Gamma(n) = 4\int_0^{\frac{\pi}{2}}\int_0^{\infty} e^{-r^2} r^{2m+2n-1} (\sin\theta)^{2n-1} (\cos\theta)^{2m-1}\,dr\,d\theta \tag{52}$$

$$= 2\int_0^{\infty} e^{-r^2} r^{2m+2n-1}\,dr \times 2\int_0^{\frac{\pi}{2}} (\sin\theta)^{2n-1}(\cos\theta)^{2m-1}\,d\theta$$

$$= \Gamma(m+n)\,\mathrm{B}(m,n), \tag{53}$$

as appears from (50), and from (271), Art. 126. Hereby the fundamental theorem connecting the Beta- and the Gamma-functions is established.

Again, take (50), and let $m = \frac{1}{2}$; then

$$\Gamma\left(\frac{1}{2}\right) = 2\int_0^{\infty} e^{-x^2} dx = 2\int_0^{\infty} e^{-y^2} dy.$$

$$\therefore \quad \left\{\Gamma\left(\frac{1}{2}\right)\right\}^2 = 4\int_0^{\infty}\int_0^{\infty} e^{-(x^2+y^2)}\,dy\,dx$$

$$= 4\int_0^{\frac{\pi}{2}}\int_0^{\infty} e^{-r^2} r\,dr\,d\theta$$

$$= \pi.$$

$$\therefore \quad \Gamma\left(\frac{1}{2}\right) = \pi^{\frac{1}{2}}; \tag{54}$$

the same result as 280, Art. 129.

Ex. 5. Double integrals of certain forms are frequently simplified by the following formulæ; viz. $x = u(1-v)$, $y = uv$. From these values we have

$$dx = (1-v)\,du - u\,dv, \qquad dy = v\,du + u\,dv.$$

$$\therefore \quad dy\,dx = u\,du\,dv. \tag{55}$$

The following is the geometrical interpretation of this substitution. In fig. 44, let P be (x, y); through P draw PN parallel to, and PS making an angle of 45° with, the axis of x, and join NS; then OS $= u$, and tan OSN $= v$: for

$$\left.\begin{aligned} x &= \text{OS}-\text{SM} & y &= \text{MP} \\ &= \text{OS}-\text{ON} & &= \text{ON} \\ &= u-uv; & &= uv. \end{aligned}\right\} \tag{56}$$

Hence, if the limits of integration of the double integral are given, it is easy to assign the limits of the transformed integral.

Thus, taking the definition of the Gamma-function given in (247), Art. 122, we have

$$\Gamma(m) = \int_0^\infty e^{-x} x^{m-1}\,dx, \qquad \Gamma(n) = \int_0^\infty e^{-y} y^{n-1}\,dy.$$

$$\begin{aligned} \therefore \quad \Gamma(m)\,\Gamma(n) &= \int_0^\infty\int_0^\infty e^{-(x+y)} x^{m-1} y^{n-1}\,dy\,dx \\ &= \int_0^1\int_0^\infty e^{-u} u^{m+n-1} v^{m-1} (1-v)^{m-1}\,du\,dv \\ &= \int_0^\infty e^{-u} u^{m+n-1}\,du \times \int_0^1 v^{n-1} (1-v)^{m-1}\,dv \\ &= \Gamma(m+n)\,\mathrm{B}(m,n). \end{aligned} \tag{57}$$

By means of this substitution the integral $\iint x^m y^n f(x+y)\,dy\,dx$ is changed into $\iint u^{m+n+1}(1-v)^m v^n f(u)\,du\,dv$.

Hence

$$\begin{aligned} \int_0^\infty\int_0^\infty x^m y^n f(x+y)\,dy\,dx &= \int_0^1\int_0^\infty u^{m+n-1}(1-v)^m v^n f(u)\,du\,dv \\ &= \int_0^1 (1-v)^m v^n\,dv \int_0^\infty u^{m+n-1} f(u)\,du \\ &= \mathrm{B}(m+1, n+1)\int_0^\infty u^{m+n-1} f(u)\,du \\ &= \frac{\Gamma(m+1)\,\Gamma(n+1)}{\Gamma(m+n+2)}\int_0^\infty u^{m+n-1} f(u)\,du. \end{aligned}$$

Or, if the given integral is $\iint f\left(\frac{x}{a}+\frac{y}{b}\right) dy\, dx$, then if we replace x and y by $a\xi$ and $b\eta$ respectively, and substitute for ξ and η in terms of u and v, the integral becomes

$$\iint a\, b\, f(u)\, u\, du\, dv. \tag{58}$$

Thus $\int_0^\infty\int_0^\infty \left(\frac{x}{a}+\frac{y}{b}-1\right)^{-3} dy\, dx = ab\int_0^\infty\int_0^\infty (\xi+\eta-1)^{-3}\, d\xi\, d\eta$

$$= ab\int_0^\infty \frac{u\, du}{(u-1)^3}. \tag{59}$$

216.] Although the limits of the first integration of a multiple integral are functions of one or more of the variables, in reference to which subsequent integrations are to be effected, yet it is possible, and frequently convenient, so to change the variables, that the limits of the first integration may be constant. The mode of effecting this transformation is explained in Art. 98.

Let us take the case of a double integral; and suppose it be

$$\text{I} = \int_{x_0}^{x_1}\int_{y_0}^{y_1} f(x,y)\, dy\, dx, \tag{60}$$

where y_1 and y_0 are functions of x.

Let $\quad y = y_0+(y_1-y_0)t, \qquad (61)$

where t is a new variable; therefore

$$dy = (y_1-y_0)\, dt;$$

and observing that $t=1$, when $y=y_1$; and $t=0$, when $y=y_0$; we have

$$\text{I} = \int_{x_0}^{x_1}\int_0^1 (y_1-y_0) f\{x, y_0+(y_1-y_0)t\}\, dt\, dx$$

$$= \int_{x_0}^{x_1}(y_1-y_0)\int_0^1 f\{x, y_0+(y_1-y_0)\, t\}\, dt\, dx; \tag{62}$$

wherein the limits of the first integration are constant.

Thus, for example, if

$$\text{I} = \int_0^\infty\int_0^{(1+x^2)^{\frac{1}{2}}} \frac{e^{-(a^2x^2+b^2y^2)}}{(1+x^2+y^2)^{\frac{1}{2}}}\, dy\, dx;$$

let $y=(1+x^2)^{\frac{1}{2}}\, t$; so that $t=1$, when $y=(1+x^2)^{\frac{1}{2}}$; and $t=0$, when $y=0$; then

$$\text{I} = \int_0^\infty\int_0^1 \frac{e^{-(a^2x^2+b^2t^2+b^2x^2t^2)}}{(1+t^2)^{\frac{1}{2}}}\, dt\, dx$$

$$= \int_0^1 \frac{e^{-b^2t^2}\, dt}{(1+t^2)^{\frac{1}{2}}}\int_0^\infty e^{-(a^2+b^2t^2)x^2}\, dx;$$

$$\text{But } \int_0^\infty e^{-(a^2+b^2t^2)x^2}\,dx = \frac{1}{2(a^2+b^2t^2)^{\frac{1}{2}}}\int_0^\infty e^{-z}\,z^{-\frac{1}{2}}\,dz$$

$$= \frac{1}{2(a^2+b^2t^2)^{\frac{1}{2}}}\,\Gamma\left(\frac{1}{2}\right)$$

$$= \frac{1}{2}\left(\frac{\pi}{a^2+b^2t^2}\right)^{\frac{1}{2}};$$

$$\therefore \quad \mathrm{I} = \frac{\pi^{\frac{1}{2}}}{2}\int_0^1 \frac{e^{-b^2t^2}}{(1+t^2)^{\frac{1}{2}}(a^2+b^2t^2)^{\frac{1}{2}}}\,dt.$$

Section 3.—*The Differentiation of a Definite Multiple Integral with respect to a Variable Parameter.*

217.] If a parameter, capable of variation, is contained in the element-function of a definite multiple integral, and also in some or all of the limits of integration, the integral is a function of that parameter; so that the value of the integral will change, if the parameter varies. Let us suppose the parameter continuously to vary, within such limits however that the element-function of the integral does not become infinite or discontinuous; then the value of the integral will also continuously vary; and it is the increment of the integral, due to the infinitesimal variation of the parameter, which it is our object to determine in the present section.

In Art. 96, we have investigated the change of value of a definite single integral, due to the infinitesimal variation of a parameter, of which the limits as well as the element-function are functions; and we have shewn that if a is the variable parameter, and D denotes the total differential,

$$\mathrm{D}\int_{x_0}^{x_n} \mathrm{F}(a,x)\,dx = \mathrm{F}(a,x_n)\,dx_n - \mathrm{F}(a,x_0)\,dx_0 + \int_{x_0}^{x_n} \frac{d.\mathrm{F}(a,x)}{da}\,da\,dx; \quad (63)$$

$$\therefore \quad \frac{\mathrm{D}}{da}\int_{x_0}^{x_n} \mathrm{F}(a,x)\,dx = \left[\mathrm{F}(a,x)\,\frac{dx}{da}\right]_{x_0}^{x_n} + \int_{x_0}^{x_n} \frac{d.\mathrm{F}(a,x)}{da}\,dx. \quad (64)$$

Now this will be the foundation of our subsequent investigations.

Let us first consider a definite double integral; viz.

$$u = \int_{x_0}^{x_n}\int_{y_0}^{y_n} \mathrm{F}(a,x,y)\,dy\,dx = \int_{x_0}^{x_n}\int_{y_0}^{y_n} \Omega\,dy\,dx, \quad (65)$$

where $$\Omega = \mathrm{F}(a,x,y); \quad (66)$$

and where x_n, x_0, y_n, y_0, as also Ω, are all functions of a; a being a parameter capable of continuous variation. We have to determine the change of value in u due to an infinitesimal variation of a.

In (64) let $F(a, x)$ be replaced by $\int_{y_0}^{y_n} F(a, x, y)\, dy$; that is, by $\int_{y_0}^{y_n} \Omega\, dy$; then

$$\frac{D}{da}\int_{x_0}^{x_n}\int_{y_0}^{y_n} \Omega\, dy\, dx = \left[\frac{dx}{da}\int_{y_0}^{y_n}\Omega\, dy\right]_{x_0}^{x_n} + \int_{x_0}^{x_n}\frac{d}{da}\int_{y_0}^{y_n}\Omega\, dy\, dx. \quad (67)$$

But by reason of (64),

$$\frac{d}{da}\int_{y_0}^{y_n}\Omega\, dy = \left[\Omega\frac{dy}{da}\right]_{y_0}^{y_n} + \int_{y_0}^{y_n}\frac{d\Omega}{da}\, dy;$$

and in the first term of the right-hand member of (67), inasmuch as $\frac{dx}{da}$ does not contain y, putting it under the sign of the y-integration for symmetry of expression, we have

$$\frac{D}{da}\int_{x_0}^{x_n}\int_{y_0}^{y_n}\Omega\, dy\, dx = \left[\int_{y_0}^{y_n}\Omega\frac{dx}{da}\, dy\right]_{x_0}^{x_n} + \int_{x_0}^{x_n}\left[\Omega\frac{dy}{da}\right]_{y_0}^{y_n} dx$$
$$+ \int_{x_0}^{x_n}\int_{y_0}^{y_n}\frac{d\Omega}{da}\, dy\, dx; \quad (68)$$

which gives the total variation of the definite double integral.

Again, if the similar variation of a definite triple integral is required, let Ω in (68) be replaced by $\int_{z_0}^{z}\Omega\, dz$, where the new Ω is $F(a, x, y, z)$; then by an exactly similar process it may be shewn that

$$\frac{D}{da}\int_{x_0}^{x_n}\int_{y_0}^{y_n}\int_{z_0}^{z_n}\Omega\, dz\, dy\, dx$$
$$= \left[\int_{y_0}^{y_n}\int_{z_0}^{z_n}\Omega\frac{dx}{da}\, dz\, dy\right]_{x_0}^{x_n} + \int_{x_0}^{x_n}\left[\int_{z_0}^{z_n}\Omega\frac{dy}{da}\, dz\right]_{y_0}^{y_n} dx$$
$$+ \int_{x_0}^{x_n}\int_{y_0}^{y_n}\left[\Omega\frac{dz}{da}\right]_{z_0}^{z_n} dy\, dx + \int_{x_0}^{x_n}\int_{y_0}^{y_n}\int_{z_0}^{z_n}\frac{d\Omega}{da}\, dz\, dy\, dx. \quad (69)$$

Now the law of formation of the several terms in the second member of this equation is sufficiently obvious, so that the variation of a definite multiple integral may be easily expressed at length; and the general law may be proved inductively; viz. by assuming the truth of it for a multiple integral of the $(n-1)$th order, and deducing from that the same law for the multiple of the nth order. This process is so easy that it is unnecessary to express it at length.

218.] Definite multiple integrals also, when one integration has been effected, and the limits of that integration are functions of a variable parameter, are subject to variation by reason of the change of that variable parameter. The consequent variable of the reduced integral may be derived from the theorem given in (64). The following are examples of the process.

In (64) let $\mathrm{F}(a, x)$ be replaced by $\Big[\Omega\Big]_{y_0}^{y_n}$; then

$$\frac{\mathrm{D}}{da}\int_{x_0}^{x_n}\Big[\Omega\Big]_{y_0}^{y_n}dx = \Big[\Big[\Omega\frac{dx}{da}\Big]_{y_0}^{y_n}\Big]_{x_0}^{x_n} + \int_{x_0}^{x_n}\frac{d}{da}\Big[\Omega\Big]_{y_0}^{y_n}dx;$$

but we have evidently,

$$\frac{d}{da}\Big[\Omega\Big]_{y_0}^{y_n} = \Big[\frac{d\Omega}{da} + \frac{d\Omega}{dy}\frac{dy}{da}\Big]_{y_0}^{y_n}; \tag{70}$$

$$\therefore \frac{\mathrm{D}}{da}\int_{x_0}^{x_n}\Big[\Omega\Big]_{y_0}^{y_n}dx = \Big[\Big[\Omega\frac{dx}{da}\Big]_{y_0}^{y_n}\Big]_{x_0}^{x_n} + \int_{x_0}^{x_n}\Big[\frac{d\Omega}{da} + \frac{d\Omega}{dy}\frac{dy}{da}\Big]_{y_0}^{y_n}dx. \tag{71}$$

Again, let Ω be replaced by $\int_{z_0}^{z_n}\Omega\,dz$, so that

$$\frac{d\Omega}{da} \text{ must be replaced by } \Big[\Omega\frac{dz}{da}\Big]_{z_0}^{z_n} + \int_{z_0}^{z_n}\frac{d\Omega}{da}dz;$$

$$\frac{d\Omega}{dy} \text{ - - - - - - - - } \Big[\Omega\frac{dz}{dy}\Big]_{z_0}^{z_n} + \int_{z_0}^{z_n}\frac{d\Omega}{dy}dz;$$

consequently

$$\begin{aligned}\frac{\mathrm{D}}{da}\int_{x_0}^{x_n}\Big[\int_{z_0}^{z_n}\Omega\,dz\Big]_{y_0}^{y_n}dx &= \Big[\Big[\int_{z_0}^{z_n}\Omega\frac{dx}{da}dz\Big]_{y_0}^{y_n}\Big]_{x_0}^{x_n} \\ &+ \int_{x_0}^{x_n}\Big[\Big[\Omega\Big(\frac{dz}{da} + \frac{dz}{dy}\frac{dy}{da}\Big)\Big]_{z_0}^{z_n}\Big]_{y_0}^{y_n}dx \\ &+ \int_{x_0}^{x_n}\Big[\int_{z_0}^{z_n}\Big(\frac{d\Omega}{da} + \frac{d\Omega}{dy}\frac{dy}{da}\Big)dz\Big]_{y_0}^{y_n}dx. \end{aligned} \tag{72}$$

Again, for another example, from (70), we have

$$\frac{d}{da}\Big[\Omega\Big]_{x_0}^{x_n} = \Big[\frac{d\Omega}{da} + \frac{d\Omega}{dx}\frac{dx}{da}\Big]_{x_0}^{x_n}.$$

Let Ω be replaced by $\int_{y_0}^{y_n}\Omega\,dy$; so that

$$\frac{d\Omega}{da} \text{ must be replaced by } \Big[\Omega\frac{dy}{da}\Big]_{y_0}^{y_n} + \int_{y_0}^{y_n}\frac{d\Omega}{da}dy;$$

$$\frac{d\Omega}{dx} \text{ - - - - - - - - } \Big[\Omega\frac{dy}{dx}\Big]_{y_0}^{y_n} + \int_{y_0}^{y_n}\frac{d\Omega}{dx}dy;$$

consequently

$$\frac{d}{da}\left[\int_{y_0}^{y_n} \Omega\, dy\right]_{x_0}^{x_n} = \left[\left[\Omega\left(\frac{dy}{da} + \frac{dy}{dx}\frac{dx}{da}\right)\right]_{y_0}^{y_n}\right]_{x_0}^{x_n}$$

$$+ \left[\int_{y_0}^{y_n}\left(\frac{d\Omega}{da} + \frac{d\Omega}{dx}\frac{dx}{da}\right) dy\right]_{x_0}^{x_n}. \quad (73)$$

The variation of all other similar integrals may be determined in a similar manner, the theorems contained in (64) and in (70) being of general application; and it is unnecessary to exemplify them further at present. We shall hereafter have occasion to apply the preceding process.

CHAPTER IX.

QUADRATURE OF SURFACES, PLANE AND CURVED.

SECTION 1.—*Quadrature of Plane Surfaces. Rectangular Coordinates.*

219.] The theory of definite multiple integration which has been developed in the preceding chapter enables us to determine the area of a plane or curved surface in such a form that it may be compared to the area of a square: hence arises the name, Quadrature: and we shall first consider the most simple case, and investigate the area of a plane superficies contained between a curve whose equation is given, the axis of x, and two ordinates parallel to the axis of y and at a finite distance apart.

Let $y = f(x)$ be the equation to the bounding curve, $P_0 P P_n$; see fig. 16; $OM_0 = x_0$, $OM_n = x_n$; and let $f(x)$ be finite and continuous for all values of x between x_0 and x_n; our object is to determine the area of $P_0 M_0 M_n P_n$.

Take any point E within the boundaries of this area, and let E be (x, y); take EF and EG infinitesimal increments of y and x, so that $EF = dy$, $EG = dx$; then the area of the element, or the area-element, as it is called, $= dy\, dx$; and the area of the superficies required is the sum of all such area-elements: thus it is evidently a double integral, of which the limits will be assigned by the given geometrical conditions, and which may be determined according to the principles of the preceding chapter. I propose however to investigate the subject from those first principles of geometry which are inherent in it.

Let other lines be drawn as in the figure; and, first considering x to be constant, let us sum the elements with respect to y from the axis of x to MP; that is, let us integrate $dx\, dy$ with respect to y from $y = 0$ to $y = f(x)$, dx being a constant factor throughout the process: the result of the operation will be the area of the differential slice PMNQ, whose sides are parallel to the axis of y, because x is the same for all the elements, and which is of the breadth dx and of the length $f(x)$; therefore

$$\text{PMNQ} = f(x)\, dx; \qquad (1)$$

and as this area is expressed in general terms of x, it is the type of all similar elemental slices; and therefore the sum of all such between assigned limits is the required area. Hence if A represents the required area,

$$A = \int_{x_0}^{x_n} \int_0^{f(x)} dy\, dx \tag{2}$$

$$= \int_{x_0}^{x_n} f(x)\, dx. \tag{3}$$

If the superficies, whose area is to be determined, is of the form OPP_nM_n of fig. 4, then the inferior limit of x is 0, and we have

$$A = \int_0^{x_n} \int_0^{f(x)} dy\, dx. \tag{4}$$

Let it not be supposed that any inaccuracy of result arises from the circumstance that the differential slice is an imperfect rectangle at the point P where it meets the curve; for though the exact value of PMNQ is intermediate to

$$f(x)\, dx \quad \text{and} \quad f(x+dx)\, dx,$$

yet the difference between these two, viz. $\{f(x+dx)-f(x)\}\, dx$, is equal to $f'(x)\, dx^2$, and is therefore an infinitesimal of a higher order, and must be neglected.

The following are examples in which the preceding formulæ are applied; but one remark must be made. If the limits of integration include a value of the variables at which the element-function changes sign, the right sign must be introduced into the integral, otherwise it may be that the sum of the elements on one side of such a critical value will exactly neutralize that of those on the other side, and the result will be nugatory.

220.] Examples of quadrature of plane surfaces.

Ex. 1. To find the area contained between the axis of x, an ordinate, and the parabola whose equation is $y^2 = 4mx$.

Let the extreme abscissa, see fig. 17, $= a$, and the extreme ordinate $= b$; so that $b^2 = 4ma$; then the equation to the parabola is $ay^2 = b^2 x$:

$$\therefore \text{ the area OAB} = \int_0^a \int_0^{b\left(\frac{x}{a}\right)^{\frac{1}{2}}} dy\, dx$$

$$= \int_0^a \Big[y\Big]_0^{b\left(\frac{x}{a}\right)^{\frac{1}{2}}} dx = \int_0^a b\left(\frac{x}{a}\right)^{\frac{1}{2}} dx$$

$$= \frac{2}{3}\frac{b}{a^{\frac{1}{2}}}\Big[x^{\frac{3}{2}}\Big]_0^a = \frac{2}{3}ab.$$

Thus the parabolic area OAB is equal to two-thirds of the rectangle OABN.

Ex. 2. To find the area of a quadrant of a circle whose equation is $x^2+y^2 = a^2$.

$$\text{the area of the quadrant} = \int_0^a \int_0^{(a^2-x^2)^{\frac{1}{2}}} dy\, dx$$

$$= \int_0^a (a^2-x^2)^{\frac{1}{2}}\, dx$$

$$= \left[\frac{x\,(a^2-x^2)^{\frac{1}{2}}}{2} + \frac{a^2}{2}\sin^{-1}\frac{x}{a}\right]_0^a$$

$$= \frac{\pi a^2}{4};$$

$\therefore$ the area of the circle $= \pi a^2$.

Hence also, see fig. 18, if OC $= x_0$ and CB $= y_0$,

$$\text{the area of BCA} = \int_{x_0}^a \int_0^{(a^2-x^2)^{\frac{1}{2}}} dy\, dx$$

$$= \int_{x_0}^a (a^2-x^2)^{\frac{1}{2}}\, dx$$

$$= \left[\frac{x\,(a^2-x^2)^{\frac{1}{2}}}{2} + \frac{a^2}{2}\sin^{-1}\frac{x}{a}\right]_{x_0}^a$$

$$= \frac{a^2}{2}\cos^{-1}\frac{x_0}{a} - \frac{x_0 y_0}{2}.$$

This result is also evident geometrically; for the area of the sector BOA $= \dfrac{a^2}{2}\cos^{-1}\dfrac{x_0}{a}$, and the area of the triangle OBC $= \dfrac{x_0 y_0}{2}$.

Ex. 3. To find the whole area of the ellipse whose equation is $y = \dfrac{b}{a}(a^2-x^2)^{\frac{1}{2}}$.

$$\text{the area of the ellipse} = 4\int_0^a \int_0^{\frac{b}{a}(a^2-x^2)^{\frac{1}{2}}} dy\, dx$$

$$= 4\int_0^a \frac{b}{a}(a^2-x^2)^{\frac{1}{2}}\, dx = \pi ab.$$

Ex. 4. To determine the whole area included between the curve and the asymptote of the cissoid of Diocles; see fig. 19.

$$\because\quad y^2 = \frac{x^3}{2a-x},$$

$$y = \frac{x^2}{(2ax-x^2)^{\frac{1}{2}}};$$

for this value of y it is convenient to have a specific symbol, and we shall denote it by Y; so that it may be distinguished from the y which is the ordinate to the area-element. Hence, as OA $= 2a$,

$$\begin{aligned} \text{the area} &= 2\int_0^{2a}\int_0^{Y} dy\,dx \\ &= 2\int_0^{2a} \frac{x^2\,dx}{(2ax-x^2)^{\frac{1}{2}}}, \text{ see Ex. 1, Art. 51,} \\ &= 2\left[-\frac{x+3a}{2}(2ax-x^2)^{\frac{1}{2}} + \frac{3a^2}{2}\text{versin}^{-1}\frac{x}{a}\right]_0^{2a} \\ &= 3\pi a^2. \end{aligned}$$

Thus the whole area = three times the area of the base-circle.

Ex. 5. To find the whole area of the cycloid.

Let the vertex be the origin; see fig. 5; then the equation to the curve is

$$y = a\,\text{versin}^{-1}\frac{x}{a} + (2ax-x^2)^{\frac{1}{2}};$$

which expression, as the limit of the definite integral, we shall denote by Y: then

$$\begin{aligned} \text{the area} &= 2\ \text{OABP} \\ &= 2\int_0^{2a}\int_0^{Y} dy\,dx \\ &= 2\int_0^{2a}\left\{a\,\text{versin}^{-1}\frac{x}{a} + (2ax-x^2)^{\frac{1}{2}}\right\}dx \\ &= 2\left[x\left\{a\,\text{versin}^{-1}\frac{x}{a} + (2ax-x^2)^{\frac{1}{2}}\right\} - \int (2ax-x^2)^{\frac{1}{2}}\,dx\right]_0^{2a} \\ &= \left[(x+a)(2ax-x^2)^{\frac{1}{2}} + a(2x-a)\,\text{versin}^{-1}\frac{x}{a}\right]_0^{2a} \\ &= 3\pi a^2. \end{aligned}$$

Thus the area = three times the area of the generating circle.

The value of the indefinite integral shews that if $x = \frac{a}{2}$, the area of the segment of the cycloid does not involve the length of a circular arc, or any circular transcendent. Hence if, in fig. 20, OM $= \frac{a}{2}$, the area of POM $= \frac{3^{\frac{3}{2}}a^2}{8} =$ the triangle QMA.

Hence also, if $x = a$, OPSC $= a^2 + \frac{\pi a^2}{4}$, CS $= a + \frac{\pi a}{2}$; therefore OQDSP $= a^2$; and the segment OPS $= \frac{a^2}{2}$, and does not involve any circular transcendent.

Also the preceding expression gives

$$\text{the area OMP} = xy - \int_0^x (2ax - x^2)^{\frac{1}{2}} dx$$

$$= \text{MN} - \text{OMQ};$$

$$\therefore \quad \text{ONP} = \text{the circular area OMQ}.$$

Ex. 6. To find the area included between the tractrix, the axis of y, and the asymptote.

The differential equation to the curve is

$$\frac{dy}{dx} = -\frac{y}{(a^2 - y^2)^{\frac{1}{2}}}.$$

Then, fig. 2, taking y to be the general value of the ordinate to the curve,

$$\text{the whole area} = \int_0^\infty \int_0^y dy\, dx$$

$$= \int_0^\infty y\, dx;$$

but $y\, dx = -dy(a^2 - y^2)^{\frac{1}{2}}$; and when $x = \infty$, $y = 0$; $x = 0$, $y = a$;

$$\therefore \quad \text{the whole area} = -\int_a^0 (a^2 - y^2)^{\frac{1}{2}} dy$$

$$= -\left[\frac{y}{2}(a^2 - y^2)^{\frac{1}{2}} + \frac{a^2}{2}\sin^{-1}\frac{y}{a}\right]_a^0 = \frac{\pi a^2}{4}.$$

Ex. 7. The whole area contained between the asymptote and the witch of Agnesi is four times the area of the base-circle.

Ex. 8. If the equation to the hyperbola is $\frac{x^2}{a^2} - \frac{y^2}{b^2} = 1$, the area included between an ordinate, the axis of x, and the curve, is

$$\frac{x_n y_n}{2} - \frac{ab}{2}\log\left\{\frac{x_n}{a} + \frac{y_n}{b}\right\}.$$

Ex. 9. If the equation to the rectangular hyperbola is $xy = k^2$, the area included between two ordinates, the axis of x, and the curve, is

$$k^2 \log\left(\frac{x_n}{x_0}\right).$$

Thus the area is expressed in terms of the Napierian logarithms of the abscissæ. Conversely Napierian logarithms are functions of the area, and for this reason they are called Hyperbolic Logarithms.

Ex. 10. The whole area of the companion to the cycloid is twice that of the generating circle.

Hence the area of the cycloid is trisected by the base circle on its axis, and the companion to the cycloid; see fig. 20.

Ex. 11. The whole area of the loop of the curve whose equation is $ay^2 = x^2(a^2-x^2)^{\frac{1}{2}}$, included between $x = a$ and $x = 0$, $= \frac{4a^2}{5}$.

Ex. 12. The area included between the axis of x, two ordinates, and the logarithmic curve $y = a^x$, is

$$\frac{a^{x_n} - a^{x_0}}{\log a};$$

and that included between the curve, the asymptote, and the axis of y is $\frac{1}{\log a}$, since $x_0 = -\infty$, $x_n = 0$.

Ex. 13. The area included between the axis of x and the curve $y = a \sin \frac{x}{a}$, for the limits $x = 0$ and $x = \pi a$, is $2a^2$.

Ex. 14. The area OAPM of the catenary in fig. 7 is equal to the rectangle contained by OA and the arc AP, and therefore is equal to twice the triangle PHM.

Ex. 15. If $\int_0^x \int_0^y dy\, dx = ay$; it is required to find the equation of the curve.

$$\begin{aligned} ay &= \int_0^x \int_0^y dy\, dx \\ &= \int_0^x y\, dx; \\ \therefore \quad a\, dy &= y\, dx; \\ y &= be^{\frac{x}{a}}. \end{aligned}$$

Ex. 16. If $\int_0^x \int_0^y dy\, dx = \frac{m}{m+n} xy$, shew that $\left(\frac{x}{a}\right)^m = \left(\frac{y}{b}\right)^n$.

221.] In all the preceding examples, the y-integration has preceded the x-integration, and we have by this process first determined the general value of a differential slice of infinitesimal breadth dx, contained between parallel ordinates, and by the summation of these determined the required area. The order of integration however, as we remarked in the preceding chapter, is indifferent; though if the order is changed, of course the limits must be changed: this we shall exemplify in a few cases.

Ex. 1. In determining the area OAB, fig. 21, where OA $= a$, AB$=b$, and the equation to the bounding curve is $ay^2 = b^2x$; if we first effect the x-integration, y being the same for all, we sum the elements along the line PK, that is, from $x = a\frac{y^2}{b^2}$ to $x = a$; and thereby obtain the area of the slice PQLK contained between two parallel abscissæ separated by the distance dy; which slices must again be summed with respect to y, the limits of integration being b and 0. Hence

$$\text{the area OAB} = \int_0^b \int_{\frac{a}{b^2}y^2}^a dx\, dy$$

$$= \int_0^b \left(a - \frac{a}{b^2}y^2\right) dy$$

$$= \left[ay - \frac{a}{3b^2}y^3\right]_0^b$$

$$= ab - \frac{ab}{3} = \frac{2ab}{3};$$

the same result as that of Ex. 1, Art. 220.

Ex. 2. The equation to the equitangential curve being

$$x = a\log\frac{a+(a^2-y^2)^{\frac{1}{2}}}{y} - (a^2-y^2)^{\frac{1}{2}},$$

and X being the general value of the abscissa to the curve, the area included between the curve, the axis of x, and the axis of y

$$= \int_0^a \int_0^{X} dx\, dy$$

$$= \int_0^a \left\{a\log\frac{a+(a^2-y^2)^{\frac{1}{2}}}{y} - (a^2-y^2)^{\frac{1}{2}}\right\} dy$$

$$= \left[y\left\{a\log\frac{a+(a^2-y^2)^{\frac{1}{2}}}{y} - (a^2-y^2)^{\frac{1}{2}}\right\} + \int (a^2-y^2)^{\frac{1}{2}} dy\right]_0^a$$

$$= \frac{\pi a^2}{4};$$

the same result as that of Ex. 6, Art. 220.

222.] If it is required to determine the area contained between two ordinates corresponding to x_n, x_0, and between two curves whose equations are $y = f(x)$, $y = \phi(x)$, the former being the equation to the upper, and the latter that to the lower curve; then, as is evident from fig. 22, the y-integration must be first effected, and for the limits $f(x)$ and $\phi(x)$; the result of which will give the area of the slice PP′Q′Q; and the subsequent definite x-integration will give the sum of all such slices between the assigned limits; and this will be the required area. Thus

$$\text{the area} = \int_{x_0}^{x_n}\int_{\phi(x)}^{f(x)} dy\,dx. \tag{5}$$

If however the superficies, whose area is required, is of a form such as that delineated in fig. 23, it is more convenient to resolve it into slices whose bounding lines are parallel to the axis of x, that is, first to effect the x-integration, for in such a case the equation to the curves will give the limits of integration: the equation to the curve AQPB giving the superior and that to AQ′P′B giving the inferior limit; which manifestly they do not, if the y-integration is first performed; in this case, if the equations to the curves are

$$x = f(y), \qquad x = \phi(y),$$

and if the ordinates to A and B are y_n and y_0,

$$\text{the area} = \int_{y_0}^{y_n}\int_{\phi(y)}^{f(y)} dx\,dy. \tag{6}$$

Sometimes also it is necessary to divide a problem of quadrature into two or more parts, and to integrate each of the double integrals in the order which is most convenient for its form and limits: such division however must be left to the ingenuity of the student, for no general rules can be given, but the principles of the calculus are of sufficient breadth to include all such cases.

223.] Examples illustrative of the preceding principles.

Ex. 1. To determine the area included between the parabola whose equation is $y^2 = 4ax$, and the straight line whose equation is $y = \beta x$; see fig. 24.

The coordinates to the point B, determined by elimination between the given equations, are $\text{OA} = \frac{4a}{\beta^2}$, $\text{AB} = \frac{4a}{\beta}$; therefore

$$\begin{aligned}\text{the area OPB} &= \int_0^{\frac{4a}{\beta^2}}\int_{\beta x}^{2(ax)^{\frac{1}{2}}} dy\,dx \\ &= \int_0^{\frac{4a}{\beta^2}} \{2(ax)^{\frac{1}{2}} - \beta x\}\,dx \\ &= \left[\frac{4(a)^{\frac{1}{2}}}{3}x^{\frac{3}{2}} - \frac{\beta x^2}{2}\right]_0^{\frac{4a}{\beta^2}} = \frac{8a^2}{3\beta^3}.\end{aligned}$$

Ex. 2. To find the area contained between an hyperbola, its transverse axis, and a central radius vector; fig. 25.

Let P_n to which the radius vector is drawn be (x_n, y_n); and let the limits of x be denoted by X_n and X_0; then the equation to OP is

$$x = \frac{x_n}{y_n} y = X_0;$$

and as the equation to the hyperbola is $\frac{x^2}{a^2} - \frac{y^2}{b^2} = 1$,

$$x = \frac{a}{b}(b^2+y^2)^{\frac{1}{2}} = X_n;$$

the area $OAP_n = \int_0^{y_n}\int_{X_0}^{X_n} dx\,dy$

$$= \int_0^{y_n}\left\{\frac{a}{b}(b^2+y^2)^{\frac{1}{2}} - \frac{x_n}{y_n}y\right\}dy$$

$$= \left[\frac{a}{b}\left(\frac{y(b^2+y^2)^{\frac{1}{2}}}{2} + \frac{b^2}{2}\log\{y+(b^2+y^2)^{\frac{1}{2}}\}\right) - \frac{x_n}{y_n}\frac{y^2}{2}\right]_0^{y_n}$$

$$= \frac{ay_n(b^2+y_n^2)^{\frac{1}{2}}}{2b} + \frac{ab}{2}\log\frac{y_n+(b^2+y_n^2)^{\frac{1}{2}}}{b} - \frac{x_n y_n}{2}$$

$$= \frac{ab}{2}\log\left\{\frac{x_n}{a} + \frac{y_n}{b}\right\}.$$

The order in which the integrations have been effected, in relation to the limits, deserves attention; as the superficies P_nOA admits of being resolved into slices by lines parallel to the axis of x, the limits of which are given by the equations to the straight line and the curve, we have effected first the x-integration, and subsequently the y-integration; but the order could have been reversed, only subject to other conditions: viz. if we had integrated first with respect to y, the limits would have been the ordinate to the straight line and zero, for all values of x from O to A, but at A, and thence on to P_n, the superior and inferior limits would have been respectively the ordinate to the straight line and the ordinate to the hyperbola; and the definite integral must have been divided into two parts corresponding to these limits.

Ex. 3. AP being the catenary, in fig. 7, whose equation is

$$y = \frac{a}{2}\left\{e^{\frac{x}{a}} + e^{-\frac{x}{a}}\right\},$$

and OA $= a$, ON $= \frac{5a}{4}$, the area APN $= \frac{a^2}{4}\{5\log 2 - 3\}$.

Ex. 4. The area included between a parabola whose equation is $y^2 = 4ax$, and a straight line through the focus inclined at 45° to the axis of x, is $\frac{16}{3} a^2 2^{\frac{1}{2}}$.

Ex. 5. The equation to a curve being $(y-x)^2 = a^2 - x^2$, the area $= \pi a^2$.

224.] The quadrature of an area is frequently facilitated by a substitution, and chiefly by putting the equation to the bounding curve into simultaneous equations by the introduction of a subsidiary angle, according to the method of Art. 193, Vol. I. The following are examples of the process.

Ex. 1. The equations to an ellipse are $x = a\cos\phi$, $y = b\sin\phi$; find the whole area.

Let $b\sin\phi$, which is the ordinate of the ellipse, be denoted by Y: then

$$\text{the area} = \int_{-a}^{a}\int_{-\text{Y}}^{\text{Y}} dy\,dx = 2\int_{-a}^{a} b\sin\phi\,dx.$$

Now $dx = -a\sin\phi\,d\phi$; and when $x = a$, $\phi = 0$; when $x = -a$, $\phi = \pi$. Consequently

$$\text{the area} = 2ab\int_{0}^{\pi}(\sin\phi)^2\,d\phi = \pi ab.$$

Ex. 2. The equations to the cycloid, fig. 6, are $x = a(\theta - \sin\theta)$, $y = a(1-\cos\theta)$. Then, using the notation of the last example,

$$\text{the area} = \int_{0}^{2\pi a}\int_{0}^{\text{Y}} dy\,dx = \int_{0}^{2\pi a} a(1-\cos\theta)\,dx$$

$$= a^2\int_{0}^{2\pi}(1-\cos\theta)^2\,d\theta = 3\pi a^2.$$

Ex. 3. The equations to the companion to the cycloid being $x = a(1-\cos\theta)$, $y = a\theta$, the area $= 2\pi a^2$.

Ex. 4. The equation to the hypocycloid is $x^{\frac{2}{3}} + y^{\frac{2}{3}} = a^{\frac{2}{3}}$. Find the area enclosed by the curve: see fig. 10.

Let $x = a(\cos\phi)^3$, $y = a(\sin\phi)^3$; then

$$\text{the area} = 4\int_{0}^{a}\int_{0}^{\text{Y}} dy\,dx$$

$$= 12a^2\int_{0}^{\frac{\pi}{2}}(\cos\phi)^2(\sin\phi)^4\,d\phi$$

$$= \frac{3\pi a^2}{8}.$$

Ex. 5. The equation to the evolute of the ellipse is

$$(ax)^{\frac{2}{3}} + (by)^{\frac{2}{3}} = (a^2-b^2)^{\frac{2}{3}};$$

find the area enclosed by the curve.

Let $ax = (a^2-b^2)(\cos\phi)^3$, $by = (a^2-b^2)(\sin\phi)^3$; then

$$\text{the area} = 12\frac{(a^2-b^2)^2}{ab}\int_0^{\frac{\pi}{2}}(\cos\phi)^2(\sin\phi)^4\,d\phi$$

$$= \frac{3\pi(a^2-b^2)^2}{8ab}.$$

225.] It is sometimes convenient to refer the area, which is to be determined, and its bounding curve, to a system of oblique coordinate axes; say, to a system whose angle of ordination is ω; then the area of the surface-element is $dy\,dx\sin\omega$, and if A represents the required area,

$$\text{A} = \iint \sin\omega\,dy\,dx; \tag{7}$$

the integral being definite, and the limits being given by the circumstances of the problem.

Ex. 1. The sides of a triangle are $x=0$, $y=0$, $\frac{x}{a}+\frac{y}{b}=1$, and the angle of ordination is ω: find the area of the triangle.

$$\text{Let}\quad \frac{b}{a}(a-x) = \text{Y},\quad\text{then}$$

$$\text{the area} = \int_0^a\int_0^{\text{Y}}\sin\omega\,dy\,dx = \frac{ab\sin\omega}{2}.$$

Ex. 2. The equation to an ellipse referred to a system of oblique axes, whose angle of ordination is ω, is $\frac{x^2}{a_1^2}+\frac{y^2}{b_1^2}=1$; find the area of the ellipse.

$$\text{Let}\quad \text{Y} = \frac{b_1}{a_1}(a_1^2-x^2)^{\frac{1}{2}};\ \text{then}$$

$$\text{the area} = \int_{-a_1}^{a_1}\int_{-\text{Y}}^{\text{Y}}\sin\omega\,dy\,dx$$

$$= \pi a_1 b_1 \sin\omega.$$

On comparing this with the value found in Ex. 3, Art. 220, it appears that $a_1 b_1 \sin\omega = ab$.

Ex. 3. Find the area of a parallelogram whose sides are $2a$ and $2b$, and which are inclined to each other at an angle ω;

$$\text{the area} = \int_{-a}^{a}\int_{-b}^{b}\sin\omega\,dy\,dx = 4ab\sin\omega.$$

Section 2.—*Quadrature of Plane Surfaces.—Polar Coordinates.*

226.] When the bounding curve of the surface, whose area is required, is referred to polar coordinates, it is obviously convenient to refer also to the same system of coordinates every point in the surface, and to express in terms of them the surface-element which abuts at that point. Now this latter expression may be determined by the method of transformation of definite integrals, which has been explained in the preceding chapter; and indeed we have therein proved, see (32), Art. 213, that if $x = r\cos\theta$, $y = r\sin\theta$, $dx\,dy = r\,dr\,d\theta$; and consequently if A denotes the required area,

$$\text{A} = \int\!\!\int r\,dr\,d\theta; \tag{8}$$

this integral being of course definite, and the limits being given by the circumstances of the problem.

It is desirable however also to investigate the preceding expression on independent geometrical principles.

Now the problem is to find the area of the plane surface contained between a plane curve and two radii vectores separated by a finite angle; see fig. 26.

Let AP_0PQP_n be the curve whose equation referred to polar coordinates is

$$r = f(\theta); \tag{9}$$

and let it be required to determine the area of P_0SP_n.

Let $\text{SP}_0 = r_0$, $\text{SP}_n = r_n$, $\text{P}_0\text{SA} = \theta_0$, $\text{P}_n\text{SA} = \theta_n$; and let E be any point within the bounding lines; draw through E the radius vector SEP, and also a consecutive one inclined to SP at an infinitesimal angle $d\theta$; from S as centre and with SE as radius draw the small circular arc EG, and also another arc at an infinitesimal distance from it: then, if the polar coordinates to E are r and θ, $\text{EF} = dr$, $\text{EG} = r\,d\theta$, and the area-element $= r\,dr\,d\theta$, the element being ultimately an infinitesimal rectangle; thus the integral of $r\,dr\,d\theta$, with respect to r, between the limits 0 and $f(\theta)$, will give the area of the triangular slice SPQ, and the integral of all such triangular slices between θ_0 and θ_n will give the required area. Consequently we have the following double integral;

$$\text{the area } \text{P}_n\text{SP}_0 = \int_{\theta_0}^{\theta_n}\int_0^{f(\theta)} r\,dr\,d\theta; \tag{10}$$

which is the same as (8).

If the r-integration is effected first, the superior limit being $f(\theta)$ or the radius vector of the curve, and the inferior limit being 0,

$$\text{the area } P_n S P_0 = \frac{1}{2}\int_{\theta_0}^{\theta_n} \{f(\theta)\}^2\, d\theta; \tag{11}$$

and replacing $f(\theta)$ by its value r given in (9), r referring to the curve,

$$\text{the area } S P_0 P_n = \frac{1}{2}\int_{\theta_0}^{\theta_n} r^2\, d\theta. \tag{12}$$

The geometrical meaning of which is that the r-integration gives the area of the sectorial slice SPQ, which is manifestly equal to $\frac{1}{2} r^2\, d\theta$; and the whole required area is equal to the integral of this quantity.

And let it not be supposed that any inaccuracy of result arises from the fact that the element of the area is not rectangular at the superior limit of the r-integration, that is, at the point P: for if two infinitesimal arcs RP, QT are described from S as a centre with radii SP and SQ respectively, then, if $SP = r = f(\theta)$, $SQ = r+dr = f(\theta+d\theta)$, the area SPQ is intermediate to SPR and SQT; that is, is intermediate to $\frac{r^2\, d\theta}{2}$ and $\frac{(r+dr)^2\, d\theta}{2}$, the difference between which is an infinitesimal of the second order, and must therefore be neglected. Hence $\frac{r^2\, d\theta}{2}$ is the correct expression for the infinitesimal sectorial area.

227.] Examples illustrative of the preceding.

Ex. 1. To find the area of a sector of a circle. See fig. 27.

Let the radius of the circle $= a$, and the arc AB subtend at the centre an angle a; then

$$\begin{aligned}\text{the area BSA} &= \int_0^{a}\int_0^{a} r\, dr\, d\theta \\ &= \int_0^{a} \frac{a^2}{2}\, d\theta = \frac{a^2 a}{2} = \frac{\text{SA} \times \text{arc AB}}{2}.\end{aligned}$$

Ex. 2. To find the area of a portion of a circle cut off by equal chords drawn through a point in its circumference. See fig. 28.

Let the radius of circle $= a$; and let $\text{BSA} = \text{B}'\text{SA} = a$; the equation to the circle is $r = 2a\cos\theta$.

$$\begin{aligned}\therefore \quad \text{the area BSB}' &= 2 \times \text{the area BSA} \\ &= 2\int_0^{a}\int_0^{2a\cos\theta} r\, dr\, d\theta \\ &= 4a^2 \int^{a} (\cos\theta)^2\, d\theta \\ &= 2a^2\{a + \sin a \cos a\}.\end{aligned}$$

Ex. 3. To find the area of a loop of the lemniscata whose equation is $r^2 = a^2 \cos 2\theta$.

$$\text{The area of a loop} = 2\int_0^{\frac{\pi}{4}}\int_0^{a(\cos 2\theta)^{\frac{1}{2}}} r\,dr\,d\theta$$

$$= a^2\int_0^{\frac{\pi}{4}} \cos 2\theta\,d\theta = \frac{a^2}{2}.$$

Ex. 4. To find the area of the loop of the folium of Descartes; see fig. 63, Art. 260, Ex. 10, Vol. I.

As the equation referred to rectangular coordinates is

$$x^3 - 3axy + y^3 = 0;$$

$$\therefore \quad r = \frac{3a\sin\theta\cos\theta}{(\sin\theta)^3 + (\cos\theta)^3} = \frac{3a\tan\theta\sec\theta}{1+(\tan\theta)^3};$$

let this value of r, which is the superior limit of the first integration, be represented by r; then

$$\text{the area} = 2\int_0^{\frac{\pi}{2}}\int_0^{\mathrm{r}} r\,dr\,d\theta$$

$$= 9a^2\int_0^{\frac{\pi}{2}} \frac{(\tan\theta)^2(\sec\theta)^2\,d\theta}{\{1+(\tan\theta)^3\}^2}$$

$$= -3a^2\left[\frac{1}{1+(\tan\theta)^3}\right]_0^{\frac{\pi}{4}}$$

$$= -3a^2\left\{\frac{1}{2} - 1\right\} = \frac{3a^2}{2}.$$

Ex. 5. If the equation of the cardioid is $r = a(1+\cos\theta)$, the whole area $= \dfrac{3\pi a^2}{2}$.

Ex. 6. If the equation to a curve is $r = a\sin 3\theta$, the area of each loop is $\dfrac{\pi a^2}{12}$; and the area of all the loops $= \dfrac{\pi a^2}{4}$.

Ex. 7. If the latus rectum of a parabola $= 4a$, the area contained between two focal radii inclined at θ_n and θ_0 to the least distance $=$

$$a^2\left\{\tan\frac{\theta_n}{2} - \tan\frac{\theta_0}{2} + \frac{1}{3}\left(\tan\frac{\theta_n}{2}\right)^3 - \frac{1}{3}\left(\tan\frac{\theta_0}{2}\right)^3\right\}.$$

Ex. 8. If $r^2 = a^2(\cos\theta)^2 + b^2(\sin\theta)^2$, the whole area

$$= \frac{\pi(a^2+b^2)}{2}.$$

Ex. 9. If $x^4 + y^4 - a^2xy = 0$, the area of a loop $= \dfrac{\pi a^2}{8}$.

228.] In all the preceding examples the r-integration has preceded the θ-integration; the effect of which order has been that the area is resolved into triangular elements with a common vertex at the polé S; and the sum of these is determined by the θ-integration. Now the areas, which are ordinarily subjects of investigation, admit of this resolution: but if the θ-integration had been first effected, r being constant, it would have determined the area of a circular annulus, the radii to whose bounding circles would have been respectively r and $r+dr$, and the subsequent r-integration would have given the sum of all similar annuli; but the areas, which are commonly the subjects of these processes, do not conveniently admit of such a resolution, and the equations of the bounding curves do not commonly give convenient values of limits; and therefore, although theoretically the order of integration is indifferent, yet we choose that which is practically most convenient, and make the r-integration precede the θ-integration. The circle, I would observe, when the centre is the pole, is adapted to both orders with the same facility, because the limits of the two integrations are constant.

229.] We proceed to the investigation of areas whose limits are of a more complex character than those considered above.

Ex. 1. To find the area of a circular annulus, the radii of whose exterior and interior bounding circles are a and b.

$$\text{The area of the annulus} = \int_0^{2\pi}\int_b^a r\,dr\,d\theta$$

$$= \frac{1}{2}\int_0^{2\pi}(a^2-b^2)\,d\theta = \pi(a^2-b^2).$$

Ex. 2. To find the area contained between the conchoid of Nicomedes, its asymptote, and two given radii vectores; fig. 29.

Let SA $= a$, AB $=$ PQ $= b$; SP $= r$, BSP $= \theta$; therefore the equation to the curve is $r = a\sec\theta + b$; also SQ $= a\sec\theta$.

Let θ_n and θ_0 be the superior and inferior limits of θ; then

$$\text{the area} = \int_{\theta_0}^{\theta_n}\int_{a\sec\theta}^{a\sec\theta+b} r\,dr\,d\theta$$

$$= \frac{1}{2}\int_{\theta_0}^{\theta}(2ab\sec\theta + b^2)\,d\theta$$

$$= \Big[ab\log\tan\Big(\frac{\pi}{4}+\frac{\theta}{2}\Big) + \frac{b^2\theta}{2}\Big]_{\theta_0}^{\theta_n}$$

$$= ab\Big\{\log\tan\Big(\frac{\pi}{4}+\frac{\theta_n}{2}\Big) - \log\tan\Big(\frac{\pi}{4}+\frac{\theta_0}{2}\Big)\Big\} + \frac{b^2}{2}(\theta_n-\theta_0).$$

Ex. 3. To determine the area contained between two successive convolutions of the spiral of Archimedes; fig. 30.

Let the general form of the equation to the spiral be $r = a\phi$, ϕ being the whole angle through which the radius vector has revolved; and let SA, SB, SC, severally be the values of the radius vector after $n-1$, n, and $n+1$ complete revolutions, so that $\text{SA} = 2(n-1)\pi a$, $\text{SB} = 2n\pi a$, $\text{SC} = 2(n+1)\pi a$; let $\text{PSA} = \theta$; therefore $\text{SP} = \{2(n-1)\pi+\theta\}a$, $\text{SP}_1 = \{2n\pi+\theta\}a$, which values it is convenient to represent by r and r_1; the problem is to determine the area of $\text{APB}_1\,\text{BP}_1\,\text{C}_1\,\text{C}$, which is expressed by the following definite integral:

$$\begin{aligned}
\text{the area} &= \int_0^{2\pi}\int_{\text{r}}^{\text{r}_1} r\,dr\,d\theta \\
&= \frac{a^2}{2}\int_0^{2\pi}\{(2n\pi+\theta)^2-(2(n-1)\pi+\theta)^2\}d\theta \\
&= \frac{a^2}{6}\Big[(2n\pi+\theta)^3-(2(n-1)\pi+\theta)^3\Big]_0^{2\pi} \\
&= \frac{a^2}{6}\{(n+1)^3-2n^3+(n-1)^3\}\,8\pi^3 \\
&= 8n\pi^3a^2;
\end{aligned}$$

therefore the area generated in the first revolution of the radius vector is $8\pi^3a^2$; and hence that generated in the nth revolution is n times that generated in the first.

230.] If the equation to the curve is given in terms of r and p, instead of finding the equivalent expression in terms of r and θ, and then integrating as in the preceding articles, it is more convenient to pursue the following course:

Let $r\,dr\,d\theta$ be integrated first in respect of r; and supposing the limits of r to be the radius vector of the curve and 0, we have, θ and θ_0 being the limits of θ,

$$\text{the area} = \frac{1}{2}\int_{\theta_0}^{\theta} r^2\,d\theta.$$

But by (24), Art. 271, Vol. I, $r^2\,d\theta = \dfrac{rp\,dr}{(r^2-p^2)^{\frac{1}{2}}}$;

$$\therefore\quad \text{the area} = \frac{1}{2}\int_{r_0}^{r}\frac{rp\,dr}{(r^2-p^2)^{\frac{1}{2}}}; \tag{13}$$

where r and r_0 are the radii vectores of the bounding curve corresponding to θ and θ_0 respectively; in which expression p must be replaced by its value in terms of r, and the r-integration then effected.

Ex. 1. To find the area contained between the involute of the circle and two limiting radii vectores; see fig. 31.

The equation to the curve is $r^2 - p^2 = a^2$.

$$\therefore \quad \text{The area ASP} = \int_0^{\theta}\int_0^{r} r\,dr\,d\theta = \frac{1}{2}\int_0^{\theta} r^2\,d\theta$$

$$= \frac{1}{2}\int_a^{r} \frac{r}{a}(r^2-a^2)^{\frac{1}{2}}\,dr = \frac{(r^2-a^2)^{\frac{3}{2}}}{6a} = \frac{p^3}{6a}.$$

Ex. 2. To find the area contained between an epicycloid and its base-circle during one revolution of the generating circle; see fig. 42, Vol. I.

By (9), Art. 268, Vol. I, the equation to the curve is

$$p = \frac{(a+2b)^2}{4b(a+b)}(r^2-a^2);$$

therefore the area contained between the pole and the curve

$$= \int_a^{a+2b} \frac{pr\,dr}{(r^2-p^2)^{\frac{1}{2}}}$$

$$= \frac{a+2b}{a}\int_a^{a+2b} \frac{r(r^2-a^2)^{\frac{1}{2}}\,dr}{\{(a+2b)^2-r^2\}^{\frac{1}{2}}}$$

$$= \frac{a+2b}{a}\int_a^{a+2b} \frac{r(r^2-a^2)^{\frac{1}{2}}\,dr}{\{4b(a+b)-(r^2-a^2)\}^{\frac{1}{2}}}$$

$$= \frac{\pi b}{a}(a+b)(a+2b);$$

and as the area of the circular sector which is included in the above expression is πab, the area included between the circle and the epicycloid $= \dfrac{\pi b^2}{a}(3a+2b)$.

In reference to the expression for the area which is given in (12) it is to be observed, that the triangle OSP, which is expressed by $\dfrac{r^2\,d\theta}{2}$, is the θ-differential of the area. But the area of this small triangle, see fig. 36, $= \dfrac{1}{2}\text{SY} \times \text{PQ} = \dfrac{p\,ds}{2}$. Hence $r^2\,d\theta = p\,ds$; and if A is a sectorial area,

$$\text{A} = \frac{1}{2}\int p\,ds; \qquad (14)$$

which is another expression for the area.

231.] The method of the present section is also applicable to the following problem:

To find the area contained between a curve, its evolute, and any two limiting radii of curvature.

In fig. 32 let OPQB be the plane curve on which P and Q are two consecutive points, P being (x, y), and PQ being an infinitesimal arc, and therefore equal to ds; let PΠ be the radius of curvature at P, and be denoted by ρ: then the area of the infinitesimal triangle PΠQ is equal to

$$\frac{\text{PQ} \times \text{P}\Pi}{2} = \frac{\rho\, ds}{2}; \tag{15}$$

and as the required area is the sum of all these, we have

$$\text{the area} = \frac{1}{2}\int \rho\, ds, \tag{16}$$

in which ρ and ds must be expressed in terms of a single variable, the limits of integration being assigned by the conditions of the problem.

Ex. 1. To determine the area contained between the parabola, whose equation is $y^2 = 4ax$, its evolute, the radius of curvature at the vertex, and any other radius of curvature.

Here $\qquad \rho = \dfrac{2}{a^{\frac{1}{2}}}(a+x)^{\frac{3}{2}}, \qquad ds = \left(\dfrac{a+x}{x}\right)^{\frac{1}{2}} dx;$

$$\therefore \quad \text{the area} = \frac{1}{a^{\frac{1}{2}}}\int_0^x \frac{(a+x)^2}{x^{\frac{1}{2}}}\, dx$$

$$= \frac{1}{a^{\frac{1}{2}}}\left\{2\, a^2 x^{\frac{1}{2}} + \frac{4}{3}\, a x^{\frac{3}{2}} + \frac{2}{5} x^{\frac{5}{2}}\right\};$$

and therefore the area contained between the curve, the evolute, and the radii of curvature at the vertex and at the extremity of the latus rectum is equal to $\dfrac{56}{15} a^2$.

Ex. 2. To find the area contained between the cycloid, its evolute, and two given radii of curvature.

In fig. 32, let O, the starting point, be the origin; then

$$x = a \operatorname{versin}^{-1}\frac{y}{a} - (2ay - y^2)^{\frac{1}{2}};$$

$$\therefore \quad \rho = 2\,(2ay)^{\frac{1}{2}}, \qquad ds = \left(\frac{2a}{2a-y}\right)^{\frac{1}{2}} dy;$$

$$\therefore \text{ the area beginning at O} = 2a\int_0^y \frac{y\, dy}{(2ay-y^2)^{\frac{1}{2}}}$$

$$= 2a\left\{-(2ay-y^2)^{\frac{1}{2}} + a \operatorname{versin}^{-1}\frac{y}{a}\right\};$$

and therefore the area OB′B $= 2\pi a^2$.

SECTION 3.—*The Quadrature of Surfaces of Revolution.*

232.] In fig. 33, let APQ be a plane curve, and suppose it to generate a surface of revolution by revolving about a line Ox in its own plane, A′P′Q′ being its position, when half a revolution has been performed; and let the equation to AP be $y = f(x)$; let OM $= x$, MP $= y$, PQ $= ds$; P and Q will, in a complete revolution, describe circles whose radii are respectively y and $y+dy$, and therefore the paths traversed severally by P and Q are $2\pi y$ and $2\pi(y+dy)$: supposing the curve to be continuous and to have no points of inflexion between P and Q, the element PQ will describe a circular band whose breadth is ds, and the circumferences of whose bounding circles are $2\pi y$ and $2\pi(y+dy)$; the area therefore of the convex surface of this band is intermediate to $2\pi y\,ds$ and $2\pi(y+dy)\,ds$; and neglecting the infinitesimal of the second order, the convex surface of the infinitesimal band is equal to $2\pi y\,ds$; and therefore, as it is an infinitesimal band-element of the surface,

$$\text{the surface} = \int 2\pi y\,ds, \tag{17}$$

the integral being of course definite, and the limits being given by the conditions of the problem.

As $y = f(x)$ is the equation to the generating plane curve, $dy = f'(x)\,dx$; and

$$\begin{aligned} ds &= (dx^2+dy^2)^{\frac{1}{2}} \\ &= \{1+(f'(x))^2\}^{\frac{1}{2}}\,dx; \end{aligned}$$

$$\therefore \quad \text{the surface} = 2\pi \int f(x)\,\{1+(f'(x))^2\}^{\frac{1}{2}}\,dx; \tag{18}$$

which is the form convenient in most cases; other processes will be explained in the sequel.

Ex. 1. To find the surface of a sphere.

The equation to the generating curve is $x^2+y^2 = a^2$, so that $y\,ds = a\,dx$.

$$\therefore \quad \text{the surface of the sphere} = 4\pi a \int_0^a dx = 4\pi a^2.$$

Hence also a zone of a sphere contained between two planes perpendicular to the axis and at distances x_n and x_0 from the centre is equal to

$$2\pi a \int_{x_0}^{x_n} dx = 2\pi a (x_n - x_0);$$

see Ex. 7, Art. 24, vol. I.

Ex. 2. To determine the surface of the paraboloid of revolution. Here let $y^2 = 4ax$; so that

$$ds = \left(\frac{a+x}{x}\right)^{\frac{1}{2}} dx.$$

$$\therefore \quad \text{the surface} = 4\pi a^{\frac{1}{2}} \int_0^x (a+x)^{\frac{1}{2}}\, dx$$

$$= \frac{8\pi a^{\frac{1}{2}}}{3}\left[(a+x)^{\frac{3}{2}}\right]_0^x = \frac{8\pi a^{\frac{1}{2}}}{3}\{(a+x)^{\frac{3}{2}} - a^{\frac{3}{2}}\}.$$

Ex. 3. To find the area of the surface described by the revolution of a cycloid about its base.

Here $x = a(\theta - \sin\theta)$, $y = a(1-\cos\theta)$; so that

$$ds = 2a\, d\theta \sin\frac{\theta}{2};$$

$$\text{the whole surface} = 2\pi \int y\, ds$$

$$= 8\pi a^2 \int_0^\pi (1-\cos\theta)\sin\frac{\theta}{2}\, d\theta$$

$$= 32\pi a^2 \int_0^\pi \left\{\left(\cos\frac{\theta}{2}\right)^2 - 1\right\} d.\cos\frac{\theta}{2}$$

$$= 32\pi a^2 \left[\frac{1}{3}\left(\cos\frac{\theta}{2}\right)^3 - \cos\frac{\theta}{2}\right]_0^\pi$$

$$= \frac{64}{3}\pi a^2.$$

Ex. 4. To determine the area of the surface described by the revolution of the tractrix about the axis of x.

The differential equation to the tractrix is

$$\frac{dy}{dx} = -\frac{y}{(a^2-y^2)^{\frac{1}{2}}};$$

$$\therefore \quad y\, ds = -a\, dy;$$

$$\therefore \quad \text{the whole surface} = \int 2\pi y\, ds$$

$$= -2\pi a \int_a^0 dy$$

$$= 2\pi a^2.$$

Ex. 5. The convex surface of a cone, whose generating line is $ay - bx = 0$, is $\pi b(a^2+b^2)^{\frac{1}{2}}$.

Ex. 6. Find the equation to the plane curve which by its revolution about the x-axis generates a surface, the area of which is proportional to (1) the extreme abscissa; (2) the extreme ordinate.

Ex. 7. The whole surface of a prolate spheroid, the equation to whose generating ellipse is

$$\frac{x^2}{a^2}+\frac{y^2}{b^2}=1,$$

and whose eccentricity is e, is $2\pi b^2+\frac{2\pi ab}{e}\sin^{-1}e$.

Ex. 8. The area of a surface generated by the revolution of a logarithmic curve, $y=e^x$, about the axis of x is equal to

$$\pi\{y(1+y^2)^{\frac{1}{2}}+\log(y+(1+y^2)^{\frac{1}{2}})\}.$$

Ex. 9. The whole area of the surface generated by the revolution of a cycloid about its axis is $8\pi a^2\left(\pi-\frac{4}{3}\right)$.

233.] If the line about which the generating plane curve revolves is the axis of y, then, see fig. 34, if $\text{OM}=x$, $\text{MP}=y$, $\text{PQ}=ds$, the convex surface of the band generated by PQ in one revolution is equal to $2\pi x\,ds$, and as this is an infinitesimal band-element of the required surface,

$$\text{the whole surface} = 2\pi\int x\,ds; \qquad (19)$$

the integral being definite, and the limits being assigned by the problem.

Ex. 1. To determine the surface of an oblate spheroid.

Let the equation to the revolving ellipse be

$$\frac{x^2}{a^2}+\frac{y^2}{b^2}=1,$$

and its eccentricity be e; then

$$x\,ds=\frac{(a^4y^2+b^4x^2)^{\frac{1}{2}}}{b^2}dy$$

$$=\frac{a^2e}{b^2}\left\{y^2+\frac{b^4}{a^2e^2}\right\}^{\frac{1}{2}}dy;$$

$$\therefore\ \text{the whole surface} = 4\pi\int_{y=0}^{y=b}x\,ds$$

$$=\frac{4\pi a^2e}{b^2}\int_0^b\left(y^2+\frac{b^4}{a^2e^2}\right)^{\frac{1}{2}}dy$$

$$=2\pi\left\{a^2+\frac{b^2}{2e}\log\frac{1+e}{1-e}\right\}.$$

Ex. 2. To determine the area of the surface generated by a given length of the catenary revolving about the axis of y, when the equation is

$$y=\frac{a}{2}\left\{e^{\frac{x}{a}}+e^{-\frac{x}{a}}\right\}.$$

Hence by Ex. 7, Art. 155, $s = \frac{a}{2}\left\{e^{\frac{x}{a}} - e^{-\frac{x}{a}}\right\}$;

$$\therefore \text{ the surface} = 2\pi\int_0^x x\,ds$$

$$= 2\pi\left[xs - \int s\,dx\right]_0^x$$

$$= 2\pi\left[xs - \frac{a^2}{2}\left(e^{\frac{x}{a}} + e^{-\frac{x}{a}}\right)\right]_0^x$$

$$= 2\pi\left\{xs - \frac{a^2}{2}\left(e^{\frac{x}{a}} + e^{-\frac{x}{a}} - 2\right)\right\}$$

$$= 2\pi\{xs - a(y-a)\}.$$

234.] If the curve, whose equation is $y = f(x)$, generates a surface by revolving about, not one of its axes of reference, but an axis parallel to, say, its axis of x, at a distance a from it, and in the plane of the curve, then the surface generated

$$= 2\pi\int_{x_0}^{x}\{a+f(x)\}\,ds \qquad (20)$$

$$= 2\pi a\int_{x_0}^{x} ds + 2\pi\int_{x_0}^{x} f(x)\,ds,$$

x and x_0 being the abscissæ corresponding to the extremities of the generating curve; and therefore if s is the length of the generating arc, and S′ is the area of the surface generated by the revolution of it about its axis of x,

the required surface $= 2\pi a s + \text{S}'$.

If the generating curve is a closed figure, such as that drawn in fig. 39, and capable of being divided into two equal and symmetrical parts by a line EBC which is its axis of x, then, if AB $= a$, and the equation to EPC is $y = f(x)$, EC being its axis of x, the surface generated by the revolution of EPC about Ox

$$= 2\pi\int_{x_0}^{x_n}\{a+f(x)\}\,ds;$$

and that generated by the revolution of EP′C about the same line

$$= 2\pi\int_{x_0}^{x_n}\{a-f(x)\}\,ds;$$

therefore the surface generated by the revolution of the closed figure EPCP′

$$= 4\pi a\int_{x_0}^{x_n} ds;$$

which is equal to $4\pi a s$; that is, to $2\pi a \times$ length of the generating curve; therefore the area of the surface generated by the revolution about an axis in its own plane of a closed curve which is symmetrical with respect to a line parallel to that about which it revolves, is equal to the product of the length of the curve and the path described by a point on the line of symmetry.

Hence if a circle of radius a revolves about an axis in its own plane at a distance c from its centre, the surface of the generated ring $= 4\pi^2 ac$.

235.] If the surface is generated by a curve referred to polar coordinates, the area of it may be determined as follows. Let the axis of revolution be the prime-radius, and from P which is (r, θ), see fig. 26, let PM be drawn at right angles to SX. Then $\text{PM} = r\sin\theta$; and the arc-element PQ, which $= ds$, will by its revolution about SX describe an infinitesimal band, whose breadth is ds, and whose circumference $= 2\pi r\sin\theta$. Consequently $2\pi r\sin\theta\, ds$ is the band-element of the surface, and

$$\text{the surface} = \int 2\pi r \sin\theta\, ds; \tag{21}$$

this integral is of course definite, and the limits are assigned by the geometrical conditions of the particular problem.

Ex. 1. Find the area of the surface of a spherical sector, 2α being the vertical angle of the sector, and a being the radius of the sphere.

$$\text{The surface} = \int_0^{\alpha} 2\pi a^2 \sin\theta\, d\theta$$
$$= 4\pi a^2 \left(\sin\frac{\alpha}{2}\right)^2.$$

Consequently, if $\alpha = \pi$, the whole surface of a sphere $= 4\pi a^2$.

Ex. 2. Find the area of the surface generated by the revolution of a loop of the lemniscata about its axis.

Since $r^2 = a^2\cos 2\theta$, $\quad r\sin\theta\, ds = a^2 \sin\theta\, d\theta$;

$$\therefore \quad \text{the surface} = 2\pi a^2 \int_0^{\frac{\pi}{4}} \sin\theta\, d$$
$$= \pi a^2 \{2 - 2^{\frac{1}{2}}\}.$$

Ex. 3. The area of the surface generated by the revolution of the cardioid, whose equation is $r = a(1+\cos\theta)$, is $\dfrac{32\pi a^2}{5}$.

Ex. 4. The area of a surface generated by the revolution of a loop of the lemniscata, whose equation is $r^2 = a^2\sin 2\theta$, is $2\pi a^2$.

SECTION 4.—*Quadrature of Curved Surfaces.*

236.] We now come to the problem of Quadrature in its most general form, only particular cases having been investigated in the preceding articles. It will be observed also that in the preceding section, the integrals, on which the quadrature depends have been single, whereas the quadrature of an area must involve a double integral: the revolution of the arc-element about the axis is however equivalent to one integration.

Let the equation to the surface, on which the area, whose quadrature is to be determined, lies, be

$$F(x, y, z) = 0; \tag{22}$$

and employing the same notation as in Art. 332, Vol. I., let

$$\left(\frac{dF}{dx}\right) = U, \qquad \left(\frac{dF}{dy}\right) = V, \qquad \left(\frac{dF}{dz}\right) = W; \tag{23}$$

$$U^2 + V^2 + W^2 = Q^2; \tag{24}$$

so that if α, β, γ are the direction angles of the normal at (x, y, z),

$$\cos\alpha = \frac{U}{Q}, \qquad \cos\beta = \frac{V}{Q}, \qquad \cos\gamma = \frac{W}{Q}. \tag{25}$$

Let P, fig. 35, be the point (x, y, z) on the surface. Through P let planes PSLN, PRJN be drawn parallel to the planes of (y, z), (x, z) respectively; and also let two other planes respectively parallel to them be drawn, and at infinitesimal distances dx, dy; so that $NL = dy$, $NJ = dx$, and PSQR is the intercepted infinitesimal element of the surface; then Q is $(x+dx, y+dy, z+dz)$; and let us imagine the whole surface by a similar process to be resolved into similar infinitesimal elements: then the area of one of these having been expressed in general terms, the area of the surface will be given by the double integral which expresses the sum of such elements, the integral of course being definite.

Let A represent the required area of the surface, and dA the area of the element PRQS. As a tangent plane to a surface at a given point contains not only the point but also an infinity of other points immediately contiguous to it, so dA being infinitesimal will be coincident with the tangent plane at P, and therefore the angle between it and any other plane is equal to the angle between the tangent plane and that plane.

Now the projection of dA on the plane of (x, y) is the rectangle NK, which $= dx\,dy$;

$$\therefore \quad dx\,dy = dA\cos\gamma = \frac{W}{Q}\,dA; \tag{26}$$

Similarly, if $d\text{A}$ is projected on the planes (y, z), and (z, x),

$$dy\,dz = d\text{A}\cos\alpha = \frac{\text{U}}{\text{Q}}\,d\text{A}; \qquad (27)$$

$$dz\,dx = d\text{A}\cos\beta = \frac{\text{V}}{\text{Q}}\,d\text{A}. \qquad (28)$$

Hence $$\text{A} = \int\int\frac{\text{Q}}{\text{U}}\,dy\,dz = \int\int\frac{\text{Q}}{\text{V}}\,dz\,dx = \int\int\frac{\text{Q}}{\text{W}}\,dx\,dy; \qquad (29)$$

each of these being a double integral; and either one being employed according as it is best suited to the equation of the surface and to the given limits.

Also squaring and adding (26) (27) and (28),

$$d\text{A}^2 = dy^2\,dz^2 + dz^2\,dx^2 + dx^2\,dy^2;$$

$$d\text{A} = \{dy^2\,dz^2 + dz^2\,dx^2 + dx^2\,dy^2\}^{\frac{1}{2}}; \qquad (30)$$

$$\therefore\quad \text{A} = \int\int\{dy^2\,dz^2 + dz^2\,dx^2 + dx^2\,dy^2\}^{\frac{1}{2}}, \qquad (31)$$

which is also the general value of the area; this form is convenient whenever the equation to the surface is given in the explicit form. Thus if $z = f(x, y)$, then from (31), we have

$$\text{A} = \int\int\left\{1 + \left(\frac{dz}{dx}\right)^2 + \left(\frac{dz}{dy}\right)^2\right\}^{\frac{1}{2}} dx\,dy. \qquad (32)$$

The formula (32) may also be deduced from the last of (29), by the theory explained in Art. 50, Vol. I; or as follows; since

$$\text{F}(x, y, z) = f(x, y) - z = 0,$$

$$\text{U} = \left(\frac{dz}{dx}\right), \quad \text{V} = \left(\frac{dz}{dy}\right), \quad \text{W} = -1;$$

and therefore (29) becomes

$$\text{A} = \int\int\left\{1 + \left(\frac{dz}{dx}\right)^2 + \left(\frac{dz}{dy}\right)^2\right\}^{\frac{1}{2}} dx\,dy. \qquad (33)$$

Now in all these cases, by means of substitution from the equation to the surface, the element-function will become a function of those two variables, whose differentials enter into the element. Thus, let us suppose the element-function to be a function of x and y, and let us consider the effects of the successive integrations.

We will suppose the surface, of which the area is required, to be closed, and to be such as is contained in the octant delineated in fig. 36; then, since PRQS is the element of the surface, the effect of a y-integration, x being constant, will be, the summation of all elements similar to PQ from L to K, that is, from $y=0$ to $y=$ MK; that is, the aggregate of the elements is the band LPK; and as the area of the band will be expressed in terms of x, and is therefore the general value of all similar bands, the effect of a subsequent x-integration will be, to sum all such elemental bands of which the surface is composed, and the limits of this latter integration must be $x=0$, and $x=$ OA. If therefore MK, as determined by the equation to the surface, $=$ Y, and OA $=a$, then

$$\text{the area} = \int_0^a \int_0^{\mathrm{Y}} \left\{1 + \left(\frac{dz}{dx}\right)^2 + \left(\frac{dz}{dy}\right)^2\right\}^{\frac{1}{2}} dy\, dx. \qquad (34)$$

If the x-integration is effected first, the effect will be to determine the band GPRI, and the limits of integration will be HI $=$ X and 0; and the subsequent y-integration, with the limits OB $=b$ and 0, will sum all such bands contained between parallel planes, and will give the area of the surface. In this latter case,

$$\text{the area} = \int_0^b \int_0^{\mathrm{X}} \left\{1 + \left(\frac{dz}{dx}\right)^2 + \left(\frac{dz}{dy}\right)^2\right\}^{\frac{1}{2}} dx\, dy. \qquad (35)$$

The above is an outline of the general method of finding the area of such surfaces: the limits of integration will of course vary according to the conditions of each problem.

237.] Examples illustrative of the preceding formulæ.

Ex. 1. The surface of the eighth part of a sphere.

Let the surface delineated in fig. 36 be that of the octant of a sphere: then, O being the centre, $x^2+y^2+z^2=a^2$.

$\therefore \quad \dfrac{\mathrm{Q}}{\mathrm{W}} = \dfrac{a}{(a^2-x^2-y^2)^{\frac{1}{2}}}$; and if $\mathrm{Y}=(a^2-x^2)^{\frac{1}{2}}$, then from (29),

$$\text{the area} = \int_0^a \int_0^{\mathrm{Y}} \frac{a\, dy\, dx}{(a^2-x^2-y^2)^{\frac{1}{2}}}$$

$$= \frac{\pi a}{2} \int_0^a dx = \frac{\pi a^2}{2}.$$

Ex. 2. A sphere is pierced by a right circular cylinder whose surface passes through the centre of the sphere, and the diameter of whose generating circle is equal to the radius of the sphere; it is required to find (1) the area of the surface of the sphere intercepted by the cylinder; (2) the area of the surface of the cylinder intercepted by the sphere.

Let the cylinder be perpendicular to the plane of (x, y): then the equations to the cylinder and to the sphere are $y^2 = ax - x^2$, and $x^2+y^2+z^2 = a^2$ respectively.

(1) If $\text{Y} = (ax-x^2)^{\frac{1}{2}}$, then, from (34),

$$\text{the area of the sphere} = 2\int_0^a \int_0^{\text{Y}} \frac{a\,dy\,dx}{(a^2-x^2-y^2)^{\frac{1}{2}}}$$

$$= 2a\int_0^a \sin^{-1}\frac{(ax-x^2)^{\frac{1}{2}}}{(a^2-x^2)^{\frac{1}{2}}}\,dx$$

$$= 2a\int_0^a \sin^{-1}\left(\frac{x}{a+x}\right)^{\frac{1}{2}} d(x+a)$$

$$= 2a\left[(x+a)\sin^{-1}\left(\frac{x}{a+x}\right)^{\frac{1}{2}} - (ax)^{\frac{1}{2}}\right]_0^a$$

$$= 2a\left\{2a\frac{\pi}{4} - a\right\} = a^2(\pi - 2).$$

(2) Eliminating y, we have $z = (a^2-ax)^{\frac{1}{2}} = \text{z}$, say; and the length-element of the trace of the cylinder on the plane of (x, y) is $\dfrac{dx}{(a^2-ax)^{\frac{1}{2}}}$; therefore

$$\text{the area of the cylinder} = 2\int_0^a \int_{-\text{z}}^{\text{z}} \frac{dz\,dx}{(a^2-ax)^{\frac{1}{2}}} = 4a^2.$$

Ex. 3. On the double ordinates of a circle, and in planes perpendicular to the plane of the circle, isosceles triangles, whose vertical angle $= 2\alpha$, are described; prove that the equation to the surface thus generated is $x^2+(y+z\tan\alpha)^2 = a^2$, and that the whole convex area $= 2a^2\{\cot\alpha + \alpha(\operatorname{cosec}\alpha)^2\}$.

238.] For another application of the preceding theory of quadrature let us consider that of the ellipsoid; for although the integrals which determine this area become elliptic transcendents, and consequently do not admit of integration, yet by the introduction of new variables, and transformation according to the principles of the preceding chapter, they assume forms deserving notice on account of the geometrical interpretations which they yield.

Let the equation to the ellipsoid be

$$\frac{x^2}{a^2} + \frac{y^2}{b^2} + \frac{z^2}{c^2} = 1, \tag{36}$$

where $a > b > c$. And let A denote the whole surface; then, if $\text{Y} = \dfrac{b}{a}(a^2-x^2)^{\frac{1}{2}}$, by the last of (29),

$$\text{A} = 8\int_0^a\int_0^Y \left\{\frac{1-\dfrac{a^2-c^2}{a^2}\dfrac{x^2}{a^2}-\dfrac{b^2-c^2}{b^2}\dfrac{y^2}{b^2}}{1-\dfrac{x^2}{a^2}-\dfrac{y^2}{b^2}}\right\}^{\frac{1}{2}} dy\,dx; \quad (37)$$

which does not admit of further integration; let us however introduce new variables; and in the first place let the equation (36) be expressed in terms of subsidiary angles a and β as follows:

$$\left.\begin{aligned} x &= a\sin a\cos\beta, \\ y &= b\sin a\sin\beta, \\ z &= c\cos a; \end{aligned}\right\} \quad (38)$$

for these equations satisfy (36).

Now we may either substitute these values in (37); or may apply to them immediately equation (30), Art. 236. In either case, if $d\text{A}$ denotes the infinitesimal surface-element,

$$d\text{A} = abc\left\{\frac{(\sin a)^2(\cos\beta)^2}{a^2}+\frac{(\sin a)^2(\sin\beta)^2}{b^2}+\frac{(\cos a)^2}{c^2}\right\}^{\frac{1}{2}}\sin a\,da\,d\beta. \quad (39)$$

But if p = the length of the perpendicular from the centre of the ellipsoid on the tangent plane,

$$\begin{aligned}\frac{1}{p^2} &= \frac{x^2}{a^4}+\frac{y^2}{b^4}+\frac{z^2}{c^4} \\ &= \frac{b^2c^2(\sin a)^2(\cos\beta)^2+c^2a^2(\sin a)^2(\sin\beta)^2+a^2b^2(\cos a)^2}{a^2b^2c^2};\end{aligned}$$

and therefore $$d\text{A} = \frac{abc\sin a\,da\,d\beta}{p}; \quad (40)$$

and if A = the whole surface of the ellipsoid,

$$\text{A} = abc\int_{-\pi}^{\pi}\int_0^{\pi}\frac{\sin a\,da\,d\beta}{p}. \quad (41)$$

Hereby we may prove the following theorem. If $d\text{A}$ = the surface-element of an ellipsoid, and s is the area of the central section of the surface which is parallel to the plane of $d\text{A}$, $\int\int\frac{d\text{A}}{\text{s}} = 4$, when the ranges of integration include the whole ellipsoid.

239.] Again, let us introduce two other subsidiary angles η and ψ, such that $\sin\eta\cos\psi$, $\sin\eta\sin\psi$, $\cos\eta$ are respectively the direction-cosines of the normal of the ellipsoid at the point (x, y, z);

so that

$$\left.\begin{aligned} \sin\eta\cos\psi &= p\frac{x}{a^2}, \\ \sin\eta\sin\psi &= p\frac{y}{b^2}, \\ \cos\eta &= p\frac{z}{c^2}; \end{aligned}\right\} \tag{42}$$

$$\therefore\quad p^2 = a^2(\sin\eta)^2(\cos\psi)^2 + b^2(\sin\eta)^2(\sin\psi)^2 + c^2(\cos\eta)^2. \tag{43}$$

And since $x = \dfrac{a^2\sin\eta\cos\psi}{p}$, $\quad y = \dfrac{b^2\sin\eta\sin\psi}{p}$,

$$\therefore\quad dy\,dx = \frac{a^2b^2c^2\sin\eta\cos\eta}{p^4}\,d\eta\,d\psi;$$

$$\therefore\quad d\mathrm{A} = \frac{dy\,dx}{\cos\eta} = \frac{a^2b^2c^2\sin\eta}{p^4}\,d\eta\,d\psi; \tag{44}$$

$$\mathrm{A} = \int_{-\pi}^{\pi}\int_0^{\pi}\frac{a^2b^2c^2\sin\eta\,d\eta\,d\psi}{\{a^2(\sin\eta)^2(\cos\psi)^2+b^2(\sin\eta)^2(\sin\psi)^2+c^2(\cos\eta)^2\}^2}. \tag{45}$$

To simplify this, let

$$\left.\begin{aligned} a^2(\sin\eta)^2 + c^2(\cos\eta)^2 &= m^2, \\ b^2(\sin\eta)^2 + c^2(\cos\eta)^2 &= n^2; \end{aligned}\right\} \tag{46}$$

$$\therefore\quad p^2 = m^2(\cos\psi)^2 + n^2(\sin\psi)^2; \tag{47}$$

$$\mathrm{A} = \int_{-\pi}^{\pi}\int_0^{\pi}\frac{a^2b^2c^2\sin\eta\,d\eta\,d\psi}{\{m^2(\cos\psi)^2+n^2(\sin\psi)^2\}^2}. \tag{48}$$

Let $\qquad n\tan\psi = m\tan\omega; \qquad (49)$

then, observing that the limits of ψ, and therefore of ω, are π and $-\pi$, we have

$$\mathrm{A} = \pi a^2b^2c^2\int_0^{\pi}\frac{\sin\eta\,(m^2+n^2)\,d\eta}{m^3n^3}. \tag{50}$$

Thus the area of the surface is expressed in terms of a single integral. Let us consider the geometrical meaning of this expression.

From (42) it appears that the relation between η and the coordinates of its corresponding point on the ellipsoid is

$$\cos\eta = \frac{pz}{c^2} = \frac{z}{c^2}\left\{\frac{x^2}{a^4}+\frac{y^2}{b^4}+\frac{z^2}{c^4}\right\}^{-\frac{1}{2}};$$

$$\therefore\quad \frac{x^2}{a^4}+\frac{y^2}{b^4}-\frac{z^2(\tan\eta)^2}{c^4} = 0; \tag{51}$$

which is the equation to an elliptical cone whose vertex is at the centre of the ellipsoid; and as η is the z-direction-angle of the normal of the ellipsoid, the axis of z is the axis of the cone, and the ratio of the semi-axes of any plane elliptical section of it perpendicular to its axis is that of $a^2 : b^2$. Now the ψ-integration, which has already been effected, between the limits π and $-\pi$, gives an annulus on the surface of the ellipsoid, the breadth of the annulus being due to the variation of η. Imagine therefore two cones, represented by equation (51), to be described corresponding to η and to $\eta+d\eta$; the lines of intersection of these cones with the ellipsoid will be two curves, infinitesimally near to each other, which contain between them the band of the ellipsoidal surface expressed by

$$\pi a^2 b^2 c^2 \frac{\sin\eta\,(m^2+n^2)\,d\eta}{m^3 n^3}; \tag{52}$$

and the sum of all these bands between the limits π and 0 will be the whole surface of the ellipsoid.

The projection on the plane of (x, y) of the intersection of the given ellipsoid with the cone (51) is an ellipse, whose equation determined by the elimination of z from (36) and (51) is

$$\frac{x^2}{a^2}\left\{1-\frac{c^2}{a^2}(\cot\eta)^2\right\}+\frac{y^2}{b^2}\left\{1-\frac{c^2}{b^2}(\cot\eta)^2\right\}=1. \tag{53}$$

Thus if the band which is expressed by (52) is projected on the plane of (x, y), one of its bounding lines is the ellipse (53), and the other is the ellipse corresponding to $\eta+d\eta$.

240.] The element-function of (37) also admits of another interpretation deserving notice.

Let $\frac{x}{a}=\xi$, $\frac{y}{b}=\eta$; so that $dy\,dx = ab\,d\eta\,d\xi$; also the equation which determines the limits of integration is $\xi^2+\eta^2=1$; let $(1-\xi^2)^{\frac{1}{2}} = \text{H}$. Also let $\frac{a^2-c^2}{a^2}=\alpha^2$, $\frac{b^2-c^2}{b^2}=\beta^2$; and let

$$\frac{1-\alpha^2\xi^2-\beta^2\eta^2}{1-\xi^2-\eta^2}=\zeta^2; \tag{54}$$

whereby we have

$$\text{A} = ab\int_{-1}^{1}\int_{-\text{H}}^{\text{H}}\zeta\,d\eta\,d\xi; \tag{55}$$

Now let us suppose ξ, η, ζ, to be rectangular coordinates; then $d\eta\,d\xi$ is an area-element of the plane (ξ, η), and as ζ is the coordinate parallel to the ζ-axis, $\zeta\,d\eta\,d\xi$ expresses the volume of a rectangular parallelepipedon whose base is $d\eta\,d\xi$, and whose height is ζ; ζ being related to ξ and η by the equation (54), which may be put into the form,

$$(\zeta^2-\alpha^2)\xi^2+(\zeta^2-\beta^2)\eta^2 = \zeta^2-1; \tag{56}$$

thus this equation represents a surface, and consequently the double definite integral, whose element is $\zeta\, d\eta\, d\zeta$, expresses the volume contained within the given limits between this surface and the plane of (ξ, η). This surface is represented in fig. 45, where $\text{OC} = 1$, for since α^2 and β^2 are both less than 1, when $\zeta = 1, \xi = \eta = 0$; and consequently the surface intersects the ζ-axis at C. For values of ζ a little less than 1, the equation gives impossible values to ξ and η; but gives possible values when ζ is greater than 1; thus the surface lies above the point C. When $\xi^2+\eta^2 = 1$, $\zeta = \infty$; consequently if a right circular cylinder, whose radius $= \text{OA} = \text{OB} = 1$, and whose axis is OC, is described as in the fig., the surface (56) is asymptotic to it. Also as (56) is unchanged when ζ is replaced by $-\zeta$, another surface equal and similar to CRQ lies below the plane of (ξ, η), to which the cylinder is also asymptotic. Thus the integral (55) expresses the volume contained within the cylinder and between those two surfaces. Hereby also we have a further interpretation of the definite integral; for if we give to ζ a value greater than 1, and take it to be constant, while ξ and η vary, that is, if we cut the surface by a plane parallel to that of (ξ, η), (56) shews that the curve of section is an ellipse; such is RQ, of which the semi-axes are

$$\left(\frac{\zeta^2-1}{\zeta^2-\alpha^2}\right)^{\frac{1}{2}} \text{ and } \left(\frac{\zeta^2-1}{\zeta^2-\beta^2}\right)^{\frac{1}{2}}, \tag{57}$$

and of which therefore the area $= \dfrac{\pi(\zeta^2-1)}{(\zeta^2-\alpha^2)^{\frac{1}{2}}(\zeta^2-\beta^2)^{\frac{1}{2}}}$. (58)

If ζ varies, the area of this ellipse will also vary by a narrow elliptical ring, of which the area $= \pi d \,.\, \dfrac{\zeta^2-1}{(\zeta^2-\alpha^2)^{\frac{1}{2}}(\zeta^2-\beta^2)^{\frac{1}{2}}}$; and consequently we may consider the element of the volume, which is expressed in the right-hand of (55), to be

$$\pi\zeta d\,.\,\frac{\zeta^2-1}{(\zeta^2-\alpha^2)^{\frac{1}{2}}(\zeta^2-\beta^2)^{\frac{1}{2}}}; \text{ so that}$$

$$\text{A} = 2\pi ab \int_1^\infty \zeta d\,.\,\frac{\zeta^2-1}{(\zeta^2-\alpha^2)^{\frac{1}{2}}(\zeta^2-\beta^2)^{\frac{1}{2}}} \tag{59}$$

$$= 2\pi ab\left[\frac{\zeta(\zeta^2-1)}{(\zeta^2-\alpha^2)^{\frac{1}{2}}(\zeta^2-\beta^2)^{\frac{1}{2}}} - \int\frac{(\zeta^2-1)\,d\zeta}{(\zeta^2-\alpha^2)^{\frac{1}{2}}(\zeta^2-\beta^2)^{\frac{1}{2}}}\right]_1. \tag{60}$$

Thus by this substitution, and mode of interpretation, due to

M. Catalan *, the double integral in (37) is reduced to the single integral in (60).

241.] In review of the processes of the two preceding articles let it be observed, that in equations (39) and (45) we have

$$d\text{A} = \{b^2c^2(\sin\alpha)^2(\cos\beta)^2+c^2a^2(\sin\alpha)^2(\sin\beta)^2+a^2b^2(\cos\alpha)^2\}^{\frac{1}{2}}\sin\alpha\, d\alpha\, d\beta,$$

$$d\text{A} = \frac{a^2b^2c^2\sin\eta\, d\eta\, d\psi}{\{a^2(\sin\eta)^2(\cos\psi)^2+b^2(\sin\eta)^2(\sin\psi)^2+c^2(\cos\eta)^2\}^{\frac{3}{2}}};$$

the former of which is irrational, and the latter is rational; so that by means of the following substitutions we have been able to transform an element-function involving irrational quantities into an equivalent in terms of rational quantities only; viz. by substituting

$$\left.\begin{aligned}\sin\alpha\cos\beta &= \frac{a\sin\eta\cos\psi}{p},\\ \sin\alpha\sin\beta &= \frac{b\sin\eta\sin\psi}{p},\\ \cos\alpha &= \frac{c\cos\eta}{p};\end{aligned}\right\} \qquad (61)$$

where $p^2 = a^2(\sin\eta)^2(\cos\psi)^2+b^2(\sin\eta)^2(\sin\psi)^2+c^2(\cos\eta)^2$; and

$$\frac{1}{p^2} = \frac{(\sin\alpha)^2(\cos\beta)^2}{a^2}+\frac{(\sin\alpha)^2(\sin\beta)^2}{b^2}+\frac{(\cos\alpha)^2}{c^2}.$$

Also equating the values of the elements given in (41) and (44), we have

$$p^3\sin\alpha\, d\alpha\, d\beta = abc\sin\eta\, d\eta\, d\psi. \qquad (62)$$

Hence by the substitutions of (61), the double integral

$$\iint\frac{\text{U}\sin\alpha\, d\alpha\, d\beta}{\{b^2c^2(\sin\alpha)^2(\cos\beta)^2+c^2a^2(\sin\alpha)^2(\sin\beta)^2+a^2b^2(\cos\alpha)^2\}^{\frac{1}{2}}},$$

in which U is a rational function of $\sin\alpha\cos\beta$, $\sin\alpha\sin\beta$, and $\cos\alpha$, may be transformed into the following, which involves only rational quantities, viz. into

$$\iint\frac{a^2b^2c^2\,\text{U}\sin\eta\, d\eta\, d\psi}{a^2(\sin\eta)^2(\cos\psi)^2+b^2(\sin\eta)^2(\sin\psi)^2+c^2(\cos\eta)^2};$$

the limits of the new variables being easily obtained from those of the former variables by means of equations (61).

Again, from (62) we have

$$\int_0^{\frac{\pi}{2}}\int_0^{\frac{\pi}{2}}p^3\sin\alpha\, d\alpha\, d\beta = abc\int_0^{\frac{\pi}{2}}\int_0^{\frac{\pi}{2}}\sin\eta\, d\eta\, d\psi$$

$$= \frac{\pi abc}{2}.$$

* See Liouville's Journal, Vol. IV, page 323. The same method is extended to integrals of higher orders and more variables; the discussion however of which is beyond the scope of the present work.

242.] As position in space may be determined by a system of polar coordinates, such as that explained in Art. 165, so may of course the equation to a surface be expressed in terms of these polar coordinates; in which case for the determination of the area it is necessary to express the area-element in terms of them. This may be found by transformation of either of the expressions given in (29) and (31). I propose however first to investigate the subject on principles purely geometrical, and subsequently by analytical transformation to exhibit the identity of the result with those thus determined.

Let P, fig. 42, be the point (r, θ, ϕ) on the surface whose equation is $F(r, \theta, \phi) = 0$; let θ and ϕ vary; and let the radii vectores due to the partial variation of θ, to the partial variation of ϕ, and to the total variation of θ and ϕ, respectively, intersect the surface in the points U, V, R, so that PURV is the surface-element. From P draw PS at right angles to OP, and cutting OU in S; then $PS = r\,d\theta$, $SU = \left(\frac{dr}{d\theta}\right) d\theta$. Also from P draw PT at right angles to OP, and cutting OV in T; then $PT = \rho\, d\phi = r \sin\theta\, d\phi$, $TV = \left(\frac{dr}{d\phi}\right) d\phi$. Also let OP be produced to Z; then from the geometry, as explained in Art. 165, it is plain that PS, PT, PZ form a system of lines at P such that each one is at right angles to the other two; and consequently they may be considered as constituting a system of rectangular coordinates originating at P; we will employ them as such, and take PS, PT, PZ to be respectively the $x-$, $y-$, $z-$, axes. In reference to this system P is $(0, 0, 0)$; U is $\left(r\,d\theta, 0, \left(\frac{dr}{d\theta}\right) d\theta\right)$; V is $\left(0, r \sin\theta\, d\phi, \left(\frac{dr}{d\phi}\right) d\phi\right)$. Hence, if dA denotes the surface-element, which is approximately a parallelogram, of which P, U, V, R are the angular points,

$$dA = \left\{r^2 (\sin\theta)^2 + (\sin\theta)^2 \left(\frac{dr}{d\theta}\right)^2 + \left(\frac{dr}{d\phi}\right)^2\right\}^{\frac{1}{2}} r\, d\theta\, d\phi. \qquad (63)$$

$$\therefore \quad A = \iint \left\{r^2 (\sin\theta)^2 + (\sin\theta)^2 \left(\frac{dr}{d\theta}\right)^2 + \left(\frac{dr}{d\phi}\right)^2\right\}^{\frac{1}{2}} r\, d\theta\, d\phi; \qquad (64)$$

this integral being definite, and the limits being assigned by the circumstances of the problem.

243.] The preceding expression for the surface-element may also be derived by transformation of coordinates from the last of (29) by means of the formulæ given in (70), Art. 165. It will be convenient to effect the transformation by first replacing x and y in terms of ρ and ϕ, and subsequently replacing ρ and z in terms of r and θ.

Since $x = \rho \cos\phi$, $y = \rho \sin\phi$, therefore $dx\, dy = \rho\, d\rho\, d\phi$: and as after substitution the equation to the surface takes the form $F(\rho, \phi, z) = 0$, then, as in (190), Art. 107, Vol. I,

$$\left.\begin{aligned} U &= \left(\frac{dF}{dx}\right) = \left(\frac{dF}{d\rho}\right)\cos\phi - \left(\frac{dF}{d\phi}\right)\frac{\sin\phi}{\rho}, \\ V &= \left(\frac{dF}{dy}\right) = \left(\frac{dF}{d\rho}\right)\sin\phi + \left(\frac{dF}{d\phi}\right)\frac{\cos\phi}{\rho}. \end{aligned}\right\} \tag{65}$$

By a similar process, since $z = r\cos\theta$, $\rho = r\sin\theta$,

$$\left.\begin{aligned} W = \left(\frac{dF}{dz}\right) &= \left(\frac{dF}{dr}\right)\cos\theta - \left(\frac{dF}{d\theta}\right)\frac{\sin\theta}{r}, \\ \left(\frac{dF}{d\rho}\right) &= \left(\frac{dF}{dr}\right)\sin\theta + \left(\frac{dF}{d\theta}\right)\frac{\cos\theta}{r}. \end{aligned}\right\}$$

Hence, substituting for $\left(\frac{dF}{d\rho}\right)$, squaring and adding,

$$Q^2 = \left(\frac{dF}{dr}\right)^2 + \left(\frac{dF}{d\theta}\right)^2\frac{1}{r^2} + \left(\frac{dF}{d\phi}\right)^2\frac{1}{r^2(\sin\theta)^2}; \tag{66}$$

and

$$A = \iint \frac{Q}{W}\, dx\, dy$$

$$= \iint \frac{\left\{\left(\frac{dF}{dr}\right)^2 + \left(\frac{dF}{d\theta}\right)^2\frac{1}{r^2} + \left(\frac{dF}{d\phi}\right)^2\frac{1}{(r\sin\theta)^2}\right\}^{\frac{1}{2}}}{\left(\frac{dF}{dr}\right)\cos\theta - \left(\frac{dF}{d\theta}\right)\frac{\sin\theta}{r}}\, r\sin\theta\, d(r\sin\theta)\, d\phi; \tag{67}$$

the integral of course being definite.

If the equation to the surface is given in the explicit form $r = f(\theta, \phi)$, so that $r - f(\theta, \phi) = F(r, \theta, \phi) = 0$, then

$$\left(\frac{dF}{dr}\right) = 1, \quad \left(\frac{dF}{d\theta}\right) = -\left(\frac{dr}{d\theta}\right), \quad \left(\frac{dF}{d\phi}\right) = -\left(\frac{dr}{d\phi}\right);$$

and (67) becomes,

$$A = \iint \left\{r^2(\sin\theta)^2 + (\sin\theta)^2\left(\frac{dr}{d\theta}\right)^2 + \left(\frac{dr}{d\phi}\right)^2\right\}^{\frac{1}{2}} r\, d\theta\, d\phi; \tag{69}$$

which is the same expression as (64).

If the surface whose area is to be determined is a surface of revolution about the z-axis, then the equation to it is independent of ϕ, and (69) becomes,

$$A = \iint \left\{r^2 + \left(\frac{dr}{d\theta}\right)^2\right\}^{\frac{1}{2}} r\sin\theta\, d\theta\, d\phi. \tag{70}$$

If the whole surface is required, the ϕ-integration may be effected for the limits 2π and 0, and

$$\text{A} = 2\pi \int \{(r\,d\theta)^2 + dr^2\}^{\frac{1}{2}} r \sin\theta. \tag{71}$$

This result is manifestly identical with that given in equations (17) and (21) of the present chapter. Indeed this process is general, and includes all the preceding.

244.] The following examples are illustrative of the preceding general formulæ.

Ex. 1. To find the area of a spherical triangle.

Let ABC be the triangle whose angles are A, B, C; and O the centre of the sphere. Let OC be the z-axis; and let the central plane containing the side AB be inclined to the plane of (x, y) at an angle $= a$, and intersect it along the x-axis. Then the equation to this plane is $\tan\theta = \cot a \operatorname{cosec}\phi$. Let ϕ_0 be the ϕ-coordinate to A, and consequently $\phi_0 + \text{C}$ is the ϕ-coordinate to B. Let $r = a$ be the equation to the sphere, and let

$$\Theta = \tan^{-1}\cot a \operatorname{cosec}\phi.$$

Then from (69),

$$\text{the area} = a^2 \int_{\phi_0}^{\phi_0+\text{C}} \int_0^{\Theta} \sin\theta\,d\theta\,d\phi$$

$$= a^2 \int_{\phi_0}^{\phi_0+\text{C}} (1 - \cos\Theta)\,d\phi$$

$$\text{Now } \frac{\sin\Theta}{1} = \frac{\cos\Theta}{\tan a \sin\phi} = \frac{1}{\{1 + (\tan a \sin\phi)^2\}^{\frac{1}{2}}}$$

$$= \frac{1}{\{(\sec a)^2 - (\tan a \cos\phi)^2\}^{\frac{1}{2}}};$$

$$\therefore \quad \cos\Theta = \frac{\sin\phi}{\{(\operatorname{cosec} a)^2 - (\cos\phi)^2\}^{\frac{1}{2}}};$$

$$\therefore \quad \text{the area} = a^2 \int_{\phi_0}^{\phi_0+\text{C}} \left\{1 - \frac{\sin\phi}{\{(\operatorname{cosec} a)^2 - (\cos\phi)^2\}^{\frac{1}{2}}}\right\} d\phi$$

$$= a^2 \Big[\phi - \cos^{-1}(\sin a \cos\phi)\Big]_{\phi_0}^{\phi_0+\text{C}}$$

$$= a^2\{\text{C} - \cos^{-1}(\sin a \cos(\phi_0 + \text{C})) + \cos^{-1}(\sin a \cos\phi_0)\};$$

But from the geometry, $\cos\text{A} = \sin a \cos\phi_0$,

$$\cos(\pi - \text{B}) = \sin a \cos(\phi_0 + \text{C});$$

$$\therefore \quad \text{the area of the triangle} = a^2\{\text{A} + \text{B} + \text{C} - \pi\}.$$

If $A = B = \frac{\pi}{2}$, the triangle becomes half of a lune; and consequently the area of the lune $= 2a^2 c$.

If $c = 2\pi$, the area of the whole sphere $= 4\pi a^2$.

Ex. 2. To find the definite integral expressing the surface of the ellipsoid in terms of polar coordinates.

Let the equation to the ellipsoid be given by the system (38); then

$$\left.\begin{aligned} x &= a \sin\alpha \cos\beta = r \sin\theta \cos\phi, \\ y &= b \sin\alpha \sin\beta = r \sin\theta \sin\phi, \\ z &= c \cos\alpha \qquad\;\; = r \cos\theta; \end{aligned}\right\} \tag{72}$$

so that eliminating r,

$$\left.\begin{aligned} a \tan\alpha \cos\beta &= c \tan\theta \cos\phi, \\ b \tan\alpha \sin\beta &= c \tan\theta \sin\phi; \end{aligned}\right\}$$

$$\therefore \quad ab (\sec\alpha)^2 \tan\alpha \, d\alpha \, d\beta = c^2 (\sec\theta)^2 \tan\theta \, d\theta \, d\phi.$$

$$\therefore \quad abc \sin\alpha \, d\alpha \, d\beta = r^3 \sin\theta \, d\theta \, d\phi. \tag{73}$$

and substituting in (41),

$$A = \int_{-\pi}^{\pi} \int_{0}^{\pi} \frac{r^3 \sin\theta \, d\theta \, d\phi}{p}; \tag{74}$$

which is the required expression. Also for the whole ellipsoid

$$\int_{-\pi}^{\pi} \int_{0}^{\pi} r^3 \sin\theta \, d\theta \, d\phi = abc \int_{-\pi}^{\pi} \int_{0}^{\pi} \sin\alpha \, d\alpha \, d\beta$$

$$= 4\pi \, abc. \tag{75}$$

Ex. 3. The lemniscata, $r^2 = a^2 \sin 2\theta$, is described in the plane of (x, y); and in planes perpendicular to that of (x, y) circles are described on the radii vectores as diameters; find the area of the surface.

The equation to the surface is $(x^2+y^2+z^2)^2 = 2a^2 xy$; and in polar coordinates $r^2 = a^2 (\sin\theta)^2 \sin 2\phi$; whence, by (64),

$$dA = a^2 (\sin\theta)^2 \, d\theta \, d\phi;$$

and taking the surface described on one loop,

$$A = 2 \int_{0}^{\frac{\pi}{2}} \int_{0}^{\frac{\pi}{2}} a^2 (\sin\theta)^2 \, d\theta \, d\phi$$

$$= \frac{\pi^2 a^2}{4}.$$

Ex. 4. If p is the perpendicular from the origin on the tangent plane at the point of the surface-element, and if the surface is closed, and has no singular points,

$$A = \int_{-\pi}^{\pi} \int_{0}^{\pi} \frac{r^3 \sin\theta \, d\theta \, d\phi}{p}.$$

SECTION 5.—*Gauss' System of Curvilinear Coordinates* *.

245.] In Art. 193, Vol. I, it is remarked that the equation to a plane curve may be expressed in terms of a subsidiary angle, or indeed in terms of any subsidiary quantity, by means of two equations; so that these two equations taken simultaneously express the curve; and the original equation to the curve is found by the elimination of this subsidiary quantity; and this system of reference has been applied in this and the preceding chapters to the rectification of curves and the quadrature of plane surfaces. Now the same principle is also applicable to curved surfaces; for as the equation to a surface involves three variables, viz., x, y, z, so in the general case each of these coordinates must be a function of two subsidiary and independent quantities; and the equation to the surface will arise from the elimination of these quantities from the given equations. Let ξ and η be these subsidiary quantities; and let the equation to the surface in terms of x, y, z be

$$F(x, y, z) = 0: \tag{76}$$

and let x, y, z be connected with ξ, η by the equations,

$$x = f_1(\xi, \eta), \qquad y = f_2(\xi, \eta), \qquad z = f_3(\xi, \eta); \tag{77}$$

where f_1, f_2, f_3 are functions such that $F(x, y, z) = 0$ arises by the elimination of ξ and η from them.

The following are particular cases of these equations.

Ex. 1. The ellipsoid may be expressed by the following equations; viz.

$$x = a \sin\eta \cos\xi, \qquad y = b \sin\eta \sin\xi, \qquad z = c\cos\eta; \tag{78}$$

because if we eliminate ξ and η, we have

$$\frac{x^2}{a^2} + \frac{y^2}{b^2} + \frac{z^2}{c^2} = 1; \tag{79}$$

which is the ordinary equation to the ellipsoid.

* The matter of the following section is in a great measure taken from Gauss' celebrated Memoir entitled "Disquisitiones generales circa superficies curvas," which is contained in Vol. VI of the Memoirs of the Royal Society of Sciences of Gottingen, 1828. It has been reprinted as an Appendix to Monge's Application d'Analyse &c., edited by M. Liouville, Bachelier, Paris, 1850; and Liouville has added some notes elucidating various parts of the system in its application to curvature and to geodesic lines. Also the student desirous of further information may consult a profound paper of M. Ossian Bonnet on the General Theory of Surfaces, in the Journal de l'Ecole Polytechnique, Cahier XXXII, Paris, 1848.

If $a=b=c$, the ellipsoid becomes a sphere: in which case (78) become

$$x = a\sin\eta\cos\xi, \qquad y = a\sin\eta\sin\xi, \qquad z = a\cos\eta; \quad (80)$$

and ξ, η are the ordinary polar coordinates.

Ex. 2. The hyperboloid of one sheet may be expressed by the following equations; viz.

$$x = a\sec\eta\cos\xi, \qquad y = b\sec\eta\sin\xi, \qquad z = c\tan\eta; \quad (81)$$

because if we eliminate ξ and η, we have

$$\frac{x^2}{a^2} + \frac{y^2}{b^2} - \frac{z^2}{c^2} = 1, \quad (82)$$

which is the ordinary equation to the surface.

Ex. 3. The skew helicoid, (88), Art. 368, Vol. I, may be expressed by the equations

$$x = \xi\cos\eta, \qquad y = \xi\sin\eta, \qquad z = ka\eta. \quad (83)$$

246.] These examples are sufficient to illustrate the method; and we return to the general equations (76) and (77). As ξ and η are independent variables, we may consider them separately. Firstly let η have a determinate value, say $\eta = c_2$; and let ξ vary; and let ξ be eliminated from the equations (77); then two equations result in terms of x, y, z and η, each of which represents a surface; and as they are simultaneous, taken together they represent a line, which is generally a curve of double curvature; and which because of the simultaneity of all the equations, evidently lies on the surface (76); this line is called $\eta = c_2$; and by giving different values to c_2, we obtain a series of such lines, all lying on the surface (76). Secondly let ξ have a determinate value, say $\xi = c_1$; and let η vary; then, as before, it is evident that the equations formed by the elimination of η will represent a line lying on the surface (76); and if different values are given to c_1, we shall have a series of lines lying on the surface. Thus the two equations

$$\xi = c_1, \qquad \eta = c_2, \quad (84)$$

taken separately, represent two different systems of curves drawn upon the surface (76). Now, as the systems are continuous, every point on the surface will be at the intersection of two curves, one of which is a member of the first, and the other is a member of the second system: and the point is determined whenever these curves are given. Suppose (x_0, y_0, z_0) to be a point on the surface, corresponding to which the values of ξ and η are ξ_0 and η_0, these latter quantities admitting of the value zero; then the intersection of ξ_0 and η_0 may conveniently be taken to be the origin; and any other point to be at the intersection of two lines corresponding to given values of ξ and η. And ξ and η may fitly be called the *curvilinear coordinates* to that point.

Thus take the ellipsoid, as given by the equations (78). Let ξ be eliminated: then we have

$$\frac{x^2}{a^2} + \frac{y^2}{b^2} = (\sin\eta)^2, \qquad z = c\cos\eta; \tag{85}$$

now for a given value of η the former of these equations represents an ellipse in the plane of (x, y), and generally an elliptical cylinder whose axis is the z-axis; and as the latter represents a plane parallel to the plane of (x, y), it appears that these equations represent an ellipse drawn on the surface of the ellipsoid, and lying in a plane parallel to that of (x, y). Thus the second system of lines as given in (84) is a series of ellipses described on the surface of the ellipsoid in planes perpendicular to the z-axis. Again, if we eliminate η from (78), we have

$$\frac{y}{x} = \frac{b}{a}\tan\xi;$$

which represents a series of planes passing through the z-axis; and as these intersect the ellipsoid in ellipses, the first system of lines as given in (84) is a series of ellipses described on the surface by planes passing through the z-axis. Thus a given point on the surface of the ellipsoid will be determined by means of the two ellipses, one of each system, which intersect at that point. The particular values of ξ and η which determine these ellipses are the curvilinear coordinates to the point.

247.] Let there be two points (x, y, z) $(x+dx,\ y+dy,\ z+dz)$ on the surface $F(x, y, z) = 0$ infinitesimally near to each other; and let us suppose the first to be at the intersection of ξ and η, and the second at the intersection of $\xi+d\xi$ and $\eta+d\eta$; and let ds be the distance between them; let us investigate the relations between these points in terms of ξ, η and their differentials. Differentiating (77) we have

$$\left.\begin{aligned} dx &= \left(\frac{df_1}{d\xi}\right)d\xi + \left(\frac{df_1}{d\eta}\right)d\eta, \\ dy &= \left(\frac{df_2}{d\xi}\right)d\xi + \left(\frac{df_2}{d\eta}\right)d\eta, \\ dz &= \left(\frac{df_3}{d\xi}\right)d\xi + \left(\frac{df_3}{d\eta}\right)d\eta; \end{aligned}\right\} \tag{86}$$

consequently, if

$$\left.\begin{array}{r} \left(\frac{df_1}{d\xi}\right)^2+\left(\frac{df_2}{d\xi}\right)^2+\left(\frac{df_3}{d\xi}\right)^2 = \text{E}, \\ \left(\frac{df_1}{d\xi}\right)\left(\frac{df_1}{d\eta}\right)+\left(\frac{df_2}{d\xi}\right)\left(\frac{df_2}{d\eta}\right)+\left(\frac{df_3}{d\xi}\right)\left(\frac{df_3}{d\eta}\right) = \text{F}, \\ \left(\frac{df_1}{d\eta}\right)^2+\left(\frac{df_2}{d\eta}\right)^2+\left(\frac{df_3}{d\eta}\right)^2 = \text{G}, \end{array}\right\} \quad (87)$$

$$\begin{aligned} ds^2 &= dx^2+dy^2+dz^2 \\ &= \text{E}\,d\xi^2+2\,\text{F}\,d\xi\,d\eta+\text{G}\,d\eta^2; \end{aligned} \quad (88)$$

and if s is the length of a curve described on the given surface between the points (ξ_1, η_1) and (ξ_0, η_0),

$$s = \int_0^1 \{\text{E}\,d\xi^2+2\,\text{F}\,d\xi\,d\eta+\text{G}\,d\eta^2\}^{\frac{1}{2}}; \quad (89)$$

1 and 0 being the subscript letters of, and thus conveniently used as abbreviated expressions for, the limits of the integral. I may observe that this is a notation which will be frequently employed hereafter.

Equation (88) enables us to give a simple geometrical interpretation to E, F and G; let $d\eta = 0$, that is, let us suppose η to be constant, and let us pass from (ξ, η) to $(\xi+d\xi, \eta)$. Then if $d\sigma$ is the distance between these points, it is plain from (88) that,

$$d\sigma = \text{E}^{\frac{1}{2}}\,d\xi; \quad (90)$$

hence $\text{E}^{\frac{1}{2}}\,d\xi$ is an infinitesimal arc-element of the curve $\eta = c_2$, intercepted between two consecutive curves of the other system.

Similarly, if $d\sigma'$ is the value of ds, when $\xi = c_1$,

$$d\sigma' = \text{G}^{\frac{1}{2}}\,d\eta; \quad (91)$$

and consequently $\text{G}^{\frac{1}{2}}\,d\eta$ is an infinitesimal arc-element of the curve $\xi = c_1$, intercepted between two consecutive curves of the other system.

Let $\omega =$ the angle at which $d\sigma$ and $d\sigma'$ are inclined to each other; let ds be the distance between (ξ, η) and $(\xi+d\xi, \eta+d\eta)$; then

$$\begin{aligned} ds^2 &= d\sigma^2+2\,d\sigma\,d\sigma'\cos\omega+d\sigma'^2 \\ &= \text{E}\,d\xi^2+2\,\text{E}^{\frac{1}{2}}\text{G}^{\frac{1}{2}}\,d\xi\,d\eta\cos\omega+\text{G}\,d\eta^2; \end{aligned} \quad (92)$$

and equating this to the value given in (88), we have

$$\cos\omega = \frac{\text{F}}{\text{E}^{\frac{1}{2}}\text{G}^{\frac{1}{2}}}. \quad (93)$$

Or thus; the direction-cosines of a curve of the first system, when

η = a constant, are proportional to $\left(\frac{df_1}{d\xi}\right), \left(\frac{df_2}{d\xi}\right), \left(\frac{df_3}{d\xi}\right)$; and those of a curve of the second system, when ξ = a constant, are proportional to $\left(\frac{df_1}{d\eta}\right), \left(\frac{df_2}{d\eta}\right), \left(\frac{df_3}{d\eta}\right)$: hence it follows that if ω is the angle at which two curves of the systems intersect at (ξ, η),

$$\cos\omega = \frac{\mathrm{F}}{\mathrm{E}^{\frac{1}{2}}\mathrm{G}^{\frac{1}{2}}}. \tag{94}$$

Hence also

$$\sin\omega = \frac{\{\mathrm{EG}-\mathrm{F}^2\}^{\frac{1}{2}}}{\mathrm{E}^{\frac{1}{2}}\mathrm{G}^{\frac{1}{2}}} = \frac{\mathrm{V}}{\mathrm{E}^{\frac{1}{2}}\mathrm{G}^{\frac{1}{2}}}, \tag{95}$$

if $\mathrm{V}^2 = \mathrm{EG}-\mathrm{F}^2$, (96)

$$= \left|\left(\frac{df_2}{d\xi}\right), \left(\frac{df_3}{d\eta}\right)\right|^2 + \left|\left(\frac{df_3}{d\xi}\right), \left(\frac{df_1}{d\eta}\right)\right|^2 + \left|\left(\frac{df_1}{d\xi}\right), \left(\frac{df_2}{d\eta}\right)\right|^2; \tag{97}$$

the last equivalent of V being expressed in the notation of determinants.

Let θ and θ' be the angles between ds and the lines at (ξ, η) of the first and second systems respectively; then

$$\cos\theta = \frac{\mathrm{E}\,d\xi + \mathrm{F}\,d\eta}{ds\,\mathrm{E}^{\frac{1}{2}}}, \tag{98}$$

$$\sin\theta = \frac{\mathrm{V}\,d\eta}{\mathrm{E}^{\frac{1}{2}}\,ds}. \tag{99}$$

$$\cos\theta' = \frac{\mathrm{F}\,d\xi + \mathrm{G}\,d\eta}{ds\,\mathrm{G}^{\frac{1}{2}}}, \tag{100}$$

$$\sin\theta' = \frac{\mathrm{V}\,d\xi}{\mathrm{G}^{\frac{1}{2}}\,ds}. \tag{101}$$

Also to determine the direction-cosines of the normal to the surface at (ξ, η) we have

$$\mathrm{U}\,dx + \mathrm{V}\,dy + \mathrm{W}\,dz = 0;$$

$$\therefore \quad \left\{\mathrm{U}\left(\frac{df_1}{d\xi}\right) + \mathrm{V}\left(\frac{df_2}{d\xi}\right) + \mathrm{W}\left(\frac{df_3}{d\xi}\right)\right\} d\xi$$
$$+ \left\{\mathrm{U}\left(\frac{df_1}{d\eta}\right) + \mathrm{V}\left(\frac{df_2}{d\eta}\right) + \mathrm{W}\left(\frac{df_3}{d\eta}\right)\right\} d\eta = 0;$$

whence, since ξ and η are independent,

$$\left.\begin{aligned} \mathrm{U}\left(\frac{df_1}{d\xi}\right) + \mathrm{V}\left(\frac{df_2}{d\xi}\right) + \mathrm{W}\left(\frac{df_3}{d\xi}\right) = 0, \\ \mathrm{V}\left(\frac{df_1}{d\eta}\right) + \mathrm{V}\left(\frac{df_2}{d\eta}\right) + \mathrm{W}\left(\frac{df_3}{d\eta}\right) = 0: \end{aligned}\right\}$$

$$\therefore \frac{\text{U}}{\left|\left(\frac{df_2}{d\xi}\right),\left(\frac{df_3}{d\eta}\right)\right|} = \frac{\text{V}}{\left|\left(\frac{df_3}{d\xi}\right),\left(\frac{df_1}{d\eta}\right)\right|} = \frac{\text{W}}{\left|\left(\frac{df_1}{d\xi}\right),\left(\frac{df_2}{d\eta}\right)\right|} = \frac{\text{Q}}{\text{V}}; \quad (102)$$

which give the direction-cosines of the normal.

If every one of either of these systems of lines intersects at right angles every one of the other system so that the two systems are orthogonal, then the surface becomes divided into a series of infinitesimal rectangles, and $\omega = 90°$; and consequently from (87), $\text{F} = 0$; that is,

$$\left(\frac{df_1}{d\xi}\right)\left(\frac{df_1}{d\eta}\right) + \left(\frac{df_2}{d\xi}\right)\left(\frac{df_2}{d\eta}\right) + \left(\frac{df_3}{d\xi}\right)\left(\frac{df_3}{d\eta}\right) = 0; \quad (103)$$

in which case, $$ds^2 = \text{E}\, d\xi^2 + \text{G}\, d\eta^2. \quad (104)$$

Also hereafter we shall find that this expression admits of further simplification, by taking such a system of lines, that either E or G is equal to unity.

248.] Although the substance of the preceding article has reference chiefly to length-elements of curves on surfaces, and thus to the rectification of curves of double curvature, yet it arises out of the necessary explanation of the system of coordinates; and it will be moreover of considerable use in the sequel. Now we return to the proper subject of the present chapter, and shall investigate the most general analytical value of the surface-element in terms of curvilinear coordinates. Imagine two systems of curves, as explained above, to be described on the surface; the surface will thereby be divided into elements, an element being contained between two curves of the first system, viz. ξ, and $\xi + d\xi$, and between two curves of the second system, viz. η, and $\eta + d\eta$. This element will be approximately a parallelogram, and the sides of it, which meet at the point (ξ, η), will be $d\sigma$ and $d\sigma'$, the length of which have been determined in (90) and (91); and ω, as determined in (93) or (94), will be the angle between them; so that if $d\text{A}$ is the surface-element,

$$d\text{A} = d\sigma\, d\sigma' \sin\omega$$

$$= \text{V}\, d\eta\, d\xi \quad (105)$$

$$= \left\{\left|\left(\frac{df_2}{d\xi}\right),\left(\frac{df_3}{d\eta}\right)\right|^2 + \left|\left(\frac{df_3}{d\xi}\right),\left(\frac{df_1}{d\eta}\right)\right|^2 + \left|\left(\frac{df_1}{d\xi}\right),\left(\frac{df_2}{d\eta}\right)\right|^2\right\}^{\frac{1}{2}} d\eta\, d\xi; \quad (106)$$

employing the value of V as given by the determinant-notation in (97).

This value of the surface-element may also be found by analytical transformation according to the principles of chap. VIII.

From (86) we have, as in (34), Art. 213,

$$\left.\begin{aligned} dy\,dz &= \left|\left(\frac{df_2}{d\xi}\right), \left(\frac{df_3}{d\eta}\right)\right| d\xi\,d\eta, \\ dz\,dx &= \left|\left(\frac{df_3}{d\xi}\right), \left(\frac{df_1}{d\eta}\right)\right| d\xi\,d\eta, \\ dx\,dy &= \left|\left(\frac{df_1}{d\xi}\right), \left(\frac{df_2}{d\eta}\right)\right| d\xi\,d\eta; \end{aligned}\right\} \tag{107}$$

whence, squaring and adding, and substituting from (30), we have

$$d\text{A} = \left\{\left|\left(\frac{df_2}{d\xi}\right),\left(\frac{df_3}{d\eta}\right)\right|^2 + \left|\left(\frac{df_3}{d\xi}\right),\left(\frac{df_1}{d\eta}\right)\right|^2 + \left|\left(\frac{df_1}{d\xi}\right),\left(\frac{df_2}{d\eta}\right)\right|^2\right\}^{\frac{1}{2}} d\xi\,d\eta; \tag{108}$$

which is the same result as (106).

As this formula is general, it includes all the preceding; and the following are examples of its application.

Ex. 1. Let $x = \xi$, $y = \eta$, $z = f(x, y) = f(\xi, \eta)$.

$$\therefore \quad \left(\frac{df_1}{d\xi}\right) = 1, \quad \left(\frac{df_2}{d\xi}\right) = 0, \quad \left(\frac{df_3}{d\xi}\right) = \left(\frac{dz}{d\xi}\right) = \left(\frac{dz}{dx}\right),$$

$$\left(\frac{df_1}{d\eta}\right) = 0; \quad \left(\frac{df_2}{d\eta}\right) = 1; \quad \left(\frac{df_3}{d\eta}\right) = \left(\frac{dz}{d\eta}\right) = \left(\frac{dz}{dy}\right);$$

$$\therefore \quad d\text{A} = \left\{\left(\frac{dz}{dx}\right)^2 + \left(\frac{dz}{dy}\right)^2 + 1\right\}^{\frac{1}{2}} dx\,dy; \tag{109}$$

which is the expression already determined.

Ex. 2. The equations to the skew helicoid are $x = \xi\cos\eta$, $y = \xi\sin\eta$, $z = ka\eta$.

$$\left(\frac{df_1}{d\xi}\right) = \cos\eta, \quad \left(\frac{df_2}{d\xi}\right) = \sin\eta, \quad \left(\frac{df_3}{d\xi}\right) = 0,$$

$$\left(\frac{df_1}{d\eta}\right) = -\xi\sin\eta; \quad \left(\frac{df_2}{d\eta}\right) = \xi\cos\eta; \quad \left(\frac{df_3}{d\eta}\right) = ka;$$

$$\therefore \quad d\text{A} = (k^2a^2 + \xi^2)^{\frac{1}{2}}\, d\xi\,d\eta.$$

Now this surface is that of the under-side of spiral staircases; thus if the radius of the cylinder which incloses the staircase is a, and the surface of one turn is required, the limits of ξ and η are a and 0, and 2π and 0 respectively: so that if $\text{A} =$ the corresponding area,

$$\begin{aligned} \text{A} &= \int_0^{2\pi}\int_0^a (k^2a^2 + \xi^2)^{\frac{1}{2}}\, d\xi\,d\eta \\ &= \pi a^2\left\{(1 + k^2)^{\frac{1}{2}} + \log\frac{1 + (1 + k^2)^{\frac{1}{2}}}{k}\right\}. \end{aligned} \tag{110}$$

249.] The preceding theory of curvilinear coordinates gives a geometrical explanation of the general transformation of a double integral. For if the double integral involves two variables x and y; and if these are connected with new variables ξ and η by the equations $x = f_1(\xi, \eta)$, $y = f_2(\xi, \eta)$; then we have, as in (34) Art. 213,

$$dx\,dy = \left|\left(\frac{df_1}{d\xi}\right), \left(\frac{df_2}{d\eta}\right)\right| d\eta\, d\xi,$$

$$= \left\{\left(\frac{dx}{d\xi}\right)\left(\frac{dy}{d\eta}\right) - \left(\frac{dx}{d\eta}\right)\left(\frac{dy}{d\xi}\right)\right\} d\eta\, d\xi; \qquad (111)$$

Now the left-hand member of (111) represents the infinitesimal area-element referred to rectangular coordinates, and therefore the right-hand member expresses the analogous element in terms of curvilinear coordinates. Suppose then in fig. 43 P to be the point of intersection of the lines corresponding to ξ and η, and P_3 to be the point of intersection of the lines corresponding to $\xi + d\xi$, $\eta + d\eta$, so that PP_2, P_1P_3 are two consecutive lines of the first system, and PP_1, P_2P_3 are two consecutive lines of the second system: then, in terms of curvilinear coordinates, P is (ξ, η), P_1 is $(\xi + d\xi, \eta)$, P_2 is $(\xi, \eta + d\eta)$, P_3 is $(\xi + d\xi, \eta + d\eta)$; and in terms of rectangular coordinates P is (x, y), P_1 is $\left(x + \left(\frac{dx}{d\xi}\right) d\xi, y + \left(\frac{dy}{d\xi}\right) d\xi\right)$, P_3 is $\left(x + \left(\frac{dx}{d\xi}\right) d\xi + \left(\frac{dx}{d\eta}\right) d\eta,\ y + \left(\frac{dy}{d\xi}\right) d\xi + \left(\frac{dy}{d\eta}\right) d\eta\right)$, and P_2 is $\left(x + \left(\frac{dx}{d\eta}\right) d\eta,\ y + \left(\frac{dy}{d\eta}\right) d\eta\right)$. Hence as the area of a quadrilateral $= \frac{1}{2}\Sigma . |x_i, y_j|$, if dA is the area of the surface-element whose angular points are P, P_1, P_3, P_2,

$$dA = \left\{\left(\frac{dx}{d\xi}\right)\left(\frac{dy}{d\eta}\right) - \left(\frac{dx}{d\eta}\right)\left(\frac{dy}{d\xi}\right)\right\} d\xi\, d\eta. \qquad (112)$$

Hence by the geometry of the figure we have

$$\left.\begin{aligned} MM_1 &= \left(\frac{dx}{d\xi}\right) d\xi, & MM_2 &= \left(\frac{dx}{d\eta}\right) d\eta, \\ N_1P_1 &= \left(\frac{dy}{d\xi}\right) d\xi, & N_2P_2 &= \left(\frac{dy}{d\eta}\right) d\eta; \end{aligned}\right\} \qquad (113)$$

$$PP_1^2 = \left\{\left(\frac{dx}{d\xi}\right)^2 + \left(\frac{dy}{d\xi}\right)^2\right\} d\xi^2; \quad PP_2^2 = \left\{\left(\frac{dx}{d\eta}\right)^2 + \left(\frac{dy}{d\eta}\right)^2\right\} d\eta^2. \quad (114)$$

If the area-element, delineated in the figure, is on a curved surface, it is evident that PP_1 and PP_2 are the lines denoted by $d\sigma$ and $d\sigma'$ in Art. 247, and by a similar process we have

$$ds = \left\{\left(\frac{dx}{d\xi}\right)^2 + \left(\frac{dy}{d\xi}\right)^2 + \left(\frac{dz}{d\xi}\right)^2\right\}^{\frac{1}{2}} d\xi$$

$$= E^{\frac{1}{2}} d\xi; \qquad (115)$$

$$ds = \left\{\left(\frac{dx}{d\eta}\right)^2 + \left(\frac{dy}{d\eta}\right)^2 + \left(\frac{dz}{d\eta}\right)^2\right\}^{\frac{1}{2}} d\eta$$

$$= G^{\frac{1}{2}} d\eta; \qquad (116)$$

which are the same results as before.

250.] The equations, which connect x, y, z with ξ and η, have been taken in (77) to be explicit. Suppose them however to be implicit, and of the forms

$$F_1(x, y, z, \xi, \eta) = 0, \quad F_2(x, y, z, \xi, \eta) = 0, \quad F_3(x, y, z, \xi, \eta) = 0; \qquad (117)$$

then by differentiation we have, making obvious substitutions,

$$\left.\begin{array}{l} a_1\, dx + b_1\, dy + c_1\, dz + \alpha_1\, d\xi + \beta_1\, d\eta = 0, \\ a_2\, dx + b_2\, dy + c_2\, dz + \alpha_2\, d\xi + \beta_2\, d\eta = 0, \\ a_3\, dx + b_3\, dy + c_3\, dz + \alpha_3\, d\xi + \beta_3\, d\eta = 0; \end{array}\right\} \qquad (118)$$

But if $F(x, y, z) = 0$ is the equation to the surface, it results by the elimination of ξ and η from (117); and as hereby x, y, z are connected by a relation, we have also

$$U\, dx + V\, dy + W\, dz = 0. \qquad (119)$$

This relation between dx dy and dz will also arise by the elimination of $d\xi$ and $d\eta$ from (118). Now from (118) we have

$$|a_1, b_2, c_3|\, dx = -\{\alpha_1\, |b_2, c_3| + \alpha_2\, |b_3, c_1| + \alpha_3\, |b_1, c_2|\}\, d\xi$$
$$- \{\beta_1\, |b_2, c_3| + \beta_2\, |b_3, c_1| + \beta_3\, |b_1, c_2|\}\, d\eta; \qquad (120)$$
$$= -|\alpha_1, b_2, c_3|\, d\xi - |\beta_1, b_2, c_3|\, d\eta; \qquad (121)$$

and also similar values for dy and dz. These equations are of the forms (86); and consequently all the theorems which have been deduced from (86) may, mutatis mutandis, be deduced from these equations.

CHAPTER X.

CUBATURE OF SOLIDS.

SECTION 1.—*Cubature of Solids of Revolution; and of similarly generated Solids.*

251.] Another important application of definite multiple integration is that whereby we are able to determine the volume, or the quantity of space, contained within given bounding surfaces, and thereby to compare it with the volume of a cube; whence arises the name Cubature of Solids. The process will generally involve triple integration; and I shall first investigate the most simple case, that viz. of the volume contained within a surface of revolution, and between two given planes which are perpendicular to the axis of revolution.

Let OPQBA, see fig. 17, be the plane surface by the revolution of which about the line OA, which is the x-axis, the volume is generated; and let the equation to OPQB, which is its bounding curve, be $y = f(x)$. Let E be the point (x, y) in the surface on which abuts the infinitesimal surface-element, the area of which is $dy\,dx$; now imagine the surface to have revolved about the line Ox through an angle ϕ; then as ϕ is increased by $d\phi$, the surface element at E will generate a volume, which is approximately a parallelepipedon, of which the base is $dy\,dx$ and the altitude is $y\,d\phi$; and consequently the volume of it $= y\,d\phi\,dy\,dx$; which is a triple differential, the definite integral of which with the proper limits will give the volume of the solid.

To consider the effects of the several integrations. The ϕ-integration with the limits 2π and 0, will give the volume of the ring which is generated by the complete revolution about Ox of the surface-element whose area is $dy\,dx$. The subsequent y-integration with the limits of the ordinate of the curve and of zero, will give the sum of all these rings which are in a plane slice perpendicular to Ox; and this sum is evidently the circular plate of thickness dx which is generated by the revolution of the slice MQ of the generating surface: and the final x-integration will give the sum of all these circular plates, which is the required volume. Thus, if V = the required volume,

$$V = \int_{x_0}^{x_n} \int_0^{f(x)} \int_0^{2\pi} y \, d\theta \, dy \, dx \tag{1}$$

$$= 2\pi \int_{x_0}^{x_n} \int_0^{f(x)} y \, dy \, dx \tag{2}$$

$$= \pi \int_{x_0}^{x_n} \{f(x)\}^2 \, dx. \tag{3}$$

This process resolves the problem of the cubature of a solid of revolution into its most simple elements, and is applicable when the generating area does not revolve through a whole circumference, and in that case the limits of the ϕ-integration will not be 2π and 0, but ϕ_n and ϕ_0 say, which will be given by the conditions of the problem. The process is also applicable when the limits of the y-integration are both ordinates to certain given curves; in which case the solid of revolution is hollow. Examples of this latter case will occur in the sequel. The final integral given in (3) may however be found by the following process without the intervention of the first two integrations. Let, as before, $y = f(x)$ be the equation to the plane curve bounding the area, by the revolution of which the solid is generated; and let the axis of x be that of revolution; see fig. 33; then, as the elemental area PQNM revolves about Ox, it generates a circular slice whose thickness is MN $= dx$, and of whose circular faces one has a radius MP $= y$, and the other has a radius NQ $= y + dy$; therefore the volume of the slice is intermediate to

$$\pi y^2 \, dx \quad \text{and} \quad \pi (y+dy)^2 \, dx;$$

whence, neglecting infinitesimals of the higher orders, as is necessary, the volume of the elemental slice is equal to $\pi y^2 \, dx$; and therefore, if x_n and x_0 are the distances from the origin of the extreme faces of the solid thus formed,

$$\text{the volume} = \pi \int_{x_0}^{x_n} y^2 \, dx \tag{4}$$

$$= \pi \int_{x_0}^{x_n} \{f(x)\}^2 \, dx. \tag{5}$$

252.] The following are examples in illustration of the preceding formula.

Ex. 1. To find the content of a right circular cone, whose altitude is a, and the radius of whose circular base is b.

The equation to the generating line is $y = \frac{b}{a}x$; therefore

$$\text{the volume of the cone} = \pi \frac{b^2}{a^2} \int_0^a x^2\, dx = \frac{\pi b^2 a}{3}.$$

Thus the volume of the cone is equal to one-third of that of the cylinder of equal altitude and equal base.

Ex. 2. To find the content of a paraboloid of revolution whose altitude is a, and the radius of whose base is b.

The equation to the generating curve is $ay^2 = b^2 x$; and

$$\text{the volume of the paraboloid} = \frac{\pi b^2}{a} \int_0^a x\, dx = \frac{\pi b^2 a}{2}.$$

Thus the volume of the paraboloid is one-half of that of the circular cylinder on the same base and of the same altitude.

Ex. 3. To find the volume generated by a circular area of radius $= a$, revolving about the diameter, the abscissæ to the extremities of which are x_n and x_0. Here, since $x^2 + y^2 = a^2$,

$$\begin{aligned}\text{the volume} &= \pi \int_{x_0}^{x_n} (a^2 - x^2)\, dx \\ &= \pi \left\{a^2 (x_n - x_0) - \frac{1}{3}(x_n^3 - x_0^3)\right\}.\end{aligned}$$

Hence the volume of a spherical segment

$$\begin{aligned}&= \pi \int_x^a (a^2 - x^2)\, dx \\ &= \pi \left\{\frac{2a^3}{3} - a^2 x + \frac{x^3}{3}\right\};\end{aligned}$$

and the volume of the sphere $= 2\pi \int_0^a (a^2 - x^2)\, dx = \frac{4}{3}\pi a^3$.

Ex. 4. The volume of a prolate spheroid $= \frac{4}{3}\pi a b^2$.

Ex. 5. Determine the volume formed by the revolution about its base of the cycloid, whose equations are

$$x = a(\theta - \sin\theta), \qquad y = a(1 - \cos\theta).$$

$$\text{the volume} = 2\pi a^3 \int_0^\pi (1 - \cos\theta)^3\, d\theta = 5\pi^2 a^3.$$

Ex. 6. The volume generated by the revolution of the cycloid about its axis is

$$\pi a^3 \left(\frac{3\pi^2}{2} - \frac{8}{3}\right).$$

Ex. 7. The volume of the solid generated by the revolution of $y = a\log x$, about the axis of x, with the limits x and 0,

$$= \pi a^2 x\{(\log x)^2 - 2\log x + 2\}.$$

Ex. 8. The volume of the solid generated by the revolution of $y = a^x$ about the axis of x, with the limits x and $-\infty$, $= \pi(\log a)^{-1} a^{2x}$.

Ex. 9. If the equation to the bounding curve of the revolving area is $xy = k^2$, the volume of the solid $= \pi k^3 (y_0 - y_n)$; and if $y_n = 0$, the volume $= \pi k^3 y_0$.

Ex. 10. Find the curve which by its revolution about the x-axis generates a solid, the volume of which varies as the nth power of (1) the extreme ordinate; (2) the extreme abscissa.

Ex. 11. The volume of the solid generated by the revolution of the equitangential curve round its x-axis $= \dfrac{\pi a^3}{3}$.

253.] If the plane area revolves about the axis of y, and thereby generates a solid, then, as is manifest from fig. 34,

$$\text{the volume of the solid} = \pi \int x^2 dy; \qquad (6)$$

$x^2 dy$ being expressed in terms of a single variable by means of the equation to the bounding curve, and the limits being assigned by the conditions of the problem.

Ex. 1. Determine the volume of an oblate spheroid. Since

$$\frac{x^2}{a^2} + \frac{y^2}{b^2} = 1,$$

$$\text{the volume} = 2\pi \int_0^b \frac{a^2}{b^2}(b^2 - y^2)\,dy = \frac{4}{3}\pi a^2 b.$$

Ex. 2. The volume formed by the revolution of the cissoid about its asymptote.

The equation to the bounding curve is

$$y^2 = \frac{x^3}{2a - x};$$

see fig. 37; OM $= x$, MP $= y$, OA $= 2a$, AM $= 2a - x$; therefore the content of the differential circular slice PQQ′P′ is $\pi(2a - x)^2 dy$;

$$\begin{aligned}\text{the volume} &= 2\pi \int_0^{2a} (2a - x)^2 dy \\ &= \pi \int_0^{2a} (3a - x)(2ax - x^2)^{\frac{1}{2}} dx \\ &= 2\pi^2 a^3.\end{aligned}$$

Ex. 3. The equation to the Witch of Agnesi being

$$y^2 = 4a^2 \frac{2a-x}{x},$$

the volume of the solid formed by its revolution about the asymptote is $4\pi^2 a^3$.

Ex. 4. The volume of the solid generated by the revolution of the right-angled triangle whose sides are a, b, c about a line passing through A and parallel to BC $= \dfrac{2\pi a b^2}{3}$.

Ex. 5. The volume of the solid generated by the revolution of the companion to the cycloid about its base $= 3\pi^2 a^3$.

254.] If the plane surface by the revolution of which the solid is formed, is referred to polar coordinates, and the prime radius is the axis of revolution, then, see fig. 26, the area of the infinitesimal area-element abutting at E is $r\,dr\,d\theta$, and the perpendicular distance of E from the prime radius is $r \sin\theta$; so that if the plane surface revolves about OX through an angle $d\phi$, it generates an approximate parallepipedon of which the base is $r\,dr\,d\theta$, and the altitude is $r\sin\theta\,d\phi$, and consequently the volume of it $= r^2 \sin\theta\,d\phi\,dr\,d\theta$; and the triple definite integral of this differential will give the volume of the generated solid. Now the ϕ-integration with the limits 2π and 0 will give the volume of the ring generated by the complete revolution of the surface-element whose area $= r\,dr\,d\theta$; the r-integration, with the limits of the radius vector to the curve and of zero, will give the sum of all these rings, which is evidently a conical shell, the thickness of which varies directly as the distance from the vertex; and the final θ-integration, with the limits assigned by the problem, will give the whole required volume. If these final limits are θ_n and 0, the solid will be full; but if the inferior limit is a finite angle, say θ_0, the solid will be hollow. Thus if V is the required volume, and $\mathrm{r} = f(\theta)$ is the equation to the bounding curve,

$$\mathrm{V} = \int_{\theta_0}^{\theta_n}\int_0^{\mathrm{r}}\int_0^{2\pi} r^2 \sin\theta\,d\phi\,dr\,d\theta, \tag{7}$$

$$= 2\pi \int_{\theta_0}^{\theta_n}\int_0^{\mathrm{r}} r^2 \sin\theta\,dr\,d\theta, \tag{8}$$

$$= \frac{2\pi}{3}\int_{\theta_0}^{\theta_n} \{f(\theta)\}^3 \sin\theta\,d\theta. \tag{9}$$

Ex. 1. The volume of the solid formed by the revolution of the cardioide whose equation is $r = a(1+\cos\theta)$, about the prime radius.

Here $r = a(1+\cos\theta) = 2a\left(\cos\frac{\theta}{2}\right)^2$.

$$\begin{aligned} \text{The volume } &= \int_0^{\pi}\int_a^{r}\int_0^{2\pi} r^2 \sin\theta\, d\phi\, dr\, d\theta \\ &= 2\pi\int_0^{\pi}\int_0^{r} r^2\sin\theta\, dr\, d\theta \\ &= \frac{16\pi a^3}{3}\int_0^{\pi}\left(\cos\frac{\theta}{2}\right)^6 \sin\theta\, d\theta \\ &= \frac{8\pi a^3}{3}. \end{aligned}$$

Ex. 2. The volume of the solid formed by the revolution of a sector of a circle, the angle of which $= \alpha$, about one of its bounding radii $= \frac{2\pi a^3}{3}(1-\cos\alpha)$.

Ex. 3. The volume of the solid formed by the revolution of the circle whose equation is $r = 2a\sin\theta$ about the prime radius $= 2\pi^2 a^3$.

Ex. 4. The volume of the solid formed by the revolution of the spiral of Archimedes, whose equation is $r = a\theta$, with the limits $\theta = \pi$, and $\theta = 0$, $= \frac{2\pi^2 a^3}{3}(\pi^2-6)$.

Ex. 5. If an ellipse revolves about its latus rectum, the volumes of the solids generated by the larger and smaller segments are respectively equal to

$$\frac{4\pi a^3}{3}(1-e^2)\left\{\frac{2+e^2}{2} + \frac{3e}{(1-e^2)^{\frac{1}{2}}}\tan^{-1}\left(\frac{1+e}{1-e}\right)^{\frac{1}{2}}\right\}, \text{ and}$$

$$\frac{4\pi a^3}{3}(1-e)^2\left\{\frac{2+e^2}{2} - \frac{3e}{(1-e^2)^{\frac{1}{2}}}\tan^{-1}\left(\frac{1-e}{1+e}\right)^{\frac{1}{2}}\right\}.$$

Hence, if an ellipse revolves about its latus rectum, the excess of volume generated by the larger segment over that generated by the smaller $= 2\pi^2 e a^2 b$.

Ex. 6. The volume formed by the revolution of the loop of the lemniscata, $r^2 = a^2\cos 2\theta$, about its prime radius

$$= \pi a^2\left\{\frac{2^{\frac{1}{2}}}{8}\log(2^{\frac{1}{2}}+1) - \frac{1}{12}\right\}.$$

Ex. 7. The vertex of a right cone, whose vertical angle is 2α, is on the surface of a sphere, and the axis of the cone passes through the centre; prove that the volume contained within the cone and the sphere $= \frac{4\pi a^3}{3}\{1-(\cos\alpha)^4\}$.

255.] The surfaces which bound the volumes investigated in the preceding articles are all surfaces of revolution, and are therefore generated by circles whose planes are parallel, and whose radii vary according to a law assigned by the equation of the bounding curve; and the preceding method of cubature consists in the summation by the Integral Calculus of the circular slices into which the solid admits of being resolved: a similar method therefore is applicable to solids whose bounding surfaces are generated by curves or lines moving according to other given laws: the following cases exemplify the process.

Ex. 1. To find the volume of the elliptical cone, of a given altitude, which is generated by a varying ellipse moving parallel to itself and perpendicular to the axis of x, along which its centre moves, and the semi-axes of which are the ordinates to two straight lines intersecting at the origin and lying respectively in the planes of (z, x) and (x, y).

Let the equations to the lines be $az = cx$, $ay = bx$; then, as the area of an ellipse whose semi-axes are α and β is $\pi\alpha\beta$, the volume of the elliptic elemental slice of the cone contained between two planes perpendicular to the axis of x, and at a distance dx apart, is $\frac{\pi bc}{a^2} x^2\, dx$;

$$\text{therefore the volume of the cone} = \frac{\pi bc}{a^2}\int_0^x x^2\, dx = \frac{\pi bc x^3}{3a^2};$$

x being the altitude of the cone.

Ex. 2. Let the generating ellipse move as in the preceding example, and let its semi-axes be the ordinates of parabolas respectively in the planes of (z, x) and (x, y) whose equations are $z^2 = 4ax$, $y^2 = 4bx$; so that the volume of the elemental slice, whose thickness is dx, $= 4\pi(ab)^{\frac{1}{2}}\, x\, dx$;

$$\begin{aligned}\text{therefore the volume of the solid} &= 4\pi(ab)^{\frac{1}{2}}\int_0^x x\, dx \\ &= 2\pi(ab)^{\frac{1}{2}} x^2.\end{aligned}$$

The bounding surface is the elliptic paraboloid.

Ex. 3. The volume of the ellipsoid may by a similar process be shewn to be equal to $\frac{4}{3}\pi abc$.

Ex. 4. Two equal quadrants of circles being described from the origin of coordinates as centre, and in the planes of (z, x) and (z, y), a variable square moving parallel to the plane of (x, y), and having the ordinates of the quadrants for sides, generates a surface called the Groin; it is required to determine the volume included between it and the coordinate planes.

The equations to the director-circles are, see fig. 38,

$$x^2+z^2 = a^2, \qquad y^2+z^2 = a^2;$$

$\therefore$ the volume of the square-element of the solid $= (a^2-z^2)\,dz$;

$$\therefore \quad \text{the volume of the solid} = \int_0^a (a^2-z^2)\,dz$$
$$= \frac{2a^3}{3}.$$

As the mode of generation may be continued into the other seven octants, the volume of the whole groin $= \frac{16}{3}a^3$.

Ex. 5. A surface is generated by a rectangle moving parallel to the plane of (x, y), one of its sides being the ordinate of a given straight line passing through the origin which is in the plane of (y, z); and the other being an ordinate of a semicircle which is in the plane of (x, z), and passes through the origin, and whose diameter is coincident with the axis of z: it is required to find the volume of the solid.

Let the equations to the director-lines be

$$y = az, \qquad x^2 = 2az-z^2;$$

$$\text{then the volume} = \frac{\pi a a^3}{2}.$$

Ex. 6. On the double ordinates of an ellipse, and in planes perpendicular to that of the ellipse, isosceles triangles, of vertical angle $= 2\alpha$, are described; prove that the volume of the surface thus generated $= \frac{4ab^2}{3\tan\alpha}$.

256.] If the generating plane area has not the axis of revolution for one of its containing sides, but is bounded by two curves whose equations are $y = \text{F}(x)$ and $y = f(x)$, then the elemental annular slice is equal to $\pi\{(\text{F}(x))^2-(f(x))^2\}\,dx$;

and the volume required $= \pi\int\{(F(x))^2-(f(x))^2\}\,dx$; (10)

the limits of integration being given by the geometry of the problem.

Of the formula (10) the following is an useful result. Suppose, as in fig. 39, that the generating plane, which for convenience sake we will call A, is such as to be divisible into two parts perfectly equal and symmetrical by a straight line parallel to Ox, the axis of revolution, in the same manner as the closed figure EPCP′ is divided by EC which is parallel to Ox: then, if the equation to EPC is $y = f(x)$, when EBC is the axis of x, and if AB $= a$, the equations to EPC and EP′C with respect to Ox as the axis of x are severally $y = a+f(x)$, and $y = a-f(x)$; and therefore by (10), the volume of the solid generated by the revolution of EPCP′ about Ox

$$= \pi\int\{(a+f(x))^2-(a-f(x))^2\}\,dx$$

$$= 4\pi a\int f(x)\,dx$$

$$= 2\pi a \times \text{A}; \qquad (11)$$

that is, the volume of the solid is equal to the product of the revolving area and the circumference of the circle whose radius is the distance between the axis of revolution and that of symmetry.

Ex. 1. The volume generated by a circle of radius a about an axis in its own plane at a distance b from its centre $= 2\pi^2a^2b$.

Ex. 2. The volume generated by an ellipse revolving about a tangent at the extremity of the major axis is $2\pi^2a^2b$.

Ex. 3. The volume generated by an isosceles triangle, whose altitude $= a$ and base $= 2b$, about a line in its plane parallel to and at a distance c from the bisector of the vertical angle $= 2\pi abc$.

SECTION 2.—*Cubature of Solids bounded by any Curved Surface.*

257.] If the bounding surface of the solid is not one of the particular forms discussed in the preceding section, but is determined by a general equation involving the coordinates of a point in space, the problem of cubature involves a triple integral, and is solved by the following process.

First, let position be determined by means of a system of rectangular coordinates fixed in space; see fig. 40; and let E be a point within the space whose volume is to be found, and let E be (x, y, z); let F be another point within the space, and infinitesimally near to E, and let F be $(x+dx, y+dy, z+dz)$; then the volume of the infinitesimal rectangular parallelepipedon of which E and F are two opposite angles is $dx\, dy\, dz$: and the aggregate of all such, for the limits of integration assigned by the problem, is the required volume. If therefore V represents the required volume,

$$\text{V} = \iiint dz\, dy\, dx; \tag{12}$$

and the volume depends on the integration of this triple integral for the assigned limits.

Let us consider the effects of these successive integrations, and the relations between the limits of integration and the geometrical data; and, to fix our thoughts, let us suppose the volume to be such as that delineated in fig. 36, and contained within three coordinate planes.

Now $dz\, dy\, dx$ is the volume of the infinitesimal parallelepipedon, one angle of which is at E, whose coordinates are x, y, z; the effect of the z-integration therefore, x and y not changing value, is the determination of the volume of a prismatic column, whose base is $dy\, dx$, and whose altitude is given by the equations to the bounding surfaces: thus, if the volume is such as that delineated in fig. 36, and the equation to the surface is $z = f(x, y)$, the limits of the z-integration are $f(x, y)$ and 0; and the volume of the elemental column, whose height is NP, is $f(x, y)\, dy\, dx$: or if the volume to be determined is that contained between two surfaces whose equations are $z = f_1(x, y)$, $z = f_2(x, y)$ respectively, then the volume of the elemental column is

$$\{f_1(x, y) - f_2(x, y)\}\, dy\, dx; \tag{13}$$

and similarly whatever the bounding surfaces may be.

Suppose that we next integrate with respect to y; the integral expressing the volume is now a double integral, and of the form

$$\text{V} = \iint f(x, y)\, dy\, dx, \tag{14}$$

$f(x, y)$ being a function of x and of y, which is introduced by the limits of z. Taking then fig. 36 to express the normal case, since $f(x, y)\, dy\, dx$ is the volume of the elemental column, the sum of all such determined by the y-integration, when x is constant, is the elemental slice LPKJR contained between two planes parallel to the coordinate plane of (y, z), and at an infinitesimal distance dx apart. In the case therefore of the volume being such as that of the figure, the limits of y are MK and 0, MK being the y to the trace of the surface on the plane of (x, y) and which may therefore be found in terms of x by putting $z = 0$ in the equation of the surface. Or if the volume is contained between two planes parallel to that of (x, z) and at distances y_n, y_0 from it, y_n and y_0 being constants, they are in that case the limits of y; and similarly must the limits be determined if the bounding surface is a cylinder whose generating lines are parallel to the axis of z; in all these cases the result of the y-integration is the volume of a slice contained between two planes at an infinitesimal distance apart, the length of which, viz. that parallel to the axis of y, is a function of its distance, from the plane of (y, z); and therefore if the length is expressed by $\mathrm{F}(x)$,

$$\mathrm{V} = \int \mathrm{F}(x)\, dx, \tag{15}$$

and the volume is the sum of all such differential slices taken between the assigned limits. Thus suppose in fig. 36 the volume contained in the octant to be required, and $\mathrm{OA} = a$, then the limits of x are a and 0: or suppose the required volume to be contained between two planes at distances x_n and x_0 from the plane of (y,z), then x_n and x_0 are the limits of x. The following examples however of such cubatures will serve best to clear up any difficulties which arise from inadequate explanations of general principles.

258.] Examples of cubature.

Ex. 1. Find the content of a rectangular parallelepipedon, three of whose edges meeting at a point are a, b, c.

Let the point at which the edges a, b, c meet be the origin, and let the axes of x, y, z severally lie along them; then if V is the volume,

$$\mathrm{V} = \int_0^a \int_0^b \int_0^c dz\, dy\, dx = \int_0^a \int_0^b c\, dy\, dx$$

$$= \int_0^a b\, c\, dx = abc.$$

Ex. 2. Determine the volume in the first octant of a right elliptical cylinder whose axis is the x-axis, and whose altitude $= a$: and the equation to the generating ellipse is

$$c^2 y^2 + b^2 z^2 = b^2 c^2.$$

Here the superior limit of z is $\frac{c}{b}(b^2-y^2)^{\frac{1}{2}}$, which I will call z, and the inferior limit is 0; the limits of y are b and 0; and of x, a and 0.

$$\begin{aligned}
\mathrm{v} &= \int_0^a \int_0^b \int_0^{\mathrm{z}} dz\, dy\, dx \\
&= \frac{c}{b}\int_0^a \int_0^b (b^2-y^2)^{\frac{1}{2}}\, dy\, dx \\
&= \frac{c}{b}\int_0^a \left[\frac{y\,(b^2-y^2)^{\frac{1}{2}}}{2} + \frac{b^2}{2}\sin^{-1}\frac{y}{b}\right]_0^b dx \\
&= \frac{c}{b}\int_0^a \frac{b^2}{2}\frac{\pi}{2}\, dx = \frac{\pi a\, bc}{4}.
\end{aligned}$$

Ex. 3. To find the volume of the ellipsoid whose equation is

$$\frac{x^2}{a^2} + \frac{y^2}{b^2} + \frac{z^2}{c^2} = 1.$$

Hence the limits of integration for the volume in the first octant are, for z, $c\left\{1-\frac{x^2}{a^2}-\frac{y^2}{b^2}\right\}^{\frac{1}{2}}$, which I will call z, and 0; for y, $b\left(1-\frac{x^2}{a^2}\right)^{\frac{1}{2}}$, which I will call Y, and 0; for x, a and 0;

$$\begin{aligned}
\therefore\ \mathrm{v} &= 8\int_0^a \int_0^{\mathrm{Y}} \int_0^{\mathrm{z}} dz\, dy\, dx \\
&= 8c\int_0^a \int_0^{\mathrm{Y}} \left(1-\frac{x^2}{a^2}-\frac{y^2}{b^2}\right)^{\frac{1}{2}} dy\, dx \\
&= \frac{8c}{b}\int_0^a \int_0^{\mathrm{Y}} (\mathrm{Y}^2-y^2)^{\frac{1}{2}}\, dy\, dx \\
&= \frac{8c}{b}\int_0^a \left[\frac{y\,(\mathrm{Y}^2-y^2)^{\frac{1}{2}}}{2} + \frac{\mathrm{Y}^2}{2}\sin^{-1}\frac{y}{\mathrm{Y}}\right]_0^{\mathrm{Y}} dx \\
&= \frac{8c}{b}\int_0^a \frac{\mathrm{Y}^2}{2}\frac{\pi}{2}\, dx \\
&= \frac{2\pi cb}{a^2}\int_0^a (a^2-x^2)\, dx \\
&= \frac{2\pi bc}{a^2}\left[a^2x-\frac{x^3}{3}\right]_0^a = \frac{4}{3}\pi abc.
\end{aligned}$$

Ex. 4. To find the volume of the Cono-Cuneus of Wallis.

Fig. 41. $\mathrm{OA} = \mathrm{CB} = \mathrm{CE} = a$; $\mathrm{OC} = \mathrm{AB} = c$; and the equation to the surface is, Vol. I, Art. 367, (89), $c^2z^2 = y^2(a^2-x^2)$; therefore for the volume in the first octant the limits are of z, $\frac{y}{c}(a^2-x^2)^{\frac{1}{2}}$, which I will call Z, and 0; of y, c and 0; of x, a and 0. Therefore

$$\text{the volume required} = 4\int_0^a\int_0^c\int_0^Z dz\,dy\,dx$$

$$= 4\int_0^a\int_0^c \frac{y}{c}(a^2-x^2)^{\frac{1}{2}}\,dy\,dx$$

$$= 4\int_0^a \frac{c}{2}(a^2-x^2)^{\frac{1}{2}}\,dx = \frac{\pi a^2 c}{2}.$$

Ex. 5. To determine the volume contained within the surface of an elliptic paraboloid whose equation is

$$\frac{y^2}{a}+\frac{z^2}{b} = 4x,$$

and a plane parallel to that of (y, z), and at a distance c from it.

Hence for the volume in the first octant the limits of integration are, for z, $\left(4bx-\frac{b}{a}y^2\right)^{\frac{1}{2}}$, which I shall call Z, and 0; for y, $2(ax)^{\frac{1}{2}}$, which I shall call Y, and 0; for x, c and 0;

$$\therefore\ \text{the volume required} = 4\int_0^c\int_0^Y\int_0^Z dz\,dy\,dx$$

$$= 4\int_0^c\int_0^Y \left(\frac{b}{a}\right)^{\frac{1}{2}}(Y^2-y^2)^{\frac{1}{2}}\,dy\,dx$$

$$= 4\left(\frac{b}{a}\right)^{\frac{1}{2}}\int_0^c \pi a x\,dx = 2\pi c^2(ab)^{\frac{1}{2}}.$$

Ex. 6. Find the volume of the solid cut from the cylinder $x^2+y^2 = a^2$ by the planes $z = 0$, and $z = x\tan\alpha$.

Here the limits of integration are, for z, $x\tan\alpha$, which I will call Z, and 0; for y, $(a^2-x^2)^{\frac{1}{2}}$, which I will call Y, and $-$Y; and for x, a and 0; so that, if V is the volume,

$$V = \int_0^a\int_{-Y}^{Y}\int_0^Z dz\,dy\,dx$$

$$= 2\tan\alpha\int_0^a x(a^2-x^2)^{\frac{1}{2}}\,dx$$

$$= \frac{2}{3}a^3\tan\alpha.$$

Ex. 7. If the general equation of the second degree represents a closed surface, explain the process by which the several limits of integration as required in Art. 214 are determined.

Corresponding to given values of x and y there are generally two values of z, say z_1 and z_0, which are the limits of the z-integration, so that the z-integration gives the volume of a right prism which is perpendicular to the plane of (x, y), whose altitude is $z_1 - z_0$, and whose base is $dy\,dx$. The subsequent y-integration gives the sum of all those which lie in a thin slice parallel to the plane of (y, z), the limits of this integration being fixed by those points in that slice at which the tangent plane of the closed surface is parallel to the z-axis, and at which consequently $\left(\frac{dF}{dz}\right) = 0$; and thus if z is eliminated between $F(x, y, z) = 0$, and $\left(\frac{dF}{dz}\right) = 0$, we shall have a quadratic in y, which will give two roots, say Y_1 and Y_0; and these are the limits of the y-integration. The subsequent x-integration will give the sum of all these slices, and thus the volume included within the closed surface; and the limits of the x-integration are determined by the values of the x-ordinate at which these slices parallel to the plane of (y, z) vanish, and at which consequently the surface is touched by planes parallel to that of (y, z). At these points then $F = 0$, $\left(\frac{dF}{dz}\right) = 0$, $\left(\frac{dF}{dy}\right) = 0$, and the elimination of y and z by means of these three equations will give a quadratic in terms of x, the two roots of which are the limits of the x-integration.

259.] If the infinitesimal element-function of a triple integral is of the form $F\left(\frac{x}{a}, \frac{y}{b}, \frac{z}{c}\right)$, and the inequality which assigns the limits of integration is also of the form $f\left(\frac{x}{a}, \frac{y}{b}, \frac{z}{c}\right) < 1$, so that, if I is the integral,

$$I = \int_{X_0}^{X_n} \int_{Y_0}^{Y_n} \int_{Z_0}^{Z_n} F\left(\frac{x}{a}, \frac{y}{b}, \frac{z}{c}\right) dz\,dy\,dx; \tag{16}$$

then, if $x = a\xi$, $y = b\eta$, $z = c\zeta$,

$$I = abc \int_{X'_0}^{X'_n} \int_{Y'_0}^{Y'_n} \int_{Z'_0}^{Z'_n} F(\xi, \eta, \zeta)\, d\zeta\, d\eta\, d\xi, \tag{17}$$

and the inequality which assigns the limits becomes $f(\xi, \eta, \zeta) < 1$; consequently if the definite integral given in (17) is known, that of (16) is also known.

Thus, in Ex. 3 of the preceding article, if x, y, z are replaced severally by $a\xi$, $b\eta$, $c\zeta$, $V = abc \iiint d\zeta\, d\eta\, d\xi$, and the equation to the ellipsoid, which assigns the limits of integration, becomes $\xi^2 + \eta^2 + \zeta^2 = 1$, which is the equation to a sphere whose radius $= 1$. Now $\iiint d\zeta\, d\eta\, d\xi$ for these limits expresses the volume of a sphere whose radius $= 1$; that is, $= \frac{4\pi}{3}$; and therefore

$$\text{v} = \frac{4\pi abc}{3}.$$

Again, since the value of a right cone of altitude $= c$, described on a circular base of radius $= a$, $= \frac{\pi a^2 c}{3}$, that of a right cone of the same altitude on an elliptical base $= \frac{\pi abc}{3}$.

260.] It is frequently convenient to refer the element of a volume to a mixed system of coordinates; as, for instance, to the z-axis, and to polar coordinates in the plane of (x, y); so that in the expression for the volume-element given in (12), $dy\, dx$ must be replaced by $r\, dr\, d\theta$; and consequently

$$\text{v} = \iiint dz\, r\, dr\, d\theta; \tag{18}$$

in this case the base of the elemental prismatic column, which results from the z-integration, is the plane area-element whose area $= r\, dr\, d\theta$; and the volume of that column $= zr\, dr\, d\theta$. Now if the r-integration is next effected, the sum of all these columns, θ being constant, will be a sectorial slice, the limits of r being given by the trace of the surface on the plane of (x, y); and if the inferior limit of r is zero, the edge of the sectorial slice will coincide with the z-axis; and the final θ-integration will give the sum of all such sectorial slices, and consequently the volume required by the conditions of the problem. The following examples illustrate the process.

Ex. 1. The axis of a right circular cylinder of radius b passes through the centre of a sphere of radius a, b being less than a; find the volume of the solid bounded by the surfaces.

Let the plane passing through the centre of the sphere and perpendicular to the axis of the cylinder be that of (x, y); then the equations of the surfaces are $x^2+y^2+z^2 = a^2$, and $x^2+y^2 = b^2$; or, in terms of polar coordinates, the equation to the cylinder is, $r = b$.

For the volume therefore contained in the first octant the limits are, of z, $(a^2-x^2-y^2)^{\frac{1}{2}}$ or $(a^2-r^2)^{\frac{1}{2}}$, which I will call z, and 0; of r, b and 0; of θ, $\frac{\pi}{2}$ and 0; therefore

$$\text{v} = 8\int_0^{\frac{\pi}{2}}\int_0^b\int_0^z dz\, r\, dr\, d\theta$$

$$= \int_0^{\frac{\pi}{2}}\int_0^b (a^2-r^2)^{\frac{1}{2}}\, r\, dr\, d\theta$$

$$= -\frac{8}{3}\int_0^{\frac{\pi}{2}}\Big[(a^2-r^2)^{\frac{3}{2}}\Big]_0^b d\theta$$

$$= \frac{4\pi}{3}\{a^3-(a^2-b^2)^{\frac{3}{2}}\}.$$

Ex. 2. Through the centre of a sphere of radius a passes the surface of a right circular cylinder the radius of whose director-circle is half that of the sphere: find the volume of the solid bounded by the surfaces.

If the centre of the sphere is the origin, and the plane through it and perpendicular to the axis of the cylinder is that of (x, y), then the equations of the sphere and the cylinder are

$$x^2+y^2+z^2 = a^2, \qquad x^2+y^2 = ax;$$

and if the angular point of the volume-element nearest to the origin is (z, r, θ), the equations to these surfaces may be put into the following forms, $z^2 = a^2-r^2$, $r = a\cos\theta$; so that if v is the required volume,

$$\text{v} = 4\int_0^{\frac{\pi}{2}}\int_0^{a\cos\theta}\int_0^{(a^2-r^2)^{\frac{1}{2}}} dz\, r\, dr\, d\theta$$

$$= \frac{2(3\pi-4)}{9}a^3.$$

Ex. 3. To determine the volume contained within a surface of revolution, and two planes perpendicular to the axis of revolution.

Let the axis of revolution be the z-axis; then, by Vol. I, Art. 369, it is evident that the general equation of such surfaces is $x^2+y^2 = f(z)$; and if r is the distance of (x, y, z) from the z-axis, the equation to these surfaces is $r^2 = f(z)$. Now if the angular point of the volume-element nearest to the origin is (z, r, θ), and if $r^2 = f(z)$, and if z_n and z_0 are the distances from the origin of the two bounding planes,

$$\text{v} = \int_{z_0}^{z_n}\int_0^{2\pi}\int_0^r r\, dr\, d\theta\, dz$$

$$= \pi\int_{z_0}^{z_n} f(z)\, dz. \qquad (19)$$

which is the same result as that obtained in Art. 251.

261.] The value of the volume-element given in (18), and the ensuing triple integration is readily applicable to those problems in which the trace of the bounding surface on the plane of (x, y) is a circle or other curve which is expressed in polar coordinates more simply than in rectangular coordinates. For other traces other systems of reference may be more convenient, and these must be left to the ingenuity of the student, the preceding principles being of sufficient breadth for all. I will however take another example, that of the cubature of the ellipsoid. Let V be the volume of the octant contained within the positive rectangular coordinate planes. Then, if z is the height of an elemental prism whose base is $dy\,dx$, $z\,dy\,dx$ is the volume of that prism, and the double integral of this volume-element is the volume required. Now if the equation to the ellipsoid is

$$\frac{x^2}{a^2}+\frac{y^2}{b^2}+\frac{z^2}{c^2} = 1,$$

we may as in (38), Art. 238, replace as follows; let $x = a\sin\alpha\cos\beta$, $y = b\sin\alpha\sin\beta$, $z = c\cos\alpha$; then $dy\,dx = ab\sin\alpha\cos\alpha\,d\alpha\,d\beta$; so that

$$\text{V} = \int\int z\,dy\,dx$$

$$= abc\int_0^{\frac{\pi}{2}}\int_0^{\frac{\pi}{2}} (\cos\alpha)^2\sin\alpha\,d\alpha\,d\beta$$

$$= \frac{\pi abc}{6}.$$

Ex. 4. The volume of the solid contained between the paraboloid $x^2+y^2 = 2az$, the cylinder $x^2+y^2-2ax = 0$, and the plane of (x, y), $= \dfrac{3\pi a^3}{4}$.

Ex. 5. The volume of the solid contained between the surface $z = ae^{-\frac{x^2+y^2}{c^2}}$ and the plane of (x, y), $= \pi a c^2$.

262.] If the system of reference is that of polar coordinates in space, which has been explained in Art. 165, the volume-element will be expressed in the following form. Let E, fig. 42, be the point (r, θ, ϕ), and let H be the point $(r+dr, \theta+d\theta, \phi+d\phi)$; this latter point having been determined from the former by the following process. Let the meridian plane, in which E is, be drawn, and let OE be increased by the infinitesimal element $\text{EJ} = dr$; let the radius OE revolve in the meridian-plane through an infinitesimal angle $\text{EOF} = d\theta$, so that $\text{EF} = r\,d\theta$, and the area

EFGJ $= r\,dr\,d\theta$. Also let the meridian-plane revolve about OZ through an infinitesimal angle NOQ $= d\phi$, whereby the rectangle EFGJ will come into the position IDHK, and each one of the angular points will have passed through a distance equal to EI or NQ, neglecting infinitesimals of the highest orders; but NQ $=$ ON $\times\, d\phi = \rho\, d\phi = r \sin\theta\, d\phi$; and therefore the volume of the elemental parallelepipedon, of which E and H are opposite angular points, is $r^2 \sin\theta\, dr\, d\theta\, d\phi$; and consequently by integration between assigned limits,

$$\text{the volume} = \int\int\int r^2 \sin\theta\, dr\, d\theta\, d\phi. \tag{20}$$

The order in which the integrations are to be effected is scarcely arbitrary; for the form of the equations to surfaces in most cases renders it necessary to integrate first with respect to r; and as the infinitesimal expression represents an element of a pyramid whose vertex is at O, and whose base is an element of the bounding surface, so does the r-integration produce the volume of the pyramid, complete or truncated, according as the inferior limit of r is 0 or a given finite quantity; and every solid admits of being resolved into such pyramids; but the order in which the θ- and ϕ-integrations are effected is arbitrary; and if the θ-integration is effected next after that of r, it produces a sectorial or cuneiform slice of the solid, complete or truncated, according to the limits of r; and the subsequent ϕ-integration will produce the aggregate of all such slices: or if the ϕ-integration is effected next after r, it produces a complete or truncated cuneiform conical slice, the aggregate of all which is found by the subsequent θ-integration.

The particular values of the limits of course may be different for every case, and are assigned by the geometrical conditions of the problem. If the origin is in the interior of a closed surface, the volume contained within which is to be determined, then, if the equation to the bounding surface is $r = f(\theta, \phi)$, this value, as the superior limit of the r-integration, it is convenient to express by R; whereby if V is the volume,

$$\text{V} = \int_0^{2\pi}\int_0^{\pi}\int_0^{\text{R}} r^2 \sin\theta\, dr\, d\theta\, d\phi \tag{21}$$

$$= \frac{1}{3}\int_0^{2\pi}\int_0^{\pi} \{f(\theta, \phi)\}^3 \sin\theta\, d\theta\, d\phi. \tag{22}$$

The preceding expression for the volume-element in terms of

polar coordinates, viz. $r^2 \sin\theta\, dr\, d\theta\, d\phi$, may also be deduced by transformation from the expression of the volume-element, viz., $dz\, dy\, dx$, which is in terms of rectangular coordinates. The equations which connect rectangular and polar coordinates have already been given in (70), Art. 165, and the transformation has been made in (2), Ex. 2, Art. 213, the result being given in (38); so that it is necessary here only to call the attention of the reader to those equivalents.

263.] Examples illustrating the preceding process.

Ex. 1. Determine the volume of a sphere.

Let the radius of the sphere $= a$; then

$$\text{the volume} = \int_0^{2\pi}\int_0^{\pi}\int_0^{a} r^2 \sin\theta\, dr\, d\theta\, d\phi$$

$$= \frac{a^3}{3}\int_0^{2\pi}\int_0^{\pi} \sin\theta\, d\theta\, d\phi$$

$$= \frac{2a^3}{3}\int_0^{2\pi} d\phi = \frac{4\pi a^3}{3}.$$

Ex. 2. Determine the volume of that part of a sphere which abuts at the centre, and is bounded by the ring contained between two parallels of latitude, for which the values of θ are θ_n and θ_0.

$$\text{V} = \int_0^{2\pi}\int_{\theta_0}^{\theta_n}\int_0^{a} r^2 \sin\theta\, dr\, d\theta\, d\phi$$

$$= \frac{2\pi a^3}{3}(\cos\theta_0 - \cos\theta_n).$$

Hence if $\theta_n = a$, and $\theta_0 = 0$, $\text{V} = \dfrac{2\pi a^3}{3}(1-\cos a)$. This is evidently the volume of the solid common to the sphere and to a right circular cone whose vertical angle is $2a$, and whose centre is at the centre of the sphere.

Ex. 3. Determine the general expression of the volume included within a conical surface and a given base.

Let the origin be at the vertex of the cone, and let the axis of z fall within the conical surface; so that the equation to the surface is

$$\frac{y}{z} = f\left(\frac{x}{z}\right); \qquad (23)$$

then this equation, in terms of polar coordinates given by (70), Art. 165, becomes

$$\tan\theta \sin\phi = f(\tan\theta\cos\phi);$$

and let us suppose this to be put into the form

$$\tan\theta = \mathrm{F}(\sin\phi). \tag{24}$$

Now suppose the required volume to be contained between the conical surface, and a plane perpendicular to the z-axis at a distance c from the origin; then the limits of the r-integration are $c\sec\theta$ and 0; of the θ-integration, $\tan^{-1}\mathrm{F}(\sin\phi)$, which I shall call Θ, and 0; and of the ϕ-integration 2π and 0; so that

$$\begin{aligned}
\mathrm{v} &= \int_0^{2\pi}\int_0^{\Theta}\int_0^{c\sec\theta} r^2\sin\theta\, dr\, d\theta\, d\phi \\
&= \frac{c^3}{3}\int_0^{2\pi}\int_0^{\Theta}\tan\theta\,(\sec\theta)^2\, d\theta\, d\phi \\
&= \frac{c^3}{6}\int_0^{2\pi}\{\mathrm{F}(\sin\phi)\}^2\, d\phi;
\end{aligned}$$

but by (24), $c\,\mathrm{F}(\sin\phi) = c\tan\theta =$ the radius vector of the plane base of the cone drawn from the point where it is pierced by the axis of z; therefore by equation (12), Art. 226,

$$\frac{c^2}{2}\int_0^{2\pi}\{\mathrm{F}(\sin\phi)\}^2\, d\phi = \text{the area of the base of the cone};$$

$$\therefore\quad \mathrm{v} = \frac{c}{3}\times\text{the area of the base of the cone};$$

consequently the volume of a cone is one-third of the right cylinder of the same base and of the same altitude.

CHAPTER IX.

ON SOME QUESTIONS IN THE CALCULUS OF PROBABILITIES, AND ON THE DETERMINATION OF MEAN VALUES.

SECTION 1.—*On the Calculus of Probabilities.*

264.] The Calculus of Probabilities contains two or three classes of problems, which cannot be solved without the Integral Calculus; and as they afford a good and apt illustration of definite integration, and are moreover in themselves curious and instructive, it is desirable to devote a few pages to their treatment. It is not of course my intention to enter on a discussion of the difficulties, metaphysical and moral, in which the calculus of probabilities is often enveloped. I shall assume and merely state the mathematical principles of it, which are few and simple, and apply them to certain problems which require definite integration.

The first principle, which is indeed the foundation of the calculus, is the following;

In a system of possible events, of which all are equally probable; that is, in which we know no reason why any one should occur rather than another; the probability in favour of any one is the ratio of the number of events favourable to its occurrence to that of all the events; and the probability against any one is the ratio of the number of events adverse to its occurrence to that of all the events.

The word Chance is synonymous with Probability; so that the preceding principle is valid when the word Probability is replaced by the word Chance.

Hence we have the following definitions;

The chance or probability in favour of an event is the ratio of the number of favourable events to that of all the events.

The chance or probability against an event is the ratio of the number of unfavourable events to that of all the events.

And thus the mathematical definitions take the form of fractions.

The probability in favour of an event is a fraction of which the numerator is the number of the favourable events, and the denominator is that of all the events.

The probability against an event is a fraction of which the numerator is the number of the unfavourable events, and the denominator is that of all the events.

The sum of these probabilities is unity; and as one or other of the possible events must (as it is assumed) occur, the result is certainty. Hence unity is the mathematical definition of certainty.

This result is also otherwise evident: in the case of certainty all the events are favourable, so that the numerator and the denominator of the fraction which defines the favourable chance are equal; and consequently the favourable chance, which is certainty, is unity.

The ratio of the number of favourable events to that of unfavourable events is called the odds in favour of an event.

The ratio of the number of unfavourable events to that of favourable events is called the odds against an event.

Hence it appears that probabilities and odds do not vary, when the number of possible events varies, if the ratio of the numbers of favourable and unfavourable events remains the same.

All these definitions may be expressed mathematically in the following manner; let A and B be two contradictory events; that is, B occurs if A does not occur; then, if $a =$ the number of possible events favourable to the occurrence of A, and $b =$ that of possible events favourable to the occurrence of B, all being equally probable,

$$\text{The chance in favour of A} = \frac{a}{a+b}. \tag{1}$$

$$\text{The chance against A} = \frac{b}{a+b}. \tag{2}$$

$$\text{The odds in favour of A} = \frac{a}{b}. \tag{3}$$

$$\text{The odds against A} = \frac{b}{a}. \tag{4}$$

265.] Now the case to which I propose first to apply these principles is that in which the number of possible events, as also that of the favourable events, is infinite; in which however the numbers of, or the areas or the extents which include these several classes of, events are not equal; so that these are to be calculated;

and they become the numerators or denominators of the fractions which express the chances and the odds. In these problems the several events are no longer discreet cases, as distinct units or integers; of which the sum is found by simple addition; but they are elements of a continuous series, and their sums are definite integrals of which the limits are assigned by the conditions of the question. From the preceding principles we have the following definitions;

The chance in favour of an event

$$= \frac{\text{The extent of favourable cases}}{\text{The extent of all possible cases}}; \quad (5)$$

The chance against an event

$$= \frac{\text{The extent of unfavourable cases}}{\text{The extent of all possible cases}}; \quad (6)$$

and similar definitions are of course true of the odds for and against an event. Although the direct solution of problems of this kind requires the integral calculus for the determination of these several extents, yet in many cases these can be compared without that process, and I propose first to give some examples of this kind.

Ex. 1. A needle is thrown on a plane horizontal table. What is the chance that the line of it is within 15° of the meridian?

In this example the whole extent of possible cases is 360°; and as the needle may lie within 15° of the meridian to the east or to the west, and either towards the north or the south, the extent of favourable cases = 60°. Hence

$$\text{The chance in favour of the event} = \frac{60}{360} = \frac{1}{6}.$$

Ex. 2. A plane surface of indefinite area is divided into equal squares, and a thin circular coin falls flat on the surface. What is the probability that it does not fall across one of the dividing lines of the surface?

Let a = the side of each square, and r = the radius of the coin. As all the squares are equal, it is evident that the estimation of the chance for any one square is sufficient for that of all. Draw lines parallel to the sides of a given square, and within it, at a distance = r from each side; then a new square is described within the original one, each side of which = $a-2r$, and of which consequently the area = $(a-2r)^2$. Now if the centre of the coin falls within this area, the coin does not intersect a dividing line of the area, and consequently the area of favourable cases $= (a-2r)^2$. But as the centre of the coin may fall anywhere within the square, the area of all possible cases $= a^2$. Hence

$$\text{The chance of the coin falling clear of a dividing line} = \frac{(a-2r)^2}{a^2}.$$

The chances may be similarly determined if the plane surface is divided into equilateral triangles or into regular hexagons.

Ex. 3. A circular coin falls flat on a circular plate; determine the chance that (1) the coin does not meet the edge of the plate; (2) the coin meets the edge of the plate and does not fall over; (3) the coin meets the edge and falls over.

Let R = the radius of the plate, r = the radius of the coin; then it is evident that we may estimate only those areas on which the centre of the coin may fall. Now in each of the cases the area of possible events $= \pi (R+r)^2$; and the areas of the favourable events are respectively $\pi(R-r)^2$, $\pi r(2R-r)$, $\pi r(2R+r)$; so that if C_1, C_2, C_3 are the chances,

$$C_1 = \frac{(R-r)^2}{(R+r)^2}; \qquad C_2 = \frac{2Rr-r^2}{(R+r)^2}; \qquad C_3 = \frac{2Rr+r^2}{(R+r)^2}.$$

As these three cases exhaust all the possible events, $C_1+C_2+C_3=1$.

Ex. 4. A heavy spherical ball falls through a grating, the bars of which are infinitesimally thin, and are equidistant. What is the chance that it falls clear of a bar?

Let $2a$ = the distance of the bars from each other, and let r = the radius of the ball; then the chance $= \dfrac{a-r}{a}$.

Ex. 5. Four points are taken at random in a plane of indefinite area. What is the chance that some one of the four points is included within the triangle of which the other three points are the vertices?

With these four points four triangles may be formed, let A, B, C be the three points which form the largest triangle, and let P be the fourth point. Through A, B, C respectively draw the lines B′C′, C′A′, A′B′ parallel to BC, CA, AB respectively; thus forming a new triangle A′B′C′, the area of which is four times that of ABC, and so that A′BC = B′CA = C′AB = ABC. Now P must fall within the area A′B′C′; for if it fell without it, the triangle ABC would not be the greatest which could be made out of the original four points. Consequently A′B′C′ is the area of all possible events;

that is, of events consistent with the asumption that ABC is the largest triangle; and ABC is the area of all the favourable events; and hence

$$\text{The required chance} = \frac{\text{the area of ABC}}{\text{the area of A}'\text{B}'\text{C}'},$$

$$= \frac{1}{4}.$$

The solution of this problem is due to Professor Cayley of Cambridge.

Similarly, if five points are taken at random in space, the chance that the fifth point may be included within the tetrahedron of which the other four points are the vertices $= \frac{1}{27}$.

266.] The following are examples in which the integral calculus is directly required.

Ex. 1. A large plane area is ruled with equidistant straight lines; a thin straight needle, the length of which is less than the distance between two consecutive lines, falls on the plane; what is the probability that it falls clear of a dividing line?

In this problem we may evidently confine our attention to the space contained between two consecutive lines, and to the middle point of the needle with reference to that space.

Let $2a =$ the distance between two consecutive lines; and let $2c$ be the length of the needle. Let the distance between two consecutive lines be bisected by a straight line parallel to each of the lines, and let $x =$ the perpendicular distance of the middle point of the needle in any position from this line; also let ϕ be the angle at which the needle is inclined to the line perpendicular to the dividing lines. Now it is evident that x admits of all values from $-a$ to a, and that ϕ admits of all values from $-\pi$ to π; and that the needle does not intersect a dividing line so long as $x+c\cos\phi$ is less than a, or $-x-c\cos\phi$ is greater than $-a$. Let dx be an infinitesimal-increment of x, which we will suppose to be coincident with the centre of the needle, and let dy be an infinitesimal increment of ϕ, and we will suppose $d\phi$ to be coincident with the needle. Then

$$\text{The extent of all possible cases} = \int_{-a}^{a}\int_{-\pi}^{\pi} d\phi\, dx$$

$$= 4\pi a.$$

Let $a-c\cos\phi = \text{X}$; then

$$\text{The extent of favourable cases} = 2\int_{-\frac{\pi}{2}}^{\frac{\pi}{2}}\int_{-\text{X}}^{\text{X}} dx\, d\phi$$

$$= 4\int_{-\frac{\pi}{2}}^{\frac{\pi}{2}} (a-c\cos\phi)\, d\phi$$

$$= 4(\pi a - 2c).$$

Hence the chance that the needle is clear of a dividing line $= 1-\dfrac{2c}{\pi a}$; and the probability that it falls across a dividing line $= \dfrac{2c}{\pi a}$.

If in the calculation of the favourable extent, the ϕ-integration had preceded the x-integration, then, if $\Phi = \cos^{-1}\dfrac{a-x}{c}$,

The extent of the favourable cases

$$= 8\int_0^{a-c}\int_0^{\frac{\pi}{2}} d\phi\, dx + 8\int_{a-c}^{a}\int_{\Phi}^{\frac{\pi}{2}} d\phi\, dx$$

$$= 4\pi a - 8c;$$

which is the same result as before.

The result of this problem suggests a curious way of finding the value of π. Let a needle of given length be thrown on a table ruled as in the question; and in $m+n$ trials let $m =$ the number of times in which the needle falls clear of a dividing line; then $\dfrac{2c}{\pi a} = \dfrac{n}{m+n}$; so that $\pi = \dfrac{(m+n)2c}{na}$.

Ex. 2. In the adjacent sides OA, OB of a rectangle, two points P and Q are taken at random; what is the probability that the rectangle contained by OP and OQ is not greater than the nth part of that contained by OA and OB?

Let OA $= a$, OB $= b$; OP $= x$, OQ $= y$; then

$$\text{The extent of all possible cases} = \int_0^a\int_0^b dy\, dx$$

$$= ab.$$

In calculating the favourable cases, the limits of x and y are given by the inequality $\dfrac{ab}{n} - xy > 0$; so that for all values of x between $\dfrac{a}{n}$ and 0, y ranges from b to 0; but for any value of x intermediate to a and $\dfrac{a}{n}$, y ranges from $\dfrac{ab}{nx}$ to 0; consequently

$$\text{the extent of favourable cases} = \int_0^{\frac{a}{n}}\int_0^b dy\,dx + \int_{\frac{a}{n}}^a\int_0^{\frac{ab}{nx}} dy\,dx$$

$$= \frac{ab}{n} + \frac{ab}{n}\log n.$$

$$\therefore \quad \text{The required chance} = \frac{1+\log n}{n}.$$

The following is another solution of this problem;

Let OA and OB be taken as rectangular axes of x and y respectively, and let the rectangular hyperbola, whose equation is $xy = \frac{ab}{n}$, be drawn; and let it intersect the sides of the rectangle opposite to OA and OB in C and D respectively; let R be the point (x, y); and from R draw OP and OQ perpendicular to OA and OB respectively; then the conditions of the problem are fulfilled if R falls within the area OADCB.

Now the probability that P falls between x and $x+dx = \frac{dx}{a}$; and that Q falls between y and $y+dy = \frac{dy}{b}$; consequently the probability of the concurrence of these two events $= \frac{dy\,dx}{ab}$; and this is the probability that R falls within the area $dy\,dx$. Hence, since BC $= \frac{a}{n}$,

$$\text{The required probability} = \frac{1}{ab}\left\{\int_0^{\frac{a}{n}}\int_0^b dy\,dx + \int_{\frac{a}{n}}^a\int_0^{\frac{ab}{nx}} dy\,dx\right\}$$

$$= \frac{1+\log n}{n}.$$

Ex. 3. A stick is broken into three parts; what is the chance that the sum of the lengths of every two is greater than the length of the third; that is, that the three parts will form a triangle?

Let the length of the rod $= 2a$; and let $x =$ the length of the first part, and $y =$ the length of the second part. Then

$$\text{The extent of all possible cases} = \int_0^{2a}\int_0^{2a-x} dy\,dx$$

$$= 2a^2.$$

In calculating the favourable cases x may range from a to 0; and since $x+y>a$, y may range from a to $a-x$; so that

$$\text{the extent of favourable cases} = \int_0^a \int_{a-x}^a dy\, dx$$

$$= \frac{a^2}{2}.$$

$$\therefore \quad \text{The required chance} = \frac{1}{4}.$$

Ex. 4. Two arrows are fixed in a circular target. What is the probability that the distance between them is greater than the radius of the target?

Let the radius of the target $= a$; and let $2r =$ the distance of one of the arrows from the centre of the target; so that $4r\,dr\,d\theta$ is the area-element of the target which we may suppose to be occupied by this arrow. Now the area of that part of the target, all of whose points are at a distance from this arrow not less than the radius of the target, $= 2r(a^2-r^2)^{\frac{1}{2}} + 2a^2 \sin^{-1}\frac{r}{a}$. Consequently the extent of the favourable cases

$$= 8\int_0^{\frac{a}{2}}\int_0^{2\pi} \left\{r(a^2-r^2)^{\frac{1}{2}} + a^2 \sin^{-1}\frac{r}{a}\right\} r\, d\theta\, dr$$

$$= 16\pi \int_0^{\frac{a}{2}} \left\{r(a^2-r^2)^{\frac{1}{2}} + a^2 \sin^{-1}\frac{r}{a}\right\} r\, dr$$

$$= \frac{3^{\frac{3}{2}}\pi a^4}{4}.$$

$$\text{The extent of all possible cases} = \int_0^{\frac{a}{2}}\int_0^{2\pi} 4\pi a^2 r\, d\theta\, dr$$

$$= \pi^2 a^4.$$

$$\therefore \quad \text{The required chance} = \frac{3^{\frac{3}{2}}}{4\pi}.$$

Ex. 5. Two arrows are fixed in a circular target. What is the probability that they and the centre of the target are at the angles of an acute-angled triangle?

Let $a =$ the radius of the target; and let P, the place of one of the arrows, be at the point (r, θ). Let O be the centre of the target; join OP, and on OP as a diameter describe a circle; at O and P draw lines perpendicular to OP; then the area of the part of the target included between these two lines

$$= r(a^2-r^2)^{\frac{1}{2}} + a^2 \sin^{-1}\frac{r}{a};$$

and if Q, the place of the other arrow, is in any point of this area, except within the small circle, the triangle OPQ is acute-angled. Hence

The extent of favourable cases

$$= \int_0^{2\pi}\int_0^{a} \left\{ r(a^2-r^2)^{\frac{1}{2}} + a^2 \sin^{-1}\frac{r}{a} - \frac{\pi r^2}{4} \right\} r\, dr\, d\theta$$

$$= \frac{\pi^2 a^4}{4}.$$

The extent of all the cases $= \int_0^{2\pi}\int_0^{a} \pi a^2 r\, dr\, d\theta$

$$= \pi^2 a^4.$$

$\therefore$ The chance required $= \frac{1}{4}$.

As this result is independent of the radius of the target, it is true when that radius is infinite, in which case the centre is an indeterminate point; and hereby we have the conclusion that in infinite circular space, if three points are taken at random, one of them being the centre of the space, the probability that they are the angles of an acute-angled triangle is $\frac{1}{4}$.

267.] The integral calculus is also required for the complete discussion of the laws of combination of events, when the number of events is infinite; and this inquiry is most important in the indirect application of the theory of probabilities to the investigation of the probabilities of the causes to which certain observed events are due; because in most physical investigations, the number of events or of, what amounts to the same thing, observations may be infinite, and it is necessary to ascertain the character which this circumstance brings into the result. Here however I must first have recourse to some elementary considerations of compound events.

Let there be a system of possible events of only two classes; one of which produces A, and the other produces B; and let a and b be the numbers of events of these two classes respectively; and let us suppose all to be equally probable, and assume the necessity that either A or B must occur. Thus $\frac{a}{a+b}$, $\frac{b}{a+b}$ are the probabilities in favour of A and B respectively. Let

$$\frac{a}{a+b} = p, \qquad \frac{b}{a+b} = q; \tag{7}$$

so that p and q are the probabilities in favour of A and B respectively. Also, it is evident that

$$p+q = \frac{a+b}{a+b} = 1, \tag{8}$$

so that $q = 1-p$; and the sum of the probabilities $= 1$, which represents certainty.

Now let us suppose m trials, or observations, to be made on this system of possible events; then the probability that

A will occur m times, - - - - - - $= p^m$,

A - - - - $(m-1)$ times, and B once $= m p^{m-1} q$,

A - - - - $(m-2)$ times, and B twice $= \frac{m(m-1)}{1.2} p^{m-2} q^2$,

. . . .

A - - - - once, and B $(m-1)$ times $= m p q^{m-1}$,

B will occur m times $= q^m$;

these are all the possible combinations of the events, and are evidently the successive terms of the development of a binomial; consequently the sum of all the probabilities $= (p+q)^m = 1$.

Hence it appears that of the several terms of $(p+q)^m$ in its expanded form each expresses the probability of an event compounded of A repeated as many times as is the index of p, and of B repeated as many times as is the index of q. Also that the sum of the first $n+1$ terms gives the probability that in m trials A will occur not less than $m-n$ times; or, what is the same thing, that B will occur not more than n times.

Since the number of terms of $(p+q)^m$ is $m+1$, the number of terms increases as m increases; but as $(p+q)^m = 1$, the sum of the terms is always the same, and consequently each term decreases as m increases. Thus the probability of a particular combination becomes less and less, although the sum of all the probabilities is the same.

These results admit of the following graphic representation. Take a straight line OA of definite length, in the axis of x, say; and divide it into m equal parts; at each of the $m+1$ points of partition draw ordinates in order, severally proportional to the several successive terms of the development of $(p+q)^m$, and join their extremities; we shall hereby have a broken line, which will ultimately become a curve under certain conditions, when $m = \infty$, the ordinates of which represent the probabilities of certain compound events which are assigned by the abscissæ.

Thus, if MP is the ordinate corresponding to the nth point of partition of OA, MP is proportional to the nth term of the developed binomial, and expresses the probability that the event will be compounded of A repeated $m-n+1$ times, and of B repeated $n-1$ times; that is, MP corresponds to the probability that the number of repetition of the events A is to the number of repetition of the events B in the ratio of AM to MO. This curve is called the curve of possibility; and an examination of the several terms of the binomial will at once shew the general course of it.

Let the origin be at the foot of the ordinate which corresponds to the first term; then, as the values of the terms of the binomial continually increase up to a maximum, so will the ordinates of this curve increase, and ultimately attain to a maximum; and afterwards they will decrease as the terms of the binomial series decrease, until the curve comes to a point whose abscissa is OA, which is nearest to the x-axis because it corresponds to the last term of the series. If $p = q$, the curve will be symmetrical on the two sides of its greatest ordinate; but such will not be the case if p and q are not equal.

268.] Now a most important application of the preceding theory is that to the combination of errors of observation. All observations of physical facts, whether made by instruments or otherwise, are subject to certain elementary errors, of which the number is indeterminate, and the causes are unknown; all are supposed to be independent of each other, and to be of the same absolute magnitude, but to be either positive or negative. The true error in each particular case is the algebraical sum of these elementary errors, and the probable true error is the object of our inquiry.

In applying to this case the theory of the preceding article, I shall take the existence of a positive elementary error to be the event A, and that of a negative elementary error to be the event B; so that the number of cases which produce A and B respectively is the same; consequently $a = b$, $p = q = \frac{1}{2}$. Let each elementary error be denoted by Δx; and let $2m =$ the number of them, each of which may be either positive or negative. Then the several terms of $\left(\frac{1}{2}+\frac{1}{2}\right)^{2m}$ give the probabilities of the combinations of the repetitions of the positive elementary errors with those of the negative elementary errors: as the two terms of the binomial are equal, the middle term of the development, which is the $(m+1)$th term, is the largest, and terms equidistant from it are equal. Thus the combination of an equal number of positive and negative elementary errors is the most probable, and the probability of the combination of $2m-n$ positive with n negative errors is equal to that of the combination of n positive errors with $2m-n$ negative errors. Hence the most probable event is that in which there is a compensation of errors, and consequently no resultant error; and the cases that are most likely to occur after this are those in which the errors, whether positive or negative, are small.

Now in this case the curve of possibility is symmetrical relatively to its greatest ordinate; and this corresponds to the middle term of the series. Let the point where this greatest ordinate intersects the axis of x be the origin; and along the x-axis in both directions take a series of lengths, each of which $= \Delta x$, and thus corresponds to an elementary error; and at each of the points on the x-axis thus determined, draw ordinates proportional to the corresponding terms of the expanded binomial. Let $y_0, y_1, y_2, \ldots$ be the ordinates thus drawn; then

$$y_0 = 2^{-2m} \frac{2m(2m-1)\ldots(m+2)(m+1)}{1.2.3\ldots(m-1)m}; \tag{9}$$

$$y_1 = 2^{-2m} \frac{2m(2m-1)\ldots(m+2)}{1.2.3\ldots(m-1)}; \tag{10}$$

. . . .

$$y_k = 2^{-2m} \frac{2m(2m-1)\ldots(m+k+1)}{1.2.3\ldots(m-k)}; \tag{11}$$

$$y_{k+1} = 2^{-3m} \frac{2m(2m-1)\ldots(m+k+2)}{1.2.3\ldots(m-k-1)}; \tag{12}$$

$$\therefore \quad \frac{y_{k+1}}{y_k} = \frac{m-k}{m+k+1}. \tag{13}$$

Let us suppose the number of causes which produce these elementary errors, and consequently the number of errors, to be infinite, so that $m = \infty$; and let us suppose the error due to each one to be infinitesimal, so that Δx becomes dx; then the curve of possibility becomes a continous curve; and if we take y_k to be a general value of the ordinate, to be y, say, $y_{k+1} = y + dy$; also $k\,dx = x$; so that multiplying the numerator and denominator of the second member of (13) by dx, (13) becomes

$$\frac{y+dy}{y} = \frac{m\,dx - x}{m\,dx + x + dx};$$

$$\therefore \quad \frac{dy}{y} = -\frac{2x+dx}{m\,dx + x + dx}$$

$$= -\frac{2x\,dx + dx^2}{m\,dx^2 + x\,dx + dx^2}. \tag{14}$$

Now with reference to the terms in the numerator and denominator of this last fraction, which must be omitted, it will be observed that $m\,dx$ represents the whole length of the x-axis, inasmuch as an infinite number of causes allows it to be extended to any distance; it is indeed that distance beyond which no error can be supposed to reach; and this distance is evidently infinite, inasmuch as we know of no limit to the number of errors; consequently $m\,dx$ will be in the general case infinitely greater than x; and thus if $m\,dx^2 = \frac{1}{h^2}$, (14) becomes

$$\frac{dy}{y} = -2h^2 x\,dx; \tag{15}$$

$$\therefore \quad \log\frac{y}{y_0} = -h^2 x^2; \tag{16}$$

since $y = y_0$, as given in (9), when $x = 0$; hence the equation of the curve of possibility in this case is

$$y = y_0 e^{-h^2 x^2}; \tag{17}$$

and this equation gives the relation between the possibility of an error and the magnitude of that error; that is, if the abscissa represents an error, the ratio of the corresponding ordinate to the greatest ordinate represents the possibility of that error.

The curve whose equation is (17), cuts the axis of y at a distance y_0 from the origin, and the tangent at the point of intersection is parallel to the axis of x, and y_0 is the maximum ordinate. The axis of x is an asymptote; the curve always lying on the positive side of it; and there are points of inflexion when $x = \pm\frac{1}{2h}$. The curve is also evidently symmetrical relatively to the y-axis.

Since the ordinates to this curve are proportional to the possibilities of the errors which correspond to the several abscissæ, a variation of y_0 will change all the ordinates in the same ratio, and consequently will produce no alteration in the relative values of the ordinates. Any variation of h however will produce an important change in the curve, for the greater h is, the more rapidly does the curve approach its asymptote, and the less do the ordinates become for a given abscissa, and consequently the less is the possibility, or the probability, of a given error. Thus h is a measure of the precision of the observations.*

269.] If the curve of possibility is also the curve of probability, then the sum of all the probabilities $= 1$: and as the probabilities are measured by the several ordinates, it is necessary that the sum of all these ordinates, that is, the area contained between the curve and the x-axis, should be equal to unity. Hence

$$1 = \int_{-\infty}^{\infty} y_0 e^{-h^2x^2} dx$$

$$= \frac{y_0 \pi^{\frac{1}{2}}}{h};$$

$$\therefore \quad y_0 = \frac{h}{\pi^{\frac{1}{2}}}; \qquad y = \frac{h}{\pi^{\frac{1}{2}}} e^{-h^2x^2}; \tag{18}$$

so that this last equation is that of the curve of probability.

270.] We can also hereby determine the probability of a given error of observation, say x; for if $p =$ the probability required, it is equal to the ratio of its particular possibility to the sum of all the possibilities. Hence

$$p = \frac{e^{-h^2x^2} dx}{\int_{-\infty}^{\infty} e^{-h^2x^2} dx}$$

$$= \frac{h}{\pi^{\frac{1}{2}}} e^{-h^2x^2} dx, \text{ by (342), Art. 138;} \tag{19}$$

$$= y\,dx;$$

if y is the ordinate of the curve of probability given in (18); so that the probability of an error x is equal to the area-element $y\,dx$ of the curve of probability.

Hence, if P is the probability of an error which is not greater than x,

$$P = \frac{h}{\pi^{\frac{1}{2}}} \int_0^x e^{-h^2x^2} dx. \tag{20}$$

If we compare two observations of different precisions, so that the

* On the subject of the measure of precision (mensura præcisionis), as also on other properties of the preceding function, see Gauss, "Motus Corporum Cœlestium," art. 178.

possibility of error is the same in both cases, that is, so that $\frac{y}{y_0}$ is the same in both cases, then $h^2 x^2 = h'^2 x'^2$; and consequently

$$\frac{x}{x'} = \frac{h'}{h}.$$

that is, the errors vary inversely as the precisions.

271.] Other very important problems in the theory of probabilities are the determinations of the probability of an event, and of a precedent cause of that event, by means of certain events which have been observed. In this case the cause and the probability of its action are supposed to be unknown, so that the problem is the determination of its probability, from a given number of observed events. When the number of possible causes or of hypotheses is finite, we have the following theorem. Let $h_1, h_2, \dots h_n$ be the probabilities of the n possible causes, of which let the probabilities as shewn by observed events be respectively $p_1, p_2, \dots p_n$; then, as the sum of the probabilities of the several possible causes $= 1$,

$$\frac{h_1}{p_1} = \frac{h_2}{p_2} = \dots\dots = \frac{h_n}{p_n} = \frac{1}{\Sigma . p}; \tag{21}$$

and thus the probability of each possible cause is assigned by means of the observed events; h and p are called respectively the a priori and a posteriori probabilities of a possible cause.

Now when this theory is applied to the probabilities of physical facts, the number of hypothetical causes which may be assigned as the fore-runners of these facts is infinite, and the probability of each cause may have any value between the limits 0 and 1; thus, in this case the denominator of (21) is the sum of a continuous series and becomes a definite integral, and the integral calculus is required for the investigation of its properties.

Suppose then we have two contradictory events A and B, of which A has already occurred m times, and B, n times; and of the producing causes of which we know nothing beyond that which these facts supply; then, if $x =$ the probability of A, $1-x$ $=$ the probability of B; and x may have all values ranging from 0 to 1. Then the a posteriori probability of an event compounded of A repeated m times and of B repeated n times is represented by an expression of the form $kx^m (1-x)^n$: and the probabilities of all the possible producing causes will be given by this formula when x varies from 0 to 1; so that as we have hereby a continuous variation, the probability, infinitesimal indeed, of the action of the cause which produces the event A, and of which the probability

$$\text{is } x, \quad = \frac{x^m (1-x)^n \, dx}{\int_0^1 x^m (1-x)^n \, dx} \tag{22}$$

$$= \frac{x^m (1-x)^n \, dx}{\mathrm{B}(m+1, n+1)}, \text{ by reason of (267), Art. 126;}$$

$$= \frac{\Gamma(m+n+2)}{\Gamma(m+1)\,\Gamma(n+1)} x^m (1-x)^n \, dx. \tag{23}$$

If $n = 0$, so that all the past events are A, then the probability of the hypothesis which assigns to the event A the particular probability x

$$= \frac{\Gamma(m+2)}{\Gamma(m+1)} x^m \, dx = (m+1)\, x^m \, dx. \tag{24}$$

272.] Hence we can easily deduce the probability of the future event A. As (22) gives the probability of the hypothesis which assigns to A the particular probability x, the probability of the occurrence of that event

$$= \frac{x^{m+1} (1-x)^n \, dx}{\int_0^1 x^m (1-x)^n \, dx};$$

and the sum of all these probabilities, as x varies from 0 to 1, will give the whole probability of the event A. Hence

$$\text{the probability of A} = \frac{\int_0^1 x^{m+1} (1-x)^n \, dx}{\int_0^1 x^m (1-x)^n \, dx} \tag{25}$$

$$= \frac{\mathrm{B}(m+2, n+1)}{\mathrm{B}(m+1, n+1)}$$

$$= \frac{\Gamma(m+2)\,\Gamma(n+1)}{\Gamma(m+n+3)} \; \frac{\Gamma(m+n+2)}{\Gamma(m+1)\,\Gamma(n+1)}$$

$$= \frac{m+1}{m+n+2}. \tag{26}$$

$$\text{Similarly the probability of B} = \frac{n+1}{m+n+2}. \tag{27}$$

The sum of these probabilities $= 1$, as the result ought to be, since the events are contradictory.

If the event A has occurred m times consecutively, and B has not occurred at all, $n = 0$; so that

$$\left.\begin{aligned}&\text{the probability of the occurrence of A} = \frac{m+1}{m+2};\\&\text{the probability of the occurrence of B} = \frac{1}{m+2}.\end{aligned}\right\} \qquad (28)$$

Thus, for example, if the Sun has already risen m times, and it has not been observed not to rise,

$$\text{the probability of the Sun's rising again} = \frac{m+1}{m+2}:$$

which = 1, that is, certainty, when $m = \infty$.

273.] When of two contradictory events A and B, A, of which the probability $= x$, has already occurred m times, and B, n times, the probability that in the next $(p+q)$ times A will occur p times and B q times may thus be found.

By reason of Art. 267 the probability of the required compound event

$$= \frac{(p+q)(p+q-1)\dots(p+1)}{1.2.3.\dots q} x^p (1-x)^q;$$

and multiplying this by the probability of the hypothesis which is given in (22); and taking the sum of all the products thus formed for the limits 0 and 1 of x, the required probability

$$= \frac{(p+q)(p+q-1)\dots(p+1)}{1.2.3.\dots q} \frac{\int_0^1 x^{m+p}(1-x)^{n+q}\,dx}{\int_0^1 x^m (1-x)^n\,dx}$$

$$= \frac{\Gamma(p+q+1)\,\Gamma(m+p+1)\,\Gamma(n+q+1)\,\Gamma(m+n+2)}{\Gamma(p+1)\,\Gamma(q+1)\,\Gamma(m+1)\,\Gamma(n+1)\,\Gamma(m+n+p+q+2)}. \qquad (29)$$

If $n = q = 0$, then the probability that A will occur p times in succession, when it has already occurred m times,

$$= \frac{\Gamma(m+p+1)\,\Gamma(m+2)}{\Gamma(m+1)\,\Gamma(m+p+2)} = \frac{m+1}{m+p+1}. \qquad (30)$$

Thus, if a coin has turned up heads four times in succession, and no tail has occurred, the probability that it will turn up heads the following three times in succession $= \frac{5}{8}$.

If $p = m$, and m is very large, the probability that A will occur m times without interruption, when it has already occurred m times without interruption, $= \frac{1}{2}$.

We may also hereby determine the value of x which gives the most probable of all the hypothetical causes which can produce A. Thus let P be the general probability; then by (22),

$$P = \frac{x^m (1-x)^n \, dx}{\int_0^1 x^m (1-x)^n \, dx}. \qquad (31)$$

Now the denominator of this fraction being a definite integral does not vary with x. Consequently if $\frac{dP}{dx} = 0$, we have,

$$x^{m-1}(1-x)^{n-1}\{m(1-x)-nx\} = 0;$$

$$\therefore \quad x = \frac{m}{m+n}; \qquad 1-x = \frac{n}{m+n}. \qquad (32)$$

Thus, the most probable of all the hypotheses is that in which the probabilities of the events A and B are equal to the ratios which the number of the favourable past events bear to the whole number of past events.

274.] From the preceding equations we can also deduce the following problem, which is of considerable interest. If in a certain number of observations in which two contradictory events A and B are possible, one, say A, has occurred more frequently than the other, we are naturally led to suspect the existence of some cause of this result. If the producing causes of the two were equal, one should in the long run occur as frequently as the other; but if this equality is interrupted, we suspect a preponderance of cause in favour of that event which more frequently occurs; that is, we suspect that the probability in favour of that event is greater than $\frac{1}{2}$; and the suspected preponderance increases according as the number of events of that particular kind increases. In this case if A is that event on the side of which the preponderance exists, then the limits of x, which is its probability, are 1 and $\frac{1}{2}$: so that if P is the probability of the existence of a cause which produces A, by (31)

$$P = \frac{\int_{.5}^1 x^m (1-x)^n \, dx}{\int_0^1 x^m (1-x)^n \, dx}. \qquad (33)$$

If $n = 0$, so that A has occurred m times without interruption, and B has not occurred at all,

$$P = \frac{\int_{.5}^1 x^m \, dx}{\int_0^1 x^m \, dx} = \frac{2^{m+1}-1}{2^{m+1}}; \qquad (34)$$

which expresses the probability of the existence of a cause, that is, of an hypothesis the probability of which is greater than $\frac{1}{2}$, which favours the repetition of A.

Thus, for example, all the eighty-eight planets which constitute the Solar system as at present (Oct. 1, 1864) known, have a direct motion; that is, move round the Sun from west to east. And, as the motion of all these in the same direction is the repetition of the same fact, whereas the contrary fact might exist, it shews the high probability of the existence of a cause which produces this fact: and if P is this probability,

$$\text{P} = \frac{2^{89}-1}{2^{89}},$$

which fraction nearly $=1$; and consequently the doctrine of chances shews the well-nigh absolute certainty of a physical cause of this fact.

The limits of my work preclude me from giving other problems in this most interesting branch of the higher mathematics, and I can only refer the reader to treatises where the subject is specially investigated.

SECTION 2.—*On the Determination of Mean Values.*

275.] When n different values are assigned to the variable of a function, so that the function thereby receives n values, the nth part of the sum of these values is called the mean or average value of the function, or in more precise terms, the arithmetical mean value of the function. This is the definition of mean value when the values of the function arise from discontinuous values of the variable; but an analogous definition is also applicable when the variable varies continuously. In this case let us suppose $f(x)$ to be a function of x which varies continuously, and does not become infinite, between the limits x_n and x_0; and let us suppose $x_n - x_0$ to be divided into n equal parts each of which $= i$; so that $x_n - x_0 = ni$; then the mean value of the functions corresponding to the several points of partition

$$= \frac{f(x_0)+f(x_0+i)+f(x_0+2i)+\ldots+f\{x_0+(n-1)i\}}{n}. \quad (35)$$

Let the numerator and denominator be multiplied by i; and let us suppose x to increase continuously from x_0 to x_n; then

$$\text{The mean value of } f(x) = \frac{\{f(x_0)+f(x_0+i)+\ldots+f(x_n-i)\}\,i}{ni}$$

$$= \frac{\int_{x_0}^{x_n} f(x)\,dx}{x_n - x_0}$$

$$= \frac{\int_{x_0}^{x_n} f(x)\,dx}{\int_{x_0}^{x_n} dx}; \qquad (36)$$

and this is the definition of mean or average value in its application to a continuous function. The numerator and denominator will be definite multiple integrals, if their elements are multiple differentials. The following examples are in illustration of their definitions.

Ex. 1. Find the mean distance of all points within a circle (1) from the centre; (2) from a point in the circumference, the radii vectores to the points being regularly distributed in each case.

The meaning of the last condition is, that the finite angle, through which the radii vectores drawn to the several points of the circular area extend, is divided into equal elements. Let a = the radius of the circle; and in both cases let the point P from which the line is drawn be coincident with and expressed by the area-element; so that $r\,dr\,d\theta$ denotes that point. Then as r is the distance of this point from the origin, $r^2\,dr\,d\theta$ is the element of the definite integral which is the numerator, and $r\,dr\,d\theta$ is the element of the definite integral which is the denominator of the fraction whereby the mean value is determined.

That is, point til that spa as also the 'ordin' taken in space

In case (1), $\quad$ the numerator $= \int_0^{2\pi}\int_0^a r^2\,dr\,d\theta = \dfrac{2\pi a^3}{3}$;

the denominator $= \int_0^{2\pi}\int_0^a r\,dr\,d\theta = \pi a^2$;

$\therefore$ the required mean value $= \dfrac{2a}{3}$.

In case (2), $\quad$ if $\mathrm{r} = 2a\cos\theta$,

$$\text{The numerator} = \int_{-\frac{\pi}{2}}^{\frac{\pi}{2}}\int_0^{\mathrm{r}} r^2\,dr\,d\theta$$

$$= \frac{8a^3}{3}\int_{-\frac{\pi}{2}}^{\frac{\pi}{2}} (\cos\theta)^3\,d\theta = \frac{32a^3}{9};$$

$$\text{the denominator} = \int_{-\frac{\pi}{2}}^{\frac{\pi}{2}}\int_0^{\mathrm{r}} r\,dr\,d\theta = \pi a^2;$$

$\therefore$ the required mean value $= \dfrac{32a}{9\pi}$.

Ex. 2. The mean distances of all points within a sphere from its centre $= \frac{3a}{4}$.

Ex. 3. Find the mean inclination to a given plane of a system of planes, whose number is infinite, regularly distributed in space.

Let the positions of these planes be determined by means of their normals, and let us suppose all these to be drawn through the centre of a given sphere: then, as these planes are regularly distributed in space, the number of normals to them contained within a given portion of the surface of the sphere varies as that portion, and consequently we may consider the surface-element of the sphere to express the point of section of the normal of a particular plane with it. Now if $a =$ the radius of the sphere, and (a, θ, ϕ) is the place of this element in reference to the system of polar coordinates explained in Art. 165, $a^2 \sin\theta \, d\theta \, d\phi$ is by (69), Art. 243, the area-element of the surface; so that if the given plane is the plane of (x, y), θ is the angle at which the other plane is inclined to it. Consequently in this problem

$$\text{The numerator of (36)} = a^2 \int_0^{2\pi} \int_0^{\frac{\pi}{2}} \theta \sin\theta \, d\theta \, d\phi$$
$$= 2\pi a^2;$$

$$\text{The denominator of (36)} = a^2 \int_0^{2\pi} \int_0^{\frac{\pi}{2}} \sin\theta \, d\theta \, d\phi$$
$$= 2\pi a^2.$$

$\therefore$ the mean angle $= 1 = 57.29578°$;

that is, is equal to the angle, the subtending arc of which is equal to the radius.

In the eight large planets, the mean value of the inclination of the planes of their orbits to the plane of the ecliptic is 2° 19′ 64″; and in case of seventy-nine small planets the mean inclination of the planes of their orbits to that of the earth's orbit is 7° 41′ 3″. Consequently as these observed results are so far below the mean a priori determined as above, we are necessarily led to infer a physical connection between these planets and the plane of the ecliptic; that is, the doctrine of chances indicates thus far a physical law which binds these several bodies to the solar system.

On the other hand in the case of the comets, so far as a general inference can be drawn from the calculated elements of 190, which have been observed between the years 1556 and 1861*; the mean inclination to the plane of the ecliptic is about 50°, which is so near to the a priori mean as calculated above that no connection can be hence inferred as to a physical law of relation. Many more comets have been seen, but the above are all which have been observed between these years, and whose elements have been calculated. Also of the 18 which have elliptic orbits, and whose periodic times are known, at least approximately, the inclinations are very small, so that hereby is shewn a very strong probability of a law of relation between them and the solar system. And if these comets are excepted, the mean inclination of the others rises considerably above 50°.

Ex. 4. Find the mean of all squares inscribed in a given square.

Let $a =$ the side of the given square, and let $x =$ the distance from one of its angles of the angle of the inscribed square; so that the side of the inscribed square $= (2x^2 - 2ax + a^2)^{\frac{1}{2}}$. Hence

$$\text{The numerator of (36)} = \int_0^a (2x^2 - 2ax + a^2)\, dx = \frac{2a^3}{3};$$

$$\text{The denominator of (36)} = \int_0^a dx = a.$$

$$\therefore \text{ The required mean value } = \frac{2a^2}{3}.$$

Ex. 5. Find the mean of all the focal radii vectores of an ellipse, which are drawn at equal angular intervals.

Let the equation to the ellipse be $r = \dfrac{a(1-e^2)}{1-e\cos\theta}$; then

$$\text{The numerator of (36)} = \int_{-\pi}^{\pi} \frac{a(1-e^2)\, d\theta}{1-e\cos\theta}$$

$$= 2a(1-e^2)^{\frac{1}{2}} \left[\tan^{-1}\left\{\left(\frac{1+e}{1-e}\right)^{\frac{1}{2}} \tan\frac{\theta}{2}\right\}\right]_{-\pi}^{\pi}$$

$$= 2\pi a (1-e^2)^{\frac{1}{2}};$$

$$\text{The denominator of (36)} = \int_{-\pi}^{\pi} d\theta = 2\pi;$$

* Chambers' Hand-book of Astronomy. Appendix III. Murray, London, 1861.

$\therefore$ The required mean value $= a(1-e^2)^{\frac{1}{2}} = b$,
that is, is equal to half the minor axis of the ellipse.

Ex. 6. The mean length of all parallels of latitude on a sphere whose radius $= a$, drawn at equal angular intervals from the centre, $= 4a$.

Ex. 7. Another interesting problem of mean value is that of the distribution of double stars in the celestial vault, and it is also important on account of the inference drawn from it by W. Struve as to the physical connection of such binary systems.

Let $n =$ the number of stars up to a given order, say, to the eighth order inclusive; in which case $n = 100{,}000$, more or less. So that the greatest number of pairs of these $= \frac{n(n-1)}{2}$; let $r =$ the number (very small) of angular seconds of separation of the two members of a double star; so that πr^2 expresses in seconds the area of the circle of which r is the radius. Let $x =$ the number of double stars which the space πr^2 occupies, when these pairs are regularly distributed; then as the surface of the whole sphere expressed in seconds $= 4\pi(206265)^2$,

$$x = \frac{n(n-1)r^2}{8(206265)^2};$$

which assigns the number of double stars which, within a radius of r'', the celestial vault ought to exhibit.

Now Struve has found 311 double stars between the north pole and 15° of south declination, the angular distance between the members of which does not exceed 4″; whereas if such stars were regularly distributed, there ought, as he finds, to be *one* at most. We are therefore obliged to infer that the distribution in pairs is not fortuitous, but that there exists a true physical law connecting the two members of such a binary combination.

CHAPTER XII.

REDUCTION OF MULTIPLE INTEGRALS.

SECTION 1.—*Reduction of Multiple Integrals by simple application of the Gamma-function.*

276.] An examination of the processes required for the complete solution of the problems of the two preceding chapters and of other similar questions shews that they depend on the determination of the value of multiple integrals; and that the problems are only solved when these integrals are evaluated. Now in many cases the element-functions are of certain special forms, the general forms of which can be determined by means of certain other integrals which have already been evaluated; such as the Gamma-function and the integral-logarithm. In other cases the order of the multiple integral can be so far reduced, that the solution depends on a single integration; when it is said to be reduced to a quadrature. There are other processes of simplification which are frequently of great importance. All these we propose to investigate in the present chapter.

277.] The limits of the definite integrals which ordinarily occur are either constant, or are determined by an equation which fixes the range of all the integration processes; and according as the limits are assigned in one or other of these two modes, so will the method for the reduction of the integral vary. I will first take the case in which all the limits are constant; and explain the process, devised by M. Cauchy, by which the value of a given multiple integral may be made to depend on that of one or more single integrals.

Let the integral, which I will denote by I, be a multiple integral of the nth order, and of the form

$$\mathrm{I} = \iiint \dots \frac{\mathrm{P}}{\mathrm{Q}^m} \dots dz\, dy\, dx; \tag{1}$$

in which n variables $x, y, z, \dots$ are involved; where m is a positive quantity; P and Q are functions of the variables of the form

$$\mathrm{P} = \mathrm{P}_x \,.\, \mathrm{P}_y \,.\, \mathrm{P}_z \dots; \tag{2}$$

$$\mathrm{Q} = \mathrm{Q}_0 + \mathrm{Q}_x + \mathrm{Q}_y + \mathrm{Q}_z + \dots, \tag{3}$$

where Q_0 is a constant; P_x and Q_x are functions of x only; P_y and Q_y are functions of y only; and so on; where Q is always a positive quantity, and may have an impossible part, if the real part is always positive; and where all the limits of integration are constant.

Now according to an artifice due to M. Cauchy, which has already been applied in Art. 131, I propose to replace Q^{-m} in (1) by a definite integral, with constant limits, in terms of a variable t which is independent of the former variables. For this purpose we have by (250), Art. 122,

$$\int_0^{\infty} e^{-Qt} t^{m-1} dt = \frac{\Gamma(m)}{Q^m};$$

$$\therefore \quad \frac{1}{Q^m} = \frac{1}{\Gamma(m)} \int_0^{\infty} e^{-Qt} t^{m-1} dt; \tag{4}$$

so that

$$I = \frac{1}{\Gamma(m)} \int_0^{\infty} \int\int\int \ldots t^{m-1} P e^{-Qt} \ldots dz\, dy\, dx\, dt. \tag{5}$$

Let P and Q be replaced by their values given in (2) and (3); and separating the variables as the limits are constant, we have

$$I = \frac{1}{\Gamma(m)} \int_0^{\infty} e^{Q_0 t} t^{m-1} dt \ldots \int P_y e^{-Q_y t} dy \int P_x e^{-Q_x t} dx. \tag{6}$$

To simplify this, let us make the following substitutions for the definite integrals, each integral being taken with its proper limits;

$$\int P_x e^{-Q_x t} dx = u, \qquad \int P_y e^{-Q_y t} dy = v, \qquad \int P_z e^{-Q_z t} dz = w, \tag{7}$$

so that $u, v, w, \ldots$ are functions of t; then

$$I = \frac{1}{\Gamma(m)} \int_0^{\infty} t^{m-1} e^{-Q_0 t} \ldots w v u\, dt; \tag{8}$$

and thus the multiple integral given in (1) is reduced to a single integral.

278.] The following is an example of this general theorem; in which however I take only three variables, as the process is the same in all cases.

Let
$$\left.\begin{aligned} P &= x^{p-1} y^{q-1} z^{r-1} e^{-(ax+by+cz)}, \\ Q &= k + \alpha x + \beta y + \gamma z, \end{aligned}\right\} \tag{9}$$

where $p, q, r, a, b, c, k, \alpha, \beta, \gamma$ are positive constants; and let us suppose ∞ and 0 to be the limits of integration for each of the variables; then

$$\text{I} = \int_0^\infty \int_0^\infty \int_0^\infty \frac{x^{p-1}\, y^{q-1}\, z^{r-1}\, e^{-(ax+by+cz)}}{(k+\alpha x+\beta y+\gamma z)^m}\, dz\, dy\, dx \qquad (10)$$

$$= \frac{1}{\Gamma(m)} \int_0^\infty \int_0^\infty \int_0^\infty \int_0^\infty t^{m-1} x^{p-1} y^{q-1} z^{r-1} e^{-(\text{Q}t+ax+by+cz)}\, dz\, dy\, dx\, dt. \; (11)$$

Let Q be replaced by its value given in (9); then since

$$\int_0^\infty x^{p-1}\, e^{-(\alpha t+a)x}\, dx = \frac{\Gamma(p)}{(\alpha t+a)^p};$$

$$\int_0^\infty y^{q-1}\, e^{-(\beta t+b)y}\, dy = \frac{\Gamma(q)}{(\beta t+b)^q};$$

$$\int_0^\infty z^{r-1}\, e^{-(\gamma t+c)z}\, dz = \frac{\Gamma(r)}{(\gamma t+c)^r};$$

substituting these values in (11) we have

$$\text{I} = \frac{\Gamma(p)\,\Gamma(q)\,\Gamma(r)}{\Gamma(m)} \int_0^\infty \frac{t^{m-1}\, e^{-kt}\, dt}{(\alpha t+a)^p\, (\beta t+b)^q\, (\gamma t+c)^r}; \qquad (12)$$

whereby the multiple integral given in (10) is reduced to the single integral in (12).

The following are particular cases of the preceding theorems.

Let $a = b = c = 0$; then

$$\int_0^\infty \int_0^\infty \int_0^\infty \frac{x^{p-1} y^{q-1} z^{r-1}}{(k+\alpha x+\beta y+\gamma z)^m}\, dz\, dy\, dx$$

$$= \frac{\Gamma(p)\,\Gamma(q)\,\Gamma(r)}{\Gamma(m)} \int_0^\infty \frac{t^{m-p-q-r-1}\, e^{-kt}\, dt}{\alpha^p\, \beta^q\, \gamma^r}$$

$$= \frac{\Gamma(p)\Gamma(q)\Gamma(r)\Gamma(m-p-q-r)}{\Gamma(m)\, \alpha^p\, \beta^q\, \gamma^r\, k^{m-p-q-r}}; \qquad (13)$$

and if $k = \alpha = \beta = \gamma = 1$,

$$\int_0^\infty \int_0^\infty \int_0^\infty \frac{x^{p-1}\, y^{q-1}\, z^{r-1}}{(1+x+y+z)^m}\, dz\, dy\, dx$$

$$= \frac{\Gamma(p)\,\Gamma(q)\Gamma(r)\Gamma(m-p-q-r)}{\Gamma(m)}. \qquad (14)$$

From (13) a more general theorem may be deduced by replacing αx by $\left(\frac{x}{a}\right)^a$, βy by $\left(\frac{y}{\beta}\right)^b$, γz by $\left(\frac{z}{\gamma}\right)^c$; but it is unnecessary to express it at length.

If in the general theorem (12) there is only one variable x,

$$\int_0^\infty \frac{x^{p-1}\, e^{-ax}\, dx}{(k+\alpha x)^m} = \frac{\Gamma(p)}{\Gamma(m)} \int_0^\infty \frac{t^{m-1}\, e^{-kt}\, dt}{(a+\alpha t)^p}. \qquad (15)$$

In this equation let $k = \alpha = 1$; and in the right-hand member let t be replaced by x; then

$$\Gamma(m)\int_0^\infty \frac{x^{p-1}e^{-x}\,dx}{(1+ax)^m} = \Gamma(p)\int_0^\infty \frac{x^{m-1}e^{-x}\,dx}{(1+ax)^p}. \tag{16}$$

Again, in (15) let $a = 0$, $k = \alpha = 1$; then

$$\int_0^\infty \frac{x^{p-1}\,dx}{(1+x)^m} = \frac{\Gamma(p)}{\Gamma(m)}\int_0^\infty t^{m-p-1}e^{-t}\,dt$$

$$= \frac{\Gamma(p)\,\Gamma(m-p)}{\Gamma(m)}; \tag{17}$$

a result which has already been found in Art. 125.

279.] Let us now consider these cases in which a reduction of a multiple integral can be made by means of the Gamma-function, and in which the limits of integration are given by an equation of condition.

The integral which I shall take is the following of n variables.

$$\text{I} = \iiint \ldots x^{p-1}y^{q-1}z^{r-1}\ldots(a-x-y-z-\ldots)^{s-1}\ldots dz\,dy\,dx; \tag{18}$$

the limits of integration of which include all positive values of the variables satisfying the inequality

$$x+y+z+\ldots\ldots < a. \tag{19}$$

Let us however confine ourselves to three variables; for we thereby fix our thoughts, and do not restrict the nature of the process of reduction which is the same, whatever is the number of variables. In this case the equation of limits gives the surface which bounds all those points in space which are included in the integral; and in this special case that surface is the plane whose equation is $x+y+z = a$. Hence

$$\text{I} = \int_0^a\int_0^{a-x}\int_0^{a-x-y} x^{p-1}y^{q-1}z^{r-1}(a-x-y-z)^{s-1}\,dz\,dy\,dx. \tag{20}$$

Now by (270), Art. 126,

$$\int_0^c u^{m-1}(c-u)^{n-1}\,du = c^{m+n-1}\frac{\Gamma(m)\,\Gamma(n)}{\Gamma(m+n)}; \tag{21}$$

so that applying this theorem to the successive integrations in (20), we have

$$\text{I} = \int_0^a\int_0^{a-x} x^{p-1}y^{q-1}(a-x-y)^{r+s-1}\frac{\Gamma(r)\,\Gamma(s)}{\Gamma(r+s)}\,dy\,dx$$

$$= \frac{\Gamma(r)\,\Gamma(s)}{\Gamma(r+s)}\int_0^a x^{p-1}(a-x)^{q+r+s-1}\frac{\Gamma(q)\,\Gamma(r+s)}{\Gamma(q+r+s)}\,dx$$

$$= \frac{\Gamma(q)\,\Gamma(r)\,\Gamma(s)}{\Gamma(q+r+s)}\,\frac{\Gamma(p)\,\Gamma(q+r+s)}{\Gamma(p+q+r+s)}\,a^{p+q+r+s-1}$$

$$= \frac{\Gamma(p)\,\Gamma(q)\,\Gamma(r)\,\Gamma(s)}{\Gamma(p+q+r+s)}\,a^{p+q+r+s-1}; \tag{22}$$

which result completely determines the value of the integral given in (20).

If $s = 1$,

$$\int_0^a \int_0^{a-x} \int_0^{a-x-y} x^{p-1} y^{q-1} z^{r-1}\, dz\, dy\, dx = \frac{\Gamma(p)\,\Gamma(q)\,\Gamma(r)}{\Gamma(p+q+r+1)} a^{p+q+r}; \quad (23)$$

all positive values of the variables being included which lie within the plane whose equation is $x+y+z = a$.

280.] This theorem is capable of extension, and in its extended form supplies the solution of many important problems. For we can by a similar process determine the value of

$$\text{I} = \iiint x^{p-1} y^{q-1} z^{r-1}\, dz\, dy\, dx; \quad (24)$$

when the limits of integration include all positive values of the variables given by the inequality

$$\left(\frac{x}{a}\right)^{\alpha} + \left(\frac{y}{b}\right)^{\beta} + \left(\frac{z}{c}\right)^{\gamma} < 1, \quad (25)$$

where $p, q, r, a, b, c, \alpha, \beta, \gamma$ are all positive quantities.

For let $$\left(\frac{x}{a}\right)^{\alpha} = \xi, \quad \left(\frac{y}{b}\right)^{\beta} = \eta, \quad \left(\frac{z}{c}\right)^{\gamma} = \zeta, \quad (26)$$

so that the inequality (25), which assigns the limits, becomes

$$\xi+\eta+\zeta = 1; \quad (27)$$

Also $$dx = \frac{a}{\alpha} \xi^{\frac{1}{\alpha}-1} d\xi, \quad dy = \frac{b}{\beta} \eta^{\frac{1}{\beta}-1} d\eta, \quad dz = \frac{c}{\gamma} \zeta^{\frac{1}{\gamma}-1} d\zeta; \quad (28)$$

and substituting these in (24),

$$\text{I} = \frac{a^p b^q c^r}{\alpha\beta\gamma} \int_0^1 \int_0^{1-\xi} \int_0^{1-\xi-\eta} \xi^{\frac{p}{\alpha}-1} \eta^{\frac{q}{\beta}-1} \zeta^{\frac{r}{\gamma}-1}\, d\zeta\, d\eta\, d\xi;$$

so that by (23),

$$\text{I} = \frac{a^p b^q c^r}{\alpha\beta\gamma} \frac{\Gamma\left(\frac{p}{\alpha}\right)\Gamma\left(\frac{q}{\beta}\right)\Gamma\left(\frac{r}{\gamma}\right)}{\Gamma\left(\frac{p}{\alpha}+\frac{q}{\beta}+\frac{r}{\gamma}+1\right)}. \quad (29)$$

This useful theorem was first given by Lejeune Dirichlet*, having been deduced by him from a general process which will be described in section 2 of the present chapter; the following are examples in which it is applied.

Ex. 1. Let there be two variables, and let $\alpha = \beta = 2$; then the equation which assigns the limits is $\left(\frac{x}{a}\right)^2 + \left(\frac{y}{b}\right)^2 = 1$; and

* Comptes Rendus, Tome VIII, p. 159; 1839.

consequently the range includes all values of the variables which correspond to points within the first quadrant of an ellipse; and if $\text{Y} = \frac{b}{a}(a^2 - x^2)^{\frac{1}{2}}$,

$$\int_0^a \int_0^{\text{Y}} x^{p-1} y^{q-1} dy\, dx = \frac{a^p b^q}{4} \frac{\Gamma\left(\frac{p}{2}\right)\Gamma\left(\frac{q}{2}\right)}{\Gamma\left(\frac{p}{2}+\frac{q}{2}+1\right)}. \tag{30}$$

If $p = q = 1$; then, since $\Gamma\left(\frac{1}{2}\right) = \pi^{\frac{1}{2}}$, and $\Gamma(2) = 1$,

$$\int_0^a \int_0^{\text{Y}} dy\, dx = \frac{\pi a b}{4}. \tag{31}$$

And as the left-hand member evidently expresses the area of a quadrant of an ellipse, the area of the ellipse $= \pi a b$.

If $p = 2$, $q = 1$; then, since $\Gamma\left(\frac{5}{2}\right) = \frac{3}{2}\Gamma\left(\frac{3}{2}\right) = \frac{3}{4}\pi^{\frac{1}{2}}$;

$$\int_0^a \int_0^{\text{Y}} x\, dy\, dx = \frac{a^2 b}{3}. \tag{32}$$

Similarly, if $p = 1$, $q = 2$,

$$\int_0^a \int_0^{\text{Y}} y\, dy\, dx = \frac{a b^2}{3}. \tag{33}$$

Again, let $p = 3$, $q = 1$; then, since $\Gamma(3) = 2\,\Gamma(2) = 2$,

$$\int_0^a \int_0^{\text{Y}} x^2\, dy\, dx = \frac{\pi a^3 b}{16}; \tag{34}$$

similarly, if $p = 1$, $q = 3$,

$$\int_0^a \int_0^{\text{Y}} y^2\, dy\, dx = \frac{\pi a b^3}{16}. \tag{35}$$

Hence by addition

$$\int_0^a \int_0^{\text{Y}} (x^2 + y^2)\, dy\, dx = \frac{\pi a b}{16}(a^2 + b^2); \tag{36}$$

and consequently for the whole ellipse,

$$\int_{-a}^a \int_{-\text{Y}}^{\text{Y}} (x^2 + y^2)\, dy\, dx = \frac{\pi a b}{4}(a^2 + b^2). \tag{37}$$

All these expressions will be of considerable use in the sequel.

Ex. 2. Let three variables be involved in the integral (29), and let $\alpha = \beta = \gamma = 2$, so that the equation which assigns the limits is

$$\frac{x^2}{a^2} + \frac{y^2}{b^2} + \frac{z^2}{c^2} = 1; \tag{38}$$

and consequently the range includes all values of the variables which correspond to points within the first octant of the ellipsoid. Also let $c\left(1-\frac{x^2}{a^2}-\frac{y^2}{b^2}\right)^{\frac{1}{2}} = \text{z}$, $b\left(1-\frac{x^2}{a^2}\right)^{\frac{1}{2}} = \text{Y}$; then

$$\int_0^a \int_0^{\text{Y}} \int_0^{\text{z}} x^{p-1} y^{q-1} z^{r-1} dz\, dy\, dx = \frac{a^p b^q c^r}{8} \frac{\Gamma\left(\frac{p}{2}\right)\Gamma\left(\frac{q}{2}\right)\Gamma\left(\frac{r}{2}\right)}{\Gamma\left(\frac{p}{2}+\frac{q}{2}+\frac{r}{2}+1\right)}. \quad (39)$$

Let $p = q = r = 1$; then

$$\int_0^a \int_0^{\text{Y}} \int_0^{\text{z}} dz\, dy\, dx = \frac{\pi abc}{6}. \quad (40)$$

As the left-hand member of this equation expresses the volume of the octant of the ellipsoid, the whole volume of the ellipsoid $= \frac{4}{3}\pi abc$.

If $p = 2$, $q = r = 1$, then

$$\int_0^a \int_0^{\text{Y}} \int_0^{\text{z}} x\, dz\, dy\, dx = \frac{\pi a^2 bc}{16}. \quad (41)$$

If $p = 3$, $q = r = 1$, then

$$\int_0^a \int_0^{\text{Y}} \int_0^{\text{z}} x^2\, dz\, dy\, dx = \frac{\pi a^3 bc}{30}. \quad (42)$$

And for the whole ellipsoid

$$\int_{-a}^a \int_{-\text{Y}}^{\text{Y}} \int_{-\text{z}}^{\text{z}} x^2\, dz\, dy\, dx = \frac{4\pi a^3 bc}{15}. \quad (43)$$

Similar equivalents are of course true for the other variables.

Ex. 3. Let the definite integral (29) contain three variables; and let $\alpha = \beta = \gamma = 4$; so that the equation which assigns the limits of integration is

$$\left(\frac{x}{a}\right)^4 + \left(\frac{y}{b}\right)^4 + \left(\frac{z}{c}\right)^4 = 1; \quad (44)$$

then, if $p = q = r = 1$, I in (29) expresses the volume of the octant of the surface whose equation is (44); and if V is the whole volume,

$$\text{V} = \frac{abc}{6 \cdot 2^{\frac{1}{2}}\pi} \left\{\Gamma\left(\frac{1}{4}\right)\right\}^4;$$

which, as shewn by (57), Art. 162, may be expressed in terms of the arc of a lemniscata.

Ex. 4. Generally if the definite integral contains three variables x, y, z, and α, β, γ are even numbers so that the equation

$$\left(\frac{x}{a}\right)^{\alpha} + \left(\frac{y}{b}\right)^{\beta} + \left(\frac{z}{c}\right)^{\gamma} = 1$$

represents a closed surface; then, if $p = q = r = 1$, the given definite integral expresses the volume of the octant of that surface, and $8v$ = the whole volume.

$$v = \frac{abc}{\alpha\beta\gamma} \frac{\Gamma\left(\frac{1}{\alpha}\right)\Gamma\left(\frac{1}{\beta}\right)\Gamma\left(\frac{1}{\gamma}\right)}{\Gamma\left(\frac{1}{\alpha}+\frac{1}{\beta}+\frac{1}{\gamma}+1\right)} \qquad (45)$$

291.] Soon after the publication of the preceding theorem by Dirichlet, M. Liouville* gave an extension of it to those cases in which the element-function involves an arbitrary function of that particular combination of the variables which is given in the following form,

$$I = \iiint \ldots f(x+y+z+\ldots)\, x^{p-1} y^{q-1} z^{r-1} \ldots \, dz\, dy\, dx \ldots; \qquad (46)$$

where the range of integration includes all positive values of the variables given by the inequality

$$x+y+z+\ldots < h, \qquad (47)$$

where h is a positive constant.

Let us first take the case of two variables only: then

$$I = \int_0^h \int_0^{h-x} f(x+y)\, x^{p-1} y^{q-1}\, dy\, dx. \qquad (48)$$

Let this integral be transformed, and let $x = uv$, $y = u(1-v)$; then, as in (55), Art. 215, $dy\, dx = u\, du\, dv$; and taking the limits of the new variables so that the same range may be included,

$$I = \int_0^1 \int_0^h f(u)\, u^{p+q-1} v^{p-1} (1-v)^{q-1}\, du\, dv. \qquad (49)$$

As the limits of both integrations are constant, the order in which the integrations are effected is indifferent; and consequently

$$I = \int_0^1 v^{p-1} (1-v)^{q-1}\, dv \int_0^h f(u)\, u^{p+q-1}\, du$$

$$= \frac{\Gamma(p)\,\Gamma(q)}{\Gamma(p+q)} \int_0^h f(u)\, u^{p+q-1}\, du; \qquad (50)$$

so that the double integral given in (48) is reduced to the product of a certain combination of the Gamma-function and of a single definite integral.

* Liouville's Journal, Tome IV, p. 230.

Next let us suppose three variables to enter into the integral (46); so that

$$\text{I} = \int_0^h \int_0^{h-x} \int_0^{h-x-y} f(x+y+z)\, x^{p-1} y^{q-1} z^{r-1}\, dz\, dy\, dx. \quad (51)$$

To reduce this, let a transformation be made of two variables similar to the preceding; and let $y = vw$, $z = v(1-w)$; then, $dy\, dz = v\, dv\, dw$; and

$$\text{I} = \int_0^h \int_0^1 \int_0^{h-x} f(x+v)\, x^{p-1} v^{q+r-1} w^{q-1} (1-w)^{r-1}\, dv\, dw\, dx$$

$$= \frac{\Gamma(q)\,\Gamma(r)}{\Gamma(q+r)} \int_0^h \int_0^{h-x} f(x+v)\, x^{p-1} v^{q+r-1}\, dv\, dx;$$

in which the definite integral involves only two variables: and transforming again as in the former cases we shall have

$$\text{I} = \frac{\Gamma(q)\,\Gamma(r)}{\Gamma(q+r)} \frac{\Gamma(p)\,\Gamma(q+r)}{\Gamma(p+q+r)} \int_0^h f(u)\, u^{p+q+r-1}\, du$$

$$= \frac{\Gamma(p)\,\Gamma(q)\,\Gamma(r)}{\Gamma(p+q+r)} \int_0^h f(u)\, u^{p+q+r-1}\, du. \quad (52)$$

This process may evidently be extended to any number of variables; and we shall have ultimately

$$\iiint \dots f(x+y+z+\dots)\, x^{p-1} y^{q-1} z^{r-1} \dots dz\, dy\, dx$$

$$= \frac{\Gamma(p)\,\Gamma(q)\,\Gamma(r)\dots}{\Gamma(p+q+r+\dots)} \int_0^h f(u)\, u^{p+q+r+\dots-1}\, du; \quad (53)$$

where the limits of integration are given by the inequality

$$x+y+z+\dots\dots < h; \quad (54)$$

so that by this theorem the multiple integral is reduced to a single integral.

If $f(u) = 1$, we have the result already given in (23).

282.] This theorem also admits of extension as to the inequality which determines the limits, similar to that of Art. 280 from Art. 279. In this case

$$\text{I} = \iiint \dots f\left\{ \left(\frac{x}{a}\right)^{\alpha} + \left(\frac{y}{b}\right)^{\beta} + \left(\frac{z}{c}\right)^{\gamma} + \dots \right\} x^{p-1} y^{q-1} z^{r-1} \dots dz\, dy\, dx;$$

where the range includes all positive values of the variables which are given by the inequality

$$\left(\frac{x}{a}\right)^{\alpha} + \left(\frac{y}{b}\right)^{\beta} + \left(\frac{z}{c}\right)^{\gamma} + \dots\dots < h. \quad (55)$$

Let us take the case of three variables, so that we have the integral

$$\text{I} = \iiint f\left\{\left(\frac{x}{a}\right)^{\alpha} + \left(\frac{y}{b}\right)^{\beta} + \left(\frac{z}{c}\right)^{\gamma}\right\} x^{p-1} y^{q-1} z^{r-1}\, dz\, dy\, dx; \qquad (56)$$

where the limits are assigned by the equation

$$\left(\frac{x}{a}\right)^{\alpha} + \left(\frac{y}{b}\right)^{\beta} + \left(\frac{z}{c}\right)^{\gamma} = h; \qquad (57)$$

and let us make the substitutions which are given in (26) and (28); so that the range includes all values of the new variables ξ, η, ζ which are given by the inequality

$$\xi + \eta + \zeta < h; \text{ and} \qquad (58)$$

$$\text{I} = \frac{a^p b^q c^r}{\alpha\beta\gamma} \int_0^h \int_0^{h-\xi} \int_0^{h-\xi-\eta} f(\xi+\eta+\zeta)\, \xi^{\frac{p}{\alpha}-1} \eta^{\frac{q}{\beta}-1} \zeta^{\frac{r}{\gamma}-1}\, d\zeta\, d\eta\, d\xi \qquad (59)$$

$$= \frac{a^p b^q c^r}{\alpha\beta\gamma} \frac{\Gamma\left(\frac{p}{\alpha}\right)\Gamma\left(\frac{q}{\beta}\right)\Gamma\left(\frac{r}{\gamma}\right)}{\Gamma\left(\frac{p}{\alpha}+\frac{q}{\beta}+\frac{r}{\gamma}\right)} \int_0^h f(u)\, u^{\frac{p}{\alpha}+\frac{q}{\beta}+\frac{r}{\gamma}-1}\, du. \qquad (60)$$

If $f(u) = 1$, and $h = 1$, we have the result already given in (29); so that this theorem includes all those previously given.

The following are examples in which it is applied.

Ex. 1. Determine the value of $\iint \left(\frac{1-x^2-y^2}{1+x^2+y^2}\right)^{\frac{1}{2}} dy\, dx$, where the limits are given by the inequality $x^2+y^2<1$.

Here we have only two variables; and $p=q=1$; $\alpha=\beta=2$, $a=b=1$; $h=1$; so that if $\text{Y} = (1-x^2)^{\frac{1}{2}}$,

$$\int_0^1 \int_0^{\text{Y}} \left(\frac{1-x^2-y^2}{1+x^2+y^2}\right)^{\frac{1}{2}} dy\, dx = \frac{\pi}{4}\int_0^1 \left(\frac{1-u}{1+u}\right)^{\frac{1}{2}} du$$

$$= \frac{\pi(\pi-2)}{8}. \qquad (61)$$

Ex. 2. Determine the value of $\iiint \frac{dz\, dy\, dx}{(1-x^2-y^2-z^2)^{\frac{1}{2}}}$, where the limits are given by the inequality $x^2+y^2+z^2<1$. In this case the range includes all points lying within the surface of a sphere whose radius $= 1$. If $\text{Z} = (1-x^2-y^2)^{\frac{1}{2}}$, $\text{Y} = (1-x^2)^{\frac{1}{2}}$,

$$\text{then} \quad \int_0^1 \int_0^{\text{Y}} \int_0^{\text{Z}} \frac{dz\, dy\, dx}{(1-x^2-y^2-z^2)^{\frac{1}{2}}} = \frac{\pi}{4}\int_0^1 \frac{u^{\frac{1}{2}}\, du}{(1-u)^{\frac{1}{2}}} = \frac{\pi^2}{8}. \qquad (62)$$

Ex. 3. Determine the value of $\iiint \left(\frac{1-x^2-y^2-z^2}{1+x^2+y^2+z^2}\right)^{\frac{1}{2}} dz\, dy\, dx$, where the limits are given by the inequality $x^2+y^2+z^2<1$.

Taking the same notation as in the preceding example,

$$\int_0^1 \int_0^{\mathrm{X}} \int_0^{\mathrm{Z}} \left(\frac{1-x^2-y^2-z^2}{1+x^2+y^2+z^2}\right)^{\frac{1}{2}} dz\, dy\, dx$$

$$= \frac{(2\pi)^{\frac{1}{2}}}{8} \left\{4\pi^2 \left\{\Gamma\left(\frac{1}{4}\right)\right\}^{-2} - \left\{\Gamma\left(\frac{1}{4}\right)\right\}^{2}\right\} \qquad (63)$$

Section 2.—*Dirichlet's Method of Reduction by means of a Factor of Discontinuity. The Application of Fourier's Integral.*

283.] An examination into the theory of multiple integrals at once shews that generally the difficulty of evaluating those whose limits are given by an inequality which involves the variables is much greater than that of evaluating those of which the limits are constant. In the latter case, as we have demonstrated in Art. 99, the order of the integrations may be changed without any change in the value of the result, and in very many cases we are able to simplify the integral by means of properties of the Gamma-function, and of other allied integrals, which have been proved heretofore. In the former case however the order of integration is prescribed, and cannot be changed without (in many cases) considerable difficulty and consequent risk of error; this fact is apparent from the difficulty of assigning the limits of integration when the integral is transformed by a change of variable. Now the knowledge of this circumstance appears to have suggested to L. Dirichlet the process of so operating on the infinitesimal element-function of a multiple integral, of which the limits are assigned by an inequality, that the limits may be constant; and indeed as he has shewn, and as it is convenient to take them, that the limits of all the several integrations may be ∞ and 0. To effect this, he introduces a factor in the form of a definite integral, into the element-function; this factor being a discontinuous function which $= 1$ for all values of the variables within the range of integration, and $= 0$ for all values of the variables beyond that range. Consequently when this factor has been introduced, we may enlarge the range of integration to any extent; and may indeed include all values from ∞ to 0; or from ∞ to $-\infty$. We have already had similar cases in which the range has been so enlarged; see Art. 197.

The mode of applying this principle is as follows; Let the given definite integral contain n variables, and be of the form

$$\text{I} = \iiint \ldots \text{F}(x, y\, z, \ldots) \ldots dz\, dy\, dx\,; \tag{64}$$

and let us suppose the integral to include all positive values of the variables within limits assigned by the inequality

$$x + y + z + \ldots\ldots < 1. \tag{65}$$

Now suppose that we have a single definite integral of the form $\int_\alpha^\beta f(t, k)\, dt$, containing the undetermined constant k; and that this integral $= 1$ for all values of k less than 1, and $= 0$ for all values of k greater than 1; then if we replace k by $x + y + z + \ldots$, this integral will be equal to 1 or 0, according as we are considering values of the variables within or beyond the range assigned by the inequality (65). Consequently if we introduce within the integration-symbol in (64) the factor

$$\int_\alpha^\beta f(t,\, x + y + z + \ldots)\, dt, \tag{66}$$

we may enlarge the limits of all the other integrations to ∞ and 0 without changing the value of the integral. Thus we have

$$\text{I} = \int_0^\infty \int_0^\infty \int_0^\infty \ldots \int_\alpha^\beta \text{F}(x, y, z, \ldots) f(t, x + y + z + \ldots) \ldots dz\, dy\, dx\, dt. \tag{67}$$

The case in which $k = x + y + z + \ldots = 1$ will only give one element of the definite integral, when the limits are extended; and consequently, assuming that the introduced factor is finite when $k = 1$, that element must be neglected in the definite integral; and it is necessary only to take account of the elements when k is greater than and less than 1. This remark is important; and is applicable to all the subsequent cases in the present section where the limits of integration are given by an equality, and the limit of the value of this inequality is expressed by a discontinuous factor.

Now a factor satisfying the preceding conditions is called a factor of discontinuity. A quantity possessing these qualities has been investigated in Ex. 4, Art. 100; in which it is shewn that $\frac{2}{\pi} \int_0^\infty \frac{\sin mt \cos kt}{t}\, dt = 1$, for all values of k less than m; $= \frac{1}{2}$, when $k = m$; and $= 0$, for all values of k greater than m; consequently if $m = 1$,

$$\left.\begin{aligned} \frac{2}{\pi} \int_0^\infty \frac{\sin t \cos kt}{t}\, dt &= 1, \text{ when } k \text{ is less than } 1\,; \\ &= 0, \text{ when } k \text{ is greater than } 1. \end{aligned}\right\} \tag{68}$$

Hence if the following integral includes all positive values of the variables within limits of integration assigned by the inequality $$x+y+z+\ldots\ldots<1, \tag{69}$$
then

$$\text{I} = \iiint \ldots \text{F}(x, y, z, \ldots) \ldots dz\, dy\, dx$$

$$= \frac{2}{\pi}\int_0^\infty\int_0^\infty\int_0^\infty \ldots \frac{\sin t}{t}\,\text{F}(x,y,\ldots)\cos(x+y+\ldots)t\ldots dy\, dx\, dt. \tag{70}$$

This may also be reduced to a form more convenient for the evaluation of the several separate integrations by means of the Gamma-function. Since $\cos kt$ is the real part of $e^{-kt\sqrt{-1}}$, (70) may be expressed in the form,

I = the real part

$$\text{of } \frac{2}{\pi}\int_0^\infty\int_0^\infty\int_0^\infty \ldots \frac{\sin t}{t}\,\text{F}(x,y,\ldots)e^{-(x+y+\ldots)t\sqrt{-1}}\ldots dy\, dx\, dt. \tag{71}$$

This last equation, as also (70), states the theorem discovered by Dirichlet for the evaluation of multiple definite integrals. One of the most useful applications of it is to a problem in attractions which occurs in a future part of this treatise. The following examples, which refer to subjects already discussed, are sufficient to exhibit its application.

284.] Ex. 1. Let us take a case of three variables, and assume $\text{F}(x, y, z) = x^{p-1}y^{q-1}z^{r-1}e^{-a(x+y+z)}$; and let the limits be assigned by the inequality $x+y+z<1$; then

$$\int_0^1\int_0^{1-x}\int_0^{1-x-y} x^{p-1}y^{q-1}z^{r-1}e^{-a(x+y+z)}\,dz\,dy\,dx = \text{the real part}$$

$$\text{of } \frac{2}{\pi}\int_0^\infty\int_0^\infty\int_0^\infty\int_0^\infty \frac{\sin t}{t}\,x^{p-1}y^{q-1}z^{r-1}e^{-(a+t\sqrt{-1})(x+y+z)}\,dz\,dy\,dx\,dt$$

$$= \frac{2}{\pi}\int_0^\infty \frac{\sin t}{t}\,dt\int_0^\infty x^{p-1}e^{-(a+t\sqrt{-1})x}\,dx\int_0^\infty y^{q-1}e^{-(a+t\sqrt{-1})y}\,dy \times\int_0^\infty z^{r-1}e^{-(a+t\sqrt{-1})z}\,dz$$

$$= \frac{2}{\pi}\int_0^\infty \frac{\sin t}{t}\,dt\,\frac{\Gamma(p)}{(a+t\sqrt{-1})^p}\,\frac{\Gamma(q)}{(a+t\sqrt{-1})^q}\,\frac{\Gamma(r)}{(a+t\sqrt{-1})^r}$$

$$= \frac{2\,\Gamma(p)\,\Gamma(q)\,\Gamma(r)}{\pi}\int_0^\infty \frac{\sin t\,dt}{t(a+t\sqrt{-1})^{p+q+r}}. \tag{72}$$

If this definite integral involves only one variable,

$$\int_0^1 x^{p-1} e^{-ax} dx = \text{the real part of } \frac{2\,\Gamma(p)}{\pi} \int_0^\infty \frac{\sin t\, dt}{t(a+t\sqrt{-1})^p};$$

therefore the real part of

$$\int_0^\infty \frac{\sin t\, dt}{t(a+t\sqrt{-1})^{p+q+r}} = \frac{\pi}{2\,\Gamma(p+q+r)} \int_0^1 t^{p+q+r-1} e^{-at} dt; \quad (73)$$

substituting which in (72),

$$\int_0^1 \int_0^{1-x} \int_0^{1-x-y} x^{p-1} y^{q-1} z^{r-1} e^{-a(x+y+z)} dz\, dy\, dx$$
$$= \frac{\Gamma(p)\,\Gamma(q)\,\Gamma(r)}{\Gamma(p+q+r)} \int_0^1 t^{p+q+r-1} e^{-at} dt; \quad (74)$$

and thus the triple integral in the left-hand member is expressed in terms of the single integral in the right-hand member.

This result is evidently a particular form of (53). If $a = 0$,

$$\int \int_0^{1-x} \int_0^{1-x-y} x^{p-1} y^{q-1} z^{r-1} dz\, dy\, dx = \frac{\Gamma(p)\,\Gamma(q)\,\Gamma(r)}{\Gamma(p+q+r+1)}. \quad (75)$$

which is the same result as (23), if a, in (23), $= 1$.

Ex. 2. Hereby also we may find the volume of the ellipsoid; and this example is a good illustration of the mode in which these functions are to be treated.

In this case the inequality which assigns the limits is

$$\frac{x^2}{a^2} + \frac{y^2}{b^2} + \frac{z^2}{c^2} < 1. \quad (76)$$

Hence if v is the volume of the octant,

$$v = \int\int\int dz\, dy\, dx = \text{the real part}$$

$$\text{of } \frac{2}{\pi} \int_0^\infty \int_0^\infty \int_0^\infty \int_0^\infty \frac{\sin t}{t} e^{-\left(\frac{x^2}{a^2}+\frac{y^2}{b^2}+\frac{z^2}{c^2}\right) t\sqrt{-1}} dz\, dy\, dx\, dt \quad (77)$$

$$= \frac{2}{\pi} \int_0^\infty \frac{\sin t}{t} dt \int_0^\infty e^{-\frac{t\sqrt{-1}}{a^2}x^2} dx \int_0^\infty e^{-\frac{t\sqrt{-1}}{b^2}y^2} dy \int_0^\infty e^{-\frac{t\sqrt{-1}}{c^2}z^2} dz \quad (78)$$

$$= \frac{2}{\pi} \int_0^\infty \frac{\sin t}{t} dt \frac{\pi^{\frac{1}{2}} a}{2(t\sqrt{-1})^{\frac{1}{2}}} \frac{\pi^{\frac{1}{2}} b}{2(t\sqrt{-1})^{\frac{1}{2}}} \frac{\pi^{\frac{1}{2}} c}{2(t\sqrt{-1})^{\frac{1}{2}}},$$

by reason of (342), Art. 138;

$$= \frac{abc\,\pi^{\frac{1}{2}}}{4} \int_0^\infty \frac{\sin t\, dt}{t^{\frac{5}{2}}(\sqrt{-1})^{\frac{3}{2}}}. \quad (79)$$

Now $\frac{1}{(\sqrt{-1})^{\frac{3}{2}}} = \cos\frac{3\pi}{4} - \sqrt{-1}\sin\frac{3\pi}{4}$; also by (349), Art. 139, and (404), Art. 146,

$$\int_0^{\infty} \frac{\sin t\, dt}{t^{\frac{5}{2}}} = \sin\left(-\frac{3\pi}{4}\right)\Gamma\left(-\frac{3}{2}\right) = -\frac{2(2\pi)^{\frac{1}{2}}}{3}; \tag{80}$$

$$\therefore \quad v = \frac{\pi abc}{6}.$$

And thus the whole volume $= \dfrac{4\pi abc}{3}$.

285.] An extension may be given to the process of reduction of a multiple integral by the employment of a discontinuous factor of a more general form than that hitherto introduced. For such a factor is suggested by Fourier's theorem, which is stated in (143), Art. 204; that particular theorem however not being sufficient for the purpose, because limits of a more general form are required. Now if λ and μ are positive quantities, λ being greater than μ, and if $f(t)$ is finite and continuous for all values of t over the range $\lambda-\mu$, then

$$\left.\begin{array}{l}\dfrac{2}{\pi f(k)}\displaystyle\int_0^{\infty}\int_{\mu}^{\lambda} f(t)\cos ku \cos tu\, dt\, du = 1, \text{ or } = 0, \\ \text{according as } k \text{ falls within or beyond the range } \lambda-\mu.\end{array}\right\} \tag{81}$$

Let

$$\begin{aligned}
\text{I} &= \int_0^{\infty}\int_{\mu}^{\lambda} f(t)\cos ku\cos tu\, dt\, du \\
&= \int_0^{\infty}\cos ku\, du\int_{\mu}^{\lambda} f(t)\cos tu\, dt \\
&= \int_0^{\infty}\cos ku\, du\left[\frac{f(t)\sin tu}{u} - \frac{1}{u}\int f'(t)\sin tu\, dt\right]_{\mu}^{\lambda} \\
&= f(\lambda)\int_0^{\infty}\frac{\sin\lambda u\cos ku}{u}du - f(\mu)\int_0^{\infty}\frac{\sin\mu u\cos ku}{u}du \\
&\qquad - \int_{\mu}^{\lambda} f'(t)\, dt\int_0^{\infty}\frac{\sin tu\cos ku}{u}du.
\end{aligned} \tag{82}$$

Now by the theorem given in Ex. 4, Art. 100,

$$\int_0^{\infty}\frac{\sin mx\cos nx}{x}dx = \frac{\pi}{2}, = \frac{\pi}{4}, = 0,$$

according as n is less than, is equal to, or is greater than m. Hence applying this theorem to the terms of (82), we have three several cases;

(1) Let $k < \mu < \lambda$, and consequently less than every value of t involved in the last term of (82); so that

$$\begin{aligned}
\text{I} &= f(\lambda)\frac{\pi}{2} - f(\mu)\frac{\pi}{2} - \int_{\mu}^{\lambda} f'(t)\, dt\frac{\pi}{2} \\
&= \frac{\pi}{2}\{f(\lambda)-f(\mu)-f(\lambda)+f(\mu)\} = 0.
\end{aligned} \tag{83}$$

(2) Let $\mu < k < \lambda$; now the values of the first two terms of (82) are evident; but as t in the last integral ranges from μ to λ, and as k falls between μ and λ, it will be convenient to divide this integral into two parts, as follows;

$$\int_\mu^\lambda f'(t)\,dt \int_0^\infty \frac{\sin tu \cos ku}{u}\,du$$

$$= \int_k^\lambda f'(t)\,dt \int_0^\infty \frac{\sin tu \cos ku}{u}\,du + \int_\mu^k f'(t)\,dt \int_0^\infty \frac{\sin tu \cos ku}{u}\,du;$$

so that

$$\text{I} = f(\lambda)\frac{\pi}{2} - f(\mu)\times 0 - \frac{\pi}{2}\{f(\lambda)-f(k)\} + \int_\mu^k f'(t)\,dt \times 0$$

$$= \frac{\pi}{2} f(k).$$

(3) Let $\mu < \lambda < k$; then

$$\text{I} = f(\lambda) \times 0 - f(\mu) \times 0 - \int_\mu^\lambda f'(t)\,dt \times 0 = 0.$$

Hence

$$\int_0^\infty \int_\mu^\lambda f(t) \cos ku \cos tu\, dt\, du = \frac{\pi}{2} f(k), \quad \text{or} \quad = 0,$$

according as k falls within or beyond the range $\lambda - \mu$.

In the same manner it may be shewn that

$$\int_0^\infty \int_\mu^\lambda f(t) \sin ku \sin tu\, dt\, du = \frac{\pi}{2} f(k), \quad \text{or} \quad = 0,$$

according as k falls within or beyond the range $\lambda - \mu$.

Hence we have generally

$$\left.\begin{aligned} &\frac{2}{\pi}\int_0^\infty \int_\mu^\lambda f(t) \cos ku \cos tu\, dt\, du \\ &\quad = \frac{2}{\pi}\int_0^\infty \int_\mu^\lambda f(t) \sin ku \sin tu\, dt\, du = f(k), \quad \text{or} \quad = 0, \\ &\text{according as } k \text{ falls within or beyond the range } \lambda - \mu. \end{aligned}\right\} \quad (84)$$

In the preceding investigation I have not calculated the values of the definite integral when $k = \lambda$, and when $k = \mu$, because such values are evidently finite, and in the use which will be made of the function each will give only one value of the infinitesimal element; and the omission of this element will not vitiate the definite integral.

If, in (84), $\lambda = \infty$ and $\mu = 0$, the theorem takes the form of Fourier's theorem as given in (145), Art. 204. Also in the first part of (84), if $f(t) = 1$, $\lambda = 1$, $\mu = 0$, we have

$$\frac{2}{\pi}\int_0^{\infty} \frac{\sin u \cos ku}{u}\, du = 1, \quad \text{or} \quad = 0,$$

according as k is less than or is greater than 1. This is the result already given in (68) of the present chapter, and is the factor of discontinuity which has been heretofore employed.

286.] I propose however now to take the more general factor of discontinuity which is given in the first part of (84); whereby we have

$$\left.\begin{array}{l}\frac{2}{\pi}\int_0^{\infty}\int_{\mu}^{\lambda} f(t) \cos ku \cos tu\, dt\, du = f(k), \quad \text{or} \quad = 0, \\ \text{according as } k \text{ falls within or beyond the range } \lambda - \mu.\end{array}\right\} \quad (85)$$

The following are examples in which this factor is employed. I have taken integrals of only two or three variables, so that the formulæ may be shorter; and as the process is the same in all cases, these will exemplify it quite as well as those in which more variables are employed.

Let an integral involve three variables, and be of the form

$$\mathrm{I} = \iiint x^{p-1} y^{q-1} z^{r-1} e^{-(ax+by+cz)} \mathrm{F}(x+y+z)\, dz\, dy\, dx, \qquad (86)$$

where a, b, c are positive quantities, or positive in at least the real parts of them, and where the limits of integration are given by the inequality,

$$\mu < x+y+z < \lambda. \qquad (87)$$

For the sake of abbreviation, let $x+y+z = k$; and by means of (85) let us replace $\mathrm{F}(k)$ in (86) by

$$\frac{2}{\pi}\int_0^{\infty}\int_{\mu}^{\lambda} \mathrm{F}(t) \cos ku \cos tu\, dt\, du,$$

which $= \mathrm{F}(k)$ for all values of k within the range $\lambda - \mu$; that is, within the range given by the inequality (87); and $= 0$, for all values of k outside of that range. Now this discontinuous equivalent of $\mathrm{F}(k)$ is the real part of $\frac{1}{\pi}\int_{-\infty}^{\infty}\int_{\mu}^{\lambda} \mathrm{F}(t)\, e^{-ku\sqrt{-1}} \cos ut\, dt\, du$;
so that substituting in (86), and extending the ranges of the $x-$, $y-$, $z-$ integrations to the limits ∞ and 0, we have

$\mathrm{I} =$ the real part

$$\text{of } \frac{1}{\pi}\int_0^{\infty}\int_0^{\infty}\int_0^{\infty} x^{p-1} y^{q-1} z^{r-1} e^{-(ax+by+cz)}\, dz\, dy\, dx$$

$$\times \int_{-\infty}^{\infty}\int_{\mu}^{\lambda} \mathrm{F}(t)\, e^{-ku\sqrt{-1}} \cos ut\, dt\, du$$

I = the real part

$$\text{of } \frac{1}{\pi}\int_{\mu}^{\lambda} \mathrm{F}(t)\, dt \int_{-\infty}^{\infty} \cos ut\, du \int_0^{\infty} x^{p-1} e^{-(a+u\sqrt{-1})x}\, dx$$

$$\times \int_0^{\infty} y^{q-1} e^{-(b+u\sqrt{-1})y}\, dy \int_0^{\infty} z^{r-1} e^{-(c+u\sqrt{-1})z}\, dz$$

$$= \frac{1}{\pi}\int_{\mu}^{\lambda} \mathrm{F}(t)\, dt \int_{-\infty}^{\infty} \cos ut\, du \frac{\Gamma(p)}{(a+u\sqrt{-1})^p} \frac{\Gamma(q)}{(b+u\sqrt{-1})^q} \frac{\Gamma(r)}{(c+u\sqrt{-1})^r}$$

$$= \frac{\Gamma(p)\Gamma(q)\Gamma(r)}{\pi}\int_{\mu}^{\lambda} \mathrm{F}(t)\, dt$$

$$\times \int_{-\infty}^{\infty} \frac{\cos ut\, du}{(a+u\sqrt{-1})^p (b+u\sqrt{-1})^q (c+u\sqrt{-1})^r}; \quad (88)$$

so that the value of the triple integral given in (86) is made to depend on that of a double integral. Similarly if I had been a multiple integral of the nth order, its value may be made to depend on that of a double integral.

287.] An important theorem may be derived from (88). Let a, b, c, and u be replaced by θa, θb, θc, θu respectively; and to shorten the formulæ, let us take only two variables: then, from (86) and (88), the limits being given by $\mu < x+y < \lambda$,

$$\iint x^{p-1} y^{q-1} e^{-(ax+by)\theta}\, \mathrm{F}(x+y)\, dy\, dx = \text{the real part of}$$

$$\frac{\Gamma(p)\,\Gamma(q)}{\pi}\int_{\mu}^{\lambda} \mathrm{F}(t)\, dt \int_{-\infty}^{\infty} \frac{\cos \theta ut\, du}{(a+u\sqrt{-1})^p (b+u\sqrt{-1})^q\, \theta^{p+q-1}}. \quad (89)$$

Let each side of this equation be multiplied by $\theta^{p+q-1} e^{-\omega\theta}\, d\theta$, and let the θ-integral of each side be taken for the limits ∞ and 0; then

$$\iint x^{p-1} y^{q-1} \mathrm{F}(x+y)\, dy\, dx \int_0^{\infty} e^{-(\omega+ax+by)\theta}\, \theta^{p+q-1}\, d\theta = \text{the real part of}$$

$$\frac{\Gamma(p)\,\Gamma(q)}{\pi}\int_{\mu}^{\lambda} \mathrm{F}(t)\, dt \int_{-\infty}^{\infty} \frac{du}{(a+u\sqrt{-1})^p (b+u\sqrt{-1})^q} \int_0^{\infty} \cos ut\theta\, e^{-\omega\theta}\, d\theta.$$

Hence, since by (250) Art. 122, and by (6), Art. 82,

$$\left.\begin{aligned} \int_0^{\infty} e^{-(\omega+ax+by)\theta}\, \theta^{p+q-1}\, d\theta &= \frac{\Gamma(p+q)}{(\omega+ax+by)^{p+q}}, \\ \text{and} \quad \int_0^{\infty} \cos ut\theta\, e^{-\omega\theta}\, d\theta &= \frac{\omega}{\omega^2+(ut)^2}; \end{aligned}\right\} \quad (90)$$

therefore for the limits assigned by the inequality, $\mu < x+y < \lambda$,

$$\iint x^{p-1}\,y^{q-1}\,\frac{F(x+y)}{(\omega+ax+by)^{p+q}}\,dy\,dx = \text{the real part of}$$

$$\frac{\omega}{\pi}\frac{\Gamma(p)\Gamma(q)}{\Gamma(p+q)}\int_{\mu}^{\lambda}F(t)\,dt\int_{-\infty}^{\infty}\frac{du}{\{\omega^2+(ut)^2\}(a+u\sqrt{-1})^p(b+u\sqrt{-1})^q}. \quad (91)$$

Now the last integral in the right-hand member admits of evaluation by the following process due to Lejeune Dirichlet.*

From (161) and (162), Art. 106, it is evident that

$$\int_{-\infty}^{\infty}\frac{\cos ku\,du}{h^2+u^2} = \frac{\pi}{h}e^{-hk}, \qquad \int_{-\infty}^{\infty}\frac{\sin ku\,du}{h^2+u^2} = 0;$$

so that
$$\int_{-\infty}^{\infty}\frac{e^{-ku\sqrt{-1}}\,du}{h^2+u^2} = \frac{\pi}{h}e^{-hk}. \quad (92)$$

Also since
$$\int_0^{\infty} v^{p-1}\,e^{-(a+u\sqrt{-1})v}\,dv = \frac{\Gamma(p)}{(a+u\sqrt{-1})^p};$$

$$\Gamma(p)\int_{-\infty}^{\infty}\frac{e^{-k\sqrt{-1}}\,du}{(h^2+u^2)(a+u\sqrt{-1})^p}$$

$$= \int_{-\infty}^{\infty}\int_0^{\infty}\frac{v^{p-1}\,e^{-(a+u\sqrt{-1})v}\,e^{-ku\sqrt{-1}}\,dv\,du}{h^2+u^2}$$

$$= \int_0^{\infty} v^{p-1}\,e^{-av}\,dv\int_{-\infty}^{\infty}\frac{e^{-(v+k)u\sqrt{-1}}\,du}{h^2+u^2}$$

$$= \int_0^{\infty} v^{p-1}\,e^{-av}\,dv\,\frac{\pi}{h}e^{-h(v+k)}$$

$$= \frac{\pi e^{-hk}}{h}\int_0^{\infty} v^{p-1}\,e^{-(a+h)v}\,dv$$

$$= \frac{\pi e^{-hk}}{h}\,\frac{\Gamma(p)}{(a+h)^p}; \quad (93)$$

so that
$$\int_{-\infty}^{\infty}\frac{e^{-ku\sqrt{-1}}\,du}{(h^2+u^2)\,(a+u\sqrt{-1})^p} = \frac{\pi e^{-hk}}{h\,(a+h)^p}. \quad (94)$$

Also by a further similar process it may be shewn that

$$\int_{-\infty}^{\infty}\frac{e^{-ku\sqrt{-1}}\,du}{(h^2+u^2)(a+u\sqrt{-1})^p(b+u\sqrt{-1})^q} = \frac{\pi e^{-hk}}{h(a+h)^p(b+h)^q}; \quad (95)$$

so that, if $k = 0$,

$$\int_{-\infty}^{\infty}\frac{du}{(h^2+u^2)(a+u\sqrt{-1})^p(b+u\sqrt{-1})^q} = \frac{\pi}{h\,(a+h)^p(b+h)^q}. \quad (96)$$

And substituting in (91), and observing that the whole expression is real, we have

* Crelle's Journal, Vol. IV, p. 94.

$$\iint x^{p-1} y^{q-1} \frac{f(x+y)}{(\omega+ax+by)^{p+q}} dy\, dx$$

$$= \frac{\Gamma(p)\,\Gamma(q)}{\Gamma(p+q)} \int_{\mu}^{\lambda} \frac{t^{p+q-1} f(t)\, dt}{(at+\omega)^p\,(bt+\omega)^q}, \quad (97)$$

where the limits of integration in the left-hand member are given by the inequality $\mu < x+y < \lambda$.

288.] This theorem also admits of extension to the case where the limits are given by the inequality,

$$\mu < \left(\frac{x}{a}\right)^{\alpha} + \left(\frac{y}{b}\right)^{\beta} + \left(\frac{z}{c}\right)^{\gamma} + \dots < \lambda, \quad (98)$$

and the definite integral is of the form

$$\iint \dots \frac{f\left\{\left(\frac{x}{a}\right)^{\alpha} + \left(\frac{y}{b}\right)^{\beta} + \dots\right\}}{\{\omega + lx^{\alpha} + my^{\beta} + \dots\}^{\frac{p}{\alpha}+\frac{q}{\beta}+\dots}} x^{p-1} y^{q-1} \dots dy\, dx. \quad (99)$$

The process of reducing this to the form (97) is exactly similar to that of Art. 280; and it is unnecessary to repeat it. The following example will illustrate it.

Let it be required to determine the value of

$$\iiint \frac{dz\, dy\, dx}{(\omega + x^2 + y^2 + z^2)^{\frac{3}{2}}}; \quad (100)$$

where the limits of integration are given by the inequality,

$$\mu^2 < \frac{x^2}{a^2} + \frac{y^2}{b^2} + \frac{z^2}{c^2} < \lambda^2. \quad (101)$$

Let $\frac{x^2}{a^2} = \xi$, $\frac{y^2}{b^2} = \eta$, $\frac{z^2}{c^2} = \zeta$; so that for limits assigned by the inequality (101),

$$\iiint \frac{dz\, dy\, dx}{(\omega + x^2 + y^2 + z^2)^{\frac{3}{2}}}$$

$$= \frac{\left\{\Gamma\left(\frac{1}{2}\right)\right\}^3}{\Gamma\left(\frac{3}{2}\right)} \frac{abc}{8} \int_{\mu^2}^{\lambda^2} \frac{t^{\frac{1}{2}}\, dt}{\{(a^2 t+\omega)\,(b^2 t+\omega)\,(c^2 t+\omega)\}^{\frac{1}{2}}};$$

$$= \frac{\pi abc}{2} \int_{\mu}^{\lambda} \frac{u^2\, du}{\{(a^2u^2+\omega)\,(b^2u^2+\omega)\,(c^2u^2+\omega)\}^{\frac{1}{2}}}; \quad (102)$$

and thus the triple integral is reduced to a single integral.

289.] Thus far, when the infinitesimal element has contained an arbitrary function, the subject of that function has been the expression by which the inequality assigning the limits of integration has been determined. Let us consider the following integral in which this is not the case; and here, as before, I will take an integral involving only three variables; viz.

$$\text{I} = \iiint \text{F}(x+y+z)\, dz\, dy\, dx,$$

where the limits are assigned by the inequality,

$$0 < \left(\frac{x}{a}\right)^2 + \left(\frac{y}{b}\right)^2 + \left(\frac{z}{c}\right)^2 < 1;$$

so that the integral includes all values of the variables which lie within the surface of an ellipsoid. Let

$$\frac{x^2}{a^2} + \frac{y^2}{b^2} + \frac{z^2}{c^2} = k; \qquad x+y+z = s. \tag{103}$$

Then since $\frac{2}{\pi}\int_0^\infty \frac{\sin t \cos kt}{t}\, dt = 1$ or 0, according as k is less than or greater than 1, so will the real part of $\frac{2}{\pi}\int_0^\infty \frac{\sin t}{t} e^{kt\sqrt{-1}}\, dt = 1$, or 0, according as k is less than or greater than 1. Hence if we introduce this factor of discontinuity into the definite integral, the ranges of the x-, y-, and z-integrations may be extended to ∞ and $-\infty$; so that with the limits assigned by the given inequality,

$$\iiint \text{F}(x+y+z)\, dz\, dy\, dx$$

$$= \text{the real part of } \int_{-\infty}^{\infty}\int_{-\infty}^{\infty}\int_{-\infty}^{\infty} \text{F}(s)\, dz\, dy\, dx\, \frac{2}{\pi}\int_0^\infty \frac{\sin t}{t} e^{kt\sqrt{-1}}\, dt.$$

$$\text{- - - - - - } \frac{2}{\pi}\int_0^\infty \frac{\sin t}{t} dt \int_{-\infty}^{\infty}\int_{-\infty}^{\infty}\int_{-\infty}^{\infty} \text{F}(s) e^{kt\sqrt{-1}} dz\, dy\, dx. \tag{104}$$

Now by Fourier's Theorem, (144), Art. 204,

$$\text{F}(s) = \frac{1}{2\pi}\int_{-\infty}^{\infty}\int_{-\infty}^{\infty} \text{F}(\omega) \cos u(s-\omega)\, du\, d\omega$$

$$= \text{the real part of } \frac{1}{2\pi}\int_{-\infty}^{\infty}\int_{-\infty}^{\infty} \text{F}(\omega)\, e^{u(s-\omega)\sqrt{-1}}\, du\, d\omega; \tag{105}$$

so that in (104), replacing $\text{F}(s)$ by this value, we have

$$\iiint \text{F}(x+y+z)\, dz\, dy\, dx = \text{the real part}$$

$$\text{of } \frac{1}{\pi^2}\int_0^\infty \frac{\sin t}{t} dt \int_{-\infty}^{\infty} \text{F}(\omega)\, d\omega \int_{-\infty}^{\infty} e^{-u\omega\sqrt{-1}} du \int_{-\infty}^{\infty}\int_{-\infty}^{\infty}\int_{-\infty}^{\infty} e^{(kt+us)\sqrt{-1}} dz\, dy\, dx$$

$$\text{- - - - - - - - - - - - - } dy \int_{-\infty}^{\infty} e^{\left(\frac{x^2}{a^2}t + xu\right)\sqrt{-1}}\, dx,$$

replacing k and s by their values given in (103).

But by (362), Art. 140,

$$\int_{-\infty}^{\infty} e^{\left(\frac{x^2}{a^2}t+x\omega\right)\sqrt{-1}}\,dx = a\left(\frac{\pi}{t}\right)^{\frac{1}{2}} e^{\left(\frac{\pi}{4}-\frac{a^2\omega^2}{4t}\right)\sqrt{-1}};$$

and similar values are true for the y- and the z-integrals; so that substituting these, and replacing $a^2+b^2+c^2$ by r^2, we have

$$\iiint \mathrm{F}(x+y+z)\,dz\,dy\,dx =$$

$$\text{the real part of } \frac{abc}{\pi^{\frac{1}{2}}} e^{\frac{3\pi}{4}\sqrt{-1}} \int_0^{\infty} \frac{\sin t}{t^{\frac{3}{2}}}\,dt \int_{-\infty}^{\infty} \mathrm{F}(\omega)\,d\omega \int_{-\infty}^{\infty} e^{-\left(\frac{r^2u^2}{4t}+\omega u\right)\sqrt{-1}}\,du$$

$$\text{- - - - - - -}\ \frac{abc}{\pi^{\frac{1}{2}}} e^{\frac{3\pi}{4}\sqrt{-1}} \int_0^{\infty} \frac{\sin t}{t^{\frac{3}{2}}}\,dt \int_{-\infty}^{\infty} \mathrm{F}(\omega)\,d\omega\, \frac{2(\pi t)^{\frac{1}{2}}}{r} e^{\left(\frac{\omega^2 t}{r^2}-\frac{\pi}{4}\right)\sqrt{-1}}$$

$$\text{- - - - - - -}\ \frac{2\,abc}{r} e^{\frac{\pi}{2}\sqrt{-1}} \int_0^{\infty} \frac{\sin t}{t}\,dt \int_{-\infty}^{\infty} \mathrm{F}(\omega)\, e^{\frac{\omega^2 t}{r^2}\sqrt{-1}}\,d\omega$$

$$= \frac{2\,abc}{r} \int_{-\infty}^{\infty} \mathrm{F}(\omega)\,d\omega \int_0^{\infty} \frac{\sin t}{t} \sin\frac{\omega^2 t}{r^2}\,dt; \qquad (106)$$

whereby the triple integral is expressed in terms of a double integral.

If the integral contains n variables, the limits being given by the inequality,

$$0 < \frac{x^2}{a^2} + \frac{y^2}{b^2} + \frac{z^2}{c^2} + \ldots\ldots < 1, \qquad \text{then}$$

$$\iiint \ldots \mathrm{F}(x+y+z+\ldots)\ldots dz\,dy\,dx$$

$$= 2\pi^{\frac{n-3}{2}} \frac{abc\ldots}{r} \int_{-\infty}^{\infty} \mathrm{F}(\omega)\,d\omega \int_0^{\infty} \frac{\sin t}{t^{\frac{n+1}{2}}} \cos\left(\frac{n-1}{4}\pi + \frac{\omega^2 t}{r^2}\right) dt; \ (107)$$

and thus the evaluation of the multiple integral depends on that of the double integral.

CALCULUS OF VARIATIONS.

CHAPTER XIII.

EXPOSITION OF THE PRINCIPLES OF THE CALCULUS OF VARIATIONS.

290.*] The subjects of investigation in the preceding parts of our treatise have been functions whose forms are known and determinate; such as those symbolized by cos, $\tan^{-1}$, log, $\log^{-1}$,

* The authors and titles of the principal works on this branch of infinitesimal calculus are the following, and from them much assistance has been derived :—

Euler, Methodus inveniendi lineas curvas maximi minimive proprietate gaudentes, Lausanne, 1744.

Lagrange, Leçons sur le Calcul des Fonctions, Paris, 1808.

Lagrange, Théorie des Fonctions Analytiques, 3me edition, par J. A. Serret, Paris, 1847.

Poisson, Mémoires de l'Institut de France, Tome XII, Paris, 1833.

Jacobi, Journal für Mathematik, Crelle, Band XVII, Berlin, 1837.

Ostrogradsky, Mémoires de l'Académie de St. Petersbourg, Tome I, St. Petersbourg, 1838.

Ibid. Tome III.

Delaunay, Journal de l'Ecole Polytechnique, XXIX Cahier, Paris, 1843.

Delaunay, Journal de Mathématiques, par Liouville, Tome VI, Paris, 1841.

Sarrus, Mémoires presentés par divers savants à l'Académie des Sciences, Tome X, Paris, 1848.

Strauch, Theorie und Anwendung des Variations-calcul, Zurich, 1849.

Jellett, Calculus of Variations, Dublin, 1850.

Schellbach, Probleme der Variationsrechnung, Crelle, Band XLI, p. 293, 1851.

Stegmann, Lehrbuch der Variationsrechnung, Cassel, 1854.

Todhunter, History of the Progress of the Calculus of Variations; Cambridge, 1861.

Moigno et Lindelöf, Calcul des Variations; Paris, 1861.

and other such like: and the inquiry has been for the most part confined to the properties of these functions, which arise from the continuous and infinitesimal variation of their subject-variables; we have had no occasion to consider the functions themselves as undergoing continuous change as to *form*; certain invariable relations have been shewn to exist between certain functions; for by the process of derivation we pass from one function to another; but these are nevertheless determinate, and the relation arises from a continuous growth of the subject-variable, and not from a continuous and infinitesimal change of the function as to *form*. This distinction is important; for there is no conceivable reason why functions should not be continuously variable as to form, as well as numbers be as to magnitude. Thus for instance suppose the subject of investigation to be $y = \sin x$; the value of y may manifestly be changed either by a change in the subject-variable x, or by a change of the functional symbol into any other, as $\tan^{-1}$; changes due to the former cause are considered in the Differential Calculus; but those arising from a continuous change in the form of the function require another mode of investigation; and whereas heretofore we have passed *per saltum* from one function to another, the new calculus requires a continuous passage: a wide extension then is opened before us, one the subject-matter of which is not number but functions: and as a functional symbol expresses the *law* of combination of its subject-variables, we shall have to consider laws, and not subjects of laws. Functions then, as they are the subject of this new calculus, are free from all concrete or applied signification, and express laws; and the proper end and object of such a calculus of functions is to investigate their origin and their principles, their growth and extent, their laws of combination, and to deduce from them properties with which they are pregnant. As differential calculus investigates properties of continuous number, so in this new calculus properties of continuous functions are discussed.

291.] Apart however from these general considerations, let us view the calculus in the light of an easy problem of that class, the attempt to solve which gave rise to it. Suppose that it is required to determine the form of the function connecting x and y which expresses the shortest distance between two given points: if the function were given, the problem would be one of rectification and would be solved by the integral calculus: also *a posteriori* we know that the required function is that which expresses a straight line: but the *direct* solution of the problem requires a different process; viz. the assumption of a general functional symbol undetermined as to form, and the expression for the distance between the two points in terms of it; so that if an infinitesimal variation of the distance due to an infinitesimal variation of the form of the function is calculated, the required form will be determined by equating to zero that variation: provided that the form so determined is such as to make the first variation change its sign: or what is equivalent, such as to make the second variation either positive or negative for all values of the determined function within the given points: for such an operation it is necessary (1) to calculate the infinitesimal change of the distance due to the infinitesimal change of the form of a function, (2) to be able to determine the form of the function by equating to zero the variation of the distance; in other words, we must be able to differentiate functions as to form, and to determine functions by means of given conditions; also if these conditions give many results, we must be able to discriminate according as one or another is taken. Such a process then requires a knowledge of functions as accurate and complete as that of number required in the differential calculus. It will be observed that, as the two points which are to terminate the line are given, the only variable quantity of the problem is the form of the function.

Suppose however that the problem is, to determine the form of the function which expresses the shortest distance between two given curves in space; let the distance be expressed by means of a general undetermined function, as in the former case, and in terms of the current coordinates of the two curves which it is to meet; then it becomes dependent on the form of the function, and on the coordinates of these two curves; and as these quantities are independent of each other, they may be considered as independent variables, and their variations may be taken separately; that arising from a change in the form of the function may be estimated as in the former case, and thence may be deduced the form that gives the least distance: and those which arise from the coordinates of the points on the given curves at which the required curve is to meet them must be calculated according to the rules of the differential calculus, and by equating them to zero we shall be able to determine the points of meeting. In the solution of this problem therefore two kinds of variations will be required, one arising from a change in the form of the function, and the other from the differentiation of the equations of the given curves.

292.] The infinitesimal variations therefore of the calculus of functions and of the differential calculus are essentially distinct in *kind*: in the former they result from a change of form of an undetermined function; in the latter from a change of the subject-variables of a determinate function: and to use language borrowed from the geometry of curves, a variation of the former kind leads from a point on a curve to a point on another curve infinitesimally near to it; a variation of the latter from a point on a curve to a point on the same curve infinitesimally near to it. It is convenient therefore to have different names for quantities so different, and to express them by different symbols: in the former calculus they are called *variations*, in the latter differentials: hence arises the name "calculus of variations," and so henceforth we shall employ the term "variation" in a technical sense, to indicate the particular infinitesimal change of this calculus: we also shall use d to express *differential*, and δ to express *variation*: consequently d indicates a passage from one system of variables to another, both of which satisfy a given determinate function; δ indicates a passage from a system which satisfies one function to a system satisfying a function infinitesimally different from the former one: thus a variation as applied to a function may be defined as *the infinitesimal change of the value of the function due to its change of form*; and variation as applied to a variable is *the infinitesimal arbitrary increment of it.*

293.] The symbols in relation to their subjects stand as follow: let u be an undetermined function; then δu is the change of value of u due to its change of form. Now let a certain operation symbolized by F be performed on u; it may be differentiation or integration; and let

$$\mathrm{V} = \mathrm{F}(u);$$

then

$$\delta \mathrm{V} = \delta . \mathrm{F}(u); \qquad (1)$$

and $\delta \mathrm{V}$ is the change in V, $= \mathrm{F}(u)$, due to a change of form of u.

As in the differential calculus there are partial and total differentials of functions of many variables, according as one or all of the variables change value; so if a function, whose variation is to be calculated, involves many undetermined and independent functions, it is susceptible of different variations according as one or more or all of these undetermined functions vary, and therefore in the present calculus there will be partial and total variations; and, by the principle of such infinitesimal changes, the total variation is equal to the sum of the several partial variations.

Thus let $u_1, u_2, \dots u_n$ denote n undetermined functions, and let F denote an operation performed on a certain combination of them; and let

$$\text{V} = \text{F}(u_1, u_2, \dots u_n); \tag{2}$$

then
$$\delta \text{V} = \left(\frac{\delta \text{V}}{\delta u_1}\right)\delta u_1 + \left(\frac{\delta \text{V}}{\delta u_2}\right)\delta u_2 + \dots + \left(\frac{\delta \text{V}}{\delta u_n}\right)\delta u_n; \tag{3}$$

using brackets to denote partial variations. Now, and this remark is important, so long as the relation between F and u_1 remains the same, the ratio of the *infinitesimal* changes of F and u_1 must be independent of the particular species of them, that is, must be the same, whether the changes are of magnitude or of form; and consequently

$$\left(\frac{\delta \text{V}}{\delta u_1}\right) = \left(\frac{d\text{V}}{du_1}\right); \tag{4}$$

and similarly for the others; and thus

$$\delta \text{V} = \left(\frac{d\text{V}}{du_1}\right)\delta u_1 + \left(\frac{d\text{V}}{du_2}\right)\delta u_2 + \dots + \left(\frac{d\text{V}}{du_n}\right)\delta u_n; \tag{5}$$

whence it follows that the variations of finite quantities and of finite functions follow the same laws as the differentials of similar quantities.

294.] Thus far as to the general principles of the Calculus of Variations: we proceed to investigate methods by which it may be applied to the solution of problems which are of the greatest importance in the present state of mathematical science, and which the Differential Calculus fails to solve.

Of functions in their integral and determinate forms our knowledge is too scanty for the attainment of the present object; but there are certain general expressions for infinitesimal elements, independent of the functions of which they are elements, and therefore the same for all, provided that the functions satisfy the law of continuity within the range for which they are considered; thus $ds = \{dx^2 + dy^2\}^{\frac{1}{2}}$ is the distance between two points (x, y) $(x+dx, y+dy)$ on a plane curve, whatever is the form of the function $y = f(x)$, which is the equation to the curve. Thus also $\{dy^2 dz^2 + dz^2 dx^2 + dx^2 dy^2\}^{\frac{1}{2}}$ is the surface-element, whatever is the form of $\text{F}(x, y, z) = c$, which is the equation to the surface:

similarly $dx\,dy\,dz$ is the volume-element referred to rectangular coordinates, and is independent of the particular form of the bounding surface.

Now these and similar general expressions for infinitesimal elements are made the subjects of investigation; and we calculate their variations according to processes which will be developed hereafter. If an integral function is the subject of inquiry, it is considered as the integral or sum of elements; and to this sum we apply the conditions, so far as they are applicable, for determining the unknown function. By this artifice we avoid the difficulty of making to vary the function in its general form. Thus, for instance, in the problem of finding the shortest line between two given points, (x_n, y_n) and (x_0, y_0); instead of assuming the distance to be $F(x_n, y_n, x_0, y_0)$, where F represents some undetermined function, and then determining F by equating to zero the change of the distance due to the variation or change of form of F, we assume $ds = \{dx^2+dy^2\}^{\frac{1}{2}}$ to be an element of the distance, so that the distance $=$* $\int_0^n \{dx^2+dy^2\}^{\frac{1}{2}}$; and we make the latter sum the subject of investigation.

And in the most general case; suppose that we have to investigate the form of the relation $y = f(x)$, where f is the symbol of some unknown function, so that a given condition should be satisfied, when that condition is the sum or integral of a series of elements, each of which is a given function of $x, dx, d^2x, \ldots d^nx, y, dy, d^2y, \ldots d^ny$, neither x nor y being equicrescent; then, if the element $= F(x, dx, d^2x, \ldots d^nx, y, dy, d^2y, \ldots d^ny)$, where F represents a given function, and if x_1, y_1, x_0, y_0, are the given limiting values of x and y, the unknown function

$$= \int_0^1 F(x, dx, d^2x, \ldots d^nx, y, dy, d^2y, \ldots d^ny); \qquad (6)$$

and the relation which exists between y and x, that is, the form of f, is determined by means of conditions to which (6) is subject. A similar method is applicable if the element of the unknown function involves more variables and their differentials.

Now the principle by which these and similar problems are solved is of the greatest importance; and is that of which the calculus of variations is the development. It may be formally stated in the following terms;

* Instead of expressing the limits of integration at length, we have merely put their subscript letters, as we have already explained in Art. 247.

If a quantity depends on a certain unknown function, and the determination of the form of that function depends on certain given conditions, which it has to satisfy; in the general case our knowledge of functions and of their laws is insufficient for the determination of the function, and especially when the conditions require an infinitesimal variation of it: but as the form of many infinitesimal elements is known, and is the same whatever is the unknown continuous function, we may consider the quantity which depends on the function to be the sum of certain elements between given limits, and may make the quantity in its latter form the subject of inquiry.

295.] When the problem has been put into the above form, the following is the most convenient method of effecting and of symbolizing the necessary operations: the unknown function is made to assume a new form by an infinitesimal change of the variables and their differentials which are involved in the given element-function, the infinitesimal variations being functions of the variables to which they are applied; and as hereby the element-function will have changed value, so will also the sum of all these; and as these infinitesimal changes are not made subject to the conditions of an original given function, they may be, and generally will be, inconsistent with it, and thus a new law will be introduced which will be expressed by a new functional symbol. Or to employ the language of geometry: suppose a certain curve to be expressed by the undetermined function: and suppose each point of the curve to be shifted, and thereby each of the length-elements and each of the successive differentials to change value; the curve in its new position will generally have taken a new form, and so will require a new function to express it. Thus, suppose the curve under consideration to be a curve of double curvature, and let the position and form of it be changed; then if δx, δy, δz are the variations of the coordinates, these being functions of x, y, z, the point (x, y, z) becomes $(x+\delta x, y+\delta y, z+\delta z)$; observe then the change; the point on the old curve infinitesimally near to (x, y, z) is $(x+dx, y+dy, z+dz)$, whereas $x+\delta x$, $y+\delta y$, $z+\delta z$ refer to the same point as x, y, z, but to it in a new position, and on a new curve, and when the form of the function has varied. Similarly also $\delta.dx$, $\delta.dy$, $\delta.dz$, $\delta.d^2x, \ldots \delta.d^n x$ express variations which the several successive differentials undergo, and which are due to the change of the form of the function.

As however the element-function which will be the subject of variation will generally contain the differentials of the variables as well as the variables themselves, it is necessary to consider with great precision the relations between variations and differentials. And as the operations of which these infinitesimal quantities are the results are evidently of the same nature though of different species, the order in which they are effected on a given subject is evidently indifferent; and consequently they are subject to the commutative law; so that we have

$$\left.\begin{aligned} \delta.dx &= d.\delta x; \\ \delta.d^2x &= d.\delta.dx = d^2.\delta x; \\ &\cdots\cdots\cdots \\ \delta.d^n x &= d^n.\delta x. \end{aligned}\right\} \qquad (7)$$

Similarly, if $y = f(x)$,

$$\left.\begin{aligned} \delta.dy &= \delta.df(x) = d.\delta f(x); \\ \delta.d^2y &= \delta.d^2f(x) = d^2.\delta f(x); \\ &\cdots\cdots\cdots\cdots \\ \delta.d^n y &= d^n.\delta f(x). \end{aligned}\right\} \qquad (8)$$

And similar results are also true for other variables.

Now as d denotes an operation subject to the index law, and which is true for negative as well as for positive values of the index, the results of the operations effected on x and on $f(x)$ being true for positive integral values of n will also be true for negative integral values; that is, as they are true for differentiation, so will they also be true for integration. Thus

$$\left.\begin{aligned} \delta.\int dx &= \int \delta.dx; \\ \delta.\int f(x)\,dx &= \int \delta.f(x)\,dx; \\ \delta.\int\int f(x,y)\,dy\,dx &= \int\int \delta.f(x,y)\,dy\,dx. \end{aligned}\right\} \qquad (9)$$

Similar results are also true for successive variations; so that we have generally

$$\left.\begin{aligned} \delta^m d^n f(x) &= d^n \delta^m f(x); \\ \delta^m.\int\int \ldots \mathrm{v} \ldots dz\,dy\,dx &= \int\int \ldots \delta^m.(\mathrm{v} \ldots dz\,dy\,dx). \end{aligned}\right\} \qquad (10)$$

Also as $\left(\frac{du}{dx}\right), \left(\frac{du}{dy}\right), \ldots \left(\frac{d^2u}{dx^2}\right), \left(\frac{d^2u}{dx\,dy}\right), \ldots$ are generally functions of $x, y, z, \ldots$, so they are susceptible of variation; and we have

$$\left.\begin{array}{ll} \delta . \frac{du}{dx} = \frac{d.\delta u}{dx}; & \delta . \frac{du}{dy} = \frac{d.\delta u}{dy}; \ldots\ldots \\ \delta . \frac{d^2u}{dx^2} = \frac{d^2.\delta u}{dx^2}; & \delta . \frac{d^2u}{dx\,dy} = \frac{d^2.\delta u}{dx\,dy}; \end{array}\right\} \quad (11)$$

and also still more generally

$$\delta . \frac{d^{m+n+r+\cdots}u}{dx^m\,dy^n\,dz^r \ldots} = \frac{d^{m+n+r+\cdots}\delta u}{dx^m\,dy^n\,dz^r \ldots}. \quad (12)$$

296.] As these results, especially (9), are of importance in the sequel, let us consider them in reference to a plane curve; see fig. 47.

Let P_0PQP_1 be a plane curve whose equation is $y = f(x)$;

$$\left.\begin{array}{l} OM_0 = x_0 \\ M_0P_0 = y_0 \end{array}\right\}; \quad \left.\begin{array}{l} OM = x \\ MP = y \end{array}\right\}; \quad \left.\begin{array}{l} MN = dx \\ GQ = dy \end{array}\right\}; \quad \left.\begin{array}{l} OM_1 = x_1 \\ M_1P_1 = y_1 \end{array}\right\};$$

Q being a point on the curve infinitesimally near to P.

Now suppose each point of the curve to have shifted, so that the points indicated by the letters in the figure assume the points indicated by the letters accented, and suppose hereby the form of the function to have changed; so that

$$\left.\begin{array}{l} MM' = \delta x \\ P'L = \delta y \end{array}\right\}; \quad \left.\begin{array}{l} NN' = \delta(x+dx) = \delta x + \delta.dx \\ KQ' = \delta(y+dy) = \delta y + \delta.dy \end{array}\right\}.$$

Also as P', Q' are points on the new curve infinitesimally near to each other, since $OM' = x + \delta x$, therefore

$$M'N' = d(x+\delta x) = dx + d.\delta x. \quad \text{Also}$$

$$\because \quad MN + NN' = MM' + M'N';$$

$$\therefore \quad dx + \delta x + \delta.dx = \delta x + dx + d.\delta x;$$

$$\therefore \quad \delta.dx = d.\delta x.$$

By a similar process we may prove that

$$\delta.dy = d.\delta y;$$

and by repetition of the process that

$$\left.\begin{array}{l} \delta.d^n x = d^n.\delta x; \\ \delta.d^n y = d^n.\delta y. \end{array}\right\} \quad (13)$$

It will be observed that we have made the limiting points of the curve, viz. P_0 and P_1, to change position, so that there are variations of x_0, y_0, x_1, y_1: in the general case this will of course occur; and the change of value in the integral, of which these are the limits, is the sum of the quantities which have been determined in (86) and (87), Art. 96. If the limits are fixed, there are of course no variations of them; and if the limits are constrained to be on certain curves, their variations are not arbitrary, but must be in agreement with the equations to these curves. This illustration indicates how the total variation of a quantity expressed in the form of a definite integral involves partial variations due to changes of the limits, as well as of those due to the change of form of the function.

297.] Problems within the range of this calculus may involve either a single infinitesimal element, or the integral of such elements between given limits of integration. In the latter case a finite function is the subject of investigation; but as it is, and is expressed as, the sum of a series of infinitesimals, and as the order in which the operations of variation and of integration are effected is indifferent in the result, here also the infinitesimal elements are the first subjects of investigation. It is necessary therefore in the first place to investigate the effects of the operations of the calculus of variations on these elements; and I shall take those which have arisen in the preceding chapters.

Let (x, y, z) $(x+dx,\ y+dy,\ z+dz)$ be two points in space infinitesimally near to each other; and let ds be the distance between them; so that

$$ds^2 = dx^2+dy^2+dz^2. \tag{14}$$

Let (x, y, z) be shifted to $(x+\delta x,\ y+\delta y,\ z+\delta z)$, where $\delta x, \delta y, \delta z$ are arbitrary functions of x, y and z; so that the displacement of the point (x, y, z) may be most general; then, taking the variation of (14) according to the theorem given in (5), we have

$$ds\,\delta.ds = dx\,\delta.dx+dy\,\delta.dy+dz\,\delta.dz;$$

$$\therefore\quad \delta.ds = \frac{dx}{ds}\delta.dx+\frac{dy}{ds}\delta.dy+\frac{dz}{ds}\delta.dz; \tag{15}$$

which is the total variation of ds, and shews that it is equal to the sum of the projections on the tangent to the curve at (x, y, z) of the several variations of the coordinates.

The following however is another proof of this theorem, and as it involves quantities which will be of great use in the sequel, it is desirable to insert it at once.

Let (ξ, η, ζ) be the *varied* place of (x, y, z); so that

$$\xi = x+\delta x, \qquad \eta = y+\delta y, \qquad \zeta = z+\delta z. \tag{16}$$

Now as δx, δy, δz are in the most general case arbitrary functions of x, y, and z*,

$$\left.\begin{aligned} d\xi &= dx + \frac{d.\delta x}{dx}dx + \frac{d.\delta x}{dy}dy + \frac{d.\delta x}{dz}dz, \\ d\eta &= dy + \frac{d.\delta y}{dx}dx + \frac{d.\delta y}{dy}dy + \frac{d.\delta y}{dz}dz, \\ d\zeta &= dz + \frac{d.\delta z}{dx}dx + \frac{d.\delta z}{dy}dy + \frac{d.\delta z}{dz}dz. \end{aligned}\right\} \quad (17)$$

But as $\delta x, \delta y, \delta z$ are variations of the $x-$, $y-$, and $z-$ coordinates respectively, it is evident in reference to δx that $\frac{d.\delta x}{dy}$ and $\frac{d.\delta x}{dz}$ are infinitesimals of a higher order than $\frac{d.\delta x}{dx}$; and analogous results are true of the differentials of δy and δz; so that taking infinitesimals of the lowest order in (17), we have

$$d\xi = \left\{\frac{d.\delta x}{dx}+1\right\}dx, \quad d\eta = \left\{\frac{d.\delta y}{dy}+1\right\}dy, \quad d\zeta = \left\{\frac{d.\delta z}{dz}+1\right\}dz. \quad (18)$$

Let $d\xi^2 + d\eta^2 + d\zeta^2 = d\sigma^2$; then $d\sigma = ds + \delta.ds$; and consequently

$$\{ds + \delta.ds\}^2 = \left\{\frac{d.\delta x}{dx}+1\right\}^2 dx^2 + \left\{\frac{d.\delta y}{dy}+1\right\}^2 dy^2 + \left\{\frac{d.\delta z}{dz}+1\right\}^2 dz^2; \quad (19)$$

and omitting infinitesimals of the highest orders, this on expansion becomes

$$ds^2 + 2\,ds\,\delta.ds = dx^2 + dy^2 + dz^2 + 2\{dx\,d.\delta x + dy\,d.\delta y + dz\,d.\delta z\};$$

$$\therefore \quad \delta.ds = \frac{dx}{ds}\delta.dx + \frac{dy}{ds}\delta.dy + \frac{dz}{ds}\delta.dz; \quad (20)$$

the order of the symbols d and δ having been changed by reason of (7). And this is the same result as (15).

Now, as $\delta.ds$ is the change in length which the arc-element undergoes by reason of the displacement of ds, it has been called the *linear dilatation* of the arc-element; and similarly, $\frac{d\xi}{dx}dx$, $\frac{d\eta}{dy}dy$, $\frac{d\zeta}{dz}dz$ are respectively the linear dilatations of dx, dy, dz. We shall have important applications of this and of similar theorems in the higher mechanics.

* Many of the brackets which are indicative of partial differentiation are omitted in this and the following Articles, that the heaviness of the formulæ may be relieved.

298.] Next let $dy\,dx$ be the infinitesimal whose variation is to be determined, which is the area-element of a plane surface. Let ξ, η, as in the last article be the varied values of x and y; so that $\xi = x+\delta x$, $\eta = y+\delta y$; then

$$\left.\begin{aligned} d\xi &= dx + \frac{d.\delta x}{dx}\,dx + \frac{d.\delta x}{dy}\,dy, \\ d\eta &= dy + \frac{d.\delta y}{dx}\,dx + \frac{d.\delta y}{dy}\,dy; \end{aligned}\right\} \qquad (21)$$

and consequently, by the method of Art. 211,

$$d\xi\,d\eta = \left\{\left(1+\frac{d.\delta x}{dx}\right)\left(1+\frac{d.\delta y}{dy}\right) - \frac{d.\delta x}{dy}\,\frac{d.\delta y}{dx}\right\}dx\,dy \quad (22)$$

$$= \left\{1 + \frac{d.\delta x}{dx} + \frac{d.\delta y}{dy}\right\}dx\,dy,$$

where infinitesimals of the higher orders are omitted. Thus, since $\delta.dx\,dy =$ the excess of the area of the displaced element over that of the original element $= d\xi\,d\eta - dx\,dy$,

$$\delta.dx\,dy = \left\{\frac{d.\delta x}{dx} + \frac{d.\delta y}{dy}\right\}dx\,dy. \qquad (23)$$

As $\delta.dx\,dy$ is the variation which the infinitesimal area-element undergoes by reason of its displacement, it is called *the superficial dilatation* of the area-element; and this is of course expressed by the right-hand member of (23).

Let us however consider this subject by the light of first principles, for we shall thereby be able to trace not only the area of the displaced element, but also its form; and this is important in a problem wherein great clearness and precision are required. Let A, B, C, D be the four angular points of the original area-element, $dx\,dy$, where A is (x, y); B is $(x+dx, y)$; C is $(x, y+dy)$; D is $(x+dx, y+dy)$; also let A′, B′, C′, D′, be the places of A, B, C, D after the displacement; so that A′ is $(x+\delta x, y+\delta y)$;

B′ is $\left(x+\delta x+dx+\dfrac{d.\delta x}{dx}\,dx,\ y+\delta y+\dfrac{d.\delta y}{dx}\,dx\right)$;

C′ is $\left(x+\delta x+\dfrac{d.\delta y}{dx}\,dx,\ y+\delta y+dy+\dfrac{d.\delta y}{dy}\,dy\right)$;

D′ is $\left(x+\delta x+dx+\dfrac{d.\delta x}{dx}\,dx+\dfrac{d.\delta x}{dy}\,dy,\right.$

$\left.\qquad y+\delta y+dy+\dfrac{d.\delta y}{dx}\,dx+\dfrac{d.\delta y}{dy}\,dy\right)$.

Thus $\text{A}'\text{B}' = dx\left\{\left(1+\frac{d.\delta x}{dx}\right)^2+\left(\frac{d.\delta y}{dx}\right)^2\right\}^{\frac{1}{2}} = \text{C}'\text{D}'$;

$$\text{A}'\text{C}' = dy\left\{\left(\frac{d.\delta x}{dy}\right)^2+\left(1+\frac{d.\delta y}{dy}\right)^2\right\}^{\frac{1}{2}} = \text{B}'\text{D}';$$

and thus the figure $\text{A}'\text{B}'\text{C}'\text{D}'$ is a parallelogram. If infinitesimals of the higher orders are neglected, these results are the same as (18).

Also let ω be the angle between the two sides $\text{A}'\text{B}'$ and $\text{A}'\text{C}'$; then since

$$\text{B}'\text{C}'^2 = \left\{\left(1+\frac{d.\delta x}{dx}\right)dx-\frac{d.\delta x}{dy}dy\right\}^2+\left\{\left(1+\frac{d.\delta y}{dy}\right)dy-\frac{d.\delta y}{dx}dx\right\}^2;$$

$$\therefore \quad \cos\omega = \frac{\text{A}'\text{B}'^2+\text{A}'\text{C}'^2-\text{B}'\text{C}'^2}{2\,\text{A}'\text{B}'\times\text{A}'\text{C}'}$$

$$= \frac{d.\delta x}{dy}+\frac{d.\delta y}{dx}, \tag{24}$$

if infinitesimals of the higher orders are neglected. As those however involved in the right-hand member are of a higher order than those retained in (23), it appears that approximately $\cos\omega = 0$; consequently $\omega = 90°$, and the displaced area-element is a rectangle as it was in its original state.

Again, let us investigate the variation of that general surface-element, which is given in (30), Art. 236; viz.

$$d\text{A}^2 = dy^2\,dz^2+dz^2\,dx^2+dz^2\,dy^2. \tag{25}$$

Let $d\text{A}'$ be the area of the displaced surface-element, so that if (ξ, η, ζ) is the *varied* place of (x, y, z),

$$d\text{A}'^2 = d\eta^2\,d\zeta^2+d\zeta^2\,d\xi^2+d\xi^2\,d\eta^2. \tag{26}$$

And substituting for $d\xi, d\eta, d\zeta$ the values given in (18), we have

$$d\text{A}'^2 = \left(1+\frac{d.\delta y}{dy}\right)^2\left(1+\frac{d.\delta z}{dz}\right)^2 dy^2\,dz^2+\ldots+\ldots$$

$$= \left\{1+2\frac{d.\delta y}{dy}+2\frac{d.\delta z}{dz}\right\}dy^2\,dz^2+\ldots+\ldots$$

$$= d\text{A}^2+2\left\{\left(\frac{d.\delta y}{dy}+\frac{d.\delta z}{dz}\right)dy^2\,dz^2+\ldots+\ldots\right\};$$

and taking the square root of both members,

$$d\text{A}' = d\text{A}+\left\{\frac{d.\delta y}{dy}+\frac{d.\delta z}{dz}\right\}\frac{dy^2\,dz^2}{d\text{A}}+\left\{\frac{d.\delta z}{dz}+\frac{d.\delta x}{dx}\right\}\frac{dz^2\,dx^2}{d\text{A}}$$

$$+\left\{\frac{d.\delta x}{dx}+\frac{d.\delta y}{dy}\right\}\frac{dx^2\,dy^2}{d\text{A}}, \tag{27}$$

all infinitesimals of the higher orders having been omitted. Now since $\delta.d\text{A}$ is the excess of the area of the surface-element in its displaced state over that of it in its original state, $\delta.d\text{A} = d\text{A}' - d\text{A}$; and consequently

$$\delta.d\text{A} = \left(\frac{d.\delta y}{dy} + \frac{d.\delta z}{dz}\right)\frac{dy^2\,dz^2}{d\text{A}} + \left(\frac{d.\delta z}{dz} + \frac{d.\delta x}{dx}\right)\frac{dz^2\,dx^2}{d\text{A}} + \left(\frac{d.\delta x}{dx} + \frac{d.\delta y}{dy}\right)\frac{dx^2\,dy^2}{d\text{A}}: \quad (28)$$

which is the required result; and expresses the areal dilatation of the surface-element. I may observe that (28) may be deduced from (25) by taking the variation of (25) according to the rules of differentiation.

If a, β, γ are the direction-cosines of the normal to the surface at the point (x, y, z), then employing the values given in (26), (27), (28), Art. 236, (28) becomes

$$\delta.d\text{A} = \left(\frac{d.\delta y}{dy} + \frac{d.\delta z}{dz}\right)\cos a\,dy\,dz + \left(\frac{d.\delta z}{dz} + \frac{d.\delta x}{dx}\right)\cos\beta\,dz\,dx + \left(\frac{d.\delta x}{dx} + \frac{d.\delta y}{dy}\right)\cos\gamma\,dx\,dy; \quad (29)$$

which shews that the variation of the general surface-element is equal to the sum of the projections on its plane of the several variations of its projections on the coordinate planes.

299.] Again, let the infinitesimal whose variation is to be determined be $dx\,dy\,dz$; that is, be the volume-element in reference to a system of rectangular coordinates in space.

Let (ξ, η, ζ) be as heretofore the varied place of (x, y, z); then by reason of (17),

$$d\xi\,d\eta\,d\zeta = \begin{vmatrix} 1+\frac{d.\delta x}{dx}, & \frac{d.\delta x}{dy}, & \frac{d.\delta x}{dz} \\ \frac{d.\delta y}{dx}, & 1+\frac{d.\delta y}{dy}, & \frac{d.\delta y}{dz} \\ \frac{d.\delta z}{dx}, & \frac{d.\delta z}{dy}, & 1+\frac{d.\delta z}{dz} \end{vmatrix} dx\,dy\,dz; \quad (30)$$

and omitting infinitesimals of the higher orders, we have

$$d\xi\,d\eta\,d\zeta = \left\{1 + \frac{d.\delta x}{dx} + \frac{d.\delta y}{dy} + \frac{d.\delta z}{dz}\right\} dx\,dy\,dz. \quad (31)$$

And as the variation of the volume-element is equal to the excess of the volume in its displaced position over that of it in its original position,

$$\delta.dx\,dy\,dz = \left\{\frac{d.\delta x}{dx} + \frac{d.\delta y}{dy} + \frac{d.\delta z}{dz}\right\} dx\,dy\,dz. \qquad (32)$$

This is called the cubical dilatation of the volume-element.

As this subject is of peculiar importance in reference to subsequent investigations, let us examine it more closely, as in the preceding article, by means of first principles, and retain infinitesimals of a higher order; so that hereby we may clear up all the obscurity which surrounds it, depending as it does on the preceding partial differentials.

Let A, B, C, D be the four angular points of that face of the elemental parallelepipedon which is parallel to and nearest to the plane of (x, y); where A is (x, y, z); B is $(x+dx, y, z)$; C is $(x, y+dy, z)$; D is $(x+dx, y+dy, z)$. Let A′, B′, C′, D′ be the places of A, B, C, D when the variation has taken place; so that

A′ is $(x+\delta x, y+\delta y, z+\delta z)$,

B′ is $\left(x+dx+\delta x+\frac{d.\delta x}{dx}dx,\ \ y+\delta y+\frac{d.\delta y}{dx}dx,\ \ z+\delta z+\frac{d.\delta z}{dx}dx\right)$,

C′ is $\left(x+\delta x+\frac{d.\delta x}{dy}dy,\ y+dy+\delta y+\frac{d.\delta y}{dy}dy,\ \ z+\delta z+\frac{d.\delta z}{dy}dy\right)$,

D′ is $\left(x+dx+\delta x+\frac{d.\delta x}{dx}dx+\frac{d.\delta x}{dy}dy, y+dy+\delta y+\frac{d.\delta y}{dx}dx+\frac{d.\delta y}{dy}dy,\right.$

$$\left. z+\delta z+\frac{d.\delta z}{dx}dx+\frac{d.\delta z}{dy}dy\right).$$

$$\therefore \quad A'B' = dx\left\{\left(1+\frac{d.\delta x}{dx}\right)^2+\left(\frac{d.\delta y}{dx}\right)^2+\left(\frac{d.\delta z}{dx}\right)^2\right\}^{\frac{1}{2}} = C'D'; \qquad (33)$$

$$A'C' = dy\left\{\left(\frac{d.\delta x}{dy}\right)^2+\left(1+\frac{d.\delta y}{dy}\right)^2+\left(\frac{d.\delta z}{dy}\right)^2\right\}^{\frac{1}{2}} = B'D'; \qquad (34)$$

so that the figure A′B′C′D′ is also a parallelogram. If infinitesimals of a higher order are neglected, these results are the same as the first two of (18).

Similarly if the variations of the other faces of the new volume-element are calculated, it will be found that they are parallelograms; so that the volume-element is a parallelepipedon, its volume being given in (31), so far as infinitesimals of the first order are involved. Also let α, β, γ be the angles between the edges $d\eta$ and $d\zeta$, $d\zeta$ and $d\xi$, $d\xi$ and $d\eta$ respectively; then, as γ is the angle between the two sides A′B′ and A′C′, and as

$$B'C'^2 = A'B'^2 - 2A'B' \times A'C' \cos\gamma + A'C'^2; \quad \text{and}$$

$$\text{B}'\text{C}'^2 = \left\{\left(1+\frac{d.\delta x}{dx}\right)dx - \frac{d.\delta x}{dy}dy\right\}^2 + \left\{\left(1+\frac{d.\delta y}{dy}\right)dy - \frac{d.\delta y}{dx}dx\right\}^2 + \left\{\frac{d.\delta z}{dx}dx - \frac{d.\delta z}{dy}dy\right\}^2; \quad (85)$$

therefore neglecting infinitesimals of the higher orders,

$$\cos\gamma = \frac{d.\delta x}{dy} + \frac{d.\delta y}{dx}. \quad (86)$$

Similarly

$$\cos\alpha = \frac{d.\delta y}{dz} + \frac{d.\delta z}{dy}; \quad (87)$$

$$\cos\beta = \frac{d.\delta z}{dx} + \frac{d.\delta x}{dz}. \quad (88)$$

And as all the quantities in the right-hand members of these last three equations are infinitesimals of an order higher than that of those retained in (81), it is evident that α, β, and γ are approximately right angles; and that the volume-element in its displaced position is approximately a rectangular parallelepipedon.

Hence it appears that if $\delta x, \delta y, \delta z$ are functions of only x, y, z respectively, we shall have rigorously $\cos\alpha = \cos\beta = \cos\gamma = 0$; so that the volume-element is a rectangular parallelepipedon in both its original and displaced states. Thus, the variation of the volume-element is, to the first order of infinitesimals, as general in this restricted case of the variations of the coordinates, as it is in the general case where $\delta x, \delta y, \delta z$ are arbitrary functions of x, y, and z.

300.] Finally, let us take the general case, and investigate the variation of a product of differentials of the form $dx_1\, dx_2\, dx_3 \dots dx_n$, $x_1, x_2, x_3, \dots x_n$ being n variables independent of each other.

Let $\xi_1, \xi_2, \xi_3, \dots$ be the values of $x_1, x_2, x_3, \dots$ in their varied states; so that

$$\xi_1 = x_1 + \delta x_1,$$
$$\xi_2 = x_2 + \delta x_2,$$
$$\dots\dots\dots$$
$$\xi_n = x_n + \delta x_n;$$

where $\delta x_1, \delta x_2, \dots$ are functions of $x_1, x_2, x_3, \dots$: then we have

$$d\xi_1 = \left\{1+\frac{d.\delta x_1}{dx_1}\right\}dx_1 + \frac{d.\delta x_1}{dx_2}dx_2 + \dots$$

$$d\xi_2 = \frac{d.\delta x_2}{dx_1}dx_1 + \left\{1+\frac{d.\delta x_2}{dx_2}\right\}dx_2 + \dots$$

$$\dots\dots\dots\dots\dots\dots$$

$$d\xi_n = \frac{d.\delta x_n}{dx_1}dx_1 + \frac{d.\delta x_n}{dx_2}dx_2 + \dots\dots + \left\{1+\frac{d.\delta x_n}{dx_n}\right\}dx_n;$$

so that, by the method of Art. 211, and neglecting infinitesimals of the higher orders, we have

$$d\xi_1 d\xi_2 d\xi_3 \ldots = \left(1+\frac{d.\delta x_1}{dx_1}\right)\left(1+\frac{d.\delta x_2}{dx_2}\right)\left(1+\frac{d.\delta x_3}{dx_3}\right)\ldots dx_1 dx_2 dx_3 \ldots$$

$$= \left\{1+\frac{d.\delta x_1}{dx_1}+\frac{d.\delta x_2}{dx_2}+\ldots\right\} dx_1 dx_2 dx_3 \ldots; \quad (39)$$

$$\therefore \quad \delta.dx_1 dx_2 dx_3 \ldots = \left\{\frac{d.\delta x_1}{dx_1}+\frac{d.\delta x_2}{dx_2}+\ldots\right\} dx_1 dx_2 dx_3 \ldots \quad (40)$$

and this gives the variation of the multiple infinitesimal element.

301.] I come now to the investigation of the variation of definite integrals, in which the element-functions involve finite variables, their differentials, and differential coefficients. Here the processes will be longer; but by a judicious employment of integration, the final results assume a practicable form.

Now the value of a definite integral may be varied in many ways; (1) by a change of form in the element-function; this is a variation which falls entirely within the scope of this Calculus: (2) by a change in the limits; and this may occur in two ways; either the limits may be assigned by certain determinate functions; in which case a change of them involves a passage from one value to another in accordance with certain given conditions, as along a given curve or a given surface; and although the variations of the variables at these limits will have to be determined, yet at those points they will not be arbitrary, but of the nature of differentials and not of variations; or the functions which assign the limits may be undetermined, and thus subject to variation; and thus there will be variations of these limits due to a change of form in the functions which assign them. In all these several cases the total variation will be the sum of the several partial variations which are due to the changes as they occur in particular problems; and these partial variations will be estimated separately.

In these problems it appears to me most desirable to maintain as far as possible symmetry of notation and symmetry of expression; for a large amount of labour, both of brain and hand, is thereby saved. With this object in view I have at first considered all the variables to be equally dependent and equally subject to variation, and have made none equicrescent; the rationale of this process being that each variable is supposed to be a function of another variable which is not involved in the expressions either explicitly or implicitly. The process is applicable with especial advantage, as the sequel will shew, when the number of variables is small. In many cases however it leads to long expressions which are capable of abbreviation, if some hypothesis is made as to the character of one or more of the variables or of their variations, as to equicrescence and variation, and the generality of the problem is not thereby abridged. Thus, suppose the element-function of the definite integral to contain two variables x and y, the functional relation between which is not fixed; then, for the variation of the element-function apart from its values at the limits, so long as the variation of one of the variables, say y, is arbitrary and indeterminate, the generality of the whole variation is not abridged if x is not subject to variation, or if a particular value, or a series of particular values, is given to δx; apart from the limits, I say; because at the limits a relation may be given between x and y, and this relation taken in connection with the general functional relation may absolutely determine the values of δx and of δy at the limits. This is also evident geometrically. Each point in a plane curve may be displaced in a direction parallel to the axis of y, and thus $\delta x = 0$. Thus in fig. 47, P may be shifted to R; in which case if all the points do not move through equal spaces, but through spaces which are functions of the coordinates of the point in its original position, the form of the equation to the curve will change, although the point has the same abscissa in both its positions. If however the extreme points P_0 and P_1 are constrained to move on given curves, at the limits generally x and y must both vary, and consistently with the equations to the limiting curves. In many cases in the sequel I shall take one or more of the variables to be equicrescent, or assume that they have no variation, or take the variation of a variable to be a function of that variable only, or make some other hypothesis, as far as it is applicable, which will shorten the operations, and not abridge the generality.

302.] In the first place let us consider a function of two variables x and y, which are connected with each other by a functional symbol, which is not known; and suppose the element to be

$$F(x, dx, d^2x, \ldots d^n x, y, dy, d^2y, \ldots d^m y),$$

where F expresses a known function; let u represent the sum of these elements, between the limits x_1, y_1 and x_0, y_0; so that

$$u = \int_0^1 \text{F}(x, dx, d^2x, \ldots d^n x, y, dy, d^2y, \ldots d^m y); \tag{41}$$

it is required to calculate the variation of u, the relation between x and y being an unknown function.

Let the variation be of the most general kind that is possible; so that not only x, y, but also $dx, d^2x, \ldots d^n x, dy, d^2y, \ldots d^m y$ receive variations; and let

$$\Omega = \text{F}(x, dx, d^2x, \ldots d^n x, y, dy, d^2y, \ldots d^m y); \tag{42}$$

and thus
$$u = \int_0^1 \Omega; \tag{43}$$

$$\therefore \quad \delta u = \delta \int_0^1 \Omega = \int_0^1 \delta\Omega; \tag{44}$$

then since Ω is a function of $x, dx, d^2x, \ldots d^n x, y, dy, d^2y, \ldots d^m y$, by virtue of equation (5) we have,

$$\delta\Omega = \left(\frac{d\Omega}{dx}\right)\delta x + \left(\frac{d\Omega}{d.dx}\right)\delta.dx + \left(\frac{d\Omega}{d.d^2x}\right)\delta.d^2x + \ldots + \left(\frac{d\Omega}{d.d^n x}\right)\delta.d^n x$$
$$+ \left(\frac{d\Omega}{dy}\right)\delta y + \left(\frac{d\Omega}{d.dy}\right)\delta.dy + \left(\frac{d\Omega}{d.d^2y}\right)\delta.d^2y + \ldots + \left(\frac{d\Omega}{d.d^m y}\right)\delta.d^m y. \tag{45}$$

To acquire a more convenient notation, let

$$\left.\begin{array}{ll} \left(\dfrac{d\Omega}{dx}\right) = \text{X}, & \left(\dfrac{d\Omega}{dy}\right) = \text{Y}, \\ \left(\dfrac{d\Omega}{d.dx}\right) = \text{X}_1, & \left(\dfrac{d\Omega}{d.dy}\right) = \text{Y}_1, \\ \left(\dfrac{d\Omega}{d.d^2x}\right) = \text{X}_2, & \left(\dfrac{d\Omega}{d.d^2y}\right) = \text{Y}_2, \\ \cdot \quad \cdot \quad \cdot \quad \cdot \quad \cdot & \cdot \quad \cdot \quad \cdot \quad \cdot \quad \cdot \\ \left(\dfrac{d\Omega}{d.d^n x}\right) = \text{X}_n; & \left(\dfrac{d\Omega}{d.d^m y}\right) = \text{Y}_m; \end{array}\right\} \tag{46}$$

therefore

$$\begin{aligned} \delta\Omega = {} & \text{X}\,\delta x + \text{X}_1\,\delta.dx + \text{X}_2\,\delta.d^2x + \ldots + \text{X}_n\,\delta.d^n x \\ & + \text{Y}\,\delta y + \text{Y}_1\,\delta.dy + \text{Y}_2\,\delta.d^2y + \ldots + \text{Y}_m\,\delta.d^m y. \end{aligned} \tag{47}$$

And similarly

$$\begin{aligned} d\Omega = {} & \text{X}\,dx + \text{X}_1\,d.dx + \text{X}_2\,d.d^2x + \ldots + \text{X}_n\,d.d^n x \\ & + \text{Y}\,dy + \text{Y}_1\,d.dy + \text{Y}_2\,d.d^2y + \ldots + \text{Y}_m\,d.d^m y; \end{aligned} \tag{48}$$

$$\begin{aligned} \therefore \quad \delta\int_0^1 \Omega &= \int_0^1 \delta\Omega \\ &= \int_0^1 \{\text{X}\,\delta x + \text{X}_1\,d.\delta x + \text{X}_2\,d^2.\delta x + \ldots + \text{X}_n\,d^n.\delta x \\ &\qquad + \text{Y}\,\delta y + \text{Y}_1\,d.\delta y + \text{Y}_2\,d^2.\delta y + \ldots + \text{Y}_m\,d^m.\delta y\}; \end{aligned} \tag{49}$$

the order of the symbols of operation having been changed in accordance with the commutative law established in Art. 295. But

$$\left.\begin{aligned}
&\int_0^1 x_1\, d.\delta x = \Big[x_1 \delta x\Big]_0^1 - \int_0^1 dx_1\, \delta x\,;\\
&\int_0^1 x_2 d^2\delta x = \Big[x_2\, d\delta x - dx_2\, \delta x\Big]_0^1 + \int_0^1 d^2 x_2\, \delta x\,;\\
&\cdot\quad\cdot\quad\cdot\quad\cdot\quad\cdot\quad\cdot\quad\cdot\quad\cdot\quad\cdot\quad\cdot\quad\cdot\quad\cdot\quad\cdot\quad\cdot\\
&\int_0^1 x_n d^n\delta x = \Big[x_n d^{n-1}\delta x - dx_n d^{n-2}\delta x + \ldots (-)^{n-1} d^{n-1} x_n \delta x\Big]_0^1\\
&\qquad\qquad\qquad\qquad\qquad (-)^n \int_0^1 d^n x_n \delta x\,;
\end{aligned}\right\} \quad (50)$$

and similar results are of course true for the integration of the Y's; therefore, substituting in (49),

$$\begin{aligned}
\delta.\int_0^1 \Omega = \Big[&x_1 \delta x\\
&+ x_2\, d\delta x - dx_2 \delta x\\
&+ x_3 d^2\delta x - dx_3 d\delta x + d^2 x_3 \delta x\\
&+ \cdot\quad\cdot\quad\cdot\quad\cdot\quad\cdot\quad\cdot\quad\cdot\quad\cdot\quad\cdot\\
&+ x_n d^{n-1}\delta x - dx_n d^{n-2}\delta x + d^2 x_n d^{n-3}\delta x - \ldots (-)^{n-1} d^{n-1} x_n \delta x\\
&+ y_1 \delta y\\
&+ y_2 d\delta y - dy_2 \delta y\\
&+ y_3 d^2\delta y - dy_3 d\delta y + d^2 y_3 \delta y\\
&+ \cdot\quad\cdot\quad\cdot\quad\cdot\quad\cdot\quad\cdot\quad\cdot\quad\cdot\\
&+ y_m d^{m-1}\delta y - dy_m d^{m-2}\delta y + d^2 y_m d^{m-3}\delta y - \ldots (-)^{m-1} d^{m-1} y_m \delta y\Big]_0^1\\
&+ \int_0^1 \{x - dx_1 + d^2 x_2 - \ldots (-)^n d^n x_n\}\, \delta x\\
&+ \int_0^1 \{y - dy_1 + d^2 y_2 - \ldots (-)^m d^m y_m\}\, \delta y\,;
\end{aligned} \quad (51)$$

which expression it will be observed consists of two parts: one of which depends on the variables at the limits and their variations; and the other involves a sign of integration, and being therefore dependent on the form of the function connecting x and y cannot be determined unless that function is known: but by means of which in many cases, as we shall see hereafter, the unknown form may be found. It is also to be noticed that the former part vanishes if the limits are fixed; and if they are constrained to fulfil certain conditions, relations will exist between their variations with which the former part of (51) must consist.

If x is assumed to be equicrescent, so that

$$d^2x = d^3x = \ldots = d^n x = 0;$$

then $\Omega = F(x, dx, y, dy, d^2y, \ldots d^m y)$,

and $X_2 = X_3 = \ldots = X_n = 0$;

$$\begin{aligned}\therefore\ \delta.\int_0^1 \Omega = \Big[& X_1 \delta x \\ & + Y_1 \delta y \\ & + Y_2 d\delta y - dY_2 \delta y \\ & + \quad \cdot \quad \cdot \quad \cdot \quad \cdot \\ & + Y_m d^{m-1}\delta y - dY_m d^{m-2}\delta y + \ldots (-)^{m-1} d^{m-1} Y_m \delta y \Big]_0^1 \\ & + \int_0^1 \{X - dX_1\} \delta x \\ & + \int_0^1 \{Y - dY_1 + d^2 Y_2 - \ldots (-)^m d^m Y_m\} \delta y. \qquad (52)\end{aligned}$$

303.] If however the element of the definite integral is expressed in terms of differential coefficients, in which x is equicrescent, so that

$$u = \int_0^1 V dx,$$

where $V = F\left(x, y, \frac{dy}{dx}, \frac{d^2y}{dx^2}, \ldots \frac{d^n y}{dx^n}\right)$,

F representing a known function, and the relation between x and y being undetermined; then to give to V the most general variation that is possible, $x, y, \frac{dy}{dx}, \frac{d^2y}{dx^2}, \ldots$ will all vary.

For convenience of notation let us substitute as follows;

$$\frac{dy}{dx} = y', \ \frac{d^2y}{dx^2} = y'', \ldots \frac{d^n y}{dx^n} = y^{(n)};$$

so that $V = F(x, y, y', y'', \ldots y^{(n)})$; therefore

$$\left.\begin{aligned}\delta V &= \left(\frac{dV}{dx}\right)\delta x + \left(\frac{dV}{dy}\right)\delta y + \left(\frac{dV}{dy'}\right)\delta y' + \ldots + \left(\frac{dV}{dy^{(n)}}\right)\delta y^{(n)}; \\ dV &= \left(\frac{dV}{dx}\right)dx + \left(\frac{dV}{dy}\right)dy + \left(\frac{dV}{dy'}\right)dy' + \ldots + \left(\frac{dV}{dy^{(n)}}\right)dy^{(n)};\end{aligned}\right\} \quad (53)$$

$$\begin{aligned}\therefore\ \delta.\int_0^1 V dx &= \int_0^1 \delta.V dx \\ &= \int_0^1 \{V\delta.dx + \delta V dx\}\end{aligned}$$

$$\delta.\int_0^1 \text{V}\,dx = \int_0^1 \{\text{V}d.\delta x + \delta\text{V}\,dx\}$$
$$= \Big[\text{V}\delta x\Big]_0^1 + \int_0^1 \{\delta\text{V}\,dx - d\text{V}\,\delta x\}$$
$$= \Big[\text{V}\delta x\Big]_0^1 + \int_0^1 \Big\{\Big(\frac{d\text{V}}{dy}\Big)\{dx\,\delta y - \delta x\,dy\} + \Big(\frac{d\text{V}}{dy'}\Big)\{dx\,\delta y' - \delta x\,dy'\} + \ldots$$
$$\ldots + \Big(\frac{d\text{V}}{dy^{(n)}}\Big)\{dx\,\delta y^{(n)} - \delta x\,dy^{(n)}\}\Big\}. \quad (54)$$

Again, let

$\Big(\frac{d\text{V}}{dy}\Big) = \text{Y}$, $\Big(\frac{d\text{V}}{dy'}\Big) = \text{Y}'$, $\Big(\frac{d\text{V}}{dy''}\Big) = \text{Y}''$, ... $\Big(\frac{d\text{V}}{dy^{(n)}}\Big) = \text{Y}^{(n)}$; therefore

$$\delta.\int_0^1 \text{V}\,dx = \Big[\text{V}\delta x\Big]_0^1 + \int_0^1 \{\text{Y}(dx\,\delta y - \delta x\,dy) + \text{Y}'(dx\,\delta y' - \delta x\,dy') + \ldots$$
$$\ldots + \text{Y}^{(n)}(dx\,\delta y^{(n)} - \delta x\,dy^{(n)})\}$$
$$= \Big[\text{V}\delta x\Big]_0^1 + \int_0^1 \{\text{Y}(\delta y - y'\,\delta x) + \text{Y}'(\delta y' - y''\,\delta x) + \ldots$$
$$\ldots + \text{Y}^{(n)}(\delta y^{(n)} - y^{(n+1)}\delta x)\}\,dx.$$

Let $\delta y - y'\delta x = \omega$;

$$\therefore\ \delta y' - y''\delta x = \delta.\frac{dy}{dx} - y''\delta x$$
$$= \frac{\delta.dy}{dx} - \frac{dy}{dx^2}\delta.dx - y''\delta x$$
$$= \frac{d.\delta y}{dx} - \frac{dy}{dx}\frac{d.\delta x}{dx} - \frac{d}{dx}\Big(\frac{dy}{dx}\Big)\delta x$$
$$= \frac{d.\delta y}{dx} - \frac{d}{dx}\Big\{\frac{dy}{dx}\delta x\Big\}$$
$$= \frac{d}{dx}\{\delta y - y'\delta x\} = \frac{d\omega}{dx}$$
$$= \omega';$$

similarly $\delta y'' - y'''\delta x = \omega''$;

. . . .

$\delta y^{(n)} - y^{(n+1)}\delta x = \omega^{(n)}$; therefore

$$\delta.\int_0^1 \text{V}\,dx = \Big[\text{V}\delta x\Big]_0^1 + \int_0^1 \{\omega\text{Y} + \omega'\text{Y}' + \omega''\text{Y}'' + \ldots + \omega^{(n)}\text{Y}^{(n)}\}\,dx. \quad (55)$$

Now

$$\int_0^1 \omega'\,\text{Y}'\,dx = \Big[\omega\text{Y}'\Big]_0^1 - \int_0^1 \frac{d\text{Y}'}{dx}\,\omega\,dx;$$
$$\int_0^1 \omega''\text{Y}''dx = \Big[\omega'\text{Y}'' - \omega\frac{d\text{Y}''}{dx}\Big]_0^1 + \int_0^1 \frac{d^2\text{Y}''}{dx^2}\,\omega\,dx;$$

. . . .

$$\int_0^1 \omega^{(n)} \mathrm{Y}^{(n)} dx = \Big[\omega^{(n-1)} \mathrm{Y}^{(n)} - \omega^{(n-2)} \frac{d\mathrm{Y}^{(n)}}{dx} + \dots (-)^{n-1} \omega \frac{d^{n-1} \mathrm{Y}^{(n)}}{dx^{n-1}}\Big]_0^1$$

$$(-)^n \int_0^1 \frac{d^n \mathrm{Y}^{(n)}}{dx^n} \omega\, dx;$$

substituting which values in (55) we have

$$\delta.\int_0^1 \mathrm{V}\, dx = \Big[\mathrm{V}\,\delta x + \Big\{\mathrm{Y}' - \frac{d\mathrm{Y}''}{dx} + \frac{d^2 \mathrm{Y}'''}{dx^2} - \dots (-)^{n-1} \frac{d^{n-1} \mathrm{Y}^{(n)}}{dx^{n-1}}\Big\}\omega$$

$$+ \Big\{\mathrm{Y}'' - \frac{d\mathrm{Y}'''}{dx} + \dots (-)^{n-2} \frac{d^{n-2} \mathrm{Y}^{(n)}}{dx^{n-2}}\Big\}\omega'$$

$$+ \quad \cdot \quad \cdot \quad \cdot \quad \cdot \quad \cdot \quad \cdot \quad \cdot \quad \cdot \quad \cdot \quad \cdot \quad \cdot$$

$$+ \Big\{\mathrm{Y}^{(n-1)} - \frac{d\mathrm{Y}^{(n)}}{dx}\Big\}\omega^{(n-2)}$$

$$+ \mathrm{Y}^{(n)} \omega^{(n-1)}\Big]_0^1$$

$$+ \int_0^1 \Big\{\mathrm{Y} - \frac{d\mathrm{Y}'}{dx} + \frac{d^2 \mathrm{Y}''}{dx^2} - \frac{d^3 \mathrm{Y}'''}{dx^3} + \dots (-)^n \frac{d^n \mathrm{Y}^{(n)}}{dx^n}\Big\}\omega\, dx; \quad (56)$$

which expression*, it will be observed, consists of two parts; the former depends on the values of the variables at the limits of integration; the latter involves an integration, which cannot generally be effected unless the function connecting x and y is known.

As to the former part, the first term is $\Big[\mathrm{V}\,\delta x\Big]_0^1 = \mathrm{V}_1 \delta x_1 - \mathrm{V}_0 \delta x_0$, where V_1 and V_0 are the values taken by V at the limits; these do not involve δy, but only involve δx at the limits; and are the variations of the definite integral due to the change of the limits, being the quantities which have been determined in (86) and (87), Art. 96. The other terms contain ω and its derived functions; and if $\delta x = 0$, $\omega = \delta y$, all these terms involve the values which δy and its derived functions take at the limits, and are independent of the general functional connection between x and y.

304.] Certain other properties of (56) require notice. If $\omega = 0$, that is if $\delta y - y' \delta x = 0$,

$$\frac{\delta y}{\delta x} = \frac{dy}{dx};$$

* In the Memoir on this Calculus by M. Poisson, which was read to the French Academy in 1831, and is printed in Vol. XII of the Memoirs of the Institute, equation (56) is deduced from first principles.

and thus the ratio of the variations and of the differentials of y and x is the same: and in this case,

$$\delta.\int_0^1 \mathrm{v}\, dx = \Big[\mathrm{v}\, dx\Big]_0^1 = \mathrm{v}_1\, dx_1 - \mathrm{v}_0\, \delta x_0; \qquad (57)$$

that is, if we make the coordinates of a point on a curve to vary, so that the ratio of the variations and of the differentials of the coordinates is the same, we do not leave the curve, but pass to a consecutive point of it, and the definite integral is increased by the value of its element-function corresponding to the superior limit, and diminished by that corresponding to the inferior limit.

Also the geometrical meaning of $\omega\, dx$ deserves notice. Let the variations of the coordinates of any point on the plane curve under consideration be δx and δy, and let the projections of the space through which the point (x, y) has moved be estimated along the tangent and normal of the original curve at the given point, and let these projections be τ and ν; then

$$\delta x = \tau\frac{dx}{ds} - \nu\frac{dy}{ds}, \qquad \delta y = \tau\frac{dy}{ds} + \nu\frac{dx}{ds}; \qquad (58)$$

$$\tau = \frac{dy}{ds}\delta y + \frac{dx}{ds}\delta x, \qquad \nu = \frac{dx}{ds}\delta y - \frac{dy}{ds}\delta x; \qquad (59)$$

$$\therefore\ \omega\, dx = (\delta y - y'\delta x)\, dx = \delta y\, dx - dy\, \delta x$$
$$= \nu\, ds; \qquad (60)$$

$$\therefore\ \omega = \nu\frac{ds}{dx}; \qquad \omega' = \frac{d}{dx}\left(\nu\frac{ds}{dx}\right); \qquad \omega^{(n)} = \frac{d^n}{dx^n}\left(\nu\frac{ds}{dx}\right);$$

substituting which values in (56), it will be seen that every term in the part at the limits except the first involves only ν, the normal displacement, and the part of that involving τ is $\Big[\mathrm{v}\,\tau\frac{dx}{ds}\Big]_0^1$, which is equivalent to $\Big[\mathrm{v}\,\delta x\Big]_0^1$, if the variation is made on the supposition that $\nu = 0$. Also the part under the sign of integration involves ν only; the reason of which is, the variation in the form of a curve due to the shifting of its several points and elements arises from the infinitesimal normal displacement only; the effect of the tangential displacement being to shift a point to another consecutive point on the curve.

305.] In reference also to the general expression (51) it is worth remarking, that if δx and δy are replaced by dx and dy, that is, if the shifting of the point takes place along the curve only, and if there is no normal displacement, then the total variation of $\int_0^1 \Omega$ is that which takes place at the limits; thus in this case

$$\delta.\int_0^1 \Omega = \Big[\mathrm{X}_1\, dx + \mathrm{Y}_1\, dy\Big]_0^1$$
$$+ \Big[\mathrm{X}_2\, d^2x - d\mathrm{X}_2\, dx + \mathrm{Y}_2\, d^2y - d\mathrm{Y}_2\, dy\Big]_0^1$$
$$+ \quad . \quad . \quad . \quad . \quad . \quad . \quad . \quad . \quad . \quad . \quad . \quad .$$
$$+ \Big[\mathrm{X}_n\, d^nx - d\mathrm{X}_n\, d^{n-1}x + \dots (-)^{n-1}\, d^{n-1}\,\mathrm{X}_n\, dx\Big]_0^1$$
$$+ \Big[\mathrm{Y}_m\, d^my - d\mathrm{Y}_m\, d^{m-1}\,y + \dots (-)^{m-1} d^{m-1}\,\mathrm{Y}_m dy\Big]_0^1$$
$$+ \int_0^1 \{\mathrm{X} - d\mathrm{X}_1 + d^2\,\mathrm{X}_2 - \dots (-)^n\, d^n\,\mathrm{X}_n\}\, dx$$
$$+ \int_0^1 \{\mathrm{Y} - d\mathrm{Y}_1 + d^2\,\mathrm{Y}_2 - \dots (-)^m\, d^m\,\mathrm{Y}_m\}\, dy\,;$$

and the last two terms, after integration by parts, become

$$\int_0^1 \{\mathrm{X}\, dx + \mathrm{Y}\, dy\}$$
$$- \Big[\mathrm{X}_1\, dx + \mathrm{Y}_1\, dy\Big]_0^1 + \int_0^1 \{\mathrm{X}_1\, d^2x + \mathrm{Y}_1\, d^2y\}$$
$$+ \Big[d\mathrm{X}_2\, dx + d\mathrm{Y}_2\, dy\Big]_0^1 - \Big[\mathrm{X}_2 d^2x + \mathrm{Y}_2 d^2y\Big]_0^1 + \int_0^1 \{\mathrm{X}_2\, d^3x + \mathrm{Y}_2\, d^3y\}$$
$$- \quad . \quad . \quad . \quad . \quad . \quad . \quad . \quad . \quad . \quad . \quad . \quad . \quad . \quad . \quad . \quad . \quad . \quad . \;;$$

and so on. Hence

$$\delta.\int_0^1 \Omega = \int_0^1 \{\mathrm{X}\, dx + \mathrm{X}_1\, d.dx + \mathrm{X}_2\, d.d^2x + \dots + \mathrm{X}_n\, d.d^nx$$
$$+ \mathrm{Y}\, dy + \mathrm{Y}_1\, d.dy + \mathrm{Y}_2\, d.d^2y + \dots + \mathrm{Y}_m\, d.d^my\}$$
$$= \int_0^1 d\Omega = \Big[\Omega\Big]_0^1;$$

by reason of equation (48); so that the total variation is reduced to the difference between the values of the infinitesimal element at the first and the last limits.

306.] We proceed now to investigate the variation of a definite integral whose element involves three independent variables and their successive differentials; and to consider the variations in their greatest generality let us suppose all the variables and differentials to receive variations. Let

$$u = \int_0^1 \Omega, \qquad \text{where}$$

$$\Omega = F(x, dx, d^2x, \dots d^n x, y, dy, d^2y, \dots d^m y, z, dz, d^2z, \dots d^k z).$$

Let us first substitute as follows:

$$\left(\frac{d\Omega}{dx}\right) = X, \qquad \left(\frac{d\Omega}{dy}\right) = Y, \qquad \left(\frac{d\Omega}{dz}\right) = Z,$$

$$\left(\frac{d\Omega}{d.dx}\right) = X_1, \qquad \left(\frac{d\Omega}{d.dy}\right) = Y_1, \qquad \left(\frac{d\Omega}{d.dz}\right) = Z_1,$$

$$\cdot \quad \cdot \quad \cdot \quad \cdot \qquad \cdot \quad \cdot \quad \cdot \quad \cdot \quad \cdot \quad \cdot \qquad \cdot \quad \cdot \quad \cdot \quad \cdot \quad \cdot$$

$$\left(\frac{d\Omega}{d.d^n x}\right) = X_n; \qquad \left(\frac{d\Omega}{d.d^m y}\right) = Y_m; \qquad \left(\frac{d\Omega}{d.d^k z}\right) = Z_k;$$

$$\therefore \ \delta u = \int_0^1 \delta\Omega$$

$$\begin{aligned} = \int_0^1 \{ & X\,\delta x + X_1\, d.\delta x + X_2\, d^2.\delta x + \dots + X_n\, d^n.\delta x \\ & + Y\,\delta y + Y_1\, d.\delta y + Y_2\, d^2.\delta y + \dots + Y_m\, d^m.\delta y \\ & + Z\,\delta z + Z_1\, d.\delta z + Z_2\, d^2.\delta z + \dots + Z_k\, d^k.\delta z \}; \end{aligned} \qquad (61)$$

and reducing these terms by partial integration, we have finally,

$$\begin{aligned} \delta.\int_0^1 \Omega = \Big[& X_1\,\delta x \\ & + X_2\, d\,\delta x - dX_2\,\delta x \\ & + \quad \cdot \quad \cdot \quad \cdot \quad \cdot \quad \cdot \quad \cdot \quad \cdot \quad \cdot \quad \cdot \\ & + X_n d^{n-1}\delta x - dX_n d^{n-2}\delta x + d^2X_n d^{n-3}\delta x - \dots (-)^{n-1} d^{n-1} X_n\,\delta x \\ & + Y_1\,\delta y \\ & + Y_2\, d\,\delta y - dY_2\delta y \\ & + \quad \cdot \quad \cdot \quad \cdot \quad \cdot \quad \cdot \quad \cdot \quad \cdot \quad \cdot \quad \cdot \\ & + Y_m d^{m-1}\delta y - dY_m d^{m-2}\delta y + d^2Y_m d^{m-3}\delta y - \dots (-)^{m-1} d^{m-1} Y_m\delta y \\ & + Z_1\,\delta z \\ & + Z_2\, d\,\delta z - dZ_2\,\delta z \\ & + \quad \cdot \quad \cdot \quad \cdot \quad \cdot \quad \cdot \\ & + Z_k d^{k-1}\delta z - dZ_k d^{k-2}\delta z + d^2Z_k d^{k-3}\delta z - \dots (-)^{k-1} d^{k-1} Z_k\delta z \Big]_0^1 \\ & + \int_0^1 \{X - dX_1 + d^2X_2 - \dots (-)^n d^n X_n\}\,\delta x \\ & + \int_0^1 \{Y - dY_1 + d^2\,Y_2 - \dots (-)^m\, d^m\, Y_m\}\,\delta y \\ & + \int_0^1 \{Z - dZ_1 + d^2Z_2 - \dots (-)^k d^k Z_k\}\delta z. \end{aligned} \qquad (62)$$

When Ω involves more than three independent variables, the expression for the variation of $\int_0^1 \Omega$ is of course similar.

307.] Suppose however that an equation of relation is given between the variables and their differentials which are involved in Ω; and, to fix our thoughts, let us take the case of three variables x, y, z; and suppose the equation to be

$$\text{L} = f\{x, dx, d^2x, \dots y, dy, d^2y, \dots z, dz, d^2z, \dots\} = 0. \quad (63)$$

If L involves only x, y, z, z may be expressed in terms of x and y, and thence $dz, d^2z, \dots$ may be found, and substituted in Ω, so that Ω will become a function of only two variables, x and y: but as L involves the differentials of the variables, such an elimination is generally impossible, and we are obliged to have recourse to the following process. Take the variation of L, viz.

$$\begin{aligned}&\left(\frac{d\text{L}}{dx}\right)\delta x + \left(\frac{d\text{L}}{d.dx}\right) d.\delta x + \left(\frac{d\text{L}}{d.d^2x}\right) d^2\delta x + \dots \\ &+ \left(\frac{d\text{L}}{dy}\right)\delta y + \left(\frac{d\text{L}}{d.dy}\right) d.\delta y + \left(\frac{d\text{L}}{d.d^2y}\right) d^2\delta y + \dots \\ &+ \left(\frac{d\text{L}}{dz}\right)\delta z + \left(\frac{d\text{L}}{d.dz}\right) d.\delta z + \left(\frac{d\text{L}}{d.d^2z}\right) d^2\delta z + \dots = \delta\text{L} = 0; \quad (64)\end{aligned}$$

and employing a convenient and abbreviating notation,

$$\begin{aligned}\delta\text{L} = &\,\xi\delta x + \xi_1 d.\delta x + \xi_2 d^2\delta x + \dots \\ &+ \eta\delta y + \eta_1 d.\delta y + \eta_2 d^2\delta y + \dots \\ &+ \zeta\delta z + \zeta_1 d.\delta z + \zeta_2 d^2\delta z + \dots = 0. \quad (65)\end{aligned}$$

Now since the equation $\text{L} = 0$ must be satisfied for all values of x, y, z which are admissible into the problem, therefore the variation of x, y, z must be subject to the condition $\delta\text{L} = 0$, that is, to equation (65); but since

$$\begin{aligned}\delta\text{V} = &\,\text{X}\delta x + \text{X}_1 d.\delta x + \text{X}_2 d^2\delta x + \dots \\ &+ \text{Y}\delta y + \text{Y}_1 d.\delta y + \text{Y}_2 d^2\delta y + \dots \\ &+ \text{Z}\delta z + \text{Z}_1 d.\delta z + \text{Z}_2 d^2\delta z + \dots, \quad (66)\end{aligned}$$

it is plain that we may add to it the right-hand member of (65) multiplied by an undetermined constant λ, without destroying the truth of the expressions; so that

$$\begin{aligned}\delta\text{V} = &\,(\text{X}+\lambda\xi)\delta x + (\text{X}_1+\lambda\xi_1) d.\delta x + (\text{X}_2+\lambda\xi_2) d^2\delta x + \dots \\ &+ (\text{Y}+\lambda\eta)\delta y + (\text{Y}_1+\lambda\eta_1) d.\delta y + (\text{Y}_2+\lambda\eta_2) d^2\delta y + \dots \\ &+ (\text{Z}+\lambda\zeta)\delta z + (\text{Z}_1+\lambda\zeta_1) d.\delta z + (\text{Z}_2+\lambda\zeta_2) d^2\delta z + \dots. \quad (67)\end{aligned}$$

Observing now the process by which (62) was deduced from

(61), a result similar to (62) will be deduced from (67), wherein instead of X will be $X+\lambda\xi$, instead of X_1, $X_1+\lambda\xi_1, \ldots$ instead of Y, $Y+\lambda\eta, \ldots$, and so on for the others; and as λ is undetermined, we may consider the variation to involve three independent quantities.

The variation will also be found in a similar manner if the original element-function involves more variables, and if these are related to each other by many equations of condition.

308.] Suppose however that the element involves three variables x, y, z; that x is equicrescent, and that y and z are two unknown functions of x, and independent of each other, and that the quantity whose variation is to be calculated is

$$u = \int_0^1 V\, dx,$$

where $V = F\left(x, y, \frac{dy}{dx}, \frac{d^2y}{dx^2}, \ldots \frac{d^my}{dx^m}, z, \frac{dz}{dx}, \frac{d^2z}{dx^2}, \ldots \frac{d^nz}{dx^n}\right)$,

F being a known function.

To give V the most general variation, let us suppose that not only x, y, z, but that also the derived functions of y and z vary: then, adopting the following substitutions,

$$\frac{dy}{dx} = y', \frac{d^2y}{dx^2} = y'', \ldots \frac{d^my}{dx^m} = y^{(m)};$$

$$\frac{dz}{dx} = z', \frac{d^2z}{dx^2} = z'', \ldots \frac{d^nz}{dx^n} = z^{(n)};$$

$$V = F(x, y, y', y'', \ldots y^{(m)}, z, z', z'', \ldots z^{(n)});$$

also let

$$\left(\frac{dV}{dy}\right) = Y, \left(\frac{dV}{dy'}\right) = Y', \left(\frac{dV}{dy''}\right) = Y'', \ldots \left(\frac{dV}{dy^{(m)}}\right) = Y^{(m)};$$

$$\left(\frac{dV}{dz}\right) = Z, \left(\frac{dV}{dz'}\right) = Z', \left(\frac{dV}{dz''}\right) = Z'', \ldots \left(\frac{dV}{dz^{(n)}}\right) = Z^{(n)};$$

$$\therefore \quad \delta V = \left(\frac{dV}{dx}\right)\delta x + Y\delta y + Y'\delta y' + Y''\delta y'' + \ldots + Y^{(m)}\delta y^{(m)}$$
$$+ Z\delta z + Z'\delta z' + Z''\delta z'' + \ldots + Z^{(n)}\delta z^{(n)}; \quad (68)$$

and following a process precisely similar to that of Art. 303, and extending it to z, and putting

$$\delta y - y'\delta x = \omega, \qquad \delta z - z'\delta x = \omega_1,$$
$$\delta y' - y''\delta x = \omega', \qquad \delta z' - z''\delta x = \omega'_1,$$
$$\cdots\cdots \qquad \cdots\cdots$$
$$\delta y^{(m)} - y^{(m+1)}\delta x = \omega^{(m)}; \qquad \delta z^{(n)} - z^{(n+1)}\delta x = \omega_1{}^{(n)};$$

we have the following result:

$$\delta.\int_0^1 \mathrm{v}\,dx = \Big[\mathrm{v}\delta x + \Big\{\mathrm{Y}' - \frac{d\mathrm{Y}''}{dx} + \frac{d^2\mathrm{Y}'''}{dx^2} - \ldots (-)^{m-1}\frac{d^{m-1}\mathrm{Y}^{(m)}}{dx^{m-1}}\Big\}\omega$$

$$+ \Big\{\mathrm{Y}'' - \frac{d\mathrm{Y}'''}{dx} + \ldots (-)^{m-2}\frac{d^{m-2}\mathrm{Y}^{(m)}}{dx^{m-2}}\Big\}\omega'$$

$$+ \Big\{\mathrm{Y}''' - \ldots (-)^{m-3}\frac{d^{m-3}\mathrm{Y}^{(m)}}{dx^{m-3}}\Big\}\omega''$$

$$\cdots\cdots\cdots\cdots$$

$$+ \Big\{\mathrm{Y}^{(m-1)} - \frac{d\mathrm{Y}^{(m)}}{dx}\Big\}\omega^{(m-2)}$$

$$+ \mathrm{Y}^{(m)}\omega^{(m-1)}$$

$$+ \Big\{\mathrm{Z}' - \frac{d\mathrm{Z}''}{dx} + \frac{d^2\mathrm{Z}'''}{dx^2} - \ldots (-)^{n-1}\frac{d^{n-1}\mathrm{Z}^{(n)}}{dx^{n-1}}\Big\}\omega_1$$

$$+ \Big\{\mathrm{Z}'' - \frac{d\mathrm{Z}'''}{dx} + \ldots (-)^{n-2}\frac{d^{n-2}\mathrm{Z}^{(n)}}{dx^{n-2}}\Big\}\omega'_1$$

$$+ \Big\{\mathrm{Z}''' - \ldots (-)^{n-3}\frac{d^{n-3}\mathrm{Z}^{(n)}}{dx^{n-3}}\Big\}\omega''_1$$

$$\cdots\cdots\cdots\cdots$$

$$+ \Big\{\mathrm{Z}^{(n-1)} - \frac{d\mathrm{Z}^{(n)}}{dx}\Big\}\omega_1^{(n-2)}$$

$$+ \mathrm{Z}^{(n)}\omega_1^{(n-1)}\Big]_0^1$$

$$+ \int_0^1 \Big\{\mathrm{Y} - \frac{d\mathrm{Y}'}{dx} + \frac{d^2\mathrm{Y}''}{dx^2} - \ldots (-)^m \frac{d^m\mathrm{Y}^{(m)}}{dx^m}\Big\}\omega\,dx$$

$$+ \int_0^1 \Big\{\mathrm{Z} - \frac{d\mathrm{Z}'}{dx} + \frac{d^2\mathrm{Z}''}{dx^2} - \ldots (-)^n \frac{d^n\mathrm{Z}^{(n)}}{dx^n}\Big\}\omega_1\,dx; \quad (69)$$

an expression consisting of two parts; of which one involves the values of the variables and their variations at the limits only; and the other involves a process of integration, and which cannot be performed unless the relations between x and y and z are given. These several parts admit of explanation similar to that which has been given of (56) in Art. 304.

Also it admits of a geometrical explanation similar to that of Art. 304; the relations between x and y and between x and z represent two cylinders which are perpendicular to the planes of (x, y), and of (x, z) respectively; and these by their intersection define a curve in space.

Let us consider the general displacement of a point on this curve to be due, (1) to two displacements perpendicular to each other in the normal plane, and (2) to one along the tangent line; now by a process exactly analogous to that of Art. 304 it may be shewn that the quantities under the signs of integration involve the normal displacements only; and that $\Big[\mathrm{V}\,\delta x\Big]_0^1$ is the only term wherein the tangential displacement appears.

If V contains any number of undetermined functions, the variation of $\int_0^1 \mathrm{V}\,dx$ will be calculated in a similar manner, and will consist of a series of terms and quantities similar to those of equation (69).

309.] In the last article y and z are considered to be independent of each other; if a relation is given connecting them and their derived-functions and x, and of the form

$$\mathrm{L} = f\left(x, y, \frac{dy}{dx}, \frac{d^2y}{dx^2}, \dots z, \frac{dz}{dx}, \frac{d^2z}{dx^2}, \dots\right) = 0\,; \qquad (70)$$

then $$\delta\mathrm{L}=0=\left(\frac{d\mathrm{L}}{dx}\right)\delta x+\left(\frac{d\mathrm{L}}{dy}\right)\delta y+\left(\frac{d\mathrm{L}}{dy'}\right)\delta y'+\left(\frac{d\mathrm{L}}{dy''}\right)\delta y''+\dots$$
$$+\left(\frac{d\mathrm{L}}{dz}\right)\delta z+\left(\frac{d\mathrm{L}}{dz'}\right)\delta z'+\left(\frac{d\mathrm{L}}{dz''}\right)\delta z''+\dots; \quad (71)$$

multiplying which by an undetermined constant λ, and adding it to $\delta\mathrm{V}$, we have

$$\delta\mathrm{V}=\left\{\left(\frac{d\mathrm{V}}{dx}\right)+\lambda\left(\frac{d\mathrm{L}}{dx}\right)\right\}\delta x+\left\{\mathrm{Y}+\lambda\left(\frac{d\mathrm{L}}{dy}\right)\right\}\delta y+\left\{\mathrm{Y}'+\lambda\left(\frac{d\mathrm{L}}{dy'}\right)\right\}\delta y'+\dots$$
$$+\left\{\mathrm{Z}+\lambda\left(\frac{d\mathrm{L}}{dz}\right)\right\}\delta z+\left\{\mathrm{Z}'+\lambda\left(\frac{d\mathrm{L}}{dz'}\right)\right\}\delta z'+\dots \quad (72)$$

comparing which expression with that of (68), and noticing the process by which (69) is deduced from (68), it is palpable that (72) will lead to a result of the form given in (69), and with quantities such that in the place of Y will be $\mathrm{Y}+\lambda\left(\frac{d\mathrm{L}}{dy}\right)$; in the place of Y', $\mathrm{Y}'+\lambda\left(\frac{d\mathrm{L}}{dy'}\right), \dots$; in the place of Z, $\mathrm{Z}+\lambda\left(\frac{d\mathrm{L}}{dz}\right)$; in the place of Z', $\mathrm{Z}'+\lambda\left(\frac{d\mathrm{V}}{dz'}\right), \dots$; and so on: and thus the variation will be reduced to the form of a definite integral, whose element-function involves x and two unknown and independent functions of x.

310.] Certain processes in the sequel will require the calculation of the variation of a variation, that is, of the second variation of a definite integral. As the principles and the method are the same as those explained and applied in the preceding articles, I will consider only one simple instance: viz. that in which it is required to find $\delta^2 u$, when $u = \int_0^1 \mathrm{v}\, dx$, and

$$\mathrm{v} = \mathrm{F}(x, y, y', y'', y''', \dots y^{(n)}).$$

$$\begin{aligned} \delta u &= \int_0^1 \delta.\mathrm{v}\, dx \\ &= \int_0^1 \{dx\, \delta\mathrm{v} + \mathrm{v}\, \delta.dx\}; \end{aligned}$$

$$\therefore \quad \delta^2 u = \int_0^1 \{dx\, \delta^2\mathrm{v} + 2\, \delta\mathrm{v}\, \delta.dx + \mathrm{v}\, \delta^2.dx\}. \qquad (73)$$

$$\text{but } \delta\mathrm{v} = \left(\frac{d\mathrm{v}}{dx}\right)\delta x + \left(\frac{d\mathrm{v}}{dy}\right)\delta y + \left(\frac{d\mathrm{v}}{dy'}\right)\delta y' + \left(\frac{d\mathrm{v}}{dy''}\right)\delta y'' + \dots$$

$$\begin{aligned} \therefore \quad \delta^2\mathrm{v} = {} & \delta\left(\frac{d\mathrm{v}}{dx}\right)\delta x + \delta\left(\frac{d\mathrm{v}}{dy}\right)\delta y + \delta\left(\frac{d\mathrm{v}}{dy'}\right)\delta y' + \delta\left(\frac{d\mathrm{v}}{dy''}\right)\delta y'' + \dots \\ & + \left(\frac{d\mathrm{v}}{dx}\right)\delta^2 x + \left(\frac{d\mathrm{v}}{dy}\right)\delta^2 y + \left(\frac{d\mathrm{v}}{dy'}\right)\delta^2 y' + \dots; \end{aligned} \qquad (74)$$

but $\left(\frac{d\mathrm{v}}{dx}\right)$, $\left(\frac{d\mathrm{v}}{dy}\right)$, $\left(\frac{d\mathrm{v}}{dy'}\right)$... are functions of $x, y, y', y'' \dots$;

$$\begin{aligned} \therefore \quad \delta.\left(\frac{d\mathrm{v}}{dx}\right) &= \left(\frac{d^2\mathrm{v}}{dx^2}\right)\delta x + \left(\frac{d^2\mathrm{v}}{dy\, dx}\right)\delta y + \left(\frac{d^2\mathrm{v}}{dy'\, dx}\right)\delta y' + \dots \\ \delta.\left(\frac{d\mathrm{v}}{dy}\right) &= \left(\frac{d^2\mathrm{v}}{dx\, dy}\right)\delta x + \left(\frac{d^2\mathrm{v}}{dy^2}\right)\delta y + \left(\frac{d^2\mathrm{v}}{dy'\, dy}\right)\delta y' + \dots \\ & \dots\dots\dots\dots\dots\dots\dots\dots\dots ; \end{aligned}$$

substituting which values in (74) we have

$$\begin{aligned} \delta^2\mathrm{v} = {} & \left(\frac{d^2\mathrm{v}}{dx^2}\right)\delta x^2 + \left(\frac{d^2\mathrm{v}}{dy^2}\right)\delta y^2 + \left(\frac{d^2\mathrm{v}}{dy'^2}\right)\delta y'^2 + \dots \\ & + 2\left\{\left(\frac{d^2\mathrm{v}}{dx\, dy}\right)\delta x\, \delta y + \left(\frac{d^2\mathrm{v}}{dx\, dy'}\right)\delta x\, \delta y' + \dots + \left(\frac{d^2\mathrm{v}}{dy\, dy'}\right)\delta y\, \delta y' + \dots\right\}; \end{aligned} \qquad (75)$$

and therefore

$$\begin{aligned} \delta^2.\int_0^1 \mathrm{v}\, dx = {} & \int_0^1 \left\{\left(\frac{d^2\mathrm{v}}{dx^2}\right)\delta x^2 + \left(\frac{d^2\mathrm{v}}{dy^2}\right)\delta y^2 + \left(\frac{d^2\mathrm{v}}{dy'^2}\right)\delta y'^2 + \dots\right\} dx \\ & + 2\int_0^1 \left\{\left(\frac{d^2\mathrm{v}}{dx\, dy}\right)\delta x\, \delta y + \left(\frac{d^2\mathrm{v}}{dx\, dy'}\right)\delta x\, \delta y' + \dots + \left(\frac{d^2\mathrm{v}}{dy\, dy'}\right)\delta y\, \delta y' + \dots\right\} dx \\ & + 2\int_0^1 \left\{\left(\frac{d\mathrm{v}}{dx}\right)\delta x + \left(\frac{d\mathrm{v}}{dy}\right)\delta y + \left(\frac{d\mathrm{v}}{dy'}\right)\delta y' + \dots\right\} \delta.dx \\ & + \int_0^1 \mathrm{v}\, \delta^2.dx. \end{aligned} \qquad (76)$$

By similar processes may $\delta^3 u, \delta^4 u, \dots$ be calculated.

311.] We come now to the investigation of the variation of a multiple integral; but instead of taking the problem in its most general form, I will consider the case of a double integral only; for the principle on which the inquiry is founded is the same in all cases; and the number of terms in the result increases so rapidly with each new integral sign, that by taking any higher order the formulæ are so complicated as to require new symbols and new modes of abbreviation, and no result useful for our present purpose is arrived at.

And I shall consider only a simple case of a double integral: that, viz. in which the element-function involves x, y, z, (z being an undetermined function of x and y), and the partial derived functions of z of the first and second orders; and in which also the limits of integration are given by an inequality in accordance with the principles explained in Art. 214: so that the range of integration includes the values of the variables corresponding to all points within a given closed surface. This case will suffice for all the examples to which I shall at present have occasion to apply the calculus; and the student who desires further information will find an investigation of the general case in those memoirs of Ostrogradsky, Sarrus, Delaunay, and Lindelöf, which are mentioned in the foot-note of page 411. In those by Sarrus and Lindelöf especially the difficulties of the variations of double definite integrals are elucidated, and to them I am under great obligation for the following investigation. A peculiar symbol, which they call the symbol of substitution, has been largely employed by them; I however have found the symbols already employed in this work sufficient for the purpose.

Let the double definite integral which is the subject of variation be

$$u = \int_{\mathrm{x}_0}^{\mathrm{x}_1} \int_{\mathrm{y}_0}^{\mathrm{y}_1} \mathrm{v}\, dy\, dx; \tag{77}$$

and let us suppose v to be a function of x, y, z, and of $\left(\frac{dz}{dx}\right)$, $\left(\frac{dz}{dy}\right)$, $\left(\frac{d^2 z}{dx^2}\right)$, $\left(\frac{d^2 z}{dx\, dy}\right)$, $\left(\frac{d^2 z}{dy^2}\right)$, where z is an undetermined function of x and y. For the sake of abbreviation let the following symbols be adopted;

$$\left(\frac{dz}{dx}\right) = z', \qquad \left(\frac{dz}{dy}\right) = z_{,}; \tag{78}$$

$$\left(\frac{d^2 z}{dx^2}\right) = z'', \qquad \left(\frac{d^2 z}{dx\, dy}\right) = z_{,}', \qquad \left(\frac{d^2 z}{dy^2}\right) = z_{,,}; \tag{79}$$

the upper and lower accents referring to the x— and y— partial derivations respectively. Also let us take the limits in their most general form; that is, as the integral stands in (77), let Y_1 and Y_0 be functions of x, and let X_1, X_0 be independent of x and y; although, if the order of integration is changed, X_1 and X_0 are functions of y, and Y_1 and Y_0 are independent of x and y. Now, taking the total variation of u as given in (77),

$$\begin{aligned}\delta u &= \int_{\text{X}_0}^{\text{X}_1}\int_{\text{Y}_0}^{\text{Y}_1} \delta.\text{V}\,dy\,dx \\ &= \int_{\text{X}_0}^{\text{X}_1}\int_{\text{Y}_0}^{\text{Y}_1} \{dy\,dx\,\delta\text{V} + \text{V}\,\delta.dy\,dx\} \qquad (80)\\ &= \int_{\text{X}_0}^{\text{X}_1}\int_{\text{Y}_0}^{\text{Y}_1} \left\{dy\,dx\,\delta\text{V} + \text{V}\left(\frac{d.\delta x}{dx} + \frac{d.\delta y}{dy}\right)dy\,dx\right\}, \qquad (81)\end{aligned}$$

substituting for $\delta.dy\,dx$ from equation (23). And integrating by parts the last terms, and remembering that the order of integration is indifferent, provided that attention is paid to the value of the limits,

$$\delta u = \int_{\text{X}_0}^{\text{X}_1}\Big[\text{V}\delta y\Big]_{\text{Y}_0}^{\text{Y}_1} dx + \int_{\text{Y}_0}^{\text{Y}_1}\Big[\text{V}\delta x\Big]_{\text{X}_0}^{\text{X}_1} dy + \int_{\text{X}_0}^{\text{X}_1}\int_{\text{Y}_0}^{\text{Y}_1}\left\{\delta\text{V} - \left(\frac{d\text{V}}{dx}\right)\delta x - \left(\frac{d\text{V}}{dy}\right)\delta y\right\}dy\,dx.$$

Now let us assume that in the general variation of the element, $\delta x = \delta y = 0$, because we shall thereby shorten an expression which is under any circumstances long, and (see Art. 301) shall not abridge the generality; and let us also replace δz by ω, so that ω is an arbitrary function of x and y; then

$$\left.\begin{aligned}\delta z' &= \delta.\frac{dz}{dx} = \frac{d.\delta z}{dx} = \frac{d\omega}{dx} = \omega'; \\ \delta z_{,} &= \delta.\frac{dz}{dy} = \frac{d.\delta z}{dy} = \frac{d\omega}{dy} = \omega_{,}; \\ \delta z'' &= \omega''; \quad \delta z_{,}' = \omega_{,}'; \quad \delta z_{,,} = \omega_{,,};\end{aligned}\right\} \qquad (83)$$

where ω', $\omega_{,}$, ω'', … are partial derived functions of ω.

Let Z, Z', $\text{Z}_{,}$, Z'', … denote the partial differential coefficients of V with respect to z, z', $z_{,}$, z'', … respectively; then

$$\begin{aligned}\delta\text{V} &= \text{Z}\delta z + \text{Z}'\delta z' + \text{Z}_{,}\delta z_{,} + \text{Z}''\delta z'' + \text{Z}_{,}'\delta z_{,}' + \text{Z}_{,,}\delta z_{,,} \\ &= \text{Z}\omega + \text{Z}'\omega' + \text{Z}_{,}\omega_{,} + \text{Z}''\omega'' + \text{Z}_{,}'\omega_{,}' + \text{Z}_{,,}\omega_{,,}; \qquad (84)\end{aligned}$$

so that

$$\begin{aligned}\delta u = &\int_{\text{X}_0}^{\text{X}_1}\Big[\text{V}\,\delta y\Big]_{\text{Y}_0}^{\text{Y}_1} dx + \int_{\text{Y}_0}^{\text{Y}_1}\Big[\text{V}\,\delta x\Big]_{\text{X}_0}^{\text{X}_1} dy \\ &+ \int_{\text{X}_0}^{\text{X}_1}\int_{\text{Y}_0}^{\text{Y}_1}(\text{Z}\omega + \text{Z}'\omega' + \text{Z}_{,}\omega_{,} + \text{Z}''\omega'' + \text{Z}_{,}'\omega_{,}' + \text{Z}_{,,}\omega_{,,})\,dy\,dx. \qquad (85)\end{aligned}$$

Now the last part of this equivalent of δu involves not only ω, which is an arbitrary function of x and y, but also its x— and y— partial derived functions of the first and second orders; and as it is to be integrated with respect to the same variable for which it is differentiated, several of these terms are capable of considerable simplification by means of integration by parts, and of the theorems for the variation of definite double integrals which are given in Art. 217.

Since by (64), Art. 217,

$$\frac{d}{dx}\int_{Y_0}^{Y_1} Z'\omega\, dy = \left[Z'\omega\frac{dy}{dx}\right]_{Y_0}^{Y_1} + \int_{Y_0}^{Y_1}\frac{d.Z'\omega}{dx}dy$$

$$= \quad - \ - \ - \ - \quad + \int_{Y_0}^{Y_1}\left(Z'\frac{d\omega}{dx} + \frac{dZ'}{dx}\omega\right)dy;$$

$$\therefore \int_{Y_0}^{Y_1} Z'\omega' dy = \frac{d}{dx}\int_{Y_0}^{Y} Z'\omega\, dy - \left[Z'\frac{dy}{dx}\omega\right]_{Y_0}^{Y_1} - \int_{Y_0}^{Y_1}\frac{dZ'}{dx}\omega\, dy. \quad (86)$$

Thus taking the x-integral of this equivalence with the limits X_1 and X_0, we have

$$\int_{X_0}^{X_1}\int_{Y_0}^{Y_1} Z'\omega' dy\, dx$$

$$= \left[\int_{Y_0}^{Y_1} Z'\omega\, dy\right]_{X_0}^{X_1} - \int_{X_0}^{X_1}\left[Z'\frac{dy}{dx}\omega\right]_{Y_0}^{Y_1} dx - \int_{X_0}^{X_1}\int_{Y_0}^{Y_1}\frac{dZ'}{dx}\omega\, dy\, dx. \quad (87)$$

Also since

$$\int_{Y_0}^{Y_1} Z_{,}\omega_{,} dy = \left[Z_{,}\omega\right]_{Y_0}^{Y_1} - \int_{Y_0}^{Y_1}\frac{dZ_{,}}{dy}\omega\, dy;$$

$$\therefore \int_{X_0}^{X_1}\int_{Y_0}^{Y_1} Z_{,}\omega_{,} dy\, dx = \int_{X_0}^{X_1}\left[Z_{,}\omega\right]_{Y_0}^{Y_1} dx - \int_{X_0}^{X_1}\int_{Y_0}^{Y_1}\frac{dZ_{,}}{dy}\omega\, dy\, dx. \quad (88)$$

Again, taking the last three terms in the latter part of the second member of (85), as they involve second derived functions of ω, two integrations will be required for their simplification. Let us first take the term whose element-function is $Z''\omega''$; then, by the theorem expressed by (87),

$$\int_{X_0}^{X_1}\int_{Y_0}^{Y_1} Z''\omega'' dy\, dx$$

$$= \left[\int_{Y_0}^{Y_1} Z''\omega' dy\right]_{X_0}^{X_1} - \int_{X_0}^{X_1}\left[Z''\frac{dy}{dx}\omega'\right]_{Y_0}^{Y_1} dx - \int_{X_0}^{X_1}\int_{Y_0}^{Y_1}\frac{dZ''}{dx}\omega' dy\, dx. \quad (89)$$

The first term of this second member does not admit of further reduction, because the element-function of a y-integral involves two x-derived functions. The second and third terms admit of the following simplification.

Since by (70), Art. 218,

$$\frac{d}{dx}\left[z''\frac{dy}{dx}\omega\right]_{Y_0}^{Y_1}=\left[z''\frac{dy}{dx}\left(\omega'+\omega_{,}\frac{dy}{dx}\right)\right]_{Y_0}^{Y_1}$$
$$+\left[\left(z''\frac{d^2y}{dx^2}+\frac{dz''}{dy}\frac{dy^2}{dx^2}+\frac{dz''}{dx}\frac{dy}{dx}\right)\omega\right]_{Y_0}^{Y_1};$$

$$\therefore\quad \left[z''\frac{dy}{dx}\omega'\right]_{Y_0}^{Y_1}=\frac{d}{dx}\left[z''\frac{dy}{dx}\omega\right]_{Y_0}^{Y_1}$$
$$-\left[\left(z''\frac{d^2y}{dx^2}+\frac{dz''}{dx}\frac{dy}{dx}+\frac{dz''}{dy}\frac{dy^2}{dx^2}\right)\omega\right]_{Y_0}^{Y_1}-\left[z''\frac{dy^2}{dx^2}\omega_{,}\right]_{Y_0}^{Y_1};$$

$$\int_{x_0}^{x_1}\left[z''\frac{dy}{dx}\omega'\right]_{Y_0}^{Y_1}dx=\left[\left[z''\frac{dy}{dx}\omega\right]_{Y_0}^{Y_1}\right]_{x_0}^{x_1}$$
$$-\int_{x_0}^{x_1}\left[\left(z''\frac{d^2y}{dx^2}+\frac{dz''}{dx}\frac{dy}{dx}+\frac{dz''}{dy}\frac{dy^2}{dx^2}\right)\omega\right]_{Y_0}^{Y_1}dx-\int_{x_0}^{x_1}\left[z''\frac{dy^2}{dx^2}\omega_{,}\right]_{Y_0}^{Y_1}dx. \quad (90)$$

Also by reason of (87),

$$\int_{x_0}^{x_1}\int_{Y_0}^{Y_1}\frac{dz''}{dx}\omega'\,dy\,dx$$
$$=\left[\int_{Y_0}^{Y_1}\frac{dz''}{dx}\omega\,dy\right]_{x_0}^{x_1}-\int_{x_0}^{x_1}\left[\frac{dz''}{dx}\frac{dy}{dx}\omega\right]_{Y_0}^{Y_1}dx-\int_{x_0}^{x_1}\int_{Y_0}^{Y_1}\frac{d^2z''}{dx^2}\omega\,dy\,dx; \quad (91)$$

so that substituting these equivalents in (89), we have

$$\int_{x_0}^{x_1}\int_{Y_0}^{Y_1}z''\omega''\,dy\,dx=-\left[\left[z''\frac{dy}{dx}\omega\right]_{Y_0}^{Y_1}\right]_{x_0}^{x_1}+\left[\int_{Y_0}^{Y_1}\left(z''\omega'-\frac{dz''}{dx}\omega\right)dy\right]_{x_0}^{x_1}$$
$$+\int_{x_0}^{x_1}\left[\left(z''\frac{d^2y}{dx^2}+2\frac{dz''}{dx}\frac{dy}{dx}+\frac{dz''}{dy}\frac{dy^2}{dx^2}\right)\omega+z''\frac{dy^2}{dx^2}\omega_{,}\right]_{Y_0}^{Y_1}dx$$
$$+\int_{x_0}^{x_1}\int_{Y_0}^{Y_1}\frac{d^2z''}{dx^2}\omega\,dy\,dx. \quad (92)$$

By a similar process we may shew that

$$\int_{x_0}^{x_1}\int_{Y_0}^{Y_1}z_{,}'\omega_{,}'\,dy\,dx=\left[\left[z_{,}'\omega\right]_{Y_0}^{Y_1}\right]_{x_0}^{x_1}-\left[\int_{Y_0}^{Y_1}\frac{dz_{,}'}{dy}\omega\,dy\right]_{x_0}^{x_1}$$
$$-\int_{x_0}^{x_1}\left[\frac{dz_{,}'}{dx}\omega+z_{,}'\frac{dy}{dx}\omega_{,}\right]_{Y_0}^{Y_1}dx+\int_{x_0}^{x_1}\int_{Y_0}^{Y_1}\frac{dz_{,}'}{dx\,dy}\omega\,dy\,dx. \quad (93)$$

Also by integration by parts

$$\int_{x_0}^{x_1}\int_{Y_0}^{Y_1}z_{,,}\omega_{,,}\,dy\,dx=-\int_{x_0}^{x_1}\left[\frac{dz_{,,}}{dy}\omega-z_{,,}\omega_{,}\right]_{Y_0}^{Y_1}dx+\int_{x_0}^{x_1}\int_{Y_0}^{Y_1}\frac{d^2z_{,,}}{dy^2}\omega\,dy\,dx. \quad (94)$$

Hence, substituting all these equivalents in (85), we have

$$\delta u = \int_{x_0}^{x_1}\Big[\text{v}\,\delta y\Big]_{y_0}^{y_1} dx + \int_{y_0}^{y_1}\Big[\text{v}\,\delta x\Big]_{x_0}^{x_1} dy + \Big[\Big[\Big(z_{,}' - z''\frac{dy}{dx}\Big)\omega\Big]_{y_0}^{y_1}\Big]_{x_0}^{x_1}$$

$$+ \Big[\int_{y_0}^{y_1}\Big\{\Big(z' - \frac{dz''}{dx} - \frac{dz_{,}'}{dy}\Big)\omega + z''\omega'\Big\}\, dy\Big]_{x_0}^{x_1}$$

$$+ \int_{x_0}^{x_1}\Big[\Big(z_{,} - z'\frac{dy}{dx} + z''\frac{d^2y}{dx^2} + 2\frac{dz''}{dx}\frac{dy}{dx} + \frac{dz''}{dy}\frac{dy^2}{dx} - \frac{dz_{,}'}{dx} - \frac{dz_{,,}}{dy}\Big)\omega$$

$$+ \Big(z''\frac{dy^2}{dx^2} - z_{,}'\frac{dy}{dx} + z_{,,}\Big)\omega_{,}\Big]_{y_0}^{y_1} dx$$

$$+ \int_{x_0}^{x_1}\int_{y_0}^{y_1}\Big(z - \frac{dz'}{dx} - \frac{dz_{,}}{dy} + \frac{d^2z''}{dx^2} + \frac{d^2z_{,}'}{dx\,dy} + \frac{d^2z_{,,}}{dy^2}\Big)\omega\, dy\, dx; \qquad (95)$$

which gives the complete value of δu.

Now this expression consists of three distinct groups and classes of terms:

(1) The first two terms in the first row of (95); these are independent of ω, and only involve the value of δx and δy at the limits; these consequently express the variation of u which is due to the deformation of the functions which assign the limits. This fact is also otherwise evident inasmuch as these terms have arisen from the variation of $dy\,dx$ and not from that of v. Those which arise from the variation of v are contained in the following groups.

(2) The last term in the first row, and the terms contained in the next three rows of (95): these do not depend on the general value of ω; but only on the values which it and its partial derived-functions ω' and $\omega_{,}$ have at the limits; and consequently on the functions which assign the limits of integration.

(3) The definite double integral which is in the last row of (95). This depends on the arbitrary function ω, and cannot be determined unless that function is given.

312.] In the calculation of the value of δu which is given in (95), the variations at the limits have been taken in the most general form, and have not been subjected to any conditions. In the application however of this expression, the limits are frequently determined by certain equations or inequalities, by reason of which the preceding value is much reduced.

Now one of the most common cases is that in which the range of integration is determined by a certain closed function or surface, of which the equation is given, so that values of the variables corresponding to all points within this surface are included within the integral, and all values outside are excluded; and thus if $\text{F}(x, y, z) = 0 = \text{L}$ is the equation to the limiting surface, the integral includes those values of the variables for which L is negative. This, it will be observed, is the case which I have considered in Art. 214, and have applied in Ex. 7, Art. 258. In a more general case the form of this surface may vary, but I shall assume it now not to be subject to any deformation. Moreover, I shall also assume that $\text{L} = 0$ gives only two real values of z; that if z is eliminated by means of $\text{L}=0$, and $\left(\frac{d\text{L}}{dz}\right) = 0$, we have only two real values of y, which are Y_1 and Y_0; and that if z and y are eliminated by means of $\text{L} = 0$, $\left(\frac{d\text{L}}{dz}\right) = 0$, $\left(\frac{d\text{L}}{dy}\right) = 0$, the result gives only two real values of x, which are X_1 and X_0. Under these circumstances of limits $\text{Y}_1 = \text{Y}_0$ when $x = \text{X}_1$, and also when $x = \text{X}_0$; so that all terms of the form $\left[\int_{\text{Y}_0}^{\text{Y}_1} \Omega\, dy\right]_{\text{X}_0}^{\text{X}_1}$ vanish, because the y-limits are equal; and consequently the second term of the second group of (95) vanishes.

If the limits of x and y are constants, their variations vanish; and consequently the first group of terms in (95) disappears.

313.] If the element-function of the double definite integral which is the subject of variation, is a function of $x, y, z, \left(\frac{dz}{dx}\right), \left(\frac{dz}{dy}\right)$, where as heretofore z is an undetermined function of x and y, so that

$$u = \int_{\text{X}_0}^{\text{X}_1}\int_{\text{Y}_0}^{\text{Y}_1} \text{V}\, dy\, dx\,; \tag{96}$$

where V is a function of x, y, z, z', $z_{,}$, according to the notation of Art. 311; and where the limits of integration are taken in the most general form; then, taking the results of that article, and putting $z'' = z'_{,} = z_{,,} = 0$; so that also $\text{Z}'' = \text{Z}'_{,} = \text{Z}_{,,} = 0$; from (95) we have

$$\delta u = \int_{\text{X}_0}^{\text{X}_1}\Big[\text{V}\,\delta y\Big]_{\text{Y}_0}^{\text{Y}_1} dx + \int_{\text{Y}_0}^{\text{Y}_1}\Big[\text{V}\,\delta x\Big]_{\text{X}_0}^{\text{X}_1} dy + \left[\int_{\text{Y}_0}^{\text{Y}_1} \text{Z}'\omega\, dy\right]_{\text{X}_0}^{\text{X}_1}$$
$$+ \int_{\text{X}_0}^{\text{X}_1}\left[\left(\text{Z}_{,} - \text{Z}'\frac{dy}{dx}\right)\omega\right]_{\text{Y}_0}^{\text{Y}_1} dx + \int_{\text{X}_0}^{\text{X}_1}\int_{\text{Y}_0}^{\text{Y}_1}\left(\text{Z} - \frac{d\text{Z}'}{dx} - \frac{d\text{Z}_{,}}{dy}\right) dy\, dx. \tag{97}$$

which is the general value of δu. The several groups of terms admit of explanation similar to that given in Art. 304.

If the district of integration is bounded by a closed surface, then, as in the preceding article, the first term of the second group in (97) vanishes.

314.] The preceding examples of variation are sufficient both to illustrate the theory and for the solution of special problems to which we shall apply the calculus. It is good however to consider a difficult subject, such as that under discussion, from another point of view. We have conceived the quantity involving the unknown function to be resolved into its elements, and the definite integral of these elements to be the finite quantity which is the subject of inquiry: and the limits have been taken to be values whose symbols have subscripts 1 and 0. Now imagine the definite integral to represent some property of a plane curve, and between the values x_1 and x_0; this restriction is convenient to fix our thoughts; and let the quantity x_1-x_0 be resolved into n elements, and $\xi_1, \xi_2, \xi_3, \ldots \xi_{n-1}$ be the values of x corresponding to the points of division, and the corresponding values of y be $y_1, y_2, \ldots y_{n-1}$: then, as the definite integral is the sum of a series of quantities, of each of which the element is a type; so if we replace the definite integral by its equivalent series, it will be a function of $x_0, \xi_1, \xi_2, \ldots \xi_{n-1}, x_n$, that is, of $n+1$ variables; and when the elements are infinitesimal, of an infinite number of variables. This then is a distinguishing mark of the calculus of variations; its immediate subjects of inquiry are functions of an *infinite* number of variables generally independent of each other; but as these functions consist of a series of terms, all of which are of the same form, the differential, or variation of the sum of them, is equal to the sum of the differentials or variations of the separate terms: hence the cause of δ and $\int$ being subject to the commutative law. The principles of the calculus of variations therefore are only different from those of the differential calculus, because its subject is a function of an infinite number instead of a finite number of variables.

It will also be observed, that if for the definite integral the equivalent series of terms involving intermediate variables is substituted, the number of variables that enter into each term will depend on the order of the highest differential which enters into the element-function; thus if the element-function involves d^2y, three consecutive values of y will enter into each term; and so for other forms of the element-function.

I will not however enter on further inquiry into this method of the calculus of variations, because the process is much longer than, and ultimately leads to the same results as, the preceding. But because the principles of the calculus become hereby resolved into their most simple elements; nay rather, because the processes of perhaps the most transcendental analysis hereby become capable of geometrical interpretation and construction, I shall take an opportunity, in the next Chapter, of solving a simple problem by this method; and the mode of application will thereby be evident.

CHAPTER XIV.

APPLICATION OF THE CALCULUS OF VARIATIONS TO PROBLEMS OF MAXIMA AND MINIMA.

SECTION 1.—*Determination of the critical values of a definite integral whose element-function involves variables and their differentials.*

315.] WE proceed to apply the principles of the preceding Chapter to a large class of problems of maxima and minima involving unknown functions.

At this part of our treatise it is superfluous to repeat the conditions and the criteria for determining maxima and minima values of known functions, which depend on particular values of the subject-variables of these functions; for the whole question has been fully discussed in Chapter VII of Vol. I, and the reader is supposed to be familiar with it. Suppose however that the problem is to determine the form of a curve or curved surface between certain limits, so that a property of it, such as its length or the area inclosed by it, may have a maximum or minimum value; the principles of Vol. I are plainly insufficient, because the form of the function is unknown; and we have recourse to the following mode of solution: let the property, whose value is critical, be resolved into its elements, the element being a known function of the variables and their differentials, and this being independent of the relation between the variables; then the sum of all these, or, in other words, their definite integral, is the quantity whose critical value is to be found, and by which means the form of the function is to be determined. The definite integral therefore is the subject of inquiry, and is such as those whose variations have been calculated in the preceding Chapter.

316.] Let u represent the definite integral, of which the critical value is to be determined; and first suppose that the variables and their differentials of which it is a function are independent of each other; that is, that there is no equation of relation amongst them: a maximum or minimum of such a kind is termed *absolute*: then, by the theory of maxima and minima, it is plain, if u has a critical value, that $\delta u = 0$ and changes its sign; and that the change of sign may be determined by the sign of $\delta^2 u$; so that if $\delta u = 0$, u has a maximum or minimum value, according as $\delta^2 u$ is negative or positive; the solution of the problem therefore requires the calculation of δu and of $\delta^2 u$; and by the condition $\delta u = 0$, the form of the functional symbol connecting the variables is to be found.

In the first place, let

$$u = \int_0^1 \Omega, \tag{1}$$

where $$\Omega = F(x, dx, d^2x, \ldots d^n x, y, dy, d^2y, \ldots d^m y), \tag{2}$$

and F is the symbol of a known function. On referring to the value of δu given in equation (51) of the preceding Chapter, it will be observed that it consists of two parts; one of which is integrated, and depends on the values of the variables, of their differentials, and of their variations at the limits; the other is under signs of integration, and cannot be further reduced, because δx and δy are unknown functions of x and y, and because the other factors in the element-functions involve the undetermined function and its differentials. What conditions therefore are requisite so that $\delta u = 0$? For convenience of reference let

$$\begin{aligned} &\{X_1 - dX_2 + d^2X_3 - \ldots (-)^{n-1} d^{n-1} X_n\}\,\delta x \\ &\quad + \{X_2 - dX_3 + \ldots (-)^{n-2} d^{n-2} X_n\}\, d\delta x \\ &\quad + \quad \cdot \quad \cdot \quad \cdot \quad \cdot \quad \cdot \quad \cdot \quad \cdot \\ &\qquad\qquad + \{X_{n-1} - dX_n\}\, d^{n-2}\delta x \\ &\qquad\qquad\qquad + X_n d^{n-1}\delta x = \alpha; \end{aligned} \tag{3}$$

and let the analogous quantity involving δy and its differentials $= \beta$: also let

$$X - dX_1 + d^2X_2 - \ldots (-)^n d^n X_n = \Xi, \tag{4}$$

$$Y - dY_1 + d^2Y_2 - \ldots (-)^m d^m Y_m = H; \tag{5}$$

so that we have

$$\delta u = \Big[\alpha + \beta\Big]_0^1 + \int_0^1 (\Xi\,\delta x + H\,\delta y). \tag{6}$$

Now, as δx and δy are arbitrary functions of x and y, δu cannot vanish unless $\Big[\alpha + \beta\Big]_0^1 = 0$; whence we have

$$\Big[\alpha + \beta\Big]_1 = 0, \tag{7}$$

$$\left[\alpha+\beta\right]^{0} = 0; \tag{8}$$

and also,

$$\Xi = 0, \tag{9}$$

$$H = 0; \tag{10}$$

and these are the conditions which are primarily necessary to the definite integral u having a maximum or a minimum value.

317.] Although it is desirable, both for symmetry and for the discussion of an expression in its most general form, thus far to retain all the terms in δu, and although in many of our subsequent examples we shall retain them throughout, yet it is necessary somewhat to abridge them, that we may point out some general properties of the above equations.

First, let the difference between (7), (8), and (9), (10) be observed: (7) and (8) involve limiting values of δx, δy, and of their differentials; whereas $\Xi = 0$, and $H = 0$, being differential expressions, will after integration give general relations between x and y, and therein the required functional connection; and the *same* function will be deduced both from $\Xi = 0$ and from $H = 0$, provided that (and this is a necessary condition) the same limiting values are taken in the integrals of both equations: for the *form* of the function involved in them will depend on the form of function of Ω, and from Ω they are derived by a similar process; and therefore the same functional form will appear in the final result of each.

Again, let us suppose that there is no variation of x, save at the limits; and that consequently the shifting of any point from a curve to the next consecutive curve is due to a variation of y only; then $\delta x = 0$ (except at the limits), $d\delta x = d^2\delta x = \ldots = 0$: so that (6) becomes

$$\begin{aligned}\delta u = &\left[\{X_1 - dX_2 + \ldots (-)^{n-1} d^{n-1} X_n\} \delta x\right]_0^1 \\ &+ \left[\beta\right]_0^1 \\ &+ \int_0^1 H\, \delta y;\end{aligned} \tag{11}$$

each of the three lines of the second member of which must separately $= 0$, if $\delta u = 0$.

Now, as we have shewn in the preceding Chapter, Art. 301, that the generality of the formulæ is not abridged by the assumption that one of the variables undergoes no variation, so the result hereby demonstrated is as general as that given in (7), ... (10). And as the solution of the problem is necessarily the same in both cases, we can infer that the same functional relation between x and y arises from $\text{Ξ} = 0$, and from $\text{H} = 0$.

318.] Let us then take this last case to be the solution of the general problem of maxima and minima, and thereby indicate some general properties of it. And let us consider the case in which Ω is not linear with respect to $d^m y$, so that Y_m is a function of $d^m y$, and consequently $d^m\text{Y}_m$ involves $d^{2m}y$; thus $\text{H} = 0$ is a differential equation containing $d^{2m}y$; and as in process of integration a constant is manifestly introduced at each successive integration, so the complete integral involves $2m$ arbitrary constants; thus, if Υ is the complete integral, it involves $c_1, c_2, c_3, \ldots c_{2m}$, that is, $2m$ unknown constants: and these must be determined by means of the former parts of equation (11), which are functions of the limits.

Now if the limits are not restricted by any given conditions, the former parts of (11) will contain $2(m+1)$ arbitrary quantities, viz.

$$\delta x_0, \delta y_0, d\delta y_0, d^2\delta y_0, \ldots d^{m-1}\delta y_0, \delta x_1, \delta y_1, d\delta y_1, d^2\delta y_1, \ldots d^{m-1}\delta y_1, \quad (12)$$

of which the coefficients must be separately equated to zero: hereby we shall have $2(m+1)$ different and independent equations to determine $2m$ arbitrary constants, and which are manifestly more than sufficient, and thus the problem is indeterminate; this is as it should be; for if there is no restriction on the limits or their variations, the definite integral may be of any magnitude, and cannot have either a maximum or a minimum value.

If however equations are given connecting the variables at the limits; that is, if equations are given in terms of x_0 and y_0, and in terms of x_1 and y_1: then, if $\Upsilon = 0$ is the integral of $\text{H} = 0$, there will be given

$$\Upsilon_0, \left(\frac{d\Upsilon}{dy}\right)_0, \left(\frac{d^2\Upsilon}{dy^2}\right)_0, \ldots \left(\frac{d^{m-1}\Upsilon}{dy^{m-1}}\right)_0, \Upsilon_1, \left(\frac{d\Upsilon}{dy}\right)_1, \ldots \left(\frac{d^{m-1}\Upsilon}{dy^{m-1}}\right)_1, \quad (13)$$

which with the $2m+2$ quantities of (12) give us $4m+2$ different quantities whereby to determine $2m$ constants $c_1, c_2, \ldots c_{2m}$, and the $2m+2$ quantities

$$x_0, y_0, dy_0, d^2y_0, \ldots d^m y_0, x_1, y_1, dy_1, d^2y_1, \ldots d^m y_1; \quad (14)$$

and the problem is thus determinate.

When Ω is linear with respect to $d^m y$, $\text{H} = 0$ will be a differential equation of an order not higher than $2m-1$, and therefore its complete integral cannot contain more than $2m-1$ arbitrary constants; and the number of equations relative to the limits of the general integral being the same as before, the problem is impossible, because the required conditions cannot be satisfied.

319.] The following are cases wherein the differential equation $\mathrm{H} = 0$ takes particular forms, which admit of integration.

(1) If Ω does not contain y, then H becomes

$$-d\mathrm{Y}_1 + d^2\mathrm{Y}_2 - \dots (-)^{m-1} d^m \mathrm{Y}_m = 0, \tag{15}$$

which admits of one integration without any determination of relation between y and x.

(2) If Ω does not contain the first k terms of $y, dy, d^2y, \dots$, then $\mathrm{H} = 0$ becomes

$$(-)^{k-1} d^k \mathrm{Y}_k (-)^k d^{k+1} \mathrm{Y}_{k+1} - \dots (-)^{m-1} d^m \mathrm{Y}_m = 0, \tag{16}$$

which admits of being integrated k times in succession.

(3) If Ω does not contain x, or any differentials of x, then from (48), Art. 302, and from $\mathrm{H} = 0$, we have

$$d\Omega = \mathrm{Y}\,dy + \mathrm{Y}_1 d.dy + \mathrm{Y}_2 d.d^2y + \mathrm{Y}_3 d.d^3y + \dots,$$
$$0 = \mathrm{Y} - d\mathrm{Y}_1 + d^2\mathrm{Y}_2 - d^3\mathrm{Y}_3 + \dots;$$

so that eliminating Y,

$$\begin{aligned} d\Omega &= \mathrm{Y}_1 d.dy + dy\, d\mathrm{Y}_1 + \mathrm{Y}_2 d.d^2y - dy\, d^2\mathrm{Y}_2 + \mathrm{Y}_3 d.d^3y + dy\, d^3\mathrm{Y}_3 + \dots \\ &= d(\mathrm{Y}_1 dy) + d(\mathrm{Y}_2 d^2y - dy\, d\mathrm{Y}_2) + d(\mathrm{Y}_3 d^3y - d\mathrm{Y}_3 d^2y + dy\, d^2\mathrm{Y}_3) + \dots \end{aligned}$$

whence by integration,

$$\Omega = c_1 + \mathrm{Y}_1 dy + \mathrm{Y}_2 d^2y - dy\, d\mathrm{Y}_2 + \mathrm{Y}_3 d^3y - d^2y\, d\mathrm{Y}_3 + dy\, d^2\mathrm{Y}_3 + \dots \tag{17}$$

320.] Thus far I have supposed the variables x and y, which are involved in the element Ω, to be independent of each other, and the maxima and minima of such definite integrals are called absolute. If however x and y are not independent, but are subject to a certain condition given by the equation, integral or differential as the case may be,

$$\mathrm{L} = 0, \tag{18}$$

then, as explained in Art. 307, we have

$$\delta \mathrm{L} = 0; \tag{19}$$

and a relation is given which the variations of the variables and their differentials must satisfy; multiplying therefore $\delta\mathrm{L}$ by an indeterminate constant multiplier λ, and adding to δu, we have

$$\delta\{u + \lambda \mathrm{L}\} = 0; \tag{20}$$

and we may operate on $u+\lambda L$ in a manner precisely the same as that by which we have determined the necessary conditions for the critical values of $u = \int_0^1 \Omega$. These are called *relative* maxima and minima, and the method of determining them is hereby reduced to that of finding absolute maxima and minima.

It is also similarly manifest that if the problem is the determination of the maximum or minimum value of u, when the variables and their differentials are subject to conditions expressed by a series of equations, which may be in the form of definite integrals or otherwise, viz.,

$$L_1 = l_1, \qquad L_2 = l_2, \ldots \qquad L_k = l_k; \tag{21}$$

then it is sufficient to determine the absolute critical value of

$$u + \lambda_1 L_1 + \lambda_2 L_2 + \ldots + \lambda_k L_k, \tag{22}$$

where $\lambda_1, \lambda_2, \ldots \lambda_k$ are undetermined constants; and these will be determined by means of the necessary equations arising from equating to zero the variation of (22), and from the equations (21).

I may also observe that the indeterminate multipliers $\lambda_1, \lambda_2, \ldots \lambda_k$ may be introduced in such a form that they may be supposed to be subject to variation, and thus to be functions of x and y. For suppose the function whose critical value is required to be

$$u + \lambda_1 (L_1 - l_1) + \lambda_2 (L_2 - l_2) + \ldots + \lambda_k (L_k - l_k);$$

then the total variation of this quantity is

$$\delta u + \lambda_1 \delta L_1 + \lambda_2 \delta L_2 + \ldots + \lambda_k \delta L_k$$
$$+ (L_1 - l_1) \delta\lambda_1 + (L_2 - l_2) d\lambda_2 + \ldots + (L_k - l_k) d\lambda_k,$$

and this, by reason of (21), $=0$, under the same conditions as those for which the variation of (22) $= 0$.

321.] The preceding principles are also applicable to the determination of the critical values of u where

$$u = \int_0^1 \Omega, \qquad \text{and}$$

$$\Omega = F(x, dx, d^2x, \ldots d^n x, y, dy, \ldots d^m y, z, dz, \ldots d^k z); \tag{23}$$

and employing substitutions similar to those of Art. 316,

$$\delta u = \Big[\alpha + \beta + \gamma\Big]_0^1$$
$$+ \int_0^1 (\Xi \delta x + H \delta y + \Psi \delta z); \tag{24}$$

and as $\delta u = 0$, and $\delta x, \delta y, \delta z$ are arbitrary functions of x, y, z, we must have

$$\Big[\alpha+\beta+\gamma\Big]_0^1 = 0, \qquad (25)$$

$$\Xi = 0, \quad \mathrm{H} = 0, \quad \Psi = 0: \qquad (26)$$

of which (25) is a series of equations in terms of the variables and their differentials at the limits, and the integrals of (26) will give the general functional relations between the variables: it is also to be observed that the same function will be given by any two of the three equations (26), for as Ξ, H, Ψ are all deduced by a similar process from Ω, the functional form of Ω will be, at least implicitly, contained in each; and therefore all the integral equations which may be deduced from them, provided that they have the same limits, will have the same functional form: of this result many examples will occur in the sequel.

In problems of critical values for which the Calculus of Variations is required, the discriminating criteria of maxima and minima are frequently as difficult of application as they are difficult of discovery; and I shall reserve the consideration of them to a future section. No practical inconvenience will be caused hereby, for I propose to investigate at present only those problems, the conditions of which will immediately indicate the nature of the critical values.

322.] To determine the shortest line joining, (1) two given points, (2) two given curves in the same plane.

(1) Let (x_0, y_0) (x_1, y_1) be the given points; and let $u =$ the length of the line joining them; then

$$u = \int_0^1 ds, \quad \text{and} \quad \delta u = \int_0^1 \delta.ds.$$

But
$$ds^2 = dx^2 + dy^2;$$

$$\therefore \quad \delta.ds = \frac{dx}{ds}d.\delta x + \frac{dy}{ds}d.\delta y;$$

$$\therefore \quad \delta u = \int_0^1 \left\{\frac{dx}{ds}d.\delta x + \frac{dy}{dx}d.\delta y\right\}$$

$$= \left[\frac{dx}{ds}\delta x + \frac{dy}{ds}\delta y\right]_0^1 - \int_0^1 \left\{d.\frac{dx}{ds}\delta x + d.\frac{dy}{ds}\delta y\right\} = 0,$$

$$\text{if } \left[\frac{dx}{ds}\delta x + \frac{dy}{ds}\delta y\right]_0^1 = 0; \qquad (27)$$

and if
$$d.\frac{dx}{ds} = 0; \qquad d.\frac{dy}{ds} = 0;$$

from these last two we have $\frac{dx}{ds} = \alpha$, $\frac{dy}{ds} = \beta$; so that

$$x = \alpha s + a, \qquad y = \beta s + b;$$

$$\therefore \quad \frac{x-a}{\alpha} = \frac{y-b}{\beta}; \tag{28}$$

this is the equation to a straight line, which is therefore the shortest line; α, β, a, b, being four arbitrary constants introduced in integration; these may be determined as follows: since the limits are fixed, $\delta x_0 = 0$, $\delta y_0 = 0$, $\delta x_1 = 0$, $\delta y_1 = 0$; and therefore equation (27) is satisfied without any relation between the constants of the straight line and the limits: but as the line is to pass through the two points, x and y must satisfy simultaneously x_0, y_0, and x_1, y_1; consequently (28) becomes

$$\frac{x-x_0}{x_1-x_0} = \frac{y-y_0}{y_1-y_0},$$

which is the equation to a straight line passing through the two given points.

(2) The process of determining the unknown function is the same in both parts of the problem; but in the second part the constants α, β, a, b must be found as follows: from (27) we have

$$\left.\begin{aligned} \left(\frac{dx}{ds}\right)_0 \delta x_0 + \left(\frac{dy}{ds}\right)_0 \delta y_0 &= 0, \\ \left(\frac{dx}{ds}\right)_1 \delta x_1 + \left(\frac{dy}{ds}\right)_1 \delta y_1 &= 0. \end{aligned}\right\} \tag{29}$$

Let the equations to the limiting curves be

$$F_0(x_0, y_0) = 0, \qquad F_1(x_1, y_1) = 0;$$

then as δx_0, δy_0 are the variations of x and y as we pass from one point on the first limiting curve to another consecutive point, and as δx_1, δy_1 are the similar quantities for the second limiting curve, and as $\left(\frac{dx}{ds}\right)_0, \left(\frac{dy}{ds}\right)_0, \left(\frac{dx}{ds}\right)_1, \left(\frac{dy}{ds}\right)_1$ are the direction-cosines of the straight line at the limiting points, (29) shew that the straight line cuts both limiting curves at right angles. Let (x_0, y_0), (x_1, y_1) be the points where it meets $F_0(x_0, y_0) = 0$, and $F_1(x_1, y_1) = 0$ respectively, then we have

$$\frac{x_1-x_0}{\alpha} = \frac{y_1-y_0}{\beta};$$

also, as this straight line is normal to both curves,

$$\frac{\left(\frac{dF_0}{dx}\right)_0}{x_1-x_0} = \frac{\left(\frac{dF_0}{dy}\right)_0}{y_1-y_0}; \quad \frac{\left(\frac{dF_1}{dx}\right)_1}{x_1-x_0} = \frac{\left(\frac{dF_1}{dy}\right)_1}{y_1-y_0};$$

thus these and the equations to the curve give four equations which are sufficient to determine the values of x_0, y_0, x_1, y_1; and thus the problem is completely solved.

323.] Determine the form of the longest or shortest line which can be drawn from one curve to another curve in space.

Let the equations to the curves be

$$\left.\begin{array}{l} y_0 = f_0(x_0), \\ z_0 = \phi_0(x_0); \end{array}\right\} \qquad \left.\begin{array}{l} y_1 = f_1(x_1), \\ z_1 = \phi_1(x_1). \end{array}\right\}$$

Then $$u = \int_0^1 ds, \quad \text{and} \quad \delta u = \int_0^1 \delta.ds;$$

$$\therefore \quad \delta u = \int_0^1 \left\{\frac{dx}{ds} d.\delta x + \frac{dy}{ds} d.\delta y + \frac{dz}{ds} d.\delta z\right\}$$

$$= \left[\frac{dx}{ds}\delta x + \frac{dy}{ds}\delta y + \frac{dz}{ds}\delta z\right]_0^1 - \int_0^1 \left\{d.\frac{dx}{ds}\delta x + d.\frac{dy}{ds}\delta y + d.\frac{dz}{ds}\delta z\right\};$$

and as $\delta u = 0$, we have

$$\left.\begin{array}{l} \left(\frac{dx}{ds}\right)_0 \delta x_0 + \left(\frac{dy}{ds}\right)_0 \delta y_0 + \left(\frac{dz}{ds}\right)_0 \delta z_0 = 0, \\ \left(\frac{dx}{ds}\right)_1 \delta x_1 + \left(\frac{dy}{ds}\right)_1 \delta y_1 + \left(\frac{dz}{ds}\right)_1 \delta z_1 = 0; \end{array}\right\} \quad (30)$$

$$d.\frac{dx}{ds} = 0, \qquad d.\frac{dy}{ds} = 0, \qquad d.\frac{dz}{ds} = 0;$$

$$\therefore \quad \frac{dx}{ds} = \alpha, \qquad \frac{dy}{ds} = \beta, \qquad \frac{dz}{ds} = \gamma;$$

$$x-a = \alpha s, \quad y-b = \beta s, \quad z-c = \gamma s;$$

$$\therefore \quad \frac{x-a}{\alpha} = \frac{y-b}{\beta} = \frac{z-c}{\gamma}. \quad (31)$$

Thus a straight line, whose equations are (31), is the longest or shortest line joining two curves in space; and as $\frac{dx}{ds}, \frac{dy}{ds}, \frac{dz}{ds}$ are the direction-cosines of the line, and $\delta x_0, \delta y_0, \delta z_0, \delta x_1, \delta y_1, \delta z_1$ are the variations of the limits, and consequently the projections of elements of the limiting curves, (30) shew that the straight line cuts both curves at right-angles; and from these conditions combined with the equations to the limiting curves the unknown constants of (31) may be determined.

The solution gives also the longest, as well as the shortest line, which can be drawn from one curve to another curve. The discriminating conditions of these two cases will be explained hereafter.

324.] Determine the critical value of $\int_0^1 \mu\, ds$, where ds is an element of a plane curve, and μ is a function of x and y.

Let $$u = \int_0^1 \mu\, ds;$$

$$\therefore \quad \delta u = \int_0^1 (ds\,\delta\mu + \mu\delta.ds)$$

$$= \int_0^1 \left\{ ds\,\delta\mu + \mu\left(\frac{dx}{ds}d.\delta x + \frac{dy}{ds}d.\delta y\right)\right\};$$

also $$\delta\mu = \left(\frac{d\mu}{dx}\right)\delta x + \left(\frac{d\mu}{dy}\right)\delta y.$$

$$\therefore \quad \delta u = \left[\mu\frac{dx}{ds}\delta x + \mu\frac{dy}{ds}\delta y\right]_0^1$$

$$+\int_0^1 \left\{\left(\frac{d\mu}{dx}\right)ds - d\left(\mu\frac{dx}{ds}\right)\right\}\delta x + \int_0^1 \left\{\left(\frac{d\mu}{dy}\right)ds - d\left(\mu\frac{dy}{ds}\right)\right\}\delta y;$$

and since $\delta u = 0$,

$$\left.\begin{aligned} \mu_0\left\{\left(\frac{dx}{ds}\right)_0\delta x_0 + \left(\frac{dy}{ds}\right)_0\delta y_0\right\} &= 0, \\ \mu_1\left\{\left(\frac{dx}{ds}\right)_1\delta x_1 + \left(\frac{dy}{ds}\right)_1\delta y_1\right\} &= 0; \end{aligned}\right\} \tag{32}$$

$$\left.\begin{aligned} \left(\frac{d\mu}{dx}\right)ds - d\left(\mu\frac{dx}{ds}\right) &= 0; \\ \left(\frac{d\mu}{dy}\right)ds - d\left(\mu\frac{dy}{ds}\right) &= 0. \end{aligned}\right\} \tag{33}$$

Therefore from (33),

$$\mu d\frac{dx}{ds} = \left(\frac{d\mu}{dx}\right)ds - \frac{dx}{ds}d\mu$$

$$= \left(\frac{d\mu}{dx}\right)ds - \frac{dx}{ds}\left\{\left(\frac{d\mu}{dx}\right)dx + \left(\frac{d\mu}{dy}\right)dy\right\}$$

$$= \left(\frac{d\mu}{dx}\right)\frac{dy^2}{ds} - \left(\frac{d\mu}{dy}\right)\frac{dx\,dy}{ds}$$

$$= \left\{\left(\frac{d\mu}{dx}\right)dy - \left(\frac{d\mu}{dy}\right)dx\right\}\frac{dy}{ds};$$

similarly

$$\mu d\frac{dy}{ds} = \left\{\left(\frac{d\mu}{dy}\right)dx - \left(\frac{d\mu}{dx}\right)dy\right\}\frac{dx}{ds};$$

whence, squaring and adding, and substituting from equation (19), Art. 285, Vol. I, we have

$$\frac{\mu}{\rho} = \left(\frac{d\mu}{dx}\right)\frac{dy}{ds} - \left(\frac{d\mu}{dy}\right)\frac{dx}{ds},$$

where ρ is the radius of curvature at the point (x, y); and therefore

$$\frac{1}{\rho} = \frac{1}{\mu}\left\{\left(\frac{d\mu}{dx}\right)\frac{dy}{ds} - \left(\frac{d\mu}{dy}\right)\frac{dx}{ds}\right\}. \tag{34}$$

This equation gives a geometrical property of the curve; but we cannot proceed further with the integration unless the form of μ is given.

If the limiting values of x and y are given, the equations (32) are satisfied without any relation between $x_0, y_0, x_1, y_1, \left(\frac{dx}{ds}\right)_0, \ldots$ because $\delta x_0 = \delta y_0 = \delta x_1 = \delta y_1 = 0$; if the limits of integration are on two given plane curves, then (32) shew that the required curve cuts both the limiting curves at right angles.

325.] If ds in the last problem is an element of a curve in space, and μ is a function of x, y, z, then the equations of limits and of the indefinite terms become

$$\left.\begin{aligned}\mu_0\left\{\left(\frac{dx}{ds}\right)_0\delta x_0 + \left(\frac{dy}{ds}\right)_0\delta y_0 + \left(\frac{dz}{ds}\right)_0\delta z_0\right\} = 0,\\ \mu_1\left\{\left(\frac{dx}{ds}\right)_1\delta x_1 + \left(\frac{dy}{ds}\right)_1\delta y_1 + \left(\frac{dz}{ds}\right)_1\delta z_1\right\} = 0;\end{aligned}\right\} \tag{35}$$

$$\left.\begin{aligned}\left(\frac{d\mu}{dx}\right)ds - d\left(\mu\frac{dx}{ds}\right) = 0,\\ \left(\frac{d\mu}{dy}\right)ds - d\left(\mu\frac{dy}{ds}\right) = 0,\\ \left(\frac{d\mu}{dz}\right)ds - d\left(\mu\frac{dz}{ds}\right) = 0.\end{aligned}\right\} \tag{36}$$

From (35) we infer, that if the curve is to be drawn between given limiting curves, it cuts both these curves at right-angles.

Also from (36) we have

$$\mu d.\frac{dx}{ds} = \left(\frac{d\mu}{dx}\right)ds - d\mu\frac{dx}{ds},$$

$$\mu d.\frac{dy}{ds} = \left(\frac{d\mu}{dy}\right)ds - d\mu\frac{dy}{ds},$$

$$\mu d.\frac{dz}{ds} = \left(\frac{d\mu}{dz}\right)ds - d\mu\frac{dz}{ds};$$

therefore, squaring and adding, and substituting by means of equation (23), Art. 377, Vol. I,

$$\mu^2\frac{ds^2}{\rho^2} = ds^2\left\{\left(\frac{d\mu}{dx}\right)^2 + \left(\frac{d\mu}{dy}\right)^2 + \left(\frac{d\mu}{dz}\right)^2\right\}$$
$$-2\,d\mu\left\{\left(\frac{d\mu}{dx}\right)dx + \left(\frac{d\mu}{dy}\right)dy + \left(\frac{d\mu}{dz}\right)dz\right\} + d\mu^2$$
$$= ds^2\left\{\left(\frac{d\mu}{dx}\right)^2 + \left(\frac{d\mu}{dy}\right)^2 + \left(\frac{d\mu}{dz}\right)^2\right\} - \left\{\left(\frac{d\mu}{dx}\right)dx + \left(\frac{d\mu}{dy}\right)dy + \left(\frac{d\mu}{dz}\right)dz\right\}^2$$
$$= \left\{\left(\frac{d\mu}{dy}\right)dz - \left(\frac{d\mu}{dz}\right)dy\right\}^2 + \left\{\left(\frac{d\mu}{dz}\right)dx - \left(\frac{d\mu}{dx}\right)dz\right\}^2$$
$$+\left\{\left(\frac{d\mu}{dx}\right)dy - \left(\frac{d\mu}{dy}\right)dx\right\}^2;$$

and this equation does not admit of further reduction unless the form of μ is given. It is to be observed that the line is straight if

$$\frac{\left(\frac{d\mu}{dx}\right)}{dx} = \frac{\left(\frac{d\mu}{dy}\right)}{dy} = \frac{\left(\frac{d\mu}{dz}\right)}{dz}.$$

The examples given in Arts. 322 and 323 are only particular cases of the preceding, viz. where $\mu = 1$. But this general case is of very great importance in subsequent physical applications in both Mechanics and Optics.

326.] To determine the form of a plane curve which passing through two points (x_1, y_1), (x_0, y_0) generates by its revolution about the axis of x a surface whose area is a minimum. In this case

$$u = 2\pi\int_0^1 y\,ds;$$

$$\therefore\quad \delta u = 0 = 2\pi\int_0^1\left\{ds\,\delta y + y\left(\frac{dx}{ds}d.\delta x + \frac{dy}{ds}d.\delta y\right)\right\}$$
$$= 2\pi\left[y\frac{dx}{ds}\delta x + y\frac{dy}{ds}\delta y\right]_0^1$$
$$+2\pi\int_0^1\left\{\left(ds - d.y\frac{dy}{ds}\right)\delta y - d.y\frac{dx}{ds}\delta x\right\};$$

$$\therefore\quad \left.\begin{aligned} y_0\left\{\left(\frac{dx}{ds}\right)_0\delta x_0 + \left(\frac{dy}{ds}\right)_0\delta y_0\right\} = 0,\\ y_1\left\{\left(\frac{dx}{ds}\right)_1\delta x_1 + \left(\frac{dy}{ds}\right)_1\delta y_1\right\} = 0;\end{aligned}\right\} \qquad (37)$$

$$\therefore \quad ds - d.y\frac{dy}{ds} = 0, \qquad d.y\frac{dx}{ds} = 0. \tag{38}$$

From the latter of which, $\quad y\dfrac{dx}{ds} = c;$ (39)

thus the projection of y on the normal of the curve is constant. Substituting in (38),

$$ds - cd.\frac{dy}{dx} = 0;$$

$$\therefore \quad \frac{d.\dfrac{dy}{dx}}{\left\{1+\left(\dfrac{dy}{dx}\right)^2\right\}^{\frac{1}{2}}} = \frac{dx}{c};$$

$$\therefore \quad \log\left\{\frac{dy}{dx} + \left(1+\frac{dy^2}{dx^2}\right)^{\frac{1}{2}}\right\} = \frac{x-a}{c};$$

$$\therefore \quad \frac{dy}{dx} + \left\{1 + \left(\frac{dy}{dx}\right)^2\right\}^{\frac{1}{2}} = e^{\frac{x-a}{c}},$$

$$-\frac{dy}{dx} + \left\{1 + \left(\frac{dy}{dx}\right)^2\right\}^{\frac{1}{2}} = e^{-\frac{x-a}{c}};$$

$$\therefore \quad 2\frac{dy}{dx} = e^{\frac{x-a}{c}} - e^{-\frac{x-a}{c}}$$

$$y - b = \frac{c}{2}\left\{e^{\frac{x-a}{c}} + e^{-\frac{x-a}{c}}\right\};$$

which is the equation to the catenary, a, b, and c being constants thus far undetermined. For the sake of symmetry, let us suppose the limiting values of y to be equal, and let the axis of y bisect the line joining the extreme points of the curve; then $y_0 = y_1$, $x_0 = -x_1$, and consequently $a = 0$; whence we have

$$y - b = \frac{c}{2}\left\{e^{\frac{x}{c}} + e^{-\frac{x}{c}}\right\}; \tag{40}$$

Also since $\dfrac{dy}{dx} = 0$, when $x = 0$, the curve cuts the axis of y at right-angles, and, as appears from (39), at a distance c from the origin; c being an arbitrary constant which we have no means of determining; and therefore, from (40), $b = 0$. Hence the final equation becomes

$$y = \frac{c}{2}\left\{e^{\frac{x}{c}} + e^{-\frac{x}{c}}\right\}. \tag{41}$$

If the curve is to be drawn between two given curves, then equations (37) shew that it cuts both at right-angles.

This example is plainly a case of Art. 324, where $\mu = y$, and consequently $\left(\frac{d\mu}{dx}\right) = 0$, $\left(\frac{d\mu}{dy}\right) = 1$; and therefore from (34)

$$\rho = -y\frac{ds}{dx};$$

that is, the radius of curvature is equal and in opposite direction to the normal; which is a known property of the catenary.

327.] Of all plane curves which can be drawn between two given points, to find that which contains between the curve, its evolute, and the radii of curvature at its extremities the least area.

Let ρ be the radius of curvature, and ds be the arc of the curve; then it is manifest that if u is the required area, and (x_1, y_1), $(x_0 y_0)$ are the limiting points,

$$u = \frac{1}{2}\int_0^1 \rho\, ds;$$

$$\therefore\ \delta u = \frac{1}{2}\int_0^1 \{ds\,\delta\rho + \rho\,\delta.ds\}.$$

Now $$\frac{1}{\rho} = \frac{d^2x\,dy - d^2y\,dx}{ds^3};$$ therefore

$$-\frac{\delta\rho}{\rho^2} = \frac{dy\,d^2\delta x - dx\,d^2\delta y + d^2x\,d\delta y - d^2y\,d\delta x}{ds^3} - \frac{3(d^2x\,dy - d^2y\,dx)\,d\delta s}{ds^4}$$

$$= \text{- - - - - - - - - - - - - - -} - \frac{3\,d\delta s}{\rho\,ds};$$

$$\therefore\ ds\,\delta\rho + \rho\,\delta.ds = -\frac{\rho^3}{ds^2}\{dy\,d^2\delta x - dx\,d^2\delta y + d^2x\,d\delta y - d^2y\,d\delta x\} + 4\rho\left\{\frac{dx}{ds}d\delta x + \frac{dy}{ds}d\delta y\right\}.$$

Hence integrating by parts,

$$2\delta u = 0 = \int_0^1 \{ds\,\delta\rho + \rho\,\delta.ds\}$$

$$= \left[\rho^3\frac{dx}{ds^2}d\delta y - \rho^3\frac{dy}{ds^2}d\,\delta x\right]_0^1$$

$$- \left[\left\{d.\frac{\rho^2\,dx}{ds^2} + \rho^3\frac{d^2x}{ds^2} - 4\rho\frac{dy}{ds}\right\}\delta y - \left\{d.\frac{\rho^3 dy}{ds^2} + \rho^3\frac{d^2y}{ds^2} + 4\rho\frac{dx}{ds}\right\}\delta x\right]_0^1$$

$$+ \int_0^1 \left\{d\left(\rho^2\frac{d^2x}{ds^2} - 4\rho\frac{dy}{ds} + d.\frac{\rho^2 dx}{ds^2}\right)\delta y - d\left(\rho^3\frac{d^2y}{ds^2} + 4\rho\frac{dx}{ds} + d.\frac{\rho^2 dy}{ds^2}\right)\delta x\right\};$$

Hence if a and b are arbitrary constants, we have

$$\left.\begin{aligned} \rho^2\frac{d^2x}{ds^2} - 4\rho\frac{dy}{ds} + d.\frac{\rho^2 dx}{ds^2} &= a, \\ \rho^2\frac{d^2y}{ds^2} + 4\rho\frac{dx}{ds} + d.\frac{\rho^2 dy}{ds^2} &= b; \end{aligned}\right\} \quad (42)$$

$$\therefore \quad \rho^2\frac{dx\,d^2x + dy\,d^2y}{ds^2} + dx\,d.\frac{\rho^2 dx}{ds^2} + dy\,d.\frac{\rho^2 dy}{ds^2} = a\,dx + b\,dy$$

$$\rho^2\frac{d^2s}{ds} + 2\rho\,d\rho + \rho^2\frac{(dx\,d^2x + dy\,d^2y)\,ds^2 - 2\,ds^3 d^2s}{ds^4} = a\,dx + b\,dy$$

$$\rho^2\frac{d^2s}{ds} + 2\rho\,d\rho - \rho^2\frac{d^2s}{ds} = a\,dx + b\,dy;$$

$$\therefore \quad \rho^2 = ax + by + c. \quad (43)$$

Also from (42) by subtraction

$$\rho^2\frac{dy\,d^2x - dx\,d^2y}{ds^2} - 4\rho\,ds + dy\,d.\frac{\rho^2 dx}{ds^2} - dx\,d.\frac{\rho^2 dy}{ds^2} = a\,dy - b\,dx;$$

whence, we have

$$\rho = \frac{1}{2}\left\{b\frac{dx}{ds} - a\frac{dy}{ds}\right\}. \quad (44)$$

Either (43) or (44) is a geometrical definition of the curve. From (43) it appears that the square of the radius of curvature is a linear function of the coordinates; and as the radius of curvature is an absolute quantity and independent both of the origin and of the particular system of coordinate axes, we may, without thereby affecting the generality of the problem, choose our system of reference such that $a = 0$, $c = 0$; whereby

$$\rho^2 = by; \qquad \rho = \frac{b}{2}\frac{dx}{ds}.$$

Whence it follows that the curve lies wholly on the positive side of the axis of x, and that the curvature is the same at all points equally distant from that axis; also that $\frac{dx}{ds} = 0$, when $y = 0$: the curve therefore meets the axis of x at right-angles. And since

$$\frac{b}{2}\frac{dx}{ds} = (by)^{\frac{1}{2}}; \qquad \therefore \quad dx = \frac{2y\,dy}{(by - 4y^2)^{\frac{1}{2}}};$$

$$\therefore \quad x = \frac{b}{8}\,\text{versin}^{-1}\frac{8y}{b} - \frac{1}{2}(by - 4y^2)^{\frac{1}{2}};$$

the equation to a cycloid of which the starting point is the origin, $\frac{b}{8}$ is the radius of the generating circle, and the constants are such that the origin is on the curve, and the axis of x is the base of the cycloid.

328.] To find the relation between x and y, so that

$$u = \int_0^1 (x^2+y^2)^{\frac{n}{2}} ds \text{ may be a minimum.}$$

Let u be expressed in polar coordinates, so that

$$u = \int_0^1 r^n \, ds.$$

$$\therefore \quad \delta u = \int_0^1 \{n r^{n-1} ds\, \delta r + r^n d\delta s\}$$

$$\text{but} \qquad ds^2 = dr^2 + r^2 d\theta^2;$$

$$\therefore \quad d.\delta s = \frac{dr}{ds} d.\delta r + \frac{r^2 d\theta}{ds} d.\delta\theta + \frac{r\, d\theta^2}{ds}\delta r; \qquad \text{then}$$

$$\delta u = 0 = \int_0^1 \left\{ r^n \frac{dr}{ds} d.\delta r + r^{n+2}\frac{d\theta}{ds} d.\delta\theta + \left(r^{n+1}\frac{d\theta^2}{ds} + n r^{n-1} ds\right)\delta r \right\}$$

$$= \left[r^n \frac{dr}{ds}\delta r + r^{n+2}\frac{d\theta}{ds}\delta\theta \right]_0^1$$

$$+ \int_0^1 \left\{ \left(r^{n+1}\frac{d\theta^2}{ds} + n r^{n-1} ds - d.\frac{r^n dr}{ds} \right)\delta r - d.\frac{r^{n+2} d\theta}{ds}\delta\theta \right\};$$

$$\therefore \quad r^{n+1}\frac{d\theta^2}{ds} + n r^{n-1} ds - d.\frac{r^n dr}{ds} = 0, \qquad \text{and } d.\frac{r^{n+2} d\theta}{ds} = 0;$$

$$\therefore \quad \frac{r^{n+2} d\theta}{ds} = a;$$

$$\text{whence} \qquad r^{n+1} = a \sec (n+1)\theta; \qquad (45)$$

If $n = 0$, (45) is the equation to a straight line, and the result is in accordance with that of Art. 322.

329.] Let us now consider some problems of *relative* maxima and minima; those namely wherein the variables are not independent of each other but are connected by some given relation, which may be integral or differential, or in the form of a definite integral. These problems are often called *isoperimetrical*, because the given condition when interpreted geometrically is frequently equivalent to the length of the curve being given between certain fixed points or limiting lines.

And although the method of introducing indeterminate multipliers, indicated in Art. 320, is most convenient for explaining the course to be adopted in the general case, yet as in the following problems only one condition or relation will be given, it is better to use a process which results from the theory of indeterminate multipliers as explained in Art. 168, Vol. I, and which consists in equating to a constant quantity the ratio of the several coefficients of the variations of the variables in both the definite and the unintegrated parts of the given equations.

To determine the form of a plane curve which being of given length revolves about a given line (the axis of x), and generates a solid whose volume is a maximum or a minimum.

Here $u = \pi\int_0^1 y^2\,dx, \qquad \int_0^1 ds = c =$ a given length;

$$\therefore \quad \delta u = 0 = \pi\int_0^1 (2y\,dx\,\delta y + y^2\,d.\delta x)$$

$$= \pi\Big[y^2\,\delta x\Big]_0^1 - \pi\int_0^1 (2y\,dy\,\delta x - 2y\,dx\,\delta y); \qquad (46)$$

$$\delta c = 0 = \int_0^1 \delta.ds$$

$$= \int_0^1 \left(\frac{dx}{ds}\,d.\delta x + \frac{dy}{ds}\,d.\delta y\right)$$

$$= \left[\frac{dx}{ds}\,\delta x + \frac{dy}{ds}\,\delta y\right]_0^1 - \int_0^1 \left(d.\frac{dx}{ds}\,\delta x + d.\frac{dy}{ds}\,\delta y\right); \quad (47)$$

whence equating to a constant λ the ratio of the coefficients of δx and δy in the unintegrated parts of (46) and (47), we have

$$\frac{2y\,dy}{d.\frac{dx}{ds}} = \frac{-2y\,dx}{d.\frac{dy}{ds}} = \lambda = 2y\rho,$$

the last term of the equality being deduced from the first two by means of equation (19), Art. 285, Vol. I;

$$\therefore \quad \rho = \frac{\lambda}{2y};$$

that is, the radius of curvature varies inversely as the ordinate.

$$\text{Also} \qquad 2y\,dy = \lambda d.\frac{dx}{ds};$$

$$\therefore \quad y^2 - b^2 = \lambda\frac{dx}{ds};$$

whence we have

$$\frac{dx}{y^2-b^2} = \pm\frac{dy}{\{\lambda^2-(y^2-b^2)^2\}^{\frac{1}{2}}} = \frac{ds}{\lambda}; \qquad (48)$$

expressions which do not admit of further integration, but are the equations of the elastic curve, the mechanical form of which will be the subject of investigation hereafter.

If the limiting points of the curve are given, then $\delta x_0 = \delta y_0 = 0$, and $\delta x_1 = \delta y_1 = 0$, and therefore in (46) and (47) the terms at the limits disappear: but if the line is to be drawn between two given curves, the arbitrary constants will be determined by means of the equations to those curves at the limits.

330.] To determine the form of the closed plane curve which is of given length c, and incloses the greatest area.

Since the length $= \int_0^1 ds$, and the area $= \int_0^1 y\, dx$,

$$c = \int_0^1 ds, \qquad u = \int_0^1 y\, dx;$$

$$\therefore \quad \delta c = 0 = \left[\frac{dx}{ds}\delta x + \frac{dy}{ds}\delta y\right]_0^1 - \int_0^1 \left(d.\frac{dx}{ds}\delta x + d.\frac{dy}{ds}\delta y\right); \quad (49)$$

$$\delta u = 0 = \Big[y\,\delta x\Big]_0^1 + \int_0^1 (dx\,\delta y - dy\,\delta x); \quad (50)$$

whence equating to a constant λ the ratio of the coefficients of δx and δy in the unintegrated parts of (49) and (50), we have

$$\frac{dx}{d.\dfrac{dy}{ds}} = \frac{-dy}{d.\dfrac{dx}{ds}} = \lambda = \rho, \quad (51)$$

where ρ is the radius of curvature, by reason of (19) Art. 285, Vol. I. Consequently the radius of curvature is constant, and the required curve is a circle.

Also from (51) we have

$$x - a = \lambda\frac{dy}{ds}; \qquad y - b = -\lambda\frac{dx}{ds};$$

$$\therefore \quad (x-a)^2 + (y-b)^2 = \lambda^2;$$

where a, b, and λ, which last = the radius, are constants to be determined by the conditions of the problems.

Thus suppose that it is required to find a curve of given length, and such that the area contained between it and the positive coordinate-axes of x and y is a maximum. Let the superior and inferior limits correspond to the points where the curve cuts the axes of x and y respectively. Then by a comparison of the coefficients of δx and δy in the integrated portions of (49) and (50), we have at the limits

$$\frac{y}{\frac{dx}{ds}} = \frac{0}{\frac{dy}{ds}} = \lambda;$$

at the superior limit $y = 0$, so that $\frac{dx}{ds} = 0$, and the curve cuts the axis of x at right-angles, and $b = 0$; at the inferior limit $\frac{dy}{ds} = 0$, consequently the curve cuts the axis of y at right-angles, and $a = 0$; also as λ is the radius, and c is the length of the quadrant, $\lambda = \frac{2c}{\pi}$; consequently the equation of the required curve is

$$x^2 + y^2 = \frac{4c^2}{\pi^2}.$$

This problem may also be conveniently solved by means of polar coordinates. See the first edition of this volume, Art. 221.

331.] Of all isoperimetrical curves joining two given points, to find that, the product of whose length-element and the square of its distance from the line joining the two points is a maximum.

Let the line joining the two points be the axis of x, and let the origin be at the middle point of this line; let $2a$ be the distance between the two points; then x, y, z being the coordinates of any point on the curve corresponding to the commencement of the element, and c being the length of the curve,

$$u = \int_0^1 (y^2+z^2)\, ds, \qquad \text{and } c = \int_0^1 ds;$$

and by a process similar to that of the preceding Articles,

$$\frac{d.\frac{dx}{ds}}{d.(y^2+z^2)\frac{dx}{ds}} = \frac{d.\frac{dy}{ds}}{d.(y^2+z^2)\frac{dy}{ds} - 2y\,ds} = \frac{d.\frac{dz}{ds}}{d.(y^2+z^2)\frac{dz}{ds} - 2z\,ds} = \frac{1}{\lambda}; \quad (52)$$

from the second and third of which terms we have

$$d(y^2+z^2-\lambda)\left(z\frac{dy}{ds} - y\frac{dz}{ds}\right) + (y^2+z^2-\lambda)\left(zd.\frac{dy}{ds} - yd.\frac{dz}{ds}\right) = 0;$$

$$\therefore \quad (y^2+z^2-\lambda)\left(z\frac{dy}{ds} - y\frac{dz}{ds}\right) = c';$$

but by the particular system of reference which we have chosen, when $z = 0$, $y = 0$; therefore $c' = 0$:

$$\therefore \quad (y^2+z^2-\lambda)\left(z\frac{dy}{ds} - y\frac{dz}{ds}\right) = 0;$$

and as $y^2+z^2-\lambda$ cannot vanish for all points of the curve, the above equation can be satisfied only by $z\,dy-y\,dz = 0$; whence $y = kz$; and consequently the curve lies wholly in one plane passing through the axis of x: let this plane be that of (x, y): then $z = 0$, and from (52) we have

$$\lambda d.\frac{dx}{ds} = d.y^2\frac{dx}{ds},$$

$$\therefore \quad (\lambda-y^2)\frac{dx}{ds} = b^2;$$

which is the differential equation of the required curve, and does not admit of further integration.

332.] The following are other examples in which the calculus of variations is required.

Ex. 1. Of all plane curves of the same length, that which generates the greatest or least surface by its revolution about a given axis is the catenary.

Ex. 2. Of all plane curves of the same length the elastic curve bounds the area which by its revolution about a given axis generates the greatest or least volume.

Ex. 3. Prove that the plane curve, which by its revolution about an axis in its plane generates a surface of given area and of which the content is a maximum or a minimum, is such that the sum of the principal curvatures at every point of the generated surface is constant.

Ex. 4. Two parallel planes being given, it is required to draw from a given point in one a curve of given length to the other, such that the surface of the cylinder, of which the curve is the director, and lines perpendicular to the planes are the generators, may be a maximum or a minimum.

333.] To find the line of constant curvature whose length is a maximum or a minimum.

In this example I propose to follow the general method for resolving problems of relative maxima and minima; and for the purpose of shortening the process and formulæ, shall suppose s to be equicrescent.

Let k = the constant radius of absolute curvature; so that

$$\frac{1}{k} = \frac{\{(d^2x)^2+(d^2y)^2+(d^2z)^2\}^{\frac{1}{2}}}{ds^2};$$

and therefore

$$u = \int_0^1 \left\{(dx^2 + dy^2 + dz^2)^{\frac{1}{2}} + \lambda \left(\frac{\{(d^2x)^2 + (d^2y)^2 + (d^2z)^2\}^{\frac{1}{2}}}{ds^2} - \frac{1}{k} \right) \right\};$$

then, applying the symbols of Arts. 306, 321, we have

$$\Omega = \{dx^2 + dy^2 + dz^2\}^{\frac{1}{2}} + \lambda \left\{ \frac{\{(d^2x)^2 + (d^2y)^2 + (d^2z)^2\}^{\frac{1}{2}}}{dx^2 + dy^2 + dz^2} - \frac{1}{k} \right\};$$

$$\text{X} = 0;$$

$$\text{X}_1 = \frac{d\Omega}{d.dx} = \frac{dx}{ds} - \frac{2\lambda\, dx \{(d^2x)^2 + (d^2y)^2 + (d^2z)^2\}^{\frac{1}{2}}}{ds^4}$$

$$= \frac{dx}{ds} - \frac{2\lambda}{k} \frac{dx}{ds^2};$$

$$\text{X}_2 = \frac{d\Omega}{d.d^2x} = \frac{\lambda\, d^2x}{ds^2 \{(d^2x)^2 + (d^2y)^2 + (d^2z)^2\}^{\frac{1}{2}}}$$

$$= \frac{k\lambda\, d^2x}{ds^4};$$

and therefore substituting in $\text{E} = 0$, and pursuing similar processes for H and Ψ, we have

$$\left.\begin{aligned} d\left\{\frac{dx}{ds} - \frac{2\lambda}{k}\frac{dx}{ds^2}\right\} - d^2\frac{k\lambda\, d^2x}{ds^4} &= 0; \\ d\left\{\frac{dy}{ds} - \frac{2\lambda}{k}\frac{dy}{ds^2}\right\} - d^2\frac{k\lambda\, d^2y}{ds^4} &= 0; \\ d\left\{\frac{dz}{ds} - \frac{2\lambda}{k}\frac{dz}{ds^2}\right\} - d^2\frac{k\lambda\, d^2z}{ds^4} &= 0; \end{aligned}\right\} \quad (53)$$

and by integration

$$\left.\begin{aligned} \frac{dx}{ds} - \frac{2\lambda}{k}\frac{dx}{ds^2} - d.\frac{k\lambda\, d^2x}{ds^4} &= \alpha; \\ \frac{dy}{ds} - \frac{2\lambda}{k}\frac{dy}{ds^2} - d.\frac{k\lambda\, d^2y}{ds^4} &= \beta; \\ \frac{dz}{ds} - \frac{2\lambda}{k}\frac{dz}{ds^2} - d.\frac{k\lambda\, d^2z}{ds^4} &= \gamma; \end{aligned}\right\} \quad (54)$$

α, β, γ being constants introduced in the integration: to determine them, let it be observed that the definite part of δu given in equation (62), Art. 306, becomes in this case

$$\Bigg[\left\{\frac{dx}{ds} - \frac{2\lambda}{k}\frac{dx}{ds^2} - k\lambda\, d.\frac{d^2x}{ds^4}\right\}\delta x$$

$$+ \left\{\frac{dy}{ds} - \frac{2\lambda}{k}\frac{dy}{ds^2} - k\lambda\, d.\frac{d^2y}{ds^4}\right\}\delta y$$

$$+ \left\{\frac{dz}{ds} - \frac{2\lambda}{k}\frac{dz}{ds^2} - k\lambda\, d.\frac{d^2z}{ds^4}\right\}\delta z$$

$$+ \frac{k\lambda\, d^2x}{ds^4} d.\delta x + \frac{k\lambda\, d^2y}{ds^4} d.\delta y + \frac{k\lambda\, d^2z}{ds^4} d.\delta z\Bigg]_0^1;$$

and this must vanish by virtue of the reasoning in Art. 316; and as no relation is given between the values of $\delta x, \delta y, \delta z, \ldots$ at the two limits, the coefficients of these quantities must separately be equal to zero;

$$\left.\begin{aligned} \therefore \quad \left[\frac{dx}{ds} - \frac{2\lambda}{k}\frac{dx}{ds^2} - k\lambda\, d.\frac{d^2x}{ds^4}\right]_0 &= 0; \\ \left[\frac{dx}{ds} - \frac{2\lambda}{k}\frac{dx}{ds^2} - k\lambda\, d.\frac{d^2x}{ds^4}\right]_1 &= 0; \end{aligned}\right\}$$

and as these are particular values of the first equation of (54), it must be consistent with them; therefore $a = 0$; for a similar reason $\beta = 0$, $\gamma = 0$: whereby, and differentiating, bearing in mind that s is equicrescent,

$$\left.\begin{aligned} \frac{dx}{ds} - \frac{2\lambda}{k}\frac{dx}{ds^2} - k\lambda\frac{d^3x}{ds^4} &= 0; \\ \frac{dy}{ds} - \frac{2\lambda}{k}\frac{dy}{ds^2} - k\lambda\frac{d^3y}{ds^4} &= 0; \\ \frac{dz}{ds} - \frac{2\lambda}{k}\frac{dz}{ds^2} - k\lambda\frac{d^3z}{ds^4} &= 0; \end{aligned}\right\} \qquad (55)$$

and employing the symbols of Art. 377, Vol. I, equations (6), multiplying the preceding equations successively by X, Y, Z, and observing that $\quad \text{X}\,dx + \text{Y}\,dy + \text{Z}\,dz = 0,$

we have $$\text{X}\,d^3x + \text{Y}\,d^3y + \text{Z}\,d^3z = 0; \qquad (56)$$

and therefore, by reason of equation (40), Art. 382, Vol. I, the radius of torsion is infinite; and therefore all points of the required curve lie in one plane.

Again, from (55), since λ is an arbitrary constant, and ds is also constant, we may replace λ by $\lambda' ds$: and also, replacing $1 - 2\frac{\lambda'}{k}$ by h, and $k\lambda'$ by h', we have

$$\left.\begin{aligned} h\frac{dx}{ds} - h'\frac{d^3x}{ds^3} &= 0, \\ h\frac{dy}{ds} - h'\frac{d^3y}{ds^3} &= 0, \\ h\frac{dz}{ds} - h'\frac{d^3z}{ds^3} &= 0; \end{aligned}\right\}$$

whence by integration,

$$\left.\begin{aligned} hx - h'\frac{d^2x}{ds^2} &= c_1, \\ hy - h'\frac{d^2y}{ds^2} &= c_2, \\ hz - h'\frac{d^2z}{ds^2} &= c_3; \end{aligned}\right\} \qquad (57)$$

also because s is equicrescent,

$$h(x\,dx+y\,dy+z\,dz) = c_1 dx + c_2 dy + c_3 dz;$$

$$\therefore \quad h(x^2+y^2+z^2) = 2c_1 x + 2c_2 y + 2c_3 z + c_4; \qquad (58)$$

which is the equation to a sphere: and therefore, combining (56) and (58), it follows that the curve is a plane section of a sphere, and therefore is a circle.

It may also thus be proved that the curve is a plane curve: from the last two equations of (57) we have

$$h(z\,d^2y - y\,d^2z) = c_3\,d^2y - c_2\,d^2z;$$

$$\therefore \quad h(z\,dy - y\,dz) = c_3\,dy - c_2\,dz + k_1;$$

therefore also $\quad h(x\,dz - z\,dx) = c_1\,dz - c_3\,dx + k_2,$

$$h(y\,dx - x\,dy) = c_2\,dx - c_1\,dy + k_3:$$

multiplying these severally by dx, dy, dz, and adding, we have

$$k_1\,dx + k_2\,dy + k_3\,dz = 0;$$

$$\therefore \quad k_1 x + k_2 y + k_3 z = k;$$

the equation to a plane: and therefore the curve required is a plane section of a sphere.

334.] In Art. 314 it has been stated that the calculus of variations may be considered as a particular form of differential calculus, wherein the number of subject-variables of any function is infinite: I propose to illustrate this mode of viewing the calculus by the following simple example: Between two given points to draw a curve of given length, so that the area contained between it, the ordinates to the two points, and the axis of x, may be a maximum or a minimum*.

Let the two points be (x_0, y_0), (x_n, y_n): and let the distance $x_n - x_0$ on the axis of x be divided into n equal parts, and the abscissæ corresponding to the points of partition be $x_1, x_2, \ldots x_{n-1}$; and let the corresponding ordinates be $y_1, y_2, \ldots y_{n-1}$; and also for convenience of notation let $y_1 - y_0 = \Delta y_0$, $y_2 - y_1 = \Delta y_1, \ldots$; $x_1 - x_0 = \Delta x_0 = x_2 - x_1 = \ldots$; and suppose the several points, to which these coordinates refer, to be joined by straight lines, of which let the lengths be $\Delta s_0, \Delta s_1, \ldots \Delta s_{n-1}$; and let the sum of these lengths be equal to the given length c; then, if $\mathrm{A} =$ the required area,

* For other examples of maxima and minima solved by this process see Schellbach, Variationsrechnung, Crelle, Band XLI, p. 293, 1851.

$$\text{A} = \frac{1}{2}(x_1-x_0)(y_1+y_0)+\frac{1}{2}(x_2-x_1)(y_2+y_1)+\dots$$
$$\dots+\frac{1}{2}(x_n-x_{n-1})(y_n+y_{n-1}),$$

$$c = \Delta s_0+\Delta s_1+\dots+\Delta s_{n-1}$$
$$= \{(x_1-x_0)^2+(y_1-y_0)^2\}^{\frac{1}{2}}+\{(x_2-x_1)^2+(y_2-y_1)^2\}^{\frac{1}{2}}+\dots$$
$$\dots+\{(x_n-x_{n-1})^2+(y_n-y_{n-1})^2\}^{\frac{1}{2}}.$$

Let $u =$ the required critical function; then, λ being an undetermined constant,

$$u = \text{A}+\lambda c;$$

and since x_n-x_0 is divided into n equal parts,

$$x_1-x_0 = x_2-x_1 = \dots = \text{a constant};$$

so that u is a function of $(n-1)$ independent variables, viz. $y_1, y_2, \dots y_{n-1}$; and therefore, taking the partial differentials of u with respect to them, and equating them to zero, we have the following series of equations:

$$\left.\begin{aligned}
x_2-x_0-2\lambda\left\{\frac{\Delta y_1}{\Delta s_1}-\frac{\Delta y_0}{\Delta s_0}\right\} &= 0;\\
x_3-x_1-2\lambda\left\{\frac{\Delta y_2}{\Delta s_2}-\frac{\Delta y_1}{\Delta s_1}\right\} &= 0;\\
\dots\dots\dots\dots\dots\dots\dots\dots\\
x_{m+2}-x_m-2\lambda\left\{\frac{\Delta y_{m+1}}{\Delta s_{m+1}}-\frac{\Delta y_m}{\Delta s_m}\right\} &= 0;\\
\dots\dots\dots\dots\dots\dots\dots\dots\\
x_n-x_{n-2}-2\lambda\left\{\frac{\Delta y_{n-1}}{\Delta s_{n-1}}-\frac{\Delta y_{n-2}}{\Delta s_{n-2}}\right\} &= 0;
\end{aligned}\right\}$$

and because

$$x_{m+2}-x_m = x_{m+1}+\Delta x_{m+1}-x_m = 2\Delta x, \text{ say},$$

each of these equations is of the form

$$2\Delta x-2\lambda\Delta\left(\frac{\Delta y_k}{\Delta s_k}\right)=0; \qquad (59)$$

now suppose the number of the points of division of x_n-x_0 to become infinite, then, taking x, y, s to be the general types of their particular values, from (59) we have

$$dx = \lambda d.\frac{dy}{ds};$$

whence by integration, a and b being arbitrary constants, we have

$$(x-a)^2+(y-b)^2 = \lambda^2. \qquad (60)$$

And this is the equation to a circle. To determine a and b and λ, we have

$$\left.\begin{aligned}(x_0-a)^2+(y_0-b)^2 &= \lambda^2,\\ (x_n-a)^2+(y_n-b)^2 &= \lambda^2;\end{aligned}\right\} \tag{61}$$

also
$$c = \int_0^n ds$$
$$= \lambda \int_{y_0}^{y_n} \frac{dy}{\{\lambda^2-(y-b)^2\}^{\frac{1}{2}}};$$
$$\therefore \quad \frac{c}{\lambda} = \sin^{-1}\frac{y_n-b}{\lambda} - \sin^{-1}\frac{y_0-b}{\lambda};$$
$$\therefore \quad (x_0-a)(y_n-b)-(x_n-a)(y_0-b) = \lambda^2 \sin\frac{c}{\lambda}; \tag{62}$$

and from (61) and (62), a, b, and λ may be determined.

335.] In the preceding Article x has been supposed to be equicrescent; but we might manifestly have supposed y to be equicrescent, in which case a similar process would have led to the equation

$$dy = \lambda d.\frac{dx}{ds}; \tag{63}$$

Also the problem might have been treated more generally; neither of the variables might have been supposed to be equicrescent; and in this case, as the coordinates would be independent of each other, we should have had two simultaneous groups of equations similar to (59); and from them, by a passage to infinitesimal subdivision, we should obtain two simultaneous equations, viz.

$$dx = \lambda d.\frac{dy}{ds}, \qquad dy = \lambda d.\frac{dx}{ds};$$

whence, integrating, squaring and adding,

$$(x-a)^2+(y-b)^2 = \lambda^2.$$

A careful examination of the process by which this example has been solved will shew that the method which has been employed in the previous cases, and which was explained in all its generality in the preceding Chapter, is precisely the same. In that form however it is concealed under symbols of integration and variation; whereas in this Article it has been resolved into its simplest elements, and has been laid bare to inspection and exact investigation. Other problems may of course be solved by the same process.

SECTION 2.—*On Geodesic Lines.*

336.] One problem which requires the Calculus of Variations deserves especial notice; it is that of the determination of the longest or shortest line which can be drawn on a given surface from one point to another, or from one curve to another; for these lines possess important properties in the theory of geodesy, and consequently in reference to an ellipsoid of three unequal axes. The name of Geodesic lines have been given to them, and it is necessary to consider their properties at considerable length, and from various points of view.

Geodesic lines, or Geodesics, are defined to be the longest or shortest lines which can be drawn on a curved surface between two given points, or between two given curved lines.

Let the equation to the surface on which the lines are drawn be

$$F(x, y, z) = 0; \tag{64}$$

and let us employ the same abbreviating symbols as in (1), (2), (3), of Art. 398, Vol. I. Let s = the length of the geodesic between the given limits; then $s = \int_0^1 ds$; and consequently as

$$ds^2 = dx^2 + dy^2 + dz^2;$$

$$\delta s = 0 = \left[\frac{dx}{ds}\delta x + \frac{dy}{ds}\delta y + \frac{dz}{ds}\delta z\right]_0^1$$
$$-\int_0^1 \left\{d.\frac{dx}{ds}\delta x + d.\frac{dy}{ds}\delta y + d.\frac{dz}{ds}\delta z\right\}. \tag{65}$$

Now δx, δy, δz are subject to the relation

$$U\,\delta x + V\,\delta y + W\,\delta z = 0,$$

and as this must consist with the part of (65), which is under the sign of integration, we have

$$\frac{d.\frac{dx}{ds}}{U} = \frac{d.\frac{dy}{ds}}{V} = \frac{d.\frac{dz}{ds}}{W}; \tag{66}$$

which are the differential equations to geodesic lines on a given surface: the complete integrals of them have never yet been found, but many properties may be deduced both in the general case and in the particular case of the ellipsoid.

If the geodesic line is drawn from one given point to another given point on the surface, then, as there are no variations at these limits, the definite part of (65) vanishes; but if the geodesic line is drawn from one curve to another curve on the surface, then, since

$$\left[\frac{dx}{ds}\delta x + \frac{dy}{ds}\delta y + \frac{dz}{ds}\delta z\right]_0^1 = 0,$$

and as $\frac{dx}{ds}, \frac{dy}{ds}, \frac{dz}{ds}$ are the direction-cosines of the tangent to the geodesic line, and $\delta x, \delta y, \delta z$ are proportional to the direction-cosines of the tangent to the limiting curve at the limit, it appears that the geodesic line cuts both the limiting curves at right-angles: this is also manifest by general reasoning.

It will be observed however that although an infinitesimal arc of a geodesic is of a minimum length between its extremities, yet if the distance between the extreme points is finite, a geodesic passing through them may be either a maximum or a minimum, or indeed only one of such critical lines, the number of which may be infinite. Of this theorem we shall hereafter have some instances.

Since $d.\frac{dx}{ds}, d.\frac{dy}{ds}, d.\frac{dz}{ds}$ are proportional to the direction-cosines of the principal normal, or of the direction of the radius of absolute curvature of a curve in space, and since U, V, W are proportional to the direction-cosines of the normal to the surface at the point (x, y, z), (66) shew that the radius of absolute curvature of a geodesic line drawn on a surface is coincident in direction with the normal to the surface; or, in other and equivalent words, that the osculating plane of a geodesic line is a normal plane to the surface.

Hence it appears that if a cylinder or other developable surface touches a given surface along a geodesic, the line of contact becomes straight when the developable surface is unwrapped into a plane.

337.] The equations to a geodesic line on a surface may be put under the following form:

Since the osculating plane of the geodesic line contains the normal to the surface, we have

$$\text{U}(dy\,d^2z - dz\,d^2y) + \text{V}(dz\,d^2x - dx\,d^2z) + \text{W}(dx\,d^2y - dy\,d^2x) = 0,$$

or $$(\text{V}d^2z - \text{W}d^2y)\,dx + (\text{W}d^2x - \text{U}d^2z)\,dy + (\text{U}d^2y - \text{V}d^2x)\,dz = 0;$$

also $$\text{U}\,dx + \text{V}\,dy + \text{W}\,dz = 0;$$

whence we have the equality

$$\frac{dx}{Q^2 d^2x - U(U d^2x + V d^2y + W d^2z)} = \ldots\ldots = \ldots\ldots$$

$$= \frac{dx^2 + dy^2 + dz^2}{Q^2 (dx\, d^2x + dy\, d^2y + dz\, d^2z)} \quad (67)$$

$$= \frac{dU\, dx + dV\, dy + dW\, dz}{Q^2 (dU\, d^2x + dV\, d^2y + dW\, d^2z - (U\, dU + V dV + W dW)(U d^2x + V d^2y + W d^2z)};$$

and since $\quad U\, dx + V\, dy + W\, dz = 0,$

$$\therefore \quad U\, d^2x + V\, d^2y + W\, d^2z = -(dU\, dx + dV\, dy + dW\, dz);$$

so that from (67) we have

$$\frac{dU\, d^2x + dV\, d^2y + dW\, d^2z}{dU\, dx + dV\, dy + dW\, dz} + \frac{U\, dU + V\, dV + W\, dW}{Q^2} - \frac{dx\, d^2x + dy\, d^2y + dz\, d^2z}{dx^2 + dy^2 + dz^2} = 0,$$

or $$\frac{dU\, d^2x + dV\, d^2y + dW\, d^2z}{dU\, dx + dV\, dy + dW\, dz} + \frac{dQ}{Q} - \frac{d.ds}{ds} = 0;$$

$$\therefore \quad \int \frac{dU\, d^2x + dV\, d^2y + dW\, d^2z}{dU\, dx + dV\, dy + dW\, dz} = \log \frac{ds}{Q}. \quad (68)$$

The element of the integral in the first member of this equation is manifestly a differential of the second order; and the integral will consequently involve two arbitrary constants one of which will depend on, say, the initial direction, and the other on the initial point of the geodesic. The complete integral of (68) has not yet been found. In the case of surfaces of the second degree the first integral can be found, as we shall presently see.

338.] Let ρ be the radius of absolute curvature of a geodesic line; then by (23), Art. 377, Vol. I,

$$\frac{ds^2}{\rho^2} = \left(d.\frac{dx}{ds}\right)^2 + \left(d.\frac{dy}{ds}\right)^2 + \left(d.\frac{dz}{ds}\right)^2;$$

therefore, from (66),

$$\frac{d.\frac{dx}{ds}}{U} = \frac{d.\frac{dy}{ds}}{V} = \frac{d.\frac{dz}{ds}}{W} = \frac{ds}{\rho Q}; \quad (69)$$

by means of either of which equations the length of the radius of absolute curvature at any point may be determined.

Also let ρ' be the radius of curvature of the normal section of the surface which contains ds; then, from (69),

$$\frac{ds}{\rho Q} = \frac{U d.\frac{dx}{ds} + V d.\frac{dy}{ds} + W d.\frac{dz}{ds}}{Q^2}, \quad (70)$$

$$\therefore \quad \frac{ds}{\rho Q} = \frac{\text{U}\,d^2x + \text{V}\,d^2y + \text{W}\,d^2z}{Q^2\,ds}, \text{ if } s \text{ is equicrescent;}$$

$$= \frac{Q\,ds^2}{\rho' Q^2\,ds}; \text{ by reason of (12), Art. 399, Vol. I.}$$

$$\therefore \quad \rho = \rho';$$

that is, the radius of absolute curvature of a geodesic line is equal to the radius of curvature of the normal section of the surface, which at their common point touches the line.

This result may also be inferred from the property stated in Art. 336; viz., the osculating plane of a geodesic line is a normal plane to the surface. Consequently two consecutive elements of the geodesic are in the normal plane, and are coincident with two consecutive elements of the normal curve of section; thus the radius of absolute curvature of both curves is the same.

339.] Also to determine the torsion of a geodesic, let us take the value of the radius of torsion which is given in (41), Art. 382, Vol. I: whereby we have

$$\frac{1}{\text{R}} = \frac{\rho^2}{ds^6}\{\text{X}\,d^3x + \text{Y}\,d^3y + \text{Z}\,d^3z\}. \qquad (71)$$

Now by (69), if s is equicrescent, we have

$$\left.\begin{aligned} \text{X} &= dy\,d^2z - dz\,d^2y = \frac{ds^2}{\rho Q}(\text{W}\,dy - \text{V}\,dz), \\ \text{Y} &= dz\,d^2x - dx\,d^2z = \frac{ds^2}{\rho Q}(\text{U}\,dz - \text{W}\,dx), \\ \text{Z} &= dx\,d^2y - dy\,d^2x = \frac{ds^2}{\rho Q}(\text{V}\,dx - \text{U}\,dy); \end{aligned}\right\} \qquad (72)$$

and substituting in (71),

$$\frac{1}{\text{R}} =$$

$$\frac{\rho}{Q\,ds^4}\{(dz\,d^3y - dy\,d^3z)\text{U} + (dx\,d^3z - dz\,d^3x)\text{V} + (dy\,d^3x - dx\,d^3y)\text{W}\}$$

$$= -\frac{\rho}{Q\,ds^4}\{\text{U}\,d\text{X} + \text{V}\,d\text{Y} + \text{W}\,d\text{Z}\}. \qquad (73)$$

And since $\text{X}\,d^2x + \text{Y}\,d^2y + \text{Z}\,d^2z = 0$, by (69) we have

$$\text{U}\text{X} + \text{V}\text{Y} + \text{W}\text{Z} = 0. \qquad (74)$$

$$\therefore \quad -(\text{U}\,d\text{X} + \text{V}\,d\text{Y} + \text{W}\,d\text{Z}) = \text{X}\,d\text{U} + \text{Y}\,d\text{V} + \text{V}\,d\text{W}. \qquad (75)$$

$$\therefore \text{ from (73),} \quad \frac{1}{\text{R}} = \frac{\rho}{Q\,ds^4}\{\text{X}\,d\text{U} + \text{Y}\,d\text{V} + \text{Z}\,d\text{W}\} \qquad (76)$$

$$\frac{1}{R} = \frac{1}{Q^2 ds^2}\{(W\,dy - V\,dz)\,dU + (U\,dz - W\,dx)\,dV + (V\,dx - U\,dy)\,dW\}$$

$$= \frac{(V\,dW - W\,dV)\,dx + (W\,dU - U\,dW)\,dy + (U\,dV - V\,dU)\,dz}{Q^2 ds^2}; \qquad (77)$$

which gives the radius of torsion at any point of a geodesic.

If the numerator of (77) $= 0$, dx, dy, dz refer to a line of curvature at the point (x, y, z), by reason of (7), Art. 398, Vol. I. Hence it follows that if a geodesic touches a line of curvature, its torsion is suspended at the point of contact. And consequently if a line of curvature is a geodesic, it is a plane curve, because every point on it is a point of suspended torsion.

Also since there are two lines of curvature passing through every point and at right angles to each other, there are also at every point two geodesics whose torsion is suspended at that point.

340.] Let us further consider geodesics at a point in reference to the lines of curvature which pass through that point.

Let the numerator of (77) be replaced by its equivalent which is given in (24), Art. 399, Vol. I; then

$$\frac{1}{R} = \frac{U(K-L)\,dy\,dz + V(L-H)\,dz\,dx + W(H-K)\,dx\,dy}{Q^2 ds^2}. \qquad (78)$$

Let (l, m, n), (l_1, m_1, n_1), (l_2, m_2, n_2), be sets of direction-cosines referring to the geodesic and to the two lines of curvature at the point (x, y, z); let θ be the angle at which the geodesic in its first element is inclined to the first line of curvature; and let ρ_1 and ρ_2 be the principal radii of curvature at the point; then

$$\frac{1}{R} = \frac{U(K-L)\,mn + V(L-H)\,nl + W(H-K)\,lm}{Q^2}; \qquad (79)$$

$$\left.\begin{aligned} U(K-L)\,m_1 n_1 + V(L-H)\,n_1 l_1 + W(H-K)\,l_1 m_1 = 0, \\ U(K-L)\,m_2 n_2 + V(L-H)\,n_2 l_2 + W(H-K)\,l_2 m_2 = 0; \end{aligned}\right\} \qquad (80)$$

and by (41), Art. 403, Vol. I,

$$\left.\begin{aligned} l &= l_1 \cos\theta + l_2 \sin\theta, \\ m &= m_1 \cos\theta + m_2 \sin\theta, \\ n &= n_1 \cos\theta + n_2 \sin\theta; \end{aligned}\right\} \qquad (81)$$

therefore substituting these in (79), and omitting terms which vanish by reason of (80),

$$\frac{1}{R} = \frac{\sin\theta\cos\theta}{Q^2}\{U(K-L)(m_1 n_2 + n_1 m_2)$$
$$+ V(L-H)(n_1 l_2 + l_1 n_2) + W(H-K)(l_1 m_2 + m_1 l_2)\};$$

but since $\quad \text{U}l_1+\text{V}m_1+\text{W}n_1 = \text{U}l_2+\text{V}m_2+\text{W}n_2 = 0,$

and $\quad l_1 l_2+m_1 m_2+n_1 n_2 = 0;$

$$\therefore \quad \frac{m_1 n_2-n_1 m_2}{\text{U}} = \frac{n_1 l_2-l_1 n_2}{\text{V}} = \frac{l_1 m_2-m_1 l_2}{\text{W}} = \frac{1}{\text{Q}};$$

$$\therefore \frac{1}{\text{R}} = \frac{\sin\theta\cos\theta}{\text{Q}}\{(\text{K}-\text{L})(m_1^2 n_2^2-n_1^2 m_2^2)$$

$$+(\text{L}-\text{H})(n_1^2 l_2^2-l_1^2 n_2^2)+(\text{H}-\text{K})(l_1^2 m_2^2-m_1^2 l_2^2)\}$$

$$= \frac{\sin\theta\cos\theta}{\text{Q}}\{\text{H}(l_1^2-l_2^2)+\text{K}(m_1^2-m_2^2)+\text{L}(n_1^2-n_2^2)\}$$

$$= \sin\theta\cos\theta\left(\frac{1}{\rho_1}-\frac{1}{\rho_2}\right), \text{ by (20), Art. 399, Vol. I;} \qquad (82)$$

and this assigns the radius of torsion in terms of the principal radii of curvature, and of the angle between the first element of the geodesic and a line of curvature. Hence we have the following theorems.

The torsion is suspended when $\theta = 0$, and when $\theta = \frac{\pi}{2}$, that is when the geodesic touches a line of curvature.

The torsion is a maximum when $\theta = 45°$, and $\theta = 135°$; that is, when the geodesics bisect the angles between the lines of curvature. Thus if lines of maximum geodesic torsion are traced on a given surface, two such will pass through every point on the surface, and they will bisect the angles between the lines of curvature at that point.

The torsions of two geodesics which intersect at right-angles are equal at the point of intersection.

If $\rho_1 = \rho_2$, that is, if the point (x, y, z) is an umbilic, the torsion at that point of every geodesic passing through it is suspended. And conversely, if at a given point the torsion of every geodesic passing through it is suspended, that point is an umbilic.

The equation (82) may also be put into the following form. Let ρ and ρ' be the radii of curvature of the normal sections coincident with and perpendicular to the geodesic; then by Euler's Theorem, (45), Art. 403, Vol. I,

$$\frac{1}{\rho} = \frac{(\cos\theta)^2}{\rho_1}+\frac{(\sin\theta)^2}{\rho_2}, \quad \frac{1}{\rho'} = \frac{(\sin\theta)^2}{\rho_1}+\frac{(\cos\theta)^2}{\rho_2};$$

$$\therefore \frac{1}{\text{R}^2} = \frac{1}{\rho\rho'}-\frac{1}{\rho_1\rho_2}. \qquad (83)$$

341.] The preceding theory of geodesics leads to some theorems concerning lines drawn on a surface which deserve attention.

Let PQR ... be a curved line drawn on a surface, of which

PQ, QR, ... are equal elements, and each = to ds. Produce PQ to T, making QT = PQ = ds. From T draw TN at right-angles to the surface, and meeting the surface in N; join QN, TR, RN. Then the plane PQTN is evidently the normal plane to the surface at P, and PQ, QN are consecutive elements of the geodesic of which PQ is the first element; consequently this geodesic touches the curve PQR at P, and has contact of the second order at least with the curve made by the normal section at P. Now the line NR indicates the deviation of the curve PQR from the geodesic PQN which touches it at P; and for this reason the deviation is called geodesic curvature. And as in the fig. thus described TQR is the angle of absolute curvature of PQR, TQN is the angle of absolute curvature of the geodesic PQN, so is the angle RQN the angle of geodesic curvature of PQR. Let ρ and ρ' be the radii of absolute curvature of PQN and PQR respectively; then by the definitions of these we have $ds = \rho \angle \text{TQN}$; $ds = \rho' \angle \text{TQR}$. Analogously let us suppose $ds = \rho'' \angle \text{NQR}$; then ρ'' is called the radius of geodesic curvature*. Now these three infinitesimal angles form at Q the vertical angle of the tetrahedron, of which the three adjacent sides QT, QR, QN are approximately equal: hence the angles are proportional to the opposite sides; so that

$$\rho \times \text{TN} = \rho' \times \text{TR} = \rho'' \times \text{NR}. \tag{84}$$

Let ϕ = the angle which the osculating plane of PQR makes with the normal plane PQN; that is, let RTN = ϕ; then

$$\text{NT} = \text{RT} \cos\phi, \quad \text{RN} = \text{RT} \sin\phi;$$

consequently

$$\rho' = \rho \cos\phi; \quad \rho'' = \frac{\rho'}{\sin\phi}; \tag{85}$$

the former of which equations is Meunier's theorem; see Art. 414, Vol. I; the latter gives the value of the radius of geodesic curvature in terms of the radius of absolute curvature.

Since for a geodesic $\phi = 0$, the radius of geodesic curvature of a geodesic is infinite.

342.] The following are also theorems of considerable importance.

(1) Let O, fig. 49, be a point (x_0, y_0, z_0) on a given surface $F(x, y, z) = 0$; and from O let a series of geodesics OP, OQ, OR... be drawn infinitesimally near to each other; and from them let

* Liouville in the appendix to Monge, Application d'Analyse, &c., pp. 568, 574; and Bonnet, Journal de l'Ecole Polytechnique, Cah. XXXII.

equal lengths OP = OQ = OR, ... = s; be cut off; and through the extremities P, Q, R ... let the curve PQR ..., of which the current coordinates are ξ, η, ζ, be drawn; this curve, which is the locus of all points the geodesic distances of which from O are equal to s, is called a geodesic circle, whose radius is s, and whose centre is O. Similarly from the same centre O a series of concentric circles may be described. Now since for any given circle the radius is constant, and that radius = the distance from (x_0, y_0, z_0) to (ξ, η, ζ), = s, the variation of s between these two points vanishes; consequently the coordinates of these points must satisfy the integrated part of (65). As one of the limits, viz. (x_0, y_0, z_0), is a fixed point, there are no variations of its coordinates; but for the other limit, viz. (ξ, η, ζ), we have

$$\frac{dx}{ds}\delta\xi + \frac{dy}{ds}\delta\eta + \frac{dz}{ds}\delta\zeta = 0; \qquad (86)$$

in which $\delta\xi$, $\delta\eta$, $\delta\zeta$ are the projections on the coordinate axes of an arc-element of the geodesic circle, and $\frac{dx}{ds}, \frac{dy}{ds}, \frac{dz}{ds}$ are the direction-cosines of that element of a geodesic radius which meets the circle at (ξ, η, ζ); consequently these two elements are perpendicular to each other, and each of the radii is cut at right-angles by the geodesic circle; and thus the geodesic circle intersects orthogonally a series of geodesic radii originating at a given point.

This theorem may also be demonstrated by the geometrical process of infinitesimals. Let OP and OQ be two consecutive geodesics of equal length, and PQ be an element of the curve which joins the extremities of the series of similar geodesics of equal length. Then PQ is perpendicular to both OP and OQ. For if PQO is not a right-angle, let us suppose it to be greater than a right-angle. From Q let QT be drawn at right-angles to PQ and intersecting OP in T; then considering the triangle PQT, as PQ is an infinitesimal and PQT is a right angle, PT is greater than QT; so that adding OT to each, OT + TP is greater than OT + TQ; but OT + TP = OP = OQ; therefore OQ is greater than OT + TQ; that is OQ is not the shortest line from O to Q, as it is assumed to be. Consequently OQP is not an obtuse angle; in the same way it may be shewn that it is not an acute angle: and therefore it is a right-angle.

(2) Hence also conversely, if every geodesic through O to

points on the curve PQR ... cuts that curve at right angles, the lengths of the geodesics are equal.

(3) By a similar process it may be shewn that if from all points of a line drawn on a given surface geodesics of equal length are drawn perpendicular to that line, the locus of their extremities is a curve which is perpendicular to the geodesics. Lines on a surface thus related are called geodesic parallels; such are PQR, P′Q′R′, ... in fig. 48; the number of them is of course infinite.

(4) If from a point O on a surface two geodesic radii vectores OP, OQ are drawn to two points P and Q on a line on the surface, and infinitesimally near to each other, then, if $PQ = ds$, and $OPQ = \phi$,

$$OP - OQ = ds \cos\phi. \qquad (87)$$

As these theorems are fundamental in the demonstration of geometrical theorems concerning straight lines and planes, they are applicable to the proof of theorems concerning lines on surfaces in reference to geodesics, when the lines are of infinitesimal length and may be considered straight, and when surface-elements are considered which are approximately planes. The truth of the following theorems is evident from these principles.

(5) If two points are taken on a surface, and a curve is drawn on it such that the sum of the lengths of the two geodesics drawn from any point of it to the two given points is constant; the tangent at any point of the locus is equally inclined to the two geodesics. Such a curve is called a geodesic ellipse, the given points being the geodesic foci, and the two geodesics drawn from any point on the curve to the foci being the geodesic focal radii.

(6) If two points are taken on a surface, and a curve is drawn on it such that the difference of the geodesic lengths from any point on it to these two points is constant, the tangent at any point of the locus bisects the angle between these two geodesics. The locus thus defined is called a geodesic hyperbola.

(7) A series of geodesic confocal ellipses is intersected orthogonally by a series of geodesic confocal hyperbolas.

343.] The following are examples in which the preceding theory of geodesics is applied.

Ex. 1. Determine the geodesics on the sphere, $x^2+y^2+z^2=a^2$.

In this case from (66) we have

$$\frac{d.\frac{dx}{ds}}{x} = \frac{d.\frac{dy}{ds}}{y} = \frac{d.\frac{dz}{ds}}{z};$$

$$\therefore \quad yd.\frac{dz}{ds} - zd.\frac{dy}{ds} = 0;$$

whence by integration, if h_1 is an arbitrary constant,

similarly
$$\left.\begin{aligned} y\frac{dz}{ds} - z\frac{dy}{ds} &= h_1; \\ z\frac{dx}{ds} - x\frac{dz}{ds} &= h_2, \\ x\frac{dy}{ds} - y\frac{dx}{ds} &= h_3; \end{aligned}\right\} \qquad (88)$$

where h_2 and h_3 are also arbitrary constants:

$$\therefore \quad h_1 x + h_2 y + h_3 z = 0,$$

which is the equation to a plane passing through the origin, which is the centre of the sphere, and consequently intersects the sphere in a great circle. Hence a great circle is a geodesic on a sphere; and the geodesic joining two given points on a sphere is the arc of a great circle. As the great circle passing through two given points will be divided at these points into two arcs, of which one is greater and the other is less than a semicircle, so the points will be joined by two different geodesics of which one is a maximum, and the other is a minimum. This illustrates the remark in Art. 337, that if a geodesic joins two given points we cannot assert absolutely that the length is either a maximum or a minimum.

Ex. 2. Determine the geodesic joining two given points on a right circular cylinder.

Let the equation to the cylinder be $x^2 + y^2 = a^2$; and let the two points be $(a, 0, 0)$, $(a \cos \alpha, a \sin \alpha, c)$. Then from (66) we have

$$\frac{d.\frac{dx}{ds}}{x} = \frac{d.\frac{dy}{ds}}{y} = \frac{d.\frac{dz}{ds}}{0}; \qquad (89)$$

from the last of which $d.\frac{dz}{ds} = 0$; therefore

$$\frac{dz}{ds} = \text{a constant} = \sin \beta, \text{ say};$$

and this shews that the geodesic cuts all the generating lines of the cylinder at a constant angle; consequently the geodesic is the helix, the inclination of the line of which to the plane of $(x, y) = \beta$. The equations to it may be found as follows. From the first two of (89)

$$x\,d.\frac{dy}{ds} - y\,d.\frac{dx}{ds} = 0;$$

$$\therefore\quad x\frac{dy}{ds} - y\frac{dx}{ds} = h = a\cos\beta,$$

because when $x = a$, $y = 0$, $\frac{dy}{ds} = \cos\beta$. Therefore integrating again we have

$$z = s\sin\beta; \qquad \tan^{-1}\frac{z}{x} = \frac{s}{a}\cos\beta. \tag{90}$$

And if $x = a\cos\theta$, $y = a\sin\theta$, then $z = ak\theta$, if $k = \tan\beta$; and these are the equations to the helix. See Art. 347, Vol. I.

To determine the quantity β, we have, by the values of the superior limit, $c = a\alpha\tan\beta$;

$$\therefore\quad \beta = \tan^{-1}\frac{c}{a\alpha}.$$

And as the number of the values of β which satisfy this equation is infinite, so the number of geodesics which can be drawn joining the two points is also infinite. This is also evident from the geometry, inasmuch as the geodesic may be drawn from the first point round the cylinder any number of times before it falls into the second point; and the greater the number of times that it is drawn round the cylinder, the smaller is β, since $c = s\sin\beta$, where $s =$ the length of the geodesic.

344.] Determine the equations of the geodesic drawn on the surface of an ellipsoid.

Let the equation to the ellipsoid be

$$\frac{x^2}{a^2} + \frac{y^2}{b^2} + \frac{z^2}{c^2} = 1;$$

so that

$$\text{U} = \frac{2x}{a^2}, \qquad \text{V} = \frac{2y}{b^2}, \qquad \text{W} = \frac{2z}{c^2};$$

$$d\text{U} = \frac{2\,dx}{a^2}, \qquad d\text{V} = \frac{2\,dy}{b^2}, \qquad d\text{W} = \frac{2\,dz}{c^2};$$

and the equations to a geodesic which are given by (68) give after integration

$$\left(\frac{dx^2}{a^2} + \frac{dy^2}{b^2} + \frac{dz^2}{c^2}\right)\left(\frac{x^2}{a^4} + \frac{y^2}{b^4} + \frac{z^2}{c^4}\right) = \frac{ds^2}{k^2}; \tag{91}$$

where k is a constant introduced in integration, and may be expressed in terms of the coordinates of the initial point, and of the initial direction of the geodesic. This equation does not admit of further integration, but gives the following geometrical theorem. Let p be the length of the perpendicular from the centre of the ellipsoid on the tangent plane at (x, y, z); then

$$\frac{1}{p^2} = \frac{x^2}{a^4} + \frac{y^2}{b^4} + \frac{z^2}{c^4}; \tag{92}$$

and let d be the central radius vector of the ellipsoid which is parallel to the tangent of the geodesic at (x, y, z); then

$$\frac{1}{d^2} = \frac{1}{a^2}\frac{dx^2}{ds^2} + \frac{1}{b^2}\frac{dy^2}{ds^2} + \frac{1}{c^2}\frac{dz^2}{ds^2}; \tag{93}$$

so that from (91) we have $\qquad pd = k^2$*; $\qquad$ (94)

and thus for all points along the same geodesic, the rectangle contained by the perpendicular from the centre on the tangent plane and the central radius parallel to the tangent of the geodesic are constant.

345.] The following are deductions from Joachimsthal's theorem.

(1) For all geodesics passing through an umbilic pd has the same value. At the umbilic p is the same for all geodesics, and $p = \frac{ac}{b}$; and as the central plane section parallel to the tangent plane at the umbilic is a circle whose radius $= b$, d is the same for all geodesics at the umbilic, and $= b$. Consequently for all geodesics passing through an umbilic, $pd = k^2 = ac$.

(2) A geodesic ellipse on an ellipsoid, of which two adjacent umbilics are the geodesic foci, is a line of curvature.

Let U, V be two adjacent umbilics, and let P be a point on a geodesic ellipse, of which U and V are the geodesic foci; let ds be an arc-element of the geodesic ellipse at P; then by (5), Art. 342, PU and PV are equally inclined to ds. Let $p_1 d_1$, $p_2 d_2$ refer to PU and PV respectively; then, by the preceding theorem, $p_1 d_1 = p_2 d_2$: and $p_1 = p_2$, as P is common to the two geodesics; consequently $d_1 = d_2$; and thus these radii are equally inclined to the principal axes of the central plane section which is parallel to the

* This theorem was first given by Joachimsthal, Crelle, Vol. XXVI, p. 158, and is now known as Joachimsthal's theorem. An elegant proof, by the geometrical infinitesimal method, is given by Prof. Charles Graves, of Dublin, Crelle, Vol. XLII, p. 279.

tangent plane at P. But d_1, d_2 are parallel to the tangents of PU and PV at the point P; and therefore ds is parallel to a principal axis of the central plane section. But the tangent of a line of curvature on an ellipsoid is always parallel to a principal axis of the plane central section which is parallel to its tangent plane; consequently ds is an arc-element of a line of curvature; and thus the geodesic ellipse is a line of curvature.

If ds' is an arc-element of the geodesic hyperbola of which U and V are the foci, and which intersects the geodesic ellipse at P, then by (7), Art. 342, ds' is perpendicular to ds, and as the two lines of curvature at P are perpendicular to each other, ds' is an arc-element of a line of curvature; consequently a geodesic hyperbola, of which the umbilics are foci, is a line of curvature.

(3) All geodesics which connect two opposite umbilics are of equal length.

Let U′, V′ be the umbilics opposite to U and V respectively. Let P be a point in which the geodesic ellipse whose foci are U and V intersects a geodesic hyperbola whose foci are V and U′; then

$$\text{PU}+\text{PV} = \text{a constant}; \quad \text{PU}'-\text{PV} = \text{a constant};$$

$$\therefore \text{ by addition} \qquad \text{PU}+\text{PU}' = \text{UPU}' = \text{a constant}.$$

And by the properties of geodesics proved in (5) and (6), Art. 342, it is evident that the angles made by PU and PU′ with a line of curvature at P are supplemental to each other; so that PU and PU′ are consecutive parts of the same geodesic. Consequently the lengths of all geodesics which join two opposite umbilics are equal.

There are many other theorems of geodesics on an ellipsoid relative to lines of curvature which are important and deserve investigation; but the method of elliptical coordinates devised by Lamè and the whole theory of confocal quadric surfaces are so well adapted to the demonstration of them, that it is not worth while to discuss them by other processes; and a sufficient and satisfactory exposition of those systems requires more space than can be allotted within the limits of the present work; I must consequently refer the student to the treatises which are especially devoted to these subjects.

346.] Again, let ρ = the radius of absolute curvature of a geodesic on an ellipsoid; then from (69) we have, if s is equicrescent,

$$\frac{d^2x}{\mathrm{U}} = \frac{d^2y}{\mathrm{V}} = \frac{d^2z}{\mathrm{W}} = \pm\frac{ds^2}{\rho\mathrm{Q}}$$

$$= \frac{\mathrm{U}\,d^2x + \mathrm{V}\,d^2y + \mathrm{W}\,d^2z}{\mathrm{Q}^2}$$

$$= -\frac{d\mathrm{U}\,dx + d\mathrm{V}\,dy + d\mathrm{W}\,dz}{\mathrm{Q}^2}$$

$$= -2\,\frac{\dfrac{dx^2}{a^2} + \dfrac{dy^2}{b^2} + \dfrac{dz^2}{c^2}}{\mathrm{Q}^2};$$

$$\therefore\quad \left(\frac{x^2}{a^4} + \frac{y^2}{b^4} + \frac{z^2}{c^4}\right)^{\frac{1}{2}} = \rho\left\{\frac{1}{a^2}\frac{dx^2}{ds^2} + \frac{1}{b^2}\frac{dy^2}{ds^2} + \frac{1}{c^2}\frac{dz^2}{ds^2}\right\};$$

and consequently by (92) and (93),

$$d^2 = \rho p; \tag{95}$$

that is, of geodesics which pass through the same point, the radii of absolute curvature are proportional to the squares of the central radii which are parallel to their tangents. Hence also

$$\rho = \frac{d^2}{p} = \frac{k^4}{p^3}; \tag{96}$$

that is, along the same geodesic the radius of absolute curvature but since E, F and G are functions of ξ and η,

$$\delta E = \left(\frac{dE}{d\xi}\right)\delta\xi + \left(\frac{dE}{d\eta}\right)\delta\eta;$$

and similar values are true for δF and δG; so that substituting in (97), and equating to zero the coefficients of $\delta\xi$ and $\delta\eta$ which are under the sign of integration, and bearing in mind equations (98) and (100), Art. 247, we have

$$\left.\begin{aligned} d\{\cos\theta\, E^{\frac{1}{2}}\} - \frac{d\xi^2}{2\,ds}\left(\frac{dE}{d\xi}\right) - \frac{d\xi\,d\eta}{ds}\left(\frac{dF}{d\xi}\right) - \frac{d\eta^2}{2\,ds}\left(\frac{dG}{d\xi}\right) = 0,\\ d\{\cos\theta' G^{\frac{1}{2}}\} - \frac{d\eta^2}{2\,ds}\left(\frac{dG}{d\eta}\right) - \frac{d\eta\,d\xi}{ds}\left(\frac{dF}{d\eta}\right) - \frac{d\xi^2}{2\,ds}\left(\frac{dE}{d\eta}\right) = 0;\end{aligned}\right\} \quad (98)$$

from which we have

$$\left.\begin{aligned} 2\nabla\, d\theta &= \frac{F}{E}\,dE + \left(\frac{dE}{d\eta}\right)d\xi - 2\left(\frac{dF}{d\xi}\right)d\xi - \left(\frac{dG}{d\xi}\right)d\eta,\\ 2\nabla\, d\theta' &= \frac{F}{G}\,dG + \left(\frac{dG}{d\xi}\right)d\eta - 2\left(\frac{dF}{d\eta}\right)d\eta - \left(\frac{dE}{d\eta}\right)d\xi;\end{aligned}\right\} \quad (99)$$

also from Art. 247, we have

$$\cot\theta = \frac{E}{\nabla}\frac{d\xi}{d\eta} + \frac{F}{\nabla}, \qquad \cot\theta' = \frac{G}{\nabla}\frac{d\eta}{d\xi} + \frac{F}{\nabla};$$

by means of which equations θ and θ' may be eliminated from (99), and thereby a differential equation found which will be that to the geodesics on the surface.

348.] The preceding formulæ are much simplified if ξ and η are so assumed that the angle at which the lines of one system intersect those of the other system is 90°; in which case, see equation (93), Art. 247, $\cos\omega = 0$, and therefore $F = 0$: and the equations to the geodesics become

$$2(EG)^{\frac{1}{2}}\,d\theta = \left(\frac{dE}{d\eta}\right)d\xi - \left(\frac{dG}{d\xi}\right)d\eta, \qquad \cot\theta = \left(\frac{E}{G}\right)^{\frac{1}{2}}\frac{d\xi}{d\eta};$$

the equations in terms of θ' being identical with these.

This also admits of further simplification: let us suppose the systems of lines, ξ and η, to be geodesic: then if $\eta =$ a constant, $\theta = 0$, and consequently $\frac{dE}{d\eta} = 0$; and E either is a constant or is independent of η; so that

$$2(EG)^{\frac{1}{2}}\,d\theta = -\left(\frac{dG}{d\xi}\right)d\eta, \qquad \cot\theta = \left(\frac{E}{G}\right)^{\frac{1}{2}}\frac{d\xi}{d\eta}; \qquad (100)$$

by means of which θ may be eliminated, and the resulting differential equation will be that to the geodesic lines.

And to take the simplest case of all: let a series of geodesic lines of equal length originate at the point O; and let it be referred to a system of geodesic polar coordinates, analogous to that of plane polar coordinates, and such as we have described in Art. 342; but to avoid the inconvenience of new symbols, let ξ be the geodesic radial-distance of any point from O, and let η be the angle between the first elements of ξ and of an originating prime radius which abut at O; then, by Art. 342, the condition of orthogonality is satisfied, and by (90), Art. 247, since $d\sigma = d\xi$, $E = 1$: hence (100) become

$$2G^{\frac{1}{2}}\, d\theta = -\left(\frac{dG}{d\xi}\right) d\eta, \qquad \cot\theta = \frac{1}{G^{\frac{1}{2}}} \frac{d\xi}{d\eta}.$$

To simplify these, let $G^{\frac{1}{2}} = m$;

$$\therefore\; d\theta = -\left(\frac{dm}{d\xi}\right) d\eta;$$

and eliminating θ, we have

$$m^2\left(\frac{dm}{d\xi}\right) + 2\left(\frac{dm}{d\xi}\right)\frac{d\xi^2}{d\eta^2} + \left(\frac{dm}{d\eta}\right)\frac{d\xi}{d\eta} - m\frac{d}{d\eta}\frac{d\xi}{d\eta} = 0; \quad (101)$$

which is the differential equation to the geodesic lines, but does not generally admit of integration.

m is generally a function of both ξ and η; and $m\,d\eta$, by reason of equation (91), Art. 247, is the element of a line of the second system; but if all the lines of the first system originate at a common point O, η may be assumed so that it $= 0$, when $\xi = 0$; and taking η to be the angle between the first elements of the originating geodesic, and of any other geodesic corresponding to ξ, the element of a line of the second system may be considered as the arc of an infinitesimal circle when ξ is infinitesimal, and equal to $\xi\, d\eta$: therefore for an infinitesimal value of ξ, $\xi = m$, and $\frac{dm}{d\xi} = 1$.

A further inquiry into the subject of geodesic lines from this point of view is beyond the scope of our work; but it has important applications in the determination of curvature of surfaces, according to the principles of the system invented by Gauss, and explained in his memoir, "Disquisitiones generales circa superficies curvas;" and for these I must refer the reader to that work. There is also much information on the same subject in the notes appended by M. Liouville to his edition of Monge's Analyse appliquée, &c., Paris 1852.

SECTION 3.—*Investigation of the critical values of a definite integral, whose element-function involves derived-functions.*

349.] In all the preceding problems of maxima and minima, the differentials contained in the element have been taken in their most general forms; no supposition has been made as to one or more being equicrescent or as subject to no variation, and they have not been put in the forms of derived-functions; and the solutions, it will be observed, have been deduced from first principles, and without the intervention of general formulæ: the results arrived at are left in their symmetrical forms; and hereby have we been able to infer geometrical properties, which are frequently the only available definition of the function which satisfies the critical property that is required. Now for elegance and symmetry nothing more can be desired: but we have not investigated any critical function whose element contains differentials above the second order; the simplest cases only have been considered, and a slight inspection of the general results of Art. 316 will shew that the complexity of the formulæ rapidly increases if higher differentials enter into the calculation: in this latter case then, it is desirable to simplify the formulæ as far as is possible, ere they become the subjects of inquiry; and as such a simplification is obtained by making one of the variables equicrescent, and by using derived-functions instead of differentials, although it is with the loss of symmetry, it is necessary to consider the conditions under which a definite integral, whose element-function involves derived-functions of different orders, may have a critical value. And there is also another reason why the subject must be investigated from this point of view: it is only when the element-function is of this form that criteria for discriminating maxima and minima have been constructed. We proceed then at once to the investigation.

Let the definite integral, whose maximum or minimum is to be determined, be

$$u = \int_0^1 \mathrm{v}\, dx;$$

where

$$\mathrm{v} = f(x, y, y', y'', \dots y^{(n)}), \tag{102}$$

using the notation of derived functions.

Now the variation of u, where v is of the given form, has already been investigated in Art. 303; and for an abridgement of notation of the results of that Article which are given in equation (56) of it, let

$$\text{V}\,\delta x + \left\{\text{Y}' - \frac{d\text{Y}''}{dx} + \frac{d^2\text{Y}'''}{dx^2} - \ldots (-)^{n-1}\frac{d^{n-1}\text{Y}^{(n)}}{dx^{n-1}}\right\}\omega$$
$$+\left\{\text{Y}'' - \frac{d\text{Y}'''}{dx} + \ldots (-)^{n-2}\frac{d^{n-2}\text{Y}^{(n)}}{dx^{n-2}}\right\}\omega'$$
$$+ \cdot \cdot \cdot \cdot \cdot \cdot \cdot \cdot \cdot \cdot \cdot \cdot \cdot$$
$$+\left\{\text{Y}^{(n-1)} - \frac{d\text{Y}^{(n)}}{dx}\right\}\omega^{(n-2)}$$
$$+\text{Y}^{(n)}\,\omega^{(n-1)} = a; \tag{103}$$

$$\text{Y} - \frac{d\text{Y}'}{dx} + \frac{d^2\text{Y}''}{dx^2} - \ldots (-)^n \frac{d^n\text{Y}^{(n)}}{dx^n} = \text{H}; \tag{104}$$

so that equation (56), Art. 303, becomes

$$\delta u = \Big[a\Big]_0^1 + \int_0^1 \text{H}\,\omega\, dx; \tag{105}$$

and as u is a maximum or minimum, $\delta u = 0$; and to satisfy this condition it is manifest that

$$\Big[a\Big]_0^1 = 0; \qquad \text{H} = 0; \tag{106}$$

of which expressions the former depends on the values of certain variables and their derived functions at the limits; the latter by integration gives the general functional relation, and thereby the form whence the required critical value may be found.

Now since V contains $y^{(n)}$ or $\frac{d^n y}{dx^n}$, H, which contains $\frac{d^n\text{Y}^{(n)}}{dx^n}$, will generally contain $\frac{d^{2n}y}{dx^{2n}}$, and therefore will be a differential equation of the $(2n)$th order: the solution of this equation will therefore contain $2n$ arbitrary constants; and the determination of these depends on the values which $a = 0$ assumes at its limiting values; the process however of finding these, being similar to that explained in Art. 318, it is unnecessary to repeat; but it is desirable to investigate one or two cases in which the equation $\text{H} = 0$ assumes particular forms analogous to those of Art. 319, and thereby admits of immediate integration.

350.] First, suppose V not to contain the first m of the quantities $y, y', y'', \ldots$; then the equation $\text{H} = 0$ becomes

$$\frac{d^m\text{Y}^{(m)}}{dx^m} - \frac{d^{m+1}\text{Y}^{(m+1)}}{dx^{m+1}} + \ldots = 0;$$

and this admits of m successive integrations; and thus we have

$$\text{Y}^{(m)}-\frac{d.\text{Y}^{(m+1)}}{dx}+\ldots = c_0+c_1 x+c_2 x^2+\ldots+c_{m-1}x^{m-1}. \quad (107)$$

Secondly, suppose V not to contain x: then

$$d\text{V} = \text{Y}\,dy+\text{Y}'\,dy'+\text{Y}''\,dy''+\ldots+\text{Y}^{(n)}\,dy^{(n)};$$

also

$$0 = \text{Y}-\frac{d\text{Y}'}{dx}+\frac{d^2\text{Y}''}{dx^2}-\ldots(-)^n\frac{d^n\text{Y}^{(n)}}{dx^n};$$

$$\therefore\quad d\text{V} = \text{Y}'\,d\text{Y}'+\frac{d\text{Y}'}{dx}dy+\text{Y}''\,dy''-\frac{d^2\text{Y}''}{dx^2}dy+\ldots$$

$$= d(\text{Y}'y')+d\Big(\text{Y}''y''-\frac{d\text{Y}''}{dx}y'\Big)+\ldots;$$

$$c+\text{Y}'y'+\Big\{\text{Y}''y''-\frac{d\text{Y}''}{dx}y'\Big\}+\Big\{\text{Y}'''y'''-\frac{d\text{Y}'''}{dx}y''+\frac{d^2y'''}{dx^2}y'\Big\}+\ldots; \quad (108)$$

which is a differential equation of an order not higher than $2n-1$; and therefore whenever V does not contain x, the equation H $= 0$ always admits of being integrated at least once.

Thirdly, let V $= f(y')$; then, by the preceding equation,

$$\text{V} = c+\text{Y}'y';$$

but as V and Y′ contain y' only, this may be put into the form

$$y' = \frac{dy}{dx} = \text{F}(c);$$

$$\therefore\quad y = \text{F}(c)x+c_1; \quad (109)$$

and thence we infer that a linear function, as (109), is such that the variation of any function of $\frac{dy}{dx}$ deduced from it vanishes.

Lastly, if

$$\text{V} = f\Big(y, \frac{dy}{dx}\Big),$$

$$\text{V} = c+\text{Y}'y'. \quad (110)$$

351.] Let us apply the preceding method to the solution of one or two problems.

Ex. 1. Let

$$u = \int_0^1 y^n\frac{y''}{y'}\,dx.$$

Here V $= y^n\frac{y''}{y'}$; and therefore equation (108) is applicable.

$$\text{Y} = \Big(\frac{d\text{V}}{dy}\Big) = n\,y^{n-1}\frac{y''}{y'}; \qquad \text{Y}' = \Big(\frac{d\text{V}}{dy'}\Big) = -y^n\frac{y''}{y'^2};$$

$$\text{Y}'' = \Big(\frac{d\text{V}}{dy''}\Big) = \frac{y^n}{y'}; \qquad \therefore\quad \frac{d\text{Y}''}{dx} = \frac{n\,y^{n-1}y'^2-y^n\,y''}{y'^2}.$$

Therefore (108) becomes

$$y^n \frac{y''}{y'} = c - \frac{y^n y''}{y'} + \frac{y^n y''}{y'} - \frac{n y^{n-1} y'^2 - y^n y''}{y'};$$

$$\therefore \quad c = n y^{n-1} y';$$

whence by integration, $y^n = cx + c'$.

Ex. 2. It is required to determine the shortest line between two given points.

$$u = \int_0^1 \{1 + y'^2\}^{\frac{1}{2}}\, dx.$$

Here also, as V does not involve x, (108) is applicable; and as

$$\text{V} = \{1+y'^2\}^{\frac{1}{2}}; \qquad \text{Y}' = \left(\frac{d\text{V}}{dy'}\right) = \frac{y'}{\{1+y'^2\}^{\frac{1}{2}}};$$

so that (108) becomes

$$\{1+y'^2\}^{\frac{1}{2}} = c + \frac{y'^2}{\{1+y'^2\}^{\frac{1}{2}}};$$

$$\therefore \quad y' = \frac{dy}{dx} = \frac{(1-c^2)^{\frac{1}{2}}}{c};$$

$$y = \frac{(1-c^2)^{\frac{1}{2}}}{c} x + c';$$

which is the equation to a straight line; and we have the same result as that arrived at in Art. 322.

Ex. 3. Let the element-function be of the form given in Art. 308; viz. $\text{V} = f(x, y, y', y'', \ldots z, z', z'', z''', \ldots)$; and let us also take a simple case, and suppose that

$$u = \int_0^1 \{1 + y'^2 + z'^2\}^{\frac{1}{2}}\, dx;$$

then, by reasoning similar to that which has been frequently employed, both the terms under the integral signs in equation (69), Art. 308, must vanish: and therefore, as V involves only y' and z', we have

$$\text{Y}' = \frac{y'}{\{1+y'^2+z'^2\}^{\frac{1}{2}}}; \qquad \text{Z}' = \frac{z'}{\{1+y'^2+z'^2\}^{\frac{1}{2}}};$$

and therefore, since $\dfrac{d\text{Y}'}{dx} = 0$; $\quad \dfrac{d\text{Z}'}{dx} = 0$;

$$\frac{y'}{\{1+y'^2+z'^2\}^{\frac{1}{2}}} = c_1, \qquad \frac{z'}{\{1+y'^2+z'^2\}^{\frac{1}{2}}} = c_2;$$

$$\therefore \quad \frac{dy}{dx} = c_1, \quad y = c_1 x + c_1'; \qquad \frac{dz}{dx} = c_2, \quad z = c_2 x + c_2';$$

which are manifestly the equations to a straight line in space. A straight line therefore is the shortest distance.

Ex. 4. To find the plane curve of given length enclosing the greatest area between itself, two given extreme ordinates, and the axis of x.

This is a problem of relative critical value. Let λ be a constant multiplier; then,

$$\text{the area} = \int_0^1 \int_0^y dy\, dx = \int_0^1 y\, dx;$$

$$\text{and the length} = \int_0^1 \{1+y'^2\}^{\frac{1}{2}} dx;$$

$$\therefore \quad u = \int_0^1 \{y+\lambda(1+y'^2)^{\frac{1}{2}}\} dx;$$

so that

$$\mathrm{v} = y+\lambda\{1+y'^2\}^{\frac{1}{2}};$$

and because v does not involve x, (108) is applicable; and

$$\mathrm{Y}' = \left(\frac{d\mathrm{v}}{dy'}\right) = \frac{\lambda y'}{\{1+y'^2\}^{\frac{1}{2}}};$$

$$\therefore \quad y+\lambda\{1+y'^2\}^{\frac{1}{2}} = c_1 + \frac{\lambda y'^2}{\{1+y'^2\}^{\frac{1}{2}}};$$

whence

$$(y-c_1)^2+(x-c_2)^2 = \lambda^2,$$

which is the equation to a circle, whose radius is equal to λ; and λ may be expressed in terms of the known length of the curve by a process similar to that of Art. 330.

A comparison of the two methods by which problems have been solved plainly shews that, although the former immediately involves first principles and from them is directly deduced; yet, as the results assume complicated forms when all the differentials are retained, it is convenient to make one of the variables equicrescent, and to express the element-function in terms of derived-functions, and then to apply the process of these latter articles.

SECTION 4.—*The discriminating conditions of Maxima and Minima.*

352.] The process which has been developed in the preceding articles of this chapter, and which has been applied to the solution of problems involving maxima and minima of definite integrals, although necessary, is yet insufficient for the object proposed, because no discriminating conditions of maxima and minima have been investigated. For the existence indeed of such critical values it is necessary that the first variation should vanish; but at the same time such vanishing is consistent with the definite integral being either a maximum or a minimum or a constant, and with being none of these: the truth of this statement is evident from the ordinary theory of maxima and minima. For a critical value it is necessary that the first variation of the definite integral should not only vanish, but also change its sign: and I know of no process immediately applicable by which to determine whether a function deduced from the differential equation $H = 0$, see Art. 349, and involving $2n$ arbitrary constants, will or will not cause the required change of sign of δu. In accordance then with the theory explained in Art. 149, Vol. I, we are obliged to have recourse to the second variation of the definite integral, with the object of determining its sign, and hereby to obtain the discriminating condition; so that when $\delta u = 0$, and if $\delta^2 u$ does not vanish, and does not become infinite or discontinuous, and does not change its sign within the limits of integration, u is a maximum or minimum according as $\delta^2 u$ is negative or positive. We proceed to the further development of these conditions.

But, to narrow the investigation as far as possible, I will take the case which has last been considered; that, namely, in which the infinitesimal element-function involves x, y and the derived functions of y, and in which also x is not only equicrescent but undergoes no variation; that is, δx is not one of the subjects of calculation, but the variation is due to a variation of y only: or, geometrically viewed, the displacement of the point on the curve is in a direction parallel to the axis of y only: for it is to this case that Jacobi*, the discoverer of the criteria, has confined himself. And first let the object of the research be clearly understood.

If the infinitesimal element-function contains a derived function of the nth order, the differential equation $H = 0$ will generally be of the $(2n)$th order, and therefore the value of y deduced from it is of the form

$$y = f(x, c_1, c_2, \ldots c_{2n}), \tag{111}$$

and contains $2n$ arbitrary constants which have been introduced

* Zur theorie der Variations-Rechnung und der Differential-Gleichungen, von C. G. J. Jacobi; Crelle, vol. XVII, p. 68.

in the process of integration: and therefore, if u is the given definite integral, it is plain that, after the substitution of y by means of the above equation, u will depend partly on the *form* of the function f, and partly on the arbitrary constants. It may seem then that the critical value of u will depend on both these quantities: as to the constants, however, it has been shewn that all their values may be determined by means of the given limiting values of the variables and of the derived functions; and hence, as these are determinate constants, the value of the definite integral cannot be made critical by any change of them: and even more than this, did u depend on such quantities it would become an integral (not differential) function of many variables, and would have its critical value determined by the ordinary rules of the differential calculus.

It is then the other question which we have to discuss; namely, whether the *form* of the function deduced from the equation $\text{H} = 0$ is such as to render the definite integral a maximum or minimum. For this purpose we must, as in Art. 310, calculate $\delta^2 u$, and determine its sign, subject to the conditions that when $\delta u = 0$,

(1) $\delta^2 u$ has the same sign for all values of the variables and their derived functions between the limits;

(2) $\delta^2 u$ does not become infinite for any values between the limits;

(3) $\delta^2 u$ does not vanish: for if so, we must, in accordance with the theory of maxima and minima, proceed to the investigation of $\delta^3 u$ and $\delta^4 u$, and so on; a work beyond our present purpose.

Let the definite integral, which is the subject of inquiry, be

$$u = \int_0^1 \text{V}\, dx;$$

where $\text{V} = f(x, y, y', y'', \dots y^{(n)})$;

then, by equation (105), since $\omega = \delta y$, because $\delta x = 0$,

$$\delta u = \Big[a\Big]_0^1 + \int_0^1 \text{H}\, \delta y\, \delta x;$$

and as $\text{H} = 0$,

$$\delta^2 u = \int_0^1 \delta \text{H}\, \delta y\, dx; \tag{112}$$

the sign of which is to be determined; and we have moreover to examine whether it can change its sign or not.

352.] For this purpose it is necessary that $\delta^2 u$ should not vanish when $\delta u = 0$; that is, δH must not vanish, when $H = 0$. Now the complete integral of $H = 0$ is

$$y = f(x, c_1, c_2, \dots c_{2n}), \tag{113}$$

the right-hand member of which contains $2n$ arbitrary constants. Let each of these constants vary, and let δy be the consequent variation of y; so that

$$\delta y = \left(\frac{dy}{dc_1}\right)\delta c_1 + \left(\frac{dy}{dc_2}\right)\delta c_2 + \dots + \left(\frac{dy}{dc_{2n}}\right)\delta c_{2n}; \tag{114}$$

and H becomes $H + \delta H$: then, as the varied value of y differs from the original value only in the arbitrary constants, it must also satisfy $H + \delta H$; and as the variations of the arbitrary constants are arbitrary, we may replace them by new constants $C_1, C_2, \dots C_{2n}$, so that the equation $\delta H = 0$ becomes satisfied by

$$\delta y = C_1\left(\frac{dy}{dc_1}\right) + C_2\left(\frac{dy}{dc_2}\right) + \dots + C_{2n}\left(\frac{dy}{dc_{2n}}\right); \tag{115}$$

and as y contains $2n$ arbitrary constants, so will δy also contain them: but $\delta H = 0$ cannot involve derived functions of an order higher than the $2n$th; and therefore the above value of δy is the *complete* integral of the equation $\delta H = 0$; consequently, if δy is of the form given in (115), H and δH vanish simultaneously; and u cannot thus far have a critical value. Hence, if $H = 0$ is satisfied by an equation of the form (113), the first thing to be done in the examination of the character of that result is to inquire whether δy satisfies the equation (115); if so, for that value of y, u is neither a maximum nor a minimum, and it is unnecessary to pursue the inquiry.

353.] If however the form of y given in (113), and which satisfies $H=0$, is not such as to satisfy (115), we must return to (112), and examine the sign of δH.

Now $H = \left(\frac{dV}{dy}\right) - \frac{d}{dx}\left(\frac{dV}{dy'}\right) + \frac{d^2}{dx^2}\left(\frac{dV}{dy''}\right) - \dots (-)^n \frac{d^n}{dx^n}\left(\frac{dV}{dy^{(n)}}\right)$;

$\therefore \quad \delta H = \frac{d}{dy}\delta V - \frac{d}{dx}\frac{d}{dy'}\delta V + \frac{d^2}{dx^2}\frac{d}{dy''}\delta V - \dots (-)^n \frac{d^n}{dx^n}\frac{d}{dy^{(n)}}\delta V$;

also $\delta V = \left(\frac{dV}{dy}\right)\delta y + \left(\frac{dV}{dy'}\right)\delta y' + \left(\frac{dV}{dy''}\right)\delta y'' + \dots + \left(\frac{dV}{dy^{(n)}}\right)\delta y^{(n)}$;

$$\delta \mathrm{H} = \left(\frac{d^2 \mathrm{V}}{dy^2}\right) \delta y + \left(\frac{d^2 \mathrm{V}}{dy\, dy'}\right) \delta y' + \left(\frac{d^2 \mathrm{V}}{dy\, dy''}\right) \delta y'' + \ldots + \left(\frac{d^2 \mathrm{V}}{dy\, dy^{(n)}}\right) \delta y^{(n)}$$
$$- \frac{d}{dx} \left\{ \left(\frac{d^2 \mathrm{V}}{dy\, dy'}\right) \delta y + \left(\frac{d^2 \mathrm{V}}{dy'^2}\right) \delta y' + \ldots\ldots + \left(\frac{d^2 \mathrm{V}}{dy^{(n)} dy'}\right) \delta y^{(n)} \right\}$$
$$+ \frac{d^2}{dx^2} \left\{ \left(\frac{d^2 \mathrm{V}}{dy\, dy''}\right) \delta y + \left(\frac{d^2 \mathrm{V}}{dy'\, dy''}\right) \delta y' + \ldots + \left(\frac{d^2 \mathrm{V}}{dy^{(n)} dy''}\right) \delta y^{(n)} \right\}$$
$$\cdots\cdots\cdots\cdots$$
$$(-)^n \frac{d^n}{dx^n} \left\{ \left(\frac{d^2 \mathrm{V}}{dy\, dy^{(n)}}\right) \delta y + \left(\frac{d^2 \mathrm{V}}{dy'\, dy^{(n)}}\right) \delta y' + \cdots + \left(\frac{d^2 \mathrm{V}}{dy^{(n)2}}\right) \delta y^{(n)} \right\}.$$

Since $$\delta y^{(k)} = \delta \frac{d^k y}{dx^k} = \frac{d^k \delta y}{dx^k},$$

it appears that the preceding expression for $\delta \mathrm{H}$ contains terms of the form
$$\frac{d^r}{dx^r} \left\{ \frac{d^2 \mathrm{V}}{dy^{(r)2}} \frac{d^r . \delta y}{dx^r} \right\}; \tag{116}$$

wherein the order of the derived of δy is the same as the index of $\frac{d}{dx}$ which affects the whole of its subordinate subject; and it appears also that the other terms may be grouped in pairs of the form,
$$\frac{d^r}{dx^r} \left(\mathrm{C} \frac{d^s . \delta y}{dx^s}\right) (-)^{s-r} \frac{d^s}{dx^s} \left(\mathrm{C} \frac{d^r . \delta y}{dx^r}\right); \tag{117}$$

where $$\mathrm{C} = \left(\frac{d^2 \mathrm{V}}{dy^{(r)} dy^{(s)}}\right); \tag{118}$$

the connecting sign in (117) being + or −, according as $s-r$ is even or odd.

Now all the terms of which such a series as (117) is composed can be put into the form (116); and consequently $\delta \mathrm{H}$ admits of being expressed in a series of terms, the type of each one of which will be
$$\frac{d^k}{dx^k} \left\{ \mathrm{A}_k \frac{d^k . \delta y}{dx^k} \right\},$$

where A_k is a determinable function of x.

By the theorem given in (57), Art. 426, Vol. I, if P and Q are two functions of x, whose derived functions are denoted by $\mathrm{P}', \mathrm{P}'', \ldots \mathrm{P}^{(n)}$, $\mathrm{Q}', \mathrm{Q}'', \ldots \mathrm{Q}^{(n)}$,

$$\mathrm{P} \frac{d^n \mathrm{Q}}{dx^n} = \frac{d^n . \mathrm{PQ}}{dx^n} - \frac{n}{1} \frac{d^{n-1} . \mathrm{P}'\mathrm{Q}}{dx^{n-1}} + \frac{n(n-1)}{1.2} \frac{d^{n-2} . \mathrm{P}''\mathrm{Q}}{dx^{n-2}} - \ldots$$
$$\ldots (-)^{n-1} \frac{n}{1} \frac{d . \mathrm{P}^{(n-1)} \mathrm{Q}}{dx} (-)^n \mathrm{P}^{(n)} \mathrm{Q}; \tag{119}$$

which theorem we shall apply to the subordinate subjects of differentiation in (117), when some convenient substitutions have been made.

Let $\delta y = \eta$; so that employing the ordinary notation of derived-functions

$$\frac{d^k.\delta y}{dx^k} = \eta^{(k)};$$

also suppose that $s = \sigma+\rho$, $r = \sigma-\rho$; whence

$$s+r = 2\sigma, \qquad s-r = 2\rho;$$

that is, we suppose first that $s-r$ is an even number, and therefore that the pair of values in (177) is connected with a positive sign; then, by reason of (119),

$$\text{c}\frac{d^s.\delta y}{dx^s} = \text{c}\frac{d^{\sigma+\rho}\eta}{dx^{\sigma+\rho}} = \text{c}\frac{d^\rho\eta^{(\sigma)}}{dx^\rho}$$

$$= \frac{d^\rho.\text{c}\eta^{(\sigma)}}{dx^\rho} - \frac{\rho}{1}\frac{d^{\rho-1}.\text{c}'\eta^{(\sigma)}}{dx^{\rho-1}} + \frac{\rho(\rho-1)}{1.2}\frac{d^{\rho-2}.\text{c}''\eta^{(\sigma)}}{dx^{\rho-2}} - \ldots(-)^\rho\text{c}^{(\rho)}\eta^{(\sigma)};$$

$$\therefore \quad \frac{d^r}{dx^r}\left\{\text{c}\frac{d^s.\delta y}{dx^s}\right\}$$

$$= \frac{d^{\sigma-\rho}}{dx^{\sigma-\rho}}\left\{\frac{d^\rho.\text{c}\eta^{(\sigma)}}{dx^\rho} - \frac{\rho}{1}\frac{d^{\rho-1}.\text{c}'\eta^{(\sigma)}}{dx^{\rho-1}} + \frac{\rho(\rho-1)}{1.2}\frac{d^{\rho-2}.\text{c}''\eta^{(\sigma)}}{dx^{\rho-2}} - \ldots(-)^\rho\text{c}^{(\rho)}\eta^{(\sigma)}\right\}$$

$$= \frac{d^\sigma.\text{c}\eta^{(\sigma)}}{dx^\sigma} - \frac{\rho}{1}\frac{d^{\sigma-1}.\text{c}'\eta^{(\sigma)}}{dx^{\sigma-1}} + \frac{\rho(\rho-1)}{1.2}\frac{d^{\sigma-2}.\text{c}''\eta^{(\sigma)}}{dx^{\sigma-2}} - \ldots(-)^\rho\frac{d^{\sigma-\rho}.\text{c}^{(\rho)}\eta^{(\sigma)}}{dx^{\sigma-\rho}} . \quad (120)$$

Also applying Leibnitz's theorem, (5), Art. 55, Vol. I,

$$\frac{d^s}{dx^s}\left\{\text{c}\frac{d^r.\delta y}{dx^r}\right\} = \frac{d^{\sigma+\rho}}{dx^{\sigma+\rho}}\left\{\text{c}\frac{d^{\sigma-\rho}\eta}{dx^{\sigma-\rho}}\right\}$$

$$= \frac{d^\sigma}{dx^\sigma}\frac{d^\rho}{dx^\rho}\left\{\text{c}\frac{d^{\sigma-\rho}\eta}{dx^{\sigma-\rho}}\right\}$$

$$= \frac{d^\sigma}{dx^\sigma}\left\{\text{c}\frac{d^\sigma\eta}{dx^\sigma} + \frac{\rho}{1}\text{c}'\frac{d^{\sigma-1}\eta}{dx^{\sigma-1}} + \frac{\rho(\rho-1)}{1.2}\text{c}''\frac{d^{\sigma-2}\eta}{dx^{\sigma-2}} + \ldots + \text{c}^{(\rho)}\frac{d^{\sigma-\rho}\eta}{dx^{\sigma-\rho}}\right\},$$

$$= \frac{d^\sigma.\text{c}\eta^{(\sigma)}}{dx^\sigma} + \frac{\rho}{1}\frac{d^\sigma.\text{c}'\eta^{(\sigma-1)}}{dx^\sigma} + \frac{\rho(\rho-1)}{1.2}\frac{d^\sigma.\text{c}''\eta^{(\sigma-2)}}{dx^\sigma} + \ldots + \frac{d^\sigma.\text{c}^{(\rho)}\eta^{(\sigma-\rho)}}{dx^\sigma}; \quad (121)$$

adding which to (120), we have

$$\frac{d^r}{dx^r}\left\{\text{c}\frac{d^s.\delta y}{dx^s}\right\} + \frac{d^s}{dx^s}\left\{\text{c}\frac{d^r.\delta y}{dx^r}\right\}$$

$$= 2\frac{d^\sigma.\text{c}\eta^{(\sigma)}}{dx^\sigma} - \frac{\rho}{1}\left\{\frac{d^{\sigma-1}.\text{c}'\eta^{(\sigma)}}{dx^{\sigma-1}} - \frac{d^\sigma.\text{c}'\eta^{(\sigma-1)}}{dx^\sigma}\right\}$$

$$+ \frac{\rho(\rho-1)}{1.2}\left\{\frac{d^{\sigma-2}.\text{c}''\eta^{(\sigma)}}{dx^{\sigma-2}} + \frac{d^\sigma.\text{c}''\eta^{(\sigma-2)}}{dx^\sigma}\right\} - \ldots \quad (122)$$

But $\frac{d^{\sigma-1}.C'\eta^{(\sigma)}}{dx^{\sigma-1}} - \frac{d^{\sigma}.C'\eta^{(\sigma-1)}}{dx^{\sigma}} = \frac{d^{\sigma-1}}{dx^{\sigma-1}}\left\{C'\eta^{(\sigma)} - \frac{d}{dx}.C'\eta^{(\sigma-1)}\right\}$

$$= \frac{d^{\sigma-1}}{dx^{\sigma-1}}\left\{C'\eta^{(\sigma)} - C'\eta^{(\sigma)} - C''\eta^{(\sigma-1)}\right\}$$

$$= -\frac{d^{\sigma-1}.C''\eta^{(\sigma-1)}}{dx^{\sigma-1}}.$$

Also by the theorem of Leibnitz, and that given in (119),

$$\frac{d^{\sigma-2}}{dx^{\sigma-2}}\left\{C''\eta^{(\sigma)} + \frac{d^2}{dx^2}.C''\eta^{(\sigma-2)}\right\}$$

$$= \frac{d^{\sigma-2}}{dx^{\sigma-2}}\left\{2C''\eta^{(\sigma)} + 2C'''\eta^{(\sigma-1)} + C''''\eta^{(\sigma-2)}\right\}$$

$$= \frac{d^{\sigma-2}}{dx^{\sigma-2}}\left\{2C''\frac{d}{dx}\eta^{(\sigma-1)} + 2C'''\eta^{(\sigma-1)} + C''''\eta^{(\sigma-2)}\right\}$$

$$= \frac{d^{\sigma-2}}{dx^{\sigma-2}}\left\{2\frac{d}{dx}.C''\eta^{(\sigma-1)} - 2C'''\eta^{(\sigma-1)} + 2C'''\eta^{(\sigma-1)} + C''''\eta^{(\sigma-2)}\right\}$$

$$= 2\frac{d^{\sigma-1}.C''\eta^{(\sigma-1)}}{dx^{\sigma-1}} + \frac{d^{\sigma-2}.C''''\eta^{(\sigma-2)}}{dx^{\sigma-2}};$$

which terms are of the form given in (116); and by a similar process the other terms of (122) may be transformed into equivalents of the same form; so that ultimately,

$$\delta H = A_0\eta + \frac{d.A_1\eta'}{dx} + \frac{d^2.A_2\eta''}{dx^2} + \frac{d^3.A_3\eta'''}{dx^3} + \ldots + \frac{d^n.A_n\eta^{(n)}}{dx^n}; \quad (123)$$

that is, δH consists of a series of terms in each of which the order of the derived-function of η or δy is equal to the order of the index of $\frac{d}{dx}$ which affects it; and in which $A_0, A_1, \ldots A_n$ are functions of x.

A process similar to that pursued above is also applicable if $r-s$ is an odd number.

It is manifest that A_n is $(-)^n\left(\frac{d^2V}{dy^{(n)2}}\right)$; but the other coefficients, viz., $A_0, A_1\ldots$, are of a form so complicated that it is useless to calculate them in the general case.

Thus substituting (123) in (112),

$$\delta^2 u = \int_0^1\left\{A_0\eta + \frac{d.A_1\eta'}{dx} + \frac{d^2.A_2\eta''}{dx^2} + \ldots + \frac{d^n.A_n\eta^{(n)}}{dx^n}\right\}\eta\, dx, \quad (124)$$

where $\eta = \delta y$, and $A_0, A_1, \ldots$ are determinate functions of x.

354.] Now we proceed to shew that, when δH is expressed in the form of the right-hand member of (123), $\delta\mathrm{H}\,\eta\,dx$ is an exact differential when η is replaced in $\delta\mathrm{H}$ by a certain value; and that consequently the second variation of the required definite integral admits of integration: hereby we shall be led to a reduction of the result to such a form as will immediately indicate its sign.

$\delta\mathrm{H}$ is manifestly a differential expression containing η and its derived-functions up to the $2n$th order inclusive.

Let z be a value of η satisfying the equation $\delta\mathrm{H} = 0$; that is, suppose z to be a function of x which substituted for η renders $\delta\mathrm{H} = 0$. Such a value is given in (114), Art. 352: so that we have

$$\mathrm{A}_0 z + \frac{d.\mathrm{A}_1 z'}{dx} + \frac{d^2.\mathrm{A}_2 z''}{dx^2} + \ldots + \frac{d^n.\mathrm{A}_n z^{(n)}}{dx^n} = 0. \qquad (125)$$

In $\delta\mathrm{H}$ let ηz be substituted for η, and let the result be multiplied by z, and be subsequently represented by U for convenience of notation: so that

$$\mathrm{U} = z\left\{\mathrm{A}_0\,\eta z + \frac{d.\mathrm{A}_1(\eta z)'}{dx} + \frac{d^2.\mathrm{A}_2(\eta z)''}{dx^2} + \ldots + \frac{d^n.\mathrm{A}_n(\eta z)^{(n)}}{dx^n}\right\}. \quad (126)$$

Then the following investigations will prove

(1) that $\mathrm{U}\,dx$ is an exact differential, whatever is the value of η:

(2) that $\int \mathrm{U}\,dx$ will have the same form as U, except that the index of $\frac{d}{dx}$ will in each term be diminished by unity: in other words, that we shall have

$$\int \mathrm{U}\,dx = \mathrm{B}_1 u' + \frac{d.\mathrm{B}_2 u''}{dx} + \frac{d^2.\mathrm{B}_3 u'''}{dx^2} + \ldots + \frac{d^{n-1}.\mathrm{B}_n u^{(n)}}{dx^{n-1}}. \quad (127)$$

Multiplying (125) by ηz, and subtracting it from U as given in (126), we have

$$\begin{aligned}\mathrm{U} = {} & z\frac{d.\mathrm{A}_1(\eta z)'}{dx} + z\frac{d^2.\mathrm{A}_2(\eta z)''}{dx^2} + \ldots + z\frac{d^n.\mathrm{A}_n(\eta z)^{(n)}}{dx^n} \\ & -\eta z\frac{d.\mathrm{A}_1 z'}{dx} - \eta z\frac{d^2.\mathrm{A}_2 z''}{dx^2} - \ldots - \eta z\frac{d^n.\mathrm{A}_n z^{(n)}}{dx^n},\end{aligned} \qquad (128)$$

which series consists of pairs of terms; of each of which the type is

$$z\frac{d^m.\mathrm{A}_m(\eta z)^{(m)}}{dx^m} - \eta z\frac{d^m.\mathrm{A}_m z^{(m)}}{dx^m};$$

to these let the theorem given in (119), and that of Leibnitz be respectively applied: then

$$z\frac{d^m.A_m(\eta z)^m}{dx^m}$$

$$= \frac{d^m}{dx^m}.A_m z\left\{z\eta^{(m)} + \frac{m}{1}z'\eta^{(m-1)} + \frac{m(m-1)}{1.2}z''\eta^{(m-2)} + \ldots + z^{(m)}\eta\right\}$$

$$-\frac{m}{1}\frac{d^{m-1}}{dx^{m-1}}.A_m z'\left\{z\eta^{(m)} + \frac{m}{1}z'\eta^{(m-1)} + \frac{m(m-1)}{1.2}z''\eta^{(m-2)} + \ldots + z^{(m)}\eta\right\}$$

$$+ \quad . \quad . \quad . \quad . \quad . \quad . \quad . \quad . \quad . \quad . \quad . \quad . \quad . \quad . \quad . \quad . \quad .$$

$$(-)^{m-1}\frac{m}{1}\frac{d}{dx}.A_m z^{(m-1)}\left\{z\eta^{(m)} + \frac{m}{1}z'\eta^{(m-1)} + \frac{m(m-1)}{1.2}z''\eta^{(m-2)} + \ldots + z^{(m)}\eta\right\}$$

$$(-)^m A_m z^{(m)}\left\{z\eta^{(m)} + \frac{m}{1}z'\eta^{(m-1)} + \ldots\ldots\ldots\ldots + z^{(m)}\eta\right\}; \quad (129)$$

also

$$\eta z\frac{d^m.A_m z^{(m)}}{dx^m}$$

$$= \frac{d^m.A_m z^{(m)}\eta z}{dx^m} - \frac{m}{1}\frac{d^{m-1}.A_m z^{(m)}(\eta z)'}{dx^{m-1}} + \frac{m(m-1)}{1.2}\frac{d^{m-2}.A_m z^{(m)}(\eta z)''}{dx^{m-2}} - \ldots$$

$$(-)^{m-1}\frac{m}{1}\frac{d.A_m z^{(m)}(\eta z)^{(m-1)}}{dx}(-)^m A_m z^{(m)}(\eta z)^m; \quad (130)$$

as the last term is the same in both expressions they disappear in the subtraction; and as these are the only integral terms, (128) consists of a series of derived-functions; and therefore U, which is made up of a system of terms satisfying the distributive law, is also a derived-function, and $U\,dx$ is integrable immediately by virtue of its form.

355.] Upon an examination of the series (129) it appears (1) that there are terms of the form

$$M.\frac{d^{m-k}.A_m z^{(k)} z^{(k)}\eta^{(m-k)}}{dx^{m-k}};$$

where M is a constant, and A^m, $z^{(k)}$ are functions of x, and in which therefore the order of the derived of η is the same as the index of $\frac{d}{dx}$, which affects the whole subordinate subject; and (2) that there are other terms, the general type of which is plainly,

$$(-)^k\frac{m(m-1)\ldots(m-k+1)}{1.2.3\ldots k}\,\frac{m(m-1)\ldots(m-k'+1)}{1.2.3\ldots k'}\,\frac{d^{m-k}.A_m z^{(k)} z^{(k')}\eta^{(m-k')}}{dx^{m-k}};$$

and of these, if k and k' correspond to any particular term, so must there also be another to which k' and k correspond, and which is therefore

$$(-)^{k'} \frac{m(m-1)\dots(m-k'+1)}{1.2.3\dots k'} \frac{m(m-1)\dots(m-k+1)}{1.2.3\dots k} \frac{d^{m-k'}.A_m z^{(k')} z^{(k)} \eta^{(m-k)}}{dx^{m-k'}};$$

so that there are pairs of terms of the form

$$M \left\{ \frac{d^{m-k}.C\eta^{(m-k')}}{dx^{m-k}} (-)^{k'-k} \frac{d^{m-k'}.C\eta^{(m-k)}}{dx^{m-k'}} \right\}; \qquad (131)$$

where $C = A_m z^{(k)} z^{(k')}$, and is therefore a function of x.

Now in Art. 353 it has been shewn that a pair of terms, such as (131), can be expressed in a series of terms of the form

$$b_0 \eta + \frac{d.b_1 \eta'}{dx} + \frac{d^2.b_2 \eta''}{dx^2} + \dots + \frac{d^m.b_m \eta^{(m)}}{dx^m};$$

and we shall suppose all the terms of (129) to be so expressed; and by a similar process all the terms of (130); so that ultimately by addition

$$U = B_0 \eta + \frac{d.B_1 \eta'}{dx} + \frac{d^2.B_2 \eta''}{dx^2} + \dots + \frac{d^n.B_n \eta^{(n)}}{dx^n};$$

but $B_0 = 0$, because it has been proved in the preceding Article that $U\,dx$ is an exact differential: and consequently

$$U = \frac{d.B_1 \eta'}{dx} + \frac{d^2.B_2 \eta''}{dx^2} + \dots + \frac{d^n.B_n \eta^{(n)}}{dx^n};$$

and therefore, finally,

$$\int U\,dx = B_1 \eta' + \frac{d.B_2 \eta''}{dx} + \frac{d^2.B_3 \eta'''}{dx^2} + \dots + \frac{d^{n-1}.B_n \eta^{(n)}}{dx^{n-1}};$$

where $B_1, B_2, B_3, \dots B_n$ are functions of x: the general form of these may be found, but in the general case it is too complicated to be available for any useful purpose; and it is better to determine them, if necessary, in each particular case. It is plain however that the only term in (128) which will give $\frac{d^n.B_n \eta^{(n)}}{dx^n}$ is $z \frac{d^n.A_n (\eta z)^{(n)}}{dx^n}$; and that the only term of this latter expression when expanded as in (129), which is of the required form, is

$$\frac{d^n.A_n z^2 \eta^{(n)}}{dx^n};$$

whence it follows that

$$B_n = A_n z^2 = (-)^n \left(\frac{d^2 V}{dy^{(n)2}}\right) z^2. \qquad (132)$$

356.] Hence it appears that

$$\delta^2 u = \int_0^1 \delta H\, \delta y\, dx$$

$$= \int_0^1 \left\{ A_0 \eta + \frac{d.A_1 \eta'}{dx} + \frac{d^2.A_2 \eta''}{dx^2} + \dots + \frac{d^n.A_n \eta^{(n)}}{dx^n} \right\} \eta\, dx; \quad (133)$$

and that if z is a function of x for which, when substituted for η in the expression for δH, the whole vanishes; then

$$z\left\{A_0\eta z+\frac{d.A_1(\eta z)'}{dx}+\frac{d^2.A_2(\eta z)''}{dx^2}+\ldots+\frac{d^n.A_n(\eta z)^{(n)}}{dx^n}\right\}dx \quad (134)$$

is an exact differential by virtue of its form, and ***independently of the value of*** η; and that its integral is of the form

$$B_1\eta'+\frac{d.B_2\eta''}{dx}+\frac{d^2.B_3\eta'''}{dx^2}+\ldots+\frac{d^{n-1}.B_n\eta^{(n)}}{dx^{n-1}}; \quad (135)$$

and from this we infer as a corollary, that as $\delta H\,\delta y$ or $\eta\,\delta H$ when expressed in the form (133) is a particular case of (134), so will the integral of $\delta H\,\eta\,dx$ be of the form (135): but as δH and (134) contain $\eta, \eta', \ldots \eta^{(n)}$, and $\eta z, (\eta z)', (\eta z)'', \ldots (\eta z)^{(n)}$ in corresponding places, so in the integral of $\delta H\,\eta\,dx$, when expressed in the form (135), $\eta, \eta', \ldots \eta^{(n)}$ must be replaced by $\frac{\eta}{z}, \left(\frac{\eta}{z}\right)', \ldots \left(\frac{\eta}{z}\right)^{(n)}$; and therefore

$$\int_0^1 \delta H\,z\,dx=\left[B_1\left(\frac{\eta}{z}\right)'+\frac{d}{dx}.B_2\left(\frac{\eta}{z}\right)''+\ldots+\frac{d^{n-1}}{dx^{n-1}}.B_n\left(\frac{\eta}{z}\right)^{(n)}\right]_0^1. \quad (136)$$

Now the process to be pursued is as follows; we must find a value of z; that is, we must investigate a certain expression, which, when substituted for η, will satisfy $\delta H=0$: this is given in Art. 352, and by (115); hereby we shall be able to integrate by parts the infinitesimal element-function of the second variation, viz. $\delta H\,\eta\,dx$, and to express it in the form (136): and in the general case, by the repetition of a similar process we shall ultimately arrive at an expression consisting of two factors, of which one will be a complete square, and the other, which is easily determined, will by its sign determine the sign of the second variation of the definite integral, and hereby give the required criterion of the critical values.

357.] For a first application of these criteria let us take the case wherein $V=f(x,y,y')$; so that by (104),

$$H = Y - \frac{dY'}{dx} = 0;$$

$$\therefore \quad \delta H = A_0\eta + \frac{d.A_1\eta'}{dx}; \quad (137)$$

where $\eta=\delta y$, $A_0=\frac{d^2V}{dy^2}-\frac{d}{dx}\left(\frac{d^2V}{dy\,dy'}\right)$, and $A_1=-\left(\frac{d^2V}{dy'^2}\right)$.

Let c_1 and c_2 be the arbitrary constants contained in the integral of $H = 0$, and let

$$z = C_1\left(\frac{dy}{dc_1}\right) + C_2\left(\frac{dy}{dc_2}\right);$$

then if η is replaced by z, the right-hand member of (137) vanishes by reason of (115), Art. 352; and therefore

$$z\left\{A_0\eta z + \frac{d.A_1(\eta z)'}{dx}\right\}dx,$$

is an exact differential; and consequently as η or δy is arbitrary, $z\,\delta H\,dx$ is also an exact differential; and its integral is, by reason of (136),

$$B_1\frac{d}{dx}\left(\frac{\eta}{z}\right).$$

Now suppose that $\eta = \delta y = z\,\delta' y$;

$$\therefore \quad \delta' y = \frac{\eta}{z},$$

where $\delta' y$ is a new function of x; then, substituting in the second variation of the definite integral, we have

$$\begin{aligned}
\delta^2 u &= \int_0^1 \delta H\,\delta y\,dx \\
&= \int_0^1 z\,\delta H\,\delta' y\,dx \\
&= \left[\delta' y\,B_1\frac{d}{dx}\left(\frac{\eta}{z}\right)\right]_0^1 - \int_0^1 B_1\frac{d}{dx}\left(\frac{\eta}{z}\right)\frac{d.\delta' y}{dx}\,dx \\
&= \;-\;-\;-\;-\;-\int_0^1 B_1\left\{\frac{d}{dx}\left(\frac{\eta}{z}\right)\right\}^2 dx. \qquad (138)
\end{aligned}$$

And, to take the case most free from difficulty, let us suppose the limiting values to be fixed, so that δy and therefore $\delta' y$ vanishes at both limits. Then, replacing B_1 by its value given in (132), (138) becomes

$$\delta^2 u = \int_0^1\left(\frac{d^2 V}{dy'^2}\right)\left(\frac{z\eta' - \eta z'}{z}\right)^2 dx.$$

Therefore the definite integral $\int_0^1 V\,dx$ will be a maximum or a minimum according as $\frac{d^2 V}{dy'^2}$ is negative or positive; provided that it does not change sign nor become infinite between the assigned limits; and provided also that the constants C_1 and C_2 and η are not such as to make $z\eta' - \eta z'$ vanish or become infinite.

It is worth remarking, that if $z\eta' - \eta z' = 0$, then $\eta = z = \delta y$;

in which case $\delta H = 0$, and therefore the second variation of the definite integral vanishes; and this is plainly inconsistent with the possibility of our deducing from it the criteria of maxima and minima.

For an application of the preceding, let us consider the case of the longest or shortest line between two given points; here

$$V = (1+y'^2)^{\frac{1}{2}};$$

$$\therefore \quad \frac{d^2V}{dy'^2} = \frac{1}{(1+y'^2)^{\frac{3}{2}}};$$

which is always positive, if the radical in V is affected with a positive sign.

Also, since the complete integral of $H=0$ is, see Ex. 2, Art. 351,

$$y = c_1 x + c_2;$$

$$\left(\frac{dy}{dc_1}\right) = x, \qquad \left(\frac{dy}{dc_2}\right) = 1;$$

$$\therefore \quad \delta y = C_1 x + C_2 = z, \quad \text{and} \quad z' = C_1;$$

$$\therefore \quad \left(\frac{z\eta' - \eta z'}{z}\right)^2 = \left\{\frac{(C_1 x + C_2)\eta' - \eta C_1}{C_1 x + C_2}\right\}^2;$$

C_1 and C_2 therefore must not be so assumed as to make $C_1 x + C_2 = 0$ for any value of x between the assigned limits.

358.] For a second example of the criteria, let

$$V = f(x, y, y', y'');$$

$$\therefore \quad H = Y - \frac{dY'}{dx} + \frac{d^2Y''}{dx^2} = 0;$$

$$\therefore \quad \delta H = A_0 \eta + \frac{d.A_1 \eta'}{dx} + \frac{d^2.A_2 \eta''}{dx^2}. \tag{139}$$

Let c_1, c_2, c_3, c_4 be the four arbitrary constants which enter into the complete integral of $H = 0$; then the value of z, which, substituted for η, satisfies (139), is

$$z = C_1\left(\frac{dy}{dc_1}\right) + C_2\left(\frac{dy}{dc_2}\right) + C_3\left(\frac{dy}{dc_3}\right) + C_4\left(\frac{dy}{dc_4}\right); \tag{140}$$

so that, as before, $z\,\delta H\,dx$ is an exact differential; and its integral is, by virtue of (136),

$$B_1 \frac{d}{dx}\left(\frac{\eta}{z}\right) + \frac{d}{dx}.B_2 \frac{d^2}{dx^2}\left(\frac{\eta}{z}\right). \tag{141}$$

Let $\qquad \frac{\eta}{z} = \frac{\delta y}{z} = \delta' y;$

where $\delta' y$ represents a new variation of y; then, integrating the expression for the second variation of the definite integral, and assuming the limits to be fixed, so that the terms at them vanish, we have

$$\begin{aligned}\delta^2 u &= \int_0^1 \delta H\, \delta y\, dx \\ &= \int_0^1 z\, \delta H\, dx\, \delta' y \\ &= -\int_0^1 \left\{ B_1 \frac{d.\delta' y}{dx} + \frac{d}{dx}.B_2 \frac{d^2.\delta' y}{dx^2} \right\} \frac{d.\delta' y}{dx}\, dx \\ &= -\int_0^1 \left\{ B_1 \delta' y' + \frac{d}{dx}.B_2 \frac{d.\delta' y'}{dx} \right\} \delta' y'\, dx \\ &= -\int_0^1 \left\{ B_1 \eta_1 + \frac{d.B_2 \eta'_1}{dx} \right\} \eta_1\, dx, \qquad (142)\end{aligned}$$

if $$\eta_1 = \delta' y' = \frac{d.\delta' y}{dx} = \frac{d}{dx}\frac{\eta}{z} = \frac{z\eta' - \eta z'}{z^2}.$$

Now let

$$z_1 = c'_1 \left(\frac{dy}{dc_1}\right) + c'_2 \left(\frac{dy}{dc_2}\right) + c'_3 \left(\frac{dy}{dc_3}\right) + c'_4 \left(\frac{dy}{dc_4}\right),$$

where c'_1, c'_2, c'_3, c'_4 are other new arbitrary constants employed like the former ones in (140) to represent arbitrary variations of the constants c_1, c_2, c_3, c_4: so that z_1 is a value of η which satisfies $\delta H = 0$.

Also since from (141)

$$\int_0^1 z\, \delta H\, dx = B_1 \left(\frac{\eta}{z}\right)' + \frac{d}{dx}.B_2 \left(\frac{\eta}{z}\right)'',$$

it appears that any value of η, which makes $\delta H = 0$, will also satisfy the right-hand member of the equation; but $\delta H = 0$, if $\eta = z_1$, therefore $\left(\frac{z_1}{z}\right)'$ is a solution of the right-hand member: also

$$\left(\frac{z_1}{z}\right)' = \frac{zz'_1 - z_1 z'}{z^2}.$$

$$\text{Let} \quad \eta_1 = \left(\frac{z_1}{z}\right)' \delta'' y;$$

and substituting this in (142) we have

$$\delta^2 u = -\int_0^1 \left\{ B_1 \eta_1 + \frac{d.B_2 \eta'_1}{dx} \right\} \left(\frac{z_1}{z}\right)' \delta'' y\, dx;$$

whence, integrating by parts, and omitting the integrated part which vanishes at the limits, we have

$$\delta^2 u = \int_0^1 c_2 \frac{d}{dx}\left\{\frac{\eta_1}{\left(\frac{z_1}{z}\right)'}\right\} \frac{d.\delta'' y}{dx}\, dx = \int_0^1 c_2 \left\{\frac{d.\delta'' y}{dx}\right\}^2 dx;$$

where, by virtue of (132),

$$c_2 = B_2 \left(\frac{z_1}{z}\right)'^2 = \left(\frac{zz_1' - z'z_1}{z}\right)^2 \left(\frac{d^2 V}{dy''^2}\right).$$

Also, $$\frac{d}{dx}.\delta'' y = \frac{d}{dx}.\eta_1 \frac{z^2}{zz_1' - z'z_1} = \frac{d}{dx}.\frac{z\eta' - \eta z'}{zz_1' - z'z_1}$$

$$= \frac{z\{(z'z_1'' - z_1'z'')\eta + (z''z_1 - zz'')\eta' + (zz_1' - z'z_1)\eta''\}}{(zz_1' - z'z_1)^2};$$

therefore

$$\delta^2 u = \int_0^1 \left(\frac{d^2 V}{dy''^2}\right) \left\{\frac{z'z_1'' - z'z_1''}{zz_1' - z'z_1}\delta y + \frac{z_1''z - zz''}{zz_1' - z'z_1}\delta y' + \delta y''\right\}^2 dx. \quad (143)$$

And therefore for a maximum or minimum value of the definite integral it is requisite that $\left(\frac{d^2 V}{dy''^2}\right)$ should be respectively negative or positive for all values of the variables between the limits; also the second factor must neither vanish nor become infinite: the arbitrary constants therefore must be so determined as to fulfil these conditions.

359.] If the infinitesimal element-function of the definite integral contains derived-functions of y up to the nth, the process to be pursued is exactly similar to those of the two preceding particular cases; and therefore I need give no more than an outline of it.

Let $z, z_1, z_2, \ldots z_{n-1}$ be n values of δy expressed in the preceding forms, and containing n different series of arbitrary constants: then the second variation is

$$\delta^2 u = \int_0^1 \left\{A_0 \eta + \frac{d.A_1\eta'}{dx} + \frac{d^2.A_2\eta''}{dx^2} + \ldots + \frac{d^n.A_n\eta^{(n)}}{dx^n}\right\}\eta\, dx,$$

of which the integral becomes, by neglecting the quantities at the limits, reduced to

$$\delta^2 u = -\int_0^1 \left\{B_1\eta_1 + \frac{d.B_2\eta'_1}{dx} + \ldots + \frac{d^{n-1}.B_n\eta_1^{(n-1)}}{dx^{n-1}}\right\}\eta_1\, dx;$$

and so on; until ultimately

$$\delta^2 u = (-)^n \int_0^1 J_n \left(\frac{d.\delta^{(n)} y}{dx}\right)^2 dx, \quad (144)$$

wherein J_n involves $\left(\frac{d^2\text{V}}{dy^{(n)2}}\right)$, and another factor which is of the form of a complete square; and where

$$z\delta' y = \delta y,$$

$$\left(\frac{z_1}{z}\right)' \delta'' y = \frac{d}{dx}\cdot\delta' y,$$

and so on. It appears therefore that the maximum and minimum value will depend on the sign of $\left(\frac{d^2\text{V}}{dy^{(n)2}}\right)$; and that it is necessary that this latter quantity should not change its sign for any value of the variables between the given limits; and the arbitrary constants must not be such as to allow the other factor in (144) to vanish or to become infinite.

360.] We need not enter at length on the determination of criteria for relative maxima and minima, because we have shewn above that such cases are by means of an indeterminate multiplier reduced to those of absolute critical values, and the criteria determined for this latter case are therefore applicable to the former one. Let us however shew that the solution given in the fourth example of Art. 351 is a maximum:

$$u = \int_0^1 \{y+\lambda(1+y'^2)^{\frac{1}{2}}\}\,dx;$$

$$\therefore\quad \text{V} = y+\lambda(1+y'^2)^{\frac{1}{2}};$$

$$\text{and}\quad \left(\frac{d^2\text{V}}{dy'^2}\right) = \frac{\lambda}{(1+y'^2)^{\frac{3}{2}}}.$$

Also, since the curve is determined by the differential equation,

$$\text{H} = 0 = \text{Y} - \frac{d\text{Y}_1}{dx};$$

$$\therefore\quad 0 = 1 - \frac{d}{dx}\,\frac{\lambda y'}{(1+y'^2)^{\frac{1}{2}}}$$

$$= 1 - \frac{\lambda y''}{(1+y'^2)^{\frac{3}{2}}};$$

$$\therefore\quad \lambda = \frac{(1+y'^2)^{\frac{3}{2}}}{y''};\quad \text{and}\quad \left(\frac{d^2\text{V}}{dy'^2}\right) = \frac{1}{y''};$$

and therefore the answer gives a maximum or minimum value according as y'' is negative or positive. Let the origin be at the centre of the circle; then, since, as shewn by the value of u, the curve is taken in the first quadrant, y'' is negative, and consequently the solution corresponds to a maximum.

SECTION 5.—*Investigation of the critical values of a double definite integral.*

361.] It only remains for us now to investigate the conditions of the critical values of a definite double integral, of which the variation has been calculated in Art. 311. On referring to equation (95), Art. 311, it appears that the expression for δu consists of three parts; viz., two partially integrated terms whose value depends on the values which ω and its derived-functions have at the limits which are assigned by the given limiting equation; and the third term, which is wholly unintegrated, and cannot be reduced unless ω receives a determinate value. Now let

$$\mathrm{z} - \frac{d\mathrm{z}'}{dx} - \frac{d\mathrm{z}_{,}}{dy} + \frac{d^2\mathrm{z}''}{dx^2} + \frac{d^2\mathrm{z}_{,}'}{dx\,dy} + \frac{d^2\mathrm{z}_{,,}}{dy^2} = \Omega. \qquad (145)$$

Then as $\delta u = 0$, by-reason of u having a critical value, it follows that $\Omega = 0$; and from this differential equation the required function is to be determined. (145) is plainly a partial differential equation of the fourth order; the general integral of which is in most cases beyond the present powers of the integral calculus: we can in many cases however deduce from it some geometrical property which is sufficient to define the required surface.

362.] To find the surface the portion of which enclosed by a given curve has a minimum area.

In this problem the limits of integration are given by the given curve: and

$$u = \int_0^1 \int_0^1 \{1 + z'^2 + z_{,}^2\}^{\frac{1}{2}}\, dy\, dx;$$

and therefore (145) becomes

$$\frac{d\mathrm{z}'}{dx} + \frac{d\mathrm{z}_{,}}{dy} = 0;$$

whence there manifestly results

$$\left(\frac{d^2z}{dx^2}\right)\left\{1 + \left(\frac{dz}{dy}\right)^2\right\} - 2\left(\frac{dz}{dx}\right)\left(\frac{dz}{dy}\right)\left(\frac{d^2z}{dx\,dy}\right) + \left(\frac{d^2z}{dy^2}\right)\left\{1 + \left(\frac{dz}{dx}\right)^2\right\} = 0;$$

and on comparing this with (94), Art. 413, Vol. I, it is seen that the two are identical; and therefore the geometrical interpretation is, "The surface of minimum area is such that the sum of the reciprocals of its principal radii of curvature at every point vanishes:" hence we infer that the principal radii of curvature at every point are equal and of opposite signs.

363.] Let the problem be "To determine the form of the surface which being of given extent, and terminated by a given curve, includes the greatest volume between it, the plane of (x, y), and the right cylinder whose director is the projection of the given curve on the plane of (x, y):" in this case

$$\text{the content} = \iiint dz\, dy\, dx = \iint z\, dy\, dx;$$

$$\text{the surface} = \iint \{1 + z'^2 + z_,^2\}^{\frac{1}{2}}\, dy\, dx;$$

therefore, if λ is an undetermined multiplier,

$$u = \iint \{z + \lambda(1 + z'^2 + z_,^2)^{\frac{1}{2}}\}\, dy\, dx;$$

and thus the equation $\Omega = 0$ becomes

$$z - \frac{dz'}{dx} - \frac{dz_,}{dy} = 0;$$

whence by development, as in the preceding example, we have

$$\left(\frac{d^2z}{dx^2}\right)\left\{1 + \left(\frac{dz}{dy}\right)^2\right\} - 2\left(\frac{dz}{dx}\right)\left(\frac{dz}{dy}\right)\left(\frac{d^2z}{dx\,dy}\right) + \left(\frac{d^2z}{dy^2}\right)\left\{1 + \left(\frac{dz}{dx}\right)^2\right\}$$

$$= \frac{1}{\lambda}\left\{1 + \left(\frac{dz}{dx}\right)^2 + \left(\frac{dz}{dy}\right)^2\right\}^{\frac{3}{2}}; \quad (146)$$

and to interpret this geometrically; let ρ_1 and ρ_2 be the principal radii of curvature at any point on a surface; then by equation (27), Art. 399, Vol. I, we have

$$\frac{1}{\rho_1} + \frac{1}{\rho_2} = \frac{\mathrm{U}^2(\mathrm{K} + \mathrm{L}) + \mathrm{V}^2(\mathrm{L} + \mathrm{H}) + \mathrm{W}^2(\mathrm{H} + \mathrm{K})}{(\mathrm{U}^2 + \mathrm{V}^2 + \mathrm{W}^2)^{\frac{3}{2}}}; \quad (147)$$

and if these symbols are expressed in terms of the derived-functions of z, it will be seen by comparison with (146) that

$$\frac{1}{\rho_1} + \frac{1}{\rho_2} = \frac{1}{\lambda};$$

and therefore the surface which under a given superficial area contains the greatest volume is such, that the sum of its principal curvatures at every point is constant: and this result is usually expressed as "The mean curvature is the same at every point of the surface."

The equations (146) and (147) have never yet been directly integrated, but Mr. Jellet has in Liouville's Journal* shewn indirectly that the sphere is the only surface which satisfies them.

* Tome XVIII, p. 163, 1853.

DIFFERENTIAL EQUATIONS,

OR

THE INTEGRATION OF DIFFERENTIAL FUNCTIONS OF TWO OR MORE VARIABLES.

CHAPTER XV.

THE INTEGRATION OF DIFFERENTIAL EQUATIONS OF THE FIRST ORDER.

SECTION 1.—*General considerations on Differential Equations.*

364.] EXPRESSIONS or equations which involve differentials, and the variables of which they are the differentials, are called differential expressions, and are distinguished from finite expressions, inasmuch as the latter contain only finite quantities. In the preceding parts of this volume differential expressions have been the subject-matter, but in that restricted form wherein the element-function of an integral is the function of only one variable for that integration. A more general case however is that in which a differential expression involves many variables, and the determination of the corresponding finite expression from that given differential expression is the problem. The process by which such differential expressions are derived from the finite equation has been explained, and fully illustrated in Section 7, Chap. III, Vol. I; and it is the inverse process of that of that section which we have here to develop. The formation or the genesis of differential equations, containing both total and partial differential coefficients, has been so largely explained, that it is unnecessary to say more on the subject; but the student is recommended to study that section, that he may ascertain not only the conditions of the problem but also the form of the solution in the most general case.

It is convenient to classify differential equations. They may contain either total differentials or derived-functions; or they may contain partial derived-functions.

In the former case they will be of either of the following forms;

$$F(x, y, \ldots dx, dy, \ldots d^2x, d^2y, \ldots d^m x, d^n y, \ldots) = 0, \qquad (1)$$

$$F\left(x, y, \frac{dy}{dx}, \frac{d^2y}{dx^2}, \ldots \frac{d^ry}{dx^r}\right) = 0; \qquad (2)$$

when they are called *total differential equations*. Or they may be of the forms,

$$F\left\{x, y, z, \left(\frac{dz}{dx}\right), \left(\frac{dz}{dy}\right), \left(\frac{d^2z}{dx^2}\right), \left(\frac{d^2z}{dx\,dy}\right), \left(\frac{d^2z}{dy^2}\right)\ldots\right\} = 0, \quad (3)$$

$$F\left\{x, y, z, \ldots\ldots \left(\frac{du}{dx}\right), \left(\frac{du}{dy}\right), \ldots \left(\frac{d^2u}{dx^2}\right), \left(\frac{d^2u}{dy^2}\right)\ldots\right\} = 0; \quad (4)$$

in which case they are called *partial differential equations*. On referring to Vol. I, Chap. III, Section 7, it will be observed that equations of the former class arise from the elimination of constants and determinate functions; whereas the latter arise from the elimination of arbitrary and undetermined functions. Much more however will be said on this subject hereafter.

These equations also require to be classified on other principles: (1) on the *order* of the highest differential or derived-function which is involved; and (2) on the *degree* or index to which the highest differential or derived-function is raised; thus *order* is predicated of a differential equation as to the former, *degree* as to the latter; and if a differential equation contains x, y, dx, dy, d^2x, d^2y, or $x, y, \frac{dy}{dx}, \frac{d^2y}{dx^2}$, it is said to be of the second order; and if the highest differentials or derived-functions enter in only *linear* forms, or to the first power, such an expression is said to be of the first degree; but an equation containing $x, y, \left(\frac{dy}{dx}\right)^3, \left(\frac{d^2y}{dx^2}\right)^2$ is of the second order and of the second degree; and so of other similar expressions.

365.] As in Chap. III, Vol. I, the subject is considered from only an analytical point of view, let us also examine the geometrical aspect of it. And to fix our thoughts let us first take a total differential equation of the first order and first degree, and suppose it to be in the form

$$\frac{dy}{dx} = f(x, y); \tag{5}$$

let x and y be the rectangular coordinates of a plane curve; and let τ be the angle between the axis of x and the tangent to the curve at the point (x, y): so that $\tan\tau = \frac{dy}{dx} = f(x, y)$. As x and y are general in (5), let x_0, y_0 be particular, although arbitrarily chosen, values of x and y; and let τ_0 be the corresponding value of τ: so that

$$\tan\tau_0 = f(x_0, y_0);$$

and through the point (x_0, y_0) let a line be drawn cutting the axis of x at the angle τ_0. On this line let there be taken a point (x_1, y_1) contiguous to (x_0, y_0), and through it let a line be drawn cutting the axis of x at an angle τ_1, so that

$$\tan\tau_1 = f(x_1, y_1);$$

on this line let there be taken a point (x_2, y_2) contiguous to (x_1, y_1), and through (x_2, y_2) let a line be drawn making an angle τ_2 with the axis of x, where

$$\tan\tau_2 = f(x_2, y_2):$$

and let a similar process be repeated n times, until at last we arrive at the point (x_n, y_n); hereby we shall have formed a series of short lines inclined to each other at different angles, and abutting at the points (x_0, y_0) and (x_n, y_n). Let now every two successive points be infinitesimally near to each other, and also let the number of times that the process is repeated be infinite; then the distance between the extreme points is still finite, and the broken line which joins them becomes a continuous curve, and the distances between each two successive points become arc-elements of the curve: and hereby the curve between the two points will have been constructed from the given differential equation. Now from the process thus conducted it is manifest that the position of each point of the curve depends on that of the immediately preceding point, the law of dependence being given by the differential equation (5); the *nature* of the curve therefore is given by the differential equation: but it is also equally manifest that the *position* of every point, and so of the curve, depends on that of the first assumed point, viz., on (x_0, y_0), and the position of this point is arbitrary: although therefore the nature of the curve remains the same, whatever are the values of x_0 and y_0, yet the position of it alters; and consequently the differential equation expresses a property common to a series of curves, the particular one of which is determined by means of the arbitrary values x_0 and y_0. But as a complete integral equation determines both the nature and position of the curve which it represents, it is plain that the coordinates of the first point must enter into the integral equation; and therefore the integral of (5) must contain these, and cannot be complete without them; the integral therefore of (5) must be definite; but it is convenient to leave the superior limits in the general form x, y, so that they may refer to any point on the curve. It is plain also, from the theory of definite integration, that if $\mathrm{F}(x, y)$ is the indefinite integral of (5), the definite integral is

$$\mathrm{F}(x,y)-\mathrm{F}(x_0,y_0) = 0; \tag{6}$$

and as x_0, y_0 are arbitrary constants, we may replace $\mathrm{F}(x_0, y_0)$ by an arbitrary constant c; and thus the integral equation of (5) is of the form

$$\mathrm{F}(x,y) = c; \tag{7}$$

or, more generally, $\mathrm{F}(x, y, c) = 0$;

that is, in the process of integration one arbitrary constant c has been introduced.

Again, suppose the given differential equation to be of the second order and of the form

$$f\left(x, y, \frac{dy}{dx}, \frac{d^2y}{dx^2}\right) = 0; \tag{8}$$

and let the inferior limits of integration correspond to the point (x_0, y_0), from which we will suppose the curve to begin: this point is of course arbitrary. Also since $\frac{d^2y}{dx^2}$ involves *three* consecutive points, see Art. 243, Vol. I, and as there is only one relation, viz. (8), between the three points, the second as also the first is arbitrary; but not so the third; its position with reference to the other two becomes fixed by means of equation (8); and similarly will every other consecutive point on the curve, and thus the whole curve, become fixed; in the complete integral therefore the coordinates to these first two points must enter; and, by the theory of definite integration, in the form of two constants, which will be arbitrary, because the first two points on the curve are arbitrary: and thus if c_1, c_2 are two arbitrary constants, the complete integral of (8) is of the form

$$F(x, y, c_1, c_2) = 0. \tag{9}$$

This is also otherwise manifest: the first indefinite integral of (8) will be a function of x, y, and $\frac{dy}{dx}$; and therefore the definite integral will be of the form

$$F\left(x, y, \frac{dy}{dx}\right) - F\left(x_0, y_0, \frac{dy_0}{dx_0}\right) = 0.$$

And replacing the second term by an arbitrary constant c_1, the first integral will be of the form

$$F\left(x, y, \frac{dy}{dx}, c_1\right) = 0:$$

and the integral of this will, by reason of what has already been said, involve another new arbitrary constant c_2. It appears therefore that the complete integral of a differential equation of the second order requires the introduction of two arbitrary constants.

By a similar process we may shew that n arbitrary constants enter into the complete integral of a differential equation of the nth order.

It may perhaps be superfluous to remark that in thus taking the definite integral of a differential equation, the differentials or derived functions must not become infinite or discontinuous for any value of the variables between the limits.

366.] And to form a correct notion as to the meaning of a partial differential equation, let us consider the following example of a partial differential equation of the first order:

$$(x-a)\left(\frac{du}{dx}\right) + (y-b)\left(\frac{du}{dy}\right) + (z-c)\left(\frac{du}{dz}\right) = 0; \tag{10}$$

or, as it may be otherwise and equivalently expressed,

$$(x-a)\left(\frac{dz}{dx}\right) + (y-b)\left(\frac{dz}{dy}\right) = z-c. \tag{11}$$

Equation (10) is the general equation of a tangent plane of a surface, which passes through a given point (a, b, c); or, what is equivalent, (10) implies that all the normals to the surface are perpendicular to straight lines which pass through a given point: and it is not for one surface only, or for one particular species of surface, that this property is true; it is not only for a given cone or for circular cones that the property holds good; but it is true of all conical surfaces of which the given point is the vertex: and thus a symbol expressing a condition equally general must enter into the final integral equation: in other words, the complete integral must contain the law of the director-curve of the conical surface; and such can be the case only when an arbitrary function is introduced: the complete integral therefore of a partial differential equation of the form (10) or (11) must contain an arbitrary functional symbol: in fact we know that the integral of (10) or (11) is either

$$\mathrm{F}\left(\frac{x-a}{z-c}, \frac{y-b}{z-c}\right) = 0; \tag{12}$$

or

$$\frac{x-a}{z-c} = f\left(\frac{y-b}{z-c}\right). \tag{13}$$

Hence it appears that the integral of a partial differential equation of the first order requires the introduction of one arbitrary function.

Thus it appears that these geometrical explanations as to the introduction of constants and arbitrary functions in the case of total and partial differential expressions are in accordance with the reverse analytical process of Section 7, Chap. III, Vol. I.

367.] We may also thus prove that the complete integral of a differential equation of the nth order and first degree involves n arbitrary constants.

Let us suppose the differential equation to be of the form

$$\mathrm{F}\left(x, y, \frac{dy}{dx}, \dots \frac{d^n y}{dx^n}\right) = 0,$$

and to admit of being put into the form

$$\frac{d^n y}{dx^n} = f\left(x, y, \frac{dy}{dx}, \dots \frac{d^{n-1} y}{dx^{n-1}}\right); \tag{14}$$

and let us suppose that it, and all its integrals up to the last, satisfy the conditions which are requisite for development in Taylor's series.

Let (14) be differentiated successively, and let the necessary eliminations be performed, so that we can determine $\frac{d^{n+1} y}{dx^{n+1}}$, $\frac{d^{n+2} y}{dx^{n+2}}, \dots$ in terms of $x, y, \frac{dy}{dx}, \dots \frac{d^{n-1} y}{dx^{n-1}}$; and let the limits of the integral of (14) be x_0, y_0 and x, y; then, by equation (84), Art. 74, Vol. I,

$$y_0 + \left(\frac{dy}{dx}\right)_0 \frac{x-x_0}{1} + \left(\frac{d^2 y}{dx^2}\right)_0 \frac{(x-x_0)^2}{1.2} + \dots + \left(\frac{d^n y}{dx^n}\right)_0 \frac{(x-x_0)^n}{1.2.3\dots n} + \dots; \tag{15}$$

where the subscript cyphers indicate particular values of the symbols; those, namely, which correspond to the inferior limit. Now from the preceding remarks it is plain that all the differential coefficients after $\left(\frac{d^{n-1}y}{dx^{n-1}}\right)_0$, may be expressed in terms of $x_0, y_0, \left(\frac{dy}{dx}\right)_0, \ldots \left(\frac{d^{n-1}y}{dx^{n-1}}\right)_0$, so that the series (15) will involve n and only n undetermined quantities, viz. the term independent of x, and the several coefficients of $x, x^2, \ldots x^{n-1}$, which are n in number and may be expressed by n constants, $c_1, c_2, \ldots c_n$; and therefore into the complete integral of (14) n arbitrary constants enter.

Of course it is supposed that none of the quantities y_0, $\left(\frac{dy}{dx}\right)_0, \ldots \left(\frac{d^{n-1}y}{dx^{n-1}}\right)_0$ is infinite or discontinuous between the limits; as however no criteria are given for determining whether these conditions are satisfied or not, the above must be taken only to establish an *a priori* probability that the theorem, as stated, is true. A rigorous proof of a particular case will be given hereafter, and might be extended generally.

As an example of this process, let us take the equation

$$\frac{dy}{dx} + ay + bx^2 = 0;$$

$$\therefore \quad \frac{d^2y}{dx^2} + a\frac{dy}{dx} + 2bx = 0;$$

$$\frac{d^3y}{dx^3} + a\frac{d^2y}{dx^2} + 2b = 0; \qquad \frac{d^4y}{dx^4} + a\frac{d^3y}{dx^3} = 0.$$

$$\cdots\cdots\cdots\cdots\cdots\cdots$$

Let the inferior limits be y_0 and $x_0 = 0$;

$$\therefore \quad \left(\frac{dy}{dx}\right)_0 = -ay_0, \ \left(\frac{d^2y}{dx^2}\right)_0 = a^2y_0, \ \left(\frac{d^3y}{dx^3}\right)_0 = -a^3y_0 - 2b, \ldots;$$

thus (15) becomes

$$y = y_0 - ay_0\frac{x}{1} + a^2y_0\frac{x^2}{1.2} - \left(a^3y_0 + 2b\right)\frac{x^3}{1.2.3} + \left(a^4y_0 - 2ab\right)\frac{x^4}{1.2.3.4} + .$$

$$= y_0\left\{1 - \frac{ax}{2} + \frac{a^2x^2}{1.2} - \ldots\right\} - \frac{2b}{a^3}\left\{\frac{a^3x^3}{1.2.3} - \frac{a^4x^4}{1.2.3.4} + \ldots\right\}$$

$$= y_0e^{-ax} + \frac{2b}{a^3}\left\{e^{-ax} - 1 + \frac{ax}{2} - \frac{a^2x^2}{1.2}\right\};$$

which involves only one arbitrary constant, viz. y_0.

368.] When the integral of a given differential equation contains n arbitrary constants, and these in their most general form, it is called the *general integral*; and conversely, if an equation in terms of x and y satisfies a given differential equation of the nth order and contains n arbitrary constants, it is the general integral. If particular values are given to one or more of these arbitrary constants, as, for instance, if any of them is zero, then the integral is called a *particular integral*. Also sometimes one or more of the arbitrary constants may be replaced by a particular function of x and y, and the equation will still satisfy the given differential equation, when at the same time such a result cannot be obtained by giving any particular constant value to one or more of the arbitrary constants of the general integral: in this case the integral is called a *singular solution*. Our capital problem is the discovery of the *general integral*, by means of which particular integrals evidently may be determined. But we shall also investigate as far as possible the general properties of singular solutions, and indicate some specific forms of differential equations which admit of such solutions.

In most cases we shall be obliged to leave the arbitrary constants undetermined; the complete integral of a differential equation requires that the integral should be *definite*, and therefore the constants ought to be expressed in terms of the limits; but it is manifest that this can be done only when the conditions of the problem are given, as in the geometrical applications of the calculus. Differential equations, however, for the most part arise in mechanics and other applied mathematics, on the investigation of which we have not yet entered: the constants therefore which are introduced in the process of integration must in most cases be left arbitrary, at least for the present.

369.] A simple form of differential equation which admits of integration immediately, or, as it is commonly said, by simple quadrature, is that where the variables are separated; in which case the expression contains the algebraical sum of several elements, each of which is a function of a single variable. The general form in the case of two variables is

$$f(x)\,dx+\phi(y)\,dy = 0; \qquad (16)$$

whence we have for the definite integral, x_0 and y_0 being corresponding values of x and y,

$$\int_{x_0}^{x} f(x)\,dx + \int_{y_0}^{y} \phi(y)\,dy = 0; \qquad (17)$$

and if the integrals are indefinite,

$$\int f(x)\,dx + \int \phi(y)\,dy = c, \tag{18}$$

where c is an arbitrary constant.

And if there are three variables, the general form of the equation is

$$f(x)\,dx + \phi(y)\,dy + \chi(z)\,dz = 0; \tag{19}$$

$$\therefore \quad \int_{x_0}^{x} f(x)\,dx + \int_{y_0}^{y} \phi(y)\,dy + \int_{z_0}^{z} \chi(z)\,dz = 0. \tag{20}$$

Ex. 1. $$\frac{dx}{(1-x^2)^{\frac{1}{2}}} + \frac{dy}{(1-y^2)^{\frac{1}{2}}} = 0;$$

$$\therefore \quad \sin^{-1} x + \sin^{-1} y = \sin^{-1} c,$$

c being an arbitrary constant;

$$\therefore \quad x(1-y^2)^{\frac{1}{2}} + y(1-x^2)^{\frac{1}{2}} = c,$$

which is the general integral; and if $c = 0$, we have a particular integral $y = -x$.

370.] Another form in which the variables immediately admit of separation is

$$\text{X}\text{Y}_1\,dx + \text{Y}\text{X}_1\,dy = 0; \tag{21}$$

where X and X_1 are functions of x only, and Y and Y_1 are functions of y only; for dividing through by X_1Y_1, we have

$$\frac{\text{X}}{\text{X}_1}\,dx + \frac{\text{Y}}{\text{Y}_1}\,dy = 0; \tag{22}$$

$$\therefore \quad \int \frac{\text{X}}{\text{X}_1}\,dx + \int \frac{\text{Y}}{\text{Y}_1}\,dy = c, \tag{23}$$

which is the general integral of (22).

Ex. 1. $y\,dx + x\,dy = 0;$

$$\therefore \quad \frac{dx}{x} + \frac{dy}{y} = 0;$$

$$\log x + \log y = \log c^2;$$

$$\therefore \quad xy = c^2;$$

c being an arbitrary constant.

Ex. 2. $dy\,x^{\frac{1}{2}} + dx\,y^{\frac{1}{2}} = 0;$

$$\frac{dy}{y^{\frac{1}{2}}} + \frac{dx}{x^{\frac{1}{2}}} = 0;$$

$$2y^{\frac{1}{2}} + 2x^{\frac{1}{2}} = 2a^{\frac{1}{2}};$$

$$\therefore \quad x^{\frac{1}{2}} + y^{\frac{1}{2}} = a^{\frac{1}{2}},$$

where a is an arbitrary constant.

Ex. 3. If $\sin x \cos y\, dx - \sin y \cos x\, dy = 0$, then $\cos x = m \cos y$.

Ex. 4. If $(\sec x)^2 \tan y\, dx + (\sec y)^2 \tan x\, dy = 0$,

then $\tan x \tan y = k^2$.

These methods however are so simple that it is unnecessary to add other examples.

SECTION 2.—*Integration of exact total differentials of two and more variables.*

371.] Let us first take the case of two variables, and suppose the differential expression to be

$$P\, dx + Q\, dy = 0, \tag{24}$$

where P and Q are functions of x and y: it may be that (24) is the exact differential of some integral function of the form $u = F(x, y) = c$; or it may be that some factor common to the two terms has been divided out, and that (24) will not be an exact differential until this factor, or some other factor, has been introduced; this latter case is reserved to Section 6 of the present Chapter.

If (24) is the exact differential of the function

$$u = F(x, y) = c; \tag{25}$$

then $$P\, dx + Q\, dy = Du = \left(\frac{du}{dx}\right) dx + \left(\frac{du}{dy}\right) dy; \tag{26}$$

and as dx and dy are arbitrary, though infinitesimal, increments of x and y, (26) can only be true when

$$\left(\frac{du}{dx}\right) = P, \quad \text{and} \quad \left(\frac{du}{dy}\right) = Q. \tag{27}$$

Hence we have a criterion whether (24) is an exact differential or not; for since $\left(\frac{d^2u}{dy\,dx}\right) = \left(\frac{d^2u}{dx\,dy}\right)$, if (27) are true,

$$\left(\frac{dP}{dy}\right) = \left(\frac{dQ}{dx}\right): \tag{28}$$

and consequently if it is not on inspection plain whether (24) is an exact differential or not, we may apply the condition (28); and, if it is fulfilled, we are assured that (24) is an exact differential.

The equation (28) is commonly called *the condition of integrability*. Let us suppose it to be fulfilled. Since $P\, dx$ is the

x-partial differential of u, the x-integral of $\text{P}\,dx$ will give the function of x which enters into the general integral; and similarly the y-integral of $\text{Q}\,dy$ will give the function of y: the addition therefore to the x-integral of $\text{P}\,dx$ of those functions of y which the y-integral of $\text{Q}\,dy$ contains and which are not in the x-integral of $\text{P}\,dx$ will give the whole variable part of the general integral of (24); and the addition of a constant, or the determination of the definite integral, when the limits are given, at last gives the general integral of the differential equation.

Hence we have
$$u = \int \text{P}\,dx + \text{Y}, \tag{29}$$
$$u = \int \text{Q}\,dy + \text{X}; \tag{30}$$

where Y and X are functions severally of y and x only, and which are added to the partial integrals of $\text{P}\,dx$ and $\text{Q}\,dy$; and where Y is the sum of all the y-functions which are in (30) and are not in (29); and where X is the sum of the x-functions which are in (29) and are not in (30).

It will be observed that, if the variables are separated, as in (22), P and Q are functions severally of x and y only, and that
$$\left(\frac{d\text{P}}{dy}\right) = 0 = \left(\frac{d\text{Q}}{dx}\right);$$
so that the criterion of integrability is satisfied; and thus the general integral is determined by two single integrations.

Also if $\text{P}\,dx + \text{Q}\,dy$ can be divided into two parts, one of which is evidently an exact differential, it is sufficient to ascertain whether the other part is also such.

The following are examples of these processes;

Ex. 1. $(2ax+by+g)\,dx+(2cy+bx+e)\,dy = 0$;
$$\text{P} = 2ax+by+g, \qquad \therefore \quad \left(\frac{d\text{P}}{dy}\right) = b;$$
$$\text{Q} = 2cy+bx+e, \qquad \left(\frac{d\text{Q}}{dx}\right) = b;$$
so that the criterion of integrability is satisfied. Let u_x and u_y denote the x- and y-partial integrals; then
$$\left.\begin{aligned} u_x &= \int (2ax+by+g)\,dx = ax^2+bxy+gx; \\ u_y &= \int (2cy+bx+e)\,dy = cy^2+bxy+ey; \end{aligned}\right\} \tag{31}$$

and let Y and X be the functions of y and x respectively, which are added to the partial integrals as above; then

$$Y = cy^2 + ey; \qquad X = ax^2 + gx;$$

by means of either of which we have from (31),

$$u = ax^2 + bxy + cy^2 + gx + ey + k;$$

where k is an arbitrary constant.

Ex. 2. $\dfrac{y\,dx - x\,dy}{x^2+y^2} = 0;$

$$P = \frac{y}{x^2+y^2}; \qquad \therefore \quad \left(\frac{dP}{dy}\right) = \frac{x^2-y^2}{(x^2+y^2)^2};$$

$$Q = \frac{-x}{x^2+y^2}; \qquad \left(\frac{dQ}{dx}\right) = \frac{x^2-y^2}{(x^2+y^2)^2};$$

and thus the criterion of integrability is satisfied.

$$\therefore \quad u_x = \int \frac{y\,dx}{x^2+y^2} = \tan^{-1}\frac{x}{y};$$

$$u_y = \int \frac{-x\,dy}{x^2+y^2} = -\tan^{-1}\frac{y}{x} = \tan^{-1}\frac{x}{y} - \frac{\pi}{2};$$

$$\therefore \quad u = \tan^{-1}\frac{x}{y} + c.$$

Ex. 3. $\{\phi(xy) + xy\,\phi'(xy)\}\,dx + x^2\phi'(xy)\,dy = 0;$

$$P = \phi(xy) + xy\,\phi'(xy); \quad \therefore \quad \left(\frac{dP}{dy}\right) = 2x\phi'(xy) + x^2y\,\phi''(xy);$$

$$Q = x^2\phi'(xy); \qquad \left(\frac{dQ}{dx}\right) = 2x\phi'(xy) + x^2y\,\phi''(xy);$$

and thus the criterion of integrability is satisfied; and we have

$$u_x = \int \{\phi(xy) + xy\,\phi'(xy)\}\,dx = x\,\phi(xy);$$

$$u_y = \int x^2\phi'(xy)\,dy = x\,\phi(xy);$$

$$u = x\,\phi(xy) + c.$$

372.] If (29) has been found in a definite form, the unknown function Y in it may be determined without the integral (30). Take the y-differential of the definite integral of (29); viz.,

$$\left(\frac{du}{dy}\right) = \frac{d}{dy}\int_{x_0}^{x} P\,dx + \frac{dY}{dy}; \tag{32}$$

Then since $\left(\dfrac{du}{dy}\right) = Q$, we have $\dfrac{dY}{dy} = Q - \displaystyle\int_{x_0}^{x} \frac{dP}{dy}\,dx;$

$$\therefore \quad \frac{d\text{Y}}{dy} = \text{Q} - \int_{x_0}^{x} \frac{d\text{Q}}{dx}\, dx, \text{ by reason of (28),}$$
$$= \text{Q} - \text{Q} + \text{Q}_0,$$

representing by Q_0 the value of Q, when $x = x_0$:

$$\therefore \quad \text{Y} = \int_{y_0}^{y} \text{Q}_0\, dy; \tag{33}$$

$$\text{and} \quad u = \int_{x_0}^{x} \text{P}\, dx + \int_{y_0}^{y} \text{Q}_0\, dy. \tag{34}$$

Or representing P and Q by $f(x, y)$ and $\phi(x, y)$ we have

$$u = \int_{x_0}^{x} f(x, y)\, dx + \int_{y_0}^{y} \phi(x_0, y)\, dy; \tag{35}$$

$$\text{or} \quad u = \int_{y_0}^{y} \phi(x, y)\, dy + \int_{x_0}^{x} f(x, y_0)\, dx. \tag{36}$$

373.] Next let us take the case of a differential equation of three variables, and of the form

$$\text{P}\, dx + \text{Q}\, dy + \text{R}\, dz = 0; \tag{37}$$

where P, Q and R are functions of x, y and z. Now of course it may be that either (37) is an exact differential; or that some factor common to all the terms has been divided out, and that the expression can be made exact only by introducing this, or some other equivalent, factor: this latter case we shall, as heretofore, reserve to Section 6 of the present Chapter, and shall first consider the case where (37) is an exact differential. If we recognise immediately the general integral of (37), it is of the form,

$$u = \text{F}(x, y, z) = c; \tag{38}$$

and we need not apply criteria of integrability: and this is manifestly the case in such an example as,

$$\frac{x}{a^2}\, dx + \frac{y}{b^2}\, dy + \frac{z}{c^2}\, dz = 0;$$

$$\text{whence} \quad \frac{x^2}{a^2} + \frac{y^2}{b^2} + \frac{z^2}{c^2} - k^2 = u = 0;$$

where k is an arbitrary constant; and in

$$(y+z)\, dx + (z+x)\, dy + (x+y)\, dz = 0;$$

$$\text{whence} \quad yz + zx + xy - k^2 = u = 0.$$

If however the integral is not discoverable at first sight, still let us suppose (37) to be the exact differential of a function of three independent variables of the form (38); then

$$
\begin{aligned}
\mathrm{P}\,dx + \mathrm{Q}\,dy + \mathrm{R}\,dz &= \mathrm{D}u \\
&= \left(\frac{du}{dx}\right) dx + \left(\frac{du}{dy}\right) dy + \left(\frac{du}{dz}\right) dz; \qquad (39)
\end{aligned}
$$

and as dx, dy, dz are arbitrary, though infinitesimal, this equation can be true only when

$$
\mathrm{P} = \left(\frac{du}{dx}\right), \quad \mathrm{Q} = \left(\frac{du}{dy}\right), \quad \mathrm{R} = \left(\frac{du}{dz}\right). \qquad (40)
$$

Hereby we have criteria whether (37) is or is not an exact differential, for since

$$
\left(\frac{d^2u}{dy\,dz}\right) = \left(\frac{d^2u}{dz\,dy}\right), \left(\frac{d^2u}{dz\,dx}\right) = \left(\frac{d^2u}{dx\,dz}\right), \text{ and } \left(\frac{d^2u}{dx\,dy}\right) = \left(\frac{d^2u}{dy\,dx}\right),
$$

if (40) are true, we have

$$
\left(\frac{d\mathrm{R}}{dy}\right) = \left(\frac{d\mathrm{Q}}{dz}\right), \quad \left(\frac{d\mathrm{P}}{dz}\right) = \left(\frac{d\mathrm{R}}{dx}\right), \quad \left(\frac{d\mathrm{Q}}{dx}\right) = \left(\frac{d\mathrm{P}}{dy}\right), \qquad (41)
$$

which equations are called the conditions of integrability of (37); and if they are fulfilled we can integrate as follows:

Since $\mathrm{P}\,dx$ is the x-partial differential of the general integral, the x-integral of $\mathrm{P}\,dx$ will give us the whole function of x which enters into the general integral; similarly the y-integral of $\mathrm{Q}\,dy$ will give the whole function of y; and the z-integral of $\mathrm{R}\,dz$, the whole function of z: if therefore we add to the x-integral of $\mathrm{P}\,dx$ those functions of y which are in the y-integral of $\mathrm{Q}\,dy$ and are not in the x-integral of $\mathrm{P}\,dx$; and if again we add to the sum those functions of z which are in the z-integral of $\mathrm{R}\,dz$, and which have not already entered, the result will evidently contain all the variable part of the general integral, and therefore by the addition of an arbitrary constant the general integral will be obtained.

Hence we have

$$
\left.
\begin{aligned}
u &= \int \mathrm{P}\,dx + \mathrm{Y} + \mathrm{Z}; \\
u &= \int \mathrm{Q}\,dy + \mathrm{Z}_1 + \mathrm{X}_1; \\
u &= \int \mathrm{R}\,dz + \mathrm{X}_2 + \mathrm{Y}_2;
\end{aligned}
\right\} \qquad (42)
$$

where X, Y, .. are severally functions of x, y, ... only; and are determined in the manner explained above.

In this case it is evident that if the variables are separated, the criteria of integrability are satisfied.

Also that if $\mathrm{P}\,dx + \mathrm{Q}\,dy + \mathrm{R}\,dz$ can be divided into portions of which some are at once perceived to be exact differentials, it is sufficient to ascertain by the criteria whether the other parts are also exact differentials.

Ex. 1. $yz\,dx+zx\,dy+xy\,dz = 0.$

$$\left.\begin{array}{l}\left(\frac{d\text{P}}{dy}\right) = z, \\ \left(\frac{d\text{P}}{dz}\right) = y;\end{array}\right\} \quad \left.\begin{array}{l}\left(\frac{d\text{Q}}{dz}\right) = x, \\ \left(\frac{d\text{Q}}{dx}\right) = z;\end{array}\right\} \quad \left.\begin{array}{l}\left(\frac{d\text{R}}{dx}\right) = y, \\ \left(\frac{d\text{R}}{dy}\right) = x;\end{array}\right\}$$

and the conditions of integrability are satisfied. Let u_x, u_y, u_z denote the several partial integrals; then, taking indefinite integrals,

$$u_x = \int \text{P}\,dx = \int yz\,dx = xyz;$$

similarly, $u_y = xyz$; $u_z = xyz$;

$$\therefore\quad u = xyz-k^3 = 0;$$

where k is an arbitrary constant; and this is the general integral.

Ex. 2. $\dfrac{x\,dx+y\,dy+z\,dz}{(x^2+y^2+z^2)^{\frac{1}{2}}}+\dfrac{z\,dx-x\,dz}{x^2+z^2}+3ax^2dx+2by\,dy+c\,dz=0.$

The several parts of this equation are so evidently exact differentials, that it is unnecessary to apply the criteria; and for the general integral, if k is an arbitrary constant, we have

$$u = (x^2+y^2+z^2)^{\frac{1}{2}}+\tan^{-1}\frac{x}{z}+ax^3+by^2+cz+k.$$

374.] When the criteria of integrability are satisfied, the general integral may be expressed in terms of definite partial integrals as follows: as the process is similar to that of Art. 372, it is unnecessary to repeat every step of it; let the differential equation be

$$f_1(x,y,z)\,dx+f_2(x,y,z)\,dy+f_3(x,y,z)\,dz = 0;$$

$$\therefore\quad u = \int_{x_0}^{x} f_1(x,y,z)\,dx+v, \tag{43}$$

where v is a function of y and z;

$$\therefore\quad \left(\frac{du}{dy}\right) = \int_{x_0}^{x}\frac{d}{dy}f_1(x,y,z)\,dx+\left(\frac{dv}{dy}\right);$$

$$\therefore\quad \left(\frac{dv}{dy}\right) = f_2(x,y,z)-\int_{x_0}^{x}\frac{d}{dx}f_2(x,y,z)\,dx$$

$$= f_2(x_0,y,z);$$

$$\therefore\quad v = \int_{y_0}^{y} f_2(x_0,y,z)\,dy+w, \tag{44}$$

where w is a function of z only;

$$\therefore \quad u = \int_{x_0}^{x} f_1(x, y, z)\,dx + \int_{y_0}^{y} f_2(x_0, y, z)\,dy + w,$$

$$\left(\frac{du}{dz}\right) = \int_{x_0}^{x} \frac{d}{dz} f_1(x, y, z)\,dx + \int_{y_0}^{y} \frac{d}{dz} f_2(x_0, y, z)\,dy + \frac{dw}{dz};$$

$$\therefore \quad f_3(x, y, z) = \int_{x_0}^{x} \frac{d}{dx} f_3(x, y, z)\,dx + \int_{y_0}^{y} \frac{d}{dy} f_3(x_0, y, z)\,dy + \frac{dw}{dz};$$

$$\therefore \quad \frac{dw}{dz} = f_3(x_0, y_0, z);$$

$$\therefore \quad w = \int_{z_0}^{z} f_3(x_0, y_0, z)\,dz;$$

$$\therefore \quad u = \int_{x_0}^{x} f_1(x, y, z)\,dx + \int_{y_0}^{y} f_2(x_0, y, z)\,dy + \int_{z_0}^{z} f_3(x_0, y_0, z)\,dz. \quad (45)$$

As an example of this form let us take the simple equation,

$$\frac{3x^2dx + 2y\,dy + dz}{x^3 + y^2 + z} = 0:$$

$$\therefore \quad u = \int_{x_0}^{x} \frac{3x^2dx}{x^3 + y^2 + z} + \int_{y_0}^{y} \frac{2y\,dy}{x_0^3 + y^2 + z} + \int_{z_0}^{z} \frac{dz}{x_0^3 + y_0^2 + z}$$

$$= \log(x^3 + y^2 + z) - \log(x_0^3 + y^2 + z)$$

$$+ \log(x_0^3 + y^2 + z) - \log(x_0^3 + y_0^2 + z)$$

$$+ \log(x_0^3 + y_0^2 + z) - \log(x_0^3 + y_0^2 + z_0)$$

$$= \log(x^3 + y^2 + z) - \log(x_0^3 + y_0^2 + z_0).$$

375.] Lastly, let us briefly consider a differential equation of n independent variables of the form

$$P_1\,dx_1 + P_2\,dx_2 + \ldots + P_n\,dx_n = 0; \quad (46)$$

where $P_1, P_2, \ldots P_n$ are functions of the n variables $x_1, x_2, \ldots x_n$. In order that this may be an exact differential of a function

$$u = F(x_1, x_2, \ldots x_n) = c, \quad (47)$$

we must have

$$\left(\frac{du}{dx_1}\right) = P_1, \quad \left(\frac{du}{dx_2}\right) = P_2, \ldots \quad \left(\frac{du}{dx_n}\right) = P_n; \quad (48)$$

and that these equations should be true, it is necessary that

$$\left.\begin{array}{r} \left(\frac{dP_1}{dx_2}\right) = \left(\frac{dP_2}{dx_1}\right), \left(\frac{dP_1}{dx_3}\right) = \left(\frac{dP_3}{dx_1}\right), \ldots \left(\frac{dP_1}{dx_n}\right) = \left(\frac{dP_n}{dx_1}\right), \\ \left(\frac{dP_2}{dx_3}\right) = \left(\frac{dP_3}{dx_2}\right), \ldots \left(\frac{dP_2}{dx_n}\right) = \left(\frac{dP_n}{dx_2}\right), \\ \ldots\ldots \left(\frac{dP_3}{dx_n}\right) = \left(\frac{dP_n}{dx_3}\right), \\ \cdot \quad \cdot \quad \cdot \quad \cdot \quad \cdot \quad \cdot \quad \cdot \\ \left(\frac{dP_{n-1}}{dx_n}\right) = \left(\frac{dP_n}{dx_{n-1}}\right); \end{array}\right\} \quad (49)$$

the number of which conditions is the same as that of n things taken two and two together, that is, is $\frac{n(n-1)}{2}$; and when these are satisfied, and the n partial-integrals are found, the general integral may be determined from them by a process similar to that employed in the cases of two and three independent variables.

We may also express the general integral in terms of definite partial integrals in the following manner. Let the coefficients of the differentials in (46) be $f_1(x_1, x_2, \dots x_n)$, $f_2(x_1, x_2, \dots x_n)$, $\dots f_n(x_1, x_2, \dots x_n)$; and let the inferior limits of integration be $\mathrm{X}_1, \mathrm{X}_2, \dots \mathrm{X}_n$: then by a process similar to that of the last Article,

$$u = \int_{\mathrm{X}_1}^{x_1} f_1(x_1, x_2, \dots x_n)\, dx_1 + \int_{\mathrm{X}_2}^{x_2} f_2(\mathrm{X}_1, x_2, \dots x_n)\, dx_2 + \dots$$
$$\dots + \int_{\mathrm{X}_n}^{x_n} f_n(\mathrm{X}_1, \mathrm{X}_2, \dots \mathrm{X}_{n-1}, x_n)\, dx_n. \quad (50)$$

SECTION 3.—*Integration of homogeneous equations of two variables.*

377.] Differential equations of the first order and degree can generally be integrated only when they satisfy the criteria of integrability; and therefore when an equation does not fulfil these conditions, our first object is to investigate, if it be possible, some mode of so transforming it that its equivalent may be in the required form: the principal means which are useful for such a transformation are (1) an introduction of new variables by way of substitution, (2) the multiplication of the equation by a factor which will render it an exact differential, and which is commonly called *an integrating factor:* these processes we go on to examine.

First, let us take the case of two variables x and y, and suppose the equation to be

$$\mathrm{P}\, dx + \mathrm{Q}\, dy = 0; \quad (51)$$

let us suppose that the criterion of integrability is not satisfied, but that P and Q are homogeneous functions of x and y of n dimensions: then dividing through by x^n, so that x^n may stand as a common factor, (51) takes the form

$$x^n f\left(\frac{y}{x}\right) dx + x^n \phi\left(\frac{y}{x}\right) dy = 0. \quad (52)$$

Let $y = xz$; therefore $dy = x\,dz + z\,dx$; and neglecting the factor x^n, (52) becomes

$$f(z)\,dx + \phi(z)\,\{x\,dz + z\,dx\} = 0;$$

$$\therefore \quad \frac{dx}{x} + \frac{\phi(z)\,dz}{f(z) + z\phi(z)} = 0; \tag{53}$$

in which equation the variables are separated; and consequently the condition of integrability is satisfied; and thus the integration depends on that of two simple differentials of one variable.

Instead of arranging the equation (51) in the form (52), wherein x^n is the common factor, it might equally as well have been put into the form

$$y^n f\left(\frac{x}{y}\right) dx + y^n \phi\left(\frac{x}{y}\right) dy = 0; \tag{54}$$

and if x is replaced by yz the variables will be separated, and the criterion of integrability will be satisfied.

Ex. 1. $y^2\,dx + (xy + x^2)\,dy = 0.$

Let $x = yz$; $\quad \therefore \quad dx = y\,dz + z\,dy$;

$$y^2(y\,dz + z\,dy) + (y^2 z + y^2 z^2)\,dy = 0;$$

$$\therefore \quad \frac{dy}{y} + \frac{dz}{z^2 + 2z} = 0;$$

$$\log\frac{y}{c} + \frac{1}{2}\log\frac{z}{z+2} = 0;$$

$$\therefore \quad y^2 = c^2\,\frac{x + 2y}{x}.$$

Ex. 2. $x^2 y\,dx - (x^3 + y^3)\,dy = 0.$

Let $x = yz$; $\quad \therefore \quad dx = y\,dz + z\,dy$;

$$\therefore \quad \frac{dy}{y} = z^2\,dz;$$

$$y = c e^{\frac{x^3}{3y^3}}.$$

Ex. 3. $x\,dy - y\,dx = (x^2 + y^2)^{\frac{1}{2}}\,dx.$

This is an homogeneous equation of one dimension.

Let $y = xz$; $\quad dy = x\,dz + z\,dx$;

$$\therefore \quad x(x\,dz + z\,dx) = xz\,dx + (x^2 + x^2 z^2)^{\frac{1}{2}}\,dx;$$

$$\therefore \quad \frac{dx}{x} = \frac{dz}{(1 + z^2)^{\frac{1}{2}}};$$

$$x^2 = 2cy + c^2;$$

which is the general integral.

Although either of the substitutions $y = xz$, or $x = yz$, will produce the result, yet a judicious choice will frequently shorten the process: the student however must in this matter be left to his own skill.

378.] Let us also consider homogeneous equations of the form (52), and the introduction of the new variable, from a geometrical point of view; (52) may evidently be put in the form

$$\frac{dy}{dx} = \text{F}\left(\frac{y}{x}\right);$$

and since $\frac{dy}{dx} = \tan \tau$, and $\frac{y}{x} = \tan \theta$, τ and θ being the angles respectively at which the tangent to a curve and the radius vector are inclined to the axis of x, the above equation, interpreted geometrically, expresses a relation between these two angles. Thus if $\tau = 2\theta$; then

$$\frac{dy}{dx} = \frac{2xy}{x^2 - y^2};$$

Let $x = yz$; $\quad \therefore \quad dx = y\,dz + z\,dy$;

$$\therefore \quad \frac{dy}{y} + \frac{2z\,dz}{z^2+1} = 0;$$

$$\therefore \quad \log \frac{y}{2c} + \log (z^2 + 1) = 0;$$

$$\therefore \quad x^2 = 2cy - y^2;$$

which is the equation to a circle, whose radius is the arbitrary constant c.

And to take another example, see fig. 50: to find the equation to a curve such that a perpendicular MS let fall from M, the foot of the ordinate, on the radius vector OP shall cut the axis of y at the point T′, where it is cut by the tangent PT′.

In this problem $\tan \text{SOM} = \tan \text{OT′M}$;

$$\therefore \quad \frac{y}{x} = \frac{x\,dx}{y\,dx - x\,dy};$$

$$\therefore \quad xy\,dy + (x^2 - y^2)\,dx = 0;$$

$$\therefore \quad x^2 = c^2 e^{\frac{y^2}{x^2}}.$$

379.] By the introduction of the new variable z the original expression (51) has been so transformed that the variables admit of separation; let us examine the process more closely: take the form (52) and compare it with (51); then

$$\left.\begin{aligned} P &= x^n f\left(\frac{y}{x}\right) = x^n f(z), \\ Q &= x^n \phi\left(\frac{y}{x}\right) = x^n \phi(z); \end{aligned}\right\} \qquad (55)$$

and (53) has been found by dividing (52) by $x^{n+1}\{f(z)+z\phi(z)\}$, which is manifestly equal to $Px+Qy$; hence the equation (51) satisfies the criteria of integrability when it is divided by $Px+Qy$; therefore $(Px+Qy)^{-1}$ is an integrating factor of $P\,dx+Q\,dy = 0$. Let us apply this process of integrating homogeneous equations.

Ex. 1. $x\,dx+y\,dy = m(x\,dy-y\,dx)$.

$$(x+my)\,dx+(y-mx)\,dy = 0;$$

therefore the integrating factor is

$$\{x^2+mxy+y^2-mxy\}^{-1} = (x^2+y^2)^{-1};$$

and the equation becomes an exact differential of the form

$$\frac{(x+my)\,dx+(y-mx)\,dy}{x^2+y^2} = 0 = \mathrm{D}u;$$

$$\therefore\quad u_x = \int\frac{x+my}{x^2+y^2}\,dx = \log(x^2+y^2)^{\frac{1}{2}}+m\tan^{-1}\frac{x}{y};$$

$$u_y = \int\frac{y-mx}{x^2+y^2}\,dy = \log(x^2+y^2)^{\frac{1}{2}}+m\tan^{-1}\frac{x}{y};$$

$$\therefore\quad u = \log(x^2+y^2)^{\frac{1}{2}}+m\tan^{-1}\frac{x}{y}+c = 0.$$

Ex. 2. Again, let us take Ex. 2, in Art. 377,

$$x^2y\,dx-(x^3+y^3)\,dy = 0;$$

the integrating factor is $-y^{-4}$: therefore

$$\frac{x^2y\,dx-(x^3+y^3)\,dy}{-y^4} = \mathrm{D}u;$$

$$u_x = -\int\frac{x^2\,dx}{y^3} = -\frac{x^3}{3y^3};$$

$$u_y = \int\left(\frac{1}{y}+\frac{x^3}{y^4}\right)dy = \log y-\frac{x^3}{3y^3};$$

$$\therefore\quad u = \log\frac{y}{c}-\frac{x^3}{3y^3} = 0.$$

380.] And that the factor $(Px+Qy)^{-1}$ renders (51) an exact differential is also evident, inasmuch as the condition of integrability becomes satisfied: for multiplying by this factor, we have

$$\frac{P\,dx+Q\,dy}{Px+Qy} = 0;$$

and the condition of integrability is

$$\frac{d}{dy}\frac{P}{Px+Qy}-\frac{d}{dx}\frac{Q}{Px+Qy}=$$

$$\frac{(Px+Qy)\left\{\left(\frac{dP}{dy}\right)-\left(\frac{dQ}{dx}\right)\right\}-P\left\{x\left(\frac{dP}{dy}\right)+y\left(\frac{dQ}{dy}\right)+Q\right\}+Q\left\{P+x\left(\frac{dP}{dx}\right)+y\left(\frac{dQ}{dx}\right)\right\}}{(Px+Qy)^2}$$

$$=\frac{Q\left\{x\left(\frac{dP}{dx}\right)+y\left(\frac{dP}{dy}\right)\right\}-P\left\{x\left(\frac{dQ}{dx}\right)+y\left(\frac{dQ}{dy}\right)\right\}}{(Px+Qy)^2}$$

$$=\frac{nPQ-nPQ}{(Px+Qy)^2}$$

$$=0,$$

since by Euler's Theorem, P and Q being homogeneous functions of n dimensions,

$$\left.\begin{aligned} x\left(\frac{dP}{dx}\right)+y\left(\frac{dP}{dy}\right)&=nP,\\ x\left(\frac{dQ}{dx}\right)+y\left(\frac{dQ}{dy}\right)&=nQ;\end{aligned}\right\}\qquad(56)$$

and thus the criterion of integrability is satisfied.

If P and Q are such that $Px+Qy=0$, the preceding process fails; in this case however another integrating factor can be found by a method which will be developed in a future section of the present Chapter.

381.] A form of differential equation which is easily reduced to the homogeneous form is

$$(a_1x+b_1y+c_1)dx+(a_2x+b_2y+c_2)dy=0:\qquad(57)$$

let $\quad a_1x+b_1y+c_1=\xi,\qquad a_2x+b_2y+c_2=\eta;$

$$\therefore\quad d\xi=a_1dx+b_1dy,\qquad d\eta=a_2dx+b_2dy;$$

$$\therefore\quad dx=\frac{b_2d\xi-b_1d\eta}{a_1b_2-a_2b_1},\qquad dy=\frac{a_2d\xi-a_1d\eta}{a_2b_1-a_1b_2};\qquad(58)$$

and substituting in (57), and reducing,

$$(b_2\xi+a_2\eta)\,d\xi-(b_1\xi+a_1\eta)\,d\eta=0;\qquad(59)$$

a homogeneous equation, which is integrable as above.

This transformation is manifestly equivalent to that of a system of rectangular coordinate axes, in which the origin and the direction of the axes are changed: and it is always possible, unless

$$\frac{a_1}{a_2}=\frac{b_1}{b_2}=k\ (\text{say}),\qquad(60)$$

for in this case $d\xi$ and $d\eta$ are infinite: but then (57) becomes

$$(ka_2x+kb_2y+c_1)\,dx+(a_2x+b_2y+c_2)\,dy = 0,$$

$$(a_2x+b_2y)(dy+k\,dx)+c_1\,dx+c_2\,dy = 0; \qquad (61)$$

in which, if we put $a_2x+b_2y=z$, and eliminate x or y, the variables will be separated, and the integration can be performed.

Also by a similar substitution may the variables be separated in the equation

$$dy = f(ax+by)\,dx. \qquad (62)$$

Section 4.—*The integration of the first linear differential equation.*

382.] Another form of differential equations in which the variables admit of separation is

$$P_1\frac{dy}{dx}+P_2y+P_3 = 0, \qquad (63)$$

where P_1, P_2, P_3 are functions of x only; and which is called the *linear* equation of the first order, because $\frac{dy}{dx}$ and y enter in only the first degree, and there is a vague analogy between it and the equation to a straight line.

Dividing through by P_1 and making obvious substitutions, the equation becomes

$$dy+f(x)\,y\,dx = F(x)\,dx. \qquad (64)$$

Let $y = zt$; $\quad\therefore\; dy = z\,dt+t\,dz;$ (65)

$$\therefore\; z\,dt+t\,dz+f(x)\,zt\,dx = F(x)\,dx$$

$$z\,dt+t\,\{dz+f(x)\,z\,dx\} = F(x)\,dx. \qquad (66)$$

As we have introduced two new variables z and t, and have made only one supposition respecting them, we may make another; let this be,

$$dz+f(x)\,z\,dx = 0; \qquad (67)$$

$$\therefore\; \log z = -\int f(x)\,dx,$$

$$z = e^{-\int f(x)\,dx}; \qquad (68)$$

thus (66) takes the form $dt = e^{\int f(x)\,dx}\,F(x)\,dx$;

$$\therefore\; t = C+\int e^{\int f(x)\,dx}\,F(x)\,dx;$$

and

$$y = zt = e^{-\int f(x)\,dx}\left\{C+\int e^{\int f(x)\,dx}\,F(x)\,dx\right\}. \qquad (69)$$

No constant has been introduced in (68), because it is desirable to keep complex formulæ in as simple a form as possible: and the generality of the final result is not affected by the omission, for such a constant would disappear in (69) by reason of the form of the result.

In terms of definite integrals (69) is

$$y = e^{-\int_{x_0}^{x} f(x)\,dx}\left\{y_0 + \int_{x_0}^{x} e^{\int_{x_0}^{x} f(x)\,dx}\,\mathrm{F}(x)\,dx\right\}.$$

Ex. 1. $dy + y\,dx = e^{-x}\,dx.$

$$y = zt; \qquad \therefore \quad dy = t\,dz + z\,dt.$$

$$t\,dz + z\,(dt + t\,dx) = e^{-x}\,dx.$$

Let $\qquad dt + t\,dx = 0; \qquad \therefore \quad t = e^{-x}:$

$$\therefore \quad z = x + c,$$

$$y = zt = (x+c)\,e^{-x}.$$

Ex. 2. $(1-x^2)\,dy + xy\,dx = a\,dx.$

$$y = ax + c\,(1-x^2)^{\frac{1}{2}}.$$

Ex. 3. $(x+1)\,dy - ny\,dx = e^x\,(x+1)^{n+1}\,dx.$

$$y = (e^x + c)\,(x+1)^n.$$

Ex. 4. $2\,dy + 2y\cos x\,dx = \sin 2x.$

383.] A form which admits of reduction to (64), and consequently of having its integral determined in the form (69), is

$$\frac{dy}{dx} + y f(x) = \mathrm{F}(x)\,y^n; \tag{70}$$

which is generally known by the name of Bernoulli's linear equation of the first order; for dividing through by y^n, we have

$$y^{-n}\,dy + y^{-n+1} f(x)\,dx = \mathrm{F}(x)\,dx. \tag{71}$$

Let $\quad y^{-n+1} = z; \qquad \therefore \quad -(n-1)\,y^{-n}\,dy = dz;$

and therefore by substitution,

$$dz - (n-1)\,z f(x)\,dx = -(n-1)\,\mathrm{F}(x)\,dx; \tag{72}$$

and by (69),

$$z = \frac{1}{y^{n-1}} = e^{(n-1)\int f(x)\,dx}\left\{C - (n-1)\int e^{-(n-1)\int f(x)\,dx}\,\mathrm{F}(x)\,dx\right\}. \tag{73}$$

The explanation of the failure of the above substitution when $n = 1$ is too obvious to require more than a passing remark.

Ex. 5. $\quad \dfrac{dy}{dx} + \dfrac{xy}{1-x^2} = xy^{\frac{1}{2}};$

$$\therefore \quad dy\,y^{-\frac{1}{2}} + \frac{x}{1-x^2}y^{\frac{1}{2}}dx = x\,dx.$$

Let $y^{\frac{1}{2}} = z$; $\therefore \quad \frac{dy}{2y^{\frac{1}{2}}} = dz$;

$$\therefore \quad dz + \frac{x\,dx}{2(1-x^2)}z = \frac{x\,dx}{2};$$

$$\therefore \quad z = y^{\frac{1}{2}} = (1-x^2)^{\frac{1}{4}}\left\{c - \frac{1}{3}(1-x^2)^{\frac{3}{4}}\right\},$$

$$y^{\frac{1}{2}} = c(1-x^2)^{\frac{1}{4}} - \frac{1-x^2}{3}.$$

Ex. 2. $dy + 2y\,x\,dx = 2a\,x^3\,y^3\,dx.$

The preceding is only a particular case of the following more general differential equation which is evidently capable of solution by the same process; viz.

$$f'(y)\frac{dy}{dx} + f(y)\,\phi(x) = \text{F}(x). \tag{74}$$

Ex. 1. $6x^4\,y^2\frac{dy}{dx} - 4x^3\,y^3 = 1.$

Section 5.—*The integration of partial differential equations of the first order and degree.*

384.] We must now consider differential expressions of another character; those, viz., wherein a relation is given between partial derived-functions and the variables: the general forms of these are (3) and (4) in Art. 364. I shall at present take the case wherein the partial derived-functions enter linearly, and where the coefficients are functions of the variables, including of course the case where they are constant.

First let it be observed, that a partial differential expression which arises from an implicit function of two variables of the form

$$u = \text{F}(x, y) = c, \tag{75}$$

and the general form of which is

$$\text{P}\left(\frac{du}{dx}\right) + \text{Q}\left(\frac{du}{dy}\right) = 0, \tag{76}$$

where P and Q are functions of x and y, although involving partial derived-functions, is in fact a total differential expression; for differentiating (75), we have

$$\left(\frac{du}{dx}\right)dx + \left(\frac{du}{dy}\right)dy = 0;$$

making which identical with (76) we have

$$\frac{\text{P}}{dx} = \frac{\text{Q}}{dy};$$

$$\therefore \quad \text{Q}dx - \text{P}dy = 0; \tag{77}$$

which is a total differential equation of the form (24).

Now, from the explanation of partial differential equations which has been given in Article 366, it follows that the integral of a partial differential equation of the first order and degree requires the introduction of an arbitrary function; and although the integral may be particular, yet it is not general without it. Since then a total differential equation of the form (77) may by an inversion of the process followed above be changed into a partial differential equation, so does its general integral require an arbitrary function: the method of determining it will be explained in Section 6 of the present Chapter: thereby also we shall be led to a solution of total differential equations still more general than that of the preceding Sections.

Let us now consider a partial differential expression of three variables x, y, z, and of the form

$$\text{P}\left(\frac{dz}{dx}\right) + \text{Q}\left(\frac{dz}{dy}\right) = \text{R}, \tag{78}$$

where P, Q, R are or may be functions of x, y and z, and in which z has been considered a variable dependent on two independent variables y and x. To discuss it however in the most general form, let us suppose the original function to be of the form

$$u = \text{F}(x, y, z) = c, \tag{79}$$

where F denotes the arbitrary function, which the complete integral requires; then, by the process of Art. 50, Vol. I,

$$\left(\frac{dz}{dx}\right) = -\frac{\left(\frac{du}{dx}\right)}{\left(\frac{du}{dz}\right)}, \qquad \left(\frac{dz}{dy}\right) = -\frac{\left(\frac{du}{dy}\right)}{\left(\frac{du}{dz}\right)}; \tag{80}$$

substituting which in (78) we have

$$\text{P}\left(\frac{du}{dx}\right) + \text{Q}\left(\frac{du}{dy}\right) + \text{R}\left(\frac{du}{dz}\right) = 0; \tag{81}$$

and this is in the most general form of a partial differential equation of the first order and degree. It is the integral of this that we shall investigate.

Now of (79), the differential is

$$\left(\frac{du}{dx}\right)dx + \left(\frac{du}{dy}\right)dy + \left(\frac{du}{dz}\right)dz = 0; \qquad (82)$$

from which and (81), we have

$$\frac{\left(\frac{du}{dx}\right)}{Q\,dz - R\,dy} = \frac{\left(\frac{du}{dy}\right)}{R\,dx - P\,dz} = \frac{\left(\frac{du}{dz}\right)}{P\,dy - Q\,dx}. \qquad (83)$$

Let us assume that

$$\frac{P}{dx} = \frac{Q}{dy} = \frac{R}{dz}; \qquad (84)$$

either of which equations, it will be observed, involves the other by reason of (83); and let us suppose two independent integrals of these equations to be found, and to contain two arbitrary constants c_1 and c_2; and to be of the form

$$f_1(x, y, z) = c_1, \qquad f_2(x, y, z) = c_2, \qquad (85)$$

where c_1 and c_2 are arbitrary constants.

Then from these we have

$$\left.\begin{aligned} \left(\frac{df_1}{dx}\right)dx + \left(\frac{df_1}{dy}\right)dy + \left(\frac{df_1}{dz}\right)dz &= 0; \\ \left(\frac{df_2}{dx}\right)dx + \left(\frac{df_2}{dy}\right)dy + \left(\frac{df_2}{dz}\right)dz &= 0; \end{aligned}\right\} \qquad (86)$$

from which, by reason of (84), we have

$$\left.\begin{aligned} P\left(\frac{df_1}{dx}\right) + Q\left(\frac{df_1}{dy}\right) + R\left(\frac{df_1}{dz}\right) &= 0, \\ P\left(\frac{df_2}{dx}\right) + Q\left(\frac{df_2}{dy}\right) + R\left(\frac{df_2}{dz}\right) &= 0; \end{aligned}\right\} \qquad (87)$$

on comparison of which with (81) it appears that either f_1 or f_2 satisfies (81): and so also will any arbitrary function of f_1 and f_2: for let F represent an arbitrary function of f_1 and f_2, that is, of c_1 and c_2; then multiplying the members of (87) severally by $\left(\frac{dF}{df_1}\right)$ and $\left(\frac{dF}{df_2}\right)$, and adding, we have

$$P\left(\frac{dF}{dx}\right) + Q\left(\frac{dF}{dy}\right) + R\left(\frac{dF}{dz}\right) = 0;$$

and therefore $F(f_1, f_2)$ satisfies (81): and therefore we have for the general integral

$$\left.\begin{aligned} u &= F(f_1, f_2) = 0, \\ &= F(c_1, c_2) = 0; \end{aligned}\right\} \qquad (88)$$

or, as it may be expressed,

$$\left.\begin{array}{l} c_1 = f(c_2), \\ f_1 = f(f_2); \end{array}\right\} \tag{89}$$

and either (88) or (89) is the general integral, because each contains an arbitrary function in its most general form.

385.] The process requires further development and illustration: but it will be better first to consider and solve some particular examples.

Suppose the given equation to be

$$\left(\frac{dz}{dx}\right) = f(x, y); \tag{90}$$

$$\text{then} \quad z = \int f(x, y)\,dx + \phi(y), \tag{91}$$

where $\phi(y)$ is the arbitrary function which enters into the general integral, and which has y only for its subject. Similarly, if

$$\left(\frac{dz}{dy}\right) = f(x, y); \qquad z = \int f(x, y)\,dy + \phi(x).$$

Thus if

$$\left(\frac{dz}{dx}\right) = \frac{z+y}{x+y};$$

$$\frac{dz}{z+y} = \frac{dx}{x+y},$$

$$\log(z+y) - \log(x+y) = \log\phi(y);$$

$$\therefore \quad \frac{z+y}{x+y} = \phi(y);$$

and this is the complete integral.

We may also thus prove (91): replacing $\left(\frac{dz}{dx}\right)$ in (90) by its value from (80), we have

$$\left(\frac{du}{dx}\right) + \left(\frac{du}{dz}\right) f(x, y) = 0;$$

and therefore from (84), $\dfrac{dx}{1} = \dfrac{dy}{0} = \dfrac{dz}{f(x, y)}$;

$$\therefore \quad dy = 0, \qquad y = c_1;$$

$$dz = dx\,f(x, y) = dx\,f(x, c_1);$$

$$\therefore \quad z = \int f(x, c_1)\,dx + c_2 = \phi(c_1) + \int f(x, c_1)\,dx$$

$$= \phi(y) + \int f(x, y)\,dx.$$

Ex. 1. $a\left(\frac{dz}{dx}\right)+b\left(\frac{dz}{dy}\right)=c;$

or, what is equivalent, by means of the substitutions (80),

$$a\left(\frac{du}{dx}\right)+b\left(\frac{du}{dy}\right)+c\left(\frac{du}{dz}\right)=0. \quad (92)$$

Now by the conditions (84) we have

$$\frac{dx}{a}=\frac{dy}{b}=\frac{dz}{c}; \quad (93)$$

$$\therefore \quad bx-ay=c_1, \qquad cx-az=c_2; \quad (94)$$

and therefore by reason of (89),

$$bx-ay=f(cx-az); \quad (95)$$

or $$u=F(bx-ay,\, cx-az)=0; \quad (96)$$

either of which is the general integral and involves an arbitrary functional symbol.

It is useful to observe the geometrical interpretation of the process:

Let (95) or (96) represent a surface: then from (92) it appears that the normal to the surface at every point of it is perpendicular to a straight line, whose direction-cosines are proportional to the quantities a, b, c: but as these determine the *direction* and not the *position* of a line, we can only conclude that every normal is perpendicular to one of a series of *parallel* straight lines: and the successive positions of these lines may vary according to any law; which law however is not given by the differential equation, but is required for the integral equation of the surface: in fact the insertion of it is absolutely necessary; for otherwise the equation cannot represent a surface; and the geometrical form of the law is the equation to the director curve along which the parallel straight line moves, and generates the surface; and this surface is cylindrical. This is also manifest from (88) and (94); (94) are the equations to two systems of parallel planes respectively parallel to the axes of z and y: and the intersection of two, viz., one of each system, will give the generating line of the surface; and the line of intersection will of course vary according to the functional relation between c_1 and c_2, the particular values of which determine the particular intersecting planes.

Ex. 2. $(x-a)\left(\frac{dz}{dx}\right)+(y-b)\left(\frac{dz}{dy}\right)=z-c.$

The equivalent of this in the most general form is

$$(x-a)\left(\frac{du}{dx}\right)+(y-b)\left(\frac{du}{dy}\right)+(z-c)\left(\frac{du}{dz}\right)=0; \qquad (97)$$

and therefore by (84),

$$\frac{dx}{x-a}=\frac{dy}{y-b}=\frac{dz}{z-c};$$

$$\therefore \quad \log(x-a) = \log(y-b)+\log c_1,$$

$$\log(x-a) = \log(z-c)+\log c_2;$$

$$\therefore \quad c_1 = \frac{x-a}{y-b}, \qquad c_2 = \frac{x-a}{z-c};$$

$$\therefore \quad \frac{x-a}{y-b} = f\left(\frac{x-a}{z-c}\right); \qquad (98)$$

$$u = \mathrm{F}\left(\frac{x-a}{y-b}, \frac{x-a}{z-c}\right)=0;$$

which may also be expressed as follows,

$$u = \mathrm{F}\left(\frac{y-b}{z-c}, \frac{z-c}{x-a}, \frac{x-a}{y-b}\right)=0.$$

Observe the geometrical meaning: (97) indicates that the normal to the surface is perpendicular to a straight line which passes through a given point (a, b, c), and therefore the surface is generated by a straight line which passes through the given point and moves according to a given law: and this is a property of conical surfaces, of which therefore (98) is the general equation, and the arbitrary functional symbol contained in it expresses the law of the director-curve.

Ex. 3. $(mz-ny)\left(\frac{dz}{dx}\right)+(nx-lz)\left(\frac{dz}{dy}\right) = ly-mx.$

The equivalent to this in its most general form is

$$(mz-ny)\left(\frac{du}{dx}\right)+(nx-lz)\left(\frac{du}{dy}\right)+(ly-mx)\left(\frac{du}{dz}\right)=0; \qquad (99)$$

$$\text{let} \quad \frac{dx}{mz-ny}=\frac{dy}{nx-lz}=\frac{dz}{ly-mx}; \qquad (100)$$

then multiplying the numerators and denominators severally by x, y, z, and adding; and again operating in the same way with l, m, n; the sum of denominators in each case is zero: therefore the sum of the numerators must also vanish: consequently

$$x\,dx+y\,dy+z\,dz = 0; \qquad l\,dx+m\,dy+n\,dz = 0; \qquad (101)$$

$$\therefore \quad x^2+y^2+z^2 = c_1, \qquad lx+my+nz = c_2; \qquad (102)$$

$$\therefore \quad x^2+y^2+z^2 = f(lx+my+nz); \tag{103}$$

$$\text{or} \quad u = F(x^2+y^2+z^2,\ lx+my+nz) = 0;$$

either of which is the equation of a surface of revolution, and in which the origin is on the axis of revolution; and equation (99) implies that all the normals of the surface pass through the axis: also from (102), which are the equations of a sphere and a plane, it follows that all plane sections of the surface, which are perpendicular to the line whose direction-cosines are proportional to l, m, n, are circles.

Ex. 4. $x^2\left(\frac{dz}{dx}\right)+y^2\left(\frac{dz}{dy}\right) = z^2.$

$$x^2\left(\frac{du}{dx}\right)+y^2\left(\frac{du}{dy}\right)+z^2\left(\frac{du}{dz}\right) = 0;$$

whence we have

$$\frac{1}{x}-\frac{1}{y} = f\left(\frac{1}{x}-\frac{1}{z}\right); \tag{104}$$

$$\text{or} \quad u = F\left(\frac{1}{x}-\frac{1}{y},\ \frac{1}{x}-\frac{1}{z}\right) = 0; \tag{105}$$

$$\text{or} \quad u = F\left(\frac{1}{y}-\frac{1}{z},\ \frac{1}{z}-\frac{1}{x},\ \frac{1}{x}-\frac{1}{y}\right) = 0. \tag{106}$$

386.] The assumption made in Art. 384, by which (84) is assumed from (83), requires further elucidation; and that our notions may be definite, I shall consider it from a geometrical point of view. Suppose the integral equation to be that to a surface; then, from (81) and (82), it appears that the normal to the surface at a certain point is perpendicular to the line whose direction-cosines are proportional to the values which P, Q, R have at that point, and also to *any* line of which the element on the surface is ds, the projections on the coordinate axes of ds being dx, dy, dz; and combining these two conditions, as in (83), it follows that the normal to the surface is coincident in direction with the normal to the plane containing these two lines (P, Q, R), (dx, dy, dz). Now the direction (P, Q, R) is fixed for any one point, and the direction of ds is indeterminate; in order therefore that we may leave the most general condition to be fulfilled hereafter, we may suppose these two directions to be the same, which fact is expressed mathematically by the equations (84): so that now $\left(\frac{du}{dx}\right), \left(\frac{du}{dy}\right), \left(\frac{du}{dz}\right)$ are indeterminate, as appears from (83), and therefore the normal is only limited to being in the plane which passes through the point under consideration, and is normal to the line (P, Q, R). Thus far it appears that two consecutive points on the line (P, Q, R) will be on the surface, but nothing is determined as to consecutive points in other directions.

Now suppose the integrals of the two equations (84) to be found, and to be (85): these are manifestly the equations to two surfaces, and, being simultaneous, express a line which is their line of intersection, and lies on the surfaces, and it is for all points along it that equations (84) are satisfied. The forms of these surfaces depend on the forms of P, Q, R; and as the equation of each of them contains an arbitrary constant, c_1 or c_2, so by the variations of thèse, systems of surfaces arise, and by a relation which is arbitrary, but which we may assume to exist between these constants, we obtain a series of lines, all of which lie on the surface $u = 0$, and therefore by which, in their several and successive positions, the surface is formed; and this relation between c_1 and c_2 may be expressed by a functional symbol which will enter into the final equation; and although this function may be arbitrary, yet for any one surface it will be determinate; and hence will the values of $\left(\frac{du}{dx}\right)$, $\left(\frac{du}{dy}\right)$, $\left(\frac{du}{dz}\right)$ become determinate, and the position of the points contiguous to (x, y, z) be fixed in other directions than along (P, Q, R); that is, in other words, the resulting equations will express a continuous and determinate surface. Although then the assumption of (84) may appear to restrict the generality of (81), inasmuch as it causes the conditions expressed by it and (82) to be satisfied along only a line on the surface, yet it leaves us free to introduce the general functional symbol of relation between c_1 and c_2, and thereby are we enabled to express the class of surfaces of the greatest extent which satisfies the condition of the given partial differential equation.

The reader will perceive the agreement between the method here explained and the process of solution applied to the examples of the preceding Article.

387.] A similar method may also be applied to the integration of partial differential equations of the first order and first degree of any number of variables.

Let the partial differential equation involve n variables, x_1, x_2, $\dots x_n$; and let us suppose the required integral to be of the form

$$u = F(x_1, x_2, \dots x_n) = 0; \tag{107}$$

and suppose the differential equation to be

$$P_1\left(\frac{du}{dx_1}\right) + P_2\left(\frac{du}{dx_2}\right) + \dots + P_n\left(\frac{du}{dx_n}\right) = 0; \tag{108}$$

where all the variables are supposed to be independent; for if such were not the case, but if one were supposed to be a function of the other $(n-1)$, the equation might be changed into the form (108) by means of equivalents analogous to (80). Now the total differential of (107) is

$$Du = \left(\frac{du}{dx_1}\right)dx_1 + \left(\frac{du}{dx_2}\right)dx_2 + \dots + \left(\frac{du}{dx_n}\right)dx_n = 0; \tag{109}$$

and let us assume that the following $(n-1)$ relations exist between (108) and (109),

$$\frac{dx_1}{P_1} = \frac{dx_2}{P_2} = \dots = \frac{dx_n}{P_n} = \mu \text{ (say)}. \tag{110}$$

Suppose now that we can determine the integrals of the $(n-1)$ different equations which are involved in (110), or can by any means, as in Ex. 3 of Article 385, determine $(n-1)$ different relations between the n variables; and suppose them to be of the forms,

$$f_1(x_1, x_2, \dots x_n) = c_1, \quad f_2(x_1, x_2, \dots x_n) = c_2,$$
$$f_{n-1}(x_1, x_2, \dots x_n) = c_{n-1}; \tag{111}$$

where $c_1, c_2, \dots c_{n-1}$ represent $(n-1)$ arbitrary constants. Then these arbitrary constants must be related to each other by a functional symbol, such as

$$\phi(c_1, c_2, \dots c_{n-1}) = 0, \tag{112}$$

or

$$\phi(f_1, f_2, \dots f_{n-1}) = 0; \tag{113}$$

where $f_1, f_2, \dots$ are used as abbreviations for $f_1(x_1, x_2, \dots x_n), \dots$ in (111), as may thus be shewn: let (111) be differentiated, and we have

$$\left.\begin{aligned}
\left(\frac{df_1}{dx_1}\right)dx_1 + \left(\frac{df_1}{dx_2}\right)dx_2 + \dots + \left(\frac{df_1}{dx_n}\right)dx_n &= 0, \\
\left(\frac{df_2}{dx_1}\right)dx_1 + \left(\frac{df_2}{dx_2}\right)dx_2 + \dots + \left(\frac{df_2}{dx_n}\right)dx_n &= 0, \\
\dots\dots\dots\dots\dots\dots & \\
\left(\frac{df_{n-1}}{dx_1}\right)dx_1 + \left(\frac{df_{n-1}}{dx_2}\right)dx_2 + \dots + \left(\frac{df_{n-1}}{dx_n}\right)dx_n &= 0;
\end{aligned}\right\} \tag{114}$$

and multiplying these severally by $\left(\frac{d\phi}{df_1}\right), \left(\frac{d\phi}{df_2}\right), \dots \left(\frac{d\phi}{df_{n-1}}\right)$, and adding, the coefficients of $dx_1, dx_2, \ldots dx_n$ in the sum are evidently $\left(\frac{d\Phi}{dx_1}\right), \left(\frac{d\Phi}{dx_2}\right), \ldots \left(\frac{d\Phi}{dx_n}\right)$; and thus we have

$$\left(\frac{d\Phi}{dx_1}\right)dx_1 + \left(\frac{d\Phi}{dx_2}\right)dx_2 + \ldots + \left(\frac{d\Phi}{dx_n}\right)dx_n = 0; \qquad (115)$$

and replacing $dx_1, dx_2, \ldots dx_n$ by their proportionals from (110), we have

$$\mathrm{P}_1\left(\frac{d\Phi}{dx_1}\right) + \mathrm{P}_2\left(\frac{d\Phi}{dx_2}\right) + \ldots + \mathrm{P}_n\left(\frac{d\Phi}{dx_n}\right) = 0; \qquad (116)$$

comparing which with (108) it is manifest that, with the exception of an added constant which is immaterial, $\Phi = u$; and therefore, from (113), the general integral is

$$u = \Phi(f_1, f_2, \ldots f_{n-1}) = 0; \qquad (117)$$

or, as it may be written,

$$f_1 = \phi(f_2, f_3, \ldots f_{n-1}). \qquad (118)$$

388.] Also if we operate on the several equations of (114) with the series of equalities (110), by comparing the results with (108) it will be manifest that the functions $f_1, f_2, \ldots f_{n-1}$ are all such as when substituted for u satisfy (108); and are therefore solutions of the given equation: each however will be less general than (117), because (117) combines them all under one other arbitrary functional symbol. I may however mention that although I have shewn that (117) is such as to *satisfy* the given equation, yet I have not proved that it is the *necessary* solution; the question is, are any and what restrictions introduced by the hypothetical assumptions (110)? But these inquiries are beyond the range and scope of the present work.

Ex. 1. $(t+y+z)\left(\frac{dt}{dx}\right) + (t+x+z)\left(\frac{dt}{dy}\right) + (t+x+y)\left(\frac{dt}{dz}\right) = x+y+z.$

Let this be changed into its equivalent,

$$(x+y+z)\left(\frac{du}{dt}\right) + (y+z+t)\left(\frac{du}{dx}\right) + (z+t+x)\left(\frac{du}{dy}\right) + (t+x+y)\left(\frac{du}{dz}\right) = 0;$$

therefore, by the assumptions (110),

$$\frac{dt}{x+y+z} = \frac{dx}{y+z+t} = \frac{dy}{z+t+x} = \frac{dz}{t+x+y}$$

$$= \frac{dt+dx+dy+dz}{3(t+x+y+z)} = \frac{dt-dx}{x-t} = \ldots\ldots;$$

$$\therefore \quad \log(t+x+y+z)^{\frac{1}{3}} = \log\frac{c_1}{x-t};$$

$$\left.\begin{aligned} c_1 &= (x-t)(t+x+y+z)^{\frac{1}{3}}, \\ c_2 &= (y-t)(t+x+y+z)^{\frac{1}{3}}, \\ c_3 &= (z-t)(t+x+y+z)^{\frac{1}{3}}; \end{aligned}\right\}$$

For convenience of notation, let $t+x+y+z = \omega^3$; so that by (117), the general integral is

$$u = \phi\{(x-t)\omega, (y-t)\omega, (z-t)\omega\} = 0.$$

Ex. 2. Determine the form of a function of n variables which will satisfy the differential equation

$$x_1\left(\frac{du}{dx_1}\right)+x_2\left(\frac{du}{dx_2}\right)+\ldots+mx_n\left(\frac{du}{dx_n}\right) = 0;$$

$$\therefore \quad \frac{dx_1}{x_1} = \frac{dx_2}{x_2} = \ldots = \frac{dx_n}{mx_n};$$

$$\therefore \quad \frac{x_2}{x_1} = c_2, \quad \frac{x_3}{x_1} = c_3, \ldots \quad \frac{x_{n-1}}{x_1} = c_{n-1}, \quad \frac{x_n}{x_1^{\,m}} = c_n;$$

therefore the general integral is

$$u = \phi\left(\frac{x_2}{x_1}, \ \frac{x_3}{x_1}, \ldots \ \frac{x_{n-1}}{x_1}, \ \frac{x_n}{x_1^{\,m}}\right) = 0; \tag{119}$$

and if $x_1, x_2, \ldots x_{n-1}$ are independent variables, and x_n is dependent, it may be expressed

$$x_n = x_1^{\,m} f\left(\frac{x_2}{x_1}, \ \frac{x_3}{x_1}, \ldots \ \frac{x_{n-1}}{x_1}\right); \tag{120}$$

the right-hand member of which is an homogeneous function of $(n-1)$ variables and of m dimensions: the above is manifestly a proof of the converse of Euler's theorem.

Section 6.—*On integrating factors of differential equations of the first order and degree.*

389.] We return to total differential equations, with the object of investigating the conditions, subject to which differential expressions of two or more variables, which are not exact differentials, and do not in themselves satisfy the criteria of integrability, may yet be made to do so by means of an integrating factor; and first we shall consider an expression of two variables of the form

$$P\,dx + Q\,dy = 0; \tag{121}$$

where P and Q are functions of x and y; and we shall shew that there is always a factor μ, which is generally a function of x and y, which will render (121) an exact differential of u, so that

$$\mu(\text{P}\,dx+\text{Q}\,dy) = \text{D}u = 0. \qquad (122)$$

Suppose the general integral of (121) to be

$$f(x, y) = c; \qquad (123)$$

where c is an arbitrary constant: then we have, by differentiation,

$$\left(\frac{df}{dx}\right)dx+\left(\frac{df}{dy}\right)dy = 0; \qquad (124)$$

comparing which with (121) we have

$$\frac{\left(\frac{df}{dx}\right)}{\text{P}} = \frac{\left(\frac{df}{dy}\right)}{\text{Q}} = \mu, \text{ say}; \qquad (125)$$

so that $\left(\frac{df}{dx}\right) = \mu\text{P}$, $\left(\frac{df}{dy}\right) = \mu\text{Q}$; consequently μP and μQ are respectively the x- and y- partial derived-functions of the same function $f(x, y) = 0$; and μ is a factor which renders (121) an exact differential.

390.] Not only is there an integrating factor of a given differential expression of the form (121), but the number of such factors is infinite.

For supposing $u = 0$ to be such that

$$\mu(\text{P}\,dx+\text{Q}\,dy) = \text{D}u, \qquad (126)$$

then multiplying both members by, say, $\phi(u)$, an arbitrary function of u, we have

$$\mu\,\phi(u)(\text{P}\,dx+\text{Q}\,dy) = \phi(u)\,\text{D}u; \qquad (127)$$

and as the right-hand member of this equation is an exact differential, the left-hand member also is; and as $\phi(u)$ is an arbitrary function of u, the number of such factors is infinite.

From the form of the integrating factors it appears, that if two integrating factors of a differential expression can be found, the integral may be found without integration.

For suppose μ to be one integrating factor; then (127) shews any other integrating factor to be of the form $\mu\,\phi(u)$; consequently dividing the latter by the former, and equating the quotient to a constant c, we have

$$\phi(u) = c: \qquad (128)$$

and this is an integral of the given differential expression, inasmuch as ϕ expresses an arbitrary function.

391.] Let us however consider the question of integrating factors of a differential expression of the form (121) from a more general point of view. Let μ be an integrating factor, so that $\mu \text{P}\, dx + \mu \text{Q}\, dy$ is an exact differential; and consequently

$$\mu \text{P}\, dx + \mu \text{Q}\, dy = \text{D}u = 0; \qquad (129)$$

then by the condition of such exact differentials,

$$\frac{d}{dy}(\mu \text{P}) = \frac{d}{dx}(\mu \text{Q}); \qquad (130)$$

whence
$$\text{Q}\left(\frac{d\mu}{dx}\right) - \text{P}\left(\frac{d\mu}{dy}\right) = \mu\left(\frac{d\text{P}}{dy}\right) - \mu\left(\frac{d\text{Q}}{dx}\right); \qquad (131)$$

and as this is a partial differential equation of the first order and degree in μ, it is to be integrated by the methods of the preceding Section.

Let the general integral of this equation be

$$\mu = \phi(x, y);$$

or in the implicit form

$$\nu = \Phi(x, y, \mu) = 0; \qquad (132)$$

then (131) becomes

$$\text{Q}\left(\frac{d\nu}{dx}\right) - \text{P}\left(\frac{d\nu}{dy}\right) + \mu\left\{\left(\frac{d\text{P}}{dy}\right) - \left(\frac{d\text{Q}}{dx}\right)\right\}\left(\frac{d\nu}{d\mu}\right) = 0; \qquad (133)$$

and therefore, by virtue of the hypothesis made in (84), we have

$$\frac{dx}{\text{Q}} = \frac{dy}{-\text{P}} = \frac{d\mu}{\mu\left\{\left(\frac{d\text{P}}{dy}\right) - \left(\frac{d\text{Q}}{dx}\right)\right\}}; \qquad (134)$$

by the integration of which equations μ must be expressed as an arbitrary function of x and y. There is no general method of finding the integrals of the form (134); although, as we shall presently see, many cases admit of integration.

Since the integrals of (134) involve an arbitrary function the number of factors which will render (121) an exact differential is infinite. This is the theorem which was proved in the preceding Article.

392.] The following are examples in which the integrating factors are determined by the preceding process.

Ex. 1. $a(x\,dy + 2y\,dx) = xy\,dy.$

$$2ay\,dx + x(a-y)\,dy = 0.$$

Let μ be the integrating factor: then

$$\frac{d}{dy}(2a\mu y) = \frac{d}{dx}\{\mu x(a-y)\};$$

$$\therefore \quad x(a-y)\left(\frac{d\mu}{dx}\right) - 2ay\left(\frac{d\mu}{dy}\right) = \mu(a+y);$$

so that in this case (134) become

$$\frac{dx}{x(a-y)} = \frac{dy}{-2ay} = \frac{d\mu}{\mu(a+y)};$$

$$\therefore \quad \left.\begin{aligned} \log x &= \log y^{-\frac{1}{2}} + \frac{y}{2a} + c_1, \\ \log \mu &= \log y^{-\frac{1}{2}} - \frac{y}{2a} + c_2; \end{aligned}\right\}$$

whence the most general value of μ may be determined; but as it takes a complicated form, let us suppose that the relation between c_1 and c_2 is such that $c_1 = c_2 = 0$: then

$$\mu = \frac{1}{xy};$$

so that we have

$$\frac{2ay\,dx + x(a-y)\,dy}{xy} = \mathrm{D}u = 0;$$

$$\therefore \quad u_x = 2a\log x, \qquad u_y = a\log y - y;$$

$$\therefore \quad u = 2a\log x + a\log y - y = c.$$

Ex. 2. $y\,dx + (ax^2y^n - 2x)\,dy = 0.$

$$\therefore \quad \frac{d}{dy}\cdot\mu y = \frac{d}{dx}\cdot\mu(ax^2y^n - 2x);$$

$$\therefore \quad \frac{dx}{ax^2y^n - 2x} = \frac{dy}{-y} = \frac{d\mu}{\mu(3 - 2axy^n)} = \frac{\frac{dy}{y} - \frac{2dx}{x}}{3 - 2axy^n}; \qquad (135)$$

$$\therefore \quad \frac{d\mu}{\mu} = \frac{dy}{y} - \frac{2dx}{x};$$

$$\log\mu = \log y - \log x^2 + \log c_1;$$

$$\therefore \quad \mu = \frac{c_1 y}{x^2};$$

and this is a particular value of the integrating factor; using which, the given differential equation becomes

$$\frac{y^2\,dx + (ax^2y^{n+1} - 2xy)\,dy}{x^2} = \mathrm{D}u = 0;$$

$$\therefore \quad u = \frac{ay^{n+2}}{n+2} - \frac{y^2}{x} = c;$$

and as this integral is that of the first two members of the equality (135), we have

$$\frac{ay^{n+2}}{n+2} - \frac{y^2}{x} = c_2;$$

$$\therefore \quad \mu = \frac{y}{x^2} c_1 = \frac{y}{x^2} \phi'(c_2);$$

$$\therefore \quad \mu = \frac{y}{x^2} \phi' \left\{ \frac{ay^{n+2}}{n+2} - \frac{y^2}{x} \right\};$$

and the general integral is

$$u = \phi \left\{ \frac{ay^{n+2}}{n+2} - \frac{y^2}{x} \right\} = c.$$

Ex. 3. $(x^2+y^2+1)\,dx - 2xy\,dy = 0.$

The integrating factor is cx^{-2}; and the integral is

$$u = \phi \left\{ x - \frac{y^2+1}{x} \right\} = c.$$

393.] Certain special cases in which the integrating factors can be determined require consideration.

Let us suppose the integrating factor μ to be a function of one variable only; say, of x only; so that $\left(\frac{d\mu}{dy}\right) = 0$; then from (131),

$$Q \frac{d\mu}{dx} = \mu \left\{ \left(\frac{dP}{dy}\right) - \left(\frac{dQ}{dx}\right) \right\};$$

$$\therefore \quad \frac{d\mu}{\mu} = \frac{\left(\frac{dP}{dy}\right) - \left(\frac{dQ}{dx}\right)}{Q} dx; \qquad (136)$$

If the right-hand member of this equation is an explicit function of x only, the equation is integrable; and we have

$$\log \mu = \int \frac{\left(\frac{dP}{dy}\right) - \left(\frac{dQ}{dx}\right)}{Q} dx. \qquad (137)$$

Now this condition may be satisfied when Q contains x only, and P is of the form $X_1 y + X_2$, where X_1 and X_2 are functions of x only; in which case the equation is of the first linear form given in Art. 382; this will be fully discussed hereafter. It will also be satisfied when the functions of y which enter into $\left(\frac{dP}{dy}\right) - \left(\frac{dQ}{dx}\right)$ and into Q are the same, so that they may be divided out in the right-hand member of (136).

Similarly if μ is a function of y only

$$\log \mu = \int \frac{\left(\frac{dQ}{dx}\right) - \left(\frac{dP}{dy}\right)}{P}\, dy; \tag{138}$$

and the requisite condition that the second member should be integrable is that $\dfrac{\left(\frac{dQ}{dx}\right) - \left(\frac{dP}{dy}\right)}{P}$ should be a function of y only.

Ex. 1. $(x^2+y^2+2x)\,dx+2y\,dy = 0.$

Here $\dfrac{\left(\frac{dP}{dy}\right) - \left(\frac{dQ}{dx}\right)}{Q} = 1; \quad \therefore \ \log \mu = x.$

$$\therefore \ e^x\{(x^2+y^2+2x)\,dx+2y\,dy\} = \mathrm{D}u;$$

$$\therefore \ u = e^x(x^2+y^2) = c.$$

Ex. 2. $(xy^2-e^{\frac{1}{x^3}})\,dx-x^2y\,dy = 0.$

394.] Again operating on the first two members of (134) in the manner indicated by the following results, we have

$$\frac{dx}{Q} = \frac{-dy}{P} = \frac{d\mu}{\mu\left\{\left(\frac{dP}{dy}\right)-\left(\frac{dQ}{dx}\right)\right\}} = \frac{y\,dx-x\,dy}{Px+Qy}, \tag{139}$$

$$= \frac{y\,dx+x\,dy}{Qy-Px}. \tag{140}$$

If $Px+Qy = 0$, then from (139) we have also

$$y\,dx-x\,dy = 0; \tag{141}$$

$$\therefore \ \frac{y}{x} = c,$$

which is a solution of the differential equation; or in an ap-
...ly more general form,

$$f\left(\frac{y}{x}\right) = c. \tag{142}$$

...ue also because simultaneously
... and $Px+Qy = 0.$
... we have also
... $= c,$ (143)
...uation; and in an ap-

(144)

Equation (143) is also true because we have simultaneously $Qy - Px = 0$, and $P\,dx + Q\,dy = 0$.

395.] If however neither $Px + Qy$ nor $Qy - Px$ vanishes, then from (139), we have

$$\frac{d\mu}{\mu} = \frac{y\,dx - x\,dy}{Px + Qy}\left\{\left(\frac{dP}{dy}\right) - \left(\frac{dQ}{dx}\right)\right\}; \tag{145}$$

now the right-hand member of this equation is an exact differential if P and Q are homogeneous functions of x and y of n dimensions. For in that case by Euler's Theorem, we have

$$x\left(\frac{dP}{dx}\right) + y\left(\frac{dP}{dy}\right) = nP,$$

$$x\left(\frac{dQ}{dx}\right) + y\left(\frac{dQ}{dy}\right) = nQ;$$

and consequently as $P\,dx + Q\,dy = 0$,

$$x\left\{\left(\frac{dP}{dx}\right)dx + \left(\frac{dQ}{dx}\right)dy\right\} + y\left\{\left(\frac{dP}{dy}\right)dx + \left(\frac{dQ}{dy}\right)dy\right\} = 0;$$

and adding this to the numerator of the right-hand member of (145), and observing that

$$dP = \left(\frac{dP}{dx}\right)dx + \left(\frac{dP}{dy}\right)dy, \quad \text{and} \quad dQ = \left(\frac{dQ}{dx}\right)dx + \left(\frac{dQ}{dy}\right)dy,$$

we have

$$\frac{d\mu}{\mu} = -\frac{x\,dP + y\,dQ}{Px + Qy}$$

$$= -\frac{d(Px + Qy)}{Px + Qy}; \qquad \because \quad P\,dx + Q\,dy = 0;$$

$$\therefore \quad \log\mu = -\log(Px + Qy);$$

$$\therefore \quad \mu = \frac{1}{Px + Qy}; \tag{146}$$

and therefore $\dfrac{P\,dx + Q\,dy}{Px + Qy}$ is an exact differential; and we have

$$\frac{P\,dx + Q\,dy}{Px + Qy} = Du; \tag{147}$$

hereby then can u be determined, and an integral be found.

Let these results be compared with Articles 379 and 380.

If both members of (147) are multiplied by $F'(u)$, then we have

$$F(u) = \int F'(u)\,du = \int \frac{P\,dx + Q\,dy}{Px + Qy}F'(u); \tag{148}$$

and therefore if one integral of (147) can be found, an infinite number may also be determined.

Hence, if P and Q are homogeneous functions of n dimensions, $(\mathrm{P}x+\mathrm{Q}y)^{-1}$ is an integrating factor of $\mathrm{P}dx+\mathrm{Q}dy = 0$. The following are examples of integration by this process.

Ex. 1. $2xy\,dx+(y^2-x^2)\,dy = 0$.

Here $\mathrm{P}x+\mathrm{Q}y = x^2y+y^3$;

$$\therefore \quad \frac{2xy\,dx+(y^2-x^2)\,dy}{x^2y+y^3} = \mathrm{D}u = 0;$$

$$\text{and} \quad u_x = \int \frac{2xy\,dx}{x^2y+y^3}; \qquad u_y = \int \frac{y^2-x^2}{x^2y+y^3}\,dy;$$

$$= \log(x^2+y^2); \qquad = \log\frac{x^2+y^2}{y};$$

$$\therefore \quad u = \log\frac{x^2+y^2}{y} = c,$$

where c is an arbitrary constant.

Therefore by reason of (148) the most general integral is

$$u = \mathrm{F}\left(\frac{x^2+y^2}{y}\right).$$

Also taking equation (140) we have

$$\frac{d\mu}{\mu} = \frac{y\,dx+x\,dy}{\mathrm{Q}y-\mathrm{P}x}\left\{\left(\frac{d\mathrm{P}}{dy}\right)-\left(\frac{d\mathrm{Q}}{dx}\right)\right\}; \tag{149}$$

the right-hand member of which is an exact differential if $\mathrm{P} = yf(xy)$, $\mathrm{Q} = x\phi(xy)$; for in this case

$$\frac{d\mu}{\mu} = -\frac{d.xy\,\{f(xy)-\phi(xy)\}}{xy\,\{f(xy)-\phi(xy)\}};$$

$$\therefore \quad \log\mu = -\log.xy\,\{f(xy)-\phi(xy)\}$$

$$\mu = \frac{1}{xy\,\{f(xy)-\phi(xy)\}}$$

$$= \frac{1}{\mathrm{P}x-\mathrm{Q}y}; \tag{150}$$

and consequently we have the following exact differential,

$$\frac{\mathrm{P}dx+\mathrm{Q}dy}{\mathrm{P}x-\mathrm{Q}y} = \mathrm{D}u = 0. \tag{151}$$

And thus if $\mathrm{P} = yf(xy)$, $\mathrm{Q} = x\phi(xy)$, $(\mathrm{P}x-\mathrm{Q}y)^{-1}$ is an integrating factor of $\mathrm{P}dx+\mathrm{Q}dy$.

Ex. 1. $(1+xy)\,y\,dx+(1-xy)\,x\,dy = 0$.

$$\therefore \quad \mathrm{D}u = \frac{(1+xy)\,y\,dx+(1-xy)\,x\,dy}{2x^2y^2};$$

$$\therefore \quad u_x = \int \frac{1+xy}{2x^2 y}\,dx, \qquad u_y = \int \frac{1-xy}{2xy^2}\,dy$$

$$= -\frac{1}{2xy} + \log x^{\frac{1}{2}}; \qquad = -\frac{1}{2xy} - \log y^{\frac{1}{2}};$$

$$\therefore \quad u = -\frac{1}{2xy} + \log\left(\frac{x}{y}\right)^{\frac{1}{2}} = c.$$

396.] For another illustration of the theory of integrating factors let us take the first linear differential equation which is given in (64), Art. 382, and a means of integrating which has already been therein discussed. The form is

$$\{y f(x) - \text{F}(x)\}\,dx + dy = 0. \tag{152}$$

On comparing this with Art 393, it appears that the condition requisite for the existence of a factor which is a function of x only is satisfied, so that we might deduce the form of the factor from that article. Let us however investigate it from first principles.

Let μ be the factor; therefore

$$\left(\frac{d\mu}{dx}\right) = \frac{d}{dy}\{\mu(yf(x) - \text{F}(x))\}$$

$$= \{yf(x) - \text{F}(x)\}\left(\frac{d\mu}{dy}\right) + \mu f(x);$$

$$\therefore \quad \left(\frac{d\mu}{dx}\right) + \{\text{F}(x) - yf(x)\}\left(\frac{d\mu}{dy}\right) = \mu f(x); \tag{153}$$

$$\text{and} \quad \frac{dx}{1} = \frac{dy}{\text{F}(x) - yf(x)} = \frac{d\mu}{\mu f(x)}; \tag{154}$$

$$\therefore \quad \frac{d\mu}{\mu} = f(x)\,dx;$$

$$\mu = c_1 e^{\int f(x)\,dx}; \tag{155}$$

multiplying (152) by this value of μ we have

$$e^{\int f(x)\,dx}\{yf(x) - \text{F}(x)\}\,dx + e^{\int f(x)\,dx}\,dy = 0; \tag{156}$$

$$\therefore \quad u_x = \int e^{\int f(x)\,dx}\{yf(x) - \text{F}(x)\}\,dx;$$

$$= y e^{\int f(x)\,dx} - \int e^{\int f(x)\,dx}\,\text{F}(x)\,dx;$$

$$u_y = \int e^{\int f(x)\,dx}\,dy$$

$$= y e^{\int f(x)\,dx};$$

$$\therefore \quad u = y e^{\int f(x)\,dx} - \int e^{\int f(x)\,dx}\,\text{F}(x)\,dx = c; \tag{157}$$

where c is an arbitrary constant, and is the second arbitrary constant introduced in the integration of the series of equations given in (154); so that as $c_1 = \phi'(c)$, where ϕ' denotes an arbitrary function, from (155) we have

$$\mu = e^{\int f(x)\,dx}\phi'\left\{y e^{\int f(x)\,dx} - \int e^{\int f(x)\,dx}\,\mathrm{F}(x)\,dx\right\}; \qquad (158)$$

And applying this most general value of μ, we have as the *general* integral of (152)

$$u = \phi\left\{y e^{\int f(x)\,dx} - \int e^{\int f(x)\,dx}\,\mathrm{F}(x)\,dx\right\} = c. \qquad (159)$$

It appears then that the equation (152) when multiplied through by $e^{\int f(x)\,dx}$ is an exact differential, and may be integrated as such; this is also otherwise evident; since

$$dy + y f(x)\,dx = \mathrm{F}(x)\,dx;$$

$$\therefore \quad e^{\int f(x)\,dx}dy + y e^{\int f(x)\,dx} f(x)\,dx = e^{\int f(x)\,dx}\,\mathrm{F}(x)\,dx; \qquad (160)$$

whence, as the left-hand member is an exact differential, by integration we have

$$y e^{\int f(x)\,dx} = \int e^{\int f(x)\,dx}\,\mathrm{F}(x)\,dx + c, \qquad (161)$$

which result is the same as that of Art. 382.

The following are examples of this process.

Ex. 1. $dy + y\,dx = ax^n\,dx.$

$$f(x) = 1; \qquad \therefore \quad \int f(x)\,dx = x;$$

$$\therefore \quad \mu = e^x;$$

$$e^x dy + y e^x dx = a e^x x^n dx;$$

$$\therefore \quad y e^x = a\int e^x x^n dx$$

$$= a\{x^n - n x^{n-1} + n(n-1)x^{n-2} - \ldots$$

$$\ldots(-)^n n(n-1)\ldots 3.2.1\} e^x + c;$$

$$\therefore \quad y = a\{x^n - n x^{n-1} + \ldots(-)^n n(n-1)\ldots 3.2.1\} + c e^{-x},$$

where c is an arbitrary constant.

Ex. 2. $dy + \dfrac{n}{(1+x^2)^{\frac{1}{2}}}\,y\,dx = \dfrac{a\,dx}{(1+x^2)^{\frac{1}{2}}}.$

$$\therefore \quad \int f(x)\,dx = \log\{x + (1+x^2)^{\frac{1}{2}}\}^n;$$

$$\therefore \quad \mu = \{x + (1+x^2)^{\frac{1}{2}}\}^n;$$

$$\therefore \quad y\{x+(1+x^2)^{\frac{1}{2}}\}^n = a\int \frac{\{x+(1+x^2)^{\frac{1}{2}}\}^n}{(1+x^2)^{\frac{1}{2}}}\, dx$$
$$= \frac{a}{n}\{x+(1+x^2)^{\frac{1}{2}}\}^n + c;$$
$$\left(y-\frac{a}{n}\right)\{x+(1+x^2)^{\frac{1}{2}}\}^n = c.$$

397.] We proceed now to a differential expression of three independent variables, of the form

$$P\,dx + Q\,dy + R\,dz = 0; \tag{162}$$

where P, Q, R are functions of x, y, and z. Suppose μ to be a factor, by which, when multiplied, it becomes the exact differential of, say, the function,

$$u = F(x, y, z) = c; \tag{163}$$

and thus
$$\mu P\,dx + \mu Q\,dy + \mu R\,dz = Du = 0, \tag{164}$$

where μ generally is a function of all the variables; then the conditions of its being an exact differential are, see equations (41),

$$\frac{d(\mu R)}{dy} = \frac{d(\mu Q)}{dz}, \quad \frac{d(\mu P)}{dz} = \frac{d(\mu R)}{dx}, \quad \frac{d(\mu Q)}{dx} = \frac{d(\mu P)}{dy};$$

$$\left.\begin{aligned}
\therefore \quad * \quad + R\left(\frac{d\mu}{dy}\right) - Q\left(\frac{d\mu}{dz}\right) &= \mu\left\{\left(\frac{dQ}{dz}\right) - \left(\frac{dR}{dy}\right)\right\};\\
-R\left(\frac{d\mu}{dx}\right) + \quad * \quad + P\left(\frac{d\mu}{dz}\right) &= \mu\left\{\left(\frac{dR}{dx}\right) - \left(\frac{dP}{dz}\right)\right\};\\
Q\left(\frac{d\mu}{dx}\right) - P\left(\frac{d\mu}{dy}\right) + \quad * \quad &= \mu\left\{\left(\frac{dP}{dy}\right) - \left(\frac{dQ}{dx}\right)\right\};
\end{aligned}\right\} \tag{165}$$

multiplying the first of which by P, the second by Q, and the third by R, we have

$$P\left\{\left(\frac{dQ}{dz}\right) - \left(\frac{dR}{dy}\right)\right\} + Q\left\{\left(\frac{dR}{dx}\right) - \left(\frac{dP}{dz}\right)\right\} + R\left\{\left(\frac{dP}{dy}\right) - \left(\frac{dQ}{dx}\right)\right\} = 0; \tag{166}$$

which condition must be satisfied, in order that (162) may admit of being made an exact differential by means of a multiplier: we shall return hereafter to the meaning of the necessity of this condition.

Now it is manifest that the three equations (165) are equivalent to any two of them together with (166): and if of these three integrals, involving three arbitrary constants, can be found, the most general integrating factor may be determined: if however we can integrate only one or only two, we may use the resulting expression as an integrating factor, although it may not be the most general.

Also from (165) in many cases, by various combinations, other forms of differential expressions may be found, whereby integrating factors may be determined. Thus one form may be obtained in the following manner: multiply the second of the group (165) by dz, and the third by dy, and then subtract the third from the second; and we have

$$\text{P}\left\{\left(\frac{d\mu}{dy}\right)dy+\left(\frac{d\mu}{dz}\right)dz\right\}-\left(\frac{d\mu}{dx}\right)\{\text{Q}\,dy+\text{R}\,dz\}$$
$$=\mu\left\{\left(\frac{d\text{Q}}{dx}\right)dy+\left(\frac{d\text{R}}{dx}\right)dz-\left(\frac{d\text{P}}{dy}\right)dy-\left(\frac{d\text{P}}{dz}\right)dz\right\};$$

$$\therefore\ \text{P}\left\{\left(\frac{d\mu}{dx}\right)dx+\left(\frac{d\mu}{dy}\right)dy+\left(\frac{d\mu}{dz}\right)dz\right\}-\left(\frac{d\mu}{dx}\right)\{\text{P}\,dx+\text{Q}\,dy+\text{R}\,dz\}$$
$$=\mu\left\{\left(\frac{d\text{P}}{dx}\right)dx+\left(\frac{d\text{Q}}{dx}\right)dy+\left(\frac{d\text{R}}{dx}\right)dz-d\text{P}\right\};$$

and therefore by (162),

$$\text{P}\,d\mu=\mu\left\{\left(\frac{d\text{P}}{dx}\right)dx+\left(\frac{d\text{Q}}{dx}\right)dy+\left(\frac{d\text{R}}{dx}\right)dz-d\text{P}\right\};$$

$$\frac{d\mu}{\mu}+\frac{d\text{P}}{\text{P}}=\frac{1}{\text{P}}\left\{\left(\frac{d\text{P}}{dx}\right)dx+\left(\frac{d\text{Q}}{dx}\right)dy+\left(\frac{d\text{R}}{dx}\right)dz\right\};$$

$$\therefore\quad \log(\mu\text{P})=\int\frac{1}{\text{P}}\left\{\left(\frac{d\text{P}}{dx}\right)dx+\left(\frac{d\text{Q}}{dx}\right)dy+\left(\frac{d\text{R}}{dx}\right)dz\right\}+c_1; \qquad (167)$$

similarly,

$$\log(\mu\text{Q})=\int\frac{1}{\text{Q}}\left\{\left(\frac{d\text{P}}{dy}\right)dx+\left(\frac{d\text{Q}}{dy}\right)dy+\left(\frac{d\text{R}}{dy}\right)dz\right\}+c_2; \qquad (168)$$

$$\log(\mu\text{R})=\int\frac{1}{\text{R}}\left\{\left(\frac{d\text{P}}{dz}\right)dx+\left(\frac{d\text{Q}}{dz}\right)dy+\left(\frac{d\text{R}}{dz}\right)dz\right\}+c_3; \qquad (169)$$

and the general form of the integrating multiplier will be determined by the equation

$$\phi(c_1, c_2, c_3)=0; \qquad (170)$$

where ϕ expresses an arbitrary function, and c_1, c_2, c_3 are to be expressed by their equivalents determined as above. The most general form of the multiplier of course requires the integrals of all three equations; there is no method of finding the integrals of all in their above general forms; in many cases however, as the following examples shew, the integration is possible.

Ex. 1. $zy\,dx-zx\,dy+y^2\,dz=0;$

in this case (166) is $-yz(x+2y)+xyz+2y^2z$, which is equal to 0, and therefore the condition is satisfied; and from (167),

$$\log(\mu y z) = -\int \frac{dy}{y} = \log\frac{1}{y} + \log c_1;$$

$$\therefore \quad \mu = \frac{c_1}{y^2 z};$$

which gives us a particular value of μ. And multiplying the given equation by it, we have

$$\frac{zy\,dx - zx\,dy + y^2\,dz}{y^2 z} = \mathrm{D}u = 0;$$

$$\therefore \quad u = \frac{x}{y} + \log z = c. \tag{171}$$

Again, from (169),

$$\log(\mu y^2) = \int \frac{y\,dx - x\,dy}{y^2} = \frac{x}{y} + \log c_3;$$

$$\therefore \quad \mu = \frac{c_3}{y^2} e^{\frac{x}{y}};$$

and multiplying the given equation by this, and integrating, we have

$$u = z e^{\frac{x}{y}} + c'; \tag{172}$$

and therefore either this or (171) is an integral of the given equation; and thus the general integral is

$$u = \mathrm{F}(z e^{\frac{x}{y}}) = 0.$$

Ex. 2. $(bz - cy)\,dx + (cx - az)\,dy + (ay - bx)\,dz = 0;$
in this case (166) becomes

$$-2a(bz - cy) - 2b(cx - az) - 2c(ay - bx),$$

which is equal to 0, and therefore the condition is satisfied.

The equations for determining μ become

$$\left.\begin{aligned} \mu(bz - cy)^2 &= c_1, \\ \mu(cx - az)^2 &= c_2, \\ \mu(ay - bx)^2 &= c_3; \end{aligned}\right\}$$

and therefore any value of μ which will satisfy the equation

$$\phi\{\mu(bz - cy)^2, \mu(cx - az)^2, \mu(ay - bx)^2\} = 0,$$

may be used as a multiplier to render the given equation an exact differential.

Let us however take one of its particular forms, say the first of the group, and we have

$$\frac{(bz - cy)\,dx + (cx - az)\,dy + (ay - bx)\,dz}{(bz - cy)^2} = \mathrm{D}u = 0;$$

$$\therefore \quad u = \frac{cx - az}{bz - cy} = 0.$$

By taking the other values of μ we might obtain other, although equivalent, values of u; and thus the most general form of the integral is

$$u = \mathrm{F}\left(\frac{cx-az}{bz-cy}\right) = 0.$$

Ex. 3. $(y^2+yz+z^2)dx+(z^2+zx+x^2)dy+(x^2+xy+y^2)dz=0.$ The condition (166) is satisfied; and to determine μ let us have recourse to first principles:

$$\frac{d}{dz}\cdot\mu(z^2+zx+x^2) = \frac{d}{dy}\cdot\mu(x^2+xy+y^2);$$

$$\therefore \quad (x^2+xy+y^2)\left(\frac{d\mu}{dy}\right)-(x^2+xz+z^2)\left(\frac{d\mu}{dz}\right) = 2\mu(z-y);$$

whence we have

$$\frac{dx}{0} = \frac{dy}{x^2+xy+y^2} = \frac{dz}{-(x^2+xz+z^2)} = \frac{d\mu}{2\mu(z-y)}$$

$$= \frac{dx+dy+dz}{y^2+xy-xz-z^2} = \frac{dx+dy+dz}{(y-z)(x+y+z)};$$

$$\therefore \quad \frac{dx+dy+dz}{x+y+z} = -\frac{d\mu}{2\mu};$$

$$\text{and } \log\frac{c_1}{\mu} = \log(x+y+z)^2;$$

$$\therefore \quad \mu = \frac{c_1}{(x+y+z)^2};$$

and multiplying the given equation by this we have

$$\frac{(y^2+yz+z^2)dx+(z^2+zx+x^2)dy+(x^2+xy+y^2)dz}{(x+y+z)^2} = \mathrm{D}u = 0;$$

$$\therefore \quad u_x = \int\frac{y^2+yz+z^2}{(x+y+z)^2}dx$$

$$= -\frac{y^2+yz+z^2}{x+y+z}$$

$$= \frac{yz+zx+xy}{x+y+z}-y+z;$$

$$u_y = \text{- - - - -} -z+x;$$

$$u_z = \text{- - - - -} -x+y;$$

$$\therefore \quad u = \frac{yz+zx+xy}{x+y+z}-c = 0;$$

and thus the general integral becomes

$$u = \mathrm{F}\left(\frac{yz+zx+xy}{x+y+z}\right) = 0,$$

the arbitrary functional symbol F including the arbitrary constant of integration.

398.] Equations (165) admit of combination into a more simple form when P, Q, R are homogeneous and of n dimensions: for multiplying the second of them by z, and the third by y, and subtracting, we have

$$\mu\left\{z\left(\frac{dR}{dx}\right)+y\left(\frac{dQ}{dx}\right)-y\left(\frac{dP}{dy}\right)-z\left(\frac{dP}{dz}\right)\right\}$$
$$= P\left\{z\left(\frac{d\mu}{dz}\right)+y\left(\frac{d\mu}{dy}\right)\right\}-(Qy+Rz)\left(\frac{d\mu}{dx}\right);$$
$$\therefore\quad \mu\left\{x\left(\frac{dP}{dx}\right)+y\left(\frac{dQ}{dx}\right)+z\left(\frac{dR}{dx}\right)-x\left(\frac{dP}{dx}\right)-y\left(\frac{dP}{dy}\right)-z\left(\frac{dP}{dz}\right)\right\}$$
$$= P\left\{x\left(\frac{d\mu}{dx}\right)+y\left(\frac{d\mu}{dy}\right)+z\left(\frac{d\mu}{dz}\right)\right\}-(Px+Qy+Rz)\frac{d\mu}{dx};$$

but
$$x\left(\frac{dP}{dx}\right)+y\left(\frac{dP}{dy}\right)+z\left(\frac{dP}{dz}\right) = nP;$$

whence we have

$$\frac{d}{dx}\{\mu(Px+Qy+Rz)\}-\mu P(1+n) = P\left\{x\left(\frac{d\mu}{dx}\right)+y\left(\frac{d\mu}{dy}\right)+z\left(\frac{d\mu}{dz}\right)\right\}.$$

Similarly,

$$\frac{d}{dy}\{\mu(Px+Qy+Rz)\}-\mu Q(1+n) = Q\left\{x\left(\frac{d\mu}{dx}\right)+y\left(\frac{d\mu}{dy}\right)+z\left(\frac{d\mu}{dz}\right)\right\};$$
$$\frac{d}{dz}\{\mu(Px+Qy+Rz)\}-\mu R(1+n) = R\left\{x\left(\frac{d\mu}{dx}\right)+y\left(\frac{d\mu}{dy}\right)+z\left(\frac{d\mu}{dz}\right)\right\};$$

and therefore multiplying by dx, dy, dz, and adding,

$$D\{\mu(Px+Qy+Rz)\}-\mu(1+n)(P\,dx+Q\,dy+R\,dz)$$
$$= (P\,dx+Q\,dy+R\,dz)\left\{x\left(\frac{d\mu}{dx}\right)+y\left(\frac{d\mu}{dy}\right)+z\left(\frac{d\mu}{dz}\right)\right\};$$
$$\therefore\quad D\{\mu(Px+Qy+Rz)\} = 0;$$
$$\mu = \frac{c}{Px+Qy+Rz}, \tag{173}$$

where c is an arbitrary constant: we subjoin an example in which the method is applied.

It is required to integrate the partial differential equation

$$y\left(\frac{dz}{dx}\right)+z\left(\frac{dz}{dy}\right) = x;$$

or, which is equivalent,

$$y\left(\frac{du}{dx}\right)+z\left(\frac{du}{dy}\right)+x\left(\frac{du}{dz}\right) = 0.$$

Hence by (84), we have

$$\frac{dx}{y} = \frac{dy}{z} = \frac{dz}{x}$$

$$= \frac{dx(z-x)+dy(x-y)+dz(y-z)}{y(z-x)+z(x-y)+x(y-z)}$$

$$= \frac{(x^2-yz)dx+(y^2-zx)dy+(z^2-xy)dz}{(x^2-yz)y+(y^2-xz)z+(z^2-xy)x};$$

and as the denominators of these last equations are equal to zero, the numerators must also vanish: and therefore

$$(x^2-yz)dx+(y^2-zx)dy+(z^2-xy)dz = 0,$$

$$\therefore \quad \frac{x^3+y^3+z^3}{3} - xyz = c_1;$$

$$(z-x)dx+(x-y)dy+(y-z)dz = 0, \tag{174}$$

and this expression satisfies the condition (166); and as the equation is homogeneous we have by means of (173),

$$\frac{(z-x)dx+(x-y)dy+(y-z)dz}{zx+xy+yz-(x^2+y^2+z^2)} = \text{D}u = 0;$$

and by integration

$$u_x = \log\{yz+zx-xy\}^{\frac{1}{2}} + \frac{1}{3^{\frac{1}{2}}}\tan^{-1}\frac{2x-y-z}{3^{\frac{1}{2}}(y-z)};$$

$$u_y = \text{- - - - - - - -} + \frac{1}{3^{\frac{1}{2}}}\tan^{-1}\frac{2y-z-x}{3^{\frac{1}{2}}(z-x)};$$

$$u_z = \text{- - - - - - - -} + \frac{1}{3^{\frac{1}{2}}}\tan^{-1}\frac{2z-x-y}{3^{\frac{1}{2}}(x-y)};$$

and as the difference between the circular functions contained in u_x, u_y, u_z respectively is a constant, it follows that either one is an integral of (174), and that therefore another particular integral of the equation is

$$\log\{yz+zx+xy\}^{\frac{1}{2}} + \frac{1}{3^{\frac{1}{2}}}\tan^{-1}\frac{2x-y-z}{3^{\frac{1}{2}}(y-z)} = c_2;$$

and therefore the general integral of the given differential equation is

$$\log(yz+zx+xy)^{\frac{1}{2}} + \frac{1}{3^{\frac{1}{2}}}\tan^{-1}\frac{2x-y-z}{3^{\frac{1}{2}}(y-z)} = \text{F}(x^3+y^3+z^3-3xyz),$$

where F is the symbol of an arbitrary function.

I may by the way observe that the solution of homogeneous equations is often facilitated by a substitution similar to that made in Art. 377. Thus we may integrate Ex. 3 in Art. 397 by assuming $x = \xi z$, $y = \eta z$; in which case the equation becomes

$$\frac{dz}{z} + \frac{(\eta^2+\eta+1)\,d\xi + (\xi^2+\xi+1)\,d\eta}{\eta^2\xi+\xi^2\eta+3\eta\xi+\eta^2+\xi^2+\eta+\xi} = 0.$$

399.] It is manifest from the examples of the preceding Articles, that the difficulty of determining the integrating factor is the chief obstacle, and is in most cases insurmountable: there is however another mode of solution less direct than that given above, but of which it is desirable to give a brief description, because it is the only one which has hitherto been generally applied.

Since the differential equation $\text{P}\,dx + \text{Q}\,dy + \text{R}\,dz = 0$ is a function of three variables, we may consider one of them to be dependent, and the other two to be independent; let the independent variables be x and y, so that the integral is assumed to be of the form $z = f(x, y)$; now we may consider x and y to vary separately, and consequently z to vary owing to the variation of x or of y, when the other does not vary: suppose that in the differential equation we consider y to be constant, and the variation of z to be partial and to be due to that of x: in which case the equation becomes

$$\text{P}\,dx + \text{R}\,dz = 0; \tag{175}$$

let μ be an integrating factor of this equation, when y is considered constant: and let us suppose

$$\int \mu(\text{P}\,dx + \text{R}\,dz) = \text{F}(x, y, z);$$

then in the integration, since y has been considered constant, a function of y must be introduced; and if Y represents an arbitrary function of y, the integral is

$$\text{F}(x, y, z) = \text{Y}: \tag{176}$$

now this is manifestly such as to give the correct value of $\left(\frac{dz}{dx}\right)$ in (175): Y however must be so determined as to give the correct value of $\left(\frac{dz}{dy}\right)$: and it is also evident that if (176) satisfies these conditions it is an integral of the given differential equation. Of (176) let the total differential be taken; then

$$\left(\frac{d\text{F}}{dx}\right)dx + \left(\frac{d\text{F}}{dy}\right)dy + \left(\frac{d\text{F}}{dz}\right)dz = \frac{d\text{Y}}{dy}\,dy;$$

$$\text{but}\quad \left(\frac{dF}{dx}\right)dx + \left(\frac{dF}{dz}\right)dz = \mu\{P\,dx + R\,dz\},$$
$$= -\mu Q\,dy;$$
$$\therefore\quad \frac{dY}{dy}dy = \left(\frac{dF}{dy}\right)dy - \mu Q\,dy,$$
$$Y = \int\left\{\left(\frac{dF}{dy}\right) - \mu Q\right\}dy + c;$$

whereby Y may be determined.

But in order that Y, as assumed in (176), should be a function of y only, it is necessary that $\left(\frac{dF}{dy}\right) - \mu Q$ should be independent of x and z: and if this is true, the x- and z-differentials of it vanish; and therefore

$$\frac{d}{dx}\left\{\left(\frac{dF}{dy}\right) - \mu Q\right\} = 0, \qquad \frac{d}{dz}\left\{\left(\frac{dF}{dy}\right) - \mu Q\right\} = 0;$$

$$\therefore\quad \left.\begin{aligned} \left(\frac{d^2F}{dx\,dy}\right) - \mu\left(\frac{dQ}{dx}\right) - Q\left(\frac{d\mu}{dx}\right) = 0; \\ \left(\frac{d^2F}{dy\,dz}\right) - \mu\left(\frac{dQ}{dz}\right) - Q\left(\frac{d\mu}{dz}\right) = 0; \end{aligned}\right\} \qquad (177)$$

$$\text{but since}\quad \mu P = \left(\frac{dF}{dx}\right), \quad\text{and}\quad \mu R = \left(\frac{dF}{dz}\right), \qquad (178)$$

$$\left(\frac{d^2F}{dx\,dy}\right) = \mu\left(\frac{dP}{dy}\right) + P\left(\frac{d\mu}{dy}\right); \qquad \left(\frac{d^2F}{dy\,dz}\right) = \mu\left(\frac{dR}{dy}\right) + R\left(\frac{d\mu}{dy}\right);$$

substituting which in (177) and from (178) we have

$$* \quad + R\left(\frac{d\mu}{dy}\right) - Q\left(\frac{d\mu}{dz}\right) = \mu\left\{\left(\frac{dQ}{dz}\right) - \left(\frac{dR}{dy}\right)\right\};$$
$$-R\left(\frac{d\mu}{dx}\right) + \quad * \quad + P\left(\frac{d\mu}{dz}\right) = \mu\left\{\left(\frac{dR}{dx}\right) - \left(\frac{dP}{dz}\right)\right\};$$
$$Q\left(\frac{d\mu}{dx}\right) - P\left(\frac{d\mu}{dy}\right) + \quad * \quad = \mu\left\{\left(\frac{dP}{dy}\right) - \left(\frac{dQ}{dx}\right)\right\};$$

and therefore we have

$$P\left\{\left(\frac{dQ}{dz}\right) - \left(\frac{dR}{dy}\right)\right\} + Q\left\{\left(\frac{dR}{dx}\right) - \left(\frac{dP}{dz}\right)\right\} + R\left\{\left(\frac{dP}{dy}\right) - \left(\frac{dQ}{dx}\right)\right\} = 0;$$

which is the condition of integrability already determined; and consequently, if this condition is satisfied, this method of integration may be employed.

Ex. 1. $(yz+xyz)dx+(zx+xyz)dy+(xy+xyz)dz = 0.$

Here the condition of integrability is satisfied: let y be constant; then $(z+xz)dx+(x+xz)dz = 0;$

$$\therefore \quad \frac{1+x}{x}dx+\frac{1+z}{z}dz = 0;$$

$$u = \log xz+x+z+\text{Y} = 0;$$

$$\therefore \quad \text{D}u = \frac{1+x}{x}dx+\frac{1+z}{z}dz+\frac{d\text{Y}}{dy}dy$$

$$= \frac{1+x}{x}dx+\frac{1+z}{z}dz+\frac{1+y}{y}dy;$$

$$\therefore \quad d\text{Y} = \frac{1+y}{y}dy; \qquad \text{Y} = \log y+y+c;$$

$$\therefore \quad u = \log xyz+x+y+z+c.$$

400.] I will now return to the consideration of the criterion of integrability given in (166), and exhibit it from a geometrical point of view, with reference to properties of surfaces.

The differential equation $\text{P}dx+\text{Q}dy+\text{R}dz = 0$ evidently expresses the condition that the line (P, Q, R) is perpendicular to a line (dx, dy, dz), joining two consecutive points; but P, Q, R are functions of x, y, z, and vary as we pass from point to point; so that if the preceding equation expresses a property of a surface, that surface cuts orthogonally the system of straight lines whose direction-cosines are proportional to P, Q, R. It is of course conceivable that straight lines (P, Q, R) may be such that no continuous surface can cut all orthogonally, or in other words that the differential equation may not express a surface; and consequently a further condition may be required when it is capable of such an interpretation. This condition is given by the criterion of integrability, and I proceed to demonstrate that it is so.

Since the criterion involves partial differentials of P, Q, R, it evidently expresses some property of these lines in a position of displacement infinitesimally near to their former position; and as by reason of the differential equation P, Q, R are proportional to the direction-cosines of the normal to the surface, the property required must depend on the position of the normals at points on the surface infinitesimally near to the point (x, y, z). Now the theorem given in Art. 340 of the present volume assigns such a relation. For since by it the radii of torsion of two geodesics on a surface intersecting at right-angles are equal at the point of intersection, so in other words, if at a point P on a surface a normal PN is drawn, and two lines PP_1 and PP_2 of equal infinitesimal length are drawn on the surface and perpendicular to each other, the normal at P_1 shall make with the plane PNP_1 an angle equal to that which the normal at P_2 makes with the plane PNP_2.

I proceed to prove that this theorem interprets the criterion of integrability: let the point (x, y, z) to which reference is made in the equation

$$P\,dx + Q\,dy + R\,dz = 0 \tag{179}$$

be supposed to be on a surface, on which the line (dx, dy, dz) lies; so that the line (P, Q, R) is normal to the surface at the point (x, y, z). On the surface let two points P_1 and P_2 be taken equidistant from P, and let these points be $(x+\xi_1, y+\eta_1, z+\zeta_1)$, $(x+\xi_2, y+\eta_2, z+\zeta_2)$ respectively, $d\sigma$ being the distance of each of these from P; also let these lines make a right angle at P. Then we have

$$\xi_1\xi_2 + \eta_1\eta_2 + \zeta_1\zeta_2 = 0; \tag{180}$$

$$\left.\begin{aligned} P\xi_1 + Q\eta_1 + R\zeta_1 &= 0, \\ P\xi_2 + Q\eta_2 + R\zeta_2 &= 0. \end{aligned}\right\} \tag{181}$$

From the last two of which we have

$$\frac{P}{\eta_1\zeta_2 - \zeta_1\eta_2} = \frac{Q}{\zeta_1\xi_2 - \xi_1\zeta_2} = \frac{R}{\xi_1\eta_2 - \eta_1\xi_2}. \tag{182}$$

Also let

$$P^2 + Q^2 + R^2 = S^2. \tag{183}$$

Then the direction-cosines of the normal at P_1, ξ_1, η_1, ζ_1 being the increments of x, y, z, are

$$\frac{P}{S} + \xi_1 \frac{d}{dx}\frac{P}{S} + \eta_1 \frac{d}{dy}\frac{P}{S} + \zeta_1 \frac{d}{dz}\frac{P}{S}, \tag{184}$$

$$\frac{Q}{S} + \xi_1 \frac{d}{dx}\frac{Q}{S} + \eta_1 \frac{d}{dy}\frac{Q}{S} + \zeta_1 \frac{d}{dz}\frac{Q}{S}, \tag{185}$$

$$\frac{R}{S} + \xi_1 \frac{d}{dx}\frac{R}{S} + \eta_1 \frac{d}{dz}\frac{R}{S} + \zeta_1 \frac{d}{dz}\frac{R}{S}, \tag{186}$$

the sum of the squares of these quantities being equal to unity, when infinitesimals are neglected.

Let ϕ_1 be the angle which the normal at P_1 makes with the plane P_1PN. Then as ξ_2, η_2, ζ_2 are proportional to the direction-cosines of the normal of this plane, and

$$\xi_1^2 + \eta_1^2 + \zeta_1^2 = \xi_2^2 + \eta_2^2 + \zeta_2^2 = d\sigma^2;$$

$$\sin\phi_1 = \frac{\xi_2(184) + \eta_2(185) + \zeta_2(186)}{d\sigma};$$

$$\sin\phi_1 = \frac{\xi_1}{d\sigma}\left\{\xi_2\frac{d}{dx}\frac{\mathrm{P}}{\mathrm{S}} + \eta_2\frac{d}{dx}\frac{\mathrm{Q}}{\mathrm{S}} + \zeta_2\frac{d}{dx}\frac{\mathrm{R}}{\mathrm{S}}\right\}$$

$$+ \frac{\eta_1}{d\sigma}\left\{\xi_2\frac{d}{dy}\frac{\mathrm{P}}{\mathrm{S}} + \eta_2\frac{d}{dy}\frac{\mathrm{Q}}{\mathrm{S}} + \zeta_2\frac{d}{dy}\frac{\mathrm{R}}{\mathrm{S}}\right\}$$

$$+ \frac{\zeta_1}{d\sigma}\left\{\xi_2\frac{d}{dz}\frac{\mathrm{P}}{\mathrm{S}} + \eta_2\frac{d}{dz}\frac{\mathrm{Q}}{\mathrm{S}} + \zeta_2\frac{d}{dz}\frac{\mathrm{R}}{\mathrm{S}}\right\}$$

$$= \frac{1}{\mathrm{S}\,d\sigma}\left\{\xi_1\xi_2\left(\frac{d\mathrm{P}}{dx}\right) + \eta_1\eta_2\left(\frac{d\mathrm{Q}}{dy}\right) + \zeta_1\zeta_2\left(\frac{d\mathrm{R}}{dz}\right)\right\}$$

$$+ \frac{1}{\mathrm{S}\,d\sigma}\left\{\zeta_1\eta_2\left(\frac{d\mathrm{Q}}{dx}\right) + \xi_1\zeta_2\left(\frac{d\mathrm{R}}{dx}\right) + \eta_1\xi_2\left(\frac{d\mathrm{P}}{dy}\right) + \eta_1\xi_2\left(\frac{d\mathrm{R}}{dy}\right)\right.$$

$$\left. + \zeta_1\xi_2\left(\frac{d\mathrm{P}}{dz}\right) + \zeta_1\eta_2\left(\frac{d\mathrm{Q}}{dz}\right)\right\}; \quad (187)$$

Similarly, if ϕ_2 is the angle contained between the normal at P_2 and the plane $\mathrm{P}_2\,\mathrm{PN}$,

$$\sin\phi_2 = \frac{1}{\mathrm{S}\,d\sigma}\left\{\xi_1\xi_2\left(\frac{d\mathrm{P}}{dx}\right) + \eta_1\eta_2\left(\frac{d\mathrm{Q}}{dy}\right) + \zeta_1\zeta_2\left(\frac{d\mathrm{R}}{dz}\right)\right\}$$

$$+ \frac{1}{\mathrm{S}\,d\sigma}\left\{\xi_2\eta_1\left(\frac{d\mathrm{Q}}{dx}\right) + \xi_2\zeta_1\left(\frac{d\mathrm{R}}{dx}\right) + \eta_2\xi_1\left(\frac{d\mathrm{P}}{dy}\right) + \eta_2\zeta_1\left(\frac{d\mathrm{R}}{dy}\right)\right.$$

$$\left. + \zeta_2\xi_1\left(\frac{d\mathrm{P}}{dz}\right) + \zeta_2\eta_1\left(\frac{d\mathrm{Q}}{dz}\right)\right\}. \quad (188)$$

Hence as $\phi_1 = \phi_2$, equating (187) and (188), we have

$$(\eta_1\zeta_2 - \zeta_1\eta_2)\left\{\left(\frac{d\mathrm{Q}}{dz}\right) - \left(\frac{d\mathrm{R}}{dy}\right)\right\} + (\zeta_1\xi_2 - \xi_1\zeta_2)\left\{\left(\frac{d\mathrm{R}}{dx}\right) - \left(\frac{d\mathrm{P}}{dz}\right)\right\}$$

$$+ (\xi_1\eta_2 - \eta_1\xi_2)\left\{\left(\frac{d\mathrm{P}}{dy}\right) - \left(\frac{d\mathrm{Q}}{dx}\right)\right\} = 0; \quad (189)$$

whence by (182),

$$\mathrm{P}\left\{\left(\frac{d\mathrm{R}}{dy}\right) - \left(\frac{d\mathrm{Q}}{dz}\right)\right\} + \mathrm{Q}\left\{\left(\frac{d\mathrm{P}}{dz}\right) - \left(\frac{d\mathrm{R}}{dz}\right)\right\} + \mathrm{R}\left\{\left(\frac{d\mathrm{Q}}{dx}\right) - \left(\frac{d\mathrm{P}}{dy}\right)\right\} = 0; \quad (190)$$

which is the condition of integrability; and therefore we infer that if (179) does not satisfy this condition, it does not express the property of a surface, and its integral cannot be of the form

$$u = \mathrm{F}(x, y, z) = c.$$

If the point P on the surface is taken for the origin, and the normal PN is taken for the z-axis, and the lines PP_1 and PP_2 are taken for the axes of x and of y respectively, then $\xi_1 = \eta_2 = d\sigma$, $\eta_1 = \zeta_1 = \xi_2 = \zeta_2 = 0$; so that

$$\sin\phi_1 = \frac{d\sigma}{\mathrm{S}}\left(\frac{d\mathrm{Q}}{dx}\right); \qquad \sin\phi_2 = \frac{d\sigma}{\mathrm{S}}\left(\frac{d\mathrm{P}}{dy}\right);$$

$$\therefore \quad \left(\frac{d\text{Q}}{dx}\right) = \left(\frac{d\text{P}}{dy}\right); \tag{191}$$

whereby we have a geometrical interpretation of the ordinary criteria that the equation $\text{P}\,dx + \text{Q}\,dy = 0$ is an exact differential. I may also observe that if the axes of coordinates, to which the surface supposed to be represented by the differential equation

$$\text{P}\,dx + \text{Q}\,dy + \text{R}\,dz = 0$$

is referred, are transformed into another rectangular system, in which the several letters are accented, it may be shewn that (190) becomes

$$\text{P}'\left\{\left(\frac{d\text{R}'}{dy'}\right) - \left(\frac{d\text{Q}'}{dz'}\right)\right\} + \text{Q}'\left\{\left(\frac{d\text{P}'}{dz'}\right) - \left(\frac{d\text{R}'}{dx'}\right)\right\} + \text{R}'\left\{\left(\frac{d\text{Q}'}{dx'}\right) - \left(\frac{d\text{P}'}{dy'}\right)\right\} = 0; \tag{192}$$

so that this condition is invariant; a result which may generally be inferred from the circumstance that it expresses a geometrical property of surfaces which is true independently of any particular system of reference.

401.] We now return to the analytical investigation. It appears that an equation of the form $\text{P}\,dx + \text{Q}\,dy = 0$ can always be rendered an exact differential by means of a multiplier, and that its integral involves an arbitrary functional symbol; and it also appears that $\text{P}\,dx + \text{Q}\,dy + \text{R}\,dz = 0$ is not always capable of being made an exact differential by means of a multiplier, and can be made so only when the condition (166) is satisfied.

If however (166) is not satisfied, but $\text{P}\,dx + \text{Q}\,dy + \text{R}\,dz$ can be separated into two parts, which are respectively exact differentials multiplied by factors, so that it is of the form

$$\mu\,\text{DU} + \mu_1\,\text{DU}_1 = 0, \tag{193}$$

it is evidently satisfied by $\text{U} = c$, $\text{U}_1 = c_1$; U and U_1 being so related, that both are simultaneously constant; and therefore

$$\text{U}_1 = \phi(\text{U}), \tag{194}$$

the form of ϕ being at present undetermined: but as from (194)

$$\text{DU}_1 = \phi'(\text{U})\,\text{DU};$$

substituting this in (193) we have

$$\mu + \mu_1\phi'(\text{U}) = 0;$$

which equations are sufficient for determining the form of ϕ; and the result becomes

$$\text{U} = c, \qquad \text{U}_1 = \phi(c);$$

each of which is the equation to a surface; and the two when taken simultaneously, as it is necessary in this case, express the curve of intersection of the two surfaces: the differential equation therefore expresses a property of a curve and not of a surface.

Or again if we cannot by inspection separate the differential expression into two parts of the form (193); yet by the following process we can shew that it expresses a property of a curve and not of a surface; that is, if (x, y, z) is a point on a surface, it is possible to draw through the point and on the surface an infinite number of lines, the consecutive points of which shall satisfy the differential equation, although the equation to the surface does not.

For suppose the equation to the surface to be $F_1(x, y, z) = c_1$; whence we have $U\,dx + V\,dy + W\,dz = 0$; then multiplying this last by ν, and adding it to the given differential equation,

$$(P + \nu U)\,dx + (Q + \nu V)\,dy + (R + \nu W)\,dz = 0; \qquad (195)$$

now suppose ν to be so determined that this shall satisfy the condition (166): let the integral of (195) be $F_2(x, y, z) = c_2$; then F_1 and F_2 taken together satisfy the differential equation; and therefore all the curves in which these two surfaces intersect satisfy the equation: now F_2 will manifestly contain an arbitrary function, and therefore there will be an infinite number of lines of intersection; although therefore no one surface satisfies the conditions of the given equation, yet through any point on the surface F_1 may an infinite number of lines be drawn along which we may pass without violating the conditions, but we are unable to pass from one line to another across the others.

Another way of considering the matter is this; assume $y = \phi(x)$, and substitute for y in the given differential equation; then we have

$$\{P + Q\,\phi'(x)\}\,dx + R\,dz = 0, \qquad (196)$$

$\phi(x)$ having been substituted for y in P, Q and R. Suppose the integral of (196) to be

$$F(x, z, c) = 0,$$

where c is an arbitrary constant: then the intersection of the cylinders whose equations are $y = \phi(x)$, and $F(x, z, c) = 0$, satisfies the requirements of the given differential equation.

Ex. 1.
$$z\,dx + x\,dy + y\,dz = 0.$$

The condition (166) becomes in this case $-x-y-z$, which is not equal to 0.

Let $y = \phi(x)$, so that $dy = \phi'(x)dx$; then the equation becomes

$$z\,dx + x\,\phi'(x)\,dx + \phi(x)\,dz = 0,$$

$$\{z + x\,\phi'(x)\}\,dx + \phi(x)\,dz = 0;$$

the integral of which and the equation $y = \phi(x)$ together satisfy the differential equation. Thus if $y = x+c$, then

$$(x+c)\,dz + z\,dx + x\,dx = 0,$$

$$(x+c)z + \frac{x^2}{2} = c_1.$$

Now this equation is that to a hyperbolic cylinder perpendicular to the plane of (x, z): and $y = x+c$ expresses a plane perpendicular to the plane of (x, y): and each of these involves an arbitrary constant; consequently the series of lines of intersections of these two surfaces satisfy the given differential equation.

I have said nothing as to the means of determining the integrating factor of a differential expression of more than three variables, because I am unwilling to enlarge the volume by investigations which are not necessary aids in our subsequent applications of pure mathematics to physics.

SECTION 7.—*On singular solutions of differential equations of the first order.*

402.] Thus far we have investigated general and particular integrals of differential equations of the first order; but in some cases there are functions of x and y which satisfy the given equation, and yet cannot be deduced from the general integral by any particular value of the arbitrary constant: such functions are called ***singular solutions***, as we have already noticed in Art. 368, and we now proceed to investigate their properties and modes of discovery. As the inquiry is one of the most difficult in this branch of our subject, the best course is to recur to first principles of definite integration, and thus to state the question and its conditions in the most elementary form.

Let us assume the differential equation, whose singular solution is required, to be exact, and to be of the form

$$\text{P}\,dx + \text{Q}\,dy = \text{D}u = 0, \tag{197}$$

where P and Q are functions of x and y; and let us replace $-\frac{\text{P}}{\text{Q}}$ by $f(x,y)$; so that

$$dy = f(x,y)\,dx; \tag{198}$$

let us suppose the integral to be definite, and the limiting values to be (x_0, y_0), (x_n, y_n); it will be convenient in some cases to replace one or the other of these sets by general symbols x, y: also let us suppose an integral of (198) to be $y = \text{F}(x)$; so that $dy = \text{F}'(x)\,dx$; then $\text{F}(x)$ is subject to these conditions,

$$y_0 = \text{F}(x_0), \quad \text{F}'(x) = f\{x, \text{F}(x)\};$$

and also to similar conditions at the superior limit.

Let us suppose the interval $x_n - x_0$ to be divided into n infinitesimal parts, and $x_1, x_2, \dots x_{n-1}$ to be the values of x at the $n-1$ points of partition; let the corresponding values of y be $y_1, y_2, \dots y_{n-1}$; and let $f(x,y)$ be finite and continuous for all these values: then

$$y_1 - y_0 = f(x_0, y_0)\,(x_1 - x_0),$$

$$y_2 - y_1 = f(x_1, y_1)\,(x_2 - x_1),$$

$$\cdot \quad \cdot \quad \cdot \quad \cdot \quad \cdot \quad \cdot \quad \cdot \quad \cdot \quad \cdot$$

$$y_n - y_{n-1} = f(x_{n-1}, y_{n-1})\,(x_n - x_{n-1});$$

and consequently

$$y_n - y_0 = (x_n - x_0) f\{x_0 + \theta\,(x_n - x_0),\ y_0 + \theta(y_n - y_0)\}; \tag{199}$$

where θ is a *general* symbol for a positive proper fraction.

Let us also express the greatest of the values of $f(x_0, y_0)$, $f(x_1, y_1), \dots f(x_n, y_n)$ by A; then (199) takes the form,

$$y_n - y_0 = (x_n - x_0)\,\theta\,\text{A};$$

and thus (199) becomes

$$y_n - y_0 = (x_n - x_0) f\{x_0 + \theta(x_n - x_0),\ y_0 + \theta\,\text{A}\,(x_n - x_0)\}: \tag{200}$$

whence we know y_n in terms of y_0; and if for x_n and y_n the general values of x and y are substituted, we have

$$y = y_0 + (x - x_0) f\{x_0 + \theta\,(x - x_0),\ y_0 + \theta\,\text{A}\,(x - x_0)\}. \tag{201}$$

It may be shewn, by a method similar to that employed in Art. 6, that the truth of (201) does not depend on the particular mode of partition of the interval $x - x_0$, provided that the parts of it are infinitesimal.

403.] There is however a condition to which (201) must be subject: we have supposed x_0 to be a constant; but as the differential equation does not assign any values at either limit to the variables, y_0, although particular, must be arbitrary; and as y_0 and y must be continuous variables, one may be considered to vary continuously with the other. Equation (201) which gives the relation between y and y_0 must be consistent with such continuous variation; and this can only be the case when the y-differential of the coefficient of $(x-x_0)$ on the right-hand side of (201) is not infinite, that is, when $\frac{d.f(x,y)}{dy}$ does not become infinite for any value of the variables between the limits.

Hence if $x-x_0 = h$, since $y = \mathrm{F}(x)$,

$$\mathrm{F}(x) = \mathrm{F}(x_0)+hf\{x_0+\theta h,\ \mathrm{F}(x_0)+\theta \mathrm{A} h\};$$

and we infer, that if $f(x,y)$ and $\frac{d.f(x,y)}{dy}$ are finite and continuous for all values of the variables between x and x_0, there is always a function of x, viz. $\mathrm{F}(x)$, capable of satisfying the given differential equation, and of becoming a definite value, viz. $y_0 = \mathrm{F}(x_0)$, when $x = x_0$. Thus the result contains a general undetermined constant. It may also be observed that as these are the circumstances requisite for a general integral, it is thus proved that every differential equation of the first order has an integral.

404.] Now consistently with these conditions we can shew that there is only one general form of function which will satisfy the given equation; for suppose $y = \mathrm{F}(x)$ to be a general form of function which satisfies the equation $dy = f(x,y)\,dx$; and suppose another form which satisfies it to be

$$y = \mathrm{F}(x)+\phi(x).$$

Then we have simultaneously

$$y_0 = \mathrm{F}(x_0), \qquad y_0 = \mathrm{F}(x_0)+\phi(x_0),$$

$$\mathrm{F}'(x) = f\{x,\ \mathrm{F}(x)\}; \qquad \mathrm{F}'(x)+\phi'(x) = f\{x,\ \mathrm{F}(x)+\phi(x)\};$$

$$\therefore\quad \phi(x_0) = 0;$$

$$\mathrm{F}'(x)+\phi'(x) = f\{x,\ \mathrm{F}(x)\}+\phi(x)\frac{d}{dy}f\{x,\ \mathrm{F}(x)+\theta\phi(x)\};$$

$$\therefore\quad \phi'(x) = \phi(x)\frac{d}{dy}f\{x,\ \mathrm{F}(x)+\theta\phi(x)\};$$

and as this is to be true for all values between the limits, it is true when $x = x_0$, in which case $\phi(x_0) = 0$, and $\phi'(x_0)$ does not generally vanish; and therefore we have

$$\phi'(x_0) = 0 \times \frac{d}{dy_0}f(x_0,\ y_0);$$

$$\therefore \quad \frac{d}{dy_0} f(x_0, y_0) = \infty;$$

which result is inconsistent with the given conditions. The general integral therefore involves only one general form of function. Consequently if we can discover a second form of function which satisfies a differential equation, besides a general integral, we may be sure that this second form is not a general integral.

405.] Let us suppose then $y = \text{F}(x)$ to be a function of x which satisfies the given differential equation $dy = f(x, y)\,dx$; then if each of the functions $f\{x, \text{F}(x)\}$, $\frac{d}{dy} f(x, y)$ is finite and continuous for all values of x, or, at least, for all values of x between certain assigned limits, we may take any one of these for that which we have represented by x_0; and thus $y = \text{F}(x)$ will be a function of x which will satisfy all the conditions stated in Art. 403, and therefore will be either the general or a particular integral of the given differential equation.

But if, on the other hand, $f\{x, \text{F}(x)\}$ or $\frac{d}{dy} f(x, y)$ is infinite or discontinuous or indeterminate for all values of x, then the conditions necessary for a general or particular integral cannot be fulfilled, although the value is such as to satisfy the given differential equation: the case of discontinuity we may at once dismiss as beyond our province; and $f\{x, \text{F}(x)\}$, which is equal to $\text{F}'(x)$, cannot be infinite for all values of x unless $\text{F}'(x)$ is, and therefore unless $\text{F}(x)$ is, and thus this is another case which we may exclude: and therefore the cases which remain are

$$(1)\ f\{x, \text{F}(x)\} = \frac{0}{0}; \quad (2)\ \frac{d}{dy} f(x,y) = \frac{0}{0}; \quad (3)\ \frac{d}{dy} f(x,y) = \infty;$$

and when $y = \text{F}(x)$ is such as to satisfy the given differential equation and at the same time to satisfy one or other of these conditions, the integral is not *general*.

Yet such a value satisfies the differential equation, and is therefore either a particular integral or a singular solution; and to determine these it is necessary to investigate the relations between x and y which will render $f(x, y)$, or $\frac{d}{dy} f(x, y)$, indeterminate, and which will render $\frac{d}{dy} f(x, y) = \infty$; if they satisfy at the same time the given differential equation, they are either singular solutions or particular integrals; if the general integral is known, there is no difficulty in determining whether any particular constant value of the arbitrary constant will reduce the general integral to the form $y = \text{F}(x)$, but if only the differential equation is given, we must apply the criterion which will be investigated in Article 407.

The last of the above-mentioned conditions may be conveniently applied in the following way. Let us use Lagrange's notation of derived functions: then, since

$$f(x, y) = \frac{dy}{dx} = y';$$

$$\therefore \quad \frac{d}{dy} f(x, y) = \frac{dy'}{dy}. \tag{202}$$

If therefore $\frac{dy'}{dy}$ is found from the given differential equation, and equated to ∞, and a functional relation between x and y is thereby determined and of the form $y = \text{F}(x)$, this is a singular solution if it satisfies the given differential equation. Of this method of discovering singular solutions some examples are added.

Ex. 1. $x\left(\frac{dy}{dx}\right)^2 - y\frac{dy}{dx} + m = 0;$

$$\therefore \quad \frac{dy}{dx} = y' = \frac{y}{2x} \pm \frac{(y^2-4mx)^{\frac{1}{2}}}{2x};$$

$$\therefore \quad \frac{dy'}{dy} = \frac{1}{2x} \pm \frac{y}{2x(y^2-4mx)^{\frac{1}{2}}} = \infty, \quad \text{if} \quad y^2 = 4mx;$$

and as this satisfies the given differential equation, it is a singular solution.

Ex. 2. $dy\,x^{\frac{1}{2}} - dx\,y^{\frac{1}{2}} = 0;$

$$\frac{dy}{dx} = y' = \left(\frac{y}{x}\right)^{\frac{1}{2}};$$

$$\therefore \quad \frac{dy'}{dy} = \frac{1}{2(yx)^{\frac{1}{2}}} = \infty, \quad \text{if} \quad y = 0, \text{ or if } x = 0;$$

and as either of these satisfies the given differential equation, they are singular solutions.

Ex. 3. $\frac{dy}{dx} = \frac{x}{(x^2+y^2-a^2)^{\frac{1}{2}}-y};$

the singular solution is $x^2+y^2-a^2 = 0$.

(2) $y^2 = 2cx + c^2 = -x^2$; so that $c = -x$; consequently (2) is a singular solution.

In the case however where the general integral is not known, an inversion of the process of Articles 405 and (406) will determine whether $y = \text{F}(x)$ is a particular integral or a singular solution; that is, by inquiring whether $\frac{dy'}{dy}$ is not or is rendered infinite by the substitution of $\text{F}(x)$ for y; or whether $\left(\frac{df}{dy'}\right)$ vanishes by the same substitution. Of this process we subjoin some examples, and shall first take that which has just been considered.

Ex. 1. The equation $yy'^2 + 2xy' - y = f(x, y, y') = 0$ is satisfied by (1) $y^2 = 2x + 1$; (2) by $y^2 + x^2 = 0$; are they singular solutions or particular integrals?

$$\left(\frac{df}{dy'}\right) = 2yy' + 2x,$$

which does not vanish for the relation (1), but does vanish for (2): therefore (1) is a particular integral, and (2) is a singular solution.

Ex. 2. The equation $y'^2 + yy' + x = 0$ is satisfied when $y^2 + (x-1)^2 = 0$; is this expression a singular solution or a particular integral?

$$\left(\frac{df}{dy'}\right) = 2y' + y,$$

and as this does not vanish when $y^2 + (x-1)^2 = 0$, this function is a particular integral.

Ex. 3. $y = ax$ satisfies the equation $(1 - x^2)y' + xy - a = 0$; prove that this is a particular integral.

Through the preceding Articles we have considered the differential equation to be a function of $x, y, \frac{dy}{dx}$, and have deduced our results on this supposition; we might however just as well have considered it to be a function of $y, x, \frac{dx}{dy}$, in which case the conditions for a singular solution would be

$$\frac{d}{dx} \cdot \frac{dx}{dy} = \infty, \quad \text{and} \quad \frac{df}{d \cdot \frac{dx}{dy}} = 0; \qquad (205)$$

and the resulting equation $x = \text{F}(y)$ must of course satisfy the differential equation.

408.] Thus far we have investigated singular solutions with reference to the differential equations of which they are solutions; we proceed now to deduce them from, and to point out their relation to, the general integrals of the differential equations; and herein we shall recur to the characteristic property of them; viz., that they are particular forms of the general integral when the arbitrary constant of integration is replaced by a function of the variables, whereby the solution becomes a function of x and y only, and also is such as to satisfy the differential equation.

Let the general integral of a differential equation be

$$\mathrm{F}(x, y, c) = 0, \tag{206}$$

where c is an arbitrary constant introduced in integration; then the theory of the formation of differential equations shews that the given differential equation has been formed by the elimination of c between (206) and

$$\left(\frac{d\mathrm{F}}{dx}\right) dx + \left(\frac{d\mathrm{F}}{dy}\right) dy = 0. \tag{207}$$

Now it is to be considered whether the same value of $\frac{dy}{dx}$ cannot be obtained from an equation of the form (206), if c is replaced by a function of x and y, say of the form $\phi(x, y)$, which we shall abbreviate into ϕ for convenience of notation; because if this is possible, the function hereby obtained is a singular solution.

Suppose then the integral to be

$$\mathrm{F}(x, y, \phi) = 0; \tag{208}$$

$$\therefore \quad \left(\frac{d\mathrm{F}}{dx}\right) dx + \left(\frac{d\mathrm{F}}{dy}\right) dy + \left(\frac{d\mathrm{F}}{d\phi}\right) d\phi = 0; \tag{209}$$

since however ϕ denotes a function of x and y,

$$d\phi = \left(\frac{d\phi}{dx}\right) dx + \left(\frac{d\phi}{dy}\right) dy;$$

and substituting this in (208), we have

$$\left\{\left(\frac{d\mathrm{F}}{dx}\right) + \left(\frac{d\mathrm{F}}{d\phi}\right)\left(\frac{d\phi}{dx}\right)\right\} dx + \left\{\left(\frac{d\mathrm{F}}{dy}\right) + \left(\frac{d\mathrm{F}}{d\phi}\right)\left(\frac{d\phi}{dy}\right)\right\} dy = 0. \tag{210}$$

This will be identical with (207), if

$$\left(\frac{d\mathrm{F}}{d\phi}\right) \frac{\left(\frac{d\phi}{dx}\right)}{\left(\frac{d\mathrm{F}}{dx}\right)} = \left(\frac{d\mathrm{F}}{d\phi}\right) \frac{\left(\frac{d\phi}{dy}\right)}{\left(\frac{d\mathrm{F}}{dy}\right)}. \tag{211}$$

Now this condition can be satisfied in three different ways;

Firstly, if $\left(\frac{d\phi}{dx}\right) = \left(\frac{d\phi}{dy}\right) = 0$; which shew that ϕ contains neither x nor y; that is, $\phi =$ a constant; and thus we have the general integral if the constant is arbitrary, and a particular integral if it has a particular value.

409.] Secondly, (211) is satisfied if $\left(\frac{d\text{F}}{d\phi}\right) = 0$; and if ϕ is eliminated by means of this condition and of (208), or, what amounts to the same thing, if we eliminate c between

$$\text{F}(x, y, c) = 0, \quad \text{and} \quad \left(\frac{d\text{F}}{dc}\right) = 0, \tag{212}$$

the resulting equation will be a relation between x and y which will satisfy the given differential equation; and will not be a particular integral, unless c should happen to be equal to a particular constant previously involved in the differential equation.

The following are examples of this theorem.

Ex. 1. The general integral of a differential equation is $y = c(x+c)^2$; it is required to find the singular solution.

$$\text{F}(x, y, c) = y - c(x+c)^2 = 0;$$

$$\therefore \quad \left(\frac{d\text{F}}{dc}\right) = -(x+c)(x+3c) = 0;$$

$$\therefore \quad x = -c; \qquad x = -3c;$$

of which values the former makes $y = 0$, and as the same result is obtained if $c = 0$, it gives a particular integral.

The second value gives

$$4x^3 + 27y = 0,$$

which is the singular solution.

Ex. 2. The general integral of a differential equation is $c^2x - cy + a = 0$; it is required to find the singular solution.

$$\text{F}(x, y, c) = c^2x - cy + a = 0;$$

$$\therefore \quad \left(\frac{d\text{F}}{dc}\right) = 2cx - y = 0, \text{ if } c = \frac{y}{2x};$$

$$\text{whence} \qquad y^2 = 4ax;$$

and as no particular value of the constant can give this equation, it is a singular solution.

Ex. 3. The general integral of a differential equation is $y - cx - (b^2 + a^2c^2)^{\frac{1}{2}} = 0$; it is required to find the singular solution.

$$\text{F}(x, y, c) = y - cx - (b^2 + a^2c^2)^{\frac{1}{2}} = 0;$$

$$\therefore \quad \left(\frac{dr}{dc}\right) = -x - \frac{a^2 c}{(b^2+a^2c^2)^{\frac{1}{2}}} = 0, \text{ if } c = -\frac{bx}{a(a^2-x^2)^{\frac{1}{2}}};$$

whence we have $$\frac{x^2}{a^2}+\frac{y^2}{b^2} = 1;$$

and this is the singular solution.

The process by which singular solutions are thus derived from the general integral is evidently identical with that by which the envelope of a family of lines, each individual of which is given by a particular value of an arbitrary constant, has been determined in Section 2, Chapter XIII, Vol. I. Thus the general integral involving the arbitrary constant represents the family of lines; the particular integral, a particular value having been given to the arbitrary constant, expresses an individual of the family; and the singular solution which is determined by the elimination of the constant between the general integral and its c-derived function, expresses the envelope of these particular lines. In illustration of these remarks let us take the preceding Ex. 2. The general integral therein given is the general integral of

$$x\left(\frac{dy}{dx}\right)^2 - y\frac{dy}{dx} + a = 0;$$

and the integral may be expressed in the form

$$y = cx + \frac{a}{c};$$

This is manifestly the equation to a straight line; and to a series of straight lines, if c is considered a variable parameter; and the envelope of all these is the singular solution, and is a parabola whose latus rectum is $4a$, as appears from the preceding example; it will be at once seen that the equation is that to the tangent of a parabola in what is sometimes called the magical form.

The comparison of the preceding differential equation and its general integral shews that $\frac{dy}{dx}$ in the first is replaced in the second by c: the c-differentiation therefore of the second produces the same result in terms of c as the $\frac{dy}{dx}$-differentiation of the first produces for $\frac{dy}{dx}$. Hence this method of deducing the singular solution is, in this form of equation at least, the same as that investigated and applied in Art. 406: we have not therefore hereby obtained a more general method.

410.] Thirdly, (211) will be satisfied if simultaneously we have

$$\left(\frac{d\text{F}}{dx}\right) = \left(\frac{d\text{F}}{dy}\right) = \infty; \tag{213}$$

which condition is the same as that determined in Art. 405. In respect of this circumstance let it be observed that it is inconsistent with the very first principles of differentiation that the derived-functions should have infinite values: if they have, the rules according to which they have been found fail. Now in differentiating a function of one variable only, say of x, it may be that its derived-function becomes infinite for a particular *constant* value of the variable: thus, for instance, if $y = (x^2-a^2)^{\frac{1}{2}}$, $y' = \infty$, if $x = \pm a$; but in a function of two variables, as, for instance, $u = (x^2+y^2-a^2)^{\frac{1}{2}} = 0$, $\left(\frac{du}{dx}\right) = \infty$, and $\left(\frac{du}{dy}\right) = \infty$, if $x^2+y^2 = a^2$: that is, the total differential of u is infinite for this relation between x and y. Here then we have met with a case which is beside the common rules of differentiation; but which is of great importance in reference to singular solutions. For suppose u to involve other functions of x and y which are not infinite for the particular relation which makes the above values infinite, and suppose it to contain a function of an arbitrary constant, and the derived-function of it with reference to this arbitrary constant not to become infinite for this relation between x and y, then all these quantities must be neglected in comparison with those which become infinite; and therefore the function of x and y which renders them infinite satisfies the differential equation, and is independent of the arbitrary constant which the general integral contains: and as this last property is characteristic of a singular solution, the function which satisfies these conditions is a singular solution. Hence we infer that if a function of x and y, which is independent of the arbitrary constant of integration, renders infinite $\left(\frac{d\text{F}}{dx}\right)$ and $\left(\frac{d\text{F}}{dy}\right)$, and at the same time satisfies the differential equation, it is a singular solution, provided that it cannot be obtained by giving any particular constant value to the general constant of integration.

With regard to this criterion of a singular solution it must be observed that $\left(\frac{d\text{F}}{d\phi}\right)$ which is the same as $\left(\frac{d\text{F}}{dc}\right)$ must not become infinite simultaneously with $\left(\frac{d\text{F}}{dx}\right)$ and $\left(\frac{d\text{F}}{dy}\right)$; for if this is the case, (211) takes an indeterminate form.

The following are examples of this process.

Ex. 1. $$u = x+2(y-x)^{\frac{1}{2}}-c = 0;$$

$$\left(\frac{du}{dx}\right) = 1-\frac{1}{(y-x)^{\frac{1}{2}}}, \qquad \left(\frac{du}{dy}\right) = \frac{1}{(y-x)^{\frac{1}{2}}};$$

and each of these $= \infty$, if $y = x$; in which case $c = x$; that is, the arbitrary constant receives a variable value, and therefore we have a singular solution.

Ex. 2. $c^2-2cy+a^2-x^2 = 0$;

$$\therefore \quad u = y+(x^2+y^2-a^2)^{\frac{1}{2}}-c = 0;$$

$$\therefore \quad \left(\frac{du}{dx}\right) = \frac{x}{(x^2-y^2-a^2)^{\frac{1}{2}}}, \qquad \left(\frac{du}{dy}\right) = 1+\frac{y}{(x^2+y^2-a^2)^{\frac{1}{2}}};$$

and each of these $= \infty$, if $x^2+y^2-a^2 = 0$; in which case $c = y$, and therefore the solution is singular.

The same singular solution may also be found by the methods previously investigated. The differential equation of which the given equation is the integral is

$$(a^2-x^2)\frac{dy^2}{dx^2}+2xy\frac{dy}{dx}+x^2 = 0;$$

$$\therefore \quad \frac{dy}{dx} = \frac{-xy\pm x\{x^2+y^2-a^2\}^{\frac{1}{2}}}{a^2-x^2} = y';$$

$$\therefore \quad \frac{dy'}{dy} = \infty, \text{ if } x^2+y^2 = a^2;$$

and as this relation satisfies the differential equation given above, it is a singular solution.

Also if we find the c-differential of the general integral, and then eliminate c according to the method of Art. 409, we have

$$c = y; \qquad \therefore \quad y^2+x^2-a^2 = 0;$$

and thus all the methods for finding singular solutions lead to the same result.

SECTION 8.—*Differential equations of the first order and of any degree.*

411.] *Order* of differential equation depends on the index of the symbol of differentiation with which the highest differential or differential coefficient is affected, and *degree* on the power to which such highest differential or differential coefficient is raised. Thus a differential expression of the first order and nth degree is that which involves $\left(\frac{dy}{dx}\right)^n$, but no higher derived-function, and no higher power of $\frac{dy}{dx}$: the general form of such an equation is

$$\left(\frac{dy}{dx}\right)^n + F_1\left(\frac{dy}{dx}\right)^{n-1} + \dots + F_{n-2}\left(\frac{dy}{dx}\right)^2 + F_{n-1}\frac{dy}{dx} + F_n = 0, \quad (214)$$

where $F_1, F_2, \dots F_n$ are symbols for functions of x and y.

Let us suppose the equation (214) to be resolved into n factors of the form,

$$\frac{dy}{dx} - f_1 = 0, \quad \frac{dy}{dx} - f_2 = 0, \quad \dots \quad \frac{dy}{dx} - f_n = 0, \quad (215)$$

where $f_1, f_2, \dots f_n$ are the roots of (214) and are generally functions of x and y: let each of these equations be integrated separately, and let their integrals be

$$\phi_1(x,y,c_1) = 0, \quad \phi_2(x,y,c_2) = 0, \dots \phi_n(x,y,c_n) = 0, \quad (216)$$

where $c_1, c_2, \dots c_n$ are arbitrary constants. Then the equation

$$\phi_1(x,y,c_1) \times \phi_2(x,y,c_2) \times \dots \times \phi_n(x,y,c_n) = 0 \quad (217)$$

will contain all the integrals of (214), because it and (214) vanish simultaneously for each of the n functions. And the truth of this final equation will not be affected if the arbitrary constants are equal, that is, if $c_1 = c_2 = \dots = c_n = c$, because c is arbitrary, and therefore will pass through the values $c_1, c_2, \dots c_n$, if it receives all the values of which it is capable.

The following are examples of this mode of integration.

Ex. 1. $\left(\frac{dy}{dx}\right)^2 - 4a^2x^2 = 0$;

$$\therefore \quad \frac{dy}{dx} = 2ax, \qquad \frac{dy}{dx} = -2ax,$$

$$y - c_1 = ax^2; \qquad y - c_2 = -ax^2.$$

and these two solutions may be combined into the single equation

$$(y - ax^2 - c_1)(y + ax^2 - c_2) = 0;$$

either of which factors satisfies the given differential equation; and if $c_1 = c_2 = c$, we have

$$(y-c)^2 - a^2x^4 = 0..$$

Now this is equally true with the former equation, as the primitive from which the differential equation is derived; for it may be derived either from this latter by the elimination of c, or from *either* of the former by the elimination of c_1 or c_2.

The singular solution of this equation is $x=0$; and considered geometrically the general integral represents two parabolas which have a common axis, viz. that of y, and a common vertex on the axis of y at a distance c from the origin; and the singular solution represents a point on the axis of y, through which all the parabolas pass, and which is consequently an envelope.

Ex. 2. $$\left(\frac{dy}{dx}\right)^3-(a+2bx+3ex^2)\left(\frac{dy}{dx}\right)^2 + (6bex^3+3aex^2+2abx)\frac{dy}{dx}-6abex^3 = 0;$$

$$\therefore \quad \frac{dy}{dx} = a; \qquad \therefore \quad y = ax+c_1;$$

$$\frac{dy}{dx} = 2bx; \qquad y = bx^2+c_2;$$

$$\frac{dy}{dx} = 3ex^2; \qquad y = ex^3+c_3;$$

and the integral is

$$(y-ax-c_1)(y-bx^2-c_2)(y-ex^3-c_3) = 0;$$

and which may be simplified if $c_1 = c_2 = c_3 = c$. In this case also the singular solution is a point on the axis of y.

Ex. 3. $$\left\{\left(\frac{dy}{dx}\right)^2 - \frac{1}{a^2-x^2}\right\} \left\{\frac{dy}{dx} - \left(\frac{y}{x}\right)^{\frac{1}{2}}\right\} = 0;$$

$$\frac{dy}{dx} = \frac{1}{(a^2-x^2)^{\frac{1}{2}}}; \qquad y = \sin^{-1}\frac{x}{a}+c_1;$$

$$\frac{dy}{dx} = \frac{-1}{(a^2-x^2)^{\frac{1}{2}}}; \qquad y = \cos^{-1}\frac{x}{a}+c_2;$$

$$\frac{dy}{dx} = \left(\frac{y}{x}\right)^{\frac{1}{2}}; \qquad y^{\frac{1}{2}} = x^{\frac{1}{2}}+c_3;$$

and the integral is

$$\left(y-\sin^{-1}\frac{x}{a}-c_1\right)\left(y-\cos^{-1}\frac{x}{a}-c_2\right)(y^{\frac{1}{2}}-x^{\frac{1}{2}}-c_3) = 0.$$

Ex. 4. $$\left(\frac{dy}{dx}\right)^2+2\frac{y}{x}\frac{dy}{dx}-1 = 0;$$

$$\frac{dy}{dx}+\frac{y}{x} = \pm\left(1+\frac{y^2}{x^2}\right)^{\frac{1}{2}};$$

$$\therefore \quad x\,dy+y\,dx = \pm(y^2+x^2)^{\frac{1}{2}}\,dx,$$

which are homogeneous, and may be integrated by the methods explained above.

412.] Certain forms of differential equations of the first order and of any degree are capable of integration without resolution into factors according to the method of the preceding Article, and these we proceed to explain.

The first form is

$$x f(y') + y\,\phi(y') + \text{F}(y') = 0; \tag{218}$$

where y' according to Lagrange's notation denotes $\frac{dy}{dx}$ and f, ϕ, and F are symbols of given functions.

Let this equation be divided by $\phi(y')$; so that making obvious substitutions it becomes

$$y = x\,\phi(y') + f(y'); \tag{219}$$

and let us in the first place take the simple case in which $\phi(y') = y'$: so that

$$y = x\,y' + f(y'); \tag{220}$$

this form is known by the name of Clairaut's equation.

Let this be differentiated; then we have, resubstituting,

$$\left\{x + f'\left(\frac{dy}{dx}\right)\right\} \frac{d^2 y}{dx^2} = 0. \tag{221}$$

Now this may be satisfied in two ways;

$$(1) \quad \frac{d^2 y}{dx^2} = 0; \qquad \therefore \quad \frac{dy}{dx} = c; \tag{222}$$

and substituting this in (220) we have

$$y = cx + f(c),$$

which is the general integral, containing the arbitrary constant c.

We might of course integrate (222) immediately; whereby we have

$$y = cx + c_1, \tag{223}$$

where c_1 is a new arbitrary constant: but as (223) is to satisfy (220), $c_1 = f(c)$. This result is also manifest from the fact that (220) is a differential equation of the first order, and therefore its integral must contain only one arbitrary constant.

$$(2) \quad x + f'\left(\frac{dy}{dx}\right) = 0; \qquad \therefore \quad \frac{dy}{dx} = \phi(x);$$

and substituting this value of $\frac{dy}{dx}$ in the equation (220), an expression results which of course satisfies the differential equation, and is independent of c the arbitrary constant, and is therefore either a particular integral or a singular solution; and it is manifestly the latter, because c which is equal to $\frac{dy}{dx}$, is replaced by a function of x, viz. $\phi(x)$.

The following are differential equations solved by this process, which generally is called Integration by means of Differentiation.

Ex. 1. $y = x\frac{dy}{dx} + a\frac{dx}{dy}$;

$$\therefore \quad \frac{dy}{dx} = \frac{dy}{dx} + \left\{x - a\left(\frac{dx}{dy}\right)^2\right\}\frac{d^2y}{dx^2};$$

which is satisfied either by

(1) $\frac{d^2y}{dx^2} = 0$; $\quad \therefore \quad \frac{dy}{dx} = c$; $\quad \therefore \quad y = cx + \frac{a}{c}$;

and this is the general integral; or by

(2) $\frac{dy}{dx} = \left(\frac{a}{x}\right)^{\frac{1}{2}}$; $\quad \therefore \quad y = (ax)^{\frac{1}{2}} + (ax)^{\frac{1}{2}}$;

$$\therefore \quad y^2 = 4ax;$$

and this is the singular solution, since it involves no arbitrary constant, and is not a particular integral, because the constant is replaced by $\left(\frac{a}{x}\right)^{\frac{1}{2}}$, which is a function of x.

Ex. 2. $y = x\frac{dy}{dx} - \left(\frac{dy}{dx}\right)^2$;

$$\therefore \quad \frac{dy}{dx} = \frac{dy}{dx} + \left\{x - 2\frac{dy}{dx}\right\}\frac{d^2y}{dx^2};$$

thus the general integral is $y = cx - c^2$;

and the singular solution is deduced from $\frac{dy}{dx} = \frac{x}{2}$, and is $x^2 = 4y$.

Ex. 3. $y = x\frac{dy}{dx} + \sin^{-1}\frac{dy}{dx}$;

therefore $\frac{dy}{dx} = \frac{dy}{dx} + \left\{x + \frac{dx}{(dx^2 - dy^2)^{\frac{1}{2}}}\right\}\frac{d^2y}{dx^2}$;

and therefore (1) $\frac{d^2y}{dx^2} = 0$, $\quad \frac{dy}{dx} = c$, $\quad y = cx + \sin^{-1}c$;

(2) $x^2dx^2 - x^2dy^2 = dx^2$;

$$\therefore \quad \frac{dy}{dx} = \frac{(x^2-1)^{\frac{1}{2}}}{x}; \quad \therefore \quad y = (x^2-1)^{\frac{1}{2}} + \sec^{-1}x,$$

and this is the singular solution.

Ex. 4. $y = x\frac{dy}{dx} + \frac{dy}{dx} - \left(\frac{dy}{dx}\right)^2$.

Ex. 5. $y = x\frac{dy}{dx} - a\frac{dy}{ds}$.

413.] The differential equation which is integrated by the above method admits easily of a geometrical interpretation. Since $\tan^{-1}\frac{dy}{dx}$ is the angle between the axis of x and the tangent to a plane curve at the point (x, y), equation (220) is that to the tangent of a curve, in which the intercept by the tangent of the axis of y is expressed as a function of $\frac{dy}{dx}$, or, in other words, $\eta_0 = f\left(\frac{dy}{dx}\right)$; thus the general integral is the equation of any tangent line, the arbitrary constant contained in which is the tangent of the angle between it and the axis of x; and the singular solution is the curve-envelope of all such tangents, and is found in fact by a method which is identical with those of the last section: viz. by eliminating either c between the general integral and its c-differential, or $\frac{dy}{dx}$ between the differential equation and its $\frac{dy}{dx}$- differential.

The most general geometrical problem which involves a differential equation of Clairaut's form is that wherein the length of the perpendicular from the origin on the tangent to a curve is a function of the angle which the perpendicular from the origin makes with the axis of x: and as problems of this kind are numerous, and often elegant, two or three are subjoined.

Ex. 1. To find the equation to the curve, the perpendicular from the origin on the tangent of which is of constant length a.

The differential equation of the curve is plainly

$$y\,dx - x\,dy = a\,ds; \tag{224}$$

$$\therefore \quad y = x\frac{dy}{dx} + a\left\{1+\left(\frac{dy}{dx}\right)^2\right\}^{\frac{1}{2}};$$

Differentiating we have

$$0 = \frac{d^2y}{dx^2}\left\{x + \frac{a\frac{dy}{dx}}{\left\{1+\left(\frac{dy}{dx}\right)^2\right\}^{\frac{1}{2}}}\right\}; \tag{225}$$

$$\therefore \quad \frac{d^2y}{dx^2} = 0, \qquad \frac{dy}{dx} = c, \qquad y = cx + a(1+c^2)^{\frac{1}{2}};$$

which is the general integral; and is the equation to a straight line inclined at $\tan^{-1}c$ to the axis of x, and the perpendicular distance from the origin on which is equal to a. This line evidently satisfies the conditions of the problem.

Also from the second factor of (225), we have

$$\frac{dy}{dx} = -\frac{x}{(a^2-x^2)^{\frac{1}{2}}};$$

substituting which in the given differential equation,

$$x^2+y^2 = a^2;$$

the equation of a circle whose radius is a, and which is the singular solution, being the envelope of all the lines whose equations are given by the general integral.

The following process is worth noticing: let us differentiate (224) taking neither x nor y to be equicrescent;

$$\begin{aligned} y\,d^2x - x\,d^2y &= a\,d^2s \\ &= a\left\{\frac{dx}{ds}d^2x + \frac{dy}{ds}d^2y\right\}; \end{aligned}$$

$$\therefore \quad \left(y-a\frac{dx}{ds}\right)d^2x - \left(x+a\frac{dy}{ds}\right)d^2y = 0;$$

and since d^2x and d^2y are arbitrary,

$$y-a\frac{dx}{ds} = 0; \qquad x+a\frac{dy}{ds} = 0;$$

$$\therefore \quad x^2+y^2 = a^2.$$

Ex. 2. The product of two ordinates of the tangent of a curve drawn at two given points on the axis of x is constant; it is required to find the equation of the curve.

Let the origin, see fig. 51, be taken at the point of bisection of the line AB which joins the two points A and B at which the ordinates AQ, BR are drawn: and let x and y be the current co-ordinates to the tangent line: let $\text{OA} = \text{OB} = a$;

$$\therefore \quad \tan \text{RTB} = \frac{dy}{dx};$$

and let $\text{AQ} \times \text{BR} = k^2$: then the equation to RQ is

$$y = x\frac{dy}{dx} + \left\{k^2 + a^2\left(\frac{dy}{dx}\right)^2\right\}^{\frac{1}{2}}; \tag{226}$$

$$y\,dx - x\,dy = \{k^2dx^2 + a^2dy^2\}^{\frac{1}{2}}. \tag{227}$$

Now from (226) by differentiation we have

$$\frac{d^2y}{dx^2}\left\{x + \frac{a^2\dfrac{dy}{dx}}{\left\{k^2+a^2\left(\dfrac{dy}{dx}\right)^2\right\}^{\frac{1}{2}}}\right\} = 0; \tag{228}$$

$$\therefore \quad \frac{d^2y}{dx^2} = 0, \qquad \frac{dy}{dx} = c, \qquad y = cx + (k^2+a^2c^2)^{\frac{1}{2}};$$

which is the general integral: also from the second factor of (228)

$$\frac{dy}{dx} = \frac{-kx}{a\{a^2-x^2\}^{\frac{1}{2}}};$$

and therefore from (226),

$$\frac{x^2}{a^2} + \frac{y^2}{k^2} = 1,$$

which is the singular solution; and is an ellipse of which AB is the major axis.

Or thus; differentiating (227), and equating to zero the coefficients of d^2x and d^2y,

$$\frac{y}{k} = \frac{k\,dx}{\{k^2dx^2+a^2dy^2\}^{\frac{1}{2}}}, \qquad \frac{x}{a} = \frac{-a\,dy}{\{k^2dx^2+a^2dy^2\}^{\frac{1}{2}}};$$

then squaring and adding,

$$\frac{x^2}{a^2} + \frac{y^2}{k^2} = 1.$$

Ex. 3. The triangle contained between the rectangular axes and the tangent of a plane curve is of constant area and $= \frac{k^2}{2}$; shew that the equation of the tangent of the curve is

$$y\,dx - x\,dy = k\,(dy\,dx)^{\frac{1}{2}};$$

and that the singular solution is $xy = \frac{k^2}{4}$.

Ex. 4. TRQ being a tangent to a curve in fig. 51, perpendiculars AY, BZ are drawn to it from the points A and B, and the included area ABZY is constant: find the equation to the line TRQ; and shew that the singular solution of it is the equation to a parabola.

Ex. 5. Find the equation to the curve the portion of whose tangent intercepted between the coordinate axes is of constant length.

Ex. 6. Determine the curve whose tangent cuts off from the coordinate axes parts the sum of which is constant.

414.] Let us now return to the more general form given in (219); and differentiate it; then we have

$$y' = \phi(y') + \{x\phi'(y') + f'(y')\}\frac{dy'}{dx}; \qquad (229)$$

$$\{\phi(y') - y'\}\frac{dx}{dy'} + x\phi'(y') + f'(y') = 0; \qquad (230)$$

which is a linear differential equation of the first order and degree, and can be integrated by the method of Art. 382; whereby we shall obtain y' in terms of x; and thus y' may be eliminated from the given differential equation, and the resulting equation in terms of x and y will be the required solution.

Ex. 1. $y+xy' = ay'^2$;

$$\therefore \quad 2y' + (x-2ay')\frac{dy'}{dx} = 0;$$

$$2y'\,dx + x\,dy' = 2ay'\,dy';$$

$$\therefore \quad xy'^{\frac{1}{2}} = \frac{2a}{3}y'^{\frac{3}{2}} + c;$$

by means of which and the original equation, y' may be eliminated, and the resulting equation will be the required solution.

Ex. 2. $y = xy'^2 + 2y'$;

$$\because \quad y' = y'^2 + 2(xy'+1)\frac{dy'}{dx};$$

$$\text{and} \quad \frac{dx}{dy'} + \frac{2x}{y'-1} = -\frac{2}{y'(y'-1)},$$

which is a linear equation, and of which the integrating factor is $(y'-1)^2$;

$$\therefore \quad (y'-1)^2 dx + 2x(y'-1)dy' = -2\frac{y'-1}{y'}dy',$$

$$x(y'-1)^2 = \log y'^2 - 2y' + c,$$

between which and the given equation we may eliminate y', and so obtain the required result.

Ex. 3. Prove that the curve in which the length of the perpendicular from the origin on the tangent is equal to the abscissa of the point of contact is the circle, the origin being on the circumference.

Ex. 4. Determine the plane curve such that the normal makes equal angles with the radius vector and the x-axis.

415.] Also if the differential equation can be put into the form

$$y = f\left(x, \frac{dy}{dx}\right) = f(x, y'), \tag{231}$$

then differentiating we have

$$y' = f'(x, y') + \frac{d.f(x,y')}{dy'}\frac{dy'}{dx},$$

which is a differential equation involving two variables x and y', the integral of which will be of the form $F(x, y', c) = 0$, where c is an arbitrary constant: and if y' can be eliminated by means of this equation and (231), the resulting equation will contain x, y, and c, and will be the integral of (231).

Similarly, if an equation is capable of being put into the form

$$x = F(y, y'); \tag{232}$$

then by differentiation

$$1 = \left(\frac{dF}{dy}\right)y' + \left(\frac{dF}{dy'}\right)y''; \tag{233}$$

and since $\quad y'' = \dfrac{dy'}{dx} = y'\dfrac{dy'}{dy},$

(233) becomes $\quad 1 = \left(\dfrac{dF}{dy}\right)y' + \left(\dfrac{dF}{dy'}\right)y'\dfrac{dy'}{dy};$

which involves only y and y'; whence the integral of it is of the form

$$\phi(y, y', c) = 0,$$

and thus if y' is eliminated by means of this and of (232), the result will be the integral of (232).

The following examples are illustrative of these processes.

Ex. 1. $\quad x(1+y'^2) = 1;$

$$\therefore \quad dx = -\frac{2y'\,dy'}{(1+y'^2)^2} = \frac{dy}{y'}, \text{ since } \frac{dy}{dx} = y';$$

$$\therefore \quad dy = -\frac{2y'^2\,dy'}{(1+y'^2)^2},$$

$$y-c = \frac{y'}{1+y'^2} - \tan^{-1} y'$$

$$= (x-x^2)^{\frac{1}{2}} - \tan^{-1}\left(\frac{1-x}{x}\right)^{\frac{1}{2}}.$$

Ex. 2. $\quad 4y = x^2 + y'^2;$

$$\therefore \quad y'\,dy' + (x-2y')\,dx = 0;$$

as this is homogeneous, the integral is, by Art. 377,

$$\log(y'-x) - \frac{x}{y'-x} = c;$$

by means of which and the given equation y' may be eliminated.

Ex. 3. $\quad y = xy' + ax^2y'^2 + bx^3y'^3 + \dots;$

Let $\quad xy' = u; \qquad \therefore \quad y = u + au^2 + bu^3 + cu^4 + \dots;$

$$\therefore \quad \frac{dy}{dx} = y' = (1 + 2au + 3bu^2 + \dots)\frac{du}{dx};$$

but $\quad x = \dfrac{u}{y'}; \qquad \therefore \quad dx = \dfrac{y'\,du - u\,dy'}{y'^2};$

$$\therefore \quad y'(y'\,du - u\,dy') = (1 + 2au + 3bu^2 + \dots)y'^2\,du;$$

$$\therefore \quad \frac{dy'}{y'} + 2a\,du + 3bu\,du + \ldots = 0;$$

$$\therefore \quad \log\frac{y'}{c} + 2au + \frac{3bu^2}{2} + \frac{4cu^3}{3} + \ldots = 0;$$

or $$\log\frac{y'}{c} + 2axy' + \frac{3bx^2y'^2}{2} + \frac{4cx^3y'^3}{3} + \ldots = 0;$$

from which and the given equation y' may be eliminated, and the resulting expression will be the integral required.

416.] Another form of differential equation of the first order and nth degree which admits of solution is

$$\text{F}_0\left(\frac{dy}{dx}\right)^n + \text{F}_1\left(\frac{dy}{dx}\right)^{n-1} + \text{F}_2\left(\frac{dy}{dx}\right)^{n-2} + \ldots + \text{F}_{n-1}\frac{dy}{dx} + \text{F}_n = 0, \quad (234)$$

where $\text{F}_1, \text{F}_2, \ldots \text{F}_n$ are homogeneous functions of x and y; so that the equation admits of being put into the form

$$\left(\frac{dy}{dx}\right)^n + f_1\left(\frac{y}{x}\right)\left(\frac{dy}{dx}\right)^{n-1} + \ldots + f_{n-1}\left(\frac{y}{x}\right)\frac{dy}{dx} + f_n\left(\frac{y}{x}\right) = 0. \quad (235)$$

Let $y = tx$; so that (235) takes the form $f(y', t) = 0$; we have also $dy = t\,dx + x\,dt$;

$$\therefore \quad y' - t = x\frac{dt}{dx}, \qquad \text{and} \qquad \frac{dx}{x} = \frac{dt}{y'-t};$$

$$\therefore \quad \log\frac{x}{c} = \int\frac{dt}{y'-t}; \quad (236)$$

and replacing y' by its value in terms of t which is given by $f(y', t) = 0$, the integral of the right-hand member of (236) may be found, and we shall thus have the general integral of 234. The following are examples of the process.

Ex. 1. $x^2\left(\frac{dy}{dx}\right)^2 + x^2 - y^2 = 0$;

Let $y = tx$; $\quad \therefore \quad \frac{dy}{dx} = t + x\frac{dt}{dx}$;

$$\therefore \quad \left(t + x\frac{dt}{dx}\right)^2 = t^2 - 1.$$

$$\therefore \quad \frac{dx}{x} = \frac{dt}{(t^2-1)^{\frac{1}{2}} - t} = -\{(t^2-1)^{\frac{1}{2}} + t\}\,dt;$$

$$\therefore \quad 2\log\frac{x}{c} = \log\{t + (t^2-1)^{\frac{1}{2}}\} - t(t^2-1)^{\frac{1}{2}} - t^2.$$

$$\log\frac{c^2}{x^3}\{y + (y^2-x^2)^{\frac{1}{2}}\} = \frac{y}{x^2}\{y + (y^2-x^2)^{\frac{1}{2}}\},$$

which is the general integral.

The geometrical interpretation of the given equation is "To find the equation to a plane curve such that the projection of its ordinate on the normal is equal to the abscissa."

We are now able to complete the theory of the relations between lengths of curves and the coordinates of their extreme points of which the more simple examples have been given in Art. 166.

Ex. 2. To find the curve the arc of which commencing at a given point is a mean proportional between the ordinate and twice the abscissa; that is, $s^2 = 2xy$.

$$ds = \left(\frac{y}{2x}\right)^{\frac{1}{2}} dx + \left(\frac{x}{2y}\right)^{\frac{1}{2}} dy;$$

$$\therefore \quad (1+y'^2)^{\frac{1}{2}} = \left(\frac{y}{2x}\right)^{\frac{1}{2}} + \left(\frac{x}{2y}\right)^{\frac{1}{2}} y'$$

$$= \left(\frac{t}{2}\right)^{\frac{1}{2}} + \left(\frac{1}{2t}\right)^{\frac{1}{2}} y';$$

$$\therefore \quad y' = \frac{t+(t-1)(2t)^{\frac{1}{2}}}{2t-1};$$

and therefore from (236),

$$\log\frac{x}{c} = -\log(t-1) - \frac{1}{2^{\frac{1}{2}}}\log\left(\frac{t^{\frac{1}{2}}-1}{t^{\frac{1}{2}}+1}\right);$$

$$\therefore \quad (y-x)^{2^{\frac{1}{2}}} = c\,\frac{y^{\frac{1}{2}}+x^{\frac{1}{2}}}{y^{\frac{1}{2}}-x^{\frac{1}{2}}}.$$

Ex. 3. To find the curve such that $s^2 = mx^2 + ny^2$.

$$\therefore \quad 1+y'^2 = \frac{(mx+nyy')^2}{mx^2+ny^2};$$

$$\therefore \quad (m+nt^2)(1+y'^2) = (m+nty')^2;$$

whence we can easily find y' in terms of t; and by substituting in (236), we can determine the equation of the required curve.

Ex. 4. $s^2 = x^2+y^2$; $\quad \therefore \quad t = y'$;

$$\frac{dy}{dx} = \frac{y}{x}; \qquad y = cx.$$

SECTION 9.—*Partial Differential Equations of the First Order and Higher Degree.*

417. Partial differential equations of the first order and higher degree frequently offer themselves for solution in problems of

solid geometry; and it is incumbent on us to consider them so far as they are subject to integration; but here we are close on the boundaries of our knowledge; and it is often necessary for the complete investigation of functions satisfying given differential expressions to have recourse to considerations which belong to integral calculus as applied in geometry, mechanics, &c., and which are therefore beyond and extraneous to the fundamental principles of the pure science. For this reason we shall in the sequel omit some subjects which are to a certain extent within our present grasp; but which I believe it to be more advantageous for the student to defer to a future part of the course, so that he may have at his disposal all the materials which are available for the complete investigation. This course too is also *historically* preferable. Such equations as I allude to have arisen in physical investigations of light, heat, &c.; and they express properties referring to peculiar constitutions of the physical material of the theories from which these phænomena result. It is consequently with reference to these suppositions that they have been made subjects of inquiry, and it is in respect of these that their integrals become interpretable. Of some few partial differential equations of the first order and higher degree it is desirable however at once to seek the integrals.

418.] In the integration of these equations according to a received notation it is convenient to represent $\left(\frac{dz}{dx}\right)$ by p, and $\left(\frac{dz}{dy}\right)$ by q; and let us suppose the equation which is proposed for integration to be of the form

$$f(x, y, z, p, q) = 0, \tag{237}$$

where z is a dependent, and x and y are two independent variables; so that the integral is of the form $z = \text{F}(x, y)$;

$$\therefore \quad dz = p\,dx + q\,dy; \tag{238}$$

but as this is an exact differential,

$$\left(\frac{dp}{dy}\right) = \left(\frac{dq}{dx}\right); \tag{239}$$

where $\left(\frac{dp}{dy}\right)$ denotes the y-derived function of p on the assumption that x is constant; as however p may involve z which will vary with y,

$$\begin{aligned}\left(\frac{dp}{dy}\right) &= \left(\frac{dp}{dy}\right) + \left(\frac{dp}{dz}\right)\left(\frac{dz}{dy}\right) \\ &= \left(\frac{dp}{dy}\right) + q\left(\frac{dp}{dz}\right);\end{aligned}$$

Also by means of the given differential equation q may be expressed in terms of x, y, z, p; wherein z is a function of x and y, and p is a function of x, y, z; and as we now require the x-derived function of q on the assumption that y is constant, we have

$$\left(\frac{dq}{dx}\right) = \left(\frac{dq}{dx}\right) + \left(\frac{dq}{dz}\right)\left(\frac{dz}{dx}\right) + \left(\frac{dq}{dp}\right)\left(\frac{dp}{dx}\right)$$

$$= \left(\frac{dq}{dx}\right) + p\left(\frac{dq}{dz}\right) + \left(\frac{dq}{dp}\right)\left\{\left(\frac{dp}{dx}\right) + \left(\frac{dp}{dz}\right)\left(\frac{dz}{dx}\right)\right\}$$

$$= \left(\frac{dq}{dx}\right) + p\left(\frac{dq}{dz}\right) + \left(\frac{dq}{dp}\right)\left(\frac{dp}{dx}\right) + p\left(\frac{dq}{dp}\right)\left(\frac{dp}{dz}\right);$$

and substituting these values in (239), we have

$$-\left(\frac{dq}{dp}\right)\left(\frac{dp}{dx}\right) + \left(\frac{dp}{dy}\right) + \left\{q - p\left(\frac{dq}{dp}\right)\right\}\left(\frac{dp}{dz}\right) - \left(\frac{dq}{dx}\right) - p\left(\frac{dq}{dz}\right) = 0; \quad (240)$$

which is a partial differential equation of the first order and degree; and consequently by the assumptions (110), Art. 387, we have

$$\frac{dx}{-\left(\frac{dq}{dp}\right)} = \frac{dy}{1} = \frac{dz}{q - p\left(\frac{dq}{dp}\right)} = \frac{dp}{\left(\frac{dq}{dx}\right) + p\left(\frac{dq}{dz}\right)}; \quad (241)$$

which is a system of three ordinary simultaneous equations, from which the integral of the given equation may be inferred according to the process of Art. 387.

It will be observed that from (241), we have

$$dx = -\left(\frac{dq}{dp}\right)dy, \qquad dz = \left\{q - p\left(\frac{dq}{dp}\right)\right\}dy;$$

$$\therefore \quad dz = p\,dx + q\,dy;$$

so that while the system (241) is derived from the criterion of an exact differential it also secures that criterion.

Accordingly, if we can determine p by means of (241), and thence q by means of (237), we may substitute these values in (238) and thus determine z in terms of x and y. The determination of p will involve one arbitrary constant, viz. c_1, and the integration of (238) will involve a second, viz. c_2, which, by virtue of the argument of Art. 387, must be a function of the other constant.

The preceding process was originated by Lagrange, but was completed by Charpit, and now commonly bears his name.

The following are examples in which it is applied.

Ex. 1. $p^2+q^2 = 1$. $\therefore\ q = (1-p^2)^{\frac{1}{2}}$;

$$\therefore\ \left(\frac{dq}{dp}\right) = -\frac{p}{(1-p^2)^{\frac{1}{2}}}; \qquad \left(\frac{dq}{dx}\right) = 0; \qquad \left(\frac{dq}{dz}\right) = 0;$$

so that (241) become

$$\therefore\ \frac{dx}{p} = \frac{dy}{(1-p^2)^{\frac{1}{2}}} = \frac{dz}{1} = \frac{dp}{0};$$

$$\therefore\ p = c_1; \qquad \therefore\ q = (1-c_1^2)^{\frac{1}{2}};$$

$$\therefore\ dz = c_1\,dx+(1-c_1^2)^{\frac{1}{2}}\,dy;$$

$$z = c_1 x+(1-c_1^2)^{\frac{1}{2}}\,y+f(c_1),$$

which is the equation to a plane, the position of which depends on the arbitrary constant c_1.

Ex. 2. $z = pq$. $\therefore\ q = \frac{z}{p}$; and

$$\left(\frac{dq}{dx}\right) = 0; \qquad \left(\frac{dq}{dz}\right) = \frac{1}{p}; \qquad \left(\frac{dq}{dp}\right) = -\frac{z}{p^2};$$

so that (241) become

$$\frac{p^2\,dx}{z} = \frac{dy}{1} = \frac{p\,dz}{2z} = \frac{dp}{1}. \tag{242}$$

From the second and fourth of these terms, we have

$$p = y+c_1;$$

$$\therefore\ q = \frac{z}{y+c_1};$$

$$\therefore\ dz = (y+c_1)\,dx+\frac{z}{y+c_1}\,dy;$$

of which the integral evidently is

$$\frac{z}{y+c_1} = x+f(c_1);$$

$$\therefore\ z = (y+c_1)\{x+f(c_1)\};$$

which is an integral of the given equation.

Also from the third and fourth of (242) we have $p = cz^{\frac{1}{2}}$; therefore $q = \frac{z^{\frac{1}{2}}}{c}$; and

$$dz = cz^{\frac{1}{2}}\,dx+\frac{z^{\frac{1}{2}}}{c}\,dy;$$

$$\therefore\ 2z^{\frac{1}{2}} = cx+\frac{y}{c}+f(c);$$

this is also another integral of the differential equation.

Ex. 3. $q = xp+p^2$;

$$\therefore\ \left(\frac{dq}{dp}\right) = x+2p; \qquad \left(\frac{dq}{dx}\right) = p; \qquad \left(\frac{dq}{dz}\right) = 0;$$

so that (241) become

$$-\frac{dx}{x+2p} = \frac{dy}{1} = -\frac{dz}{p^2} = \frac{dp}{p}.$$

$$\therefore \quad p = ce^y; \qquad q = cxe^y + c^2e^{2y};$$

$$dz = ce^y dx + (cxe^y + c^2e^{2y})dy.$$

$$\therefore \quad z = cxe^y + \frac{c^2}{2}e^{2y} + f(c),$$

where f denotes an arbitrary function.

Ex. 4. Find the equation of a tubular surface generated by a sphere of radius a, the centre of which moves along a director-curve in the plane of (x, y).

The differential equation of the surface is, see (124), Art. 373, Vol. I,

$$1+p^2+q^2 = \frac{a^2}{z^2};$$

so that (241) become

$$\frac{dx}{p} = \frac{dy}{q} = \frac{z^2dz}{a^2-z^2} = \frac{-z^3dp}{a^2p}; \tag{243}$$

from the last two of which $\dfrac{p^2z^2}{a^2-z^2} = c^2;$

$$\therefore \quad p^2 = \frac{c^2(a^2-z^2)}{z^2}; \qquad q^2 = \frac{(1-c^2)(a^2-z^2)}{z^2};$$

$$\therefore \quad \frac{-z\,dz}{(a^2-z^2)^{\frac{1}{2}}} = c\,dx + (1-c^2)^{\frac{1}{2}}dy;$$

$$(a^2-z^2)^{\frac{1}{2}} = cx + (1-c^2)^{\frac{1}{2}}y + f(c);$$

which is the equation to a right circular cylinder, whose axis is in the plane of (x, y).

Also taking other terms of (243), we have

$$dx = \frac{pz^2dz}{a^2-z^2} \qquad dy = \frac{qz^2dz}{a^2-z^2}$$

$$= \frac{cz\,dz}{(a^2-z^2)^{\frac{1}{2}}}; \qquad = \frac{(1-c^2)^{\frac{1}{2}}z\,dz}{(a^2-z^2)^{\frac{1}{2}}};$$

$$\therefore \quad x-b = -c(a^2-z^2)^{\frac{1}{2}}; \quad y-f(b) = -(1-c^2)^{\frac{1}{2}}(a^2-z^2)^{\frac{1}{2}};$$

squaring and adding which

$$(x-b)^2+\{y-f(b)\}^2 = a^2-z^2;$$

$$\therefore \quad \{x-b\}^2+\{y-f(b)\}^2+z^2 = a^2;$$

which is the general equation to tubular surfaces, and wherein $f(b)$ is determined by the equation to the plane director-curve.

Ex. 5. Find the equation to the surface in which the part of the normal intercepted between the surface and the plane of (x, y) is constant and equal to unity.

SECTION 10.—*Various Theorems and Applications of Differential Equations of the First Order and of any Degree.*

419.] The form of a differential equation will frequently be so much modified by a judicious substitution, or by a change of coordinates when it is presented in a geometrical form, that it is desirable to exhibit this method of integration at a greater length. No general rules however can be given, and the particular substitution must be left to the judgment of the student. The following example indicates the kind of substitution.

Ex. 1. $$\frac{xy\,dy+y^2\,dx}{x^2y^2+a^4} = \frac{d.f(y)}{a^2};$$

$$\therefore \quad \frac{a^2 y\, d(xy)}{x^2y^2+a^4} = d.f(y),$$

$$\therefore \quad \tan^{-1}\frac{xy}{a^2} = \int \frac{d.f(y)}{y};$$

$$\therefore \quad xy = a^2 \tan \int \frac{d.f(y)}{y}.$$

Ex. 2. $$\left(\frac{dy}{dx}\right)^3 + y^3\frac{dy}{dx} + y^4 = 0.$$

Let $\frac{dy}{dx} = uy$; so that we have

$$u^3+uy+y = 0; \tag{244}$$

$$\therefore \quad y = -\frac{u^3}{1+u}, \qquad \frac{dy}{du} = -\frac{3+2u}{(1+u)^2}u^2;$$

$$\therefore \quad dx = \frac{dy}{uy} = \frac{3+2u}{u^2(1+u)}du;$$

$$\therefore \quad x = c - \frac{3}{u} + \log\frac{1+u}{u};$$

from which u is to be eliminated by means of (244), and the resulting equation will be the required solution of the given differential equation.

420.] The substitution however which is frequently very convenient is that which consists in the transformation from rectangular to polar coordinates, especially when the differential equation is a function of $x\,dy-y\,dx$, $x\,dx+y\,dy$, $(x^2+y^2)^{\frac{1}{2}}$; in this case x and y are replaced by $r\cos\theta$ and $r\sin\theta$ respectively, where r and θ are new and independent variables. A differential expression is often much simplified hereby. Take the problem, To find the curve such that the perpendicular distance from the origin on the tangent is equal to the abscissa. In rectangular coordinates the differential equation is

$$y\,dx - x\,dy = x\,ds; \tag{245}$$

$$\therefore \quad (x^2-y^2)\,dx + 2xy\,dy = 0;$$

which is a homogeneous equation, and of which the integral is $y^2 = 2ax - x^2$, the equation to a circle, whose radius is a. If however (245) is expressed in polar coordinates, we have

$$\left.\begin{array}{ll} x = r\cos\theta, & dx = dr\cos\theta - r\sin\theta\,d\theta; \\ y = r\sin\theta; & dy = dr\sin\theta + r\cos\theta\,d\theta; \end{array}\right\} \tag{246}$$

$$\therefore \quad \frac{dr}{r} = -\frac{\sin\theta\,d\theta}{\cos\theta},$$

$$r = 2a\cos\theta.$$

which result has already been demonstrated in Ex. 12, Art. 176.

It is indeed this process of transformation which has been applied in the preceding solution of homogeneous differential equations, where we have replaced y by xz.

The following are other examples.

421.] Sometimes the integrals of the sum of two or more expressions can be found in finite algebraical terms, although the integral of each separately would involve an elliptical or other transcendental function; the reason of course being that the transcendental parts neutralize each other: of this we have had instances in Fagnani's theorem as to elliptic arcs, and in Ex. 1, Art. 369. The following example is a remarkable illustration of this. It is required to integrate

$$\frac{dx}{\{a_0+a_1x+a_2x^2+a_3x^3+a_4x^4\}^{\frac{1}{2}}} = \frac{dy}{\{a_0+a_1y+a_2y^2+a_3y^3+a_4y^4\}^{\frac{1}{2}}}. \quad (247)$$

Let

$$a_0+a_1x+a_2x^2+a_3x^3+a_4x^4 = \text{X},$$

$$a_0+a_1y+a_2y^2+a_3y^3+a_4y^4 = \text{Y};$$

and let each term of (247) $= dt$: so that

$$\frac{dx}{\text{X}^{\frac{1}{2}}} = \frac{dy}{\text{Y}^{\frac{1}{2}}} = dt;$$

$$\therefore \quad \text{X} = \frac{dx^2}{dt^2}; \qquad \text{Y} = \frac{dy^2}{dt^2};$$

and consequently, if the new variable t is equicrescent,

$$d\text{X} = \frac{2\,dx\,d^2x}{dt^2}, \qquad d\text{Y} = \frac{2\,dy\,d^2y}{dt^2};$$

but

$$d\text{X} = (a_1+2a_2x+3a_3x^2+4a_4x^3)\,dx,$$

$$d\text{Y} = (a_1+2a_2y+3a_3y^2+4a_4y^3)\,dy,$$

therefore

$$\frac{dx^2-dy^2}{dt^2} = a_1(x-y)+a_2(x^2-y^2)+a_3(x^3-y^3)+a_4(x^4-y^4); \quad (248)$$

$$\frac{d^2x+d^2y}{dt^2} = a_1+a_2(x+y)+\frac{3a_3}{2}(x^2+y^2)+2a_4(x^3+y^3). \quad (249)$$

Let $\quad x-y=z, \qquad x+y=s;$

therefore (248) and (249) become

$$\frac{dz\,ds}{z\,dt^2} = a_1+a_2s+\frac{a_3}{4}(3s^2+z^2)+\frac{a_4}{2}s(s^2+z^2);$$

$$\frac{d^2s}{dt^2} = a_1+a_2s+\frac{3a_3}{4}(s^2+z^2)+\frac{a_4}{2}(s^3+3sz^2);$$

whence by subtraction,

$$\frac{dz\,ds-z\,d^2s}{z\,dt^2} = -\frac{a_3z^2}{2}-a_4sz^2;$$

$$\therefore \quad \frac{(2\,dz\,ds-2z\,d^2s)\,ds}{z^3\,dt^2} = (-a_3-2a_4s)\,ds;$$

$$\therefore \quad \frac{ds^2}{z^2\,dt^2} = a_3 s + a_4 s^2 + \text{C};$$

where C is an arbitrary constant;

$$\therefore \quad \frac{ds}{dt} = z\{a_3 s + a_4 s^2 + \text{C}\}^{\frac{1}{2}};$$

and therefore by substitution

$$\text{X}^{\frac{1}{2}} + \text{Y}^{\frac{1}{2}} = (x-y)\{a_3(x+y) + a_4(x+y)^2 + \text{C}\}^{\frac{1}{2}};$$

$$\{a_0 + a_1 x + a_2 x^2 + a_3 x^3 + a_4 x^4\}^{\frac{1}{2}} + \{a_0 + a_1 y + a_2 y^2 + a_3 y^3 + a_4 y^4\}^{\frac{1}{2}}$$
$$= (x-y)\,\{a_3(x+y) + a_4(x+y)^2 + \text{C}\}^{\frac{1}{2}};$$

and this is an integral equation satisfying the given equation (247).

Another and equivalent form of the same equation is

$$\frac{d\theta}{\{1-e^2(\sin\theta)^2\}^{\frac{1}{2}}} = \frac{d\phi}{\{1-e^2(\sin\phi)^2\}^{\frac{1}{2}}} = dt \text{ (say)}, \qquad (250)$$

t in this and in the former case being an elliptic function: hence

$$\frac{d\theta}{dt} = \{1-e^2(\sin\theta)^2\}^{\frac{1}{2}}; \qquad \frac{d\phi}{dt} = \{1-e^2(\sin\phi)^2\}^{\frac{1}{2}};$$

$$\frac{d^2\theta}{dt^2} = -e^2 \sin\theta\cos\theta; \qquad \frac{d^2\phi}{dt^2} = -e^2\sin\phi\cos\phi;$$

$$\therefore \quad \frac{d^2\theta}{dt^2} + \frac{d^2\phi}{dt^2} = -e^2(\sin\theta\cos\theta + \sin\phi\cos\phi)$$
$$= -e^2\sin(\theta+\phi)\cos(\theta-\phi);$$

$$\frac{d\theta^2}{dt^2} - \frac{d\phi^2}{dt^2} = -e^2\{(\sin\theta)^2 - (\sin\phi)^2\}$$
$$= -e^2\sin(\theta-\phi)\sin(\theta+\phi).$$

Let $\quad \theta+\phi = \sigma, \qquad \theta-\phi = \delta;$

$$\frac{d^2\sigma}{dt^2} = -e^2\sin\sigma\cos\delta; \qquad \frac{d\sigma}{dt}\frac{d\delta}{dt} = -e^2\sin\sigma\sin\delta;$$

$$\therefore \quad \sin\delta\frac{d^2\sigma}{dt^2} - \cos\delta\frac{d\sigma}{dt}\frac{d\delta}{dt} = 0:$$

$$\operatorname{cosec}\delta\frac{d\sigma}{dt} = \text{C}: \qquad \frac{d\sigma}{dt} = \text{C}\sin\delta;$$

$$\frac{d\theta}{dt} + \frac{d\phi}{dt} = \text{C}\sin(\theta-\phi);$$

$$\{1-e^2(\sin\theta)^2\}^{\frac{1}{2}} + \{1-e^2(\sin\phi)^2\}^{\frac{1}{2}} = \text{C}\sin(\theta-\phi),$$

which is an integral expression satisfying (250).

The preceding equation and some more general cases of a similar form are discussed by Prof. Richelot of Königsberg, in Crelle's Journal. Vol. XXIII. p. 354. The following expressions may also be integrated by a similar process:

$$\frac{dx}{\{a_0+a_2x^2+a_4x^4+a_6x^6\}^{\frac{1}{2}}} = \frac{dy}{\{a_0+a_2y^2+a_4y^4+a_6y^6\}^{\frac{1}{2}}};$$

by assuming $x^2 = \xi$, $y^2 = \eta$; and more generally

$$\frac{dx}{x\{a_0+a_1x^n+a_2x^{2n}\}^{\frac{1}{2}}} = \frac{dy}{y\{a_0+a_1y^n+a_2y^{2n}\}^{\frac{1}{2}}},$$

by assuming $x^n = \xi$, $y^n = \eta$.

These questions however are only particular cases of the general theory of the new elliptic and other functions; which is a subject requiring distinct discussion, but lying beyond that proposed in the present work.

422.] Some functional equations also are conveniently solved by means of integration and differentiation, as the following examples shew.

Ex. 1. Determine the form of $z = f(x)$, so that for all values of x and y $f(x)+f(y) = f(x+y)$.

Taking the x-differential, $f'(x) = f'(x+y)$; whence, y being independent of x, we infer that $f'(x)$ is constant whatever value x has; therefore

$$f'(x) = c,$$
$$f(x) = cx+c_1;$$

substituting which in the given equation,

$$cx+c_1+cy+c_1 = c(x+y)+c_1;$$
$$\therefore \quad c_1 = 0;$$

and therefore the most general form of $f(x)$ which satisfies the equation is

$$f(x) = cx.$$

Ex. 2. Find the form of f, so that $f(x)+f(y) = f(xy)$.

Taking the x-differential, $f'(x) = yf'(xy)$; again taking the y-differential $f'(y) = xf'(xy)$;

$$\therefore \quad xf'(x) = yf'(y);$$

consequently $xf'(x)$ is a constant: let us suppose $xf'(x) = a$;

$$\therefore \quad f(x) = a\log\frac{x}{c};$$

substituting which in the given equation,

$$a\log\frac{x}{c} + a\log\frac{y}{c} = a\log\frac{xy}{c};$$
$$\therefore \quad c = 1;$$
$$\therefore \quad f(x) = a\log x.$$

Ex. 3. If $f(x)f(y) = f(x+y)$, $\qquad f(x) = e^{ax}$.

423.] The differential equation $P_1 y' + P_2 y + P_3 = 0$, where P_1, P_2, P_3 are functions of x, has been integrated in Art. 382. The form which next suggests itself is $P_1 y' + P_2 y + P_3 y^2 + P_4 = 0$, where $P_1 \dots P_4$ are functions of x: but this has never yet been completely integrated, and will not be, until the properties of certain transcendents, which are in the form of definite integrals, have been more completely investigated: a particular form however of it is

$$\frac{dy}{dx} + ay^2 = bx^m, \tag{251}$$

which is known by the name of Riccati's Equation, having been discussed by Riccati in the year 1775 in the Acta Eruditorum, and of which, in particular cases, solutions can be found: these I proceed to investigate.

First suppose $m = 0$; then (251) becomes

$$\frac{dy}{ay^2 - b} + dx = 0,$$

in which the variables are separated.

Again, let $y = z^n$; (251) becomes

$$nz^{n-1}dz + (az^{2n} - bx^m)\,dx = 0;$$

and this will be homogeneous if $n-1 = 2n = m$; that is, if $n = -1$, $m = -2$; thus the equation $\frac{dy}{dx} + ay^2 = \frac{b}{x^2}$ becomes homogeneous, if y is replaced by z^{-1}; and the integration can be performed.

Now to investigate general conditions of integrability; let

$$y = Ax^p + x^q z; \tag{252}$$

then the equation becomes

$$x^q dz + (qx^{q-1} + 2Aax^{p+q} + ax^{2q}z)\,z\,dx$$
$$+ (pAx^{p-1} + aA^2x^{2p})\,dx = bx^m dx; \tag{253}$$

in which, let

$$p-1 = 2p, \quad pA + aA^2 = 0, \quad q-1 = p+q, \quad q + 2Aa = 0;$$

$$\therefore \quad p = -1, \qquad A = \frac{1}{a}, \qquad q = -2;$$

so that (252) becomes

$$y = \frac{1}{ax} + \frac{z}{x^2},$$

and (251) becomes

$$x^2 dz + az^2 dx = bx^{m+4} dx; \tag{254}$$

in which equation the variables are separated if $m = -4$; and we have

$$\frac{dz}{az^2 - b} + \frac{dx}{x^2} = 0.$$

Again, in (254) let $x = \frac{1}{u}$; then

$$dz - az^2\,du = -bu^{-m-4}\,du,$$

which is of the same form as (251); and therefore if (251) is integrable for any particular value of m, say $\mu = m$, it is also integrable when

$$m = -\mu - 4.$$

424.] Again, in (251) let $y = \frac{1}{z}$; then

$$dz = a\,dx - b\,z^2 x^m\,dx. \tag{255}$$

Let $(m+1)x^m\,dx = dv$; then (255) becomes

$$\frac{dz}{dv} + \frac{b}{m+1}z^2 = \frac{a}{m+1}v^{-\frac{m}{m+1}};$$

which is of the same form as (251); and therefore if (251) is integrable for any particular value, say μ, of m, it will be integrable also when

$$m = -\frac{\mu}{\mu+1}. \tag{256}$$

Now we have seen above that (251) is integrable, when $\mu = -4$, therefore the equation is also integrable, when

$$m = -\frac{4}{3}.$$

Also from the conclusion of Art. 423 we infer that the equation is integrable when

$$m = \frac{4}{3} - 4 = -\frac{8}{3};$$

and therefore from (256) it appears that $m = -\frac{8}{5}$;

and thus substituting successively in the two formulæ

$$m = -\mu - 4, \qquad m = \frac{-\mu}{\mu+1};$$

we have the following series of values:

$$-4, \qquad -\frac{8}{3}, \qquad -\frac{12}{5}, \qquad -\frac{16}{7}, \ldots\ldots; \tag{257}$$

$$-\frac{4}{3}, \qquad -\frac{8}{5}, \qquad -\frac{12}{7}, \qquad -\frac{16}{9}, \ldots\ldots; \tag{258}$$

the types of the general terms being respectively $-\frac{4n}{2n-1}$ and $\frac{-4n}{2n+1}$; and in which if $n = 0$, and if $n = \infty$, we have the two values of m, viz. 0, and -2, which on inspection render (251) integrable.

Ex. 1. $dy + y^2 dx = x^{-\frac{4}{3}} dx.$

As this form is one of those which fall under the series (258), we must put $y = z^{-1}$;

$$\therefore \quad dz - dx = -z^2 x^{-\frac{4}{3}} dx.$$

Let $\quad x^{-\frac{4}{3}} dx = dv; \qquad \therefore \quad -3x^{-\frac{1}{3}} = v;$

$$\therefore \quad dz + z^2 dv = \left(\frac{3}{v}\right)^4 dv.$$

Let
$$z = \frac{1}{v} + \frac{u}{v^2};$$

$$\therefore \quad dz = -\frac{dv}{v^2} + \frac{du}{v^2} - \frac{2u\,dv}{v^3};$$

$$\frac{du}{v^2} + \frac{u^2 dv}{v^4} = \frac{81}{v^4} dv;$$

$$\therefore \quad \frac{du}{u^2 - 81} = -\frac{dv}{v^2}; \qquad \frac{1}{18} \log \frac{u-9}{u+9} = \frac{1}{v};$$

and substituting for u and for v,

$$u = 3\,\frac{3 + x^{\frac{1}{3}} y}{y x^{\frac{2}{3}}}, \qquad v = -\frac{3}{x^{\frac{1}{3}}},$$

$$\frac{y(x^{\frac{1}{3}} + 3x^{\frac{2}{3}}) + 3}{y(x^{\frac{1}{3}} - 3x^{\frac{2}{3}}) + 3} = e^{6x^{\frac{1}{3}}}.$$

425.] This example, and it is one of the easiest, sufficiently indicates the tediousness of the process, and the succession of the substitutions. If m has a value corresponding to the first term of the series (257) the method is of course that of Article 423: but if m has any other value, then we shall have to pass successively by alternate processes from one series to the other, until at last we arrive at a form wherein m has the value -4.

The above process is unsatisfactory, because although it points out certain cases where the variables are separable, still the number of them is limited; and they are obtained by particular artifices, and the investigation does not prove that they are the only possible ones. M. Liouville, however, in the VIth volume of his Mathematical Journal, has proved by a rigorous investigation that the cases comprised in the above series are the only ones where the integral can be expressed in an algebraical, logarithmic, or exponential form.

There are also other forms which are capable of reduction to Riccati's Equation. Thus, if

$$dy + ay^2 x^n dx = bx^m dx;$$

let $\quad x^n dx = dz; \qquad \therefore \quad x^{n+1} = (n+1)z;$

$$dy + ay^2 dz = b(n+1)^{\frac{m-n}{n+1}} z^{\frac{m-n}{n+1}} dz,$$

which is of the form (251).

The Equation of Riccati also admits of transformation into a differential equation of the second order, under which it is often convenient to consider it.

$$\frac{dy}{dx} + ay^2 = bx^m.$$

Let $\quad y = -\frac{1}{az}\frac{dz}{dx}; \qquad \therefore \quad \frac{dy}{dx} = \frac{1}{az^2}\frac{dz^2}{dx^2} - \frac{1}{az}\frac{d^2z}{dx^2};$

$$\therefore \quad \frac{d^2z}{dx^2} = -ab\,x^m z.$$

Let $\quad -ab = k; \qquad \therefore \quad \frac{d^2z}{dx^2} = kx^m z. \qquad (259)$

Again, in (259) let $\quad x^{\frac{m+2}{2}} = t$, and we have

$$\frac{d^2z}{dt^2} + \frac{m}{m+2}\frac{1}{t}\frac{dz}{dt} = \frac{4k}{(m+2)^2}z; \qquad (260)$$

and if we substitute $n = \frac{m}{2(m+2)}$, $l = \frac{4k}{(m+2)^2}$,

we have $$\frac{d^2z}{dt^2} + \frac{2n}{t}\frac{dz}{dt} = lz;$$

and if $z = ut^{-n}$,

$$\frac{d^2u}{dt^2} - \frac{n(n-1)}{t^2}u = lu.$$

All these therefore are equivalents of Riccati's Equation; and the properties which are true of any one are also true of each of the others. If therefore we can determine either a particular or a general integral of either, that of Riccati's equation will be determined by the equation

$$\log z = \int y\,dx.$$

A Memoir by M. Malmsten of the University of Upsala, and inserted in Vol. XXXIX. of Crelle's Journal, p. 108, on the various forms and properties of Riccati's Equation, may be consulted with advantage by the reader who is desirous of further information.

SECTION 11.—*Solution of Geometrical Problems involving Differential Equations.*

426.] Although we have generally introduced in the course of our work the solution of those geometrical problems which depend on Differential Equations, yet some remain which require special treatment. Of these the first class is that of Trajectories; a trajectory being a line or a surface which cuts a series of lines or surfaces according to a given law, the series of lines or surfaces so cut being generally formed by the variation of a parameter contained in their general equation. And let us first consider the trajectory to cut a given series of plane curves at a constant angle.

Let $f(x, y, a) = 0$ be the equation to any one of the curves, the series being formed by the variation of the arbitrary parameter a; and let $\text{F}(x', y') = 0$ be the equation to the required trajectory, and $\frac{m}{n}$ be the tangent of the constant angle contained between the curves at their point of intersection; then

$$\frac{m}{n} = \frac{\dfrac{dy'}{dx'} - \dfrac{dy}{dx}}{1 + \dfrac{dy'}{dx'}\dfrac{dy}{dx}}; \tag{261}$$

$$\therefore \quad m + m\frac{dy'}{dx'}\frac{dy}{dx} = n\frac{dy'}{dx'} - n\frac{dy}{dx}. \tag{262}$$

Let $\frac{dy}{dx}$ be found from the equation of the given family of curves and be substituted in (261), and let a be eliminated by means of (261) and of the given equation; then if for y' and x', y and x are substituted, because they refer to the same point, the integral of (262) will be that of the required trajectory.

If the angle between the two curves is a right-angle, the trajectory is said to be *orthogonal;* in which case $n = 0$, and (262) becomes

$$1 + \frac{dy'}{dx'}\frac{dy}{dx} = 0. \tag{263}$$

Ex. 1. To find the equation to the curve which cuts at a constant angle all circles passing through a given point, and at that point touching a given straight line.

Let the point be taken for the origin and the given line for the axis of y; then the equation to the circles is

$$y^2 = 2ax - x^2;$$

$$\frac{dy}{dx} = \frac{a-x}{y} = \frac{y^2-x^2}{2xy};$$

therefore from (262),

$$m\left\{1 + \frac{dy}{dx}\frac{y^2-x^2}{2xy}\right\} = n\left\{\frac{dy}{dx} - \frac{y^2-x^2}{2xy}\right\}$$

$$\therefore \quad (ny^2+2mxy-nx^2)\,dx + (my^2-2nxy-mx^2)\,dy = 0,$$

which is homogeneous, and of the second degree. Therefore by the method of Art. 395,

$$\mathrm{D}u = \frac{(ny^2+2mxy-nx^2)\,dx + (my^2-2nxy-mx^2)\,dy}{(my-nx)(x^2+y^2)} = 0;$$

$$\therefore \quad u_x = \log(x^2+y^2) - \log(my-nx);$$

$$u_y = \log(x^2+y^2) - \log(my-nx);$$

$$\therefore \quad x^2+y^2 = 2c(my-nx),$$

where $2c$ is the arbitrary constant of integration. The equation is manifestly that to a circle.

If the trajectory is orthogonal, $n = 0$; and the equation becomes $x^2+y^2 = 2cmy$, the equation to a circle passing through the origin, and whose centre is on the axis of y, and radius $= cm$.

It will be observed that the arbitrary constant of integration leaves the particular curve undetermined, although the general integral determines the species of it.

Ex. 2. Find the trajectory of a series of parallel straight lines.

Let the equation to the lines be $x\cos a + y\sin a = p$, where a is constant, and p is the variable parameter;

$$\therefore \quad \frac{dy}{dx} = -\cot a;$$

therefore equation (262) becomes

$$m - m\cot a\frac{dy}{dx} = n\frac{dy}{dx} + n\cot a,$$

$$mx - my\cot a = ny + nx\cot a + c,$$

$$(m\sin a - n\cos a)x - (m\cos a + n\sin a)y = c\sin a.$$

The equation to a series of parallel straight lines.

Ex. 3. Find the orthogonal trajectory of a series of parabolas expressed by the equation $y^2 = 4ax$:

$$\therefore \quad \frac{dy}{dx} = \frac{2a}{y} = \frac{y}{2x};$$

so that by equation (263) we have

$$1+\frac{dy}{dx}\frac{y}{2x} = 0; \qquad 2x\,dx+y\,dy = 0;$$

$$x^2+\frac{y^2}{2} = c^2;$$

where c^2 is an arbitrary constant.

$$\therefore \quad \frac{x^2}{c^2}+\frac{y^2}{2c^2} = 1.$$

Ex. 4. Find the orthogonal trajectory of the series of hyperbolas expressed by $xy = k^2$.

$$\therefore \quad \frac{dy}{dx} = -\frac{y}{x}; \qquad 1-\frac{y}{x}\frac{dy}{dx} = 0; \qquad x^2-y^2 = c^2.$$

427.] The trajectory, orthogonal or other, of a series of curves referred to polar coordinates may be determined in a similar manner; thus

$$\tan^{-1}\frac{m}{n} = \tan^{-1}\frac{r\,d\theta}{dr}-\tan^{-1}\frac{r'd\theta'}{dr'}; \tag{264}$$

$$\therefore \quad \frac{m}{n} = \frac{\dfrac{r\,d\theta}{dr}-\dfrac{r'd\theta'}{dr'}}{1+\dfrac{rr'd\theta\,d\theta'}{dr\,dr'}},$$

$$m+mrr'\frac{d\theta}{dr}\frac{d\theta'}{dr'} = nr\frac{d\theta}{dr}-nr'\frac{d\theta'}{dr'}; \tag{265}$$

and if the trajectory is orthogonal, $n = 0$; and

$$1+rr'\frac{d\theta}{dr}\frac{d\theta'}{dr'} = 0. \tag{266}$$

Ex. 1. Find the orthogonal trajectory of a series of logarithmic spirals expressed by the equation $r = a^{\theta}$, where a varies.

$$\therefore \quad \frac{dr}{d\theta} = a^{\theta}\log a = \frac{r}{\theta}\log r;$$

therefore (266) becomes $1+\dfrac{r\,d\theta}{dr}\dfrac{\theta}{\log r} = 0;$

$$\frac{dr}{r}\log r+\theta\,d\theta = 0; \qquad (\log r)^2+\theta^2 = c^2;$$

$$\log r = (c^2-\theta^2)^{\frac{1}{2}}, \qquad r = e^{(c^2-\theta^2)^{\frac{1}{2}}}.$$

Ex. 2. Determine the orthogonal trajectory of a series of lemniscata expressed by the equation $r^2 = a^2\cos 2\theta$.

$$\therefore \quad \frac{r\,d\theta}{dr} = -\frac{\cos 2\theta}{\sin 2\theta}; \qquad 1-\frac{r\,d\theta}{dr}\frac{\cos 2\theta}{\sin 2\theta} = 0;$$

$$r^2 = c^2\sin 2\theta,$$

which is the equation to another lemniscata whose axis is inclined at 45° to that of the given one.

Ex. 3. Find the equation to the orthogonal trajectory of a series of confocal and coaxal parabolas.

$$r = \frac{2a}{1+\cos\theta}; \qquad \frac{r\,d\theta}{dr} = \frac{1+\cos\theta}{\sin\theta};$$

$$\therefore \quad 1 + \frac{r\,d\theta}{dr}\,\frac{1+\cos\theta}{\sin\theta} = 0; \qquad \frac{dr}{r} + \cot\frac{\theta}{2}\,d\theta = 0;$$

$$\therefore \quad r = \frac{2c}{1-\cos\theta},$$

the equation to a series of the confocal and coaxal parabolas.

428.] Trajectories with reference to families of curves may also be drawn according to other laws. The following examples illustrate the kind of problem.

Ex. 1. A series of cycloids, see fig. 52, have a common vertex O, and a common axis Ox; it is required to find the equation to the curve which cuts off from all of them an equal length of arc OP.

Let the length of the arc be k; and let the equation of one of the cycloids be

$$y = a \operatorname{versin}^{-1}\frac{x}{a} + (2ax - x^2)^{\frac{1}{2}};$$

$$\therefore \quad \frac{dx}{x^{\frac{1}{2}}} = \frac{dy}{(2a-x)^{\frac{1}{2}}} = \frac{ds}{(2a)^{\frac{1}{2}}};$$

$$s = 2(2ax)^{\frac{1}{2}} = k; \qquad a = \frac{k^2}{8x};$$

$$\therefore \quad y = \frac{k^2}{8x}\operatorname{versin}^{-1}\frac{8x^2}{k^2} + \left(\frac{k^2}{4} - x^2\right)^{\frac{1}{2}}.$$

Ex. 2. Many circles touch each other at a common point: find the curve which cuts them at an angle proportional to the vectorial angle at the point of section, the common point being the pole and their common diametral line being the prime radius.

$$r = 2a\cos\theta;$$

$$\therefore \quad \frac{r\,d\theta}{dr} = -\cot\theta = \tan(90° + \theta).$$

Let $k\theta =$ the angle of intersection;

$$\therefore \quad k\theta = \tan^{-1}\frac{r\,d\theta}{dr} - \tan^{-1}\frac{r'\,d\theta'}{dr'}$$

$$= 90° + \theta - \tan^{-1}\frac{r\,d\theta}{dr};$$

when the arcs intercepted betwee
of x are of a constant length. Se
Let OA $= a$, AOP $= \theta$: therefor

$$\therefore \quad a =$$

$$\therefore \quad r = a =$$

the equation to a reciprocal spiral.

429.] By a similar process ma
surfaces which are trajectories, or
surfaces of a given family. Supp
family to be $\qquad F(x, y, z) =$

and this equation to involve an ar
equation to the trajectory to be

$$f(x, y, z) =$$

and then if this second surface cuts
an angle whose cosine is m,

$$\left(\frac{dF}{dx}\right)\left(\frac{df}{dx}\right) + \left(\frac{dF}{dy}\right)\left(\frac{df}{dy}\right) + \left(\frac{dF}{dz}\right)\left(\frac{df}{dz}\right.$$

$$= m\left\{\left(\frac{dF}{dx}\right)^2 + \left(\frac{dF}{dy}\right)^2 + \left(\frac{dF}{dz}\right)^2\right\}^{\frac{1}{2}}\left\{\left(\frac{df}{dx}\right.\right.$$

and if the trajectory is orthogonal,

$$\left(\frac{dF}{dx}\right)\left(\frac{df}{dx}\right) + \left(\frac{dF}{dy}\right)\left(\frac{df}{dy}\right) +$$

wherein $\left(\frac{dF}{dx}\right)$, $\left(\frac{dF}{dy}\right)$, $\left(\frac{dF}{dz}\right)$ are to

Ex. 1. To find the orthogonal trajectory of a series of spheres touching a given plane at a given point.

Let the given point be taken as the origin, and the given plane for the plane of (y, z); then the equation to the spheres is

$$x^2-2ax+y^2+z^2 = 0 = F(x, y, z),$$

where a is variable; and therefore (270) becomes

$$\left(\frac{df}{dx}\right)\frac{x^2-y^2-z^2}{2x} + \left(\frac{df}{dy}\right) y + \left(\frac{df}{dz}\right) z = 0;$$

so that by (84), Art. 384,

$$\frac{2x\,dx}{x^2-y^2-z^2} = \frac{dy}{y} = \frac{dz}{z}; \qquad \therefore \quad \frac{y}{z} = c_1;$$

$$\frac{2xz\,dx - x^2\,dz}{z^2} + (1+c_1^2)\,dz = 0,$$

$$\frac{x^2}{z} + (1+c_1^2)z = c_2 = f(c_1);$$

$$\therefore \quad x^2+y^2+z^2 = zf\left(\frac{y}{z}\right),$$

where f expresses an arbitrary function.

Ex. 2. Find the equation of the orthogonal trajectory of

$$\frac{x^2}{a^2} + \frac{y^2}{b^2} + \frac{z^2}{c^2} = k^2,$$

where k is a variable parameter.

In this case (270) becomes

$$\left(\frac{df}{dx}\right)\frac{x}{a^2} + \left(\frac{df}{dy}\right)\frac{y}{b^2} + \left(\frac{df}{dz}\right)\frac{z}{c^2} = 0,$$

$$\frac{a^2\,dx}{x} = \frac{b^2\,dy}{y} = \frac{c^2\,dz}{z};$$

$$a^2 \log x - b^2 \log y = c_1;$$

$$b^2 \log y - c^2 \log z = c_2;$$

$$\therefore \quad a^2 \log x - b^2 \log y = f(b^2 \log y - c^2 \log z),$$

is the equation to the trajectory, where f represents an arbitrary function.

430.] The following geometrical problems involve total differentials of three variables.

Ex. 1. Find the equation to the surface whose tangent plane is

$$(\xi - x)\frac{x}{a^2} + (\eta - y)\frac{y}{b^2} + (\zeta - z)\frac{z}{c^2} = 0;$$

comparing this with the general equation of the tangent plane

$$(\xi-x)\left(\frac{dF}{dx}\right)+(\eta-y)\left(\frac{dF}{dy}\right)+(\zeta-z)\left(\frac{dF}{dz}\right)=0,$$

we have

$$\frac{\left(\frac{dF}{dx}\right)}{\frac{x}{a^2}}=\frac{\left(\frac{dF}{dy}\right)}{\frac{y}{b^2}}=\frac{\left(\frac{dF}{dz}\right)}{\frac{z}{c^2}}=\frac{\left(\frac{dF}{dx}\right)dx+\left(\frac{dF}{dy}\right)dy+\left(\frac{dF}{dz}\right)dz}{\frac{x\,dx}{a^2}+\frac{y\,dy}{b^2}+\frac{z\,dz}{c^2}};$$

and as the numerator of this last fraction $= 0$, we have also

$$\frac{x\,dx}{a^2}+\frac{y\,dy}{b^2}+\frac{z\,dz}{c^2}=0;$$

$$\therefore\quad \frac{x^2}{a^2}+\frac{y^2}{b^2}+\frac{z^2}{c^2}=k^2,$$

the equation to an ellipsoid.

Ex. 2. Find the equation to the surface whose tangent plane is

$$\frac{\xi-x}{x}+\frac{\eta-y}{y}+\frac{\zeta-z}{z}=0;$$

$$xyz=k^3.$$

431.] The following geometrical problems also involve partial differential equations of the first order.

Ex. 1. Determine the surface whose tangent planes pass through the same point.

$$(x-a)\left(\frac{dF}{dx}\right)+(y-b)\left(\frac{dF}{dy}\right)+(z-c)\left(\frac{dF}{dz}\right)=0;$$

$$\therefore\quad \frac{dx}{x-a}=\frac{dy}{y-b}=\frac{dz}{z-c};$$

$$\frac{x-a}{y-b}=c_1,\qquad \frac{x-a}{z-c}=c_2;$$

$$\therefore\quad \frac{x-a}{y-b}=f\left(\frac{x-a}{z-c}\right),$$

the general equation to conical surfaces.

Ex. 2. To determine the surface such that the intercept of the axis of x by the tangent plane is proportional to x.

The differential equation which expresses this property is

$$x\left(\frac{dF}{dx}\right)+y\left(\frac{dF}{dy}\right)+z\left(\frac{dF}{dz}\right)=nx\left(\frac{dF}{dx}\right);$$

$$\therefore\quad (1-n)x\left(\frac{dF}{dx}\right)+y\left(\frac{dF}{dy}\right)+z\left(\frac{dF}{dz}\right)=0;$$

$$\therefore\quad x=z^{1-n}f\left(\frac{y}{z}\right).$$

Ex. 2. Determine the equation to the surface in which the coordinates of the point where the normal meets the plane of (x, y) are to each other as the corresponding coordinates.

As the equations to the normal are

$$\frac{\xi - x}{\left(\frac{dF}{dx}\right)} = \frac{\eta - y}{\left(\frac{dF}{dy}\right)} = \frac{\zeta - z}{\left(\frac{dF}{dz}\right)};$$

$$\therefore \quad \left(\frac{dF}{dz}\right)\xi_0 = x\left(\frac{dF}{dz}\right) - z\left(\frac{dF}{dx}\right),$$

$$\left(\frac{dF}{dz}\right)\eta_0 = y\left(\frac{dF}{dz}\right) - z\left(\frac{dF}{dy}\right);$$

but $\frac{\xi_0}{\eta_0} = \frac{x}{y}$; $\quad\therefore \quad y\left(\frac{dF}{dx}\right) - x\left(\frac{dF}{dy}\right) = 0$;

$$z = c_1; \qquad x^2 + y^2 = c_2;$$

$$\therefore \quad z = f(x^2 + y^2),$$

where f represents an arbitrary function.

432.] Also let us consider the differential equation of the first order and of the second degree, which expresses the lines of curvature of an ellipsoid.

Let the equation to the ellipsoid be

$$\frac{x^2}{a^2} + \frac{y^2}{b^2} + \frac{z^2}{c^2} = 1; \tag{271}$$

then by the general equation (64) Art. 409, Vol. I, the equation to the lines of curvature is

$$(b^2 - c^2)\frac{x}{dx} + (c^2 - a^2)\frac{y}{dy} + (a^2 - b^2)\frac{z}{dz} = 0. \tag{272}$$

Let $\qquad \frac{x^2}{a^2} = \xi, \qquad \frac{y^2}{b^2} = \eta, \qquad \frac{z^2}{c^2} = \zeta, \qquad (273)$

so that (271) and (272) become

$$\xi + \eta + \zeta = 1, \tag{274}$$

$$(b^2 - c^2)\xi d\eta d\zeta + (c^2 - a^2)\eta d\zeta d\xi + (a^2 - b^2)\zeta d\xi d\eta = 0. \tag{275}$$

Let ζ and $d\zeta$ be eliminated from these equations; then we have

$$(c^2 - a^2)\eta d\xi^2 + \{(a^2 - c^2)\xi - (b^2 - c^2)\eta - (a^2 - b^2)\} d\xi d\eta + (b^2 - c^2)\xi d\eta^2 = 0;$$

which may be expressed in the form

$$(\eta d\xi - \xi d\eta)\{(c^2 - a^2)d\xi - (b^2 - c^2)d\eta\} = (a^2 - b^2)d\xi d\eta;$$

$$\therefore \quad \eta = \xi\frac{d\eta}{d\xi} - \frac{(a^2 - b^2)d\eta}{(a^2 - c^2)d\xi + (b^2 - c^2)d\eta}; \tag{276}$$

which is an equation of Clairaut's form; and of which, if k is an arbitrary constant, the general integral is

$$\xi k-\eta = \frac{(a^2-b^2)k}{a^2-c^2+k(b^2-c^2)};$$

$$\therefore \quad k\frac{x^2}{a^2}-\frac{y^2}{b^2} = \frac{(a^2-b^2)k}{(a^2-c^2)+k(b^2-c^2)}; \qquad (277)$$

which represents a cylinder of the second degree, the axis of which is perpendicular to the plane of (x, y); and consequently the lines of curvature are determined by the intersection of these cylindrical surfaces with the ellipsoid.

CHAPTER XVI.

INTEGRATION OF DIFFERENTIAL EQUATIONS OF ORDERS HIGHER THAN THE FIRST.

SECTION 1.—*General Properties of Differential Equations of Higher Orders.*

433.] WE are now just on the outskirts of our science, and are unable to give any general theory for the integration of differential equations of higher orders; almost all that deserves the name of philosophical treatment has been exhausted; and it only remains for us to insert such discussions on isolated topics as are useful either in the way of extending the boundaries of our knowledge, or for the purposes of subsequent application.

The most general forms of differential equations of the nth order are (1) (2) (3) (4) in Art. 364; the last two of these are partial, and the discussion of them is reserved to a future Section of the present Chapter: we shall confine our researches at present to an equation of the form

$$F\left(x, y, \frac{dy}{dx}, \frac{d^2y}{dx^2}, \dots \frac{d^ny}{dx^n}\right) = 0, \tag{1}$$

which contains only two variables, and wherein one of these is equicrescent. Of such equations we have in Art. 365 pointed out the geometrical meaning; and in Art. 367 have shewn that the general integral involves n arbitrary constants. If a function satisfies (1) and does not contain n arbitrary constants, it may be either a particular integral or a singular solution; but it is not the general integral. And it will be either a particular integral or a singular solution according as one or more of the arbitrary constants has been replaced by particular constant values or by functions of the variables: and it is manifest that such substitutions may take place, at any one, or at more than one, of the successive integrations.

434.] Now with reference to general properties of differential equations of the form (1), if (1) admits of being expressed explicitly in the form

$$\frac{d^n y}{dx^n} = f\left(x, y, \frac{dy}{dx}, \dots \frac{d^{n-1}y}{dx^{n-1}}\right); \qquad (2)$$

and, the limits of integration being (x_0, y_0) (x_1, y_1), if (2) as well as all its derived-functions remain finite and continuous for all values of the variables within the limits, then (2) can be integrated in a series, by the method of Art. 367: and its general integral will contain n arbitrary constants.

Also it is evident that a differential expression such as (1) may admit of integration by reason of the *form* of the expression, and independently of any specific relation between x and y: the conditions that this should be the case have received much consideration from Euler, Lagrange, Lexell, Poisson; and lastly from M. J. Bertrand*, and M. J. Binet, as quoted in Moigno's Calcul Integral, Vol. II. p. 551: and it is to Euler and to the last two that we are indebted for most of our knowledge of the subject. In the following articles the conditions requisite for such a case are investigated by means of the Calculus of Variations.

Suppose the integral of (1) to be definite, and the limits of integration to be those particular values of the variables which carry the subscripts 0 and 1: and let the definite integral be expressed according to the notation of Art. 247. Now our object is to determine conditions which (1) must satisfy, so as to be the x-derived function of some other function of the form,

$$\Phi\left(x, y, \frac{dy}{dx}, \frac{d^2y}{dx^2}, \dots \frac{d^{n-1}y}{dx^{n-1}}\right), \qquad (3)$$

independently of any relation between x and y; that is, so that

$$\int_0^1 \mathrm{F}(x, y, y', \dots y^{(n)})\, dx = \Big[\Phi(x, y, y', \dots y^{(n-1)})\Big]_0^1, \qquad (4)$$

and so that this equation may subsist independently of the functional connexion of x and y.

Suppose this functional relation to undergo a small variation, and the values of the variables and of the $(n-1)$ derived functions at the limits not to change; then by reason of (4) the value of the integral will not be altered, and therefore

$$\delta.\int_0^1 \mathrm{F}(x, y, y', y'', \dots y^{(n)})\, dx = 0; \qquad (5)$$

then employing the notation introduced in Article 303, it is manifest, that if we replace the left-hand member of (5) by its

* See Journal de l'Ecole Royale Polytechnique, Cahier 28, Paris 1841, p. 249.

value given in equation (56) of Art. 303, (5) cannot be true unless

$$\text{Y} - \frac{d\text{Y}'}{dx} + \frac{d^2\text{Y}''}{dx^2} - \dots (-)^n \frac{d^n \text{Y}^{(n)}}{dx^n} = 0; \tag{6}$$

and this therefore is the condition requisite that (1) should be an exact differential independently of any relation between y and x.

It will be observed that $\text{Y}, \text{Y}', \text{Y}'' \dots$ are partial derived functions; but that the subsequent x-differentiations are made on the supposition that all these quantities are implicit functions of x: and that they do not vanish, although x may not enter explicitly into them.

435.] Let us now pass to the converse of the above. Suppose that $\text{F}(x, y, y', \dots y^{(n)})$ satisfies the condition (6); then its integral is capable of being expressed in the form (4), and independently of any relation between x and y; or what is tantamount, if (6) is satisfied, the integral can be expressed in terms of the limiting values of the variables and of their derived functions; and this is what we mean by definite integration. For in this case, by virtue of equation (56), Art. 303, the variation of the integral on the left-hand side of (5) will be expressed in terms of the limiting values of the variables and of their derived-functions; and in terms of these alone; and consequently the integral will be a function of these quantities only. Hence also, if these limits are fixed, their variations disappear, and the variation of the definite integral also vanishes. Some examples are subjoined.

Ex. 1. Let V be a function of x and y: it is required to determine the condition that $\text{V}\,dx$ should be integrable independently of any relation between x and y.

In this case (6) becomes $\text{Y} = \left(\frac{d\text{V}}{dy}\right) = 0$; consequently V must not contain y.

Ex. 2. Under what circumstances is $(\text{P} + \text{Q}y')\,dx$, where P and Q are functions of x and y, integrable, independently of any relation between x and y?

In this case (6) becomes

$$\text{Y} - \frac{d\text{Y}'}{dx} = 0; \tag{7}$$

$$\text{Y} = \left(\frac{d\text{P}}{dy}\right) + \left(\frac{d\text{Q}}{dy}\right) y', \qquad \text{Y}' = \text{Q};$$

$$\therefore \quad \frac{d\text{Y}'}{dx} = \left(\frac{d\text{Q}}{dx}\right) + \left(\frac{d\text{Q}}{dy}\right) y';$$

therefore (7) becomes

$$\left(\frac{dP}{dy}\right) - \left(\frac{dQ}{dx}\right) = 0,$$

which is the same condition as (28), Art. 371; hence also we may infer that the complete integral of the differential equation of the first order and degree contains an undetermined functional symbol.

Ex. 3. Prove that $y'y'' - x^2 y y' - xy^2 = 0$ satisfies the condition of integrability.

436.] It is good also to exhibit *a posteriori* the criterion given in (6) in a particular case. Let us suppose $\frac{yy'}{x}$ to be the integral of a given differential expression, when no functional relation is given between x and y; then the x-differential of $\frac{yy'}{x}$ is

$$\frac{yy''}{x} + \frac{y'^2}{x} - \frac{yy'}{x^2} = F(x, y, y', y'');$$

and this must satisfy (6); now

$$Y = \left(\frac{dF}{dy}\right) = \frac{y''}{x} - \frac{y'}{x^2}, \quad Y' = \left(\frac{dF}{dy'}\right) = \frac{2y'}{x} - \frac{y}{x^2}, \quad Y'' = \left(\frac{dF}{dy''}\right) = \frac{y}{x};$$

$$\frac{dY'}{dx} = -\frac{2y'}{x^2} + \frac{2y}{x^3} + \frac{2y''}{x} - \frac{y'}{x^2}; \qquad \frac{d^2Y''}{dx^2} = \frac{y''}{x} - \frac{2y'}{x^2} + \frac{2y}{x^3}.$$

$$\therefore \quad Y - \frac{dY'}{dx} + \frac{d^2Y''}{dx^2} = 0.$$

437.] We may also by a similar process determine the conditions that $F\,dx^m$ should be integrable m times successively, and independently of any particular relation between x and y; m being not greater than n which is the index of the highest derived-function contained in F. Let

$$V = F(x, y, y', \dots y^{(n)}); \tag{8}$$

then it is manifest by the principles enuntiated above that, in accordance with the notation of Art. 150, the variation of the definite integral of

$$\int^m F(x, y, y', \dots y^{(n)})\, dx^m$$

must not involve terms containing signs of integration. Now using the symbols of Art. 303, and supposing $\delta x = 0$,

$$\delta . \int^m \text{v}\, dx^m = \int^m \delta \text{v}\, dx^m$$

$$= \int^m \{\text{Y}\,\delta y + \text{Y}'\delta y' + \text{Y}''\delta y'' + \ldots + \text{Y}^{(n)}\delta y^{(n)}\}\, dx^m. \quad (9)$$

Of this series let us take a typical term, say $\text{Y}^{(k)}\,\delta y^{(k)}$, which we may write in the form

$$\text{Y}^{(k)}\,\delta . \frac{d^k y}{dx^k} = \text{Y}^{(k)}\frac{d^k\,\delta y}{dx^k}. \quad (10)$$

Now, by the theorem proved in (57), Art. 426, Vol. I,

$$\text{Y}^{(k)}\frac{d^k\delta y}{dx^k} = \frac{d^k}{dx^k}.\text{Y}^{(k)}\delta y - \frac{k}{1}\frac{d^{k-1}}{dx^{k-1}}.\frac{d\text{Y}^{(k)}}{dx}\delta y + \frac{k(k-1)}{1.2}\frac{d^{k-2}}{dx^{k-2}}.\frac{d^2\text{Y}^{(k)}}{dx^2}\delta y - \ldots$$

$$\ldots(-)^{k-1}\frac{k}{1}\frac{d}{dx}.\frac{d^{k-1}\text{Y}^{(k)}}{dx^{k-1}}\delta y\,(-)^k\frac{d^k\text{Y}^{(k)}}{dx^k}\delta y. \quad (11)$$

$$\therefore \int^m \text{Y}^{(k)}\frac{d^k\delta y}{dx^k}dx^m = \int^{m-k}\text{Y}^{(k)}\delta y\, dx^{m-k} - \frac{k}{1}\int^{m-k+1}\frac{d\text{Y}^{(k)}}{dx}\delta y\, dx^{m-k+1} + \ldots$$

$$\ldots(-)^{k-1}\frac{k}{1}\int^{m-1}\frac{d^{k-1}\text{Y}^{(k)}}{dx^{k-1}}\delta y\, dx^{m-1}(-)^k\int^m\frac{d^k\text{Y}^{(k)}}{dx^k}dx^m; \quad (12)$$

and therefore the right-hand member of (9) consists of a series of terms of which (12) is the type; and wherein k receives all integral values from $k = 0$ to $k = n$, both inclusive; and where $\text{Y}^0 = \text{Y}$.

Now it is evident that, if $\delta . \int^m \text{v}\, dx^m$ is to be free from terms under signs of integration, the coefficients of δy under the several signs $\int^m, \int^{m-1}, \ldots \int^2, \int$ must vanish of themselves; whence we have

$$\left.\begin{array}{l}
\text{Y} - \dfrac{d\text{Y}'}{dx} + \dfrac{d^2\text{Y}''}{dx^2} - \ldots\ldots\ldots(-)^n\dfrac{d^n\text{Y}^{(n)}}{dx^n} = 0; \\
\text{Y}' - 2\dfrac{d\text{Y}''}{dx} + 3\dfrac{d^2\text{Y}'''}{dx^2} - \ldots\ldots(-)^{n-1}n\dfrac{d^{n-1}\text{Y}^{(n)}}{dx^{n-1}} = 0; \\
\text{Y}'' - \dfrac{3.2}{1.2}\dfrac{d\text{Y}'''}{dx} + \dfrac{4.3}{1.2}\dfrac{d^2\text{Y}''''}{dx^2} - \ldots(-)^{n-2}\dfrac{n(n-1)}{1.2}\dfrac{d^{n-2}\text{Y}^{(n)}}{dx^{n-2}} = 0. \\
\ldots\ldots\ldots\ldots\ldots\ldots\ldots\ldots
\end{array}\right\} \quad (13)$$

This series of conditions must be continued so long as the integration-signs have positive indices; for when the indices are negative, and when they vanish, the corresponding terms have their limiting values: of the general form (12) therefore we must take the last m terms; that is, the terms corresponding to values of the indices of the integration-signs until $k = m-1$; in which case we have

$$\text{Y}^{(m-1)} - m\frac{d\text{Y}^{(m)}}{dx} + \frac{m(m+1)}{1.2}\frac{d^2\text{Y}^{(m+1)}}{dx^2} - \dots$$

$$\dots (-)^{n-m+1}\frac{m(m+1)\dots n}{1.2\dots(n-m+1)}\frac{d^{n-m+1}\text{Y}^{(n)}}{dx^{n-m+1}} = 0; \quad (14)$$

so that we have m equations of condition; and if these are satisfied the given differential expression will be integrable m times successively.

438.] A similar process enables us to determine the conditions necessary that

$$\text{F}(x, y, y', y'', \dots y^{(n)}, z, z', z'', \dots z^{(n)})\,dx, \quad (15)$$

in which we have used the notation of Art. 308, should be integrable independently of any relation between x, y, and z: for if the variation of the integral of (15) does not contain a quantity under the sign of integration and depends only on the limiting values of the variable quantities, then

$$\left.\begin{aligned} \text{Y} - \frac{d\text{Y}'}{dx} + \frac{d^2\text{Y}''}{dx^2} - \dots (-)^n\frac{d^n\text{Y}^{(n)}}{dx^n} &= 0; \\ \text{Z} - \frac{d\text{Z}'}{dx} + \frac{d^2\text{Z}''}{dx^2} - \dots (-)^n\frac{d^n\text{Z}^{(n)}}{dx^n} &= 0; \end{aligned}\right\} \quad (16)$$

and similar conditions must be fulfilled if the element-function contains any number of variables; and also conditions similar to (13) and (14), if such an element-function is capable of m successive integrations: thus suppose $\text{V}\,dx^n$ to involve m variables besides x, then the number of conditions requisite that $\text{V}\,dx^n$ should be integrable n times successively is mn.

It is beyond the scope of our work to investigate the corresponding condition in the case of a multiple integral: the student, however, desirous of pursuing the inquiry will obtain the necessary aid from Jellett's Calculus of Variations, and from Moigno et Lindelöf, Calcul des Variations, referred to in the foot-note of page 411.

439.] Of a particular form of differential equations of the nth order, which is called the *linear*, many properties will be investigated in the following sections; but it is convenient to consider it at once in reference to the conditions (13) and (14). The equation is

$$\text{P}_n \frac{d^n y}{dx^n} + \text{P}_{n-1} \frac{d^{n-1} y}{dx^{n-1}} + \ldots + \text{P}_1 \frac{dy}{dx} + \text{P}_0 y + \text{Q} = 0, \tag{17}$$

where $\text{P}_n, \text{P}_{n-1}, \ldots \text{P}_2, \text{P}_1, \text{P}_0, \text{Q}$ are functions of x and y.

If this equation is integrable once without any specific relation between x and y, it must satisfy (13); and consequently

$$\text{P}_0 - \frac{d\text{P}_1}{dx} + \frac{d^2\text{P}_2}{dx^2} - \ldots (-)^n \frac{d^n \text{P}_n}{dx^n} = 0. \tag{18}$$

If it is integrable twice, it must also satisfy the condition

$$\text{P}_1 - 2\frac{d\text{P}_2}{dx} + 3\frac{d^2\text{P}_3}{dx^2} - \ldots (-)^{n-1} \frac{d^{n-1}\text{P}_n}{dx^{n-1}} = 0; \tag{19}$$

and so on. Thus if P_1 is a function of x only, the equation $\text{P}_1 \frac{dy}{dx} + \frac{d\text{P}_1}{dx} y = 0$ satisfies (18), and is evidently integrable immediately.

Again, if $\text{P}_2, \text{P}_1, \text{P}_0$ are functions of x only, it is required to determine the value of P_0 in the equation

$$\text{P}_2 \frac{d^2 y}{dx^2} + \text{P}_1 \frac{dy}{dx} + \text{P}_0 y = 0, \tag{20}$$

so that the equation should be integrable once; in this case (18) becomes

$$\text{P}_0 - \frac{d\text{P}_1}{dx} + \frac{d^2\text{P}_2}{dx^2} = 0; \tag{21}$$

$$\therefore \quad \text{P}_2 \frac{d^2 y}{dx^2} + \text{P}_1 \frac{dy}{dx} + \left(\frac{d\text{P}_1}{dx} - \frac{d^2\text{P}_2}{dx^2}\right) y = 0$$

is an exact differential expression; and of it the integral is

$$\text{P}_2 \frac{dy}{dx} + \left(\text{P}_1 - \frac{d\text{P}_2}{dx}\right) y = c_1; \tag{22}$$

suppose again that (20) is integrable twice; then in addition to (21) we must have from (19)

$$\text{P}_1 - 2\frac{d\text{P}_2}{dx} = 0; \tag{23}$$

and this condition might also have been deduced from (22), by applying to it the criterion (18), that (22) should be integrable once.

There is also one other point that deserves notice. Suppose that (20) does not satisfy (18), but can be made to do so by the introduction of a factor; let μ be the factor, then we have

$$\mu \text{P}_2 \frac{d^2 y}{dx^2} + \mu \text{P}_1 \frac{dy}{dx} + \mu \text{P}_0 y = 0; \tag{24}$$

so that (18) becomes

$$\mu P_0 - \frac{d.\mu P_1}{dx} + \frac{d^2.\mu P_2}{dx^2} = 0; \qquad (25)$$

and if from this any value of μ, general or particular, can be found, then (20) may be integrated directly. It will be observed however that (25) is a differential equation of the second order in terms of μ, and that the difficulty of solution, as far as the order is concerned, is not lessened.

SECTION 2.—*Investigation of Properties of Linear Differential Equations.*

440.] As there is no general method of solving differential expressions of the second and higher orders, we are obliged to have recourse to such particular forms of them as have yielded to the powers of analysis; and amongst these the most remarkable is that known by the name of the *linear equation*, and of which the solution is of the form $y = f(x)$; into which the independent variable and its derived-function enter in only the first degree, and where the coefficients are functions of the variable x only. Thus the most general form is

$$\frac{d^n y}{dx^n} + P_1 \frac{d^{n-1} y}{dx^{n-1}} + P_2 \frac{d^{n-2} y}{dx^{n-2}} + \ldots + P_{n-1} \frac{dy}{dx} + P_n y = X, \qquad (26)$$

where $P_1, P_2, \ldots P_n$, X are functions of x only. Of this equation we shall prove some general properties, and then proceed to the solution of particular examples.

It will be observed that two forms of this equation have already been integrated; (1) in Art. 150, where $P_1 = P_2 = \ldots = P_n = 0$; and thus

$$\frac{d^n y}{dx^n} = X;$$

(2) the general linear equation of the first order in Art. 382, viz.

$$P_1 \frac{dy}{dx} + P_2 y + P_3 = 0.$$

441.] THEOREM I.*—The integral of (26) depends on the integral of the left-hand member of the equation; that is, on the integral of the equation when $X = 0$.

* The first of the following theorems is due to Lagrange: the others are the original investigations of M. G. Libri, and are taken from Crelle's Journal, Vol. X, page 185.

Let $y = u_1 \int v_1 dx$, where u_1 and v_1 are two undetermined functions of x: then by Leibnitz's Theorem

$$\frac{d^m}{dx^m} \cdot u_1 \int v_1 dx = \int v_1 dx \frac{d^m u_1}{dx^m} + m v_1 \frac{d^{m-1} u_1}{dx^{m-1}} + \frac{m(m-1)}{1.2} \frac{dv_1}{dx} \frac{d^{m-2} u_1}{dx^{m-2}} + \dots$$

$$\dots + m \frac{d^{m-2} v_1}{dx^{m-2}} \frac{du_1}{dx} + \frac{d^{m-1} v_1}{dx^{m-1}} u_1; \quad (27)$$

and substituting the specific values of this in the several terms of (26) we have

$$\left\{ \frac{d^n u_1}{dx^n} + P_1 \frac{d^{n-1} u_1}{dx^{n-1}} + \dots + P_{n-1} \frac{du_1}{dx} + P_n u_1 \right\} \int v_1 dx$$

$$+ u_1 \left\{ \frac{d^{n-1} v_1}{dx^{n-1}} + Q_1 \frac{d^{n-2} v_1}{dx^{n-2}} + \dots + Q_{n-2} \frac{dv_1}{dx} + Q_{n-1} v_1 \right\} = X; \quad (28)$$

where $Q_1, Q_2, \dots Q_{n-1}$ are determinate functions of x and u_1. Suppose now u_1 to be a function of x which makes the left-hand member of (26) to vanish; that is, suppose u_1 to be a particular integral of (26), when $X = 0$; then the coefficient of $\int v_1 dx$ in (28) vanishes, and we have

$$\frac{d^{n-1} v_1}{dx^{n-1}} + Q_1 \frac{d^{n-2} v_1}{dx^{n-2}} + \dots + Q_{n-2} \frac{dv_1}{dx} + Q_{n-1} v_1 = \frac{X}{u_1}; \quad (29)$$

which is an equation of the same form as (26), and of the $(n-1)$th order; in this equation let

$$v_1 = u_2 \int v_2 dx,$$

and let substitutions be made in (29) according to the preceding process: then if u_2 is an integral of (29), when the right-hand member is equal to zero, the resulting equation will be of the $(n-2)$th order, and of the form

$$\frac{d^{n-2} v_2}{dx^{n-2}} + R_1 \frac{d^{n-3} v_2}{dx^{n-3}} + \dots + R_{n-3} \frac{dv_2}{dx} + R_{n-2} v_2 = \frac{X}{u_1 u_2};$$

then continuing the same process we shall finally have an equation of the first order which may be integrated by the methods of the preceding Chapter; and the function which satisfies the given equation will be determined by the successive integration of a multiple integral of the nth order. The problem will hereby become reduced to that of a multiple integral, and of simple quadrature.

442.] And to indicate more clearly the form which by this

process the last integral assumes, let us consider the case of a differential equation of the third order,

$$\frac{d^3 y}{dx^3} + \text{P}_1 \frac{d^2 y}{dx^2} + \text{P}_2 \frac{dy}{dx} + \text{P}_3 y = \text{X}. \qquad (30)$$

Let $\qquad y = u_1 \int v_1 \, dx;$

$$\therefore \quad \frac{d^3 y}{dx^3} = \frac{d^3 u_1}{dx^3} \int v_1 \, dx + 3 \frac{d^2 u_1}{dx^2} v_1 + 3 \frac{du_1}{dx} \frac{dv_1}{dx} + u_1 \frac{d^2 v_1}{dx^2},$$

$$\frac{d^2 y}{dx^2} = \frac{d^2 u_1}{dx^2} \int v_1 \, dx + 2 \frac{du_1}{dx} v_1 + u_1 \frac{dv_1}{dx},$$

$$\frac{dy}{dx} = \frac{du_1}{dx} \int v_1 \, dx + u_1 v_1;$$

then substituting in (30), we have

$$\left\{ \frac{d^3 u_1}{dx^3} + \text{P}_1 \frac{d^2 u_1}{dx^2} + \text{P}_2 \frac{du_1}{dx} + \text{P}_3 u_1 \right\} \int v_1 \, dx$$
$$+ \left\{ 3 \frac{d^2 u_1}{dx^2} + 2 \text{P}_1 \frac{du_1}{dx} + \text{P}_2 u_1 \right\} v_1$$
$$+ \left\{ 3 \frac{du_1}{dx} + \text{P}_1 u_1 \right\} \frac{dv_1}{dx} + u_1 \frac{d^2 v_1}{dx^2} = \text{X}. \qquad (31)$$

Now if u_1 is, according to our supposition, an integral of (30) when $\text{X} = 0$, the first term of (31) vanishes: also let

$$3 \frac{d^2 u_1}{dx^2} + 2 \text{P}_1 \frac{du_1}{dx} + \text{P}_2 u_1 = u_1 \text{Q}_2,$$

$$3 \frac{du_1}{dx} + \text{P}_1 u_1 = u_1 \text{Q}_1;$$

then (31) becomes

$$\frac{d^2 v_1}{dx^2} + \text{Q}_1 \frac{dv_1}{dx} + \text{Q}_2 v_1 = \frac{\text{X}}{u_1}. \qquad (32)$$

Again, let u_2 be an integral of this equation without its second member: and let

$$v_1 = u_2 \int v_2 \, dx;$$

then if $2 \frac{du_2}{dx} + \text{Q}_1 u_2 = \text{R}_1 u_2$, (32) becomes

$$\frac{dv_2}{dx} + \text{R}_1 = \frac{\text{X}}{u_1 u_2}. \qquad (33)$$

Again, let u_3 be a particular integral of (33) without its second member; and let

$$v_2 = u_3 \int v_3 \, dx;$$

then $$u_3 v_3 = \frac{X}{u_1 u_2};$$

$$\therefore \quad v_3 = \frac{X}{u_1 u_2 u_3}; \tag{34}$$

and retracing our steps we have

$$y = u_1 \int u_2 dx \int u_3 dx \int \frac{X dx}{u_1 u_2 u_3}; \tag{35}$$

where u_1, u_2, u_3 are integrals of the several equations found as above and without their second members; and thus the general integral is found in terms of a triple integral whose element-function contains one variable; and therefore by the process of integration three arbitrary constants will be introduced, and the integral will be in its most general form.

And thus to generalize the process, the integral of (26) will be

$$y = u_1 \int u_2 dx \int u_3 dx \dots \int \frac{X dx}{u_1 u_2 \dots u_n}. \tag{36}$$

443.] Now these quantities $u_1, u_2, \dots u_n$ may be expressed in terms of particular integrals of (26), when $X = 0$; so that if these latter quantities can be determined, the complete solution of the given differential equation will depend on only a single integral. To limit the extent of the investigation, let us confine our attention to an equation of the third order, viz.

$$\frac{d^3 y}{dx^3} + P_1 \frac{d^2 y}{dx^2} + P_2 \frac{dy}{dx} + P_3 y = X; \tag{37}$$

and let η_1, η_2, η_3 be three particular integrals of this equation, when $X = 0$, so that

$$\frac{d^3 \eta_1}{dx^3} + P_1 \frac{d^2 \eta_1}{dx^2} + P_2 \frac{d\eta_1}{dx} + P_3 \eta_1 = 0; \tag{38}$$

then employing u_1, u_2, u_3, v_1, v_2, v_3 in the same signification as in the preceding Article, let $u_1 = \eta_1$; and if for η_1 in (38) $u_1 \int u_2 dx$ is substituted, it will on expansion be seen that

$$\frac{d^3}{dx^3} . u_1 \int u_2 dx + P_1 \frac{d^2}{dx^2} . u_1 \int u_2 dx + P_2 \frac{d}{dx} . u_1 \int u_2 dx + P_3 u_1 \int u_2 dx = 0; \tag{39}$$

so that $u_1 \int u_2 dx$ is a particular integral of (37) when $X = 0$; suppose this integral to be η_2; then

$$\eta_2 = u_1 \int u_2 dx = \eta_1 \int u_2 dx;$$

$$\therefore \quad u_2 = \frac{d}{dx} . \frac{\eta_2}{\eta_1}. \tag{40}$$

But u_2 is a particular integral of (32) without its second member; so that

$$\frac{d^2 u_2}{dx^2} + Q_1 \frac{du_2}{dx} + Q_2 u_2 = 0. \tag{41}$$

Again, let η_3 be another particular integral of (38); and let u'_2 be another particular integral of (32) without its second member; then, pursuing the same process as before,

$$u'_2 = \frac{d}{dx} \cdot \frac{\eta_3}{\eta_1}; \tag{42}$$

so that u_2 and u'_2 are two particular integrals of (41): and employing u_3 as above, it will be seen that

$$\frac{d^2}{dx^2} \cdot u_2 \int u_3 dx + Q_1 \frac{d}{dx} \cdot u_2 \int u_3 dx + Q_2 u_2 \int u_3 dx = 0; \tag{43}$$

so that $u_2 \int u_3 dx$ is a particular integral of (41); let this be equal to u'_2; so that

$$u'_2 = u_2 \int u_3 dx,$$

$$\therefore \quad u_3 = \frac{d}{dx} \cdot \frac{u'_2}{u_2} = \frac{d}{dx} \cdot \frac{\frac{d}{dx} \cdot \frac{\eta_3}{\eta_1}}{\frac{d}{dx} \cdot \frac{\eta_2}{\eta_1}}; \tag{44}$$

so that now u_1, u_2, u_3 are expressed in terms of η_1, η_2, η_3, that is, in terms of three particular integrals of the given equation, when its right-hand member vanishes; and these may be substituted in (35); and the final value of y thus obtained will be

$$y = \eta_1 \int \frac{d}{dx} \cdot \frac{\eta_2}{\eta_1} dx \int \frac{d}{dx} \cdot \frac{\frac{d}{dx} \cdot \frac{\eta_3}{\eta_1}}{\frac{d}{dx} \cdot \frac{\eta_2}{\eta_1}} dx \int X \left\{ \eta_1 \frac{d}{dx} \cdot \frac{\eta_2}{\eta_1} \frac{d}{dx} \cdot \frac{\frac{d}{dx} \cdot \frac{\eta_3}{\eta_1}}{\frac{d}{dx} \cdot \frac{\eta_2}{\eta_1}} \right\}^{-1} dx. \tag{45}$$

The same process may manifestly be extended to equations of the order n; the final result however is of a form too complicated to be inserted: it will however involve n signs of integration, and therefore n arbitrary constants.

444.] Some examples of the above process are subjoined. Let us first consider the linear equation of the first order, viz.

$$\frac{dy}{dx} + Py = X. \tag{46}$$

Now of this equation, when $X = 0$, an integral may be found as follows:

$$\frac{dy}{y} + \text{P}dx = 0,$$

$$y = ce^{-\int \text{P}dx}, \tag{47}$$

which is η_1; and therefore substituting this value in the generalized form of (45), we have

$$y = e^{-\int \text{P}dx}\int \text{X}e^{\int \text{P}dx}\,dx; \tag{48}$$

and which is the general integral as before expressed in equation (69), Art. 382.

For a second example, let us consider

$$\frac{d^3y}{dx^3} - 6a\frac{d^2y}{dx^2} + 11a^2\frac{dy}{dx} - 6a^3y = e^{mx}; \tag{49}$$

particular integrals of the left-hand member of which are, when the right-hand member vanishes,

$$\eta_1 = e^{ax}, \quad \eta_2 = e^{2ax}, \quad \eta_3 = e^{3ax};$$

$$\therefore \quad \frac{d}{dx}\cdot\frac{\eta_2}{\eta_1} = \frac{d}{dx}e^{ax} = ae^{ax}; \qquad \frac{d}{dx}\cdot\frac{\eta_3}{\eta_1} = 2ae^{2ax};$$

substituting which in (45) we have

$$y = e^{ax}\int ae^{ax}dx\int 2ae^{ax}dx\int e^{mx}\{2a^2e^{ax}e^{ax}e^{ax}\}^{-1}dx$$

$$= e^{ax}\int e^{ax}dx\int e^{ax}dx\int e^{(m-3a)x}dx$$

$$= e^{ax}\int e^{ax}dx\int e^{ax}\left\{\frac{e^{(m-3a)x}}{m-3a}+c_1\right\}dx$$

$$= \frac{e^{mx}}{(m-a)(m-2a)(m-3a)} + \frac{c_1}{2a^2}e^{3ax} + \frac{c_2}{a}e^{2ax} + c_3e^{ax}; \tag{50}$$

and this is the general integral of (49); (49) in fact having been deduced from it by the elimination of c_1, c_2, and c_3.

Another example is

$$\frac{d^2y}{dx^2} - y = x;$$

the particular integrals of which without the second member are $\eta_1 = e^x$, $\eta_2 = e^{-x}$; and the general integral is

$$y = c_1e^x + c_2e^{-x} - x.$$

445.] The process which has been explained and illustrated above also gives the following Theorems.

THEOREM II.—If m particular integrals of a linear differential equation of the nth order without the second member are known, the integration of the equation with the second member will depend on the integration of a new linear equation of the $(n-m)$th order.

Let $\eta_1, \eta_2, \ldots \eta_m$ be m particular integrals of (26), when the right-hand member vanishes; and let us, in Art. 443, assume

$$y = \eta_1 \int v_1 dx. \tag{51}$$

Then substituting as in that Article, the coefficient of $\int v_1 dx$ as exhibited in (28) vanishes; and we have

$$\frac{d^{n-1}v_1}{dx^{n-1}} + Q_1 \frac{d^{n-2}v_1}{dx^{n-2}} + Q_2 \frac{d^{n-3}v_1}{dx^{n-3}} + \ldots + Q_{n-1}v_1 = \frac{X}{\eta_1}. \tag{52}$$

Now of this equation, without its second member, according to the process pursued in Art. 443, $(m-1)$ particular integrals are

$$\frac{d}{dx}\cdot\frac{\eta_2}{\eta_1}, \quad \frac{d}{dx}\cdot\frac{\eta_3}{\eta_1}, \ldots \frac{d}{dx}\cdot\frac{\eta_m}{\eta_1}; \tag{53}$$

let these severally be symbolized by $\zeta_1, \zeta_2, \ldots \zeta_{m-1}$; then in (52) let

$$v_1 = \zeta_1 \int v_2 dx; \tag{54}$$

and substituting according to Art. 443, the term involving $\int v_2 dx$ will vanish, and we shall have

$$\frac{d^{n-2}v_2}{dx^{n-2}} + R_1 \frac{d^{n-3}v_2}{dx^{n-3}} + \ldots + R_{n-2}v_2 = \frac{X}{\eta_1 \zeta_1}; \tag{55}$$

and of this equation again without its second member $(m-2)$ particular integrals are

$$\frac{d}{dx}\cdot\frac{\zeta_2}{\zeta_1}, \quad \frac{d}{dx}\cdot\frac{\zeta_3}{\zeta_1}, \ldots \frac{d}{dx}\cdot\frac{\zeta_{m-1}}{\zeta_1}; \tag{56}$$

which we may conveniently symbolize by $\theta_1, \theta_2, \ldots \theta_{m-2}$; and by a similar process we may make the integral of (55) without its second member dependent on the integration of an equation of the $(n-3)$th order: and in a continuance of the process it is manifest that each of the given particular integrals of (26) enables us to reduce by unity the order of the differential equation; and finally therefore the order of the equation will be the $(n-m)$th.

Hence, if a particular integral of a linear differential equation of the nth order can be found, the order of the equation may be depressed by unity.

446.] THEOREM III. If $\eta_1, \eta_2, \ldots \eta_n$ are n particular integrals of a linear differential equation of the nth order without the second member, and if y_1 is a particular integral of it with the second member, then the general integrals of the equation with and without the second member are respectively

$$\left.\begin{aligned} y &= c_1\eta_1+c_2\eta_2+\ldots+c_n\eta_n+y_1, \\ y &= c_1\eta_1+c_2\eta_2+\ldots+c_n\eta_n. \end{aligned}\right\} \tag{57}$$

The truth of the proposition is evident from the form of the equations; because each satisfies its corresponding differential equation, and each contains n constants: these however must be independent of each other; and the particular integrals must also be independent of each other: for suppose that $\eta_3 = a\eta_1+b\eta_2$, then

$$\begin{aligned} y &= c_1\eta_1+c_2\eta_2+c_3(a\eta_1+b\eta_2)+c_4\eta_4+\ldots \\ &= (c_1+ac_3)\eta_1+(c_2+bc_3)\eta_2+c_4\eta_4+\ldots \end{aligned}$$

and which contains only $n-1$ arbitrary constants, and consequently is not the complete and general integral.

447.] M. Libri has in the Memoir above referred to traced many analogies between the formation and properties of algebraical and differential equations: some of these are given in the following Theorems.

THEOREM IV.—A differential equation, linear in at least the first two terms, may be transformed into another linear equation of the same order, and without the second term.

Let (26) be the typical equation of a linear equation of the nth order: and let

$$y = uv, \tag{58}$$

where u and v are two undetermined functions of x: then, expressing $\frac{d^ny}{dx^n}$, $\frac{d^{n-1}y}{dx^{n-1}}$, ... by means of Leibnitz's theorem, (26) after substitution will become

$$\begin{aligned} v\frac{d^nu}{dx^n}+n\frac{dv}{dx}\frac{d^{n-1}u}{dx^{n-1}}+\frac{n(n-1)}{1.2}\frac{d^2v}{dx^2}\frac{d^{n-2}u}{dx^{n-2}}+\ldots \\ +\mathrm{P}_1\left\{v\frac{d^{n-1}u}{dx^{n-1}}+(n-1)\frac{dv}{dx}\frac{d^{n-2}u}{dx^{n-2}}+\ldots\right\} \\ +\mathrm{P}_2\left\{v\frac{d^{n-2}u}{dx^{n-2}}+\ldots\right\} \\ \ldots\ldots\ldots\ldots \\ +\mathrm{P}_nuv = \mathrm{X}. \end{aligned} \tag{59}$$

Consequently if v is such that

$$n\frac{dv}{dx}+\mathrm{P}_1v = 0, \tag{60}$$

the second term of (59) vanishes ; and from (60) we have

$$v = e^{-\int \frac{P_1}{n} dx}; \tag{61}$$

whence, theoretically at least, v may be found; and (59) will be a linear equation without the second term.

And more generally: A differential equation of which the first $m+1$ terms are linear may be transformed into another linear equation of the same order, and without the $(m+1)$th term, by means of the solution of a linear equation of the mth order. Thus, let it be required to deprive of its second term the equation

$$\frac{d^2y}{dx^2} - 3a\frac{dy}{dx} + 2a^2y = \text{X}; \tag{62}$$

substituting $y = uv$, we have

$$v\frac{d^2u}{dx^2} + \left(2\frac{dv}{dx} - 3av\right)\frac{du}{dx} + \left(\frac{d^2v}{dx^2} - 3av\frac{dv}{dx} + 2a^2v\right)u = \text{X}; \tag{63}$$

let $\quad 2\frac{dv}{dx} - 3av = 0; \quad$ so that $\quad v = e^{\frac{3ax}{2}};$

then (63) becomes $\frac{d^2u}{dx^2} - \frac{a^2}{4}u = \text{X}e^{-\frac{3ax}{2}}.$

448.] THEOREM V.—If a relation is given between two particular integrals of a linear differential equation of the nth order, the order of the equation may be diminished by unity.

Let η_1 and η_2 be two particular integrals of (26); and let us suppose them to be related by the equation $\eta_2 = \phi(\eta_1)$; then, if in (26) we substitute for y, first η_1, and then η_2, or, which is equivalent, $\phi(\eta_1)$, there will be two equations from which $\frac{d^n\eta_1}{dx^n}$ may be eliminated, and the order of the resulting equation will be only the $(n-1)$th. Thus suppose η_1 and η_2 to be two particular integrals of

$$\frac{d^2y}{dx^2} - a^2y = 0;$$

and suppose them to be related by the condition $\eta_1\eta_2 = 1$; then we have

$$\frac{d^2\eta_1}{dx^2} - a^2\eta_1 = 0; \qquad \frac{d^2}{dx^2}\cdot\frac{1}{\eta_1} - \frac{a^2}{\eta_1} = 0;$$

$$\therefore \quad \frac{d^2\eta_1}{dx^2} - \frac{2}{\eta_1}\left(\frac{d\eta_1}{dx}\right)^2 + a^2\eta_1 = 0;$$

$$\therefore \quad \left(\frac{d\eta_1}{dx}\right)^2 = a^2\eta_1^2, \quad \text{and} \quad \eta_1 = e^{\pm ax};$$

$$\therefore \quad \eta_1 = e^{ax}, \qquad \eta_2 = e^{-ax}.$$

449.] THEOREM VI.—If n particular integrals of a differential equation, which is without the second member, are known, the coefficients of the several terms are functions of these integrals, and may be found by a process analogous to that of forming an algebraical equation whose roots are given.

Let the differential equation be of the nth order, and of the form

$$\frac{d^n y}{dx^n}+P_1\frac{d^{n-1}y}{dx^{n-1}}+P_2\frac{d^{n-2}y}{dx^{n-2}}+\ldots+P_{n-1}\frac{dy}{dx}+P_n y=0; \qquad (64).$$

and let the n particular integrals be $\eta_1, \eta_2, \ldots \eta_n$.

Substitute in (64) for y, $y=\eta_1\int v_1 dx$; then we have

$$\eta_1\frac{d^{n-1}v_1}{dx^{n-1}}+n\frac{d\eta_1}{dx}\frac{d^{n-2}v_1}{dx^{n-2}}+\frac{n(n-1)}{1.2}\frac{d^2\eta_1}{dx^2}\frac{d^{n-3}v_1}{dx^{n-3}}+\ldots+\frac{d^n\eta_1}{dx^n}\int v_1 dx$$
$$+P_1\left\{\eta_1\frac{d^{n-2}v_1}{dx^{n-2}}+\frac{n-1}{1}\frac{d\eta_1}{dx}\frac{d^{n-3}v_1}{dx^{n-3}}+\ldots+\frac{d^{n-1}\eta_1}{dx^{n-1}}\int v_1 dx\right\}$$
$$+\ldots\ldots\ldots\ldots\ldots\ldots$$
$$+P_{n-1}\left\{\eta_1 v_1+\frac{d\eta_1}{dx}\int v_1 dx\right\}$$
$$+P_n\eta_1\int v_1 dx=0. \quad (65)$$

Now the coefficient of $\int v_1 dx = 0$; consequently, dividing through by η_1, we have

$$\frac{d^{n-1}v_1}{dx^{n-1}}+\left\{\frac{n}{\eta_1}\frac{d\eta_1}{dx}+P_1\right\}\frac{d^{n-2}v_1}{dx^{n-2}}+\ldots=0. \qquad (66)$$

Let $\frac{n}{\eta_1}\frac{d\eta_1}{dx}+P_1=Q_1$, and let the coefficients of the succeeding terms be $Q_2, Q_3, \ldots Q_{n-1}$: then (66) becomes

$$\frac{d^{n-1}v_1}{dx^{n-1}}+Q_1\frac{d^{n-2}v_1}{dx^{n-2}}+Q_2\frac{d^{n-3}v_1}{dx^{n-3}}+\ldots+Q_{n-1}v_1=0. \qquad (67)$$

Now of this equation the $(n-1)$ particular integrals are

$$\frac{d}{dx}\cdot\frac{\eta_2}{\eta_1},\quad \frac{d}{dx}\cdot\frac{\eta_3}{\eta_1},\ \ldots\ \frac{d}{dx}\cdot\frac{\eta_n}{\eta_1}; \qquad (68)$$

let us therefore repeat in (67) the same process as that to which (64) has been subjected; then if the successive coefficients of the transformed equation, which will be of the $(n-2)$th order, are $R_1, R_2, \ldots R_{n-2}$, we shall have

$$\text{R}_1 = \frac{n-1}{\frac{d}{dx}\cdot\frac{\eta_2}{\eta_1}}\frac{d}{dx}\frac{d}{dx}\cdot\frac{\eta_2}{\eta_1} + \text{Q}_1$$

$$= \;\text{-}\;\text{-}\;\text{-}\;\text{-}\;\text{-}\;\text{-}\; + \frac{n}{\eta_1}\frac{d\eta_1}{dx} + \text{P}_1,$$

$$\therefore \quad \text{P}_1 = -\frac{n}{\eta_1}\frac{d\eta_1}{dx} - \frac{n-1}{\frac{d}{dx}\cdot\frac{\eta_2}{\eta_1}}\frac{d^2}{dx^2}\cdot\frac{\eta_2}{\eta_1} + \text{R}_1; \tag{69}$$

and by continuing a similar process in the equation which involves $\text{R}_1, \text{R}_2, \ldots,$ we shall find an equation whose order is the $(n-3)$th, and shall be able to express P_1 in terms of others of the original particular integrals: and so on, until finally we arrive at a value of P_1 expressed wholly in terms of the η's.

By a process exactly similar, the other coefficients of (64) may be found in terms of the particular integrals. And thus in general, if $\text{F}_1(x), \text{F}_2(x), \ldots \text{F}_n(x)$ are n functions of x, and it is required to determine a linear differential equation of which these are n particular integrals, we can determine the coefficients of it in terms of the particular integrals. This case is plainly analogous to that of the formation of an algebraical equation of which the roots are given.

In illustration of this process let it be required to form the differential equation of the third order, of which three particular integrals are

$$\eta_1 = e^{a_1 x}, \quad \eta_2 = e^{a_2 x}, \quad \eta_3 = e^{a_3 x}.$$

Let the equation be

$$\frac{d^3y}{dx^3} + \text{P}_1\frac{d^2y}{dx^2} + \text{P}_2\frac{dy}{dx} + \text{P}_3 y = 0. \tag{70}$$

Let $\qquad y = e^{a_1 x}\int v_1\,dx;$

then, as $e^{a_1 x}$ is a particular integral of (70), the coefficient of $\int v_1\,dx$ vanishes, and the transformed equation is, after division by $e^{a_1 x}$,

$$\frac{d^2v_1}{dx^2} + (3a_1 + \text{P}_1)\frac{dv_1}{dx} + (3a_1{}^2 + 2\text{P}_1 a_1 + \text{P}_2)v_1 = 0, \tag{71}$$

of which two particular integrals are, by reason of (53),

$$(a_2 - a_1)e^{(a_2-a_1)x}, \qquad (a_3 - a_1)e^{(a_3-a_1)x}; \tag{72}$$

let therefore $\qquad v_1 = (a_2 - a_1)e^{(a_2-a_1)x}\int v_2\,dx; \tag{73}$

then substituting in (71), and observing that the coefficient of $\int v_2 dx$ vanishes, we have

$$\frac{dv_2}{dx} + (P_1 + a_1 + 2a_2)v_2 = 0; \qquad (74)$$

of which, by reason of equation (44), $e^{(a_3-a_2)x}$ is a particular integral; therefore substituting we have

$$P_1 = -(a_1 + a_2 + a_3), \qquad (75)$$

substituting which in (71) we have

$$\frac{d^2 v_1}{dx^2} + (2a_1 - a_2 - a_3)\frac{dv_1}{dx} + (a_1^2 - 2a_1 a_2 - 2a_1 a_3 + P_2)v_1 = 0; \quad (76)$$

and of this $e^{(a_3-a_1)x}$ is a particular integral: therefore substituting,

$$P_2 = a_2 a_3 + a_3 a_1 + a_1 a_2, \qquad (77)$$

and substituting in (70) for P_1 and P_2, and noticing that $e^{a_1 x}$ is a particular integral of (70), we have after substitution

$$P_3 = -a_1 a_2 a_3;$$

therefore equation (70) finally becomes

$$\frac{d^3 y}{dx^3} - (a_1 + a_2 + a_3)\frac{d^2 y}{dx^2} + (a_2 a_3 + a_3 a_1 + a_1 a_2)\frac{dy}{dx} - a_1 a_2 a_3 y = 0.$$

And this equation might also have been found as follows: Since $e^{a_1 x}$, $e^{a_2 x}$, $e^{a_3 x}$ are particular integrals, we might substitute these in it, and thereby obtain these equations,

$$\left.\begin{array}{l} a_1^3 + P_1 a_1^2 + P_2 a_1 + P_3 = 0, \\ a_2^3 + P_1 a_2^2 + P_2 a_2 + P_3 = 0, \\ a_3^3 + P_1 a_3^2 + P_2 a_3 + P_3 = 0; \end{array}\right\} \qquad (78)$$

of which three cubic equations a_1, a_2, a_3 are evidently the roots: therefore

$$P_1 = -(a_1 + a_2 + a_3),$$
$$P_2 = a_2 a_3 + a_3 a_1 + a_1 a_2,$$
$$P_3 = -a_1 a_2 a_3.$$

Similarly it may be shewn that the equation, of which particular integrals are x^{-1} and x^2, is

$$x^2 \frac{d^2 y}{dx^2} - 2y = 0.$$

SECTION 3.—*Integration of Linear Differential Equations of the nth Order, whose Coefficients are Constants, with or without Second Members.*

450.] The investigations of the preceding section shew that the integration of an equation of the linear form with the second member depends on that of the same equation without the second member, and on a multiple integral the element-function of which involves the second member: in the present and the future sections therefore we shall, if it is convenient, consider properties of linear differential equations, without the second members, and the reader will observe that the generality of the investigation is not diminished thereby. There are many processes of solution, which will be considered consecutively. The general type I shall take to be

$$\frac{d^n y}{dx^n}+\text{A}_1\frac{d^{n-1}y}{dx^{n-1}}+\text{A}_2\frac{d^{n-2}y}{dx^{n-2}}+\ldots+\text{A}_{n-1}\frac{dy}{dx}+\text{A}_n y=\text{X}, \quad (79)$$

where $\text{A}_1, \text{A}_2, \ldots \text{A}_n$ are constants, and X is a function of x.

First Method.—Let (79) be expressed by means of Lagrange's notation of derived functions; then we have

$$y^{(n)}+\text{A}_1 y^{(n-1)}+\text{A}_2 y^{(n-2)}+\ldots+\text{A}_{n-1}y'+\text{A}_n y=\text{X}; \quad (80)$$

and introducing certain undetermined constants $\theta', \theta'', \theta''', \ldots \theta^{(n-1)}$, we may put (80) in the form

$$\frac{d}{dx}\{y^{(n-1)}+\theta' y^{(n-2)}+\theta'' y^{(n-3)}+\ldots+\theta^{(n-1)}y\}$$

$$+(\text{A}_1-\theta')y^{(n-1)}+(\text{A}_2-\theta'')y^{(n-2)}+\ldots+(\text{A}_{n-1}-\theta^{(n-1)})y'+\text{A}_n y=\text{X};$$

and let us make the following substitutions;

$$y^{(n-1)}+\theta' y^{(n-2)}+\ldots+\theta^{(n-1)}y=\eta; \quad (82)$$

$$\left.\begin{array}{l}\text{A}_1-\theta'=\theta;\quad \text{A}_2-\theta''=\theta\theta',\quad \text{A}_3-\theta'''=\theta\theta''\ldots\ldots,\\ \text{A}_{n-1}-\theta^{(n-1)}=\theta\theta^{(n-2)},\quad \text{A}_n=\theta\theta^{(n-1)};\end{array}\right\} \quad (83)$$

so that (81) becomes

$$\frac{d\eta}{dx}+\theta\eta=\text{X}; \quad (84)$$

$$\therefore \quad \eta=e^{-\theta x}\left\{\int e^{\theta x}\text{X}\,dx+\text{C}\right\}; \quad (85)$$

and for θ let $-a$ be substituted: then from (83) we have

$$a^n+\text{A}_1 a^{n-1}+\text{A}_2 a^{n-2}+\ldots+\text{A}_{n-1}a+\text{A}_n=f(a)=0; \quad (86)$$

the resemblance of which to (79) in its powers and its coefficients is evident; and as we shall hereafter refer to this equation, it is convenient for it to bear a particular name: let it therefore according to a received nomenclature be called the *characteristic equation* of (79).

Now suppose the n roots of this equation to be unequal and to be $a_1, a_2, \ldots a_n$; then there are n different values of (85), viz.

$$e^{a_1 x}\left\{\int e^{-a_1 x}\text{X}\,dx + \text{C}_1\right\}; \qquad e^{a_2 x}\left\{\int e^{-a_2 x}\text{X}\,dx + \text{C}_2\right\}; \ldots \tag{87}$$

which may be denoted by $\eta_1, \eta_2, \ldots \eta_n$; also let the values of $\theta', \theta'', \ldots \theta^{(n-1)}$ corresponding to these roots be $\theta_1', \theta_1'', \ldots \theta_1^{(n-1)}$, $\theta_2', \theta_2'', \ldots \theta_2^{(n-1)}, \ldots \theta_n', \theta_n'', \ldots \theta_n^{(n-1)}$; then from (82) we have the following series:

$$\left.\begin{array}{l} y^{(n-1)} + \theta_1' y^{(n-2)} + \theta_1'' y^{(n-3)} + \ldots + \theta_1^{(n-1)} y = \eta_1; \\ y^{(n-1)} + \theta_2' y^{(n-2)} + \theta_2'' y^{(n-3)} + \ldots + \theta_2^{(n-1)} y = \eta_2; \\ \cdot \quad \cdot \quad \cdot \quad \cdot \quad \cdot \quad \cdot \quad \cdot \quad \cdot \quad \cdot \quad \cdot \\ y^{(n-1)} + \theta_n' y^{(n-2)} + \theta_n'' y^{(n-3)} + \ldots + \theta_n^{(n-1)} y = \eta_n; \end{array}\right\} \tag{88}$$

so that, employing the symbols of determinants,

$$y = \frac{\Sigma . \pm \eta_1 \theta_2^{(n-2)} \theta_3^{(n-3)} \ldots \theta'_{(n-1)} 1}{\Sigma . \pm \theta_1^{(n-1)} \theta_2^{(n-2)} \theta_3^{(n-3)} \ldots \theta'_{(n-1)} 1}; \tag{89}$$

the values of $y', y'', \ldots y^{(n-2)}, y^{(n-1)}$ are evidently similar in form.

Now the value of y given in (89) is of the form

$$y = \lambda_1 \eta_1 + \lambda_2 \eta_2 + \ldots + \lambda_n \eta_n, \tag{90}$$

where $\lambda_1, \lambda_2, \ldots \lambda_n$ are constants and functions of the θ's; and these are assigned by (89); but it is easier to discover them by the following method.

451.] Let us employ a concise notation and represent (90) thus;

$$y = \Sigma . \lambda_m \eta_m, \tag{91}$$

where Σ indicates the sum of a series of terms found by giving successive values to m from 1 to n inclusively; then

$$y = \Sigma . \lambda_m e^{a_m x}\left\{\int e^{-a_m x}\text{X}\,dx + \text{C}_m\right\}; \tag{92}$$

$$y' = \Sigma . \lambda_m a_m e^{a_m x}\left\{\int e^{-a_m x}\text{X}\,dx + \text{C}_m\right\} + \Sigma . \lambda_m \text{X}, \tag{93}$$

$$= \Sigma . a_m \lambda_m \eta_m + \Sigma . \lambda_m \text{X}; \tag{94}$$

and observing the remark made in the sentence following equation (89), that y' must be of the same form as y, and as this can be the case only when $\Sigma . \lambda_m \text{X} = 0$, and therefore when $\Sigma . \lambda_m = 0$, we have

$$y' = \Sigma . a_m \lambda_m \eta_m;$$

and therefore after differentiation

$$y'' = \Sigma . a_m^2 \lambda_m \eta_m + \Sigma . a_m \lambda_m \text{X}; \tag{95}$$

and as y'' must also be of the same form as y, $\Sigma . a_m \lambda_m \text{X} = 0$;

$$y'' = \Sigma . a_m^2 \lambda_m \eta_m;$$

and so on, until ultimately

$$y^{(n-1)} = \Sigma . a_m^{n-1} \lambda_m \eta_m + \Sigma . a_m^{n-2} \lambda_m \mathrm{X}; \tag{96}$$

whence we have $\Sigma . a^{n-2} \lambda_m \mathrm{X} = 0;$

and

$$y^{(n)} = \Sigma . a_m^{n} \lambda_m \eta_m + \Sigma . a_m^{n-1} \lambda_m \mathrm{X}; \tag{97}$$

and as these conditions must consist with equation (79), we have after substitution

$$\Sigma . \lambda_m \eta_m \{a_m^n + \mathrm{A}_1 a_m^{n-1} + \mathrm{A}_2 a_m^{n-2} + \dots + \mathrm{A}_{n-1} a_m + \mathrm{A}_n\} + \Sigma . a_m^{n-1} \lambda_m \mathrm{X} = \mathrm{X}; \tag{98}$$

but each term of the series comprehended within the symbol of aggregation vanishes, because $a_1, a_2, \dots a_n$ are the n roots of the characteristic equation; and therefore we have

$$\Sigma . \lambda_m a_m^{n-1} \mathrm{X} = \mathrm{X},$$

$$\therefore \quad \Sigma . \lambda_m a_m^{n-1} = 1. \tag{99}$$

Hence we have the following equations for the determination of $\lambda_1, \lambda_2, \dots \lambda_n$;

$$\left.\begin{aligned} \lambda_1 + \lambda_2 + \lambda_3 + \dots + \lambda_n &= 0, \\ a_1 \lambda_1 + a_2 \lambda_2 + a_3 \lambda_3 + \dots + a_n \lambda_n &= 0, \\ a_1^2 \lambda_1 + a_2^2 \lambda_2 + a_3^2 \lambda_3 + \dots + a_n^2 \lambda_n &= 0, \\ \dots\dots\dots\dots & \\ a_1^{n-1} \lambda_1 + a_2^{n-1} \lambda_2 + a_3^{n-1} \lambda_3 + \dots + a_n^{n-1} \lambda_n &= 1. \end{aligned}\right\} \tag{100}$$

Now consider the derived function of the characteristic equation (86);

$$f'(a) = (a-a_2)(a-a_3)\dots(a-a_n) + (a-a_1)(a-a_3)\dots(a-a_n) + \dots + (a-a_1)(a-a_2)\dots(a-a_{n-1}); \tag{101}$$

$$\left.\begin{aligned} \therefore \quad f'(a_1) &= (a_1-a_2)(a_1-a_3)\dots(a_1-a_n); \\ f'(a_2) &= (a_2-a_1)(a_2-a_3)\dots(a_2-a_n); \\ \dots\dots\dots & \\ f'(a_n) &= (a_n-a_1)(a_n-a_2)\dots(a_n-a_{n-1}). \end{aligned}\right\} \tag{102}$$

Of these equations let us take the first to be the type: it is plainly of $n-1$ dimensions in a_1, so that

$$f'(a_1) = a_1^{n-1} + \mathrm{C}_1 a_1^{n-2} + \mathrm{C}_2 a_1^{n-3} + \dots + \mathrm{C}_{n-2} a_1 + \mathrm{C}_{n-1}, \tag{103}$$

where $\mathrm{C}_1, \mathrm{C}_2, \dots \mathrm{C}_{n-1}$ are functions of $a_2, a_3, \dots a_n$; and let us multiply equations (100) severally by $\mathrm{C}_{n-1}, \mathrm{C}_{n-2}, \dots \mathrm{C}_1, 1$ and add them: then the coefficient of λ_1 is $f'(a_1)$, and the coefficients of $\lambda_2, \lambda_3, \dots \lambda_n$ vanish, because (103) vanishes by virtue of the first of (102) when a_1 is replaced by a_2 or $a_3 \dots$ or a_n; and therefore ultimately we have $\lambda_1 f'(a_1) = 1$; so that

$$\lambda_1 = \frac{1}{f'(a_1)}; \tag{104}$$

similarly may it be shewn that

$$\lambda_2 = \frac{1}{f'(a_2)}, \quad \lambda_3 = \frac{1}{f'(a_3)}, \ldots \quad \lambda_n = \frac{1}{f'(a_n)}; \qquad (105)$$

and thus the general integral of (79) is

$$y = \Sigma . \frac{1}{f'(a_m)} e^{a_m x} \{ \mathrm{C}_m + \int e^{-a_m x} \mathrm{X}\, dx \};$$

and including the constant factor in the arbitrary constant C_m we have

$$y = \mathrm{C}_1 e^{a_1 x} + \mathrm{C}_2 e^{a_2 x} + \ldots + \mathrm{C}_n e^{a_n x}$$
$$+ \frac{e^{a_1 x}}{f'(a_1)} \int e^{-a_1 x} \mathrm{X}\, dx + \frac{e^{a_2 x}}{f'(a_2)} \int e^{-a_2 x} \mathrm{X}\, dx + \ldots + \frac{e^{a_n x}}{f'(a_n)} \int e^{-a_n x} \mathrm{X}\, dx. \qquad (106)$$

452.] This is the general integral of the differential equation (79), when all the roots of the characteristic are unequal. And if $\mathrm{X} = 0$, that is, if the right-hand member of (79) $= 0$, then

$$y = \mathrm{C}_1 e^{a_1 x} + \mathrm{C}_2 e^{a_2 x} + \ldots + \mathrm{C}_n e^{a_n x}, \qquad (107)$$

an expression which is easily verified by means of substitution in (79), and each of the terms of which is a particular integral; and as all are different, n different arbitrary constants are contained in it, and the integral is therefore general; and the form of (106) indicates that the general integral is the sum of n particular integrals, each of which involves or may involve a different arbitrary constant.

If there are pairs of impossible roots in the characteristic of (79) they enter as conjugates: suppose a pair to be a_i, a_j: so that

$$a_i = a + b\sqrt{-1}, \qquad a_j = a - b\sqrt{-1};$$

$$\begin{aligned} \therefore \quad \mathrm{C}_i e^{a_i x} + \mathrm{C}_j e^{a_j x} &= e^{ax} \{ \mathrm{C}_i e^{b\sqrt{-1}x} + \mathrm{C}_j e^{-b\sqrt{-1}x} \} \\ &= e^{ax} \{ (\mathrm{C}_i + \mathrm{C}_j) \cos bx + (\mathrm{C}_i - \mathrm{C}_j)\sqrt{-1} \sin bx \} \\ &= k e^{ax} \cos(\gamma + bx), \qquad (108) \end{aligned}$$

if $\mathrm{C}_i + \mathrm{C}_j = k \cos\gamma$, $(\mathrm{C}_i - \mathrm{C}_j)\sqrt{-1} = -k \sin\gamma$; and where of course k and γ are possible quantities. In the case therefore of a pair of imaginary roots, two terms of (107) will in combination produce a trigonometrical function of the form (108), and instead of the arbitrary constants C_i and C_j we have the new constants, equally arbitrary, k and γ. And a similar process of combination is also applicable to the latter unintegrated terms of the general expression (106).

453.] The following are examples of the preceding process of integration;

Ex. 1. $\frac{d^3y}{dx^3}-6a\frac{d^2y}{dx^2}+11a^2\frac{dy}{dx}-6a^3y=e^{mx}$.

$$\frac{d}{dx}\left\{\frac{d^2y}{dx^2}+\theta'\frac{dy}{dx}+\theta''y\right\}-(6a+\theta')\frac{d^2y}{dx^2}+(11a^2-\theta'')\frac{dy}{dx}-6a^3y=e$$

Let $$\frac{d^2y}{dx^2}+\theta'\frac{dy}{dx}+\theta''y=\eta;$$

$$-(6a+\theta')=\theta, \qquad 11a^2-\theta''=\theta\theta', \qquad -6a^3=\theta\theta'';$$

whence we have $$\frac{dy}{dx}+\theta\eta=e^{mx};$$

$$\theta^3+6a\theta^2+11a^2\theta+6a^3=0;$$

$$\therefore \quad \eta=e^{-\theta x}\{c+\int e^{(\theta+m)x}dx\};$$

$$\theta=-a, \quad =-2a, \quad =-3a;$$

and therefore in accordance with equation (86),

$$f(a)=(a-a)(a-2a)(a-3a),$$

$$f'(a)=(a-2a)(a-3a)+(a-3a)(a-a)+(a-a)(a-2a),$$

$$f'(a)=2a^2, \quad f'(2a)=-a^2, \quad f'(3a)=2a^2;$$

$$\therefore \quad y=c_1e^{ax}+c_2e^{2ax}+c_3e^{3ax}+\frac{e^{mx}}{(m-a)(m-2a)(m-3a)}.$$

Ex. 2. $\frac{d^2y}{dx^2}+a^2y=\cos nx$.

$$\frac{d}{dx}\left\{\frac{dy}{dx}+\theta'y\right\}-\theta'\frac{dy}{dx}+a^2y=\cos nx.$$

Let $\frac{dy}{dx}+\theta'y=\eta, \qquad -\theta'=\theta, \qquad a^2=\theta\theta';$

$$\therefore \quad \frac{d\eta}{dx}+\theta\eta=\cos nx, \qquad \theta^2+a^2=0;$$

$$\therefore \quad \eta=e^{-\theta x}\{c+\int e^{\theta x}\cos nx\,dx\};$$

$$\theta=a\sqrt{-1}, \quad =-a\sqrt{-1};$$

$$f(a)=\theta^2+a^2, \qquad f'(a)=2a;$$

$$\therefore \quad y=c_1e^{a\sqrt{-1}x}+c_2e^{-a\sqrt{-1}x}$$
$$+\frac{e^{a\sqrt{-1}x}}{2a\sqrt{-1}}\int e^{-a\sqrt{-1}x}\cos nx\,dx-\frac{e^{-a\sqrt{-1}x}}{2a\sqrt{-1}}\int e^{a\sqrt{-1}x}\cos nx\,dx;$$

$$\therefore \quad y=k\cos(ax+\gamma)+\frac{\cos nx}{a^2-n^2};$$

where k and γ are arbitrary constants.

454.] In the preceding investigations we have, at least tacitly, supposed all the roots of the characteristic to be unequal: for if two or more of them are equal, the value of y, as expressed in (89) and found by elimination from the group of equations (88), becomes indeterminate, and the subsequent processes of Art. 451 fail. Now, to take a particular case, let us suppose two roots to be equal, say $a_2 = a_1$; then the terms corresponding to these two roots become

$$C_1 e^{a_1} + C_2 e^{a_1 x} = (C_1 + C_2) e^{a_1 x} = C' e^{a_1 x};$$

and thus the two particular integrals will introduce only *one* arbitrary constant, and the general integral will contain only $n-1$ different constants: and consequently its generality is lost. Let us return then, and suppose m roots, $a_1, a_2, \ldots a_m$, of the characteristic to be equal, that is,

$$a_1 = a_2 = \ldots = a_m;$$

and, for the sake of simplicity, consider a differential equation which has no second member, observing that the generality of the process is not lost by the restriction.

First, let us suppose

$$a_2 = a_1 + i_2, \qquad a_3 = a_1 + i_3, \ldots \qquad a_m = a_1 + i_m,$$

so that the roots are thus made to be unequal; but they will become equal if $i_2 = i_3 = \ldots = i_m = 0$. Then

$$C_1 e^{a_1 x} + C_2 e^{a_2 x} + \ldots + C_m e^{a_m x}$$
$$= e^{a_1 x} \{C_1 + C_2 e^{i_2 x} + C_3 e^{i_3 x} + \ldots + C_m e^{i_m x}\}$$
$$= e^{a_1 x} \{C' + C'' x + C''' x^2 + \ldots + C^{(m)} x^m + C^{(m+1)} x^{m+1} + \ldots\},$$

if
$$C' = C_1 + C_2 + C_3 + \ldots + C_m,$$
$$C'' = C_2 i_2 + C_3 i_3 + \ldots + C_m i_m,$$
$$\ldots\ldots\ldots\ldots$$
$$C^{(m)} = \frac{1}{1.2\ldots m} \{C_2 i_2^m + C_3 i_3^m + \ldots + C_m i_m^m\},$$
$$C^{(m+1)} = \frac{1}{1.2\ldots(m+1)} \{C_2 i_2^{m+1} + C_3 i_3^{m+1} + \ldots + C_m i_m^{m+1}\}.$$
$$\ldots\ldots\ldots\ldots\ldots\ldots\ldots\ldots$$

Of these equations let us take the first m to determine the new constants $C', C'', \ldots C^{(m)}$; and then let us suppose $i = 0$, so that all the subsequent terms vanish, and the m roots of the characteristic become equal; and thus ultimately for the general integral we have

$$y = \{C' + C'' x + \ldots + C^{(m)} x^m\} e^{a_1 x} + C_{m+1} e^{a_{m+1} x} + \ldots + C_n e^{a_n x}.$$

Thus if two roots of the characteristic are equal

$$y = \{C'+C''x\}e^{a_1x}+C_2e^{a_2x}+\ldots+C_ne^{a_nx}.$$

Also we may consider the case of equal roots in the following manner: and this is perhaps more direct.

Let the equation be

$$y^{(n)}+A_1y^{(n-1)}+\ldots+A_{n-1}y'+A_ny = 0; \qquad (109)$$

and let $\qquad y = ue^{ax}; \qquad (110)$

where a is a constant and u is a function of x; then substituting in (109), and assuming

$$a^n+A_1a^{n-1}+A_2a^{n-2}+\ldots+A_{n-1}a+A_n = f(a),$$

we have

$$uf(a)+\frac{du}{dx}f'(a)+\frac{d^2u}{dx^2}f''(a)+\ldots+\frac{d^{n-1}u}{dx^{n-1}}f^{(n-1)}(a)+\frac{d^nu}{dx^n}=0. \qquad (111)$$

Now this equation is satisfied if $u =$ a constant, and $f(a) = 0$; that is, if for a we substitute one of the roots of the characteristic: let then a_1 be substituted for a, and C_1 for u, so that (110) becomes

$$y = C_1e^{a_1x};$$

which is a particular integral; and in the same way may the other particular integrals, and consequently the general integral be found. If however two roots of the characteristic are equal, say $a_2 = a_1$, then $f(a_1) = 0$, $f'(a_1) = 0$; and (111) is satisfied when

$$\frac{d^2u}{dx^2} = 0; \qquad \therefore \quad u = C'+C''x;$$

$$\therefore \quad y = (C'+C''x)e^{a_1x}.$$

Similarly if m roots of the characteristic are equal, it is necessary that

$$\frac{d^mu}{dx^m} = 0;$$

$$\therefore \quad u = C'+C''x+\ldots+C^{(m)}x^m;$$

and thus we have the form of the general integral when m roots of the characteristic are equal.

455.] SECOND METHOD.—We may also apply to the solution of linear equations with constant coefficients the process of successive reduction which has been investigated in the preceding Section. Taking (80) to be the type equation, let

$$y = e^{ax}\int u_1\,dx, \qquad (112)$$

where a is an undetermined constant, and u_1 is a function of x, and let us as heretofore suppose

$$a^n+A_1a^{n-1}+A_2a^{n-2}+\ldots+A_{n-1}a+A_n = f(a); \qquad (113)$$

then substituting (112) in (80) we have

$$f(a)\int u_1 dx + \frac{f'(a)}{1}u_1 + \frac{f''(a)}{1.2}\frac{du_1}{dx} + \dots$$
$$\dots + \frac{f^{n-1}(a)}{1.2.3\dots(n-1)}\frac{d^{n-2}u_1}{dx^{n-2}} + \frac{f^n(a)}{1.2\dots n}\frac{d^{n-1}u_1}{dx^{n-1}} = \text{X}e^{-ax}. \quad (114)$$

Now as a is undetermined in (112) and (114), let us suppose it to be a root of (113), say $a = a_1$, so that $f(a_1) = 0$; then the first term of the left-hand member of (114) vanishes, and there remains a differential equation of the $(n-1)$th order: and observing that $f^n(a) = 1.2.3\dots(n-1)n$, it is of the form

$$\frac{d^{n-1}u_1}{dx^{n-1}} + \frac{f^{n-1}(a_1)}{1.2.3\dots(n-1)}\frac{d^{n-2}u_1}{dx^{n-2}} + \dots + \frac{f''(a_1)}{1.2}\frac{du_1}{dx} + \frac{f'(a_1)}{1}u_1 = \text{X}e^{-a_1x}. \quad (115)$$

Now supposing all the roots of the characteristic to be unequal, there are n different equations of this form, corresponding to these roots, $a_1, a_2, \dots a_n$; also to solve (115) let

$$u_1 = e^{\beta x}\int u_2 dx; \quad (116)$$

substituting which in (115) we have

$$\left\{\beta^{n-1} + \beta^{n-2}\frac{f^{n-1}(a_1)}{1.2\dots(n-1)} + \dots + \beta\frac{f''(a_1)}{1.2} + f'(a_1)\right\}\int u_2 dx$$
$$+ \text{B}_1 u_2 + \text{B}_2\frac{du_2}{dx} + \dots + \text{B}_{n-3}\frac{d^{n-2}u_2}{dx^{n-2}} = \text{X}e^{-(\beta+a_1)x}; \quad (117)$$

and expressing the first term of the left-hand member in the following form, and adding $f(a_1)$, which is equal to zero, because a_1 is a root of (113), we have

$$\frac{\int u_2 dx}{\beta}\left\{\beta^n\frac{f^n(a_1)}{1.2.3\dots n} + \beta^{n-1}\frac{f^{n-1}(a_1)}{1.2\dots(n-1)} + \dots + \beta^2\frac{f''(a_1)}{1.2} + \beta\frac{f'(a_1)}{1} + f(a_1)\right\}$$
$$+ \text{B}_1 u_2 + \dots + \text{B}_{n-3}\frac{d^{n-2}u_2}{dx^{n-2}} = \text{X}e^{-(\beta+a_1)x}; \quad (118)$$

$$\text{let} \quad \beta^n\frac{f^n(a_1)}{1.2\dots n} + \beta_{n-1}\frac{f^{n-1}(a_1)}{1.2.3\dots(n-1)} + \dots + \beta\frac{f'(a_1)}{1} + f(a_1) = 0; \quad (119)$$

$$\therefore \quad f(a_1+\beta) = 0,$$

by reason of the form of (113); and therefore $a_1+\beta$ is a root of (113): let this root be a_2, then $a_1+\beta = a_2$, and $\beta = a_2 - a_1$; and as (119) is an algebraical equation of $(n-1)$ dimensions, the other roots are $a_3-a_1, \dots a_n-a_1$; let these be represented by $\beta_1, \beta_2, \dots \beta_{n-1}$; and as B_{n-3} is evidently unity in (118), (118) becomes

$$\frac{d^{n-2}u_2}{dx^{n-2}}+B_{n-2}\frac{d^{n-3}u_2}{dx^{n-3}}+\ldots+B_2\frac{du_2}{dx}+B_1u_2 = Xe^{-a_2x}.$$

Again, let

$$u_2 = e^{\gamma x}\int u_3 dx;$$

and pursuing the same process, $\gamma = \beta_2-\beta_1 = a_3-a_2$; and as the equation for determining γ will be of $n-2$ dimensions, the other roots will be a_4-a_2, a_5-a_2, ... a_n-a_2; and the differential equation for determining u_3 will be of the form

$$\frac{d^{n-3}u_3}{dx^{n-3}}+C_{n-3}\frac{d^{n-4}u_3}{dx^{n-4}}+\ldots+C_2\frac{du_3}{dx}+C_1u_3 = Xe^{-a_3x}.$$

Again, let

$$u_3 = e^{\delta x}\int u_4 dx;$$

and the equation for the determination of δ will be of $n-3$ dimensions, and its roots will be a_4-a_3, a_5-a_3, ... a_n-a_3; and we shall continue the processes until we ultimately arrive at

$$u_{n-1} = e^{(a_n-a_{n-1})x}\int u_n dx;$$

$$u_n = Xe^{-a_nx};$$

and thus, returning through the several steps,

$$y = e^{a_1x}\int e^{(a_2-a_1)x}dx\int e^{(a_3-a_2)x}dx\int\ldots e^{(a_n-a_{n-1})x}dx\int Xe^{-a_nx}dx; \quad (120)$$

and as a constant is to be introduced at each successive integration, it is manifest that in the course of the process n such will be introduced, and therefore that the integral is general. And the general form of it is

$$y = C'e^{a_1x}+C''e^{a_2x}+\ldots+C^{(n)}e^{a_nx}$$
$$+e^{a_1x}\int e^{(a_2-a_1)x}dx\int\ldots e^{(a_n-a_{n-1})x}dx\int Xe^{-a_nx}dx. \quad (121)$$

If $X = 0$, that is, if the given differential equation has no second member, then

$$y = C'e^{a_1x}+C''e^{a_2x}+\ldots+C^{(n)}e^{a_nx}. \quad (122)$$

An examination of the form of the constant which will be introduced at the several integrations of the multiple integral in (120) shews that the result is of a form precisely the same as that indicated in equation (106).

I may observe that this method of solution is the same as that investigated in Art. 443; but the general form of that Article is too complicated to be of useful employment, and therefore I have chosen to give a special inquiry.

Should there be a pair of imaginary and conjugate roots in the characteristic, the corresponding result may be reduced to a circular function.

This process is also applicable when two or more of the roots of the characteristic are equal; also the general result in equation (120) holds good. Thus suppose all the roots to be equal, then

$$y = e^{ax}\int^n e^{-ax}\text{X}\,dx^n;$$

and the several integrations will plainly introduce n arbitrary constants.

456.] The following are examples solved by this process.

Ex. 1. $$\frac{d^5y}{dx^5} - 13\frac{d^3y}{dx^3} + 26\frac{d^2y}{dx^2} + 82\frac{dy}{dx} + 104y = 0.$$

The characteristic is $a^5 - 13a^3 + 26a^2 + 82a + 104 = 0$; of which the roots are $-1+\sqrt{-1}$, $-1+\sqrt{-1}$, $3+2\sqrt{-1}$, $3-2\sqrt{-1}$, -4: therefore by (122)

$$\begin{aligned} y &= e^{-x}\{C_1 e^{\sqrt{-1}x} + C_2 e^{-\sqrt{-1}x}\} + e^{3x}\{C_3 e^{2\sqrt{-1}x} + C_4 e^{-2\sqrt{-1}x}\} + C_5 e^{-4x} \\ &= k_1 e^{-x}\cos(x+\gamma_1) + k_2 e^{3x}\cos(2x+\gamma_2) + C_5 e^{-4x}. \end{aligned}$$

Ex. 2. $$\frac{d^3y}{dx^3} - 7a\frac{d^2y}{dx^2} + 16a^2\frac{dy}{dx} - 12a^3y = 0.$$

The characteristic is $a^3 - 7a\,a^2 + 16a^2\,a - 12a^3 = 0$; of which the roots are $2a$, $2a$, $3a$; and therefore by (121)

$$y = (C' + C''x)e^{2ax} + C_3 e^{3ax}.$$

Ex. 3. $$\frac{d^ny}{dx^n} + na\frac{d^{n-1}y}{dx^{n-1}} + \frac{n(n-1)}{1.2}a^2\frac{d^{n-2}y}{dx^{n-2}} + \dots$$

$$+ \dots \frac{n(n-1)}{1.2}a^{n-2}\frac{d^2y}{dx^2} + \frac{n}{1}a^{n-1}\frac{dy}{dx} + a^ny = 0.$$

Of which the characteristic is

$$a^n + na\,a^{n-1} + \frac{n(n-1)}{1.2}a^2a^{n-2} + \dots + \frac{n}{1}a^{n-1}a + a^n = 0;$$

of which the n roots are equal, and each is equal to $-a$; so that (121) gives

$$y = \{C_1 + C_2x + \dots + C_nx^n\}e^{-ax}.$$

Ex. 4. $$\frac{d^3y}{dx^3} - 7a^2\frac{dy}{dx} + 6a^3y = e^{mx}.$$

The roots of the characteristic are $-3a$, a, $2a$; therefore by (120),

$$y = e^{-3ax}\int e^{4ax}\,dx\int e^{ax}\,dx\int e^{(m-2a)x}\,dx$$

$$= e^{-3ax}\int e^{4ax}\,dx\int e^{ax}\left\{c_1+\frac{e^{(m-2a)x}}{m-2a}\right\}dx$$

$$= e^{-3ax}\int e^{4ax}\,dx\left\{c_2+\frac{c_1}{a}e^{ax}+\frac{e^{(m-a)x}}{(m-a)(m-2a)}\right\}dx$$

$$= e^{-3ax}\left\{c_3+\frac{c_2}{4a}e^{4ax}+\frac{c_1}{5a^2}e^{5ax}+\frac{e^{(m+3a)x}}{(m+3a)(m-a)(m-2a)}\right\}$$

$$= c'e^{2ax}+c''e^{ax}+c'''e^{-3ax}+\frac{e^{mx}}{(m+3a)(m-a)(m-2a)}.$$

Ex. 5. $\dfrac{d^2y}{dx^2}-2\dfrac{dy}{dx}+y = x.$

The roots of the characteristic are equal, and each is equal to 1. Therefore (120) becomes

$$y = e^x\iint xe^{-x}\,dx\,dx$$

$$= x+2+(c_1x+c_2)e^x.$$

Ex. 6. $\dfrac{d^2y}{dx^2}+a^2y = \mathrm{X}.$

The characteristic is $a^2+a^2=0$, and therefore the roots of the characteristic are $a\sqrt{-1}$, $-a\sqrt{-1}$.

$$y = e^{-a\sqrt{-1}x}\int e^{2a\sqrt{-1}x}\,dx\int e^{-a\sqrt{-1}x}\mathrm{X}\,dx,$$

whereby when X is given the general integral can be found.

Let $\mathrm{X} = \cos nx = \frac{1}{2}\{e^{nx\sqrt{-1}}+e^{-nx\sqrt{-1}}\}$;

$$y = \frac{1}{2}e^{-a\sqrt{-1}x}\int e^{2a\sqrt{-1}x}\,dx\int e^{-ax\sqrt{-1}}(e^{nx\sqrt{-1}}+e^{-nx\sqrt{-1}})\,dx$$

$$= \frac{1}{2}e^{-a\sqrt{-1}x}\int e^{2a\sqrt{-1}x}\left\{c_1+\frac{e^{(n-a)x\sqrt{-1}}}{(n-a)\sqrt{-1}}-\frac{e^{-(n+a)x\sqrt{-1}}}{(n+a)\sqrt{-1}}\right\}dx$$

$$= \frac{1}{2}e^{-a\sqrt{-1}x}\left\{\frac{c_1e^{2a\sqrt{-1}x}}{2a\sqrt{-1}}-\frac{e^{(n+a)x\sqrt{-1}}}{n^2-a^2}-\frac{e^{-(n-a)x\sqrt{-1}}}{n^2-a^2}+c_2\right\}$$

$$= k\cos(ax+\gamma)-\frac{\cos nx}{n^2-a^2}.$$

Let $\mathrm{X} = \cos ax = \frac{1}{2}\{e^{ax\sqrt{-1}}+e^{-ax\sqrt{-1}}\}$;

$$y = e^{-a\sqrt{-1}x}\int e^{2a\sqrt{-1}x}\left\{c_1+\frac{x}{2}-\frac{e^{-2ax\sqrt{-1}}}{4a\sqrt{-1}}\right\}dx$$

$$y = e^{-ax\sqrt{-1}}\left\{c_2 + \frac{c_1}{2a\sqrt{-1}}e^{2ax\sqrt{-1}} + \frac{xe^{2a\sqrt{-1}x}}{4a\sqrt{-1}} + \frac{e^{2a\sqrt{-1}x}}{8a^2} - \frac{x}{4a\sqrt{-a}}\right\}$$

$$= k\cos(ax+\gamma) + \frac{x\sin ax}{2a}.$$

Ex. 7. It is required to determine the form of the function, when $f(x+y)+f(x-y) = f(x)f(y)$; x and y being two variables independent of each other.

Taking the x-differential twice, we have

$$f''(x+y)+f''(x-y) = f''(x)f(y).$$

Again, taking the y-differential twice, we have

$$f''(x+y)+f''(x-y) = f(x)f''(y).$$

Consequently equating the right-hand members,

$$\frac{f''(x)}{f(x)} = \frac{f''(y)}{f(y)} = \pm a^2, \text{ say,}$$

where a^2 is an arbitrary constant, since x and y are independent variables;

$$\therefore \quad f''(x) \pm a^2 f(x) = 0. \qquad (123)$$

$$\therefore \quad f(x) = c_1 e^{ax} + c_2 e^{-ax}; \qquad (124)$$

$$f(x) = A\cos(ax+\alpha); \qquad (125)$$

(124) or (125) being the solution according as the lower or the upper sign is taken in (123). Taking (124) to be the solution, where c_1 and c_2 are undetermined constants, and substituting in the given equation we have $c_1 = c_2 = 1$, and a is undetermined. Also taking (125) to be the solution $A=2$, $\alpha=0$; a being undetermined; so that

$$f(x) = e^{ax}+e^{-ax};$$

$$f(x) = 2\cos ax;$$

either of which equations satisfies the proposed functional equation.

457.] If the right-hand member of the linear equation contains a constant only, so that the equation is of the form

$$\frac{d^n y}{dx^n} + A_1\frac{d^{n-1}y}{dx^{n-1}} + \ldots + A_{n-1}\frac{dy}{dx} + A_n y = A;$$

then it may be expressed as follows:

$$\frac{d^n y}{dx^n} + A_1\frac{d^{n-1}y}{dx^{n-1}} + \ldots + A_{n-1}\frac{dy}{dx} + A_n\left(y - \frac{A}{A_n}\right) = 0. \qquad (126)$$

Now replace y by $y + \frac{A}{A_n}$; then (126) becomes

$$\frac{d^n y}{dx^n} + A_1 \frac{d^{n-1} y}{dx^{n-1}} + \ldots + A_{n-1} \frac{dy}{dx} + A_n y = 0, \qquad (127)$$

and is therefore of the form which has already been discussed. In the final result we shall have to replace y by $y - \frac{A}{A_n}$; and therefore if $a_1, a_2, \ldots a_n$ are the n roots of the characteristic of (127),

$$y = \frac{A}{A_n} + C_1 e^{a_1 x} + C_2 e^{a_2 x} + \ldots + C_n e^{a_n x}.$$

Also from (120) the same result will be derived. Let $X = A$; then

$$y = e^{a_1 x} \int e^{(a_2 - a_1)x} dx \int e^{(a_3 - a_2)x} dx \int \ldots e^{(a_n - a_{n-1})x} dx \int A e^{-a_n x} dx$$

$$= \frac{A}{(-)^n a_1 a_2 \ldots a_n} + C_1 e^{a_1 x} + \ldots + C_n e^{a_n x}$$

$$= \frac{A}{A_n} + C_1 e^{a_1 x} + \ldots + C_n e^{a_n x}.$$

Also if the characteristic has impossible roots, or has equal roots, the results are similar in form to those investigated above.

Ex. 1. $\frac{d^2 y}{dx^2} - (a_1 + a_2) \frac{dy}{dx} + a_1 a_2 y = k^2$;

$$\therefore \quad y = e^{a_1 x} \int e^{(a_2 - a_1)x} dx \int k^2 e^{-a_2 x} dx$$

$$= e^{a_1 x} \int e^{(a_2 - a_1)x} \left\{ C_1 - \frac{k^2}{a_2} e^{-a_2 x} \right\} dx$$

$$= e^{a_1 x} \left\{ C' + \frac{C_1}{a_2 - a_1} e^{(a_2 - a_1)x} + \frac{k^2}{a_1 a_2} e^{-a_1 x} \right\}$$

$$= \frac{k^2}{a_1 a_2} + C' e^{a_1 x} + C'' e^{a_2 x}.$$

458.] The investigations of the preceding articles will have shewn that the solution of a linear differential equation with a second member depends in a great measure on the solution of it without the second member; for if the general integral of the latter can be determined, that of the former will also be determined if a certain given function of a single variable can be successively integrated, and added to it. A method for this purpose, other than the preceding, was devised by Lagrange, and is commonly called Lagrange's method of variation of parameters. This I proceed to explain.

Take the two following linear differential equations expressed in terms of derived functions:

$$y^{(n)} + A_1 y^{(n-1)} + A_2 y^{(n-2)} + \ldots + A_{n-1} y' + A_n y = X, \qquad (128)$$

$$z^{(n)} + A_1 z^{(n-1)} + A_2 z^{(n-2)} + \ldots + A_{n-1} z' + A_n z = 0; \qquad (129)$$

and suppose $z_1, z_2, \ldots z_n$ to be n particular integrals of (129), so that the general integral z is

$$z = C_1 z_1 + C_2 z_2 + \ldots + C_n z_n; \qquad (130)$$

then it is always possible to determine n functions of x, viz., $u_1, u_2, \ldots u_n$, so that the general integral of (128) may be

$$y = u_1 z_1 + u_2 z_2 + \ldots + u_n z_n, \qquad (131)$$

$$= \Sigma . uz; \qquad (132)$$

that is, the integrals of (128) and (129) are of the same form; but the arbitrary quantities $C_1, C_2, \ldots C_n$, which are constant in the integral of (129), are functions of x in that of (128).

To exemplify the process before I enter on the general theory, I will take first the case of the linear differential equation of the first order, viz.

$$\frac{dy}{dx} + f(x)\, y = F(x), \qquad (133)$$

where the coefficient of y and also the right-hand member are functions of x. Now omitting the second member we have

$$\frac{dy}{dx} + f(x)\, y = 0.$$

$$\therefore \quad \frac{dy}{y} + f(x)\, dx = 0;$$

$$\therefore \quad y = c e^{-\int f(x)\, dx}, \qquad (134)$$

...n arbitrary constant.

...re in the method of variation of parameters sup-... of the equation with its second member to ...that of the equation without the second ...integration in the latter case however ...e by a determinate function of x; ...al of (133), when c is a func-... ...entiate (134), and substi-...

$$\qquad (135)$$

$$\therefore \quad c = \text{C} + \int \text{F}(x)\, e^{\int f(x)dx}\, dx; \tag{136}$$

so that the integral of (133) is

$$y = e^{-\int f(x)dx} \left\{ \text{C} + \int \text{F}(x)\, e^{\int f(x)dx}\, dx \right\};$$

which is the same solution as that given in (69), Art. 382. The method is called variation of parameters, because c, which is a constant parameter in (134), is made to be that function of a variable which is given in (136).

Let us now apply the same theory to the general linear differential equation (128). Let us suppose (132) to be its integral; then differentiating,

$$\frac{dy}{dx} = \Sigma . u \frac{dz}{dx} + \Sigma . z \frac{du}{dx}; \tag{137}$$

and let us suppose $\frac{dy}{dx}$ to be of the same form in (130) and (131), so that

$$\Sigma . z \frac{du}{dx} = 0.$$

Differentiate again (137) subject to this condition: and we have

$$\frac{d^2 y}{dx^2} = \Sigma . u \frac{d^2 z}{dx^2} + \Sigma . \frac{dz}{dx} \frac{du}{dx};$$

and again suppose $\frac{d^2 y}{dx^2}$ to be of the same form in (130) and (131); then

$$\Sigma . \frac{dz}{dx} \frac{du}{dx} = 0;$$

and continuing the same process, and making similar substitutions up to $\frac{d^{n-1} y}{dx^{n-1}}$, we have

$$\Sigma . \frac{d^2 z}{dx^2} \frac{du}{dx} = 0, \quad \Sigma . \frac{d^3 z}{dx^3} \frac{du}{dx} = 0, \ldots\ldots \Sigma . \frac{d^{n-2} z}{dx^{n-2}} \frac{du}{dx} = 0;$$

$$\frac{d^{n-1} y}{dx^{n-1}} = \Sigma . u \frac{d^{n-1} z}{dx^{n-1}};$$

$$\therefore \quad \frac{d^n y}{dx^n} = \Sigma . u \frac{d^n z}{dx^n} + \Sigma . \frac{d^{n-1} z}{dx^{n-1}} \frac{du}{dx};$$

and substituting these values throughout in (128), we have

$$\Sigma . u \{ z^{(n)} + \text{A}_1 z^{(n-1)} + \ldots + \text{A}_{n-1} z' + \text{A}_n z \} + \Sigma . \frac{d^{n-1} z}{dx^{n-1}} \frac{du}{dx} = \text{X}.$$

Now of the expression on the left-hand side of this equation, the first part vanishes, because $z_1, z_2, \dots z_n$ are particular integrals of (129); and therefore

$$\text{x}.\frac{d^{n-1}z}{dx^{n-1}}\frac{du}{dx} = \text{x}.$$

Hence (131) is the general integral of (128), the values of the u's being found from the following system of equations:

$$\left.\begin{aligned} z_1 u_1' + z_2 u_2' + z_3 u_3' + \dots + z_n u_n' &= 0, \\ z_1' u_1' + z_2' u_2' + z_3' u_3' + \dots + z_n' u_n' &= 0, \\ \dots\dots\dots\dots\dots\dots \\ z_1^{(n-2)} u_1' + z_2^{(n-2)} u_2' + z_3^{(n-2)} u_3' + \dots + z_n^{(n-2)} u_n' &= 0, \\ z_1^{(n-1)} u_1' + z_2^{(n-1)} u_2' + z_3^{(n-1)} u_3' + \dots + z_n^{(n-1)} u_n' &= \text{x}. \end{aligned}\right\} \quad (138)$$

These equations will of course in general give n different values of $u_1', u_2', \dots u_n'$ in terms of the z's and of x, and each value will have x as a factor; suppose the other factors to be $v_1, v_2, \dots v_n$: then

$$\left.\begin{aligned} u_1' &= v_1 \text{x}; & \therefore\quad u_1 &= \text{c}_1 + \int v_1 \text{x}\, dx; \\ u_2' &= v_2 \text{x}; & u_2 &= \text{c}_2 + \int v_2 \text{x}\, dx; \\ &\dots\dots & &\dots\dots \\ u_n' &= v_n \text{x}; & u_n &= \text{c}_n + \int v_n \text{x}\, dx; \end{aligned}\right\} \quad (139)$$

and substituting these in (131), we have the general integral of (128).

In this process we have made no restriction as to the coefficients of the given differential equation; they may be either constants or functions of x: if however they are constants, and $a_1, a_2, \dots a_n$ are the roots of the characteristic of (129),

$$z_1 = e^{a_1 x}, \qquad z_2 = e^{a_2 x}, \dots\dots z_n = e^{a_n x};$$

and these must be substituted in the series (138); and thence may be deduced the values of $v_1, v_2, \dots v_n$ which are required for (139).

The following are examples of this process.

Ex. 1. $$\frac{d^2 y}{dx^2} - 5a\frac{dy}{dx} + 6a^2 y = e^{mx};$$

$$\therefore\quad \frac{d^2 z}{dx^2} - 5a\frac{dz}{dx} + 6a^2 z = 0;$$

$$z = \text{c}_1 e^{2ax} + \text{c}_2 e^{3ax}; \qquad y = u_1 e^{2ax} + u_2 e^{3ax};$$

$$\frac{dy}{dx} = 2au_1 e^{2ax} + 3au_2 e^{3ax} + e^{2ax}\frac{du_1}{dx} + e^{3ax}\frac{du_2}{dx};$$

let $$e^{2ax}\frac{du_1}{dx}+e^{3ax}\frac{du_2}{dx} = 0;$$

$$\therefore \quad \frac{d^2y}{dx^2} = 4a^2u_1e^{2ax}+9a^2u_2e^{3ax}+2ae^{2ax}\frac{du_1}{dx}+3ae^{3ax}\frac{du_2}{dx};$$

and substituting in the given differential equation, we have

$$2ae^{2ax}\frac{du_1}{dx}+3ae^{3ax}\frac{du_2}{dx} = e^{mx};$$

from which, combined with the supposition made above, we have

$$\frac{du_1}{dx} = -\frac{1}{a}e^{(m-2a)x}; \qquad \therefore \quad u_1 = \text{C}_1-\frac{e^{(m-2a)x}}{a(m-2a)};$$

$$\frac{du_2}{dx} = \frac{1}{a}e^{(m-3a)x}; \qquad u_2 = \text{C}_2+\frac{e^{(m-3a)x}}{a(m-3a)};$$

$$\therefore \quad y = \text{C}_1e^{2ax}+\text{C}_2e^{3ax}+\frac{e^{mx}}{(m-2a)(m-3a)}.$$

Ex. 2. $\frac{d^2y}{dx^2}-a^2y = e^{nx}$.

So that the equation without the second member is

$$\frac{d^2y}{dx^2}-a^2y = 0;$$

of which the integral by the processes of the preceding Articles is

$$y = c_1e^{ax}+c_2e^{-ax},$$

where c_1 and c_2 are arbitrary constants of integration.

Let us suppose c_1 and c_2 to be functions of x: then

$$\frac{dy}{dx} = ac_1e^{ax}-ac_2e^{-ax}+e^{ax}\frac{dc_1}{dx}+e^{-ax}\frac{dc_2}{dx};$$

let $$e^{ax}\frac{dc_1}{dx}+e^{-ax}\frac{dc_2}{dx} = 0; \tag{140}$$

$$\therefore \quad \frac{d^2y}{dx^2} = a^2c_1e^{ax}+a^2c_2e^{-ax}+ae^{ax}\frac{dc_1}{dx}-ae^{-ax}\frac{dc_2}{dx};$$

Let these values be substituted in the given differential equation; and we have

$$ae^{ax}\frac{dc_1}{dx}-ae^{-ax}\frac{dc_2}{dx} = e^{nx}; \tag{141}$$

$$\therefore \quad \frac{dc_1}{dx} = \frac{e^{(n-a)x}}{2a}, \qquad \frac{dc_2}{dx} = -\frac{e^{(n+a)x}}{2a};$$

$$\therefore \quad c_1 = \text{C}_1+\frac{e^{(n-a)x}}{2a(n-a)}; \qquad c_2 = \text{C}_2-\frac{e^{(n+a)x}}{2a(n+a)};$$

so that the integral of the given differential equation is

$$y = \text{C}_1e^{ax}+\text{C}_2e^{-ax}+\frac{e^{nx}}{n^2-a^2}.$$

Ex. 3. $\dfrac{d^2y}{dx^2} - a^2 y = e^{ax}$.

Taking the same process as in the preceding example,

$$y = c_1 e^{ax} + c_2 e^{-ax} + \frac{x e^{ax}}{2a}.$$

SECTION 4.—*Integration of some Particular Forms of Linear Differential Equations with Variable Coefficients.*

459.] The linear differential equation of the following form admits of being reduced to one with constant coefficients by means of a change of variable, and therefore its integral may be determined by one or other of the preceding methods.

$$(a+bx)^n \frac{d^n y}{dx^n} + A_1 (a+bx)^{n-1} \frac{d^{n-1}y}{dx^{n-1}} + A_2 (a+bx)^{n-2} \frac{d^{n-2}y}{dx^{n-2}} + \dots$$
$$\dots + A_{n-1}(a+bx)\frac{dy}{dx} + A_n y = 0; \quad (142)$$

and I may at once remark that if the equation admits of integration when the right-hand member vanishes, it may also be integrated when the right-hand member is a function of x.

Let $a+bx = z$; then as x is equicrescent in (142), so will also z be; and therefore after the substitution the equation is

$$b^n z^n \frac{d^n y}{dz^n} + b^{n-1} z^{n-1} A_1 \frac{d^{n-1}y}{dz^{n-1}} + \dots + bz A_{n-1} \frac{dy}{dz} + A_n y = 0; \quad (143)$$

so that the form of the equation is

$$x^n \frac{d^n y}{dx^n} + A_1 x^{n-1} \frac{d^{n-1}y}{dx^{n-1}} + \dots + A_{n-1} x \frac{dy}{dx} + A_n y = 0. \quad (144)$$

$$\text{Let } \frac{dx}{x} = dt; \qquad \therefore \quad x = e^t; \qquad \frac{dy}{dt} = x\frac{dy}{dx};$$

$$\frac{d^2y}{dt^2} = x\frac{d^2y}{dx^2}\frac{dx}{dt} + \frac{dy}{dx}\frac{dx}{dt} = x^2\frac{d^2y}{dx^2} + \frac{dy}{dt};$$

$$\frac{d^3y}{dt^3} = x^2\frac{d^3y}{dx^3}\frac{dx}{dt} + 2x\frac{d^2y}{dx^2}\frac{dx}{dt} + \frac{d^2y}{dt^2} = x^3\frac{d^3y}{dx^3} + 3\frac{d^2y}{dt^2} - 2\frac{dy}{dt};$$

and so on; thus $x\dfrac{dy}{dx}$, $x^2\dfrac{d^2y}{dx^2}$, ... may be expressed in terms of $\dfrac{dy}{dt}$, $\dfrac{d^2y}{dt^2}$ ...; and (144) will become a linear differential equation with constant coefficients.

Ex. 1. $$x^2\frac{d^2y}{dx^2}+x\frac{dy}{dx}-y = x^m.$$

Let $$x = e^t; \qquad x\frac{dy}{dx} = \frac{dy}{dt}; \qquad x^2\frac{d^2y}{dx^2} = \frac{d^2y}{dt^2}-\frac{dy}{dt};$$

$$\therefore \quad \frac{d^2y}{dt^2}-y = e^{mt};$$

so that by the methods of the preceding section

$$y = c_1x+c_2x^{-1}+\frac{x^m}{m^2-1}.$$

Ex. 2. $$x^2\frac{d^2y}{dx^2}-3x\frac{dy}{dx}+4y = x^m.$$

Let $$x = e^t; \qquad \frac{d^2y}{dt^2}-4\frac{dy}{dt}+4y = e^{mt};$$

$$\therefore \quad y = \frac{x^m}{(m-2)^2}+x^2\{c_1+c_2\log x\}.$$

Ex. 3. $$\frac{d^2y}{dx^2}-\frac{y}{x^2} = 0.$$

460.] The form of linear differential equation which I shall consider next is

$$(a_n+b_nx)\frac{d^ny}{dx^n}+(a_{n-1}+b_{n-1}x)\frac{d^{n-1}y}{dx^{n-1}}+\dots$$
$$\dots+(a_1+b_1x)\frac{dy}{dx}+(a_0+b_0x)y = 0, \qquad (145)$$

where the a's and the b's are constant. Now let us suppose a particular integral of this equation to be of a more general form than any heretofore assumed, and to be the definite integral of the form

$$y = \int_{u_0}^{u_1} e^{ux}\,\mathrm{v}\,du, \qquad (146)$$

where u is a new variable independent of x, v is a function of u, and u_1 and u_0 are the limits of integration and are independent of x; and let us consider the result of the substitution of this quantity in the given differential equation; differentiating (146),

$$\frac{dy}{dx} = \int_{u_0}^{u_1} ue^{ux}\,\mathrm{v}\,du, \quad \frac{d^2y}{dx^2} = \int_{u_0}^{u_1} u^2e^{ux}\,\mathrm{v}\,du,\dots \quad \frac{d^ny}{dx^n} = \int_{u_0}^{u_1} u^ne^{ux}\,\mathrm{v}\,du;$$

and moreover let us substitute as follows:

$$a_nu^n+a_{n-1}u^{n-1}+\dots+a_1u+a_0 = \mathrm{U}_0,$$
$$b_nu^n+b_{n-1}u^{n-1}+\dots+b_1u+b_0 = \mathrm{U}_1;$$

so that (145) becomes

$$\int_{u_0}^{u_1} \{\mathrm{U}_0+\mathrm{U}_1 x\} e^{ux} \mathrm{V}\, du = 0;$$

and integrating by parts,

$$\Big[e^{ux}\mathrm{U}_1\mathrm{V}\Big]_{u_0}^{u_1} + \int_{u_0}^{u_1} \{\mathrm{U}_0\mathrm{V}\, du - d.\mathrm{U}_1\mathrm{V}\} e^{ux} = 0. \qquad (147)$$

Now as V is an undetermined function of u, let us assume that

$$\mathrm{U}_0\mathrm{V}\, du - d.\mathrm{U}_1\mathrm{V} = 0;$$

$$\therefore \quad \frac{\mathrm{U}_0}{\mathrm{U}_1} du = \frac{d.\mathrm{U}_1\mathrm{V}}{\mathrm{U}_1\mathrm{V}}, \qquad \mathrm{V} = \frac{\mathrm{C}}{\mathrm{U}_1} e^{\int \frac{\mathrm{U}_0}{\mathrm{U}_1} du}.$$

And in consequence of this assumption (147) becomes

$$\Big[e^{ux}\mathrm{U}_1\mathrm{V}\Big]_{u_0}^{u_1} = 0;$$

$$\therefore \quad \Big[\mathrm{C}e^{ux+\int \frac{\mathrm{U}_0}{\mathrm{U}_1} du}\Big]_{u_0}^{u_1} = 0; \qquad (148)$$

and

$$y = \int_{u_0}^{u_1} \frac{\mathrm{C}}{\mathrm{U}_1} e^{ux+\int \frac{\mathrm{U}_0}{\mathrm{U}_1} du} du;$$

but in this expression u_1 and u_0 are undetermined; they must however satisfy (148); and as there will in general be no relation between them, each separately must satisfy it: and therefore we must discover the roots of the equation

$$e^{ux+\int \frac{\mathrm{U}_0}{\mathrm{U}_1} dx} = 0;$$

let us suppose them to be $u_0, u_1, u_2, \ldots u_k$; then if we take u_0 to be the inferior limit in all cases, and the others in turn to be superior limits, we have the following k values of y, viz.

$$\left.\begin{array}{l} y = \int_{u_0}^{u_1} \mathrm{C}_1 e^{ux+\int \frac{\mathrm{U}_0}{\mathrm{U}_1} du} \frac{du}{\mathrm{U}_1}; \\ y = \int_{u_0}^{u_2} \mathrm{C}_2 e^{ux+\int \frac{\mathrm{U}_0}{\mathrm{U}_1} du} \frac{du}{\mathrm{U}_1}; \\ \cdot\quad\cdot\quad\cdot\quad\cdot\quad\cdot\quad\cdot\quad\cdot\quad\cdot \\ y = \int_{u_0}^{u_k} \mathrm{C}_k e^{ux+\int \frac{\mathrm{U}_0}{\mathrm{U}_1} du} \frac{du}{\mathrm{U}_1}; \end{array}\right\} \qquad (149)$$

and from the form of the equation it is plain that the sum of these also satisfies the equations. If therefore it is possible to find $n+1$ such values of u, the resulting expression of the form (149) is the general integral of the given equation; in other cases it may be only a particular integral.

And I must observe that the definite integrals which enter into the final result generally do not admit of further reduction; and hence we infer that the integral of a differential equation of the form (145) is a transcendent of a higher order than any of the commonly tabulated functions.

461.] The following are examples of which the integral is expressed as a definite integral.

Ex. 1. $(a_2+b_2x)\frac{d^2y}{dx^2}+(a_1+b_1x)\frac{dy}{dx}+(a_0+b_0x)y=0.$

For x write $x-\frac{a_2}{b_2}$; then we have an equation of the form

$$x\frac{d^2y}{dx^2}+(a_1+b_1x)\frac{dy}{dx}+(a_0+b_0x)y=0;$$

$$\therefore \quad \mathrm{U}_0=a_1u+a_0, \qquad \mathrm{U}_1=u^2+b_1u+b_0,$$

$$\int\frac{\mathrm{U}_0}{\mathrm{U}_1}du=\int\frac{a_1u+a_0}{u^2+b_1u+b_0}du$$

$$=\mathrm{A}\log(u-\alpha)+\mathrm{B}\log(u-\beta),$$

if α and β are the roots of the denominators, and A and B are determinate constants dependent on a_1, a_0, b_1, b_0: so that from (148) we have

$$e^{ux}(u-\alpha)^{\mathrm{A}}(u-\beta)^{\mathrm{B}}=0,$$

and this equation is satisfied by $u=\alpha$, $u=\beta$, $u=-\infty$; and therefore from (149)

$$y=\mathrm{C}_1\int_{-\infty}^{\alpha}e^{ux}(u-\alpha)^{\mathrm{A}-1}(u-\beta)^{\mathrm{B}-1}du+\mathrm{C}_2\int_{-\infty}^{\beta}e^{ux}(u-\alpha)^{\mathrm{A}-1}(u-\beta)^{\mathrm{B}-1}du;$$

which is the general integral.

Ex. 2. $x\frac{d^2y}{dx^2}+a\frac{dy}{dx}-b^2xy=0;$

$$\mathrm{U}_0=au, \qquad \mathrm{U}_1=u^2-b^2;$$

$$\therefore \quad \int\frac{\mathrm{U}_0}{\mathrm{U}_1}du=\int\frac{au\,du}{u^2-b^2}=\frac{a}{2}\log(u^2-b^2);$$

therefore from (148)

$$e^{ux}(u^2-b^2)^{\frac{a}{2}}=0;$$

$$\therefore \quad u=-\infty, \qquad =b, \qquad =-b;$$

$$\therefore \quad y=\mathrm{C}_1\int_{-\infty}^{b}e^{ux}(u^2-b^2)^{\frac{a}{2}-1}du+\mathrm{C}_2\int_{-\infty}^{-b}e^{ux}(u^2-b^2)^{\frac{a}{2}-1}du;$$

these definite integrals do not admit of further reduction.

Ex. 3. $\dfrac{d^2y}{dx^2}+(a+bx)y=0.$

In this example (148) becomes

$$e^{ux+\frac{u^3}{3b}+\frac{au}{b}}=0;$$

which is satisfied by $u^3=-\infty$; therefore $u^3+a^3=0$, if in the result $a=\infty$; and of this equation, if r is a primitive cube root of -1; the roots are

$$-a, \quad ar, \quad ar^2,$$

and therefore

$$y=\mathrm{C}_1\int_{-a}^{ar} e^{ux+\frac{au}{b}+\frac{u^3}{3b}}du+\mathrm{C}_2\int_{-a}^{ar^2} e^{ux+\frac{au}{b}+\frac{u^3}{3b}}du;$$

and in the final result $a=\infty$.

Ex. 4. $\dfrac{d^ny}{dx^n}+axy=0;$

$$\therefore \int\frac{\mathrm{U}_0}{\mathrm{U}_1}du=\frac{1}{a}\int u^n du=\frac{u^{n+1}}{a(n+1)};$$

Therefore from (148), $e^{ux+\frac{u^{n+1}}{a(n+1)}}=0$; and $u^{n+1}=-\infty$. Consequently if the primitive roots of $(-1)^{\frac{1}{n+1}}$ are $-1, r, r^2, \ldots r^n$, and a is a quantity which, in the result, is infinite, the roots of this equation are $-a, ar, ar^2, \ldots ar^n$;

$$\therefore\ y=\mathrm{C}_1\int_{-a}^{ar} e^{ux+\frac{u^{n+1}}{a(n+1)}}du+\mathrm{C}_2\int_{-a}^{ar^2} e^{ux+\frac{u^{n+1}}{a(n+1)}}du+\ldots$$

$$\ldots+\mathrm{C}_n\int_{-a}^{ar^n} e^{ux+\frac{u^{n+1}}{a(n+1)}}du.$$

Ex. 5. As the last example of this method let us take equation (260) in Art. 425, which is equivalent to Riccati's equation, and exhibit the function, which satisfies it, in the form of a definite integral. The equation may be put into the form

$$x\frac{d^2y}{dx^2}+2n\frac{dy}{dx}-b^2xy=0;$$

$$\therefore\ \mathrm{U}_0=2nu; \qquad \mathrm{U}_1=u^2-b^2.$$

$$\therefore\ \int\frac{\mathrm{U}_0}{\mathrm{U}_1}du=n\log(u^2-b^2);$$

and (148) becomes $\quad e^{ux}(u^2-b^2)^n=0;$

$$\therefore\ u=-\infty, \quad =+b \quad =-b;$$

$$\therefore\ y=\mathrm{C}_1\int_{-\infty}^{b} e^{ux}(u^2-b^2)^{n-1}du+\mathrm{C}_2\int_{-\infty}^{-b} e^{ux}(u^2-b^2)^{n-1}du. \qquad (150)$$

Now these integrals admit of integration in finite terms whenever n is a positive whole number: and since, see Art. 425,

$$n = \frac{m}{2(m+2)}, \quad \text{whenever} \quad m = \frac{-4n}{2n-1},$$

which is one of the conditions determined in Art. 424. And if n is a whole negative number, then

$$-n = \frac{m}{2(m+2)}; \qquad \therefore \quad m = \frac{-4n}{2n+1};$$

and this is the other condition found in that Article. Hence arises a reason why Riccati's equation can be integrated for these values of m.

The reader desirous of further information on the integration of linear differential equations by means of definite integrals is referred to the large work of Petzval on the subject; viz. Integration der Linearen Differentialgleichungen, von Joseph Petzval, Wien, 1853.

Section 5.—*Integration of certain Differential Equations of higher Orders and Degrees.*

462.] As no general theory exists for the integration of differential equations of all orders and degrees, we are obliged to have recourse to artifices, which analysts have from time to time devised, for the integration of particular examples; I propose therefore to examine the most useful of these processes as concisely as possible and in order. And firstly I shall take differential equations of higher orders, where the highest derived function is a function of either the one next, or the two next, inferior to it.

Let $f^n(x)$ be the highest derived function; then the problem is, to discover the integral of the equation

$$f^n(x) = \text{F}\{f^{n-1}(x), f^{n-2}(x)\}.$$

Let $f^{n-2}(x) = z$; $\quad \therefore \; f^{n-1}(x) = \dfrac{dz}{dx}$; $\quad f^n(x) = \dfrac{d^2 z}{dx^2}$;

$$\therefore \quad \frac{d^2 z}{dx^2} = \text{F}\left(\frac{dz}{dx}, z\right);$$

and the equation is a differential equation of the second order; of this let the integral be $z = \phi(x)$;

$$\therefore \quad f(x) = \int^{n-2} \phi(x)\, dx^{n-2};$$

so that the final value of $f(x)$ depends on a function of x which is to be integrated $(n-2)$ times in succession. Some examples are subjoined.

Ex. 1. $a\frac{d^2y}{dx^2} = \frac{dy}{dx}$; $\qquad \therefore\ a\frac{dy}{dx} = y-b,$

where b is an undetermined constant.

$$\therefore\ \frac{dy}{y-b} = \frac{dx}{a}; \qquad \log(y-b) = \frac{x-c}{a};$$

where c is another undetermined constant.

Ex. 2. $d^2y = dx\,(dx^2+dy^2)^{\frac{1}{2}}$, where x is equicrescent.

$$\frac{d^2y}{(dx^2+dy^2)^{\frac{1}{2}}} = dx; \qquad \therefore\ \log\left\{\frac{dy}{dx} + \left(1+\frac{dy^2}{dx^2}\right)^{\frac{1}{2}}\right\} = x-a;$$

$$\therefore\ \frac{dy}{dx} = \frac{1}{2}\{e^{x-a}-e^{-(x-a)}\};$$

$$y-b = \frac{1}{2}\{e^{x-a}+e^{-(x-a)}\}.$$

Or we may integrate as follows: the equation is

$$\frac{d^2y}{dx^2} = \left\{1+\frac{dy^2}{dx^2}\right\}^{\frac{1}{2}}; \qquad \therefore\ 2dy\frac{d^2y}{dx^2} = 2dy\left\{1+\frac{dy^2}{dx^2}\right\}^{\frac{1}{2}};$$

$$\frac{d.\frac{dy^2}{dx^2}}{\left\{1+\frac{dy^2}{dx^2}\right\}^{\frac{1}{2}}} = 2dy;$$

$$\therefore\ \left\{1+\frac{dy^2}{dx^2}\right\}^{\frac{1}{2}} = y-b; \qquad \frac{dy}{dx} = \{(y-b)^2-1\}^{\frac{1}{2}}$$

$$\frac{dy}{\{(y-b)^2-1\}^{\frac{1}{2}}} = dx;$$

$$\log\left(y-b+\{(y-b)^2-1\}^{\frac{1}{2}}\right) = x-a;$$

$$y-b = \frac{1}{2}\{e^{x-a}+e^{-(x-a)}\}.$$

It will be observed that in the former of the two methods we have integrated first with respect to x, and in the latter first with respect to y. The final integral also might have been found by eliminating $\frac{dy}{dx}$ by means of the two first integrals.

Ex. 3. $a\frac{d^3y}{dx^3}\frac{d^2y}{dx^2} = \left\{1+\left(\frac{d^2y}{dx^2}\right)^2\right\}^{\frac{1}{2}}.$

Let $\frac{d^2y}{dx^2} = z$; $\qquad \therefore\ az\frac{dz}{dx} = (1+z^2)^{\frac{1}{2}}$;

$$\therefore \quad a(1+z^2)^{\frac{1}{2}} = x-c; \qquad z^2 = \frac{(x-c)^2-a^2}{a^2};$$

$$\therefore \quad a\frac{d^2y}{dx^2} = \{(x-c)^2-a^2\}^{\frac{1}{2}};$$

$$2a\frac{dy}{dx} = (x-c)\{(x-c)^2-a^2\}^{\frac{1}{2}}-a^2\log(x-c+\{(x-c)^2-a^2\}^{\frac{1}{2}})+C_1,$$

whence y may be found by integration.

Ex. 4. $\dfrac{d^2y}{dx^2} = \dfrac{a}{x}$;

$$\frac{dy}{dx} = a\log\frac{x}{c}; \qquad y = a\{x\log\frac{x}{c}-x\}+C_1.$$

Ex. 5. $\dfrac{d^4y}{dx^4} - a^2\dfrac{d^2y}{dx^2} = 0.$

$$\therefore \quad \frac{d^3y}{dx^3}-a^2\frac{dy}{dx}-C_1 = 0;$$

$$\frac{d^2y}{dx^2}-a^2y-C_1x-C_2 = 0,$$

which is linear of the second order, and with constant coefficients.

Ex. 6. $\dfrac{d^2y}{dx^2} = a^2+k^2\left(\dfrac{dy}{dx}\right)^2.$

$$\therefore \quad \frac{\dfrac{d^2y}{dx^2}}{a^2+k^2\dfrac{dy^2}{dx^2}} = 1; \qquad \frac{d.\dfrac{dy}{dx}}{a^2+k^2\dfrac{dy^2}{dx^2}} = dx,$$

$$\therefore \quad \frac{dy}{dx} = \frac{a}{k}\tan ak(x-c); \qquad y-b = \frac{1}{k^2}\log\sec ak(x-c).$$

Ex. 7. $\dfrac{d^2y}{dx^2} = \dfrac{1}{(ay)^{\frac{1}{2}}}.$

Multiplying both sides by $2dy$, $\dfrac{2\,dy\,d^2y}{dx^2} = \dfrac{2\,dy}{(ay)^{\frac{1}{2}}}$;

$$\therefore \quad \left(\frac{dy}{dx}\right)^2 = 4\left(\frac{y}{a}\right)^{\frac{1}{2}}-c = \frac{4}{a^{\frac{1}{2}}}\{y^{\frac{1}{2}}-k^{\frac{1}{2}}\};$$

$$\therefore \quad \frac{dy}{(y^{\frac{1}{2}}-k^{\frac{1}{2}})^{\frac{1}{2}}} = \frac{2}{a^{\frac{1}{4}}}dx;$$

$$\therefore \quad \frac{2}{3}(y^{\frac{1}{2}}-k^{\frac{1}{2}})^{\frac{1}{2}}(y^{\frac{1}{2}}+2k^{\frac{1}{2}}) = \frac{x-b}{a^{\frac{1}{4}}}.$$

All equations of the form $\frac{d^2y}{dx^2} = f(y)$ may be integrated by this process; for multiplying both sides by $2\,dy$,

$$\frac{2\,dy\,d^2y}{dx^2} = 2\,dy\,f(y);$$

$$\therefore \quad \left(\frac{dy}{dx}\right)^2 = 2\int f(y)\,dy + \text{c};$$

and of this the root is to be extracted, and a subsequent integration is to be effected.

Ex. 8. Determine the curve whose curvature is constant.

Let the radius of curvature $= c$; so that

$$\pm \frac{\left(1+\frac{dy^2}{dx^2}\right)^{\frac{3}{2}}}{\frac{d^2y}{dx^2}} = c;$$

$$\therefore \quad \frac{d.\frac{dy^2}{dx^2}}{\left(1+\frac{dy^2}{dx^2}\right)^{\frac{3}{2}}} = \pm\frac{2\,dy}{c}; \qquad \frac{1}{\left(1+\frac{dy^2}{dx^2}\right)^{\frac{1}{2}}} = \pm\frac{y-b}{c};$$

$$\frac{(y-b)\,dy}{\{c^2-(y-b)^2\}^{\frac{1}{2}}} = \pm\,dx; \qquad \{c^2-(y-b)^2\}^{\frac{1}{2}} = \pm\,(x-a),$$

$$\therefore \quad (x-a)^2+(y-b)^2 = c^2.$$

463.] Let us also examine differential equations of the second order which involve $\frac{d^2y}{dx^2}$, $\frac{dy}{dx}$, and either x or y; and which are therefore of the form

$$\text{F}\left(\frac{d^2y}{dx^2}, \frac{dy}{dx}, x\right) = 0, \quad \text{or} \quad \text{F}\left(\frac{d^2y}{dx^2}, \frac{dy}{dx}, y\right) = 0.$$

Ex. 1. $$\frac{d^2y}{(dx^2+dy^2)^{\frac{1}{2}}} = a\frac{dx}{x};$$

$$\therefore \quad \log\left\{\frac{dy}{dx} + \left(1+\frac{dy^2}{dx^2}\right)^{\frac{1}{2}}\right\} = a\log\frac{x}{c};$$

$$\therefore \quad \frac{dy}{dx} + \left(1+\frac{dy^2}{dx^2}\right)^{\frac{1}{2}} = \left(\frac{x}{c}\right)^a;$$

$$\therefore \quad 2(y-b) = c^a\frac{x^{-(a-1)}}{a-1} + \frac{x^{a+1}}{c(a+1)}.$$

Ex. 2. $y\dfrac{d^2y}{dx^2}+\dfrac{dy^2}{dx^2}+1 = 0.$

$$\frac{d}{dx}\left(y\frac{dy}{dx}\right)+1 = 0;$$

$$y\frac{dy}{dx}+x-c = 0; \quad \therefore \quad y^2+(x-c)^2 = k^2,$$

where c and k are the arbitrary constants of integration.

Ex. 3. $a^2 d^2y(a^2+x^2)^{\frac{1}{2}}+a^2 dx\,dy = x^2 dx^2.$

$$\frac{d^2y}{dx^2}+\frac{dy}{dx}\frac{1}{(a^2+x^2)^{\frac{1}{2}}} = \frac{x^2}{a^2(a^2+x^2)^{\frac{1}{2}}};$$

which is a linear equation of the first order in terms of $\dfrac{dy}{dx}$, and therefore may be integrated.

Ex. 4. $(1+x^2)\dfrac{d^2y}{dx^2}+1+\dfrac{dy^2}{dx^2} = 0.$

$$\frac{\frac{d^2y}{dx^2}}{1+\frac{dy^2}{dx^2}}+\frac{1}{1+x^2} = 0. \qquad \therefore \quad \frac{d.\frac{dy}{dx}}{1+\frac{dy^2}{dx^2}}+\frac{dx}{1+x^2} = 0;$$

$$\tan^{-1}\frac{dy}{dx}+\tan^{-1}x = \tan^{-1}c;$$

$$\frac{dy}{dx} = \frac{c-x}{1+cx}.$$

whence y may be found.

Ex. 5. $dx^3dy-x\,ds^2d^2y = a\,dx\,ds\,\{(d^2x)^2+(d^2)\,^2\}^{\frac{1}{2}},$

where s is the equicrescent variable.

$$\because \quad ds^2 = dx^2+dy^2, \qquad 0 = dx\,d^2x+dy\,d^2y;$$

$$\therefore \quad dx^3dy-x\,ds^2d^2y = a\,ds^2d^2y;$$

$$\frac{dy}{dx} = (x+a)\frac{d^2y}{dx^2}\left(1+\frac{dy^2}{dx^2}\right);$$

$$\frac{dx}{x+a} = \frac{d.\frac{dy}{dx}}{\frac{dy}{dx}}+\frac{dy}{dx}d.\frac{dy}{dx};$$

$$\log(x+a) = \log\frac{dy}{dx}+\frac{1}{2}\left(\frac{dy}{dx}\right)^2+c;$$

but as $\dfrac{dy}{dx}$ is a transcendental function of x, the next integration cannot be effected.

Ex. 6. Determine the curve of which the radius of curvature is proportional to the normal.

$$\frac{\left(1+\frac{dy^2}{dx^2}\right)^{\frac{3}{2}}}{\frac{d^2y}{dx^2}} = ky\left(1+\frac{dy^2}{dx^2}\right)^{\frac{1}{2}}; \qquad \therefore \quad \frac{d.\frac{dy^2}{dx^2}}{1+\frac{dy^2}{dx^2}} = \frac{2dy}{ky};$$

$$1+\frac{dy^2}{dx^2} = \left(\frac{y}{c}\right)^{\frac{2}{k}};$$

where k may be either positive or negative;

$$dx = \left\{\left(\frac{y}{c}\right)^{\frac{2}{k}} - 1\right\}^{-\frac{1}{2}} dy.$$

(1) Let $k = 1$; that is, the radius of curvature is equal to the normal;

$$\therefore \quad \frac{dy}{(y^2-c^2)^{\frac{1}{2}}} = \pm\frac{dx}{c}; \qquad y = \frac{c}{2}\left\{e^{\frac{x}{c}} + e^{-\frac{x}{c}}\right\};$$

the equation of the catenary.

(2) Let $k = -1$;

$$\frac{y\,dy}{(c^2-y^2)^{\frac{1}{2}}} = \pm\, dx; \qquad (c^2-y^2)^{\frac{1}{2}} = \pm\,(x-a);$$

$$y^2+(x-a)^2 = c^2;$$

the equation of a circle, whose centre is on the axis of x.

(3) Let $k = 2$; that is, the radius of curvature is equal to twice the normal.

$$(x-a)^2 = 4c\,(y-c);$$

the equation of a parabola whose axis is the y-axis.

(4) Let $k = -2$;

$$dx = \frac{y\,dy}{(cy-y^2)^{\frac{1}{2}}};$$

$$\therefore \quad x = \frac{c}{2}\,\text{versin}^{-1}\frac{2y}{c} - (cy-y^2)^{\frac{1}{2}};$$

the equation of a cycloid, whose starting point is the origin, and whose base is the axis of x.

Ex. 7. Determine the curve whose radius of curvature varies inversely as the abscissa.

$$\frac{\left(1+\frac{dy^2}{dx^2}\right)^{\frac{3}{2}}}{\frac{d^2y}{dx^2}} = \pm\frac{k}{x}; \qquad \frac{d.\frac{dy}{dx}}{\left(1+\frac{dy^2}{dx^2}\right)^{\frac{3}{2}}} = \pm\frac{x\,dx}{k};$$

$$\frac{\frac{dy}{dx}}{\left(1+\frac{dy^2}{dx^2}\right)^{\frac{1}{2}}} = \pm \frac{x^2-a^2}{2k};$$

$$\therefore \quad dy = \pm \frac{(x^2-a^2)\,dx}{\{4k^2-(x^2-a^2)^2\}^{\frac{1}{2}}};$$

an equation which does not admit of further integration, but which represents the elastic curve. Also see Art. 329.

Ex. 8. Determine the equation of the curve of which the radius of curvature varies as the cube of the normal.

$$\frac{\left(1+\frac{dy^2}{dx^2}\right)^{\frac{3}{2}}}{\frac{d^2y}{dx^2}} = -\frac{y^3}{k^2}\left(1+\frac{dy^2}{dx^2}\right)^{\frac{3}{2}}; \qquad \frac{d^2y}{dx^2} = -\frac{k^2}{y^3};$$

$$\left(\frac{dy}{dx}\right)^2 = \frac{k^2}{y^2} - \frac{k^2}{a^2}; \qquad \therefore \quad \frac{y\,dy}{(a^2-y^2)^{\frac{1}{2}}} = \frac{k}{a}\,dx;$$

$$\therefore \quad (a^2-y^2)^{\frac{1}{2}} = \frac{k}{a}(x-c), \qquad y^2 = a^2 - \frac{k^2}{a^2}(x-c)^2,$$

$$a^2y^2 + k^2(x-c)^2 = a^4;$$

the equation of an ellipse. The curve is a hyperbola if k^2 is replaced by $-k^2$; and is a parabola if no constant is introduced at the first integration.

Ex. 9. Find the equation to the curve in which

$$\int_0^x \int_0^y dy\,dx = a\int_0^s ds,$$

where ds is the length-element of the curve.

Ex. 10. A form of differential equation which frequently occurs in subsequent investigations is

$$\frac{d^2y}{dx^2} \pm k^2 y = 0.$$

Let both terms be multiplied by dy; then integrating,

$$\frac{dy^2}{dx^2} \pm k^2(y^2-b^2) = 0;$$

and according as the upper or the lower sign is taken, we have

$$y = b\cos k(x-a),$$

$$y = \frac{b}{2}\{e^{k(x-a)} + e^{-k(x-a)}\}.$$

464.] Next, let us consider homogeneous equations of the second order: the principle of homogeneity being estimated in the following manner; the variables x, y, and their differentials dx, dy, d^2y are considered to be factors of the first degree; and each term of the equation is of the same degree in respect of them; thus the equation, $x^3d^2y-(y\,dx-x\,dy)^2 = 0$, is homogeneous and of the fourth degree. Now in such an equation let the following substitutions be made; viz.

$$y = xz; \qquad \therefore \quad dy = x\,dz+z\,dx; \tag{151}$$

also let
$$\frac{d^2y}{dx^2} = \frac{v}{x}; \tag{152}$$

and it is manifest that x will enter in the same power into all the terms, and therefore may be divided out; this property in fact is the characteristic of the equation; and thus the resulting equation will contain z, v and $\frac{dy}{dx}$; for convenience of notation, let $\frac{dy}{dx} = p$: so that from (151),

$$p\,dx = x\,dz+z\,dx; \qquad \therefore \quad \frac{dx}{x} = \frac{dz}{p-z};$$

and as
$$\frac{dp}{dx} = \frac{d^2y}{dx^2} = \frac{v}{x}; \qquad \therefore \quad v\,dx = x\,dp;$$

hence
$$\frac{dx}{x} = \frac{dz}{p-z} = \frac{dp}{v}; \tag{153}$$

and v may be expressed in terms of z and p by means of the given equation, and therefore by the last two members of the equality we shall have a differential equation of the first order in terms of p and z, whereby p may be expressed in terms of z: and therefore from the first two members of (153) we shall obtain a differential equation of the first order in terms of x and z; and this after resubstitution will give the required integral.

Ex. 1. $x^3\,d^2y = (y\,dx-x\,dy)^2$.

$$x^2v = x^2(z-p)^2; \qquad \therefore \quad v = (z-p)^2;$$

therefore from (153) $\frac{dz}{p-z} = \frac{dp}{(p-z)^2}$;

$$\therefore \quad p = z+1+\text{c}e^z; \qquad \therefore \quad \frac{dx}{x} = \frac{dz}{1+\text{c}e^z} = \frac{e^{-z}dz}{\text{c}+e^{-z}};$$

$$\log\frac{x}{\text{c}_1} = -\log(\text{c}+e^{-z}); \qquad \therefore \quad \frac{\text{c}_1}{x} = \text{c}+e^{-\frac{y}{x}};$$

and this is the required integral.

Also differential equations which become homogeneous, if we consider x to be of one dimension, y of n, $\frac{dy}{dx}$ of $n-1$, and $\frac{d^2y}{dx^2}$ of $n-2$ dimensions respectively, may be integrated by a similar process by assuming

$$y = zx^n, \qquad p = ux^{n-1}, \qquad \frac{d^2y}{dx^2} = vx^{n-2}.$$

It is to Euler that we are indebted for these processes; other examples will be found in his works, and in the ordinary collections of such problems; and particularly in the Integral Calculus of M. Moigno.

465.] The following differential equation is also capable of integration by either of the two following processes;

$$\frac{d^2y}{dx^2} + \mathrm{X}\frac{dy}{dx} + \mathrm{Y}\left(\frac{dy}{dx}\right)^2 = 0, \qquad (154)$$

where X and Y are functions respectively of x and y only: divide through by $\frac{dy}{dx^2}$, and integrate; then we have

$$\log \mathrm{c}\frac{dy}{dx} + \int \mathrm{X}\,dx + \int \mathrm{Y}\,dy = 0;$$

$$\therefore \quad \mathrm{c}e^{\int \mathrm{Y}\,dy}\,dy = e^{-\int \mathrm{X}\,dx}\,dx;$$

$$\therefore \quad \int \mathrm{c}e^{\int \mathrm{Y}\,dy}\,dy = \int e^{-\int \mathrm{X}\,dx}\,dx + \mathrm{C}_1. \qquad (155)$$

Also the integration may be effected by the variation of parameters. Omitting the last term, the equation is

$$\frac{d^2y}{dx^2} + \mathrm{X}\frac{dy}{dx} = 0;$$

$$\therefore \quad \frac{dy}{dx} = \mathrm{c}e^{-\int \mathrm{X}\,dx}; \qquad (156)$$

so that taking c to be a function of x,

$$\frac{d^2y}{dx^2} = e^{-\int \mathrm{X}\,dx}\left\{\frac{d\mathrm{c}}{dx} - \mathrm{c}\mathrm{X}\right\};$$

and substituting this in (154), we have

$$\frac{d\mathrm{c}}{\mathrm{c}} + \mathrm{Y}\,dy = 0;$$

$$\therefore \quad \mathrm{c} = \mathrm{A}e^{-\int \mathrm{Y}\,dy},$$

where A is an arbitrary constant; so that from (156),

$$\frac{dy}{dx} = \mathrm{A}e^{-\{\int \mathrm{Y}\,dy + \int \mathrm{X}\,dx\}};$$

$$\therefore \int e^{\int Y dy} dy = A \int e^{-\int X dx} dx + C_1; \qquad (157)$$

the form of which is the same as that of (155).

Ex. 1. $\dfrac{d^2y}{dx^2} + \dfrac{2}{x}\dfrac{dy}{dx} + \dfrac{1}{y}\left(\dfrac{dy}{dx}\right)^2 = 0.$

$$\therefore \; y^2 = a + \frac{b}{x}.$$

466.] One other property of differential equations of a higher order, to which allusion has already been made, deserves explanation in this place.

Let there be a differential equation of the form

$$f(x, y, y', y'', \dots y^{(n)}) = 0; \qquad (158)$$

and suppose its integral to be

$$y = F(x, C_1, C_2, \dots C_n); \qquad (159)$$

then y and its derived-functions depend not only on x, but also on the values of the n undetermined constants; but x may be considered independent of them. Let us suppose any one, say C, of these constants to vary; then the variation of (158) is

$$\left(\frac{df}{dy}\right)\frac{dy}{dC} + \left(\frac{df}{dy'}\right)\frac{dy'}{dC} + \dots + \left(\frac{df}{dy^{(n)}}\right)\frac{dy^{(n)}}{dC} = 0. \qquad (160)$$

Let $\dfrac{dy}{dC} = z$; then $\dfrac{dy'}{dC} = \dfrac{dz}{dx}$, $\dfrac{dy''}{dC} = \dfrac{d^2z}{dx^2}$, and so on; so that (160) becomes

$$z\left(\frac{df}{dy}\right) + \frac{dz}{dx}\left(\frac{df}{dy'}\right) + \frac{d^2z}{dx^2}\left(\frac{df}{dy''}\right) + \dots + \frac{d^nz}{dx^n}\left(\frac{df}{dy^{(n)}}\right) = 0; \quad (161)$$

now by reason of (159) $y, y', \dots y^{(n)}$ are functions of x and of $C_1, C_2, \dots C_n$: if then we substitute these in (158), $\left(\dfrac{df}{dy}\right), \left(\dfrac{df}{dy'}\right), \dots$ become functions of $x, C_1, C_2, \dots C_n$; and therefore the coefficients of z and of its derived-functions in (161) are variable, and the equation is linear; and we know that a particular integral of it is $z = \dfrac{dy}{dC}$, because the equation was found by making $\dfrac{dy}{dC} = z$: and as for the general value C we may substitute each of the C's, so the general integral of (161) is

$$z = C'\frac{dy}{dC_1} + C''\frac{dy}{dC_2} + \dots + C^{(n)}\frac{dy}{dC_n}. \qquad (162)$$

Let this process be compared with Art. 352.

Section 6.—*Integration of Partial Differential Equations of Higher Orders.*

467.] The integration of partial differential equations of the higher orders is surrounded with difficulties; and only some few cases have at present yielded to the powers of Analysis; of those, which are integrable, most arise in the more abstruse branches of Physical Mathematics, and therefore the discussion of them would be undertaken with inadequate means at this stage of our treatise: we shall therefore pass them by; and only introduce in the two following Articles Monge's method of integrating those of a simple class; and afterwards prove some properties of the most simple forms.

First let us consider Monge's method of solving linear partial differential equations of the second order, which are of the form

$$\text{R}\left(\frac{d^2z}{dx^2}\right)+\text{S}\left(\frac{d^2z}{dx\,dy}\right)+\text{T}\left(\frac{d^2z}{dy^2}\right)=\text{V}; \tag{163}$$

where R, S, T, V are functions of x, y, z, and the partial derived-functions $\left(\frac{dz}{dx}\right)$ and $\left(\frac{dz}{dy}\right)$; and let us employ the symbols given in Art. 418, Vol. I; so that (163) is

$$\text{R}r+\text{S}s+\text{T}t=\text{V}; \tag{164}$$

where R, S, T, V are functions of x, y, z, p, q. From the symbols we have

$$dp = r\,dx+s\,dy,$$
$$dq = s\,dx+t\,dy:$$

and by means of these eliminating r and t from (164), we have

$$\text{R}\,dp\,dy-\text{V}\,dy\,dx+\text{T}\,dq\,dx-s\,\{\text{R}\,dy^2-\text{S}\,dy\,dx+\text{T}\,dx^2\}=0. \tag{165}$$

Now suppose that

$$\text{R}\,dp\,dy-\text{V}\,dy\,dx+\text{T}\,dq\,dx = 0; \tag{166}$$

$$\text{R}\,dy^2-\text{S}\,dy\,dx+\text{T}\,dx^2 = 0; \tag{167}$$

then as we have also

$$dz = p\,dx+q\,dy, \tag{168}$$

we may suppose that it is possible to satisfy these last three equations by equations of the form

$$\left.\begin{array}{l} f_1(x,y,z,p,q) = \text{C}_1, \\ f_2(x,y,z,p,q) = \text{C}_2; \end{array}\right\} \tag{169}$$

and assuming this to be so, then if F is the symbol of an undetermined function,

$$f_1 = \text{F}(f_2) \tag{170}$$

will be the general first integral of the proposed equation.

To prove this statement; let $\frac{dy}{dx} = y'$; and let y' be the general symbol of the roots of (167), so that (166) becomes

$$\text{R}\,dp\,y' - \text{V}y'\,dx + \text{T}\,dq = 0. \qquad (171)$$

Now taking the total differential of the first of (169), we have

$$\left(\frac{df_1}{dx}\right)dx + \left(\frac{df_1}{dy}\right)dy + \left(\frac{df_1}{dz}\right)dz + \left(\frac{df_1}{dp}\right)dp + \left(\frac{df_1}{dq}\right)dq = 0; \qquad (172)$$

and substituting for dy, and for dz and dq from (168) and (171), we have

$$\left\{\left(\frac{df_1}{dx}\right) + \left(\frac{df_1}{dy}\right)y' + \left(\frac{df_1}{dz}\right)(p+qy') + \left(\frac{df_1}{dq}\right)\frac{\text{V}}{\text{T}}y'\right\}dx$$
$$+\left\{\left(\frac{df_1}{dp}\right) - \frac{\text{R}}{\text{T}}y'\left(\frac{df_1}{dq}\right)\right\}dp = 0; \qquad (173)$$

and this equation must be identical, because it satisfies each of the three equations (166), (167), and (168); therefore

$$\left.\begin{aligned}\left(\frac{df_1}{dx}\right) + \left(\frac{df_1}{dy}\right)y' + \left(\frac{df_1}{dz}\right)(p+qy') + \left(\frac{df_1}{dq}\right)\frac{\text{V}y'}{\text{T}} &= 0;\\ \left(\frac{df_1}{dp}\right) - \left(\frac{df_1}{dq}\right)\frac{\text{R}y'}{\text{T}} &= 0;\end{aligned}\right\} \qquad (174)$$

and we have similar equations in terms of f_2. Also from (170),

$$\left(\frac{df_1}{dx}\right)dx + \left(\frac{df_1}{dy}\right)dy + \ldots + \left(\frac{df_1}{dq}\right)dq$$
$$= \text{F}'(f_2)\left\{\left(\frac{df_2}{dx}\right)dx + \left(\frac{df_2}{dy}\right)dy + \ldots + \left(\frac{df_2}{dq}\right)dq\right\}; \qquad (175)$$

and replacing dz by its value (168), and $\left(\frac{df_1}{dp}\right)$ and $\left(\frac{df_1}{dx}\right)$ by their values from (174); and similarly replacing $\left(\frac{df_2}{dp}\right)$ and $\left(\frac{df_2}{dx}\right)$, (175) becomes

$$\{\text{R}y'dp + \text{T}dq - \text{V}y'dx\}\left\{\left(\frac{df_1}{dq}\right) - \text{F}'\left(\frac{df_2}{dq}\right)\right\}$$
$$+(dy - y'dx)\left\{\left(\frac{df_1}{dy}\right) - \text{F}'\left(\frac{df_2}{dy}\right) + q\left(\frac{df_1}{dz}\right) - \text{F}'q\left(\frac{df_2}{dz}\right)\right\} = 0; \qquad (176)$$

which we may conveniently express

$$\text{R}y'dp + \text{T}dq - \text{V}y'dx = \Upsilon(dy - y'dx); \qquad (177)$$

and replacing dp and dq by their values,

$$(\text{R}y'r + \text{T}s - \text{V}y' + \Upsilon y')dx + (\text{R}y's + \text{T}t - \Upsilon)dy = 0; \qquad (178)$$

and as x and y are independent variables, this equation must be identical;

$$\therefore \quad \left.\begin{aligned} \text{R}y'r+\text{T}s-\text{V}y'+\text{T}y' &= 0; \\ \text{R}y's+\text{T}t-\text{T} &= 0; \end{aligned}\right\} \qquad (179)$$

and therefore eliminating r and t, and replacing in the proposed equation (164), we have

$$s\{\text{R}y'^2-\text{S}y'+\text{T}\} = 0, \qquad (180)$$

and this equation is satisfied because y' is a root of (167): T therefore has disappeared, and as that alone in (178) and (179) involves F, the result is true whatever is the form of F: and therefore (170) involves an arbitrary functional symbol and is a general first integral of (163). Let us consider the above process, when it is applied to the solution of some examples.

468.] Ex. 1. Let R, S, T be constant, and $\text{V} = 0$; and suppose the equation to be

$$\left(\frac{d^2 z}{dx^2}\right) + 5a\left(\frac{d^2 z}{dx\,dy}\right) + 6a^2\left(\frac{d^2 z}{dy^2}\right) = 0.$$

In this case (166) and (167) become

$$dp\,dy + 6a^2 dq\,dx = 0,$$

$$dy^2 - 5a^2 dy\,dx + 6a^2 dx^2 = 0;$$

$$\therefore \quad \frac{dy}{dx} = 2a; \qquad \frac{dy}{dx} = 3a;$$

from the former of which, $y = 2ax + \text{C}_1$;

$$\text{also} \qquad dp + 3a\,dq = 0;$$

$$p + 3aq = \text{C}_2 = f_1(\text{C}_1)$$

$$= f_1(y-2ax);$$

$$\text{similarly} \qquad p + 2aq = f_2(y-3ax):$$

$$\therefore \quad p = 3f_2(y-3ax) - 2f_1(y-2ax);$$

$$\therefore \quad z = \phi_1(y-2ax) + \phi_2(y-3ax);$$

and this is the general integral of the given equation.

Ex. 2. $\left(\frac{d^2 z}{dx^2}\right) - a^2\left(\frac{d^2 z}{dy^2}\right) = 0.$ Here (166) and (167) become

$$dp\,dy - a^2 dq\,dx = 0; \qquad dy^2 - a^2 dx^2 = 0;$$

$$\frac{dy}{dx} = a; \qquad \frac{dy}{dx} = -a;$$

$$y = ax + c_1; \qquad y = -ax + c_2;$$
$$p - aq = c_1'; \qquad p + aq = c_2';$$
$$\therefore\; p - aq = f_1(y - ax); \qquad p + aq = f_2(y + ax);$$
$$\therefore\; 2p = f_1(y - ax) + f_2(y + ax);$$
$$z = \phi_1(y - ax) + \phi_2(y + ax),$$

and this is the complete integral.

Ex. 3. $q^2 r - 2pqs + p^2 t = 0$. Here (166) and (167) become

$$q^2\,dp\,dy + p^2\,dq\,dx = 0; \qquad q^2\,dy^2 + 2pq\,dx\,dy + p^2\,dx^2 = 0;$$
$$\therefore\; q\,dy + p\,dx = 0; \qquad -pq\,dp + p^2\,dq = 0;$$
$$p = 0; \qquad \frac{dp}{p} - \frac{dq}{q} = 0;$$
$$z = c_1; \qquad \frac{p}{q} = c = \phi(c_1) = \phi(z);$$
$$\therefore\; z = f(y + cx) = f\{y + x\phi(z)\}.$$

This problem is the converse of that discussed in Art. 368, Vol. I.

469.] We can also find the integral of the following linear partial differential equation of the nth order

$$A\left(\frac{d^n z}{dx^n}\right) + B\left(\frac{d^n z}{dx^{n-1}\,dy}\right) + \ldots + K\left(\frac{d^n z}{dx\,dy^{n-1}}\right) + L\left(\frac{d^n z}{dy^n}\right) = 0; \quad (181)$$

where A, B, … K, L are constant. Let $z = f(y + mx)$;

$$\therefore\; \left(\frac{d^n z}{dx^n}\right) = m^n f^n(y + mx);$$
$$\left(\frac{d^n z}{dx^{n-1}\,dy}\right) = m^{n-1} f^n(y + mx);$$
$$\cdot \quad \cdot \quad \cdot \quad \cdot \quad \cdot \quad \cdot \quad \cdot \quad \cdot \quad \cdot \quad \cdot \quad \cdot \quad \cdot$$
$$\left(\frac{d^n z}{dy^n}\right) = f^n(y + mx);$$

then (181) becomes

$$A m^n + B m^{n-1} + \ldots + K m + L = 0. \quad (182)$$

Let the n roots of this equation be $m_1, m_2, \ldots m_n$, and be unequal; then

$$z = f_1(y + m_1 x) + f_2(y + m_2 x) + \ldots + f_n(y + m_n x), \quad (183)$$

where $f_1, f_2, \ldots f_n$ express arbitrary functions.

If two roots of (182) are equal, say $m_2 = m_1$, then

$$z = f_1(y+m_1 x) + x f_2(y+m_1 x) + f_3(y+m_3 x) + \ldots$$
$$\ldots + f_n(y+m_n x); \quad (184)$$

and the result is analogous, if three or more roots are equal.

Ex. 1. $b^2 r - 2ab\,s + a^2 t = 0.$

$$\therefore \quad b^2 m^2 - 2ab\,m + a^2 = 0; \qquad m = \frac{a}{b};$$

$$\therefore \quad z = f_1(by+ax) + x f_2(by+ax).$$

470.] The following are geometrical problems of considerable interest, the solution of which depends on partial differential equations.

Ex. 1. To find the equation of a surface, every point of which is an umbilic.

By (95), Art. 413, Vol. I, the condition is

$$\frac{r}{1+p^2} = \frac{s}{pq} = \frac{t}{1+q^2};$$

$$\therefore \quad \frac{p}{1+p^2}\frac{dp}{dx} = \frac{1}{q}\frac{dq}{dx}; \qquad \frac{q}{1+q^2}\frac{dq}{dy} = \frac{1}{p}\frac{dp}{dy};$$

$$\therefore \quad 1+p^2 = \text{Y} q^2; \qquad 1+q^2 = \text{X} p^2;$$

where Y and X are undetermined functions of y and x respectively, introduced in the x- and y-partial integrations; hence we have

$$p = \left(\frac{1+\text{Y}}{\text{XY}-1}\right)^{\frac{1}{2}}; \qquad q = \left(\frac{1+\text{X}}{\text{XY}-1}\right)^{\frac{1}{2}};$$

$$\text{but since} \quad \left(\frac{dp}{dy}\right) = \left(\frac{dq}{dx}\right),$$

$$\text{we have} \qquad (1+\text{X})^{-\frac{3}{2}}\frac{d\text{X}}{dx} = (1+\text{Y})^{-\frac{3}{2}}\frac{d\text{Y}}{dy};$$

now this equation shews that X and Y are of the same form, and as there is no relation between them this identity can subsist only when each side is a constant: let therefore

$$(1+\text{X})^{-\frac{3}{2}}\frac{d\text{X}}{dx} = -\frac{2}{k} = (1+\text{Y})^{-\frac{3}{2}}\frac{d\text{Y}}{dy};$$

$$\therefore \quad \frac{k}{(1+\text{X})^{\frac{1}{2}}} = x-a; \qquad \frac{k}{(1+\text{Y})^{\frac{1}{2}}} = y-b;$$

whence

$$p = \left(\frac{dz}{dx}\right) = \frac{x-a}{\{k^2-(x-a)^2-(y-b)^2\}^{\frac{1}{2}}};$$

$$q = \left(\frac{dz}{dy}\right) = \frac{y-b}{\{k^2-(x-a)^2-(y-b)^2\}^{\frac{1}{2}}};$$

$$z-c = -\{k^2-(x-a)^2-(y-b)^2\}^{\frac{1}{2}};$$

$$\therefore \quad (x-a)^2+(y-b)^2+(z-c)^2 = k^2,$$

which is the equation of a sphere. Whence we conclude that a sphere is a surface, every point of which is an umbilic.

Ex. 2. To determine the surface of revolution at every point of which the principal radii of curvature are equal and of opposite signs.

The differential equation which expresses the stated property is, see (94), Art. 413, Vol. I,

$$(1+q^2)\,r-2pqs+(1+p^2)\,t = 0. \qquad (185)$$

Let the surface required have the axis of z for its axis of revolution, so that its equation is,

$$z = f(x^2+y^2),$$

where f expresses the arbitrary function which is to be determined;

$$\therefore \quad p = 2xf'(x^2+y^2); \qquad q = 2yf'(x^2+y^2),$$

$$r = 4x^2f''+2f'; \qquad s = 4xyf''; \qquad t = 4y^2f''+2f';$$

and (185) becomes

$$(x^2+y^2)f''+2(x^2+y^2)f'^3+f' = 0;$$

let $x^2+y^2 = \zeta$; $\quad \therefore \; z = f(\zeta)$; and we have

$$\zeta\frac{d^2z}{d\zeta^2}+2\zeta\left(\frac{dz}{d\zeta}\right)^3+\frac{dz}{d\zeta} = 0;$$

and making z to be equicrescent instead of ζ,

$$\frac{1}{\zeta}\frac{d^2\zeta}{dz^2}-\frac{1}{\zeta^2}\left(\frac{d\zeta}{dz}\right)^2 = \frac{2}{\zeta};$$

$$\text{that is,} \qquad \frac{d}{dz}\left(\frac{1}{\zeta}\frac{d\zeta}{dz}\right) = \frac{2}{\zeta};$$

$$\therefore \quad \left(\frac{1}{\zeta}\frac{d\zeta}{dz}\right)^2 = 4\frac{(\zeta-c^2)}{c^2\zeta}; \qquad \frac{d\zeta}{(\zeta^2-c^2\zeta)^{\frac{1}{2}}} = \frac{2}{c}dz;$$

$$\therefore \quad \zeta-\frac{c^2}{2} = \frac{c^2}{4}\{e^{\frac{2z}{c}}+e^{-\frac{2z}{c}}\};$$

the constant being determined so that $\zeta = c^2$, when $z = 0$:

$$\therefore \quad \zeta = \frac{c^2}{4}\{e^{\frac{2z}{c}}+2+e^{-\frac{2z}{c}}\};$$

$$\therefore \quad (x^2+y^2)^{\frac{1}{2}} = \frac{c}{2}\{e^{\frac{z}{c}} + e^{-\frac{z}{c}}\};$$

which is the equation of the surface required: and the equation to the generating curve is

$$x = \frac{c}{2}\{e^{\frac{z}{c}} + e^{-\frac{z}{c}}\}.$$

which is that of the catenary, the axis of revolution being the directrix.

CHAPTER XVII.

THE SOLUTION OF DIFFERENTIAL EQUATIONS BY THE CALCULUS OF OPERATIONS.

SECTION 1.—*The Solution of Total Differential Equations by Symbolical Methods.*

471.] I propose to apply in the present Chapter certain theorems in the Calculus of Operations to the solution of differential equations. These theorems have been demonstrated in Chapter XIX, Vol. I, and I refer to that chapter for the proof of them; as also for their character in that they are subject to the commutative, the distributive, and the index laws. This last characteristic is most important in the present applications, because it gives to the symbols that interrogative quality, as it is called by the late Professor Boole, whence their power is derived for these enquiries.

Let us first take the case of a linear differential equation, which has constant coefficients, and is of the form (79), Art. 450. Now we may separate the subject-function from the constants and the symbols of operation, inasmuch as these are subject to the distributive law; thus we have

$$\left\{\frac{d^n}{dx^n} + A_1 \frac{d^{n-1}}{dx^{n-1}} + A_2 \frac{d^{n-2}}{dx^{n-2}} + \ldots + A_{n-1}\frac{d}{dx} + A_n\right\} y = X. \quad (1)$$

The function of the operating symbol in the left-hand member is evidently of the form of an algebraical expression of the nth degree; and consequently, as such symbols are in accordance with the laws of algebraical multiplication, it may be resolved into n simple factors, the roots corresponding to which may be real or imaginary, and may be all unequal, or two or more may be equal. Let us in the first place assume all the roots, whether real or imaginary, to be unequal; so that the function of the operating symbol may be expressed in the form

$$\left(\frac{d}{dx} - a_1\right)\left(\frac{d}{dx} - a_2\right) \ldots\ldots \left(\frac{d}{dx} - a_n\right), \quad (2)$$

and consequently

$$\left(\frac{d}{dx}-a_1\right)\left(\frac{d}{dx}-a_2\right)\dots\left(\frac{d}{dx}-a_n\right)y = \mathrm{X}. \qquad (3)$$

Now the meaning of this equation is evidently as follows; the several factors, which enter into the left-hand member of the equation, symbolise certain operations; and the order in which these factors are placed shews that the operations are to be performed in order, and one on the back of the other, the subject-variable or the subject-function on which they are to be performed being placed last; and the whole or resultant effect is, that they change that function into X. This is the direct process, and it yields the direct result. Our present problem however is the inverse, and is this; What is that function which is the subject of the operations symbolized by these factors, so that when thus operated upon it becomes X? Evidently to obtain this information it is necessary to make both members of the equation subjects of a series of operations inverse of the former direct operations, so that those in the left-hand member may be neutralized. As these symbols obey the index law, the operation inverse to, say, $\frac{d}{dx}-a_j$, is $\left(\frac{d}{dx}-a_j\right)^{-1}$; and n required operations of this kind having been performed on the left-hand member of (3), it will become y; and as the same operations are to be performed on X, we have

$$y = \left(\frac{d}{dx}-a_n\right)^{-1}\left(\frac{d}{dx}-a_{n-1}\right)^{-1}\dots\left(\frac{d}{dx}-a_1\right)^{-1}\mathrm{X}. \qquad (4)$$

As the operations denoted by the factors in the left-hand member of (3) are subject to the commutative law, the result is the same, whatever is the order in which the operations are performed. I have retained in (4) the order in which the operations are arranged in (3), but the factors may be arranged in any order.

These inverse symbols are manifestly of an interrogative character. The question is, What is $\left(\frac{d}{dx}-a_j\right)^{-1}\mathrm{X}$? This question is answered in (53), Art. 425, Vol. I, and also in (20), Art. 153, of the present volume;

$$\left(\frac{d}{dx}-a\right)^{-1}\mathrm{X} = e^{ax}\int \mathrm{X}\,e^{-ax}\,dx; \qquad (5)$$

and this theorem completes the solution of the problem: for the right-hand member of (4) requires only a series of these operations, which are identical in form of result, and are to be effected consecutively one on the back of the other. Thus since

$$\left(\frac{d}{dx}-a_1\right)^{-1}\mathrm{X} = e^{a_1x}\int \mathrm{X}e^{-a_1x}\,dx,$$

therefore, taking an arrangement of symbols analogous to that of (9), Art. 208,

$$\left(\frac{d}{dx}-a_2\right)^{-1}\left(\frac{d}{dx}-a_1\right)^{-1}\mathrm{X} = e^{a_2x}\int e^{(a_1-a_2)x}\,dx\int \mathrm{X}\,e^{-a_1x}\,dx; \quad (6)$$

$$\left(\frac{d}{dx}-a_3\right)^{-1}\left(\frac{d}{dx}-a_2\right)^{-1}\left(\frac{d}{dx}-a_1\right)^{-1}\mathrm{X}$$

$$= e^{a_3x}\int e^{(a_2-a_3)x}\,dx\int e^{(a_1-a_2)x}\,dx\int \mathrm{X}e^{-a_1x}\,dx; \quad (7)$$

and so on, until finally

$$y = e^{a_nx}\int e^{(a_{n-1}-a_n)x}\,dx\int e^{(a_{n-2}-a_{n-1})x}\,dx\,\ldots\ldots\int \mathrm{X}e^{-a_1x}\,dx; \quad (8)$$

which gives y, the required function, in terms of x; and thus the problem is solved.

As each integration introduces an arbitrary constant, n constants will be introduced in the processes of the n integrations: and these will evidently in the answer take the form

$$\mathrm{C}_1\,e^{a_1x}+\mathrm{C}_2\,e^{a_2x}+\ldots+\mathrm{C}_n\,e^{a_nx}. \quad (9)$$

This process gives the result already obtained in (120), Art. 448; and the shortness of it in comparison of that at once indicates its advantage in the solution of such equations.

The general value of y given in (8) holds good, whether the a's are real or imaginary. I may remark however that if some are imaginary, the imaginary roots enter in pairs, and that the corresponding part of the value of y may be conveniently expressed in terms of circular functions.

If two or more of the factors in (3) and consequently in (4) are equal, say $a_1 = a_2 = a_3$, then by (36), Art. 422, Vol. I, and by (19), Art. 153 of the present volume,

$$\left(\frac{d}{dx}-a\right)^{-3}\mathrm{X} = e^{ax}\int^3 \mathrm{X}\,e^{-ax}\,dx^3; \quad (10)$$

and the constants introduced in integration will enter in the form

$$e^{ax}\,(c_1+c_2\,x+c_3\,x^2). \quad (11)$$

The following are examples in which the process is applied.

Ex. 1. $\frac{dy}{dx} - ay = \text{X}.$

$$\therefore \quad y = \left(\frac{d}{dx} - a\right)^{-1} \text{X} = e^{ax}\left\{\int \text{X} e^{-ax} dx + \text{C}\right\}.$$

Let $\text{X} = x^n$;

$$\therefore \quad y = \text{C}e^{ax} + e^{ax}\int e^{-ax} x^n dx$$

$$= \text{C}e^{ax} - \left\{\frac{x^n}{a} + \frac{n}{a^2} x^{n-1} + \ldots + \frac{n(n-1)\ldots 3.2.1}{a^{n+1}}\right\}.$$

Let $\text{X} = e^{mx}$; $\quad \therefore \quad y = \text{C}e^{ax} + \frac{e^{mx}}{m-a}.$

Let $\text{X} = e^{ax}$; $\quad y = \text{C}e^{ax} + xe^{ax}.$

Let $\text{X} = e^{mx} \cos nx = \frac{e^{(m+n\sqrt{-1})x} + e^{(m-n\sqrt{-1})x}}{2}$;

$$\therefore \quad y = e^{mx}\frac{(m-a)\cos nx + n \sin nx}{(m-a)^2 + n^2} + \text{C}e^{ax}.$$

Ex. 2. $\frac{d^3y}{dx^3} - 6a\frac{d^2y}{dx^2} + 11a^2\frac{dy}{dx} - 6a^3 = e^{mx}.$

$$\left(\frac{d}{dx} - a\right)\left(\frac{d}{dx} - 2a\right)\left(\frac{d}{dx} - 3a\right)y = e^{mx};$$

$$\therefore \quad \left(\frac{d}{dx} - 2a\right)\left(\frac{d}{dx} - 3a\right)y = e^{ax}\int e^{(m-a)x} dx$$

$$= \frac{e^{mx}}{m-a} + c_1 e^{ax}.$$

$$\therefore \quad \left(\frac{d}{dx} - 3a\right)y = e^{2ax}\int e^{-2ax}\left\{\frac{e^{mx}}{m-a} + c_1 e^{ax}\right\}dx$$

$$= \frac{e^{mx}}{(m-2a)(m-a)} - \frac{c_1}{a} e^{ax} + c_2 e^{2ax};$$

$$\therefore \quad y = \frac{e^{mx}}{(m-3a)(m-2a)(m-a)} + \text{C}_1 e^{ax} + \text{C}_2 e^{2ax} + \text{C}_3 e^{3ax}.$$

Ex. 3. $\frac{d^2y}{dx^2} - 4\frac{dy}{dx} + 4y = x^2$;

$$\left(\frac{d}{dx} - 2\right)^2 y = x^2;$$

$$\therefore \quad y = \left(\frac{d}{dx} - 2\right)^{-2} x^2$$

$$y = e^{2x}\int^2 e^{-2x}x^2 dx^2$$

$$= e^{2x}\left\{e^{-2x}\left(\frac{x^2}{2^2}+\frac{4x}{2^3}+\frac{6}{2^4}\right)+c_1 x+c_2\right\}$$

$$= \frac{x^2}{2^2}+\frac{4x}{2^3}+\frac{6}{2^4}+e^{2x}\{c_1 x+c_2\}.$$

Ex. 4. $\dfrac{d^2y}{dx^2}+a^2y = \cos ax.$

$$\therefore\ y = \left(\frac{d}{dx}+a\sqrt{-1}\right)^{-1}\left(\frac{d}{dx}-a\sqrt{-1}\right)^{-1}\frac{e^{ax\sqrt{-1}}+e^{-ax\sqrt{-1}}}{2}$$

$$= \left(\frac{d}{dx}+a\sqrt{-1}\right)^{-1} e^{ax\sqrt{-1}}\int\frac{1+e^{-2ax\sqrt{-1}}}{2}dx$$

$$= \left(\frac{d}{dx}+a\sqrt{-1}\right)^{-1}\left\{\frac{(x+c)\,e^{ax\sqrt{-1}}}{2}-\frac{e^{-ax\sqrt{-1}}}{4a}\right\}$$

$$= e^{-ax\sqrt{-1}}\int\left\{\frac{(x+c)\,e^{2ax\sqrt{-1}}}{2}-\frac{1}{4a}\right\}dx$$

$$= \frac{x\sin ax}{2a}+A\cos(ax-\alpha);$$

where A and α are arbitrary constants.

472.] The integral value of the right-hand member of (4) may also be determined in the following manner. The function in terms of the operating symbols is evidently a rational fraction in powers of $\frac{d}{dx}$; it may consequently be decomposed into a series of simple fractions by the process explained in Chap. II, Section 2, of the present volume.

Firstly, let us suppose all the roots to be unequal; then by (27), Art. (19), if

$$\left(\frac{d}{dx}-a_1\right)^{-1}\left(\frac{d}{dx}-a_2\right)^{-1}\cdots\left(\frac{d}{dx}-a_n\right)^{-1} = f\left(\frac{d}{dx}\right), \qquad (12)$$

$$\frac{1}{f\left(\frac{d}{dx}\right)} = \frac{1}{f'(a_1)}\left(\frac{d}{dx}-a_1\right)^{-1}+\frac{1}{f'(a_2)}\left(\frac{d}{dx}-a_2\right)^{-1}+\cdots$$
$$\cdots+\frac{1}{f'(a_n)}\left(\frac{d}{dx}-a_n\right)^{-1}.$$

Therefore, introducing the subject of the operating symbols,

$$\left\{f\left(\frac{d}{dx}\right)\right\}^{-1}X = \frac{1}{f'(a_1)}\left(\frac{d}{dx}-a_1\right)^{-1}X+\frac{1}{f'(a_2)}\left(\frac{d}{dx}-a_2\right)^{-1}X+\cdots$$
$$\cdots+\frac{1}{f'(a_n)}\left(\frac{d}{dx}-a_n\right)^{-1}X.$$

And consequently by reason of the theorem cited in (5),

$$y = \frac{1}{f'(a_1)} e^{a_1 x} \int e^{-a_1 x} \mathrm{X}\, dx + \frac{1}{f'(a_2)} e^{a_2 x} \int e^{-a_2 x} \mathrm{X}\, dx + \dots$$

$$\dots + \frac{1}{f'(a_n)} e^{a_n x} \int e^{-a_n x} \mathrm{X}\, dx; \qquad (13)$$

an expression which involves n signs of integration, and consequently n arbitrary constants; and, if these are introduced, the result becomes

$$y = \mathrm{C}_1 e^{a_1 x} + \mathrm{C}_2 e^{a_2 x} + \dots + \mathrm{C}_n e^{a_n x}$$

$$+ \frac{1}{f'(a_1)} e^{a_1 x} \int e^{-a_1 x} \mathrm{X}\, dx + \frac{1}{f'(a_2)} e^{a_2 x} \int e^{-a_2 x} \mathrm{X}\, dx + \dots$$

$$\dots + \frac{1}{f'(a_n)} e^{a_n x} \int e^{-a_n x} \mathrm{X}\, dx; \qquad (14)$$

which result is identical with (106), Art. 451.

If there is one pair, or are many pairs, of imaginary roots, we may transform the expression by the process of Article 452. Thus if a_i and a_j are a pair of conjugate imaginary roots,

$$a_i = a + b\sqrt{-1}, \qquad a = a - b\sqrt{-1};$$

then $$\mathrm{C}_i e^{a_i x} + \mathrm{C}_j e^{a_j x} = e^{ax}\{\mathrm{C}_i e^{b\sqrt{-1}x} + \mathrm{C}_j e^{-b\sqrt{-1}x}\}$$

$$= e^{ax} k \cos(bx + \gamma); \qquad (15)$$

where k and γ are two new undetermined constants; and if

$$f'(a_i) = \mathrm{M} + \mathrm{N}\sqrt{-1}, \qquad f'(a_j) = \mathrm{M} - \mathrm{N}\sqrt{-1};$$

then $$\frac{e^{a_i x}}{f'(a_i)} \int e^{-a_i x} \mathrm{X}\, dx + \frac{e^{a_j x}}{f'(a_j)} \int e^{-a_j x} \mathrm{X}\, dx$$

$$= \frac{2e^{ax}(\mathrm{L}\cos bx + \mathrm{M}\sin bx)}{\mathrm{L}^2 + \mathrm{M}^2} \int e^{-ax} \cos bx\, \mathrm{X}\, dx$$

$$+ \frac{2e^{ax}(\mathrm{L}\sin bx - \mathrm{M}\cos bx)}{\mathrm{L}^2 + \mathrm{M}^2} \int e^{-ax} \sin bx\, \mathrm{X}\, dx;$$

and this again may be further simplified if L and M are replaced by $r\cos\theta$ and $r\sin\theta$ respectively.

If however m roots of $f\left(\frac{d}{dx}\right)$ are equal to each other; that is, if $a_1 = a_2 = \dots = a_m$; then, according to Art. 21, if $\psi(x)$ is equal to the reciprocal of $\phi(x)$, where $\phi(x)$ is the product of all the factors of $f\left(\frac{d}{dx}\right)$ short of the equal factors,

$$\left\{f\left(\frac{d}{dx}\right)\right\}^{-1} = \psi(a_1)\left(\frac{d}{dx}-a_1\right)^{-m} + \frac{\psi'(a_1)}{1}\left(\frac{d}{dx}-a_1\right)^{-m+1}$$

$$\ldots + \frac{\psi^{(m-1)}(a_1)}{1.2.3\ldots(m-1)}\left(\frac{d}{dx}-a_1\right)^{-1} + \frac{1}{f'(a_{m+1})}\left(\frac{d}{dx}-a_{m+1}\right)^{-1} + \ldots$$

$$\ldots + \frac{1}{f'(a_n)}\left(\frac{d}{dx}-a_n\right)^{-1};$$

to all these terms let the subject x be affixed; then since

$$\left(\frac{d}{dx}-a\right)^{-r} \mathrm{x} = e^{ax}\int^r e^{-ax}\mathrm{x}\,dx^r,$$

$$y = \left\{f\left(\frac{d}{dx}\right)\right\}^{-1}\mathrm{x}$$

$$= \psi(a_1)e^{a_1x}\int^m e^{-a_1x}\mathrm{x}\,dx^m + \frac{\psi'(a_1)}{1}e^{a_1x}\int^{m-1} e^{-a_1x}\mathrm{x}\,dx^{m-1} + \ldots$$

$$+ \frac{\psi^{(m-1)}(a_1)}{1.2.3\ldots(m-1)}e^{a_1x}\int e^{-a_1x}\mathrm{x}\,dx$$

$$+ \frac{e^{a_{m+1}x}}{f'(a_{m+1})}\int e^{-a_{m+1}x}\mathrm{x}\,dx + \ldots + \frac{e^{a_nx}}{f'(a_n)}\int e^{-a_nx}\mathrm{x}\,dx; \qquad (16)$$

and as the constants introduced by integration are arbitrary, in the first m terms, m and only m constants will be brought in; and the remaining $n-m$ constants will arise in the other integrations. If the roots corresponding to the sets of equal factors are imaginary, the process of integration is the same; the result however is so complicated that it is not worth while to express it at length.

Ex. 1. $\frac{d^3y}{dx^3} - 6a\frac{d^2y}{dx^2} + 11a^2\frac{dy}{dx} - 6a^3y = x^2.$

$$\therefore \left(\frac{d}{dx}-a\right)\left(\frac{d}{dx}-2a\right)\left(\frac{d}{dx}-3a\right)y = x^2;$$

$$y = \frac{1}{2a^2}\left(\frac{d}{dx}-a\right)^{-1}x^2 - \frac{1}{a^2}\left(\frac{d}{dx}-2a\right)^{-1}x^2 + \frac{1}{2a^2}\left(\frac{d}{dx}-3a\right)^{-1}x^2$$

$$y = \frac{e^{ax}}{2a^2}\int e^{-ax}x^2dx - \frac{e^{2ax}}{a^2}\int e^{-2ax}x^2dx + \frac{e^{3ax}}{2a^2}\int e^{-3ax}x^2dx;$$

whereby the result depends on simple integrations.

Ex. 2. $\frac{d^2y}{dx^2} - 4a\frac{dy}{dx} + 4a^2y = \sin nx.$

$$\therefore \left(\frac{d}{dx}-2a\right)^2 y = \sin nx;$$

$$y = \left(\frac{d}{dx} - 2a\right)^{-2} \sin nx = e^{2ax} \int^2 e^{-2ax} \frac{e^{nx\sqrt{-1}} - e^{-nx\sqrt{-1}}}{2\sqrt{-1}} dx^2,$$

$$= \frac{e^{2ax}}{2\sqrt{-1}} \left\{ \frac{e^{(n\sqrt{-1}-2a)x}}{(n\sqrt{-1}-2a)^2} - \frac{e^{-(n\sqrt{-1}+2a)x}}{(n\sqrt{-1}+2a)^2} + \text{C}_1 x - \text{C}_2 \right\}$$

$$= e^{2ax}\{\text{C}'x + \text{C}''\} + \frac{(4a^2-n^2)\sin nx + 4an\cos nx}{(n^2+4a^2)^2}.$$

Ex. 3. $\frac{d^2y}{dx^2} + a^2 y = \text{X};$

$$y = \left(\frac{d^2}{dx^2} + a^2\right)^{-1} \text{X}$$

$$= \frac{1}{2a\sqrt{-1}} \left(\frac{d}{dx} - a\sqrt{-1}\right)^{-1} \text{X} - \frac{1}{2a\sqrt{-1}} \left(\frac{d}{dx} + a\sqrt{-1}\right)^{-1} \text{X}$$

$$= \frac{e^{a\sqrt{-1}x}}{2a\sqrt{-1}} \int e^{-a\sqrt{-1}x} \text{X}\, dx - \frac{e^{-a\sqrt{-1}x}}{2a\sqrt{-1}} \int e^{a\sqrt{-1}x} \text{X}\, dx,$$

and a constant must be added at each integration.

Let $\text{X} = 0;\quad y = \text{C}_1 e^{a\sqrt{-1}x} + \text{C}_2 e^{-a\sqrt{-1}x}$

$$= k\cos(ax+\gamma).$$

Let $\text{X} = \cos mx;\quad y = \frac{\cos mx}{a^2-m^2} + k\cos(ax+\gamma).$

Let $\text{X} = \cos ax;$

$$y = \frac{x\sin ax}{2a} + \frac{\cos ax}{4a^2} + k\cos(ax+\gamma)$$

$$= \frac{x\sin ax}{2a} + \text{A}\cos(ax+\text{B}).$$

473.] The preceding process, it will be observed, involves operations represented by symbols of the general forms

$$\left(\frac{d}{dx} - a\right)^{-r} \text{X},\ \left(\frac{d^2}{dx^2} - a^2\right)^{-r} \text{X}, \ldots \left(\frac{d^m}{dx^m} - a^m\right)^{-r} \text{X},$$

where r is unity or some other positive and integral number; and as the operation which such a symbol represents is subject to the laws of distribution and of repetition, we may expand the operative symbol, and operate on X with the several and successive terms of the expansion: but $\left(\frac{d}{dx} + a\right)^{-r}$ may be expressed in either of the following forms,

$$\left(\frac{d}{dx}\right)^{-r} - \frac{r}{1} a \left(\frac{d}{dx}\right)^{-r-1} + \frac{r(r+1)}{1.2} a^2 \left(\frac{d}{dx}\right)^{-r-2} - \ldots;$$

$$\text{or} \quad a^{-r} - \frac{r}{1} a^{-r-1} \frac{d}{dx} + \frac{r(r+1)}{1.2} a^{-r-2} \left(\frac{d}{dx}\right)^2 - \dots;$$

the former of which involves integration only, and the latter differentiation only: and as integration introduces arbitrary constants, and differentiation does not, it may be thought that the latter expansion is inapplicable; it may however always be employed, if we take care to introduce the arbitrary constants, or the supplementary function which they are involved in; and this we may do as follows:

$$\left(\frac{d}{dx} + a\right)^{-r} \mathrm{X} = \left(\frac{d}{dx} + a\right)^{-r} (\mathrm{X} + 0)$$

$$= \left(\frac{d}{dx} + a\right)^{-r} \mathrm{X} + \left(\frac{d}{dx} + a\right)^{-r} 0.$$

$$\text{But} \quad \left(\frac{d}{dx} + a\right)^{-r} 0 = e^{-ax} \int^r 0\, dx^r$$

$$= e^{-ax} \{\mathrm{C}_1 + \mathrm{C}_2 x + \mathrm{C}_3 x^2 + \dots + \mathrm{C}_r x^{r-1}\}. \quad (17)$$

Similarly,

$$\left\{\frac{d^2}{dx^2} + a^2\right\}^{-1} \mathrm{X} = \left\{\frac{d^2}{dx^2} + a^2\right\}^{-1} (\mathrm{X} + 0)$$

$$= \left\{a^{-2} - a^{-4} \frac{d^2}{dx^2} + a^{-6} \frac{d^4}{dx^4} - \dots\right\} \mathrm{X}$$

$$+ \left\{\left(\frac{d}{dx}\right)^{-2} - a^2 \left(\frac{d}{dx}\right)^{-4} + a^4 \left(\frac{d}{dx}\right)^{-6} - \dots\right\} 0.$$

Of which expression the latter part is equal to

$$\left\{1 - a^2 \left(\frac{d}{dx}\right)^{-2} + a^4 \left(\frac{d}{dx}\right)^{-4} - \dots\right\} (\mathrm{C}_1 + \mathrm{C}_2 x)$$

$$= \mathrm{C}_1 \left\{1 - \frac{a^2 x^2}{1.2} + \frac{a^4 x^4}{1.2.3.4} - \dots\right\} + \mathrm{C}_2 \left\{x - \frac{a^2 x^3}{1.2.3} + \frac{a^4 x^5}{1.2.3.4.5} - \dots\right\}$$

$$= \mathrm{C}_1 \cos ax + \frac{\mathrm{C}_2}{a} \sin ax. \quad (18)$$

Other forms of these operating symbols may also be expressed in terms of differentiation; and as that to which X is affixed always admits of such an expansion, we infer that if we can integrate a linear differential equation when the right-hand member is zero, we can by means of differentiation only find the integral when the right-hand member is a function of x.

Ex. 1. $\dfrac{d^3 y}{dx^3} - 3a \dfrac{d^2 y}{dx^2} + 3a^2 \dfrac{dy}{dx} - a^3 y = e^{ax}.$

$$y = \left(\frac{d}{dx} - a\right)^{-3} e^{nx}$$

$$= \left\{\left(\frac{d}{dx}\right)^{-3} + 3a\left(\frac{d}{dx}\right)^{-4} + \frac{3.4}{1.2}a^2\left(\frac{d}{dx}\right)^{-5} + \frac{3.4.5}{1.2.3}a^3\left(\frac{d}{dx}\right)^{-6} + \dots\right\}e^{nx}$$

$$+ \left(\frac{d}{dx} - a\right)^{-3} 0$$

$$= \left\{\frac{1}{n^3} + \frac{3a}{n^4} + \frac{3.4}{1.2}\frac{a^2}{n^5} + \dots\right\}e^{nx} + e^{ax}\int^3 0\,dx^3$$

$$= \frac{e^{nx}}{(n-a)^3} + e^{ax}\{c_1 + c_2 x + c_3 x^2\}.$$

Ex. 2. $\frac{d^2y}{dx^2} + n^2 y = \cos ax.$

$$y = \left(\frac{d^2}{dx^2} + n^2\right)^{-1} \cos ax$$

$$= \left\{\left(\frac{d}{dx}\right)^{-2} - n^2\left(\frac{d}{dx}\right)^{-4} + n^4\left(\frac{d}{dx}\right)^{-6} + \dots\right\}(\cos ax + 0)$$

$$= \left\{-\frac{1}{a^2} - \frac{n^2}{a^4} - \frac{n^4}{a^6} - \dots\right\}\cos ax + c_1 \cos nx + c_2 \sin nx\,;$$

the last part being concluded from (18);

$$\therefore \quad y = \frac{\cos ax}{n^2 - a^2} + k\cos(nx + \gamma).$$

The form of each example however will generally suggest the process most convenient for this solution.

474.] Differential equations of the form treated of in Art. 459 are most easily reduced to the linear form with constant coefficients by the theorem given in (46), Art. 424, Vol. I; viz. if $e^\theta = x$, we have

$$\left(x\frac{d}{dx}\right)^n y = \left(\frac{d}{d\theta} - n + 1\right)\left(\frac{d}{d\theta} - n + 2\right)\dots\left(\frac{d}{d\theta} - 1\right)\frac{d}{d\theta}y. \qquad (19)$$

whereby we have

$$x\frac{dy}{dx} = \frac{dy}{d\theta},$$

$$x^2\frac{d^2y}{dx^2} = \left(\frac{d}{d\theta} - 1\right)\frac{d}{d\theta}y = \frac{d^2y}{d\theta^2} - \frac{dy}{d\theta},$$

$$x^3\frac{d^3y}{dx^3} = \left(\frac{d}{d\theta} - 2\right)\left(\frac{d}{d\theta} - 1\right)\frac{d}{d\theta}y$$

$$= \frac{d^3y}{d\theta^3} - 3\frac{d^2y}{d\theta^2} + 2\frac{dy}{d\theta};$$

and so on for differentials of higher orders; and as the right-hand members of these equivalents involve coefficients which are constants, so will a differential equation, whose terms are of the form in the left-hand members, after substitution become a linear differential equation with constant coefficients.

The following are examples in which the process is applied.

Ex. 1. $x^2 \dfrac{d^2 y}{dx^2} - nx\dfrac{dy}{dx} + ny = ax^m.$

Let $x = e^\theta$, and let us substitute by means of (19); then we have

$$\frac{d^2 y}{d\theta^2} - \frac{dy}{d\theta} - n\frac{dy}{d\theta} + ny = ae^{m\theta};$$

$$\therefore \quad \left(\frac{d}{d\theta} - n\right)\left(\frac{d}{d\theta} - 1\right) y = ae^{m\theta};$$

$$\therefore \quad y = c_1 e^\theta + c_2 e^{n\theta} + \frac{ae^{m\theta}}{(m-1)(m-n)}$$

$$= c_1 x + c_2 x^n + \frac{ax^m}{(m-1)(m-n)}.$$

Ex. 2. $x^2 \dfrac{d^2 y}{dx^2} + 3x\dfrac{dy}{dx} + y = \dfrac{1}{(1-x)^2};$

$$\therefore \quad \frac{d^2 y}{d\theta^2} + 2\frac{dy}{d\theta} + y = \frac{1}{(1-e^\theta)^2};$$

$$\left(\frac{d}{d\theta} + 1\right)^2 y = \frac{1}{(1-e^\theta)^2};$$

$$y = e^{-\theta}\int^2 \frac{e^\theta\, d\theta^2}{(1-e^\theta)^2};$$

$$= e^{-\theta}\int\left\{\frac{1}{1-e^\theta} + c_1\right\} d\theta$$

$$= e^{-\theta}\{-\log(1-e^{-\theta}) + c_1\theta + c_2\}$$

$$= \frac{1}{x}\log\frac{x}{x-1} + \frac{c_1}{x}\log x + \frac{c_2}{x}.$$

SECTION 2.—*The Solution of Partial Differential Equations by Symbolical Methods.*

475.] The process by which the theorems of operating symbols are applied to the solution of partial differential equations is in principle the same as that by which they have been already applied to the solution of total differential equations; and the method is for the most part only applicable to two classes of equations, which are analogous to those similarly treated in the preceding section. One peculiar extension is required; and this is given by the symbolical form which Taylor's theorem takes as it is exhibited in (59), Art. 427, Vol. I, where we have

$$e^{h\frac{d}{dx}} f(x) = f(x+h); \tag{20}$$

$$\therefore \quad e^{h\frac{d}{dx}+k\frac{d}{dy}} f(x, y) = f(x+h, y+k). \tag{21}$$

In (20) if $f(x)$ is a constant, it remains the same constant, although it is the subject of the operation denoted by $e^{h\frac{d}{dx}}$. This process of integration is exemplified in the following examples.

Ex. 1. $a\left(\frac{dz}{dx}\right) + b\left(\frac{dz}{dy}\right) = c.$

$$\therefore \quad \left(a\frac{d}{dx} + b\frac{d}{dy}\right)z = c;$$

$$z = \left(a\frac{d}{dx} + b\frac{d}{dy}\right)^{-1} c.$$

As the operations symbolized by $\frac{d}{dx}$ and $\frac{d}{dy}$ are independent of each other, when one is taking effect, the other is inoperative, that is, if $\frac{d}{dx}$ is variable, $\frac{d}{dy}$ is constant, and *vice versa*. Suppose that $\frac{d}{dx}$ is taking effect; then we have

$$z = \frac{1}{a}\left(\frac{d}{dx} + \frac{b}{a}\frac{d}{dy}\right)^{-1} c,$$

when $\frac{d}{dy}$ is constant; therefore by (5), Art. 471,

$$z = \frac{1}{a} e^{-\frac{b}{a}x\frac{d}{dy}} \int e^{\frac{bx}{a}\frac{d}{dy}} c\,dx;$$

and remembering that the operation symbolized by $e^{\frac{b}{a}x\frac{d}{dy}}$, when performed on a constant, produces the same constant; and that an arbitrary function of y must be added, because an x-partial integration is being performed,

$$z = \frac{1}{a} e^{-\frac{bx}{a}\frac{d}{dy}} \{cx + \phi(y)\} = \frac{cx}{a} + e^{-\frac{bx}{a}\frac{d}{dy}} \phi(y)$$

$$= \frac{cx}{a} + \phi\left(y - \frac{bx}{a}\right) = \frac{cx}{a} + \phi(ay - bx).$$

If the operation expressed by $\frac{d}{dy}$ had been first effected, we should have had

$$z = \frac{cy}{b} + \phi(bx - ay);$$

and this is equivalent in form to the former result.

Ex. 2. $\left(\frac{d^2 z}{dx^2}\right) - a^2\left(\frac{d^2 z}{dy^2}\right) = 0.$

$$\left(\frac{d^2}{dx^2} - a^2\frac{d^2}{dy^2}\right)z = 0.$$

$$z = \frac{1}{2a}\frac{d}{dy}\left\{\left(\frac{d}{dx} - a\frac{d}{dy}\right)^{-1}0 - \left(\frac{d}{dx} + a\frac{d}{dy}\right)^{-1}0\right\}$$

$$= \frac{1}{2a}\int dy\, \phi(y+ax) - \frac{1}{2a}\int dy\, \psi(y-ax)$$

$$= \Phi(y+ax) + \Psi(y-ax).$$

476.] There is another class of partial differential equations the integral of which can be determined by the symbolical method; as the process however will be best understood by means of examples, I take the two following.

Ex. 1. $x\left(\frac{dz}{dx}\right) + y\left(\frac{dz}{dy}\right) - nz = 0.$

$$\therefore \left\{x\frac{d}{dx} + y\frac{d}{dy} - n\right\}z = 0.$$

Let $x = e^{\theta}, \quad y = e^{\phi};$

$$\therefore \left\{\frac{d}{d\theta} + \frac{d}{d\phi} - n\right\}z = 0;$$

$$z = \left\{\frac{d}{d\theta} + \frac{d}{d\phi} - n\right\}^{-1}0$$

$$= e^{\left(n - \frac{d}{d\phi}\right)\theta}\int e^{\left(\frac{d}{d\phi} - n\right)\theta}\, 0\, d\theta$$

$$= e^{\left(n - \frac{d}{d\phi}\right)\theta} f(\phi) = e^{n\theta} f(\phi - \theta)$$

$$= x^n f\left(\frac{y}{x}\right),$$

where f denotes an arbitrary function.

Ex. 2. $x^2\left(\frac{d^2 z}{dx^2}\right) + 2xy\left(\frac{d^2 z}{dx\, dy}\right) + y^2\left(\frac{d^2 z}{dy^2}\right) - n\left\{\left(\frac{dz}{dx}\right) + \left(\frac{dz}{dy}\right)\right\} + nz$
$= x^2 + y^2.$

Let $x = e^\theta$, $y = e^\phi$; also let

$$x\frac{d}{dx} + y\frac{d}{dy} = \frac{d}{d\theta} + \frac{d}{d\phi} = \nabla;$$

consequently by the theorem in (46), Art. 424, Vol. I,

$$\left(x\frac{d}{dx} + y\frac{d}{dy}\right)^2 = (\nabla - 1)\nabla.$$

$$\{(\nabla-1)\nabla - n\nabla + n\}z = e^{2\theta} + e^{2\phi};$$

$$(\nabla - n)(\nabla - 1)z = e^{2\theta} + e^{2\phi}.$$

Now we have the following theorem;

$$(\nabla - n)(e^{m\theta} + e^{m\phi}) = (m-n)(e^{m\theta} + e^{m\phi});$$

so that if this operation is effected k times consecutively,

$$(\nabla - n)^k(e^{m\theta} + e^{m\phi}) = (m-n)^k(e^{m\theta} + e^{m\phi});$$

therefore if $k = -1$,

$$(\nabla - n)^{-1}(e^{m\theta} + e^{m\phi}) = \frac{e^{m\theta} + e^{m\phi}}{m-n}.$$

It is also evident from the definition of ∇, that if u_m is a homogeneous function of m dimensions in terms of x and y,

$$(\nabla - m)u_m = 0;$$

$$\therefore\ u_m = \frac{0}{\nabla - m}:$$

and $\qquad (\nabla - m)^{-1}0 = u_m.$

Let us apply these theorems to the preceding equation;

$$\begin{aligned}(\nabla - 1)z &= (\nabla - n)^{-1}(e^{2\theta} + e^{2\phi}) \\ &= \frac{e^{2\theta} + e^{2\phi}}{2-n} + u_n;\end{aligned}$$

$$\begin{aligned}\therefore\ z &= \frac{e^{2\theta} + e^{2\phi}}{(2-1)(2-n)} + u_n + u_1 \\ &= \frac{x^2 + y^2}{(2-1)(2-n)} + u_n + u_1,\end{aligned}$$

where u_n and u_1 are homogeneous functions of x and y of m and 1 dimensions respectively.

CHAPTER XVIII.

INTEGRATION OF SIMULTANEOUS DIFFERENTIAL EQUATIONS.

477.] A system of n different equations involving the independent variable t, n dependent variables $x, y, z, \ldots$ and their t-derived functions, viz. $x', y', z', \ldots x'', y'', z'', \ldots$, is called a system of simultaneous differential equations; the order of which depends on that of the highest derived-functions; and the problem of integration consists in deducing from them integral functions of $x, y, z, \ldots t$, which also contain a sufficient number of arbitrary constants. Many most important problems in mathematical physics depend on the integration of a system of equations of this form.

Let us consider first a system of simultaneous differential equations of the first order: and suppose that they are n in number, and of the form

$$\left.\begin{aligned} \frac{dx}{dt} &= f_1(t, x, y, z, \ldots), \\ \frac{dy}{dt} &= f_2(t, x, y, z, \ldots), \\ \frac{dz}{dt} &= f_3(t, x, y, z, \ldots); \\ &\cdots\cdots \end{aligned}\right\} \qquad (1)$$

we have to eliminate, by means of these n equations, $n-1$ variables $y, z, \ldots$ and thus to obtain an equation in terms of t and of the other variable x: for this purpose we differentiate the first of the above equations n times, and substitute each time for $\frac{dy}{dt}, \frac{dz}{dt}, \ldots$ their values given in the other $n-1$ equations: by this process we obtain $n-1$ equations of the forms,

$$\left.\begin{aligned} \frac{d^2x}{dt^2} &= \phi_2(t, x, y, z, \ldots), \\ \frac{d^3x}{dt^3} &= \phi_3(t, x, y, z, \ldots), \\ &\cdots\cdots \\ \frac{d^nx}{dt^n} &= \phi_n(t, x, y, z, \ldots); \end{aligned}\right\} \qquad (2)$$

which added to f_1 give us n equations, from which $y, z, \ldots$ may be eliminated, and an equation of the nth order in terms of x and

t will result. Now this when integrated will contain n arbitrary constants; and from it we shall be able to derive the several equations (2), by means of which and the remaining equations of (1) we shall obtain the other integrals.

Ex. 1. $\frac{dx}{dt} - y = 0; \qquad \frac{dy}{dt} - x = 0.$

$$\therefore \quad \frac{d^2 x}{dt^2} = \frac{dy}{dt} = x;$$

$$x = c_1 e^t + c_2 e^{-t}; \qquad \therefore \quad y = c_1 e^t - c_2 e^{-t}.$$

Ex. 2. $a\frac{dx}{dt} + (c-b)\,yz = 0, \qquad b\frac{dy}{dt} + (a-c)\,zx = 0,$

$$c\frac{dz}{dt} + (b-a)\,xy = 0.$$

Multiply the first by x, the second by y, and the third by z; and let $\frac{d\phi}{dt} = xyz$; then we have $ax\frac{dx}{dt} = (b-c)\frac{d\phi}{d_t}$; and similar values for the others;

$$\therefore \quad ax^2 = 2(b-c)\,\phi + h_1;$$
$$by^2 = 2\,(c-a)\,\phi + h_2;$$
$$cz^2 = 2\,(a-b)\,\phi + h_3;$$

$\therefore \quad abc\,x^2 y^2 z^2$

$$= \{2(b-c)\,\phi + h_1\}\,\{2(c-a)\,\phi + h_2\}\,\{2(a-b)\,\phi + h_3\},$$

$\therefore \quad (abc)^{\frac{1}{2}}\frac{d\phi}{dt}$

$$= \{2(b-c)\,\phi + h_1\}^{\frac{1}{2}}\,\{2(c-a)\,\phi + h_2\}^{\frac{1}{2}}\,\{2(a-b)\,\phi + h_3\}^{\frac{1}{2}};$$

the integral of which will give us t in terms of ϕ. Also multiplying the three equations severally through (1) by x, y, z; (2) by ax, by, cz; and adding in each case, we have

$$ax\,dx + by\,dy + cz\,dz = 0;$$
$$a^2 x\,dx + b^2 y\,dy + c^2 z\,dz = 0;$$
$$\therefore \quad ax^2 + by^2 + cz^2 = h^2;$$
$$a^2 x^2 + b^2 y^2 + c^2 z^2 = k^2;$$

which are two relations between x, y, z, which the differential equations by their forms import.

478.] *In general* the final equation will be of the nth order, and therefore its complete integral will contain n arbitrary constants: in particular cases however we shall arrive at an equation between, say, x and t of an order lower than n; and in this case it would seem that a sufficient number of arbitrary constants will not have been introduced; but it will in the result be found that one or more of the equations in terms of t and of one of the other variables will be of the nth order, and that the full number of n arbitrary constants will thus be made complete.

Thus suppose the following equations to be given,

$$\frac{dx}{dt} = y+z, \qquad \frac{dy}{dt} = x+z, \qquad \frac{dz}{dt} = y-x;$$

$$\begin{aligned} \frac{d^2x}{dt^2} &= \frac{dy}{dt}+\frac{dz}{dt} & \qquad \frac{d^2y}{dt^2} &= \frac{dx}{dt}+\frac{dz}{dt} \\ &= y+z & &= y-c_1; \\ &= \frac{dx}{dt}; & \therefore\ y &= c_1+c_2e^t+c_3e^{-t}; \\ x &= c_1+c_2e^t; & \therefore\ z &= -c_1-c_3e^{-t}: \end{aligned}$$

and as y contains three arbitrary constants, the result has the required generality.

479.] Linear simultaneous equations are those which offer themselves for solution with the best hope of success. Suppose that there are n variables $x_1, x_2, \dots x_n$, and that t is another variable which is dependent; and that there are given n equations of the form,

$$\left.\begin{aligned} \frac{dx_1}{dt}+\text{P}_1x_1+\text{P}_2x_2+\dots+\text{P}_nx_n &= \text{T}_1, \\ \frac{dx_2}{dt}+\text{Q}_1x_1+\text{Q}_2x_2+\dots+\text{Q}_nx_n &= \text{T}_2, \\ \dots\dots\dots\dots \\ \frac{dx_n}{dt}+\text{S}_1x_1+\text{S}_2x_2+\dots+\text{S}_nx_n &= \text{T}_n; \end{aligned}\right\} \qquad (3)$$

where the P's, Q's, ... S's, T's are functions of t only. These equations are integrable in certain cases.

Thus, suppose that there are two equations

$$\left.\begin{aligned} \text{M}_1\frac{dx}{dt}+\text{N}_1\frac{dy}{dt}+\text{P}_1x+\text{Q}_1y &= \text{T}_1, \\ \text{M}_2\frac{dx}{dt}+\text{N}_2\frac{dy}{dt}+\text{P}_2x+\text{Q}_2y &= \text{T}_2. \end{aligned}\right\} \qquad (4)$$

Multiply the second by θ and add to the first; put

$$\left.\begin{aligned} \text{M}_1+\theta\text{M}_2 = m, \qquad \text{N}_1+\theta\text{N}_2 = n, \qquad \text{P}_1+\theta\text{P}_2 = p, \\ \text{Q}_1+\theta\text{Q}_2 = q, \qquad \text{T}_1+\theta\text{T}_2 = \text{T}; \end{aligned}\right\} \qquad (5)$$

then $$m\frac{dx}{dt}+n\frac{dy}{dt}+px+qy = \text{T}. \tag{6}$$

Let $$dx+\frac{n}{m}dy = du, \qquad x+\frac{q}{p}y = u; \tag{7}$$

then (6) becomes $$m\frac{du}{dt}+pu = \text{T}, \tag{8}$$

which is linear of the first degree, and therefore u may be found in terms of t. Now from (7)

$$dx+\frac{n}{m}dy = d(x+\frac{q}{p}y)$$

$$= dx+\frac{q}{p}dy+y\frac{p\,dq-q\,dp}{p^2}; \tag{9}$$

and this is satisfied by

$$\frac{n}{m} = \frac{q}{p}, \quad \text{and} \quad d.\frac{q}{p} = 0; \tag{10}$$

if the coefficients involved in (4) satisfy these conditions, then we may substitute for these quantities from (5), and hereby determine θ, and hereby, by means of (8), determine a relation between t and u.

If the coefficients in the left-hand members of (4) are constant, then from (10)

$$\frac{\text{M}_1+\theta\,\text{M}_2}{\text{N}_1+\theta\,\text{N}_2} = \frac{\text{P}_1+\theta\,\text{P}_2}{\text{Q}_1+\theta\,\text{Q}_2},$$

which is a quadratic in θ; and consequently if its two roots are unequal, and are, say, θ_1 and θ_2, we shall have two equations of the form (8), viz.,

$$m_1\,du+p_1\,u\,dt = \text{T}_1\,dt,$$
$$m_2\,du+p_2\,u\,dt = \text{T}_2\,dt; \tag{11}$$

and from these we shall obtain two equations between u and t, and thus two arbitrary constants.

Ex. 1. $$\frac{dx}{dt}+2\frac{dy}{dt}-2x+2y = 3e^t,$$

$$3\frac{dx}{dt}+\frac{dy}{dt}+2x+y = 4e^{2t}.$$

Multiply the second of these by θ and add to the first; then

$$(1+3\theta)\frac{d}{dt}\left(x+\frac{2+\theta}{1+3\theta}y\right)+(2\theta-2)\left(x+\frac{2+\theta}{2\theta-2}y\right)$$
$$= 3e^t+4\theta e^{2t}. \tag{12}$$

Let $\frac{2+\theta}{1+3\theta} = \frac{2+\theta}{2\theta-2}$; $\therefore$ $\theta = -2,$ $= -3$;

and (12) becomes

$$\frac{dx}{dt} + \frac{6}{5}x = -\frac{1}{5}(3e^t - 8e^{2t});$$

and $$\frac{d}{dt}\left(x+\frac{y}{3}\right) + \left(x+\frac{y}{3}\right) = -\frac{1}{3}(3e^t - 12e^{2t}); \qquad (13)$$

which are two linear equations of the first degree, and are easily integrated.

Ex. 2. $\frac{dx}{dt} + ax + a_1 y = 0,$ $\frac{dy}{dt} + bx + b_1 y = 0.$

$$\therefore \quad d\{x+\theta y\} + (a+b\theta)\left(x + \frac{a_1+b_1\theta}{a+b\theta}y\right) = 0. \qquad (14)$$

Let $\frac{a_1+b_1\theta}{a+b\theta} = \theta$; which is a quadratic in θ; let the two roots be α and β, so that (14) becomes

$$\frac{d}{dt}(x+\alpha y) + (a+b\alpha)(x+\alpha y) = 0;$$

$$\frac{d}{dt}(x+\beta y) + (a+b\beta)(x+\beta y) = 0;$$

$$\therefore \quad x+\alpha y = C_1 e^{-(a+b\alpha)t}; \qquad x+\beta y = C_2 e^{-(a+b\beta)t};$$

which are the two integrals: and x and y may be separately determined, and each will contain two arbitrary constants.

480.] Let us next consider the case of n linear equations with constant coefficients; and let us suppose them to be of the form

$$\left.\begin{aligned} \frac{dx_1}{dt} + a_1 x_1 + a_2 x_2 + \ldots + a_n x_n &= T_1, \\ \frac{dx_2}{dt} + b_1 x_1 + b_2 x_2 + \ldots + b_n x_n &= T_2, \\ \cdot \quad \cdot \quad \cdot \quad \cdot \quad \cdot \quad \cdot \quad \cdot \quad \cdot \quad \cdot & \\ \frac{dx_n}{dt} + k_1 x_1 + k_2 x_2 + \ldots + k_n x_n &= T_n; \end{aligned}\right\} \qquad (15)$$

or, according to the calculus of operations,

$$\left.\begin{aligned} \left(\frac{d}{dt} + a_1\right)x_1 + a_2 x_2 + \ldots + a_n x_n &= T_1, \\ b_1 x_1 + \left(\frac{d}{dt} + b_2\right)x_2 + \ldots + b_n x_n &= T_2, \\ \cdot \quad \cdot \quad \cdot \quad \cdot \quad \cdot \quad \cdot \quad \cdot \quad \cdot \quad \cdot \quad \cdot & \\ k_1 x_1 + k_2 x_2 + \ldots + \left(\frac{d}{dt} + k_n\right)x_n &= T_n; \end{aligned}\right\} \qquad (16)$$

whence, according to the notation of determinants,

$$x_1 = \frac{\Sigma \cdot \pm \left(\frac{d}{dt}+b_2\right)\left(\frac{d}{dt}+c_3\right)\cdots\left(\frac{d}{dt}+k_n\right)T_1}{\Sigma \cdot \pm \left(\frac{d}{dt}+a_1\right)\left(\frac{d}{dt}+b_2\right)\left(\frac{d}{dt}+c_3\right)\cdots\left(\frac{d}{dt}+k_n\right)}; \qquad (17)$$

so that x_1 will consist of a series of terms formed by operating, with the several factors of which the denominator of (17) consists, on the quantities which are contained in the numerator.

Ex. 1. $\frac{dx}{dt}+ax+a_1y=0, \qquad \frac{dy}{dt}+bx+b_1y=0.$

$$\left(\frac{d}{dt}+a\right)x+a_1y=0, \qquad bx+\left(\frac{d}{dt}+b_1\right)y=0;$$

$$\left\{\left(\frac{d}{dt}+a\right)\left(\frac{d}{dt}+b_1\right)-a_1b\right\}x=0;$$

$$x=\left\{\frac{d^2}{dt^2}+(a+b_1)\frac{d}{dt}+ab_1-a_1b\right\}^{-1}0.$$

Let the roots of the operating factor be α and β; so that

$$x=\left(\frac{d}{dt}-\alpha\right)^{-1}\left(\frac{d}{dt}-\beta\right)^{-1}0$$

$$=c_1e^{\alpha t}+c_2e^{\beta t};$$

$$\therefore\quad y=\frac{b_1+\beta}{a_1}c_1e^{\alpha t}+\frac{a+b_1}{a_1}c_2e^{\beta t}.$$

Ex. 2. $\left(\frac{d}{dt}+4\right)x+3y=t, \qquad 2x+\left(\frac{d}{dt}+5\right)y=e^t.$

$$\left\{\left(\frac{d}{dt}+4\right)\left(\frac{d}{dt}+5\right)-6\right\}x=\left(\frac{d}{dt}+5\right)t-3e^t,$$

$$\left(\frac{d^2}{dt^2}+9\frac{d}{dt}+14\right)x=1+5t-3e^t,$$

$$\left(\frac{d}{dt}+2\right)\left(\frac{d}{dt}+7\right)x=1+5t-3e^t;$$

$$\therefore\quad x=c_1e^{-2t}+c_2e^{-7t}-\frac{31}{196}+\frac{5}{14}t-\frac{1}{8}e^t;$$

$$y=-\frac{2}{3}c_1e^{-2t}+c_2e^{-7t}+\frac{9}{98}-\frac{1}{7}t+\frac{5}{24}e^t.$$

Ex. 3. Let there be three equations

$$\left(\frac{d}{dt}+a_1\right)x+b_1y+c_1z = 0,$$

$$a_2x+\left(\frac{d}{dt}+b_2\right)y+c_2z = 0,$$

$$a_3x+b_3y+\left(\frac{d}{dt}+c_3\right)z = 0;$$

$$\therefore \left\{\left(\frac{d}{dt}+a_1\right)\left(\frac{d}{dt}+b_2\right)\left(\frac{d}{dt}+c_3\right)-c_2b_3\left(\frac{d}{dt}+a_1\right)-a_3c_1\left(\frac{d}{dt}+b_2\right)\right.$$
$$\left.-b_1a_2\left(\frac{d}{dt}+c_3\right)+c_1a_2b_3+b_1c_2a_3\right\}x = 0;$$

$$\therefore \left\{\frac{d^3}{dt^3}+(a_1+b_2+c_3)\frac{d^2}{dt^2}+(b_2c_3-b_3c_2+c_3a_1-c_1a_3+a_1b_2-a_2b_1)\frac{d}{dt}\right.$$
$$\left.+a_1b_2c_3+c_1a_2b_3+b_1c_2a_3-a_1c_2b_3-c_1b_2a_3-b_1a_2c_3\right\}x = 0;$$

of this cubic let the roots be α, β, γ;

$$\therefore \quad x = \left\{\left(\frac{d}{dt}-\alpha\right)\left(\frac{d}{dt}-\beta\right)\left(\frac{d}{dt}-\gamma\right)\right\}^{-1}0;$$

$$\therefore \quad x = c_1e^{\alpha t}+c_2e^{\beta t}+c_3e^{\gamma t},$$

from which the values of y and z may be found.

If the three roots of the cubic are equal, then

$$x = \left(\frac{d}{dt}-\alpha\right)^{-3}0 = e^{\alpha t}\{c_1+c_2t+c_3t^2\}.$$

481.] Simultaneous differential equations of higher orders with constant coefficients may also be solved in a similar manner by the calculus of operating factors: the following are examples of the process.

Ex. 1. $\frac{d^2x}{dt^2}-ax-by = c, \qquad \frac{d^2y}{dt^2}-a_1x-b_1y = c_1.$

$$\therefore \quad \left(\frac{d^2}{dt^2}-a\right)x-by = c, \qquad -a_1x+\left(\frac{d^2}{dt^2}-b_1\right)y = c_1;$$

$$\therefore \quad \left\{\left(\frac{d^2}{dt^2}-a\right)\left(\frac{d^2}{dt^2}-b_1\right)-a_1b\right\}x = \left(\frac{d^2}{dt^2}-b_1\right)c+bc_1;$$

$$\therefore \quad \left\{\frac{d^4}{dt^4}-(a+b_1)\frac{d^2}{dt^2}+ab_1-a_1b\right\}x = bc_1-b_1c.$$

Let the roots of the operating factor be real and unequal, and be $+\alpha_1$, $-\alpha_1$, $+\alpha_2$, $-\alpha_2$: and let $bc_1-b_1c = k$,

$$x = \left\{\left(\frac{d^2}{dt^2} - a_1^2\right)\left(\frac{d^2}{dt^2} - a_2^2\right)\right\}^{-1} k;$$

$$\therefore \quad x = c_1 e^{a_1 x} + c_2 e^{-a_1 x} + c_3 e^{a_2 x} + c_4 e^{-a_2 x} + \frac{k}{a_1^2 a_2^2};$$

if the roots are impossible, the exponential expressions in x may be replaced by the equivalent circular functions; also by a similar process may the value of y be found.

Ex. 2. $$\frac{d^2 x}{dt^2} - 2\frac{dx}{dt} - 2\frac{dy}{dt} + x = e^{2t};$$

$$\frac{d^2 y}{dt^2} + 2\frac{dy}{dt} + \frac{dx}{dt} + 6y + 5x = e^t.$$

$$\therefore \quad \left(\frac{d^2}{dt^2} - 2\frac{d}{dt} + 1\right)x - 2\frac{d}{dt}y = e^{2t};$$

$$\left(\frac{d}{dt} + 5\right)x + \left(\frac{d^2}{dt^2} + 2\frac{d}{dt} + 6\right)y = e^t;$$

eliminating y

$$\left(\frac{d^2}{dt^2} + 2\right)\left(\frac{d^2}{dt^2} + 3\right)x = 14e^{2t} + 2e^t;$$

$$x = c_1 \sin(2^{\frac{1}{2}} t + \gamma_1) + c_2 \sin(3^{\frac{1}{2}} t + \gamma_2) + \frac{1}{3}e^{2t} - \frac{1}{6}e^t:$$

and by a similar process we may find y.

482.] A similar method is applicable also to linear partial differential equations. Thus if

$$\left(\frac{d^2 z}{dx\,dy}\right) + a\left(\frac{dz}{dy}\right) = 0, \qquad \left(\frac{d^2 z}{dx\,dy}\right) + c\left(\frac{d^2 z}{dx^2}\right) = 0;$$

$$\therefore \quad \left(\frac{d^3 z}{dx^2\,dy}\right) - ac\left(\frac{d^2 z}{dx^2}\right) = 0;$$

so that integrating twice with respect to x, and introducing two arbitrary functions of y,

$$\left(\frac{dz}{dy}\right) - acz = x f(y) + \phi(y),$$

which is a linear differential equation of the first order: consequently

$$z = x\,\mathrm{F}(y) + \Phi(y) + e^{acy}\,\Psi(y),$$

where F, Φ and Ψ are symbols of arbitrary functions.

483.] There is no general method of integrating simultaneous equations which are not linear, and therefore we are obliged to have recourse to such artifices as are suggested by the forms of the equations: of these we have already had numerous examples in the problems of Chapter XIV; for the sake of further illustration we insert two more; others must be deferred until they arise in the course of the treatise, because their constants of integration for the most part depend on certain conditions of the problem which at present we have not means of determining.

Ex. 1. $$\frac{d^2x}{dt^2} + \frac{\mu x}{r^3} = 0, \qquad \frac{d^2y}{dt^2} + \frac{\mu y}{r^3} = 0; \tag{18}$$

where $r^2 = x^2+y^2$; multiplying these by y and x respectively, and subtracting,

$$y\frac{d^2x}{dt^2} - x\frac{d^2y}{dt^2} = 0; \tag{19}$$

therefore integrating, $$y\frac{dx}{dt} - x\frac{dy}{dt} = h, \tag{20}$$

h being an arbitrary constant. Again, from (18) and (20),

$$h\frac{d^2x}{dt^2} = \mu\frac{r\,dy - y\,dr}{r^2\,dt};$$

$$h\frac{dx}{dt} = \mu\frac{y}{r} + c_1; \tag{21}$$

similarly $$h\frac{dy}{dt} = -\mu\frac{x}{r} - c_2; \tag{22}$$

therefore, multiplying (21) by y, and (22) by x, and subtracting,

$$\mu r + c_1 y + c_2 x = h^2. \tag{23}$$

Again, from (18),

$$\frac{2\,dx\,d^2x + 2\,dy\,d^2y}{dt^2} + \frac{2\mu(x\,dx + y\,dy)}{r^3} = 0;$$

$$\therefore \quad \frac{dx^2+dy^2}{dt^2} - \frac{2\mu}{r} = 2k, \tag{24}$$

where k is an arbitrary constant. And (20), (21), (22), (24) are the integrals of the equations.

Ex. 2. Another example of simultaneous differential equations has been solved by M. Binet, and as the form has important applications it is desirable to insert it *.

Let there be four variables t, x, y, z, whereof t is independent, and the three others are dependent: also let $R = f(r)$, where

$$r^2 = x^2+y^2+z^2; \tag{25}$$

* Liouville's Journal, Vol. II, p. 457.

and let there be a system of equations

$$\frac{d^2x}{dt^2} = \left(\frac{dR}{dx}\right), \quad \frac{d^2y}{dt^2} = \left(\frac{dR}{dy}\right), \quad \frac{d^2z}{dt^2} = \left(\frac{dR}{dz}\right); \tag{26}$$

it is required to integrate them.

From (25) it follows that

$$\left(\frac{dR}{dx}\right) = \frac{x}{r}\left(\frac{dR}{dr}\right), \quad \left(\frac{dR}{dy}\right) = \frac{y}{r}\left(\frac{dR}{dr}\right), \quad \left(\frac{dR}{dz}\right) = \frac{z}{r}\left(\frac{dR}{dr}\right); \tag{27}$$

so that (26) become

$$\frac{d^2x}{dt^2} = \frac{x}{r}\left(\frac{dR}{dr}\right), \quad \frac{d^2y}{dt^2} = \frac{y}{r}\left(\frac{dR}{dr}\right), \quad \frac{d^2z}{dt^2} = \frac{z}{r}\left(\frac{dR}{dr}\right). \tag{28}$$

From the last two of these we have

$$y\frac{d^2z}{dt^2} - z\frac{d^2y}{dt^2} = 0; \tag{29}$$

and therefore integrating,

$$\left.\begin{aligned} & y\frac{dz}{dt} - z\frac{dy}{dt} = c_1; \\ \text{similarly} \qquad & z\frac{dx}{dt} - x\frac{dz}{dt} = c_2; \\ & x\frac{dy}{dt} - y\frac{dx}{dt} = c_3; \end{aligned}\right\} \tag{30}$$

$$\therefore \quad \left(y\frac{dz}{dt} - z\frac{dy}{dt}\right)^2 + \left(z\frac{dx}{dt} - x\frac{dz}{dt}\right)^2 + \left(x\frac{dy}{dt} - y\frac{dx}{dt}\right)^2$$

$$= (x^2+y^2+z^2)\left(\frac{dx^2}{dt^2} + \frac{dy^2}{dt^2} + \frac{dz^2}{dt^2}\right) - \left(x\frac{dx}{dt} + y\frac{dy}{dt} + z\frac{dz}{dt}\right)^2$$

$$= c_1^2 + c_2^2 + c_3^2$$

$$= A^2, \text{ where } A \text{ is an undetermined constant.} \tag{31}$$

Again, multiplying (26) severally by $2\,dx$, $2\,dy$, $2\,dz$, and integrating and adding,

$$\frac{dx^2+dy^2+dz^2}{dt^2} = 2(R+B), \tag{32}$$

where B is an arbitrary constant; so that (31) becomes

$$\left(x\frac{dx}{dt} + y\frac{dy}{dt} + z\frac{dz}{dt}\right)^2 = 2r^2(R+B) - A^2;$$

$$\therefore \quad r^2\frac{dr^2}{dt^2} = 2r^2(R+B) - A^2; \tag{33}$$

$$\therefore \quad dt = \frac{r\,dr}{\{2r^2(\mathrm{R}+\mathrm{B})-\mathrm{A}^2\}^{\frac{1}{2}}}; \tag{34}$$

$$\frac{dr^2}{dt^2} = 2(\mathrm{R}+\mathrm{B}) - \frac{\mathrm{A}^2}{r^2},$$

$$\frac{d^2r}{dt^2} = \frac{d\mathrm{R}}{dr} + \frac{\mathrm{A}^2}{r^3}. \tag{35}$$

Therefore combining this with the first of (28) we have

$$r\frac{d^2x}{dt^2} - x\frac{d^2r}{dt^2} = -\frac{\mathrm{A}^2 x}{r^3},$$

$$\frac{d}{dt}\left(r^2\frac{d}{dt}.\frac{x}{r}\right) = -\frac{\mathrm{A}^2}{r^2}\,\frac{x}{r};$$

$$\therefore \quad \frac{r^2}{\mathrm{A}}\,\frac{d}{dt}\left(\frac{r^2}{\mathrm{A}}\,\frac{d}{dt}.\frac{x}{r}\right) = -\frac{x}{r}. \tag{36}$$

Let
$$\frac{\mathrm{A}\,dt}{r^2} = d\phi = \frac{\mathrm{A}\,dr}{r\{2r^2(\mathrm{R}+\mathrm{B})-\mathrm{A}^2\}^{\frac{1}{2}}}; \tag{37}$$

$$\therefore \quad \frac{d^2}{d\phi^2}.\frac{x}{r} + \frac{x}{r} = 0; \tag{38}$$

integrating which,

similarly
$$\left.\begin{aligned} x &= r(a_1\cos\phi + b_1\sin\phi); \\ y &= r(a_2\cos\phi + b_2\sin\phi); \\ z &= r(a_3\cos\phi + b_3\sin\phi); \end{aligned}\right\} \tag{39}$$

also from (34) and (37)

$$t - a = \int \frac{r\,dr}{\{2r^2(\mathrm{R}+\mathrm{B})-\mathrm{A}^2\}^{\frac{1}{2}}}; \tag{40}$$

$$\phi - \beta = \int \frac{\mathrm{A}\,dr}{r\{2r^2(\mathrm{R}+\mathrm{B})-\mathrm{A}^2\}^{\frac{1}{2}}}; \tag{41}$$

from the latter of which ϕ may be expressed in terms of r, and therefore by means of the former in terms of t; and therefore in (39) x, y, z may be expressed in terms of t.

Now it will be observed that there are at present ten arbitrary constants, viz., six in (39), and $\mathrm{A}, \mathrm{B}, a, \beta$: but there are relations connecting them, so that all are not independent; for firstly,

$$\begin{aligned} r^2 &= x^2+y^2+z^2 \\ &= r^2\{(\cos\phi)^2\,\Sigma.a^2 + 2\sin\phi\cos\phi\,\Sigma.ab + (\sin\phi)^2\,\Sigma.b^2\}; \end{aligned}$$

and in order that this equation should be true for all values of ϕ, we must have

$$\Sigma . a^2 = 1, \quad \Sigma . ab = 0, \quad \Sigma . b^2 = 1,$$

which are three equations of relation. Also again the constant β in (41) will merely change the values of $a_1, b_1, \dots$ in (39), and therefore it is not independent of them: hence the number of constants is finally reduced to six.

It may also be observed that the integrals determining t and ϕ are not independent, but may be referred to a common origin. Thus let

$$S = \int \frac{dr}{r} \{2r^2(E+B) - A^2\}^{\frac{1}{2}};$$

$$t - a = \left(\frac{dS}{dB}\right), \quad \phi - \beta = -\left(\frac{dS}{dA}\right).$$

CHAPTER XIX.

INTEGRATION OF DIFFERENTIAL EQUATIONS BY SERIES.

484.] When all other means of integrating differential equations fail, we are obliged to have recourse to integration by series; a process in which we assume the dependent variable to be expanded in a series of terms of powers of the independent variable, and determine the coefficients and powers of these terms by means of the differential equation. It is a method therefore manifestly to be employed with great caution and reserve, because the assumption that the dependent variable is capable of expansion in an algebraical series may be undue; and if it is capable of such expansion, the dependent variable and the series can be used as equivalents only when the series is convergent: and the difficulty of determining the convergence may be insuperable.

The first method of integration by series is that alluded to in Art. 367 for the purpose of proving that the general integral of a differential equation of the nth order involves n arbitrary constants: and it has been therein applied to a particular example: I intend to make some other remarks on the method, and to express by means of it the integrals of some equations.

Let the limits of integration be x, y, and x_0, y_0; so that y becomes y_0 when $x = x_0$: and let $y_0', y_0'', \dots y_0^{(n)}$ be the values of $y', y'', \dots y^{(n)}$, when $x = x_0$; then by equation (84), Art. 74, Vol. I,

$$y = y_0 + y_0' \frac{x-x_0}{1} + y_0'' \frac{(x-x_0)^2}{1.2} + \dots + y_0^{(n)} \frac{(x-x_0)^n}{1.2.3\dots n} + \dots \quad (1)$$

Let the differential equation be

$$y^{(n)} = f_n(x, y, y' \dots y^{(n-1)}); \quad (2)$$

now if the series which expresses y in terms of x, and which is deduced from this equation, is convergent, and if also $y = y_0$, when $x = x_0$, then it must be of the form (1): and if therefore we deduce from (2), by successive differentiation and by elimination, the quantities following, viz.:

$$\left.\begin{array}{l} y^{(n+1)} = f_{n+1}(x, y, y', \dots y^{(n-1)}), \\ y^{(n+2)} = f_{n+2}(x, y, y', \dots y^{(n-1)}), \\ \cdot \quad \cdot \quad \cdot \quad \cdot \quad \cdot \quad \cdot \quad \cdot \quad \cdot \quad \cdot \quad \cdot \quad \cdot \end{array}\right\} \qquad (3)$$

and then in these expressions replace x by x_0, we shall obtain the several values of $y_0^{(n)}$, $y_0^{(n+1)}$, ... which we shall substitute in (1), and shall hereby obtain the general integral; for the series is by hypothesis convergent, and therefore adequately represents y; it manifestly satisfies the differential equation, because it is deduced from it; and it contains n arbitrary constants, viz., the term independent of x, and the several coefficients of $x, x^2, \dots x^{n-1}$.

Ex. 1. $\dfrac{d^2y}{dx^2} - a^2 y = 0;$

$$y''' = a^2 y', \qquad y'''' = a^2 y'', \dots$$

$$y_0''' = a^2 y_0', \qquad y_0'''' = a^4 y_0, \dots$$

$$\therefore \; y = y_0 + y_0' \frac{x-x_0}{1} + a^2 y_0 \frac{(x-x_0)^2}{1.2} + a^2 y_0' \frac{(x-x_0)^3}{1.2.3} + a^4 y_0 \frac{(x-x_0)^4}{1.2.3.4} + \dots$$

$$= \frac{y_0}{2}\{e^{a(x-x_0)} + e^{-a(x-x_0)}\} + \frac{y_0'}{a}\{e^{a(x-x_0)} - e^{-a(x-x_0)}\};$$

and replacing the arbitrary quantities $y_0, y_0', \dots$ by other arbitrary constants, we have

$$y = c_1 e^{ax} + c_2 e^{-ax}.$$

Ex. 2. $\dfrac{d^2y}{dx^2} - xy = 0;$

$$y''' = y + xy'; \qquad y'''' = 2y' + xy''; \dots$$

$$y_0'' = x_0 y_0, \quad y_0''' = y_0 + x_0 y_0', \quad y_0'''' = 2y_0' + x_0^2 y_0; \dots$$

$$\therefore \; y = y_0 + y_0' \frac{x-x_0}{1} + x_0 y_0 \frac{(x-x_0)^2}{1.2} + (y_0 + x_0 y_0') \frac{(x-x_0)^3}{1.2.3} + \dots$$

which is the solution: if $x_0 = 0$, then

$$y = y_0 + y_0' \frac{x}{1} + y_0 \frac{x^3}{1.2.3} + 2y_0' \frac{x^4}{1.2.3.4} + \dots$$

485.] The change of form which the solution of this last example has undergone in the replacement of x_0 by 0, is equivalent to the use of Maclaurin's series instead of Taylor's as the fundamental series in (1); in this case however we must be careful that neither y_0' nor y_0'' ... becomes infinite when $x_0 = 0$; thus, to solve the equation, $xy'' + y = 0$, if $x = 0$, $y_0 = 0$;

$$\therefore \quad xy''' + y'' + y' = 0; \qquad y_0'' = -y_0',$$

$$xy'''' + 2y''' + y'' = 0; \qquad y_0''' = \frac{y_0'}{2},$$

$$\ldots\ldots\ldots \qquad y_0'''' = -\frac{y_0'}{1.2.3};$$

$$\therefore \quad y = y_0' \frac{x}{1} - y_0' \frac{x^2}{1.2} + \frac{y_0'}{2} \frac{x^3}{1.2.3} - \frac{y_0'}{1.2.3} \frac{x^4}{1.2.3.4} + \ldots$$

$$= y_0' \left\{ x - \frac{1}{1} \frac{x^2}{1.2} + \frac{1}{1.2} \frac{x^3}{1.2.3} - \frac{1}{1.2.3} \frac{x^4}{1.2.3.4} + \ldots \right\}; \quad (4)$$

but as this solution contains only one arbitrary constant, viz. y_0', it is only a particular integral: but we may obtain the general integral by the following process.

Let the function of x in the right-hand member of (4) be expressed by η, and y'_0 by c; so that

$$y = c\eta; \quad (5)$$

and let us suppose that c is no longer a constant, but a variable, say u, according to the method of Art. 458, so that

$$y = u\eta; \quad (6)$$

then differentiating

$$\frac{dy}{dx} = u\frac{d\eta}{dx} + \eta\frac{du}{dx};$$

$$\frac{d^2y}{dx^2} = u\frac{d^2\eta}{dx^2} + 2\frac{du}{dx}\frac{d\eta}{dx} + \eta\frac{d^2u}{dx^2};$$

and substituting these in the given differential equation, and bearing in mind that η is a particular integral of it, we have

$$\eta\frac{d^2u}{dx^2} + 2\frac{du}{dx}\frac{d\eta}{dx} = 0; \qquad \therefore \quad \frac{d.\frac{du}{dx}}{\frac{du}{dx}} + \frac{2\,d\eta}{\eta} = 0;$$

$$\log \eta^2 \frac{du}{dx} = \log c_1; \qquad u = \int \frac{c_1\,dx}{\eta^2} + c_2;$$

so that (6) becomes

$$y = c_2\eta + \eta\int\frac{c_1\,dx}{\eta^2}; \quad (7)$$

and this contains two arbitrary constants, and is therefore the general integral of the given differential equation.

486.] Often, instead of the series of Taylor and Maclaurin, it is convenient to assume a series with undetermined indices and coefficients, and to determine these by means of the differential equation; it is in fact the only available process when the required series involves negative or fractional powers of the independent variable.

Ex. 1. $$\frac{d^2y}{dx^2}+\frac{2}{x}\frac{dy}{dx}+n^2y = 0. \tag{8}$$

Let $$y = ax^{\alpha}+bx^{\beta}+cx^{\gamma}+\ldots \tag{9}$$

and let us suppose α, β, γ, ... arranged in the order of ascending magnitude: then substituting (9) in (8) we have

$$a\alpha(\alpha+1)x^{\alpha-2}+n^2ax^{\alpha}+b\beta(\beta+1)x^{\beta-2}+n^2bx^{\beta}$$
$$+c\gamma(\gamma+1)x^{\gamma-2}+n^2cx^{\gamma}+\ldots = 0; \tag{10}$$

now $\alpha-2$ is the lowest index, and as (10) cannot be satisfied for all values of x unless the coefficients of the different powers of x vanish, we have

$$\alpha(\alpha+1) = 0, \tag{11}$$
$$\therefore \quad \alpha = 0, \quad \text{and} \quad \alpha = -1.$$

First, let $\alpha = -1$; then the next two lowest indices are α and $\beta-2$; these may be equal or unequal; if they are unequal the term $b\beta(\beta+1)x^{\beta-2}$ cannot be compounded with any other, and must therefore vanish of itself; and β cannot be equal to -1, because it is, by hypothesis, greater than α, therefore $\beta = 0$: and thus (10) becomes

$$n^2ax^{-1}+n^2b+c\gamma(\gamma+1)x^{\gamma-2}+n^2cx^{\gamma}+\ldots = 0; \tag{12}$$
$$\therefore \quad \gamma-2 = -1, \qquad \gamma = 1;$$
$$n^2a+c\gamma(\gamma+1) = 0, \qquad c = -\frac{n^2a}{1.2};$$
$$\delta-2 = 0, \qquad \delta = 2;$$
$$n^2b+d\delta(\delta+1) = 0, \qquad d = -\frac{n^2b}{1.2.3};$$
$$\therefore \quad y = a\left\{\frac{1}{x}-\frac{n^2x}{1.2}+\frac{n^4x^4}{1.2.3.4}-\ldots\right\}$$
$$+b\left\{1-\frac{n^2x^2}{1.2.3}+\frac{n^4x^3}{1.2.3.4}-\ldots\right\}, \tag{13}$$
$$= \frac{a}{x}\cos nx+\frac{b\sin nx}{nx} = \frac{c_1\cos nx+c_2\sin nx}{x}; \tag{14}$$

which contains two arbitrary constants, and is therefore the general integral.

Secondly, let $a = 0$, then (10) becomes

$$n^2 a + b\beta(\beta+1)x^{\beta-2} + n^2 b x^{\beta} + c\gamma(\gamma+1)x^{\gamma-2} + \dots; \quad (15)$$

and as β must be greater than a, that is, greater than 0, $\beta = 2$; therefore $6b + n^2 a = 0$:

$$\therefore \quad b = -\frac{n^2 a}{1.2.3}; \qquad \gamma - 2 = 2, \qquad \therefore \quad \gamma = 4;$$

$$n^2 b + c\gamma(\gamma+1) = 0, \qquad \therefore \quad c = \frac{n^4 a}{1.2.3.4.5};$$

$$\therefore \quad y = a\left\{1 - \frac{n^2x^2}{1.2.3} + \frac{n^4x^4}{1.2.3.4.5} + \dots\right\}, \qquad y = \frac{a \sin nx}{nx};$$

which is only a particular integral: and the general integral may be determined by a process similar to that of the last Article, by assuming $a = nu$, $\sin nx = \eta x$.

We should also have found a particular integral, if in the former case we had considered $\beta - 2 = a$.

487.] For a second example of the process let us consider the equation (259), Art. 425, which is deduced by a substitution from Riccati's equation; viz.

$$\frac{d^2y}{dx^2} - kx^m y = 0. \quad (16)$$

Let
$$y = a_1 x^{a_1} + a_2 x^{a_2} + a_3 x^{a_3} + \dots \quad (17)$$
$$= \Sigma . ax^a;$$

then substituting in (16) from (17), we have

$$a_1 a_1 (a_1 - 1) x^{a_1 - 2} + a_2 a_2 (a_2 - 1) x^{a_2 - 2} + a_3 a_3 (a_3 - 1) x^{a_3 - 2} + \dots$$
$$= k a_1 x^{m + a_1} + k a_2 x^{m + a_2} + k a_3 x^{m + a_3} + \dots \quad (18)$$

Now if this equation is satisfied for all values of x, we must have

$$a_1(a_1 - 1) = 0; \qquad \therefore \quad a_1 = 0, \qquad a_1 = 1; \quad (19)$$

and if n corresponds to the general term of (17),

$$a_n - 2 = m + a_{n-1}, \quad (20)$$

$$a_n a_n (a_n - 1) = k a_{n-1}. \quad (21)$$

Now from (19) let $a_1 = 0$; therefore from (20) and (21),

$$a_2 = 2 + m, \qquad a_2 = \frac{k a_1}{(m+2)(m+1)};$$

$$a_3 = 2(2+m), \quad a_3 = \frac{k^2 a_1}{(2m+4)(2m+3)(m+2)(m+1)};$$

therefore

$$y = a_1 + \frac{k a_1 x^{m+2}}{(m+1)(m+2)} + \frac{k^2 a_1 x^{2m+4}}{(2m+3)(2m+4)(m+1)(m+2)} + \dots \quad (22)$$

Again from (19) let $\alpha_1 = 1$; therefore from (20) and (21),

$$\alpha_2 = 3 + m, \qquad a_2 = \frac{k a_1}{(m+3)(m+2)};$$

$$\alpha_3 = 5 + 2m, \qquad a_3 = \frac{k^2 a_1}{(m+3)(m+2)(2m+5)(2m+4)};$$

. . . .

$$\therefore\; y = a_1 x + \frac{k a_1 x^{m+3}}{(m+3)(m+2)} + \frac{k^2 a_1 x^{2m+5}}{(m+3)(m+2)(2m+5)(2m+4)} + \dots ($$

Now each of the series (22) and (23) involves one undetermined constant, viz. a_1, each therefore is a particular integral of (16); and the general integral is the sum of the two: it is plain also that the undetermined constant is not necessarily the same in both: replacing it therefore by c_1 and c_2 respectively, we have

$$y = c_1 \left\{ 1 + \frac{k x^{m+2}}{(m+1)(m+2)} + \frac{k^2 x^{2m+4}}{(m+1)(m+2)(2m+3)(2m+4)} + \dots \right\}$$

$$+ c_2 x \left\{ 1 + \frac{k x^{m+2}}{(m+2)(m+3)} + \frac{k^2 x^{2m+4}}{(m+2)(m+3)(2m+4)(2m+5)} + \dots \right\}, \; (2$$

[redacted]

489.] Now on referring to Art. 425, it will be seen that Riccati's equation

$$\frac{dz}{dx} + az^2 = bx^m, \tag{28}$$

is transformed into the equation (16) by putting

$$z = -\frac{1}{ay}\frac{dy}{dx}, \qquad \text{and } -ab = k; \tag{29}$$

by the simple differentiation therefore of (24) we can find z, and thereby obtain the integral of (28) in the form of a series. And by a similar process and the obvious substitutions which are given in Art. 425, we may find the integrals of

$$\frac{d^2y}{dx^2} + \frac{2n}{x}\frac{dy}{dx} = by, \tag{30}$$

and of

$$\frac{d^2y}{dx^2} - \frac{n(n-1)}{x^2}y = by. \tag{31}$$

The last equation however occurs in some future investigations, and requires an independent discussion. Let us, for the sake of greater convenience, express it in the form,

$$\frac{d^2y}{dx^2} - \frac{n(n-1)}{x^2}y + b^2y = 0; \tag{32}$$

and assume

$$y = a_1 x^{a_1} + a_2 x^{a_2} + a_3 x^{a_3} + \dots; \tag{33}$$

this, when substituted in (32), gives

$$\begin{aligned} a_1 a_1(a_1-1)x^{a_1-2} + a_2 a_2(a_2-1)x^{a_2-2} + a_3 a_3(a_3-1)x^{a_3-2} + \dots \\ = n(n-1)\{a_1 x^{a_1-2} + a_2 x^{a_2-2} + a_3 x^{a_3-2} + \dots\} \\ -\{b^2 a_1 x^{a_1} + b^2 a_2 x^{a_2} + b^2 a_3 x^{a_3} + \dots\}; \end{aligned} \tag{34}$$

and if m is a general value of the index,

$$\left.\begin{aligned} a_1(a_1-1) = n(n-1), \qquad a_m = a_1 + 2m, \\ a_m\{a_m(a_m-1) - n(n-1)\} + b^2 a_{m-1} = 0; \end{aligned}\right\} \tag{35}$$

the last of which becomes, by means of the first two,

$$2m(2a_1 + 2m - 1)a_m + b^2 a_{m-1} = 0; \tag{36}$$

and this expresses the relation between two successive coefficients. Thus for the complete integral we have

$$\begin{aligned} y = \mathrm{C}_1 x^n \left\{1 - \frac{b^2x^2}{2(2n+1)} + \frac{b^4x^4}{2.4(2n+1)(2n+3)} - \dots\right\} \\ + \mathrm{C}_2 x^{1-n}\left\{1 - \frac{b^2x^2}{2(3-2n)} + \frac{b^4x^4}{2.4(3-2n)(5-2n)} - \dots\right\}. \end{aligned} \tag{37}$$

Now this series admits of being expressed in the following

form. Integrating by parts, and taking the definite integral between the assigned limits, we have

$$\int_0^\pi (\sin\theta)^{2n-1}(\cos\theta)^p d\theta = \frac{2n+2p+1}{2p+1}\int_0^\pi (\sin\theta)^{2n-1}(\cos\theta)^{p+2}d\theta, \quad (38)$$

$$= \frac{(2n+2p+1)(2n+2p+3)}{(2p+1)(2p+3)}\int_0^\pi (\sin\theta)^{2n-1}(\cos\theta)^{p+4}d\theta, \quad (39)$$

$$= \frac{(2n+2p+1)(2n+2p+3)(2n+2p+5)}{(2p+1)(2p+3)(2p+5)}\int_0^\pi (\sin\theta)^{2n-1}(\cos\theta)^{p+6}d\theta, \quad (40)$$

. . . .

Let $p = 0$; then

$$\left.\begin{aligned}\int_0^\pi (\sin\theta)^{2n-1} d\theta &= \frac{2n+1}{1}\int_0^\pi (\sin\theta)^{2n-1}(\cos\theta)^2 d\theta \\ &= \frac{(2n+1)(2n+3)}{1.3}\int_0^\pi (\sin\theta)^{2n-1}(\cos\theta)^4 d\theta \\ &= \frac{(2n+1)(2n+3)(2n+5)}{1.3.5}\int_0^\pi (\sin\theta)^{2n-1}(\cos\theta)^6 d\theta \\ &= \ldots\ldots\ldots\end{aligned}\right\} \quad (41)$$

Also in these several terms, replacing n by $1-n$, we have

$$\left.\begin{aligned}\int_0^\pi (\sin\theta)^{1-2n} d\theta &= \frac{3-2n}{1}\int_0^\pi (\sin\theta)^{1-2n}(\cos\theta)^2 d\theta \\ &= \frac{(3-2n)(5-2n)}{1.3}\int_0^\pi (\sin\theta)^{1-2n}(\cos\theta)^4 d\theta \\ &= \ldots\ldots\ldots\end{aligned}\right\} \quad (42)$$

Now let the arbitrary constants in (37) be replaced as follows:

$$C_1 = C'\int_0^\pi (\sin\theta)^{2n-1} d\theta, \qquad C_2 = C''\int_0^\pi (\sin\theta)^{1-2n} d\theta, \quad (43)$$

then, after substitution from (41) and (42), (37) becomes

$$y = C'x^n\int_0^\pi \left\{1 - \frac{b^2x^2}{1.2}(\cos\theta)^2 + \frac{b^4x^4}{1.2.3.4}(\cos\theta)^4 - \ldots\right\}(\sin\theta)^{2n-1}d\theta$$

$$+ C''x^{1-n}\int_0^\pi \left\{1 - \frac{b^2x^2}{1.2}(\cos\theta)^2 + \frac{b^4x^4}{1.2.3.4}(\cos\theta)^4 - \ldots\right\}(\sin\theta)^{1-2n}d\theta; \quad (44)$$

or $$y = C'x^n\int_0^\pi \cos(bx\cos\theta)(\sin\theta)^{2n-1}d\theta$$

$$+ C''x^{1-n}\int_0^\pi \cos(bx\cos\theta)(\sin\theta)^{1-2n}d\theta; \quad (45)$$

and this is the general integral of (32).

If $n = 0$, the differential equation becomes

$$\frac{d^2y}{dx^2} + b^2 y = 0; \tag{46}$$

and (37) becomes

$$y = \text{c}' \cos bx + \text{c}'' \sin bx;$$

and (45) becomes

$$y = \text{c}' \int^{\pi} \cos(bx\cos\theta) \frac{d\theta}{\sin\theta} + \text{c}'' x \int_0^{\pi} \cos(bx\cos\theta) \sin\theta \, d\theta, \tag{47}$$

$$= \text{c}' \int_0^{\pi} \cos(bx\cos\theta) \frac{d\theta}{\sin\theta} + \frac{2\text{c}''}{b} \sin bx.$$

Similarly, if $n = 1$, the differential equation becomes of the form (46), and (37) becomes

$$y = \frac{\text{c}_1}{b} \sin bx + \text{c}_2 \cos bx;$$

and (45) becomes

$$y = \text{c}' x \int_0^{\pi} \cos(bx\cos\theta) \sin\theta \, d\theta + \text{c}'' \int_0^{\pi} \cos(bx\cos\theta) \frac{d\theta}{(\sin\theta)^5} \tag{48}$$

$$= \frac{2\text{c}'}{b} \sin bx + \text{c}'' \int_0^{\pi} \cos(bx\cos\theta) \frac{d\theta}{\sin\theta}. \tag{49}$$

A particular form of the differential equation (32) occurs in physical mathematics; viz., when $n = 3$;

$$\frac{d^2y}{dx^2} - \frac{6y}{x^2} + b^2 y = 0; \tag{50}$$

so that (45) becomes

$$y = \text{c}' x^3 \int_0^{\pi} \cos(bx\cos\theta)(\sin\theta)^5 d\theta + \text{c}'' x^{-2} \int_0^{\pi} \cos(bx\cos\theta) \frac{d\theta}{(\sin\theta)^5}, \tag{51}$$

of which the integral is

$$y = \text{c}\left\{\sin(bx+\beta)\left(1 - \frac{3}{b^2x^2}\right) + \frac{3}{bx}\cos(bx+\beta)\right\}, \tag{52}$$

where c and β are arbitrary constants.

Further researches into these subjects may be deferred until the differential equations arise in subsequent physical investigations. Indeed the whole subject is too incomplete for perfect theoretical discussion; and the particular problems are more advantageously studied with the light which they derive from the physical laws expressed by them.

CORRECTIONS.

Page 60, line 14, *for* $-\frac{n}{n+1}$ *read* $+\frac{n}{n+1}$.

83, — 30, *for* "them" *read* "others"

86, — 10, *for* $\Big]^{x_n}$ *read* $\Big]^{x_n}_{x_0}$

90, — 2, *insert* n *after* $n-2$

92, — 23, *for* Section 3 *read* Section 4

96, — 30, *for* $-\text{F}'(x)$ *read* $\text{F}'(x)$

112, — 24, *for* $-n$ *read* $-n-1$

115, — 22, *for* dx *read* dn

177, — 2, *for* $\frac{b^2}{a^2}$ *read* $e^{\frac{b^2}{a^2}}$

178, — 1, *omit* "greater than unity"

184, — 17, *insert* + *before* $\frac{r(r+1)}{1.2}$

215, — 6, *for* NP *read* NE

215, — 14, *for* y *read* r

220, — 28, *for* ϕ *read* ψ

240, — 29, *transpose* "less" *and* "greater"

286, — 23, *for* $\Sigma \pm a_2 \beta_3 \ldots$ *read* $\Sigma \pm a_1 \beta_2 \ldots$

366, — 1, *for* Chapter IX *read* Chapter XI.

The process in page 167 admits of the following simplification; fr (293) we have

$$\frac{d.\log\Gamma(n)}{dn} = \int_0^\infty \frac{e^{-z}dz}{z} - \int_0^\infty \frac{dz}{2(1+z)^n};$$

in the second integral of the right-hand member of this equation, let

$$1+z = e^y; \quad \therefore \quad dz = e^y dy;$$

$$\therefore \quad \frac{d.\log\Gamma(n)}{dn} = \int_0^\infty \frac{e^{-z}dz}{z} - \int_0^\infty \frac{e^{-ny}dy}{1-e^{-y}}$$

$$= \int_0^\infty \left\{\frac{e^{-y}}{y} - \frac{e^{-ny}}{1-e^{-y}}\right\} dy;$$

which is (297) in that page.

BOOKS

PRINTED AT

THE CLARENDON PRESS, OXFORD,

And Published for the University

BY MACMILLAN AND CO.,

29, 30, BEDFORD STREET, COVENT GARDEN, LONDON.

LEXICONS, GRAMMARS, &c.

A Greek-English Lexicon, by Henry George Liddell, D.D., and Robert Scott, D.D. *Sixth Edition, Revised and Augmented.* 1870. 4to. *cloth,* 1*l.* 16*s.*

A Greek-English Lexicon, abridged from the above, chiefly for the use of Schools. *Fifteenth Edition,* carefully revised throughout. 1872. square 12mo. *cloth,* 7*s.* 6*d.*

A copious Greek-English Vocabulary, compiled from the best authorities. 1850. 24mo. *bound,* 3*s.*

Graecae Grammaticae Rudimenta in usum Scholarum. Auctore Carolo Wordsworth, D.C.L. *Seventeenth Edition.* 1870. 12mo. *bound,* 4*s.*

A Practical Introduction to Greek Accentuation, by H. W. Chandler, M.A. 1862. 8vo. *cloth,* 10*s.* 6*d.*

Scheller's Lexicon of the Latin Tongue, with the German explanations translated into English by J. E. Riddle, M.A. 1835. fol. *cloth,* 1*l.* 1*s.*

A Sanskrit-English Dictionary, Etymologically and Philologically arranged, with special reference to Greek, Latin, German, Anglo-Saxon, English, and other cognate Indo-European Languages, by Monier Williams, M.A. 1872. 4to. *cloth,* 4*l.* 14*s.* 6*d.*

A Practical Grammar of the Sanskrit Language, arranged with reference to the Classical Languages of Europe, for the use of English Students, by Monier Williams, M.A. *Third Edition.* 1864. 8vo. *cloth,* 15*s.*

An Icelandic-English Dictionary. By the late R. Cleasby. Enlarged and completed by G. Vigfússon. Parts I. and II. 4to. 21*s.* *each.*

GREEK AND LATIN CLASSICS.

Aeschylus: Tragoediae et Fragmenta, ex recensione Guil. Dindorfii. *Second Edition.* 1851. 8vo. *cloth,* 5*s.* 6*d.*

Sophocles: Tragoediae et Fragmenta, ex recensione et cum commentariis Guil. Dindorfii. *Third Edition.* 2 vols. 1860. fcap. 8vo. *cloth,* 1*l.* 1*s.*

Each Play separately, *limp,* 2*s.* 6*d.*

The Text alone, square 16mo. *cloth,* 3*s.* 6*d.*

Each Play separately, *limp,* 6*d.*

Sophocles: Tragoediae et Fragmenta, ex recensione Guil. Dindorfii. *Second Edition.* 1849. 8vo. *cloth,* 5*s.* 6*d.*

Euripides: Tragoediae et Fragmenta, ex recensione Guil. Dindorfii. Tomi II. 1834. 8vo. *cloth,* 10*s.*

Aristophanes: Comoediae et Fragmenta, ex recensione Guil. Dindorfii. Tomi II. 1835. 8vo. *cloth,* 11*s.*

Aristoteles: ex recensione Immanuelis Bekkeri. Accedunt Indices Sylburgiani. Tomi XI. 1837. 8vo. *cloth,* 2*l.* 10*s.*
Each volume separately, 5*s.* 6*d.*

Catulli Veronensis Liber: recognovit, apparatum criticum prolegomena appendices addidit, Robinson Ellis, A.M. 1867. 8vo. *cloth,* 16*s.*

Catulli Veronensis Carmina Selecta: secundum recognitionem Robinson Ellis, A.M. Extra fcap. 8vo. *cloth,* 3*s.* 6*d.* *Just published.*

Demosthenes: ex recensione Guil. Dindorfii. Tomi IV. 1846. 8vo. *cloth. Price reduced from* 2*l.* 2*s.* to 1*l.* 1*s.*

Homerus: Ilias, ex rec. Guil. Dindorfii. 1856. 8vo. *cloth,* 5*s.* 6*d.*

Homerus: Odyssea, ex rec. Guil. Dindorfii. 1855. 8vo. *cloth,* 5*s.* 6*d.*

Plato: The Apology, with a revised Text and English Notes, and a Digest of Platonic Idioms, by James Riddell, M.A. 1867. 8vo. *cloth,* 8*s.* 6*d.*

Plato: Philebus, with a revised Text and English Notes, by Edward Poste, M.A. 1860. 8vo. *cloth,* 7*s.* 6*d.*

Plato: Sophistes and Politicus, with a revised Text and English Notes, by L. Campbell, M.A. 1866. 8vo. *cloth,* 18*s.*

Plato: Theaetetus, with a revised Text and English Notes, by L. Campbell, M.A. 1861. 8vo. *cloth,* 9*s.*

Plato: The Dialogues, translated into English, with Analyses and Introductions, by B. Jowett, M.A., Master of Balliol College, and Regius Professor of Greek. 4 vols. 1871. 8vo. *cloth,* 3*l.* 6*s.*

Xenophon: Historia Graeca, ex recensione et cum annotationibus L. Dindorfii. *Second Edition.* 1852. 8vo. *cloth,* 10*s.* 6*d.*

Xenophon: Expeditio Cyri, ex rec. et cum annotatt. L. Dindorfii. *Second Edition.* 1855. 8vo. *cloth,* 10*s.* 6*d.*

Xenophon: Institutio Cyri, ex rec. et cum annotatt. L. Dindorfii. 1857. 8vo. *cloth,* 10*s.* 6*d.*

Xenophon: Memorabilia Socratis, ex rec. et cum annotatt. L. Dindorfii. 1862. 8vo. *cloth,* 7*s.* 6*d.*

Xenophon: Opuscula Politica Equestria et Venatica cum Arriani Libello de Venatione, ex rec. et cum annotatt. L. Dindorfii. 1866. 8vo. *cloth,* 10*s.* 6*d.*

THE HOLY SCRIPTURES, &c.

The Holy Bible in the earliest English Versions, made from the Latin Vulgate by John Wycliffe and his followers: edited by the Rev. J. Forshall and Sir F. Madden. 4 vols. 1850. royal 4to. *cloth. Price reduced from* 5*l.* 15*s.* 6*d.* *to* 3*l.* 3*s.*

The Holy Bible: an exact reprint, page for page, of the Authorized Version published in the year 1611. Demy 4to. *half bound*, 1*l.* 1*s.*

Vetus Testamentum Graece secundum exemplar Vaticanum Romae editum. Accedit potior varietas Codicis Alexandrini. Tomi III. 1848. 12mo. *cloth*, 14*s.*

Novum Testamentum Graece. Accedunt parallela S. Scripturae loca, necnon vetus capitulorum notatio et canones Eusebii. Edidit Carolus Lloyd, S.T.P.R., necnon Episcopus Oxoniensis. 1869. 18mo. *cloth*, 3*s.*

The same on writing paper, with large margin, small 4to. *cloth*, 10*s.* 6*d.*

Novum Testamentum Graece juxta exemplar Millianum. 1868. 12mo. *cloth*, 2*s.* 6*d.*

The same on writing paper, with large margin, small 4to. *cloth*, 6*s.* 6*d.*

Evangelia Sacra Graece. *The Text of Mill.* 1870. fcap. 8vo. *limp*, 1*s.* 6*d.*

The New Testament in Greek and English, on opposite pages, arranged and edited by E. Cardwell, D.D. 2 vols. 1837. crown 8vo. *cloth*, 6*s.*

Novum Testamentum Graece. Antiquissimorum Codicum Textus in ordine parallelo dispositi. Accedit collatio Codicis Sinaitici. Edidit E. H. Hansell, S.T.B. Tomi III. 1864. 8vo. *half morocco*, 2*l.* 12*s.* 6*d.*

Diatessaron; sive Historia Jesu Christi ex ipsis Evangelistarum verbis apte dispositis confecta. Ed. J. White. 1856. 12mo. *cloth*, 3*s.* 6*d.*

Canon Muratorianus. The earliest Catalogue of the Books of the New Testament. Edited with Notes and a Facsimile of the MS. in the Ambrosian Library at Milan, by S. P. Tregelles, LL.D. 1868. 4to. *cloth*, 10*s.* 6*d.*

Horae Hebraicae et Talmudicae, a J. Lightfoot. *A new Edition*, by R. Gandell, M.A. 4 vols. 1859. 8vo. *cloth*. *Price reduced from* 2*l.* 2*s.* *to* 1*l.* 1*s.*

ECCLESIASTICAL HISTORY, &c.

Baedae Historia Ecclesiastica. Edited, with English Notes, by G. H. Moberly, M.A., Fellow of C.C.C., Oxford. 1869. crown 8vo. *cloth*, 10*s.* 6*d.*

Bingham's Antiquities of the Christian Church, and other Works. 10 vols. 1855. 8vo. *cloth*. *Price reduced from* 5*l.* 5*s.* *to* 3*l.* 3*s.*

Burnet's History of the Reformation of the Church of England. *A new Edition*. Carefully revised, and the Records collated with the originals, by N. Pocock, M.A. With a Preface by the Editor. 7 vols. 1865. 8vo. *cloth*, 4*l.* 4*s.*

Councils and Ecclesiastical Documents relating to Great Britain and Ireland. Edited, after Spelman and Wilkins, by A. W. Haddan, B.D., and William Stubbs, M.A. Vol. I. 1869. medium 8vo. *cloth*, 1*l.* 1*s.*

Vol. II. *in the Press.*

Vol. III. medium 8vo. *cloth*, 1*l.* 1*s.*

Records of the Reformation. The Divorce, 1527–1533. Mostly now for the first time printed from MSS. in the British Museum, and other Libraries. Collected and arranged by N. Pocock, M.A. 2 vols. 8vo. *cloth*, 1*l.* 16*s.*

Eusebius' Ecclesiastical History, according to the text of Burton. With an Introduction by William Bright, D.D. 1872. Crown 8vo. *cloth,* 8*s.* 6*d.*

Fuller's Church History of Britain. Edited by J. S. Brewer, M.A. 6 vols. 1845. 8vo. *cloth,* 1*l.* 19*s.*

Hussey's Rise of the Papal Power traced in three Lectures. *Second Edition.* 1863. fcap. 8vo. *cloth,* 4*s.* 6*d.*

Le Neve's Fasti Ecclesiae Anglicanae. *Corrected and continued from* 1715 *to* 1853 by T. Duffus Hardy. 3 vols. 1854. 8vo. *cloth. Price reduced from* 1*l.* 17*s.* 6*d. to* 1*l.* 1*s.*

Noelli (A.) Catechismus sive prima institutio disciplinaque Pietatis Christianae Latine explicata. Editio nova cura Guil. Jacobson, A.M. 1844. 8vo. *cloth,* 5*s.* 6*d.*

Patrum Apostolicorum, S. Clementis Romani, S. Ignatii, S. Polycarpi, quae supersunt. Edidit Guil. Jacobson, S.T.P.R. Tomi II. *Fourth Edition.* 1863. 8vo. *cloth,* 1*l.* 1*s.*

Prideaux's Connection of Sacred and Profane History. 2 vols. 1851. 8vo. *cloth,* 10*s.*

Shuckford's Sacred and Profane History connected (in continuation of Prideaux). 2 vols. 1848. 8vo. *cloth,* 10*s.*

Reliquiae Sacrae secundi tertiique saeculi. Recensuit M. J. Routh, S.T.P. Tomi V. 1846–1848. 8vo. *cloth. Price reduced from* 2*l.* 11*s. to* 1*l.* 5*s.*

Scriptorum Ecclesiasticorum Opuscula. Recensuit M. J. Routh, S.T.P. Tomi II. *Third Edition.* 1858. 8vo. *cloth. Price reduced from* 1*l. to* 10*s.*

Stubbs's (W.) Registrum Sacrum Anglicanum. An attempt to exhibit the Course of Episcopal Succession in England. 1858. small 4to. *cloth,* 8*s.* 6*d.*

ENGLISH THEOLOGY.

Butler's Works, with an Index to the Analogy. 2 vols. 1849. 8vo. *cloth,* 11*s.*

Greswell's Harmonia Evangelica. *Fifth Edition.* 1856. 8vo. *cloth,* 9*s.* 6*d.*

Hall's (Bp.) Works. *A new Edition,* by Philip Wynter, D.D. 10 vols. 1863. 8vo. *cloth. Price reduced from* 5*l.* 5*s. to* 3*l.* 3*s.*

Heurtley's Collection of Creeds. 1858. 8vo. *cloth,* 6*s.* 6*d.*

Homilies appointed to be read in Churches. Edited by J. Griffiths, M.A. 1859. 8vo. *cloth. Price reduced from* 10*s.* 6*d. to* 7*s.* 6*d.*

Hooker's Works, with his Life by Walton, arranged by John Keble, M.A. *Fifth Edition.* 1865. 3 vols. 8vo. *cloth,* 1*l.* 11*s.* 6*d.*

Hooker's Works; the text as arranged by John Keble, M.A. 2 vols. 1865. 8vo. *cloth,* 11*s.*

Jackson's (Dr. Thomas) Works. 12 vols. 1844. 8vo. *cloth,* 3*l.* 6*s.*

Jewel's Works. Edited by R. W. Jelf, D.D. 8 vols. 1847. 8vo. *cloth. Price reduced from* 2*l.* 10*s. to* 1*l.* 10*s.*

Patrick's Theological Works. 9 vols. 1859. 8vo. *cloth. Price reduced from 3l. 14s. 6d. to 1l. 1s.*

Pearson's Exposition of the Creed. Revised and corrected by E. Burton, D.D. *Fifth Edition.* 1864. 8vo. *cloth,* 10s. 6d.

Pearson's Minor Theological Works. Now first collected, with a Memoir of the Author, Notes, and Index, by Edward Churton, M.A. 2 vols. 1844. 8vo. *cloth. Price reduced from 14s. to 10s.*

Sanderson's Works. Edited by W. Jacobson, D.D. 6 vols. 1854. 8vo. *cloth. Price reduced from 1l. 19s. to 1l. 10s.*

South's Sermons. 5 vols. 1842. 8vo. *cloth. Price reduced from 2l. 10s. 6d. to 1l. 10s.*

Stanhope's Paraphrase and Comment upon the Epistles and Gospels. *A new Edition.* 2 vols. 1851. 8vo. *cloth. Price reduced from 18s. to 10s.*

Wall's History of Infant Baptism, with Gale's Reflections, and Wall's Defence. *A new Edition,* by Henry Cotton, D.C.L. 2 vols. 1862. 8vo. *cloth,* 1l. 1s.

Waterland's Works, with Life, by Bp. Van Mildert. *A new Edition,* with copious Indexes. 6 vols. 1857. 8vo. *cloth,* 2l. 11s.

Waterland's Review of the Doctrine of the Eucharist, with a Preface by the present Bishop of London. 1868. crown 8vo. *cloth,* 6s. 6d.

Wheatly's Illustration of the Book of Common Prayer. *A new Edition,* 1846. 8vo. *cloth,* 5s.

Wyclif. Select English Works. By T. Arnold, M.A. 3 vols. 1871. 8vo. *cloth,* 2l. 2s.

ENGLISH HISTORY.

Two of the Saxon Chronicles parallel, with Supplementary Extracts from the Others. Edited, with Introduction, Notes, and a Glossarial Index, by J. Earle, M.A. 1865. 8vo. *cloth,* 16s.

Burnet's History of His Own Time, with the suppressed Passages and Notes. 6 vols. 1833. 8vo. *cloth,* 2l. 10s.

Carte's Life of James Duke of Ormond. *A new Edition,* carefully compared with the original MSS. 6 vols. 1851. 8vo. *cloth. Price reduced from 2l. 6s. to 1l. 5s.*

Clarendon's (Edw. Earl of) History of the Rebellion and Civil Wars in England. To which are subjoined the Notes of Bishop Warburton. 7 vols. 1849. medium 8vo. *cloth,* 2l. 10s.

Clarendon's (Edw. Earl of) History of the Rebellion and Civil Wars in England. 7 vols. 1839. 18mo. *cloth,* 1l. 1s.

Freeman's (E. A.) History of the Norman Conquest of England: its Causes and Results. Vols. I. and II. *A new Edition,* with Index. 8vo. *cloth,* 1l. 16s.

Vol. III. The Reign of Harold and the Interregnum. 1869. 8vo. *cloth,* 1l. 1s.
Vol. IV. The Reign of William. 8vo. *cloth,* 1l. 1s.

May's History of the Long Parliament. 1854. 8vo. *cloth*, 6*s*. 6*d*.

Rogers's History of Agriculture and Prices in England, A.D. 1259–1400. 2 vols. 1866. 8vo. *cloth*, 2*l*. 2*s*.

Whitelock's Memorials of English Affairs from 1625 to 1660. 4 vols. 1853. 8vo. *cloth*, 1*l*. 10*s*.

PHILOSOPHICAL WORKS, AND GENERAL LITERATURE.

A Course of Lectures on Art, delivered before the University of Oxford in Hilary Term, 1870. By John Ruskin, M.A., Slade Professor of Fine Art. Demy 8vo. *cloth*, 6*s*.

A Critical Account of the Drawings by Michel Angelo and Raffaello in the University Galleries, Oxford. By J. C. Robinson, F.S.A. Crown 8vo. *cloth*, 4*s*.

Bacon's Novum Organum, edited, with English notes, by G.W. Kitchin, M.A. 1855. 8vo. *cloth*, 9*s*. 6*d*.

Bacon's Novum Organum, translated by G. W. Kitchin, M.A. 1855. 8vo. *cloth*, 9*s*. 6*d*.

The Works of George Berkeley, D.D., formerly Bishop of Cloyne; including many of his writings hitherto unpublished. With Prefaces, Annotations, and an Account of his Life and Philosophy, by Alexander Campbell Fraser, M.A., 4 vols. 1871. 8vo. *cloth*, 2*l*. 18*s*.

Also separately. The Works. 3 vols. *cloth*, 2*l*. 2*s*.
The Life and Letters, &c. 1 vol. *cloth*, 16*s*.

Smith's Wealth of Nations. *A new Edition*, with Notes, by J. E. Thorold Rogers, M.A. 2 vols. 8vo. *cloth*, 1*l*. 1*s*.

MATHEMATICS, PHYSICAL SCIENCE, &c.

Treatise on Infinitesimal Calculus. By Bartholomew Price, M.A., F.R.S., Professor of Natural Philosophy, Oxford.

Vol. I. Differential Calculus. *Second Edition*. 1858. 8vo. *cloth*, 14*s*. 6*d*.

Vol. II. Integral Calculus, Calculus of Variations, and Differential Equations. *Second Edition*. 1865. 8vo. *cloth*, 18*s*.

Vol. III. Statics, including Attractions; Dynamics of a Material Particle. *Second Edition*. 1868. 8vo. *cloth*, 16*s*.

Vol. IV. Dynamics of Material Systems; together with a Chapter on Theoretical Dynamics, by W. F. Donkin, M.A., F.R.S. 1862. 8vo. *cloth*, 16*s*.

Vesuvius. By John Phillips, M.A., F.R.S., Professor of Geology, Oxford. 1869. crown 8vo. *cloth*, 10*s*. 6*d*.

Rigaud's Correspondence of Scientific Men of the 17th Century, with Index by A. de Morgan. 2 vols. 1841–62. 8vo. *cloth*, 18*s*. 6*d*.

Clarendon Press Series.

The Delegates of the Clarendon Press having undertaken the publication of a series of works, chiefly educational, and entitled the **Clarendon Press Series**, have published, or have in preparation, the following.

Those to which prices are attached are already published; the others are in preparation.

I. GREEK AND LATIN CLASSICS, &c.

A Greek Primer in English, for the use of beginners. By the Right Rev. Charles Wordsworth, D.C.L., Bishop of St. Andrews. *Second Edition.* Ext. fcap. 8vo. *cloth*, 1*s*. 6*d*.

Greek Verbs, Irregular and Defective; their forms, meaning, and quantity; embracing all the Tenses used by Greek writers, with reference to the passages in which they are found. By W. Veitch. *New Edition.* Crown 8vo. *cloth*, 10*s*. 6*d*.

The Elements of Greek Accentuation (for Schools): abridged from his larger work by H. W. Chandler, M.A., Waynflete Professor of Moral and Metaphysical Philosophy, Oxford. Ext. fcap. 8vo. *cloth*, 2*s*. 6*d*.

Aeschines in Ctesiphontem and **Demosthenes de Corona.** With Introduction and Notes. By G. A. Simcox, M.A., and W. H. Simcox, M.A., Fellows of Queen's College, Oxford. 8vo. *cloth*, 12. *Just published.*

Aristotle's Politics. By W. L. Newman, M.A., Fellow and Lecturer of Balliol College, and Reader in Ancient History, Oxford.

The Golden Treasury of Ancient Greek Poetry; being a Collection of the finest passages in the Greek Classic Poets, with Introductory Notices and Notes. By R. S. Wright, M.A., Fellow of Oriel College, Oxford. Extra fcap. 8vo. *cloth*, 8*s*. 6*d*.

A Golden Treasury of Greek Prose, being a Collection of the finest passages in the principal Greek Prose Writers, with Introductory Notices and Notes. By R. S. Wright, M.A., Fellow of Oriel College, Oxford; and J. E. L. Shadwell, M.A., Senior Student of Christ Church. Extra fcap. 8vo. *cloth*, 4*s*. 6*d*.

Homer. Iliad. By D. B. Monro, M.A., Fellow and Tutor of Oriel College, Oxford.

Homer. Odyssey, Books I–XII (for Schools). By W. W. Merry, M.A., Fellow and Lecturer of Lincoln College, Oxford. *Second Edition.* Extra fcap. 8vo. *cloth*, 4*s*. 6*d*.

Homer. Odyssey, Books I–XII. By W. W. Merry, M.A., Fellow and Lecturer of Lincoln College, Oxford; and the late James Riddell, M.A., Fellow of Balliol College, Oxford.

Homer. Odyssey, Books XIII–XXIV. By Robinson Ellis, M.A., Fellow of Trinity College, Oxford.

Plato. Selections (for Schools). With Notes, by B. Jowett, M.A., Regius Professor of Greek; and J. Purves, M.A., Fellow and Lecturer of Balliol College, Oxford.

Sophocles. The Plays and Fragments. With English Notes and Introductions by Lewis Campbell, M.A., Professor of Greek, St. Andrews, formerly Fellow of Queen's College, Oxford. In Two Volumes.

Vol. I. Oedipus Tyrannus, Oedipus Coloneus, Antigone. 8vo. *cloth*, 14*s*. *Just published.*

Sophocles. Oedipus Rex: Dindorf's Text, with Notes by the Ven. Archdeacon Basil Jones, M.A., formerly Fellow of University College, Oxford. *Second Edition.* Ext. fcap. 8vo. *limp cloth*, 1*s*. 6*d*.

Theocritus (for Schools). With Notes, by H. Snow, M.A., Assistant Master at Eton College, formerly Fellow of St. John's College, Cambridge. Extra fcap. 8vo. *cloth*, 4*s*. 6*d*.

Xenophon. Selections (for Schools). With Notes and Maps, by J. S. Phillpotts, B.C.L., Assistant Master in Rugby School, formerly Fellow of New College, Oxford. *Second Edition.* Extra fcap. 8vo. *cloth*, 3*s*. 6*d*.

Arrian. Selections (for Schools). By the same Editor.

Caesar. The Commentaries (for Schools). Part I. The Gallic War, with Notes and Maps, by Charles E. Moberly, M.A., Assistant Master in Rugby School; formerly Scholar of Balliol College, Oxford. Extra fcap. 8vo. *cloth*, 4*s*. 6*d*.

Part II. The Civil War: Bk. 1. By the same Editor. Extra fcap. 8vo. *cloth*, 2*s*.

Cicero's Philippic Orations. With Notes, by J. R. King, M.A., formerly Fellow and Tutor of Merton College, Oxford. Demy 8vo. *cloth*, 10*s*. 6*d*.

Cicero pro Cluentio. With Introduction and Notes. By W. Ramsay, M.A. Edited by G. G. Ramsay, M.A., Professor of Humanity, Glasgow. Extra fcap. 8vo. *cloth*, 3*s*. 6*d*.

Cicero. Selection of interesting and descriptive passages. With Notes. By Henry Walford, M.A., Wadham College, Oxford, Assistant Master at Haileybury College. In three Parts. *Second Edition.* Extra fcap. 8vo. *cloth*, 4*s*. 6*d*.

Each Part separately, *limp*, 1*s*. 6*d*.

Part I. Anecdotes from Grecian and Roman History.
Part II. Omens and Dreams: Beauties of Nature.
Part III. Rome's Rule of her Provinces.

Cicero. Select Letters. With English Introductions, Notes, and Appendices. By Albert Watson, M.A., Fellow and Tutor of Brasenose College, Oxford. 8vo. *cloth*, 18*s*.

Cicero. Selected Letters (for Schools). With Notes. By the late C. E. Prichard, M.A., formerly Fellow of Balliol College, Oxford, and E. R. Bernard, M.A., Fellow of Magdalen College, Oxford. Extra fcap. 8vo. *cloth*, 3*s*.

Cicero de Oratore. With Introduction and Notes. By A. S. Wilkins, M.A., Professor of Latin, Owens College, Manchester.

Cornelius Nepos. With Notes, by Oscar Browning, M.A., Fellow of King's College, Cambridge, and Assistant Master at Eton College. Extra fcap. 8vo. *cloth*, 2*s*. 6*d*.

Horace. With Introduction and Notes. By Edward C. Wickham, M.A., Fellow and Tutor of New College, Oxford.

Also a small edition for Schools.

Livy, Book I. By J. R. Seeley, M.A., Fellow of Christ's College, and Regius Professor of Modern History, Cambridge. 8vo. *cloth*, 6*s*.

Also a small edition for Schools.

Ovid. Selections for the use of Schools. With Introductions and Notes, and an Appendix on the Roman Calendar. By W. Ramsay, M.A. Edited by G. G. Ramsay, M.A., Professor of Humanity, Glasgow. Ext. fcap. 8vo. *cloth*, 5*s*. 6*d*.

Persius. The Satires. With a Translation and Commentary. By John Conington, M.A., late Corpus Professor of Latin in the University of Oxford. Edited by Henry Nettleship, M.A., formerly Fellow of Lincoln College, Oxford. 8vo., *cloth*, 7*s*. 6*d*.

Pliny. Selected Letters (for Schools). By the late C. E. Prichard, M.A., formerly Fellow of Balliol College, Oxford, and E. R. Bernard, M.A., Fellow of Magdalen College, Oxford. Extra fcap. 8vo., *cloth*, 3*s*.

Fragments and Specimens of Early Latin. With Introduction, Notes, and Illustrations. By John Wordsworth, M.A., Fellow of Brasenose College, Oxford.

Selections from the less known Latin Poets. By North Pinder, M.A., formerly Fellow of Trinity College, Oxford. Demy 8vo. *cloth*, 15*s*.

Passages for Translation into Latin. For the use of Passmen and others. Selected by J. Y. Sargent, M.A., Tutor, formerly Fellow, of Magdalen College, Oxford. *Second Edition*. Ext. fcap. 8vo. *cloth*, 2*s*. 6*d*.

II. MENTAL AND MORAL PHILOSOPHY.

The Elements of Deductive Logic, designed mainly for the use of Junior Students in the Universities. By T. Fowler, M.A., Fellow and Tutor of Lincoln College, Oxford. *Fourth Edition*, with a Collection of Examples. Extra fcap. 8vo. *cloth*, 3*s*. 6*d*.

The Elements of Inductive Logic, designed mainly for the use of Students in the Universities. By the same Author. Extra fcap. 8vo. *cloth*, 6*s*.

A Manual of Political Economy, for the use of Schools. By J. E. Thorold Rogers, M.A., formerly Professor of Political Economy, Oxford. *Second Edition*. Extra fcap. 8vo. *cloth*, 4*s*. 6*d*.

III. MATHEMATICS, &c.

Acoustics. By W. F. Donkin, M.A., F.R.S., Savilian Professor of Astronomy, Oxford. Crown 8vo. *cloth*, 7*s*. 6*d*.

An Elementary Treatise on Quaternions. By P. G. Tait, M.A., Professor of Natural Philosophy in the University of Edinburgh; formerly Fellow of St. Peter's College, Cambridge. Demy 8vo. *cloth*, 12*s*. 6*d*.

Book-keeping. By R. G. C. Hamilton, Accountant to the Board of Trade, and John Ball (of the Firm of Messrs. Quilter, Ball, and Co.), Examiners in Book-keeping for the Society of Arts' Examination. *Third Edition.* Extra fcap. 8vo. *limp cloth,* 1*s.* 6*d.*

Figures made Easy: a first Arithmetic Book. (Introductory to 'The Scholar's Arithmetic,' by the same Author.) By Lewis Hensley, M.A., formerly Fellow and Assistant Tutor of Trinity College, Cambridge. Crown 8vo. *stiff covers,* 6*d.* *Just published.*

Answers to the Examples in Figures made Easy, together with two thousand additional Examples, formed from the Tables in the same, with Answers. By the same Author. Crown 8vo. *cloth,* 1*s.* *Just published.*

The Scholar's Arithmetic. By the same Author. *Nearly ready.*

A Course of Lectures on Pure Geometry. By Henry J. Stephen Smith, M.A., F.R.S., Fellow of Balliol College, and Savilian Professor of Geometry in the University of Oxford.

A Treatise on Electricity and Magnetism. By J. Clerk Maxwell, M.A., F.R.S., formerly Professor of Natural Philosophy, King's College, London. *In the Press.*

A Series of Elementary Works is being arranged, and will shortly be announced.

IV. HISTORY.

A Manual of Ancient History. By George Rawlinson, M.A., Camden Professor of Ancient History, formerly Fellow of Exeter College, Oxford. Demy 8vo. *cloth,* 14*s.*

Select Charters and other Illustrations of English Constitutional History; from the Earliest Times to the Reign of Edward I. Arranged and edited by W. Stubbs, M.A., Regius Professor of Modern History in the University of Oxford. Crown 8vo. *cloth,* 8*s.* 6*d.*

A Constitutional History of England. By the same Author.

A History of Germany and of the Empire, down to the close of the Middle Ages. By J. Bryce, B.C.L., Fellow of Oriel College, Oxford.

A History of Germany, from the Reformation. By Adolphus W. Ward, M.A., Fellow of St. Peter's College, Cambridge, Professor of History, Owens College, Manchester.

A History of British India. By S. J. Owen, M.A., Lee's Reader in Law and History, Christ Church, and Teacher of Indian Law and History in the University of Oxford.

A History of Greece. By E. A. Freeman, M.A., formerly Fellow of Trinity College, Oxford.

A History of France. By G. W. Kitchin, M.A., formerly Censor of Christ Church. *In the Press.*

V. LAW.

Gaii Institutionum Juris Civilis Commentarii Quatuor; or, Elements of Roman Law by Gaius. With a Translation and Commentary, by Edward Poste, M.A., Barrister-at-Law, and Fellow of Oriel College, Oxford. 8vo. *cloth*, 16*s*.

Elements of Law, considered with reference to principles of General Jurisprudence. By William Markby, M.A., Judge of the High Court of Judicature, Calcutta. Crown 8vo. *cloth*, 6*s*. 6*d*.

The Elements of Jurisprudence. By Thomas Erskine Holland, M.A., Barrister at Law, and formerly Fellow of Exeter College, Oxford.

The Institutes of Justinian as a recension of the Institutes of Gaius. By the same Editor. *In the Press.*

Commentaries on Roman Law; from the original and the best modern sources. By H. J. Roby, M.A., formerly Fellow of St. John's College, Cambridge; Professor of Law at University College, London.

VI. PHYSICAL SCIENCE.

Natural Philosophy. In four Volumes. By Sir W. Thomson, LL.D., D.C.L., F.R.S., Professor of Natural Philosophy, Glasgow; and P. G. Tait, M.A., Professor of Natural Philosophy, Edinburgh; formerly Fellows of St. Peter's College, Cambridge. Vol. I. 8vo. *cloth*, 1*l*. 5*s*.

Elements of Natural Philosophy. By the same Authors; being a smaller Work on the same subject, and forming a complete Introduction to it, so far as it can be carried out with Elementary Geometry and Algebra. *Just ready.*

Descriptive Astronomy. A Handbook for the General Reader, and also for Practical Observatory work. With 224 illustrations and numerous tables. By G. F. Chambers, F.R.A.S., Barrister-at-Law. Demy 8vo. 856 pp., *cloth*, 1*l*. 1*s*.

Chemistry for Students. By A. W. Williamson, Phil. Doc., F.R.S., Professor of Chemistry, University College, London. *A new Edition, with Solutions.* Extra fcap. 8vo. *cloth*, 8*s*. 6*d*.

A Treatise on Heat, with numerous Woodcuts and Diagrams. By Balfour Stewart, LL.D., F.R.S., Director of the Observatory at Kew. *Second Edition.* Ext. fcap. 8vo. *cloth*, 7*s*. 6*d*.

Forms of Animal Life. By G. Rolleston, M.D., F.R.S., Linacre Professor of Physiology, Oxford. Illustrated by Descriptions and Drawings of Dissections. Demy 8vo. *cloth*, 16*s*.

Exercises in Practical Chemistry. By A. G. Vernon Harcourt, M.A., F.R.S., Senior Student of Christ Church, and Lee's Reader in Chemistry; and H. G. Madan, M.A., Fellow of Queen's College, Oxford.

Series I. Qualitative Exercises. Crown 8vo. *cloth*, 7*s*. 6*d*.
Series II. Quantitative Exercises.

Geology of Oxford and the Valley of the Thames. By John Phillips, M.A., F.R.S., Professor of Geology, Oxford. 8vo. *cloth*, 21*s*.

Geology. By J. Phillips, M.A., F.R.S., Professor of Geology, Oxford.

Mechanics. By Bartholomew Price, M.A., F.R.S., Sedleian Professor of Natural Philosophy, Oxford.

Optics. By R. B. Clifton, M.A., F.R.S., Professor of Experimental Philosophy, Oxford; formerly Fellow of St. John's College, Cambridge.

Electricity. By W. Esson, M.A., F.R.S., Fellow and Mathematical Lecturer of Merton College, Oxford.

Crystallography. By M. H. N. Story-Maskelyne, M.A., Professor of Mineralogy, Oxford; and Deputy Keeper in the Department of Minerals, British Museum.

Mineralogy. By the same Author.

Physiological Physics. By G. Griffith, M.A., Jesus College, Oxford, Assistant Secretary to the British Association, and Natural Science Master at Harrow School.

Magnetism.

VII. ENGLISH LANGUAGE AND LITERATURE.

A First Reading Book. By Marie Eichens of Berlin; and edited by Anne J. Clough. Extra fcap. 8vo. *stiff covers, 4d.*

Oxford Reading Book, Part I. For Little Children. Extra fcap. 8vo. *stiff covers, 6d.*

Oxford Reading Book, Part II. For Junior Classes. Extra fcap. 8vo. *stiff covers, 6d.*

On the Principles of Grammar. By E. Thring, M.A., Head Master of Uppingham School. Extra fcap. 8vo. *cloth, 4s. 6d.*

Grammatical Analysis, designed to serve as an Exercise and Composition Book in the English Language. By E. Thring, M.A., Head Master of Uppingham School. Extra fcap. 8vo. *cloth, 3s. 6d.*

An English Grammar and Reading Book. For Lower Forms in Classical Schools. By O. W. Tancock, M.A., Assistant Master of Sherborne School. Extra fcap. 8vo. *cloth, 3s. 6d.*

The Philology of the English Tongue. By J. Earle, M.A., formerly Fellow of Oriel College, and Professor of Anglo-Saxon, Oxford. Extra fcap. 8vo. *cloth, 6s. 6d.*

Specimens of Early English. Part II. A.D. 1298 to A.D. 1393. *A new and revised Edition.* With Introduction, Notes, and Glossarial Index. By R. Morris, LL.D., and W. W. Skeat, M.A. Extra fcap. 8vo. *cloth, 7s. 6d.*
Part I. *In the Press.*

Specimens of English Literature, from the 'Ploughmans Crede' to the 'Shepheardes Calender' (A.D. 1394 to A.D. 1579). With Introduction, Notes, and Glossarial Index, by W. W. Skeat, M.A., formerly Fellow of Christ's College, Cambridge. Extra fcap. 8vo. *cloth, 7s. 6d.*

The Vision of William concerning Piers the Plowman, by William Langland. Edited, with Notes, by W. W. Skeat, M.A., formerly Fellow of Christ's College, Cambridge. Extra fcap. 8vo. *cloth, 4s. 6d.*

Typical Selections from the best English Authors from the Sixteenth to the Nineteenth Century, (to serve as a higher Reading Book,) with Introductory Notices and Notes, being a contribution towards a History of English Literature. Extra fcap. 8vo. *cloth, 4s. 6d.*

Specimens of the Scottish Language; being a Series of Annotated Extracts illustrative of the Literature and Philology of the Lowland Tongue from the Fourteenth to the Nineteenth Century. With Introduction and Glossary. By A. H. Burgess, M.A.

See also XII. below for other English Classics.

VIII. FRENCH LANGUAGE AND LITERATURE.

An Etymological Dictionary of the French Language, with a Preface on the Principles of French Etymology. By A. Brachet. Translated by G. W. Kitchin, M.A., formerly Censor of Christ Church. *In the Press.*

Brachet's Historical Grammar of the French Language. Translated into English by G. W. Kitchin, M.A., formerly Censor of Christ Church. *A new Edition, with a full Index.* Extra fcap. 8vo. *cloth*, 3*s.* 6*d.*

Corneille's Cinna, and **Molière's** Les Femmes Savantes. Edited, with Introduction and Notes, by Gustave Masson. Extra fcap. 8vo. *cloth*, 2*s.* 6*d.*

Racine's Andromaque, and **Corneille's** Le Menteur. With Louis Racine's Life of his Father. By the same Editor. Extra fcap. 8vo. *cloth*, 2*s.* 6*d.*

Molière's Les Fourberies de Scapin, and **Racine's** Athalie. With Voltaire's Life of Molière. By the same Editor. Extra fcap. 8vo. *cloth*, 2*s.* 6*d.*

Selections from the Correspondence of **Madame de Sévigné** and her chief Contemporaries. Intended more especially for Girls' Schools. By the same Editor. Extra fcap. 8vo. *cloth*, 3*s.*

Voyage autour de ma Chambre, by **Xavier de Maistre**; Ourika, by **Madame de Duras**; La Dot de Suzette, by **Fievée**; Les Jumeaux de l'Hôtel Corneille, by **Edmond About**; Mésaventures d'un Écolier, by **Rodolphe Töpffer**. By the same Editor. Extra fcap. 8vo. *cloth*, 2*s.* 6*d.*

A French Grammar. A complete Theory of the French Language, with the Rules in French and English, and numerous Examples to serve as first Exercises in the Language. By Jules Bué, Honorary M.A. of Oxford; Taylorian Teacher of French, Oxford; Examiner in the Oxford Local Examinations from 1858.

A French Grammar Test. A Book of Exercises on French Grammar; each Exercise being preceded by Grammatical Questions. By the same Author.

Exercises in Translation No. 1, from French into English, with general rules on Translation; and containing Notes, Hints, and Cautions, founded on a comparison of the Grammar and Genius of the two Languages. By the same Author.

Exercises in Translation No. 2, from English into French, on the same plan as the preceding book. By the same Author.

IX. GERMAN LANGUAGE AND LITERATURE.

Goethe's Egmont. With a Life of Goethe, &c. By Dr. Buchheim, Professor of the German Language and Literature in King's College, London; and Examiner in German to the University of London. Extra fcap. 8vo. *cloth*, 3*s.*

Schiller's Wilhelm Tell. With a Life of Schiller; an historical and critical Introduction, Arguments, and a complete Commentary. By the same Editor. Extra fcap. 8vo. *cloth*, 3*s.* 6*d.*

Lessing's Minna von Barnhelm. A Comedy. With a Life of Lessing, Critical Commentary, &c. By the same Editor. *Just ready.*

X. ART, &c.

A Handbook of Pictorial Art. By R. St. J. Tyrwhitt, M.A., formerly Student and Tutor of Christ Church, Oxford. With coloured Illustrations, Photographs, and a chapter on Perspective by A. Macdonald. 8vo. *half morocco*, 18*s.*

A Treatise on Harmony. By Sir F. A. Gore Ouseley, Bart., M.A., Mus. Doc., Professor of Music in the University of Oxford. 4to. *cloth*, 10*s.*

A Treatise on Counterpoint, Canon, and Fugue, based upon that of Cherubini. By the same Author. 4to. *cloth*, 16*s.*

The Cultivation of the Speaking Voice. By John Hullah. Crown 8vo. *cloth*, 3*s.* 6*d.*

XI. MISCELLANEOUS.

Outlines of Textual Criticism applied to the New Testament. By C. E. Hammond, M.A., Fellow and Tutor of Exeter College, Oxford. Extra fcap. 8vo. *cloth*, 3*s.* 6*d.*

A System of Physical Education: Theoretical and Practical. By Archibald Maclaren, The Gymnasium, Oxford. Extra fcap. 8vo. *cloth*, 7*s.* 6*d.*

The Modern Greek Language in its relation to Ancient Greek. By E. M. Geldart, B.A., formerly Scholar of Balliol College, Oxford. Extra fcap. 8vo. *cloth*, 4*s.* 6*d.*

XII. A SERIES OF ENGLISH CLASSICS.

Designed to meet the wants of Students in English Literature: under the superintendence of the Rev. J. S. Brewer, M.A., *of Queen's College, Oxford, and Professor of English Literature at King's College, London.*

It is also especially hoped that this Series may prove useful to Ladies' Schools and Middle Class Schools; in which English Literature must always be a leading subject of instruction.

A General Introduction to the Series. By Professor Brewer, M.A.

1. **Chaucer.** The Prologue to the Canterbury Tales; The Knightes Tale; The Nonne Prestes Tale. Edited by R. Morris, Editor of 'Specimens of Early English,' &c., &c. *Third Edition.* Extra fcap. 8vo. *cloth*, 2*s.* 6*d.*

2. **Spenser's Faery Queene.** Books I and II. Designed chiefly for the use of Schools. With Introduction, Notes, and Glossary. By G. W. Kitchin, M.A., formerly Censor of Christ Church. Extra fcap. 8vo. *cloth*, 2*s.* 6*d.* each.

3. **Hooker.** Ecclesiastical Polity, Book I. Edited by R. W. Church, M.A., Rector of Whatley; formerly Fellow of Oriel College, Oxford. Extra fcap. 8vo. *cloth*, 2*s.*

4. **Shakespeare.** Select Plays. Edited by W. G. Clark, M.A., Fellow of Trinity College, Cambridge; and W. Aldis Wright, M.A., Trinity College, Cambridge.

   I. The Merchant of Venice. Extra fcap. 8vo. *stiff covers*, 1*s.*
   II. Richard the Second. Extra fcap. 8vo. *stiff covers*, 1*s.* 6*d.*
   III. Macbeth. Extra fcap, 8vo. *stiff covers*, 1*s.* 6*d.*
   IV. Hamlet. Extra fcap. 8vo. *stiff covers*, 2*s.*

5. **Bacon.** Advancement of Learning. Edited by W. Aldis Wright, M.A. Extra fcap. 8vo. *cloth*, 4*s.* 6*d.*

6. **Milton.** Poems. Edited by R. C. Browne, M.A., and Associate of King's College, London. *Second edition.* 2 vols. extra fcap. 8vo. *cloth*, 6*s.* 6*d.*
   Also separately, Vol. I. 4*s.*, Vol. II. 3*s.*

7. **Dryden.** Stanzas on the Death of Oliver Cromwell; Astraea Redux; Annus Mirabilis; Absalom and Achitophel; Religio Laici; The Hind and the Panther. Edited by W. D. Christie, M.A., Trinity College, Cambridge. Extra fcap. 8vo. *cloth*, 3*s.* 6*d.*

8. **Bunyan.** Grace Abounding; The Pilgrim's Progress. Edited by E. Venables, M.A., Canon of Lincoln.

9. **Pope.** With Introduction and Notes. By Mark Pattison, B.D., Rector of Lincoln College, Oxford.

   I. Essay on Man. Extra fcap. 8vo. *stiff covers*, 1*s.* 6*d.*
   II. Satires and Epistles. Extra fcap. 8vo. *stiff covers*, 2*s.*

10. **Johnson.** Rasselas; Lives of Pope and Dryden. Edited by C. H. O. Daniel, M.A., Fellow and Tutor of Worcester College, Oxford.

11. **Burke.** Thoughts on the Present Discontents; the two Speeches on America; Reflections on the French Revolution. By Mark Pattison, B.D., Rector of Lincoln College, Oxford.

12. **Cowper.** The Task, and some of his minor poems. Edited by H. G. Griffith, M.A., Pembroke College, Oxford.

Published for the University by

MACMILLAN AND CO., LONDON.

The DELEGATES OF THE PRESS *invite suggestions and advice from all persons interested in education; and will be thankful for hints, &c. addressed to either the* Rev. G. W. KITCHIN, *St. Giles's Road East, Oxford, or the* SECRETARY TO THE DELEGATES, *Clarendon Press, Oxford.*